HANDBOOK OF RESEARCH ON BIOENERGY AND BIOMATERIALS

Consolidated and Green Processes

HANDBOOK OF RESEARCH ON BIOENERGY AND BIOMATERIALS

Consolidated and Green Processes

Edited by
Leopoldo Javier Ríos González
Jośe Antonio Rodríguez-De La Garza
Miguel Ángel Medina Morales
Cristóbal Noé Aguilar

AAP | APPLE ACADEMIC PRESS

First edition published 2022

Apple Academic Press Inc.
1265 Goldenrod Circle, NE,
Palm Bay, FL 32905 USA

4164 Lakeshore Road, Burlington,
ON, L7L 1A4 Canada

CRC Press
6000 Broken Sound Parkway NW,
Suite 300, Boca Raton, FL 33487-2742 USA

4 Park Square, Milton Park,
Abingdon, Oxon, OX14 4RN UK

Library and Archives Canada Cataloguing in Publication

Title: Handbook of research on bioenergy and biomaterials : consolidated and green processes / edited by Leopoldo Javier Ríos González, Jose Antonio Rodríguez-De La Garza, Miguel Ángel Medina Morales, Cristóbal Noé Aguilar.

Names: Ríos González, Leopoldo Javier, editor. | Rodríguez-de la Garza, Jose Antonio, editor. | Medina Morales, Miguel Ángel, editor. | Aguilar, Cristóbal Noé, editor.

Description: First edition. | Includes bibliographical references and index.

Identifiers: Canadiana (print) 20210163666 | Canadiana (ebook) 20210163887 | ISBN 9781771889551 (hardcover) | ISBN 9781774639351 (softcover) | ISBN 9781003105053 (ebook)

Subjects: LCSH: Biomass conversion. | LCSH: Biomass energy. | LCSH: Biomass. | LCSH: Biomass—Refining. | LCSH: Biomolecules. | LCSH: Green chemistry.

Classification: LCC TP248.B55 H36 2022 | DDC 662/.88—dc23

Library of Congress Cataloging-in-Publication Data

Names: Ríos González, Leopoldo Javier, editor.

Title: Handbook of research on bioenergy and biomaterials : consolidated and green processes / edited by Leopoldo Javier Ríos González, José Antonio Rodríguez-de La Garza, Miguel Ángel Medina Morales, Cristóbal Noé Aguilar.

Description: First edition. | Palm Bay, FL : Apple Academic Press, 2021. | Includes bibliographical references and index. | Summary: "The Handbook of Research on Bioenergy and Biomaterials: Consolidated and Green Processes provides an understanding of consolidated processing and biorefinery systems for the production of bio-based chemicals and value-added bioproducts from renewable sources. The chapters look at a variety of bioenergy technological advances and improvements in the energy and materials sectors that aim to lower our dependence of fossil fuels and consequently reduce greenhouse gas (GHG) emissions. The volume looks at a selection of processes for the production of energy and biomaterials, including the Fischer-Tropsch process, gasification, pyrolysis, combustion, fermentation from renewable sources (such as, plants, animals and their byproducts), and others. Applications that are explored include transportation fuels, biodiesel production, wastewater treatment, edible packaging, bioplastics, physical rehabilitation, tissue engineering, biomedical applications, thermal insulation, industrial value compounds, and more. All of the topics covered in this publication address consolidated processes that play a pivotal role in the production of bioenergy and biomaterials because these processes require fewer unitary operations needed in the process, leading to a more direct method of production. This type of production system contributes to decreasing negative effects on the environment, lowering costs, saving energy and time, and improving profitability and efficiency. This volume will be valuable for the industrial sector, for researchers and scientists, as well as for faculty and advanced students"-- Provided by publisher.

Identifiers: LCCN 2021011437 (print) | LCCN 2021011438 (ebook) | ISBN 9781771889551 (hardback) | ISBN 9781774639351 (paperback) | ISBN 9781003105053 (ebook)

Subjects: LCSH: Biotechnology. | Biomass energy.

Classification: LCC TP248.15 .H36 2021 (print) | LCC TP248.15 (ebook) | DDC 660.6--dc23

LC record available at https://lccn.loc.gov/2021011437

LC ebook record available at https://lccn.loc.gov/2021011438

ISBN: 978-1-77188-955-1 (hbk)
ISBN: 978-1-77463-935-1 (pbk)
ISBN: 978-1-00310-505-3 (ebk)

About the Editors

Leopoldo Javier Ríos González, PhD

Leopoldo J. Ríos-González, PhD, is a full-time Professor at the School of Chemistry, Autonomous University of Coahuila, México. He is biologist chemist specializing in microbiology and holds a PhD in biotechnology. He is expert in the production of liquid and gas biofuels from biomass and is a member of the Bioenergy Thematic Network (REMBIO). In 2016, he was granted the Environmental Merit Award by the Autonomous University of Coahuila, and in 2017, he received the Environmental Award of Coahuila. Dr. Rios-Gonzalez has developed research funded by the Secretariat of Agricultural, National Council of Research and Technology (CONACyT) through the International Cooperation Programs CONACyT/CONICyT, Secretariat of Economy and different projects with industry. Dr Rios-Gonzalez has published over 25 original research papers in indexed journals and is member of National Researchers System in México (SNI) from 2018 (Level I).

Jośe Antonio Rodríguez-De La Garza, PhD

Jose A. Rodríguez-De la Garza, PhD, has been a full-time Professor since 2008 at the School of Chemistry, Autonomous University of Coahuila, Mexico. in the area Environmental Biotechnology. He currently is head of the Biotechnology PhD program at the same institution. He is member of the Mexican Society of Biotechnology and Bioengineering. Dr. Rodriguez-De la Garza currently develops a research project funded by the State Council of Science and Technology of Coahuila He has been a Thesis Director of PhD, MSc, and BSc theses. He is became a mechanical engineer at the Autonomous University of Nuevo Leon, Mexico. He earned his MSc and PhD degrees in biotechnology from the Autonomous University of Coahuila. He has been in two academic stays, in the USA (Washington State University, Pullman) and in Chile (Pontificia Universidad Católica de Valparaíso). Dr Rodriguez-De la Garza has published over 15 original research papers in indexed journals and five book chapters and has participated and contributed in over 40 scientific meetings.

Miguel Ángel Medina Morales, PhD

Miguel Ángel Medina Morales, PhD, is a full-time Professor at the School of Chemistry, Autonomous University of Coahuila, Mexico, in the area environmental biotechnology. He is a member of National Researchers System (SNI) from 2017 (candidate level). In 2017 he was part of a research team that were honored with the Tritio Award given by the Autonomous University of Coahuila. He has published over 14 papers in scientific journals and several book chapters. He graduated as Biologist Chemist at the Autonomous University of Coahuila. He obtained his MSc Degree in Food Science and Technology by the same institution. In 2012 he obtained his PhD in Food Science and Technology at the Autonomous University of Coahuila.

Cristóbal Noé Aguilar, PhD

Cristóbal Noé Aguilar, PhD, is Director of Research and Postgraduate Programs at the Universidad Autonoma de Coahuila, Mexico. He is also a member of the Bioprocesses and Bioproducts Research Group and Professor in the Food Research Department in the School of Chemistry at the Universidad Autónoma de Coahuila, Saltillo, Mexico. Dr. Aguilar is Associate Editor of *Heliyon* (Microbiology) and *Frontiers in Sustainable Food Systems* (Food Processing) and has published more than 330 papers in indexed journals, more than 40 articles in Mexican journals, and 250 contributions in scientific meetings. He has also published many book chapters, several Mexican books, four editions of international books, and more. He has been awarded several prizes and awards, the most important of which are the National Prize of Research 2010 from the Mexican Academy of Sciences; the Prize "Carlos Casas Campillo 2008" from the Mexican Society of Biotechnology and Bioengineering; National Prize AgroBio–2005; and the Mexican Prize in Food Science and Technology. Dr. Aguilar is a member of the Mexican Academy of Science, the International Bioprocessing Association, Mexican Academy of Sciences, Mexican Society for Biotechnology and Bioengineering, and the Mexican Association for Food Science and Biotechnology. He has developed more than 21 research projects, including six international exchange projects.

Contents

Contributors

Karina Reyes Acosta
Chemical Engineering Department, Faculty of Chemical Sciences, Autonomous University of Coahuila, Saltillo, Coahuila, Mexico

Erika Acosta-Cruz
Departamento de Biotecnología. Facultad de Ciencias Químicas. Universidad Autónoma de Coahuila, Blvd. Venustiano Carranza, 25280 Saltillo, Coahuila, México

Cristóbal N. Aguilar
Bioprocesses and Bioproducts Research Center, FCQ-UAdeC, 25280 Saltillo, México

F. J. Alonso-Montemayor
PhD Program in Materials Science and Technology, School of Chemistry (FCQ), Autonomous University of Coahuila (UAdeC), 25280 Saltillo, México

Christian Javier Cabello Alvarado
CONACYT—Consorcio de Investigación Científica, Tecnológica y de Innovación del Estado de Tlaxcala, Calle 1 de mayo No. 22, Col. Centro, C.P. 90000, Tlaxcala de Xicothenccatl Tlaxcala, México

Olga B. Álvarez Pérez
Greencorp Biorganiks de México S.A. de C.V., Blvd Luis Donaldo Colosio, Col. San Patricio, Saltillo 25204, México

Paola Angulo-Bejarano
Tecnologico de Monterrey, Centre of Bioengineering, School of Engineering and Sciences, Campus Queretaro, Querétaro, Qro, Mexico

Diana Sofía Segovia Arévalo
Centro de Investigación y de Estudios Avanzados del Instituto Politécnico Nacional Unidad Saltillo, México

F. Ávalos-Belmontes
Departament of Polymers, FCQ-UAdeC, 25280 Saltillo, Mexico

B. A. Ayil-Gútierrez
CONACYT—Centro de Biotecnología Genómica, Instituto Politécnico Nacional, Blvd. del Maestro, s/n, Esq. Elías Piña 88710. Reynosa, Tamaulipas, México

R. Baeza-Jiménez
Research Center in Food and Development, A.C. Av. 4ta Sur 3820, Fracc. Vencedores del Desierto, 33089, Cd. Delicias, Chihuahua, México

J. J. Buenrostro-Figueroa
Research Center in Food and Development, A.C. Av. 4ta Sur 3820, Fracc. Vencedores del Desierto, 33089, Cd. Delicias, Chihuahua, México

Guadalupe Bustos-Vázquez
Departamento de Ingeniería Bioquímica, Universidad Autónoma de Tamaulipas, Unidad Académica Multidisciplinaria Mante, Blvd. Enrique Cárdenas González No. 1201 Pte. Col. Jardín, C.P. 89840 Ciudad Mante, Tamaulipas, México

Denis A. Cabrera-Munguia
Advanced Materials, School of Chemistry, Autonomous University of Coahuila, México

Mauricio Carrillo-Tripp
Laboratorio de la Diversidad Biomolecular, Centro de Investigación y de Estudios Avanzados del Instituto Politécnico Nacional Unidad Monterrey, Apodaca, Nuevo León, México

Sandra L. Castañón-Alonso
Chemistry Department, Universidad Autónoma Metropolitana-Iztapalapa, Ciudad de México, México

Salvador Castell-González
Colegio de Posgraduados en Ciencias Ambientales y Biotecnología del Sureste, A.C. Mérida, Yucatán, México

E. Cázares-Sánchez
Instituto Tecnológico de la Zona Maya, carretera Chetumal Escárcega Km. 21.5. C.P. 77960 Ejido Juan Sarabia, México

Ana V. Charles-Rodríguez
Department of Food Science and Technology, Universidad Autónoma Agraría Antonio Narro, Colonia Buenavista, Saltillo, México

M. L. Chávez-González
Nanobioscience Group, School of Chemistry, Universidad Autónoma de Coahuila, Blvd. V. Carranza esquina con José Cárdenas Valdés s/n Col, República Oriente, 25280 Saltillo, México
Bioprocesses & Bioproducts Group, Food Research Department, School of Chemistry, Universidad Autónoma de Coahuila, Blvd. V. Carranza esquina con José Cárdenas Valdés s/n Col, República Oriente, 25280 Saltillo, México

Jesús A. Claudio-Rizo
Advanced Materials Department, School of Chemical Sciences, University Autonomous of Coahuila, Saltillo, Mexico

J. C. Contreras-Esquivel
Glycobiology Research Center, FCQ-UAdeC, 25280 Saltillo, Mexico

A. V. Córdova-Quiroz
Facultad de Química, Dependencia Académica de Ciencias Química y Petrolera, Universidad Autónoma del Carmen, Carmen, México

E. A. De la Cruz-Arguijo
Centro de Biotecnología Genómica, Instituto Politécnico Nacional, Blvd. del Maestro, s/n, Esq. Elías Piña, Reynosa 88710, México

Brenda R. Cruz-Ortiz
Ceramic Materials Department, School of Chemical Sciences, University Autonomous of Coahuila, Saltillo, Mexico

Marisol Cruz-Requena
Departamento de Investigación en Alimentos, Facultad de Ciencias Químicas, Universidad Autónoma de Coahuila, 25280 Saltillo, México

Leopoldo María Antonia Cruz-Hernández
Laboratorio de Interacción Ambiente-Microorganismo, Centro de Biotecnología Genómica, Instituto Politécnico Nacional, Reynosa, Tamaulipas, México

Ben Hur Espinosa-Ramírez
División de Electromecánica Industrial, Universidad Tecnológica de Tecámac, Tecámac, Estado de México, México

Lorena Farías-Cepeda
Chemical Engineering Department, Faculty of Chemical Sciences, Autonomous University of Coahuila, Saltillo, Coahuila, Mexico

María L. Flores-López
Department of Research and Development, Biocampo S.A. de C.V., Blvd. Dr. Jesús Valdés Sánchez Km. 10, Fracc. Presa de las Casas, Arteaga 25350, México

J. J. Fuentes-Avilés
Facultad de Ciencias Químicas, Universidad Autónoma de Coahuila. Blvd., Venustiano Carranza y José Cárdenas Valdés, C.P. 25280 Saltillo, Coahuila, México

Adolfo Romero Galarza
Chemical Engineering Department, Faculty of Chemical Sciences, Autonomous University of Coahuila, Saltillo, Coahuila, Mexico

Horacio González
Faculty of Chemical Engineering, Universidad Michoacana de San Nicolás de Hidalgo, México

A. Iliná
Nanobioscience Group, School of Chemistry, Universidad Autónoma de Coahuila, Blvd. V. Carranza esquina con José Cárdenas Valdés s/n Col, República Oriente, 25280 Saltillo, México

Fernando Jasso-Juarez
Departamento de Biotecnología, Facultad de Ciencias Químicas, Universidad Autónoma de Coahuila, Blvd. Venustiano Carranza, 25280 Saltillo, México

Arturo I. Martínez Enríquez
Centro de Investigación y Estudios Avanzados del Instituto Politécnico Nacional, CINVESTAV-Saltillo, 25900 Ramos Arizpe, México

J. L. Martínez-Hernández
Nanobioscience Group, School of Chemistry, Universidad Autónoma de Coahuila, Blvd. V. Carranza esquina con José Cárdenas Valdés s/n Col, República Oriente, 25280 Saltillo, México

Marlene Lariza Andrade Guel
Department of Advanced Materials, Centro de Investigación en Química Aplicada, Blvd.

Salvador Carlos Hernández
Centro de Investigación y Estudios Avanzados del Instituto Politécnico Nacional, CINVESTAV-Saltillo, 25900 Ramos Arizpe, México Enrique Reyna Hermosillo No. 140 C.P. 25294 Saltillo, Coahuila, México

V. M. Interián-Ku
Instituto Tecnológico de la Zona Maya, carretera Chetumal Escárcega Km. 21.5, Ejido Juan Sarabia, C.P. 77960 Quintana Roo, México

Alma Berenice Jasso-Salcedo
CONACYT—Centro de Investigación en Química Aplicada, Blvd. Enrique Reyna Hermosillo No. 140 C.P. 25294 Saltillo, Coahuila, México

Lourdes Díaz Jiménez
Centro de Investigación y de Estudios Avanzados del Instituto Politécnico Nacional Unidad Saltillo, México

C. Leyva
Centro de Investigación en Ciencia Aplicada y Tecnología Avanzada-IPN-Unidad Legaria, Ciudad de México, México

Miguel A. De León-Zapata
Research Center for the Conservation of Biodiversity and Ecology of Coahuila (CICBEC), Autonomous University of Coahuila, Saltillo 25280, Mexico

Claudia M. López-Badillo
Ceramic Materials Department, School of Chemical Sciences, University Autonomous of Coahuila, Saltillo, Mexico

L. Margarita López-Castillo
Centro de Biotecnología Femsa. Escuela de Ingeniería y Ciencias, Tecnológico de Monterrey, Monterrey, Nuevo León, México

Claudia M. López-Badillo
Ceramic Materials Department, School of Chemical Sciences, University Autonomous of Coahuila, Saltillo, Mexico

H. A. Luna-García
Nanobioscience Group, School of Chemistry, Universidad Autónoma de Coahuila, Blvd. V. Carranza esquina con José Cárdenas Valdés s/n Col, República Oriente, 25280 Saltillo, México
Bioprocesses & Bioproducts Group, Food Research Department, School of Chemistry, Universidad Autónoma de Coahuila, Blvd. V. Carranza esquina con José Cárdenas Valdés s/n Col, República Oriente, 25280 Saltillo, México

Lucero Rosales Marines
Chemical Engineering Department, Faculty of Chemical Sciences, Autonomous University of Coahuila, Saltillo, Coahuila, Mexico

S. Y. Martínez-Amador
Botany Department, Agronomic Division, Universidad Autónoma Agraria Antonio Narro, Mexico

Humberto Martínez Montoya
Laboratory of Genetics and Comparative Genomics, UAM Reynosa Aztlán—Universidad Autónoma de Tamaulipas, Reynosa, Tamaulipas, México

Fernando Méndez-González
Departamento de Biotecnología, Universidad Autónoma Metropolitana-Iztapalapa, Ciudad de México, México

Miguel A. Medina-Morales
Department of Biotechnology, School of Chemistry, Autonomous University of Coahuila, Mexico

Romeo Rojas Molina
School of Agronomy, Universidad Autónoma de Nuevo León, Monterrey, México

Thelma K. Morales-Martínez
Bioprocess and Microbial Biochemistry Group, School of Chemistry, Autonomous University of Coahuila, Mexico

I. M. M. Moreno-Davila
Laboratorio de Biotecnología Ambiental. Departamento de Biotecnología.
Facultad de Ciencias Químicas. Universidad Autónoma de Coahuila, Ing J. Cardenas Valdez S/N, República, 25280 Saltillo, Coahuila, México

Mayela Moreno-Dávila
Department of Biotechnology, School of Chemistry, Autonomous University of Coahuila, Mexico

L. F. Mora-Cortes
PhD Program in Materials Science and Technology, School of Chemistry (FCQ), Autonomous University of Coahuila (UAdeC), 25280 Saltillo, Mexico

R. I. Narro-Céspedes
Departament of Polymers, FCQ-UAdeC, 25280 Saltillo, México

D. Navarro-Rodríguez
Department of Polymers Synthesis, Research Center for Applied Chemistry (CIQA) 25294, Saltillo, Coahuila, México

M. G. Neira-Velázquez
Department of Polymers Synthesis, Research Center for Applied Chemistry (CIQA) 25294, Saltillo, Coahuila, México

E. Ochoa-Reyes
Research Center in Food and Development, A.C. Av. 4ta Sur 3820, Fracc. Vencedores del Desierto, 33089, Cd. Delicias, Chihuahua, México

Carlos Alberto Ávila Orta
Department of Advanced Materials, Centro de Investigación en Química Aplicada, Blvd. Enrique Reyna Hermosillo No. 140 C.P. 25294 Saltillo, Coahuila, México

R. Parra-Saldívar
Tecnologico de Monterrey, Escuela de Ingeniería y Ciencias, Ave. Eugenio Garza Sada 2501, 64849 Monterrey, NL, México

Alejandra Pichardo-Sánchez
Departamento de Biotecnología, Universidad Autónoma Metropolitana-Iztapalapa, Ciudad de México, México

Araceli Martínez Ponce
Escuela Nacional de Estudios Superiores, Unidad Morelia, Universidad Nacional Autónoma de México, Antigua Carretera a Pátzcuaro No. 8701, Col. Ex Hacienda de San José de la Huerta, 58190 Morelia, Michoacán, México

W. Poot-Poot
Laboratorio de Biotecnología, Facultad de Ingeniería y Ciencias, Universidad Autónoma de Tamaulipas, Centro Universitario, 87120 Ciudad Victoria, Tamaulipas, México

P. Villarreal Quintero
Nanobioscience Group, School of Chemistry, Universidad Autónoma de Coahuila, Blvd. V. Carranza esquina con José Cárdenas Valdés s/n Col, República Oriente, 25280 Saltillo, México
Bioprocesses & Bioproducts Group, Food Research Department, School of Chemistry, Universidad Autónoma de Coahuila, Blvd. V. Carranza esquina con José Cárdenas Valdés s/n Col, República Oriente, 25280 Saltillo, México

V. H. Ramos-Garcia
Biotecnologia Vegetal. Centro de Biotecnología Genómica, Instituto Politécnico Nacional, Blvd. del Maestro, s/n, Esq. Elías Piña, 88710 Reynosa, México

A. G. Reyes
CONACYT—CIBNOR, Instituto Politécnico Nacional 195, Playa Palo de Santa Rita Sur, 23096 La Paz, BCS, México

Leopoldo J. Ríos-González
Department of Biotechnology, School of Chemistry, Autonomous University of Coahuila, Mexico

C. Rivera-Pérez
CONACYT—CIBNOR, Instituto Politécnico Nacional 195, Playa Palo de Santa Rita Sur, 23096 La Paz, BCS, México

Armando Robledo-Olivo
Department of Food Science and Technology, Universidad Autónoma Agraría Antonio Narro, Colonia Buenavista, Saltillo, México

José A. Rodríguez-de la Garza
Department of Biotechnology, School of Chemistry, Autonomous University of Coahuila, Mexico

Luis Víctor Rodríguez-Durán
Departamento de Ingeniería Bioquímica, Universidad Autónoma de Tamaulipas, Unidad Académica Multidisciplinaria Mante, Blvd. Enrique Cárdenas González No. 1201 Pte. Col. Jardín, C.P. 89840 Ciudad Mante, Tamaulipas, México

Nubia R. Rodríguez-Durán
Departamento de Ingeniería Bioquímica, Universidad Autónoma de Tamaulipas, Unidad Académica Multidisciplinaria Mante, Blvd. Enrique Cárdenas González No. 1201 Pte. Col. Jardín, C.P. 89840 Ciudad Mante, Tamaulipas, México

M. M. Rodríguez-Garza
Department of Biotechnology, School of Chemistry, Universidad Autonoma de Coahuila, Mexico

Adolfo Romero-Galarza
Department of Chemical Engineering, School of Chemistry, Autonomous University of Coahuila, México

Lucero Rosales-Marines
Department of Chemical Engineering, School of Chemistry, Autonomous University of Coahuila, México

Anilú Rubio Ríos
Chemical Engineering Department, Faculty of Chemical Sciences, Autonomous University of Coahuila, Saltillo, Coahuila, Mexico

A. Ruiz-Marín
Facultad de Química, Dependencia Académica de Ciencias Química y Petrolera, Universidad Autónoma del Carmen, Carmen, México

A. Sáenz-Galindo
Facultad de Ciencias Químicas, Universidad Autónoma de Coahuila. Blvd., Venustiano Carranza y José Cárdenas Valdés, C.P. 25280 Saltillo, Coahuila, México

C. Salinas-Salazar
Tecnologico de Monterrey, Escuela de Ingeniería y Ciencias, Ave. Eugenio Garza Sada 2501, 64849 Monterrey, NL, México

Thalía A. Salinas-Jasso
Department of Chemical Metrology, Universidad Politécnica de Ramos Arizpe (UPRA), Ramos Arizpe, México

Rodolfo Torres-de los Santos
Unidad Académica Multidisciplinaria Mante Centro. Universidad Autónoma de Tamaulipas, Cd. Mante, Tamaulipas, México

Gerardo de Jesús Sosa-Santillán
Departamento de Biotecnología. Facultad de Ciencias Químicas, Universidad Autónoma de Coahuila, Ing J. Cardenas Valdez S/N, República, 25280 Saltillo, Coahuila, México

E. P. Segura-Ceniceros
Nanobioscience Group, School of Chemistry, Universidad Autónoma de Coahuila, Blvd. V. Carranza esquina con José Cárdenas Valdés s/n Col, República Oriente, 25280 Saltillo, México

Leonardo Sepúlveda
Grupo de Bioprocesos y Bioquímica Microbiana, Facultad de Ciencias Químicas, Universidad Autónoma de Coahuila, 25280 Saltillo, México

L. E. Serrato-Villegas
Department of Chemical Engineering, School of Chemistry, Autonomous University of Coahuila, México

J. C. Tafolla-Arellano
Departamento de Ciencias Básicas, Laboratorio de Biotecnología y Biología Molecular. Núcleo Básico del Doctorado en Ciencias en Recursos Fitogenéticos para Zonas Áridas, Universidad Autónoma Agraria Antonio Narro, 25315. Buenavista, Saltillo, Coahuila, México

M. C. Tamayo-Ordoñez
Laboratorio de Ingeniería Genética, Departamento de Biotecnología. Facultad de Ciencias Químicas. Universidad Autónoma de Coahuila, Ing J. Cardenas Valdez S/N, República, 25280 Saltillo, Coahuila, México

Y. J. Tamayo-Ordoñez
Estancia Posdoctoral Nacional-CONACyT, Posgrado en Ciencia y Tecnología de Alimentos. Facultad de Ciencias Químicas. Universidad Autónoma de Coahuila, Ing J. Cardenas Valdez S/N, República, 25280 Saltillo, Coahuila, México

Luis Fernando Sánchez Terán
Centro de Investigación y Estudios Avanzados del Instituto Politécnico Nacional, CINVESTAV-Saltillo, 25900 Ramos Arizpe, México

J. M. Tirado-Gallegos
School of Animal Science and Ecology, Universidad Autónoma de Chihuahua. 31453, Chihuahua, Chihuahua, México

J. M. Vieira
Department of Food Engineering, Faculty of Food Engineering, University of Campinas (UNICAMP), Campinas, Brazil

Abbreviations

AA	alendronic acid
ABE	acetone–butanol–ethanol pathway
ABP	activity-based probes
AC	adenylate cyclase
ACC	acetyl-CoA carboxylase
ACL	ATP citrate lyase
ACP	acyl carrier proteins
AD	anaerobic digestion
AF	animal fat
ASF	Anderson–Schulz–Flory
BCP	bicalcium phosphate
BES	bioelectrochemical systems
BFB	bubbling fluidized bed
BG	bioglass
BMP	biochemical methane potential
BTL	biomass to liquid
cAMP	cyclic adenosine monophosphate
CAP	catabolite gene activator protein
CBP	consolidated bioprocessing
CCD	central composite design
CCM	central carbon metabolism
CCMV	Cowpea chlorotic mottle virus
CcpA	catabolite control protein
CCR	carbon catabolite repression
CD	circular dichroism
CE	carbon economy
CFB	circulating fluidized bed
CFs	carbon fibers
CHA	chabazite-type
CM	citrate malate
CMPL	carboxymethyl pullulan
CMV	cucumber mosaic virus
CNF	cellulose nanofibers
CNTs	carbon nanotubes

COD	chemical oxygen demand
CP	calcium phosphate
CP	coat protein
CPMV	Cowpea mosaic virus
CRISPR	clustered regulatory interspaced short palindromic repeats (CRISPR)
CS	calcium silicate
CSs	carbon spheres
CSTR	continuously stirred tank
CV	calorific values
DAG	diacylglycerol
DCMC	dialdehyde carboxymethyl cellulose
DES	deep eutectic solvents
DGAT	diacylglycerol acyltransferase
DGs	diglycerides
DHA	docosahexaenoic acid
DMA	dynamic mechanical analysis
DMSO	dimethyl sulfoxide
DSC	differential scanning calorimetry
DTH	delayed-type hypersensitivity
DVB	divinylbenzene
ECM	extracellular matrix
EI	enzyme I
EII	enzyme II
ELPs	ELASTIN-like polypeptides
EMY	effective mass yield
EPA	eicosapentaenoic acid
EPS	expanded polystyrene
EPSs	extracellular polymeric substances
FA	furfuryl alcohol
FADh	formaldehyde dehydrogenase
FAME	fatty acid methyl ester
FAS	fatty acid synthase system
FASI	type I fatty acid synthases (FAS I)
FB	fibrinogen
FCC	face-centered cubic
FDH	formate dehydrogenase
FFAs	free fatty acids
FT	Fischer–Tropsch

FTIR	Fourier transform infrared spectroscopy
G3P	glycerate-3-phosphate (G3P)
GA	glutamic acid
GAGs	glycosaminoglycans
GAP	glyceraldehyde phosphate
GCTs	green chemical technologies
GE	genetic engineering
GFP	green fluorescent protein
GH	glycoside hydrolase
GME	genetically modified enzymes
GMM	genetically modified microbes
GMO	genetically modified organisms
GMP	genetically modified plants
GO	graphene oxide
GPAT	glycerol-sn-3-phosphate acyl-transferase
GPDH	glycerol-3-phosphate dehydrogenase
GRAS	Generally Regarded As Safe
GW	global warming
HA	hyaluronic acid
HAp	hydroxylapatite
HBV	hepatitis B virus
HCP	hexagonal close-packed
HCP	hyper-cross-linked polymer
HDPE	high density polyethylene
HIV	human immunodeficiency virus
HLA	hyaluronic acid
HMF	5-(hydroxymethyl)furfural
hMSCs	human mesenchymal stem cells
HPAs	heteropolyacids
HPr	histidine protein
HPS	3-hexulose-6-phosphate synthase
HPW	tungstophosphoric acid
HRP	horseradish peroxidase
HSCCC	high-speed counter-current chromatography
HTFT	high-temperature Fischer–Tropsch
IBs	inclusion bodies
ILs	ionic liquids
IPNs	interpenetrated networks
IR	ischemia-reperfusion

KAS	3-ketoacyl-ACP synthase
KF	Kevlar fibers
L/S	liquid-to-solid ratio
LA	lactic acid
LAB	lactic acid bacteria
LCA	life cycle assessment
LPA	lysophosphatidate
LPAT	lysophosphatidate acyl-transferase
LPS	lipopolysaccharide
LTFT	low-temperature Fischer–Trospsch
MAE	microwave assisted-extraction
MAPK	mitogen-activated protein-kinases
MD	molecular dynamics
MDC	microbial desalination cells
MDH	methanol dehydrogenase
MEC	microbial electrolysis cell
MFC	microbial fuel cell
MI	mass intensity
MIBK	methyl isobutyl ketone
MP	mass productivity
MPT	malonyl/palmitoyl transferase
MRI	magnetic resonance imaging
MTHF	2-methyltetrahydrofuran
NEOs	nonedible oils
NGFs	nerve growth factors
nHA	natural hydroxyapatite
OLR	organic loading rate
OMPs	osseous morphogenetic proteins
PA	phosphatidate
PAA	polyacrylic acid
PADs	polyamides
PBAT	poly-(butylene adipate-co-terephthalate)
PCL	poly(ε-caprolactone)
PDA	polydopamine
PDLA	poly-D-lactic acid
PE	polyethylene
PECVD	PLASMA enhanced chemical vapor deposition
PEEK	polyether ketone
PEG	polyethylene glycol

PEM	proton exchange membrane
PET	polyethylene terephthalate
PFE	polytetrafluoroethylene
PFR	plug flow reactor
PGA	polyglycolic acid
PGM	phosphoglucomutase
PGMA	polyglutamic acid
PHA	polyhydroxyalkanoates
PHEMA	poly(2-hydroxyethyl methacrylate)
PHI	6-phospho-3-hexuloisomerase
PHMHM	poly(1-hydroxymethyl-ethylene hydroxymethyl-formal)
PL	pullulan
PLA	poly(lactic acid)
PLGA	polylactic-*co*-glycolic acid
PMI	process mass intensity
PP	pentose phosphate
PPEs	polyphosphoesters
PRDs	PTS regulatory domains
PTS	phosphotransferase system
PU	polyurethane
PVA	polyvinyl alcohol
PVC	polyvinyl chloride
PVX	potato virus X
RME	reaction mass efficiency
rMSCs	researchers cultured mesenchymal stem cells
rPMs	recycled plastics materials
RS	rice starch
RSF	regenerated silk fibroin
RT	room temperature
RTD	resident time distribution
RuBP	ribulose bisphosphate pathway
RuMP	ribulose monophosphate
SAPOs	silicoaluminophosphates
SC	chondroitin sulfate
SCO	single cell oil
SDR	spinning disc reactors
SEM	scanning electron microscope
SF	silk fibroin
SFE	supercritical fluid extraction

SHF	separate method of hydrolysis and fermentation
SS	silk sericin
SS	solid state
SS	stainless steel
SSF	accharification and fermentation
SWE	subcritical water extraction
TAGs	triacylglycerols
TALEN	transcription activator-like effector nucleases
TCP	tricalcium phosphate
TEM	transmission electron microscopy
TGA	thermogravimetric analysis
TGs	triglycerides
THF	tetrahydrofuran
THFA	tetrahydrofurfuryl alcohol
TMV	tobacco mosaic virus
TNF	tumor nuclear factor
TPP	three phases partitioning
UAE	ultrasound-assisted extraction
UHMWPE	ultra-high molecular weight polyethylene
VFA	volatile fatty acids
VLPs	virus-like particles
WCOs	waste cooking oils
WGS	water–gas shift
WW	wood waste
WWI	waste water intensity
XRD	X-ray diffraction
ZEB	zero energy building
ZFNs	zinc-finger nucleases

Preface

In recent years, an increasing interest in products of natural origin has taken root in the population. Several processes are being developed to address the need for these products and, in particular, biotechnology and green chemistry are the leading areas due to its considerably lower implications of harming the environment. The specific cases of bioenergy, biomaterials, and biomolecules extraction and production have a strong relation with biotechnology, which along with green chemistry, contribute to develop technology that leads to the creation of useful compounds.

The *Handbook of Research on Bioenergy and Biomaterials: Consolidated and Green Processes* will give an insight and an understanding of consolidate processing-biorefinery systems for biofuels production using tools of biotechnology, chemical engineering, among other branches of science.

Bioenergy produced by Fischer–Tropsch, gasification, pyrolysis, combustion, and fermentation from renewable sources (such as plants, animals, and their byproducts) is considered a great technological improvement in the energy sector in regard to lower our dependence of fossil fuels and consequently the greenhouse gas (GHG) emissions. In addition to produce biofuels, researchers are looking for the biorefinery concept perceiving agro-industrial value chains. As a result, the subject of biorefinery engineering science may need to grow as a discipline in regard of converting biomass into useful liquid fuels in an attempt to replace totally or partially the fossil fuels consumption.

Biomolecules comprise a wide array of compounds with a wide area of applications. These biomolecules can be included with those with bioactivities such as antioxidants, antimicrobials, anticancer, among many others. Enzymes are among these biomolecules and are important to many processes such as biofuels production, high added-value compounds release, enzymatic synthesis of useful compound, degradation of residual materials, and along many other processes. Production of these types of biomolecules are significant factors for quality improvement in the way of life of humans and to increase production in the meat and agricultural industry sector.

Biomaterials are defined as any material that can interact with biological systems and is also an interesting topic that will be covered in our book. There are several examples that have been used as a treatment, augmentation, and replacement in certain type of tissue or surface. Organic molecules of biological origin such as polysaccharides or proteins have been used to interact with other compounds or polymers to give way to a different material with new or different features. Also, inorganic molecules have been applied to interact with living tissues or biological molecules to be applied in several fields ranging from medicine to industrial processes. Another interesting aspect in this regard is the fact that microorganisms are able to produce high molecular weight compounds that can be applied in biomaterials and also produce precursors that apply in the same manner.

In all the topics that are covered in this publication, the term "consolidated process" plays a pivotal role due to the fact that it means that fewer unitary operations will be used in a process and in obtaining a more direct method of production. This type of production systems can contribute to decrease the negative effects on the environment, lower costs, energy and time, improving profitability and efficiency.

The editors

PART I
Green Chemistry and Biorefinery

CHAPTER 1

Merging Green Chemistry and Biorefinery: Consolidating Processes

A. G. REYES[1*], C. RIVERA-PÉREZ[1], A. SÁENZ-GALINDO[2],
J. J. FUENTES-AVILÉS[2], C. SALINAS-SALAZAR[3], and R. PARRA-SALDÍVAR[3]

[1]*CONACYT-CIBNOR, Instituto Politécnico Nacional 195,
Playa Palo de Santa Rita Sur, 23096 La Paz, BCS, México*

[2]*Facultad de Ciencias Químicas, Universidad Autónoma de Coahuila.
Blvd., Venustiano Carranza y José Cárdenas Valdés, C.P. 25280 Saltillo,
Coahuila, México*

[3]*Tecnologico de Monterrey, Escuela de Ingeniería y Ciencias,
Ave. Eugenio Garza Sada 2501, 64849 Monterrey, NL, México*

Corresponding author. E-mail: agalvarado@cibnor.mx

ABSTRACT

Green chemistry, as a work philosophy, has contributed to the design and application of safer and green processes and products. However, there are still challenges to achieve sustainability. In this regard, it is necessary to adopt a circular model, where the wastes and by-products can be used. Therefore the merging between green chemistry and biorefinery as consolidated processes can be the next step to real sustainability.

Interestingly, there are new tools and technologies to predict and extract commercially attractive compounds, thus positioning to biorefinery on a realistic and achievable stage. Currently, academia, society, industry, and government are concerned about the application of green chemistry principles. There is a joint effort to pursuit sustainability. Therefore the actual moment is envisaged as the settable moment for integrated technologies directed to a sustainable world.

1.1 INTRODUCTION

The current demand in the design of clean processes has addressed to the adoption of green philosophies. Among the main disciplines included under this necessity is chemistry, and consequently biorefineries as an application. Over the past decade, green chemistry has gained the attention of the industrial sectors, such as agriculture, medicine, cosmetic, aerospace, electronics, energy, and others (Fiorentino et al., 2018; López-Velázquez et al., 2019; Shaikh et al., 2019; Celeiro et al., 2019), because the economic and environmental benefits obtained from its application are financially attractive. Moreover, it contributes to the innovative design of technologies, positioning in high levels to the user enterprises. For its part, the biorefinery concept is the engine for green chemistry through a circular model. Both processes, when consolidated, are the merging way to sustainability. In this direction, the aim of this chapter is to introduce the basic concepts of green chemistry and biorefinery, as consolidated processes, with an emphasis on sustainability, and a circular model. In the same manner to describe modern and innovative tools to predict high valuable molecules content in biomass suitable for biorefinery processes.

1.2 GREEN CHEMISTRY: DEFINITION AND A BRIEF HISTORY

Green chemistry is considered a work philosophy, which can be implemented in any chemical transformation process, with the primary objective to reduce or eliminate the use or generation of dangerous toxic chemicals in the design, procurement, manufacture, and application of chemical products. According to de Marco et al. (2019), the green chemistry term was used for the first time by Cathcart, in the year 1990, at a paper discussion. However, it was until 1998 when Paul Anastas and John Warner proposed the definition in the book "Green Chemistry: Theory and Practice." In this book, the authors presented 12 principles of green chemistry, as criteria that must be covered in any chemical transformation process, to respect the environment (Anastas and Warner, 1988). Since that moment, the book has become a manual for the industries that deal with environmental issues. For its part, also the academic concern is drive through the green chemistry principles and application improvement. Currently, it is the Green and Sustainable Chemistry Conference, space where professionals, entrepreneurs, researchers, and students expose and discuss the advances and challenges (Green and

Sustainable Chemistry Conference, 2019).

On the other hand, a recent Nature Comment proposed an updated definition for green chemistry, where a circular economy is considered in a holistic perspective. The new definition is motivated because the classical concept has driven the performing of linear processes in chemistry, diminishing important aspects, such as remaining wastes. Therefore due to the complexity of the green philosophies, it is essential to consider a circular economy, including people, planet, and profit levels. Therefore the new concept for green chemistry must be addressed with economically viable applications. Then, having circular chemistry, to replace the common linear green chemistry (Keijer et al., 2019).

1.2.1 PRINCIPLES OF GREEN CHEMISTRY: TODAY AND TOMORROW

Green chemistry has an impact on different processes, and biorefinery is not the exception. In fact, according to Aristizábal-Marulanda and Cardona-Alzate (2019), the biorefinery is a complex system, where biomass is integrally processed or fractionated to obtain more than one product including bioenergy, biofuels, chemicals, and high value-added compounds that only can be extracted from bio-based sources. Therefore under the complex nature of the procedures, the principles of green chemistry are attached and must be applied, to control and improve biorefinery context.

- **Principle 1. It is better to prevent waste than to treat or clean up waste after it has been created. Prevention**

This principle refers to prevent contamination because of the chemical transformation process. Mainly pointing, the designs stage in a chemical transformation process, to prevent the least possible contamination, avoiding the passage of purification or disposal of hazardous waste, that increases the cost of the process. Principle 1 can be applied in different areas, here are some representative examples; implemented in different disciplines. Chemical transformation type "one-pot" or single-stage processes are an excellent alternative to avoid the use of waste. Specifically, these types of procedures are suitable in organic synthesis, achieving this type of strategy by designing catalysts, reducing the use of solvents, using reagents of natural origin. Singh and Chowdhury (2012) reported the study of solvent-free multicomponent reactions, demonstrating the importance of the design of these types of processes. Multicomponent reactions have important applications in reactions, such as Mannich, Ugi, Passerini, Gewald, and Pauson-Khand, to obtain organic compounds, precisely the compounds of the type heterocycle. Another advantage of multicomponent reaction is that it can be carried under the solvent-free conditions (Singh and Chowdhury, 2012). Szőllősi described another example of reaction one-pot in 2018. It was reported a study about heterogeneous chemical catalysis, considered sustainable processes. This condition makes efficient the one-pot reactions to obtain a stereoselective compound with at least one new chiral center, which has a significant application in biocatalysts (Szőllősi, 2018). Koutinas et al. (2014) reported the study of valorization of industrial waste and by-product streams via fermentation for the production of chemicals and biopolymers, concluding that biorefinery plays a vital role in industrial development, which by means of fermentations chemical compounds are obtained, highlighting the formation of outstanding C–C covalent bonds, to get important organic compounds as obtaining of alcohols such as ethanol, 1,3-propanediol, 2,3-butanediol, also of different types of organic acid such as propionic acid, lactic acid, 3-hydroxy propionic acid, succinic acid, malic acid, fumaric acid, butyric acid, among others different compounds (Koutinas et al., 2014).

• **Principle 2: Synthetic methods should be designed to maximize the incorporation of all materials used in the process into the final product. Atomic economic.**

The principle includes the evaluation of all reagents incorporation in a chemical transformation process, by their embodiment degree into the final product. The anatomic yield of 100% is achieved when all reagents are incorporated into the final product. In 2015, the analysis of biorefining processes of marine macroalgal for the production of biofuel and commodity chemicals demonstrated an integrated strategy that facilitates the sequential extraction of the main components of red algae biomass as primary products, such as colorants, lipids, minerals, and high energy density substrate. As a conclusion of this report, it is essential to remark that the main advantage of the process is the significant co-production of value-added by-products from feedstock (Baghel, et al., 2015). Mariscal et al., reported in 2016 a study on furfural, considering it a renewable and versatile organic molecule derived directly from biomass. In this report, the furfural is presented as useful for synthesizing various chemical products and organic fuels. Besides, finding that the analyzed reactions show zero toxic residues, mostly with reactions with the atomic economy and reactions one-pot, this due to its functionalities present in its chemical structure, important characteristics that make furfural a promising natural organic reactant (Mariscal et al., 2016).

- **Principle 3. Wherever practicable, synthetic methods should be designed to use and generate substances that possess little or no toxicity to human health and the environment. Less dangerous synthesis**

The synthesis of organic and inorganic compounds is a process that involucre various risks for the person who manipulates the process. Within green chemistry, this point is addressed to design green-sustainable processes. Fortunately, to the biorefinery Principle 3 is well adapted; most of the processes within the biorefinery are synthesis and obtaining when using renewable raw materials of low or toxicity for the person who carries out the process.

Recent research on sustainable polygeneration industrial process via fast pyrolysis of *Spiraea Japonica* combined with the Brayton cycle demonstrates that this type of transformation is safe, low or no danger. This study presents a design based on experimentation involving a fast reactor system fluid bed pyrolysis reactor and Bryton energy system in an, energy-wise, nearly self-sustainable system, considerably reducing the utilization of fossil fuel-derived utilities (Brigljević et al., 2019). In 2016, Morales et al. evaluated different sustainable technologies of succinic acid production from biomass through a metabolic engineering process. In this research two sustainable technologies were combined with metabolic engineering with the most studied technologies to produce biosuccinic acid from sugar beet and lignocellulose residues, finding that the appropriate process is assisted by metabolic engineering to obtain succinic acid (Morales et al., 2016).

- **Principle 4: Chemical products should be designed to affect their desired function while minimizing their toxicity. Designing safer chemicals**

One of the objectives of green chemistry is to design the products or chemical substances with a low or no toxicity, which does not represent a risk for the human begin and the environment, considering that a balance must be taken between the reactivity of the reactants and the toxicity of the final products. In biorefinery, this principle is generally accepted, because most of the reactants come from natural renewable sources and not represent a health risk, as far as the products obtained are usually compounds or chemical substances of low toxicity, the most common with an organic acid, alcohols, ethers, biopolymers, biofuels, etc. (Baghel et al., 2015; Shen, 2014; Brandt-Talbot et al., 2017; Van den Bosch et al., 2015; Esposito and Antonietti, 2015). This principle involves specialists in chemistry and toxicologists, who must analyze the process of chemical transformation and propose an alternative in the processes to change those reagents that are a disadvantage. Theoretical research is an excellent alternative to design safe chemical transformation processes. Solatia et al. reported the study of crude protein yield and theoretical extractable pure protein of potential biorefinery feedstocks, finding that the results can be used to propose and support a sustainable economic process. This process is useful to obtain proteins, involving different factors that affect the production of the raw material (Solatia et al., 2018). In 2018, it was published the theoretical study of the analysis of biorefineries using micro- and macro-algae feedstock, through a life cycle assessment (LCA). LCA is a tool to evaluate the potential environmental impact of a product, process, or activity throughout its life cycle. This type of method integrates green chemistry, biorefineries, low environmental impact technologies, future sustainable production chains of biofuels, and high-value chemicals from biomass. (Assacuete et al., 2018)

- **Principle 5: The use of auxiliary substances (e.g., solvents, separation agents, etc.) should be made unnecessary wherever possible and innocuous when used. Safe solvents**

The solvents are essential in a chemical transformation process, as a reaction medium, and in the purification process. Most organic solvents present a risk of human health and the environment, mainly because they are lipophilic and have variable volatility. Therefore one of the most critical areas of interest of green chemistry is the elimination, reduction, or replacement of nonhazardous solvents that have a limited environmental impact. Among the traditional solvents that meet these characteristics are the following: acetone, ethanol, methanol, isopropanol, hexane, acetic acid, and ethyl acetate. However, reactions have been developed that are carried out without solvents, free-solvents, or green solvents or alternative solvents like water (universal and nontoxic). Although, there are processes where water is not useful and have to be used nonpolar solvents, ionic liquids (ILs) such as quaternary nitrogen salts, supercritical fluids such as CO_2, biodegradable solvents, and mix green solvents. Xu et al. (2017) reported the study of pretreatment structural evolution and enzymatic hydrolysis of eucalyptus used recycled ILs for low-cost biorefinery, ILs used in this study were 1-allyl-3- methylimidazolium chloride ([amim]Cl) and 1-butyl-3-methylimidazolium acetate ([bmim]OAc), concluding that the ILs presented a good efficiency to be recycled up to 3 times, which leads to conclude that it is a green, economic, and sustainable process. Lê et al. (2018) published a study on the process of chemical recovery of a de γ-valerolactone/water biorefinery optimized the process to reduce the proportion of wood pulp, as a matter of producing up to 3 L/kg while maintaining the properties of wood pulp in order to produce textile fibers through sustainable technologies, a nontoxic mixture based on water with lignin was involved in this chemical transformation process, acting as a solvent, in addition to involving the process a distillation under reduced pressure and the extraction of liquid CO2.

- **Principle 6: Energy requirements of chemical processes should be recognized for their environmental and economic impacts and should be minimized. If possible, synthetic methods should be conducted at ambient temperature and pressure. Energy efficiency**

Energy is a topic of great interest at the laboratory, pilot, and industrial level, due to the environmental effect it presents. Currently, countless studies for collecting, conserving, and generating different types of energy are used. In chemical transformation processes, heat energy is usually used and provided by an electrical heating source. However, it is one of the most expensive. Nonetheless, there is a significant advance in this aspect, the use of emerging energies, such as ultrasound, microwave, UV radiation, infrared, solar radiation, and the combination of these. Thus to generate energy that can help to take the process through faster chemical transformation, each of these types of energies, has advantages and disadvantages, in terms of cost, operation, and requirements of raw materials, should be affordable. Here some examples are exposed that support the use of alternative energies in the biorefinery process. In 2017, Tabah et al. reported an interesting study on converting solar-energy-driven biomass to bioethanol, considering a sustainable process and highlighting that bioethanol is an attractive and promising fuel. Through this methodology of solar-energy-driven biomass, the reported research demonstrates that a better energy-efficient economy and sustainable alternative to the cutting-edge process to produce bioethanol from biomass is achieved (Tabah et al., 2017). On the other hand, Ranjan et al. reported a study on ultrasound-assisted bio-alcohol synthesis, highlighting that ultrasound is a viable alternative that intensifies various physical/chemical/biological processes, and bio-alcohol synthesis is no exception. They conclude that this type of operation is feasible as it is scaled at the industrial level for its commercialization (Ranjan et al., 2016).

• **Principle 7: A raw material or feedstock should be renewable rather than depleting whenever technically and economically practicable. Use of renewable raw materials**

The use of renewable materials is currently a viable alternative to carry out some chemical transformation process and is more attractive when that raw material is a waste of little value, this is due to the excessive use of nonrenewable or limited resources. The biorefinery is an area that supports this principle. There are innumerable reports about studies where renewable raw materials were obtained from organic raw material residues such as cellulose and derivatives, chitosan, glucose ant its derivatives, etc. (Brandt-Talbot et al., 2017; Van den Bosch et al., 2015; Gillet et al., 2017; Cárdenas-Fernández et al., 2017; Liu and Li 2017; Dai et al., 2017)

Wahlström et al. reported the study of lignocellulosic enzymatic hydrolysis of polysaccharides in the presence of IL, where the raw material is of the natural original, verifying that certain biofuels and chemical products can be produced from lignocellulosic biomass, through a sustainable process and even more if green solvents are involved, such as ILs acting as a viable alternative in enzymatic hydrolysis within

the process of obtaining biofuels and chemicals (Wahlström and Suurnäkki, 2015). Other research works related to cellulose were recently reported by Shaghaleh et al., highlighting that cellulose is the natural raw material that contains the most biological products. Currently, obtaining biopolymers bases on cellulose derivatives has been reported, it is the report that exposes different routes for the production and obtaining of biopolymers to cellulose base, the uses of cellulose and nanocellulose fibers. The use of cellulose-derived monomers (glucose and other chemicals) in the synthesis of sustainable cellulose-derived biopolymer, and functional polymeric materials not only provide viable replacements for most petroleum-based polymers. Emphasizing that sustainable cellulosic biopolymers have potential applications not only when replacing synthetic polymers from petroleum, they also have varied applications in electrochemical and energy stronger devices to biomedical applications (Shaghaleh et al., 2018). The foregoing check proves the importance of the use of renewable materials where biorefinery 100% supports this principle by using mostly renewable raw materials of natural origin.

• **Principle 8: Unnecessary derivatization (use of blocking groups, protection/deprotection, temporary modification** of physical/chemical processes) should be minimized or avoided if possible, because such steps require additional reagents and can generate waste. Derivative reduction

In any chemical transformation process, the formation of by-chemical products or unwanted products that affect the percentage of the final product can be presented. A strategy to minimize this problem is to use auxiliary or blocking materials at the end of the process, it is necessary to separate them from the final product, needing another stage of purification to it. For this reason, it is necessary to have a good process design (Principle 4) to avoid as much as possible the use of this type of reactants. Chen et al. reported a detailed study on the transformation of shrimp chitin to low molecular weight chitosan with 90% purity, this process was base-catalyzed, one-step mechanochemical, through this methodology no toxic residues are obtained, concluding that this via is an excellent alternative for obtaining low molecular weight chitosan, carrying out the process directly without using intermediate reagents during the process, is highlighted the importance of chitosan because it is the second most abundant biopolymer in nature, for this reason it is important to study it, chitosan has various physicochemical properties, including solubility which depends on molecular weight, it has been shown that molecular weight is an essential factor involved in its solubility, for that reason, it is of interest to obtain low molecular weight chitosan (Chen et al., 2017).

- **Principle 9: Catalytic reagents are superior to stoichiometric reagents. Catalysis**

Green chemistry proposes that when working within a chemical process, an adequate stoichiometric analysis between the reactants must be considered to avoid the production of the unwanted product. However, in most of the chemical processes there is a limiting reactant, which drives and intervenes in the kinetics of the process, in an ideal process all reactant reactions do not produce by-products. Nonetheless, not all processes obey this behavior. For this reason, the use of selective catalysts is applied to solve this problem. In a chemical transformation process, the uses of catalysis are significant due to the multiple advantages that it presents, from the time of the process to the percentage of yields of the final product, without forgetting the purity of the final product. This principle suggests the use of reusable selective catalysts rather than stoichiometric reagents, in the biorefinery several publications support this principle, and it is associated with Principle 4, in designing a sustainable process. Kuznetsov et al. (2018) reported green catalytic processing of native and organosolv lignins, where explain two routes of catalytic depolymerization of native wood lignin and isolated wood lignin for the designation of wood peroxide, TiO_2 catalyst was used in the medium acetic acid-water with mil condition ($\leq$100 °C, atmospheric pressure) and the thermal dissolution of organosolv lignins (ethanol-lingini and acetone-lignin) in supercritical alcohols (ethanol and butanol) on solid catalysts containing Ni. Finding the case of TiO_2 catalysts, with rutile phase, it was possible to obtain cellulosic products with a lower residual lignin content and a higher cellulose content, found for the phase identified as anatase, in the case of catalysts with Ni, found that the influence is notorious, reducing, the content of phenol and benzene derivatives and increasing the content of esters, aldehydes and ketones in hexane-soluble products obtained from organosolv-lignins, as well the as the greater conversion of lignin into supercritical alcohols 93% by weight (Kuznetsov et al., 2018).

• **Principle 10: Chemical products should be designed so that at the end of their function they break down into innocuous degradation products and do not persist in the environment. Clean degradation**

In any chemical transformation process, it is important to design the final product, analyzing each of its stages, the selection of the reagents, chemical nature, toxicity, reactivity, as well as the reaction conditions, solvents used, type of sources of energies, use of catalysts, purification processes if required, and each of the stages must be as sustainable as possible. This principle refers to having a clean degradation, once the material or product covers the need for which it was created, it should not be toxic to the environment when it is in disuse, it is of great importance that it has a time of degradation when it is already available. Usually, materials that come from natural origin have a faster degree of degradation compared to materials of synthetic origin. In this specific case, the biorefinery plays an important role in obtaining a compound, materials, substances, and chemical products, since most of its raw materials are of natural origin (Maity, 2015).

• **Principle 11: Analytical methodologies need to be further developed to allow for real-time, in-process monitoring, and controlling the formation of hazardous substances. Continuous contamination analysis**

Green chemistry studies all the factors involved in a chemical transformation process, considering the contamination that can be generated when the process is carried out. It is necessary to continue the analysis of contamination during the process in real time to monitor the stage where the generation of contamination begins and, in this way, to reduce the problem, here participle analytic chemists. Regarding this principle within the biorefinery, there are reported studies on the process analysis for obtaining and extracting compounds and chemical substance from biomass. Alzate et al. (2018) reported the results of the fermentation process, thermochemical and catalytic processes in the transformation of biomass through efficient biorefineries demonstrating that the biomass contains high percentages of cellulose and derivatives such as lignocellulosic and from this raw material through different extraction and transformation processes numerous compounds of interest are obtained. The extraction and transformation processes are usually assisted by catalysts or by alternative energy sources that help to minimize the times of obtaining, in this particular case it is concluded that catalytic processes are a viable alternative for obtaining compounds of interest from biomass, proposing that heterogeneous catalysis can represent a techno-economically viable route to be analyzed (Alzate et al., 2018).

- **Principle 12: Substances and the form of a substance used in chemical processes should be chosen to minimize the potential for chemical accidents, including releases, explosions, and fires. Intrinsic safety and accident prevention**

The prevention of accident in chemical transformation processes is never exaggerated, at all stages of the process. The hazard characteristic of chemical substances and compounds, such as toxicity, explosiveness, flammability, irritability, and others, must be considered, from the stage of the design of chemicals and processes (Principles 1 and 4). One of the objectives of green chemistry is to include all the risks of substances and compounds in chemical processes, also considering pollution. The design of a process must present a detailed analysis of accident prevention and pollution control. In some cases, the solvents are the ones that present thermo insignificant risk when the temperature or the system pressure increases, so it is crucial to use green solvents that have low or no toxicity (Principle 5).

On the other hand, there are 12 modified principles, based on a circular economy, and a holistic approach, under the concept of Circular Chemistry (Keijer et al., 2019). These new principles are the base for the green chemistry movement forward, through the sustainability of worldwide demands. Therefore it is expected that innovative industries work in the design of policies and processes that include the current modifications.

1. **Principle 1. Collect and use waste.** Waste is a valuable resource that should be transformed into marketable products.
2. **Principle 2. Maximize atom circulation.** Circular processes should aim to maximize the utility of all atoms in existing molecules.
3. **Principle 3. Optimize resource efficiency.** Resource conservation should be targeted, promoting reuse and preserving finite feedstocks.
4. **Principle 4. Strive for energy persistence.** Energy efficiency should be maximized.
5. **Principle 5. Enhance process efficiency.** Innovations should continuously improve in- and post-process reuse and recycling, preferably on-site.
6. **Principle 6. No out-of-plant toxicity.** Chemical processes should not release any toxic compounds into the environment.
7. **Principle 7. Target optimal design.** The Design should be based on the highest end-of-life options, accounting for separation, purification, and degradation.
8. **Principle 8. Assess sustainability.** Environmental assessments (typified by the Life Cycle Assessment) should become prevalent to identify inefficiencies in chemical processes.
9. **Principle 9. Apply ladder of circularity.** The end of life options for a product should

strive for the highest possibilities on the ladder of circularity.
10. **Principle 10. Sell service, not product.** Producers should employ service-based business models such as chemical leasing, promoting efficiency over production rate.
11. **Principle 11. Reject lock-in.** Business and regulatory environment should be flexible to allow the implementation of innovations.
12. **Principle 12. Unify industry and provide a coherent policy framework.** The industry and policy should be unified to create an optimal environment to enable circularity in chemical processes.

1.2.2 GREEN CHEMISTRY METRICS

The measure is the only way to get qualitative and quantitative data to manage progress and efficiency, through a verifiable parameter. For its part, a metric is a quantifiable measure that includes variables, according to the phenomena been studied. Therefore considering the holistic nature of green chemistry philosophy, and the new improvements regarding circular chemistry concept, it is crucial to design proper metrics. The lack of metrics is limiting the adoption of the green chemistry, worldwide. However, in the effort to measure the advances and impacts, the Green Chemistry and Commerce Council (GC3) propose five possible levels to measure green chemistry: the molecular level, the product/material level, the firm level, the sector level, and the societal level (Blake, 2015).

The GC3 found that the molecular level is the most developed and applicable for Green Chemistry design processes due to its link with the 12 principles. In this level, the most common metrics are carbon footprint, CO_2 production, water usage, and life cycle impacts, E-factor (Total mass of weight/mass of final product), and atom economy (molecular weight of product/sum of molecular weight of reactants) × 100) (Blake, 2015). Some other mass-based metric can be divided into two, the first group representing a percentage (%) of the ideal analogous to atom economy and the second group based on kg/kg analogous to the E factor (Sheldon, 2017).

1. Group 1, analogous to the E factor: Mass intensity (MI), process mass intensity (PMI), waste water intensity (WWI), solvent intensity (SI)
2. Group 2, analogous to atom economy: Reaction mass efficiency (RME), mass productivity (MP), effective mass yield (EMY), and carbon economy (CE).

For its part, there are some companies applying metrics at the product/material level, considering safer ingredients and supplies. Approaches and tools include the use of selected green chemistry principles as measures for specific products and processes to evaluate movement away from chemicals of concern (Blake, 2015). An example is the US EPA Design for Environment's Safer Chemical List or the Cradle-to-Cradle Material Assessment.

On the other hand, for the firm and sector level metrics, some tools have been developed. There is a checklist designed for business by the Michigan Green Chemistry Roundtable (2014). The list includes four areas: support and communication, design and innovation, education, and hiring. Its purpose is to help into the progress monitoring toward green chemistry implementation. For apart, regarding sector level, there is the Blue Finder, for the textile sector.

Finally, the societal level is directed to the human health and environmental monitoring. The tracking tools limit metrics for these levels. Nevertheless, there are some platforms and programs developing approaches and tools into the green chemistry application. That is the case of the United Nations Environment Program, and the Centers for Disease Control and Prevention (CDC).

1.3 SUSTAINABILITY AND GREEN CHEMISTRY

Clean and continuous processes are desirable for industry; they hold as a priority the residue minimizing principle and the constant optimization needs. The achievement of optimum conditions to meet these purposes can be named sustainability. Sustainability has been defined as the capacity to satisfy the current generation needs, without compromising the growth and development of future generations (WCED, S. W. S, 1987). Under this direction, green chemistry is the tool that contains the necessary principles to achieve sustainability, in an industrialized world. Therefore the innovative design of processes must consider the green chemistry as the bedrock of sustainability.

On this concern, many efforts have been made by academic instances, industries, and government. There are products and processes derived from green chemistry application, considered as sustainable. For example, in the agricultural field, a sustainable strategy for pest control is the use of botanical insecticides, obtained by the encapsulation of essential oils (Campos et al., 2019). For medical uses, it has been proposed the production of vitamin K by its extraction from microalgae biomass, with the application of a green extraction method (Tarento et al., 2019). In the aeronautical research, the corrosion of stain steel is treated with citric acid from fermentation, instead of nitric acid from synthetic chemistry, to make the passivation process, as sustainable as possible (Parsons et al., 2019). Most of the industrial fields are developing sustainable processes, to contribute to the green revolution. The natural sources, such as plants, microalgae, bacteria, fungi and organic by-products, are the most suitable substrates for biomolecules production, in the green sustainable chemistry criteria. However, based on the raw material renewable principle for green chemistry, it is essential to consider the full productivity potential from each source. The use of residual biomass for production is an important tool to achieve sustainability through green chemistry. In this case, the biorefinery production system is the best option to accomplish the full potential from a natural raw material, through a circular model of production, where biomass is integrally processed or fractionated to obtain more than one product.

1.4 BIOREFINERY CONCEPT AND CLASSIFICATION

According to the IEA Bioenergy Task 42, a biorefinery is defined as the sustainable processing of biomass into a spectrum of marketable products and energy (IEA, Task

42; 2009). Nevertheless, it is essential to mention in concordance with Aristizabal-Marulanda and Cardona-Alzate (2019) that biorefinery is a complex system that needs a comprehensive study of the raw materials to be used, and the creation of a sustainable design based on the latest state-of-the-art technologies and approaches, to reach the main goals of the biorefineries, maximizing the value of the products obtained from the biomass, to increase competitiveness and prosperity in industry, to reduce the dependence of many countries on fossil fuels, to reduce the emission of greenhouse gases, and to stimulate regional and rural development (Moncada et al., 2016). Biorefinery is defined in a graphical way in Figure 1.1.

FIGURE 1.1 Graphic representation of biorefinery concept. *Modification from Kumar et al., 2020.*

On the other hand, in regard to the kind of biorefinery systems, there is a common classification based on four different criteria (VDI, 2016): (1) platform, referring to links between feedstocks and products, the most important platforms in biorefineries are Syngas, biogas, C6 sugar, and C5/C6 sugars, plant-based oil and algae oil, organic solutions, lignin,

and pyrolysis oil; (2) feedstocks, referring to the raw material and starting point of a biorefinery, hence their provision should be renewable, consistent, and regular; (3) bioproducts or bio-based products, chemicals and energy derived from renewable biological resources; and (4) conversion processes, mainly considered and described in four groups: thermochemical, biochemical, chemical, and mechanical. Under this classification, there are many configurations for a biorefinery. Some examples are (1) based on platform classification, there is the pyrolysis oil biorefinery, which represents a source of organic acids, phenol and furfural; (2) based on the conversion process, there is the biochemical biorefinery where fermentation is used to transform a substrate into different compounds by the action of specific microorganisms; (3) based on products, it can be mentioned the bioethanol biorefinery; and (4) based on feedstocks, the biorefinery of agricultural wastes is into this classification. Figure 1.2 is a graphical example of biorefinery classification.

1.4.1 BIOREFINERY AS AN ENGINE FOR GREEN CHEMISTRY

The waste and by-products accumulation of processes and economic activities must be a major concern for governments, industries, academia, and society. Even more, when there

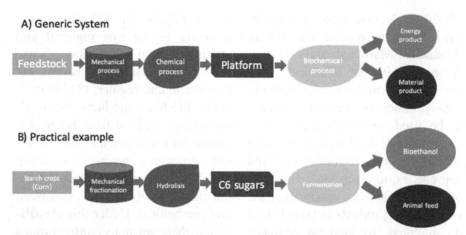

FIGURE 1.2   Graphical representation of biorefinery classification and a practical example. *Modification from Cherubini et al., 2009.*

are agricultural, municipal, industrial, forestry and animal wastes accumulation that need to be diminished. Until now, research communities and industries have applied a linear model to converse primary resources to final products, following the green chemistry principles. This mode is effective in making greener products and processes, but still need to be improved to achieve sustainability. Therefore a circular model where, remain biomass, waste and by-products can be used as a source for new products is and will be "the virtuous chain" of production.

The biorefinery is defined as a process consisting of the total valorization of a natural resource, enabling the production of a wide range of products from a unique raw material source. Under this concept, every waste is considered as a new resource that can be extracted another time

(Vernès et al., 2019). Even though, the solei concept of biorefinery strives in the fact that there is no undesirable material accumulation, the success in the application is only due to the correct use of green chemistry principles. Therefore biorefinery can be considered as the engine for green chemistry through cleaner technologies, where the result is the sustainability of processes and products (Figure 1.3).

The pursuit of zero waste processes is the objective for biorefineries. The main biorefinery technologies can be classified as extraction, biochemical, and thermochemical processes (Clark et al., 2012). On this concern, the extraction processes are, in most cases, the central point for high yields during biomolecules production, and at the same time, the point where the most wastes and by-products are obtained. Therefore the innovative

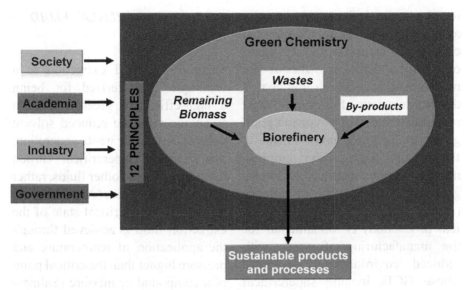

FIGURE 1.3 Merging green chemistry and biorefinery.

design of green extraction processes is relevant to biorefineries

Regarding this, there are a set of green extraction processes that have shown effectiveness. An example is the valorization of soybean residues by flavonoids extraction, using supercritical fluid extraction technology (Alvarez et al., 2019). Or the valorization of the main coproduct of the chocolate industry, cacao pod husk, using supercritical fluid extraction to obtain phenolic compounds with antioxidant activity (Valdez-Carmona et al., 2018).

On the other hand, bioprospecting is a tool where genetic information is used to search high-valuable molecules in nature. This tool is a step before of extraction processes and full of biochemical information, to decide suitable conditions in the design of green extraction technologies. Hence, it is a valuable application for biorefineries.

Consequently, the junction of innovative green technologies with the green chemistry principles into biorefineries is conducting the sustainability improvement of processes and products, consolidating the green chemistry and biorefinery.

1.5 GREEN EXTRACTION PROCESSES AS TOOLS FOR BIOREFINERY

The constant demand of materials for productive activities has reached a critical point in a consumption-controlled market (Clark, 2007). One of the main current challenges is to achieve the shift toward a sustainable society that promotes the efficient use of natural resources, based on realistic demands. One of the measures to accomplish this goal is the reassessment of chemical-based manufacturing of consumption goods to use more environmentally friendly processes. The employment of green chemical technologies (GCTs) within the biorefinery (from drying, to extraction, to chemical production) is fundamental for the manufacturing of goods with reduced environmental footprint. These GCTs include supercritical fluid extraction (SFE), microwave assisted-extraction (MAE), clean catalysis and synthesis, pressurized liquid extraction, ultrasound-assisted extraction (UAE), and pressing (Clark et al., 2006; Gallego et al., 2018; Chemat et al., 2019; Rombaut et al., 2014).

Natural sources, such as plant biomass, marine organisms (seaweeds and microalgae), foods, and agricultural by-products and wastes are the usual raw materials of biorefineries (Herrero and Ibañez, 2017). The goals of green extraction implementation are to obtain the highest productivity with reduced use of energy while maximizing efficiency and reducing the release residues that are harmful to human health and the environment (Esquivel-Hernandez et al., 2017).

1.5.1 SUPER CRITICAL FLUID EXTRACTION

Supercritical fluid extraction is a technique characterized for being a high-efficiency green extraction technique with the reduced solvent expense and lower waste generation, that employs supercritical carbon dioxide (CO_2) and other fluids, rather than organic solvents (Armenta et al., 2019). The supercritical state of the extraction fluid is achieved through the application of temperature and pressure higher than the critical point of a compound or mixture (Salinas-Salazar et al., 2019). In these conditions, the fluid has a viscosity similar to a gas, whereas maintaining a density of a liquid (Gallego et al., 2018). The combination of relatively high diffusivity and low viscosity enhances the extraction properties of the fluid, allowing shorter extraction periods, due to the penetration of the fluid throughout solid matrices. By modifying pressure and temperature, the density of a fluid can be modified, and in turn, changes its solvent strength (da Silva et al., 2016)

The main parameters to consider while designing an SFE process are temperature, pressure, solvent, use of modifiers, time of extraction, solvent flow rate, particle size, sample amount, and solvent mass ratio (Sharif et al., 2014; De Melo et al., 2014a, b). Design of experiments is a tool that enables the methodic optimization of the variables involved in a process, discriminating between significant and nonsignificant factors. The choice of an experimental design depends on the aim of the study, feasibility, time, among other related factors (da Silva et al., 2016). Experimental designs are categorized in screening designs or optimization designs. For screening purposes, full factorial, fractional factorial, and Plackett–Burman can be applied.

On the other hand, optimization processes start with a screening stage to discover the optimal parameter combination, and then the optimization can be performed through Taguchi method, central composite design (CCD), and Box-Behnken design (da Silva et al., 2016; De Melo et al. 2014; Venskutonis and Jonušaite, 2014).

The advantages of carbon dioxide SFE (SFE-CO$_2$) include the low critical temperature and pressure of CO$_2$ (30.9 °C and 73.9 bar, respectively) is an eco-friendly process that can be applied for the processing of food ingredients since the Food and Drug Administration considers CO$_2$ as generally recognized as safe (Esquivel-Hernández et al., 2016).

Furthermore, CO$_2$ separates easily from the sample after depressurization without leaving residues. This property can represent an economic advantage, since it is possible to recycle the CO$_2$ spent in the process, given the appropriate equipment setup

(Salinas-Salazar et al., 2019). Extraction of high-valuable products through SFE has been reported, like lipids, vitamins, and pigments (Esquivel-Hernández et al., 2016; Kerton et al., 2013; Esquivel-Hernández et al., 2017).

1.5.2 PRESSING

Pressing is a mechanical extraction method that consists of applying pressure over the material subject to be extracted (Rombaut et al., 2014). Within this technique, there are membrane presses, screw presses, and twin-screw extruders. The expression refers to the emanation of a fluid from a porous material, by means compression (Schwartzberg, 1997).

This technology can be improved through modification of the pressure applied, strain, the configuration of the press and pistons. Centrifugation can be used as a previous operation to separate liquids from solids, followed by porous silica filtration and concentration by evaporation (Salinas-Salazar et al., 2019).

Shear-based pressing (e.g., French and Hughes presses) forces a biomass solution or slurry through a small opening by applying pressure. This process has been reported to recover from 70% to 75% of oil, and the addition of mechanical crushing coupled with chemical pretreatment improves product recovery. The disadvantages of this technique are the high cost of maintenance and lower efficiency against other methods (Ranjith Kumar et al., 2015).

Among the products extracted through pressing we can mention lipids, proteins, and pigments, like astaxanthin and phycocyanin from microalgae biomass (Grimi et al., 2014; Moraes et al., 2010; Olaizola, 2000; Karemore and Sen, 2016)

1.5.3 INSTANT CONTROLLED PRESSURE DROP

The instant controlled pressure drop is regarded as a high-temperature short-time technology (HTST) and relies on the transformation of matter by mechanical and thermal means (Rombaut et al., 2014). In this technique, a damp product (30% dry basis) is set in an extraction vessel at atmospheric pressure. Then the pressure is lowered until near-vacuum and steam is introduced to increase the pressure to process level (< 1 MPa, 180 °C). After a set exposure time (< 60 s), an instant pressure drop valve is opened, causing a sudden drop pressure (< 5 KPa) in the vessel. Next, the steam is purged from the system to return to atmospheric pressure. Last, the mixture of the contents (extract) and water from condensation is recovered (Pingret et al., 2012). Sample

depressurization induces autovaporization of water and volatile molecules, instant cooling, and swelling that can be followed by cell rupture and secretion of contents (Allaf et al., 2012).

1.5.4 ULTRASOUND-ASSISTED EXTRACTION

Ultrasound-based extraction methods use different pressure cycles to mechanically disrupt cells by cavitation. In the low-pressure cycle, small vacuum bubbles are formed within the liquid, while on the high-pressure cycle, when the bubbles grow into a greater size, they burst, producing a high-pressure and high-speed zone in the liquid, with a resulting shear force that induces the breakage of cell membranes (Wei et al., 2008). Also, the produced high pressure, favors the diffusion of solvents into the biomass cell walls, aiding the extraction process (Cravotto et al., 2008).

In UAE, the modification of polar:nonpolar solvent ratio and the increase of time are the main factors to consider when designing a UAE process. Additionally, this technique allows the use of low temperature for the extraction, an advantage for the processing of thermally sensitive compounds. Because this technique is based on the release of cellular contents into the extraction solution, through the enhanced mass transfer and solvent penetration within the cell, it has been extensively used as a pretreatment operation (Ghasemi Naghdi et al., 2016).

1.5.5 *MICROWAVE-ASSISTED EXTRACTION*

The use of microwaves for the extraction of molecules of interest is based on the contact of a high oscillating electric field dielectric with a dielectric polar substance, such as water. This interaction generates heat because of the friction produced by inter- and intramolecular movements. Vapor is produced therefore, which in turn produces the rupture and spillage of intracellular contents, in a process known as electroporation. For the design of an MAE process extraction, time, temperature, solution dielectric properties, concentration and type of solvent and solid–liquid ratio should be considered (Ghasemi Naghdi et al., 2016). Extraction with microwaves is regarded as a convenient method since it does not require the previous removal of moisture from the sample. These methods have been regarded as economic, rapid, safe, simple, and effective for the recovery of valuable products, including oils from microalgae biomass or the transesterification of lipids to biodiesel (Ranjith Kumar et al., 2015; Šoštarič et al., 2012).

1.6 OMICS AND BIOINFORMATICS AS PREDICTIVE TOOLS FOR BIOREFINERY PRODUCTS

Biorefineries transform the residual biomass into useful products using a combination of technologies and processes (Sandun et al., 2006). There are different methods used in biorefineries: (1) thermal chemical methods, (2) chemical methods, and (3) biocatalytic methods (Cherubini, 2010). Biocatalysis encompasses whole-organism processes, such as fermentation, isolated enzymes used in chemical reactions, and chemically or genetically modified enzymes. Biocatalysis is a key step in biorefinery processes due to the high efficiency and selectivity of catalysis. However, many of these processes require enzymes that are stable at high temperatures to allow easy mixing, better substrate solubility, high mass transfer rate, and lower risk of contamination (Turner et al., 2007). These enzymes perform a dual role in the biorefinery systems that to generate metabolic building blocks for further conversion and to achieve specific catalysts in the conversion of building blocks into the desired products (Turner et al., 2007).

According to a report from the Business Communications Company Inc., the global market for industrial enzymes was estimated to reach

US$5.6 billion in 2018. Food and beverages constitute 26% that is equivalent to US$1.4 billion in 2017, followed by biofuels and detergents with 18% (US$969.3 million) and 14% (US$754.4 million) respectively in the same year. Industrial enzymes in biofuels are expected to grow 73% by 2024 (BBC, 2018).

Enzymes used in the industrial process should comply with the following industrial criteria: harsh and broad reaction conditions such as high substrate load, a broad range of temperatures, a broad range of pHs, water-deficient reaction conditions, very high solvent concentrations and process stability, and the high stereoselectivity and turnover rates (Spickermann et al., 2014; Zuhse et al., 2015; Singh, 2010). Hydrolases are the most common enzymes used in biorefineries based on biocatalytic methods, as they usually do not require cofactors (Saai Anugraha et al., 2016). There are three general classes of the enzyme that are essential to hydrolyze plant cell walls: cellulases, hemicellulases, and accessory enzymes (Himmel et al., 2007). The latter is an enzyme cocktail of hydrolytic enzymes which allow achieving cellulose hydrolysis more efficiently. Cocktail formulations require to be standardized to each biorefinery process (Jung et al., 2015; Hu et al., 2014; Harris et al., 2010). The most usual enzymes used are thermostable since high temperatures often promote better enzyme

penetration and cell-wall disorganization of the raw materials (Paes and O'Donohue, 2006).

Currently, enzymes used in biorefineries cannot meet the growing demand due to their high cost, low activity, and poor stability under the required operating conditions (Barnard et al., 2010; Fang et al., 2009). Thus the development of novel enzymes for the biorefinery industry is required. The traditional methods to discover novel enzymes is growing and screening microbial strains approach (Ogawa and Shmizu, 1999). Collections of microbial strains from different environments are used as a source of material to isolate enzymes. However, the culturable microorganisms represent a small portion of the total microbial biodiversity (~0.1%) (Alain and Querellou, 2009), and limit the potential discovery of novel enzymes, which could improve the biorefinery industry. However, with the advent of new technologies, the omics approach facilitates the discovery of enzymes with unpredictable sequences and helps to extend our knowledge about enzyme diversity that may have potential in the industry.

1.6.1 *OMICS STRATEGIES FOR METABOLITE AND ENZYME BIOPROSPECTING*

Bioprospecting, a systematic and organized search for useful products derived from bioresources, has led to increased interest in potential biotechnological applications (Pyne et al., 2016). Three novel sources for enzyme bioprospecting are metagenomics, metatranscriptomic, and metaproteomics. Metagenomics and metatranscriptomics apply a suite of genomic technologies and bioinformatics tools to directly access the genetic content from environmental samples, providing access to the functional gene composition of microbial communities and thus gives a broader description of the genome without the technically challenging culture of individual species (Lee et al., 2010), whereas metaproteomics is the large-scale identification and quantification of protein from environmental samples (Kleiner, 2019).

Metagenomics involves two basic approaches for screening biomolecules from environmental samples: function-based and sequence-based screening of metagenomic libraries (Simon and Daniel, 2011). Three different types of function-based approaches have been employed for screening of metagenomics libraries: (1) direct detection of specific phenotypes of individual clones, (2) heterologous complementation of host strains or mutants, and (3) induced gene expression (Simon and Daniel, 2009). The heterologous complementation of host strains or mutants involves the use of complemented host strains that require the presence

of the target gene for growth under selective conditions. The third type of functional screening is induced gene expression, which is a high-throughput screening that contains a reporter gene (green fluorescent protein, GFP). The approach is based on the production of GFP by the presence of a specific metabolite (Madhavan et al., 2017). However, this method is useful only for those substrates that can migrate to the cytoplasm (Yun and Ryu, 2005). All functional approaches do not depend on the availability of prior sequence information to detect enzymes, and therefore offers great potential to discover novel gene.

Sequence-based metagenomics approach is based on targeting genes using an oligonucleotide primer or probe using the colony hybridization technique. The target gene could be amplified by PCR with specific primers and cloned into an expression vector. Target gene is sequenced through a sequencing platform (e.g., 454 genome sequencer from Roche Applied Science, SOLiD platform from Applied Biosystems, the Illumina Genome Analyzer from Illumina and Ion Torrent/Ion proton platform) (Loman et al., 2012) followed by bioinformatics analysis. This strategy allows the identification of novel sequences; however, some of those may have similarity to the available preexisting sequences (Lee et al., 2010).

Metatranscriptomics and proteomics are the breakthroughs of the next-generation sequence technologies since they provide information about the functional gene from a microbial community. Metatranscriptomics is the global gene expression in the microbial community of environmental samples under certain environmental conditions (Aguiar-Pulido et al., 2016). This strategy is based on the construction of cDNA libraries from RNA. One of the challenging steps of the technique is the isolation of high quality of RNA and the removal of rRNA to enrich the mRNA for cDNA library construction (McGrath et al., 2008). In contrast to metagenomics, metatranscriptomics provides an increased number of base pairs sequenced per run since it is only focused on the expression profile of the sample. This strategy has been useful to identify unknown transcripts from complex environments such as marine and soil microbial communities (McGrath et al., 2008).

Metaproteomics allows the study of the presence and abundance of proteins in any environmental sample. The technique is based on the recovery of proteins from a sample, for example, microbial community, under certain conditions and followed by high-resolution mass spectrometry and bioinformatics analyses. Shotgun 2D-LC/MS proteomics is now the dominating

technique because of the increasing availability of genomic sequence data and development of 2D liquid chromatography and advanced mass-spectrometry (Speda et al., 2017). Proteins provide structure and activity to cells, knowing their abundances provide a cellular phenotype on the molecular level, and provides insights of their importance under a given condition (Kleiner, 2019). Discovery of novel enzymes can be achieved using activity-based probes (ABP) by metaproteomic analysis. Addition of ABPs permits systematic quantitation of individual classes of proteins that may be in lower abundance than the detectable limit and therefore missed by mass spectrometry (Mayers et al., 2017). ABPs probes are available for several enzyme classes (e.g., hydrolases, proteases, kinases, phosphatases, and glycosidases) (Cravatt et al., 2008).

1.6.2 APPLICATIONS OF METAGENOMICS, METATRANSCRIPTOMICS, AND METAPROTEOMICS FOR NOVEL ENZYME DISCOVERY

A review from the studies published over the last two decades using functional-based metagenomic approach from several habitats, such as terrestrial, marine, and freshwater habitats; nonmarine saline and alkaline lakes, acid mine drainage systems, wastewater treatment sludges, compost, and eukaryotic-associated microbiomes, revealed 5827 described enzymes, including 4034 esterase-lipase, 861 glycosidases, 859 oxid-reductase, and 73 others, such as protease, glycosyltransferase, amidase, nitrilase, phosphatase, trehalose synthase, and dehalogenase, whereas sequence-based approach revealed 211 sequences, composed by 96 esterase-lipase, 77 glycosidases, 1 oxid-reductase, and 37 represented by other enzymes, such as dehalogenase (Wang et al., 2019; Ferrer et al., 2016; Montella et al., 2016). Warnecke et al. (2007) performed the first large-scale metagenomic analysis of a wood-feeding higher termite, leading to the identification of 700 domains of different glycoside hydrolase (GH)-encoding genes, more than 100 genes related to lignocellulose hydrolysis, with (putative) pentose-acting enzymes from families GH10, GH11, GH26, GH43, and GH51. Comparative analysis from 19 metagenomic analysis searching for glycosyl hydrolases from different environmental samples reveals that the GH3 family is the most abundant in all environmental samples among all 133 GHase, followed by GH13, GH2, GH5, and GH29 (Tiwari et al., 2018). However, the sugarcane bagasse samples appear to be a good source of GHases with a rare property,

such as glucanases and xylanases (Kanokratana et al., 2013; Alvarez, 2013). Also, a deep sequencing from cow rumen revealed 27,755 putative glycoside hydrolase-encoding genes (Hess et al., 2011). The Metagenome of the gut of a wood-boring beetle *Anoplophora glabripennis* revealed bacterial enzymes involved in lignin degradation such as DyP-type peroxidase, copper oxidase, β-etherases, glutathione-*S*-transferases, and aldo-keto reductases (Scully et al., 2013).

A functional-based metagenomics approach is an attractive approach because it has the potential to study the unknown enzymes and guarantees that selected clones will harbor functional proteins, thus reducing the volume of high-throughput sequencing (Li, 2009). This strategy has been used to search for hemicellulases in microbiomes from wood, compost, rumen, and soil (Nacke et al., 2012; Li et al., 2011) using chromogenic substrates. However, low hit rates are produced due to technical problems of the libraries such as improper codon usage and/ or promoter recognition, inclusion body formation, the toxicity of the gene product or inability of the host to induce gene expression (Ekkers et al., 2012). To overcome this issue, high-throughput methods for rapid screening of cosmid metagenomics libraries have been performed allowing the identification of novel enzymes such as hemicellulose

(Maruthamuthu et al., 2016), xyla-nase enzymes (Matsuzawa et al., 2015), cellulose, protease, and lipase (Ngara and Zhang, 2018).

Metagenomics allowed the identification of thermostable enzymes from different environmental samples. For example, Pang et al. (2009) described endoglucanase from compost soil that is active in the low-temperature range (10–40 °C) with an optimum of 25 °C. Recently, Maruthamuthu et al. (2016) described thermoalkaliphilic cellulases from wheat straw degrading microbial consortia, using a multisubstrate approach for functional metagenomics. Also, thermophilic β-glucosidases and endoglucanase with activity at high temperature (90–105 °C) have been isolated from the termite gut (Wang et al., 2012), the hot spring (Schroder et al., 2014), and the compost soil (Verma et al., 2016).

Metatranscriptomics has been widely employed to investigate microbiomes in marine or aquatic environments, soils, rhizosphere, and human (Sheik et al., 2014; Maurice et al., 2013; Turner et al., 2013; de Menezes et al., 2012). In contrast to metagenomics, metatranscriptomics evaluates the expression profile and experimentally verify the activities of the enzymes encoded by the gene of interest (Mello et al., 2017; Hess et al., 2011). Carbohydrates such as cellulases are the major enzyme target since cellulosic biomass can be converted to a variety of products, including pulp and paper, textile and animal feed (Bhat et al., 2000; Sheperd et al., 1995). Metatranscriptomics of six Hu sheep rumen microbiomes revealed 2.65% unigenes encoded for GHs, which represent to 111 different carbohydrate-active enzymes (CAZymes). In total, 14,489 unigenes were annotated to 15 cellulases-containing GH families, with GH3, GH5, and GH9 being predominant. Heterologous expression from the GH5 family protei showed that proteins had endoglucanase or exoglucanase activity, with optimal activity at 40–50 °C (He et al., 2019). Studies of the rumen of dairy cow revealed 12,237 CAZymes, accounting for 1% of the transcripts. The CAZyme profile was dominated by GH94 (cellobiose-phosphorylase), GH13 (amylase), GH43 and GH10 (hemicellulases), GH9 and GH48 (cellulases), PL11 (pectinase), as well as GH2 and GH3 (oligosaccharides) (Comtet-Marre et al., 2017).

Metaproteomics studies for targeted bioprospecting of novel enzymes in microbial communities is limited, because of the magnitude of the problems and limitations of the available techniques, for example, sample preparation (Heyer et al., 2013; Leary et al., 2013). In natural environments, microbial organisms secrete hydrolases to make available the required nutrients for survival. These enzymes are the main target for industrial application. However, usually it is not possible to analyze extracellular proteins from natural environments by metaproteomics since protein concentration is low and its recovery is not reproducible (Speda et al., 2017). Therefore metaproteomics studies are focused on intracellular proteins from microbial communities (Jimenez et al., 2015). One of the attempts to study the extracellular proteins from microbial communities is by induction of protein synthesis from low baseline level that allowed the identification of 39 proteins, from which 28 were noncytosolic and 17 were carbohydrate-active enzymes (Speda et al., 2017). Metaproteome analysis has been useful to identify lipases by fluorogenic substrates using a 2D-gel zymogram (Sukul et al., 2017), and a wide range of xylan degradation-related enzymes, including xylanases, β-xylosidases, α-L-arabinofuranosidases, α-glucuronidases, and acetyl xylan esterases from corn stover (Zhu et al., 2016).

The advances in the molecular biology field (metagenomics, metatranscriptomics, and metaproteomics) promises to take this field to another unmatched level. Although enzyme discovery and optimization are challenging, the available resources are making possible the identification of novel enzymes for industries with greener ways of production.

1.7 BIOREFINERY FOR GREEN CHEMISTRY PRODUCTS: FUTURE AND CHALLENGES

The implementation of green extraction processes is one of the current challenges in the industry. The biorefinery approach offers an attractive solution looking for sustainable economic growth, based on the efficient production processes that are eco-friendly and safe for the consumers, also involving the use of innovative and efficient technologies. The development of energetically efficient biorefineries, the global implementation of sustainable government policies, the creation of incentives that promote the adoption of green technologies, either for domestic or industrial sectors, still are hurdles to overcome to exploit the full potentialities of the biorefineries to accomplish a circular economy.

Besides, an opportunity area exists in regard to the sensitization of society about the choice of more sustainable goods, either through consumer awareness or the diversification of product offer in the market, one that includes sustainable-sourced commodities. For its part, academia can contribute to the research and development of better metrics for green chemistry and biorefineries, to seek progress in the application of the principles.

On the other hand, even when the challenges are enormous, the advances in green chemistry and biorefinery application are visible today. There are companies, organizations, and institutions applying tools and designing new approaches to contribute to the progress of sustainability. Therefore it is envisaged that in the short future, the new generations will be naturally aware of the need to apply the principles of green Chemistry in a circular way, to achieve a more sustainable world.

KEYWORDS

- **green chemistry**
- **biorefinery**
- **sustainability**
- **processes**
- **circular economy**

REFERENCES

Aguiar-Pulido, V., Huang, W., Suarez-Ulloa, V., Cickovski, T., Mathee, K., and Narasimhan, G. Metagenomics, metatranscriptomics, and metabolomics approaches for microbiome analysis: supplementary issue: bioinformatics methods and applications for big metagenomics data. Evolut. Bioinform., **2016**, vol. 12, p. S36436.

Alain, K. and Querellou, J. Cultivating the uncultured: limits, advances and future challenges. Extremophiles, **2009**, vol. 13, no 4, pp. 583–594.

Allaf, T., Mounir, S., Tomao, V., and Chemat, F. Instant controlled pressure drop combined to ultrasounds as innovative extraction process combination: fundamental aspects. Proc. Eng., **2012**, vol. 42, pp. 1061–1078.

Alvarez, M.V., Cabred, S., Ramirez, C.L., and Fanovich, M. A. Valorization of an agroindustrial soybean residue by supercritical fluid extraction of phytochemical compounds. J. Supercrit. Fluids, **2019**, vol. 143, pp. 90–96.

Alvarez, T.M., Paiva, J.H., Ruiz, D.M., Cairo, J.P.L.F., Pereira, I.O., and Paixao, D.A.A. Structure and function of a novel cellulase 5 from sugarcane soil metagenome. PLoS One, **2013**, vol. 8, no 12.

Alzate, C.A.C., Toro, J.C.S., and Peña, Á.G. Fermentation, thermochemical and catalytic processes in the transformation of biomass through efficient biorefineries. Catal. Today, **2018**, vol. 302, pp. 61–72.

Anastas, P., and Warner, J. Green Chemistry: Theory and Practice. Oxford University Press, **1988**, p. 30.

Aristizábal-Marulanda, V., and Cardona Alzate, C.A. Methods for designing and assessing biorefineries. Biofuel. Bioprod. Biorefin., **2019**, vol. 13, no 3, pp. 789–808.

Armenta, S., Garrigues, S., Esteve-Turrillas, F.A., and de la Guardia, M. Green extraction techniques in green analytical chemistry. Trends Anal. Chem., **2019**.

Assacuete, l., Romagnoli, F., Cappelli, A., and Ciocci, C. Algae-based biorefinery concept: an LCI analysis for a theoretical plant. Energy Proc., **2018**, vol. 147, pp. 15–24.

BBC. Research. In Report Industrial enzymes—a global market overview, **2018**, 398 pp.

Baghel, R., Trivedi, N., Gupta, V., Neori, A., Reddy, C., Lalid, A., and Jha, B. Biorefining of marine macroalgal biomass for production of biofuel and commodity chemicals. Green Chem., **2015**, vol. 17, no 4, pp. 2436–2443.

Barnard, D., Casanueva, A., Tuffin, M., and Cowan, D. Extremophiles in biofuel synthesis. Environ. Technol., **2010**, vol. 31, no 8–9, pp. 871–888.

Bhat, M.K. Cellulases and related enzymes in biotechnology. Biotechnol. Adv., **2000**, vol. 18, no 5, pp. 355–383.

Blake, A. Measuring Progress towards Green Chemistry. Green Chemistry and Commerce Council, **2015**.

Brandt-Talbot, A.G., Fennell, P., Lammens, T., Tan, B., Wealea, J., and Hallet., J. An economically viable ionic liquid for the fractionation of lignocellulosic biomass. Green Chem., **2017**, vol. 19, no 13, pp. 3078–3102.

Brigljević, B., Liu, J., and Lim, H. Green energy from brown seaweed: Sustainable polygeneration industrial process via fast pyrolysis of *S. Japonica* combined with the Brayton cycle. Energy Convers. Manage., **2019**, vol. 195, pp. 1244–1254.

Campos, E.V., Proença, P.L., Oliveira, J.L., Bakshi, M., Abhilash, P.C., and Fraceto, L.F. Use of botanical insecticides for sustainable agriculture: future perspectives. Ecol. Indic., **2019**, vol. 105, pp. 483–495.

Cárdenas-Fernández, M., Bawn, M., Hamley-Bennett, C., Bharat, P., Fabiana Subrizi, F., Suhaili, N., Warda, J. M. An integrated biorefinery concept for conversion of sugar beet pulp into value-added chemicals and pharmaceutical intermediates. Faraday Discuss., **2017**, vol. 202, pp. 415–431.

Celeiro, M., Vazquez, L., Lamas, J.P., Vila, M., Garcia-Jares, C., and Llompart, M. Miniaturized matrix solid-phase dispersion for the analysis of ultraviolet filters and other cosmetic ingredients in personal care products. Separations, **2019**, vol. 6, no 2, p. 30.

Chemat, F., Abert-Vian, M., Fabiano-Tixier, A.S., Strube, J., Uhlenbrock, L., Gunjevic, V., and Cravotto, G. Green extraction of natural products. Origins, current status, and future challenges. Trends Anal. Chem., **2019**.

Chen, X., Yang, H., Zhong, Z., and Yan, N. Base-catalysed, one-step mechano-chemical conversion of chitin and shrimp shells into low molecular weight chitosan.

Green Chem., **2017**, vol. 19, no 12, pp. 2783–2792.

Cherubini, F., Jungmeier, G., Wellisch, M., Willke, T., Skiadas, I., Van Ree, R., and de Jong, E. Toward a common classification approach for biorefinery systems. Biofuel. Bioprod. Biorefin., **2009**, vol. 3, no 5, pp. 534–546.

Cherubini, F. The biorefinery concept: using biomass instead of oil for producing energy and chemicals. Energy Convers. Manage., **2010**, vol. 51, no 7, pp. 1412–1421.

Clark, J. H. Green chemistry for the second generation biorefinery—sustainable chemical manufacturing based on biomass. J. Chem. Technol. Biotechnol., **2007**, vol. 82, no 7, pp. 603–609.

Clark, J.H., Budarin, V., Deswarte, F.E., Hardy, J.J., Kerton, F.M., Hunt, A.J., et al. Green chemistry and the biorefinery: a partnership for a sustainable future. Green Chem., **2006**, vol. 8, no 10, pp. 853–860.

Clark, J.H., Luque, R., and Matharu, A. S. Green chemistry, biofuels, and biorefinery. Annu. Rev. Chem. Biomol. Eng., **2012**, vol. 3, pp. 183–207.

Comtet-Marre, S., Parisot, N., Leperq, P., Chaucheryras, F., Mosoni, P., Peyretaillade, E., Bayat, A.R., Shinfield, K.J., Peyret, P., and Forano, E. Metatranscriptomics reveals the active bacterial and eukaryotic fibrolytic communities in the rumen of dairy cow fed a mixed diet. Front. Microbiol., **2017**, vol. 8, p. 67.

Cravatt, B.F., Wright, A.T., and Kozarich, J.W. Activity-based protein profiling: from enzyme chemistry to proteomic chemistry. Annu. Rev. Biochem., **2008**, vol. 77, pp. 383–414.

Cravotto, G., Boffa, L., Mantegna, S., Perego, P., Avogadro, M., and Cintas, P. Improved extraction of vegetable oils under high-intensity ultrasound and/or microwaves. Ultrason. Sonochem., **2008**, vol. 15, no 5, pp. 898–902.

da Silva, R.P., Rocha-Santos, T.A., and Duarte, A.C. Supercritical fluid extraction of bioactive compounds. Trends Anal. Chem., **2016**, vol. 76, pp. 40–51.

Dai, J., Gozaydin, G., Hu, C., and Yan, N. Catalytic conversion of chitosan to glucosaminic acid by tandem hydrolysis and oxidation. ACS Sust. Chem. Eng., **2019**, vol. 7, no 14, pp. 12399–12407.

De Marco, B.A., Rechelo, B.S., Tótoli, E.G., Kogawa, A.C., and Salgado, H.R.N. Evolution of green chemistry and its multidimensional impacts: a review. Saudi Pharm. J., **2019**, vol. 27, no. 1, pp. 1–8.

De Melo, M.M., Barbosa, H.M., Passos, C.P., and Silva, C.M. Supercritical fluid extraction of spent coffee grounds: measurement of extraction curves, oil characterization and economic analysis. J. Supercrit. Fluids, **2014a**, vol. 86, pp. 150–159.

De Melo, M.M., Silvestre, A.J., and Silva, C. M. Supercritical fluid extraction of vegetable matrices: applications, trends and future perspectives of a convincing green technology. J. Supercrit. Fluids, **2014b**, vol. 92, pp. 115–176.

De Menezes, A., Clipson, N., and Doyle, E. Comparative metatranscriptomics reveals widespread community responses during phenanthrene degradation in soil. Environ. Microbiol., **2012**, vol. 14, no 9, pp. 2577–2588.

Ekkers, D.M., Cretoiu, M.S., Kielak, A.M., and van Elsas, J.D. The great screen anomaly—a new frontier in product discovery through functional metagenomics. Appl. Microbiol. Biotechnol., **2012**, vol. 93, no 3, pp. 1005–1020.

Esposito, D., and Antonietti, M. Redefining biorefinery: the search for unconventional building blocks for materials. Chem. Soc. Rev., **2015**, vol. 44, no 16, pp. 5821–5835.

Esquivel-Hernández, D.A., López, V.H., Rodríguez-Rodríguez, J., Alemán-Nava, G.S., Cuéllar-Bermúdez, S.P., Rostro-Alanis, M., and Parra-Saldívar, R. Supercritical carbon dioxide and microwave-assisted extraction of functional lipophilic compounds from Arthrospira platensis. Int. J. Mol. Sci., **2016**, vol. 17, no 5, p. 658.

Esquivel-Hernández, D.A., Rodríguez-Rodríguez, J., Cuéllar-Bermúdez, S.P., García-Pérez, J.S., Mancera-Andrade, E.I., Núñez-Echevarría, J.E., et al. Effect of supercritical carbon dioxide extraction parameters on the biological activities and metabolites present in extracts from Arthrospira platensis. Mar. Drugs, **2017**, vol. 15, no 6, p. 174.

Esquivel-Hernandez, D.A., Ibarra-Garza, I.P., Rodriguez-Rodríguez, J., Cuellar-Bermudez, S.P., Rostro-Alanis, M. d., Aleman-Nava, G.S., Parra-Saldivar, R. Green extraction technologies for high-value metabolites from algae: a review. Biofuel. Bioprod. Biorefin., **2017**, vol. 11, no 1, pp. 215–231.

Fang, X., Yano, S., Inoue, H., and Sawayama, S. Strain improvement of Acremonium cellulolyticus for cellulase production by mutation. J. Biosci. Bioeng., **2009**, vol. 107, no 3, pp. 256–261.

Ferrer, M., Martinez-Martinez, M., Bargiela, R., Streit, W.R., Golyshina, O.V., and Golyshin, P.N. Estimating the success of enzyme bioprospecting through metagenomics: current status and future trends. Microb. Biotechnol., **2016**, vol. 9, no 1, pp. 22–34.

Fiorentino, G., Zucaro, A., and Ulgiati, S. Towards an energy efficient chemistry. Switching from fossil to bio-based products in a life cycle perspective. Energy, **2019**, vol. 170, pp. 720–729.

Gallego, R., Montero, L., Cifuentes, A., Ibáñez, E., and Herrero, M. Green extraction of bioactive compounds from microalgae. J. Anal. Test., **2018**, vol. 2, no 2, pp. 109-123.

Ghasemi Naghdi, F., González, L.M., Chan, W., and Schenk, P. M. Progress on lipid extraction from wet algal biomass for biodiesel production. Microb. Biotechnol., **2016**, vol. 9, no 6, pp. 718–726.

Gillet, S., Aguedo, M., Petitjean, L., Morais, A., da Costa Lopes, A., Łukasik, R., and

Anastas, P. Lignin transformations for high value applications: towards targeted modifications using green chemistry. Green Chem., **2017**, vol. 19, no 18, pp. 4200–4233.

Green and Sustainable Chemistry Conference, **2019**.

Grimi, N., Dubois, A., Marchal, L., Jubeau, S., Lebovka, N.I., and Vorobiev, E. Selective extraction from microalgae Nannochloropsis sp. using different methods of cell disruption. Bioresource Technology, **2014**, vol. 153, pp. 254–259.

Harris, P.V., Welner, D., McFarland, K.C., Re, E., Navarro Poulsen, J.C., Brown, K., Salbo, R., Ding, H., Vlasenko, E., Merino, S., Xu, F., Cherry, J., Larsen, S., and Lo Leggio, L. Stimulation of lignocellulosic biomass hydrolysis by proteins of glycoside hydrolase family 61: structure and function of a large, enigmatic family. Biochemistry, **2010**, vol. 49, no 15, pp. 3305–3316.

He, B., Jin, S., Cao, J., Mi, L., and Wang, J. Metatranscriptomics of the Hu sheep rumen microbiome reveals novel cellulases. Biotechnol. Biofuels, **2019**, vol. 12, no 1, p. 153.

Herrero, M., and Ibañez, E. Green extraction processes, biorefineries and sustainability: Recovery of high added-value products from natural sources. J. Supercrit. Fluids, **2018**, vol. 134, pp. 252–259.

Hess, M., Sczyrba, A., Egan, R., Kim, T.W., Chokhawla, H., Scroth, G., Luo, S., Clark, D.S., Chen, F., Zhang, T., Mackie, R.I., Pennachio, L.A., Tringe,S.G., Visel, A., Woyke, T., Wang, Z., and Rubin, E.M. Metagenomic discovery of biomass-degrading genes and genomes from cow rumen. Science, **2011**, vol. 331, no 6016, pp. 463–467.

Heyer, R., Kohrs, F., Benndorf, D., Rapp, E., Kausmann, R., Heiermann, M., Klocke, M., and Reichl, U. Metaproteome analysis of the microbial communities in agricultural biogas plants. New Biotechnol., **2013**, vol. 30, no 6, pp. 614–622.

Himmel, M.E., Ding, S.Y., Johnson, D.K., Adney, W.S., Nimlos, M.R., Brady, J.W., and Foust, T.D. Biomass recalcitrance: engineering plants and enzymes for biofuels production. science, **2007**, vol. 315, no 5813, pp. 804–807.

Hu, J., Arantes, V., Pribowo, A., and Gourlay, K., Substrate factors that influence the synergistic interaction of AA9 and cellulases during the enzymatic hydrolysis of biomass. Energy Environ. Sci., **2014**, vol. 7, no 7, pp. 2308–2315.

Jimenez, D.J., Maruthamuthu, M., and van Elsas, J.D. Metasecretome analysis of a lignocellulolytic microbial consortium grown on wheat straw, xylan and xylose. Biotechnol. Biofuels, **2015**, vol. 8, no 1, p. 199.

Jung, S., Song, Y., Kim, H.M., and Bae, H.J. Enhanced lignocellulosic biomass hydrolysis by oxidative lytic polysaccharide monooxygenases (LPMOs) GH61 from Gloeophyllum trabeum. Enzyme Microb. Technol., **2015**, vol. 77, pp. 38–45.

Kanokratana, P., Mhuantong, W., Laothanachareon, T., Tangphatsornruang, S., Eurwilaichitr, L., and Pootanakit, K. Phylogenetic analysis and metabolic potential of microbial communities in an industrial bagasse collection site. Microb. Ecol., **2013**, vol. 66, no 2, pp. 322–334.

Karemore, A., and Sen, R. Downstream processing of microalgal feedstock for lipid and carbohydrate in a biorefinery concept: a holistic approach for biofuel applications. RSC Adv., **2016**, vol. 6, no 35, pp. 29486–29496.

Keijer, T., Bakker, V., and Slootweg, J. C. Circular chemistry to enable a circular economy. Nat. Chem., **2019**, vol. 11, no 3, pp. 190–195.

Kerton, F.M., Liu, Y., Omari, K.W., and Hawboldt, K. Green chemistry and the ocean-based biorefinery. Green Chem., **2013**, vol. 15, no 4, pp. 860–871.

Kleiner, M. Metaproteomics: Much More than Measuring Gene Expression in

Microbial Communities. MSystems, **2019**, vol. 4, no 3, pp. e00115–19.

Koutinas, A., Vlysidis, A., Pleissner, D., Kopsahelis, N., Lopez Garcia, I., Kookos, I.K., Lin, C. S. Valorization of industrial waste and by-product streams via fermentation for the production of chemicals and biopolymers. Chem. Soc. Rev., **2014**, vol. 43, no 8, pp. 2587–2627.

Kumar, P.S., and Yaashikaa, P.R. Sources and operations of waste biorefineries. In: Refining Biomass Residues for Sustainable Energy and Bioproducts. Academic Press, **2020**. pp. 111–133.

Kuznetsov, B.N., Chesnokov, N.V., Sudakova, I., Garyntseva, N.V., Kuznetsova, S.A., Malyara, Y.N., Djakovit, L. Green catalytic processing of native and organosolv lignins. Catal. Today, **2018**, vol. 309, pp. 18–30.

Lê, H.Q., Pokki, J.-P., Borrega, M., Uusi-Kyyny, P., Alopaeus, V., and Sixta, H. Chemical recovery of γ-valerolactone/ water biorefinery. Ind. Eng. Chem. Res., **2018**, vol. 57, no 44, pp. 15147–15158.

Leary, D.H., Hervey, W.J., Deschamps, J.R., Kusterbeck, A.W., and Vora, G.J. Which metaproteome? The impact of protein extraction bias on metaproteomic analyses. Mol. Cell. Probes, **2013**, vol. 27, no 5-6, pp. 193–199.

Lee, H.S., Kwon, K.K., Kang, S.G., Cha, S.S., Kim, S.J., and Lee, J.H. Approaches for novel enzyme discovery from marine environments. Curr. Opin. Biotechnol., **2010**, vol. 21, no 3, pp. 353–357.

Li, L.L., Taghavi, S., McCorkle, S.M., Zhang, Y.B., Blewitt, M.G., Brunecky, R., Adney, W.S., Himmel, M.E., Brumm, P., Drinkwater, C., Mead, D.A., Tringe, S.G., and van der Lelie, D. Bioprospecting metagenomics of decaying wood: mining for new glycoside hydrolases. Biotechnol. Biofuels, **2011**, vol. 4, no 1, p. 23.

Li, W. Analysis and comparison of very large metagenomes with fast clustering and functional annotation. BMC Bioinform., **2009**, vol. 10, no 1, p. 359.

Liu, T., and Li, Z. An electrogenerated base for the alkaline oxidative pretreatment of lignocellulosic biomass to produce bioethanol. RSC Adv., **2017**, vol. 7, no 75, pp. 47456–47463.

Loman, N.J., Constantinidou, C.; Chan, J.Z., Halachev, M., Sergent, M., Penn, C.W., Robinson, E.R., and Pallen, M.J. High-throughput bacterial genome sequencing: an embarrassment of choice, a world of opportunity. Nat. Rev. Microbiol., **2012**, vol. 10, no 9, pp. 599–606.

López-Velázquez, J.C., Rodríguez-Rodrí-guez, R., Espinosa-Andrews, H., Qui-Zapata, J.A., García-Morales, S., Navarro-López, D.E., and García-Carvajal, Z. Y. Gelatin–chitosan–PVA hydrogels and their application in agriculture. J. Chem. Technol. Biotechnol., **2019**, vol. 94, no 11, pp. 3495–3504.

Madhavan, A., Sindhu, R., Parameswaran, B., Sukumaran, R.K., and Pandey, A. Metagenome analysis: a powerful tool for enzyme bioprospecting. Appl. Biochem. Biotechnol., **2017**, vol. 183, no 2, pp. 636–651.

Maity, S. K. Opportunities, recent trends and challenges of integrated biorefinery: Part I. Renew. Sust. Energy Rev., **2015**, vol. 43, pp. 1427–1445.

Mariscal, R., Mairales-Torres, P., Ojeda, M., Sadaba, I., and Granados, M. Furfural: a renewable and versatile platform molecule for the synthesis of chemicals and fuels. Energy Environ. Sci., **2016**, vol. 9, no 4, pp. 1144–1189.

Maruthamuthu, M., Jimenez, D.J., Stevens, P., and van Elsas, J.D. A multi-substrate approach for functional metagenomics-based screening for (hemi) cellulases in two wheat straw-degrading microbial consortia unveils novel thermoalkaliphilic enzymes. BMC Genom., **2016**, vol. 17, no 1, p. 86.

Matsuzawa, T., Kaneko, S., and Yaoi, K. Screening, identification, and characterization of a GH43 family β-xylosidase/

α-arabinofuranosidase from a compost microbial metagenome. Appl. Microbiol. Biotechnol., **2015**, vol. 99, no 21, pp. 8943–8954.

Maurice, C.F., Haiser, H.J., and Turnbaugh, P.J. Xenobiotics shape the physiology and gene expression of the active human gut microbiome. Cell, **2013**, vol. 152, no 1–2, pp. 39–50.

Mayers, M.D., Moon, C., Stupp, G.S., Su, A.I., and Wolan, D.W. Quantitative metaproteomics and activity-based probe enrichment reveals significant alterations in protein expression from a mouse model of inflammatory bowel disease. J. Proteome Res., **2017**, vol. 16, no 2, pp. 1014–1026.

McGrath, K.C., Thomas-Hall, S.R., Cheng, C.T., Leo, L., Alexa, A., Schmidt, S., and Schenk, P.M. Isolation and analysis of mRNA from environmental microbial communities. J. Microbiol. Methods, **2008**, vol. 75, no 2, pp. 172–176.

Mello, B.L., Alessi, A.M., Riano-Pachon, D.M., deAzevedo, E.R., Guimaraes, F.E.G., Santo, M.C.E., McQueen-Mason, S., Bruce, N.C., and Polikarpov, I. Targeted metatranscriptomics of compost-derived consortia reveals a GH11 exerting an unusual exo-1, 4-β-xylanase activity. Biotechnol. Biofuels, **2017**, vol. 10, no 1, p. 254.

Michigan Green Chemistry Roundtable, Green Chemistry Checklist, **2014**

Moncada J.B., Aristizábal M.V., Carlos A., and Cardona A, Design strategies for sustainable biorefineries. Biochem. Eng. J., **2016**, vol. 116, pp. 122–134.

Montella, S., Amore, A., and Faraco, V. Metagenomics for the development of new biocatalysts to advance lignocellulose saccharification for bioeconomic development. Crit. Rev. Biotechnol., **2016**, vol. 36, no 6, pp. 998–1009.

Moraes, C.C., De Medeiros Burkert, J.F., and Kalil, S.J. C-phycocyanin extraction process for large-scale use. J. Food Biochem., **2010**, vol. 34, pp. 133–148.

Morales, M., Ataman, M., Badr, S., Linster, S., Kourlimpinis, I., Papadokonstantakis, S., Hungerbühler, K. Sustainability assessment of succinic acid production technologies from biomass using metabolic engineering. Energy Environ. Sci., **2016**, vol. 9, no 9, pp. 2794–2805.

Nacke, H., Engelhaupt, M., Brady, S., Fischer, C., Tautzt, J., and Daniel, R. Identification and characterization of novel cellulolytic and hemicellulolytic genes and enzymes derived from German grassland soil metagenomes. Biotechnol. Lett., **2012**, vol. 34, no 4, pp. 663–675.

Ngara, T.R. and Zhang, H. Recent advances in function-based metagenomic screening. Genomics Proteomics Bioinform., **2018**, vol. 16, no 6, pp. 405–415.

Ogawa, J. and Shmizu, S. Microbial enzymes: new industrial applications from traditional screening methods. Trends Biotechnol., **1999**, vol. 17, no 1, pp. 13–20.

Olaizola, M. Commercial production of astaxanthin from Haematococcus pluvialis using 25,000-liter outdoor photobioreactors. J. Appl. Phycol., **2000**, vol. 12, no 3–5, pp. 499–506.

Paes, G. and O'Donohue, M.J. Engineering increased thermostability in the thermostable GH-11 xylanase from *Thermobacillus xylanilyticus*. J. Biotechnol., **2006**, vol. 125, no 3, pp. 338–350.

Pang, H., Zhang, P., Duan, C.-J., Mo, X.-C., Tang, J.-L., and Feng, J.-X. Identification of cellulase genes from the metagenomes of compost soils and functional characterization of one novel endoglucanase. Curr. Microbiol., **2009**, vol. 58, no 4, p. 404.

Parsons, S., Poyntz-Wright, O., Kent, A., and McManus, M. C. Green chemistry for stainless steel corrosion resistance: life cycle assessment of citric acid versus nitric acid passivation. Mater. Today Sustain., **2019**, vol. 3, p. 100005.

Pingret, D., Fabiano-Tixier, A.S., and Chemat, F. Accelerated Methods for Sample Preparation in Food, **2012**.

Pyne, M.E., Narcross, L., Fossati, E., Bourgeois, L., Burton, E., Gold, N.D., and Martin, V.J.J. Reconstituting plant secondary metabolism in *Saccharomyces cerevisiae* for production of high-value benzylisoquinoline alkaloids. In: Methods in Enzymology. Academic Press, **2016**. pp. 195–224.

Ranjan, A., Singh, S., Malani, R., and Moholkar, V. Ultrasound-assisted bioalcohol synthesis: review and analysis. RSC Adv., **2016**, vol. 6, no 70, pp. 65541–65562.

Ranjith Kumar, R., Hanumantha Rao, P., and Arumugam, M. Lipid extraction methods from microalgae: a comprehensive review. Front. Energy Res., **2015**, vol. 2, p. 61.

Rombaut, N., Tixier, A.-S., Bily, A., and Chemat, F. Green extraction processes of natural products as tools for biorefinery. Biofuel. Bioprod. Biorefin., **2014**, vol. 8, no 4, p. 530–544.

Saai Anugraha, T.S., Swaminathan, T., Swaminathan, D., Meyyappan, N., and Parthiban, R. Enzymes in platform chemical biorefinery. In: Platform Chemical Biorefinery. Elsevier, **2016**. pp. 451–469.

Salinas-Salazar, C., Saul Garcia-Perez, J., Chandra, R., Castillo-Zacarias, C., Iqbal, H.M., and Parra-Saldívar, R. Methods for extraction of valuable products from microalgae biomass. In: Microalgae Biotechnology for Development of Biofuel and Wastewater Treatment. Springer, Singapore, **2019**. pp. 245–263.

Sandun, F., Adhikari, S., Chandrapal, C., and Murali, N. Biorefineries: current status, challenges, and future direction. Energy Fuels, **2006**, vol. 20, no 4, pp. 1727–1737.

Schroder, C., Elleuche, S., Blank, S., and Antranikian, G. Characterization of a heat-active archaeal β-glucosidase from a hydrothermal spring metagenome. Enzyme Microb. Technol., **2014**, vol. 57, pp. 48–54.

Schwartzberg, H. G. Expression of fluid from biological solids. Sep. Purif. Methods, **1997**, vol. 26, no 1, pp. 1–213.

Scully, E.D., Geib, S.M., Hoover, K., Tien, M., Tringe, S.G., Barry, K.W., del Rio, T.G., Chovatia, M., Herr, J., and Carlson, J.E. Metagenomic profiling reveals lignocellulose degrading system in a microbial community associated with a wood-feeding beetle. PLoS One, **2013**, vol. 8, no 9.

Shaghaleh, H., Xu, X., and Wang, S. Current progress in production of biopolymeric materials based on cellulose, cellulose nanofibers, and cellulose derivatives. RSC Adv., **2018**, vol. 8, no 2, pp. 825–842.

Shaikh, R., Zainuddin Syed, I., and Bhende, P. Green synthesis of silver nanoparticles using root extracts of *Cassia toral* L. and its antimicrobial activities. Asian J. Green Chem., **2019**, vol. 3, no 1. pp. 1–124, pp. 70–81.

Sharif, K.M., Rahman, M.M., Azmir, J., Mohamed, A., Jahurul, M.H., Sahena, F., and Zaidul, I. S. Experimental design of supercritical fluid extraction—a review. J. Food Eng., **2014**, vol. 124, pp. 105–116.

Sheik, C.S., Jain, S., and Dick, G.J. Metabolic flexibility of enigmatic SAR 324 revealed through metagenomics and metatranscriptomics. Environ. Microbiol., **2014**, vol. 16, no 1, pp. 304–317.

Sheldon, R. A. Metrics of green chemistry and sustainability: past, present, and future. ACS Sustain. Chem. Eng., **2018**, vol. 6, no 1, pp. 32–48.

Shen, Y. Carbon dioxide bio-fixation and wastewater treatment via algae photochemical synthesis for biofuels production. RSC Adv., **2014**, vol. 4, no 91, pp. 49672–49722.

Sheperd, A.C., Maslanka, M., Quinn, D., and Kung, L. Additives containing bacteria and enzymes for alfalfa silage. J. Dairy Sci., **1995**, vol. 78, no 3, pp. 565–572.

Simon, C., and Daniel, R. Achievements and new knowledge unraveled by metagenomic approaches. Appl. Microbiol. Biotechnol., **2009**, vol. 85, no 2, pp. 265–276.

Simon, C., and Daniel, R. Metagenomic analyses: past and future trends. Appl.

Environ. Microbiol., **2011**, vol. 77, no 4, pp. 1153–1161.

Singh, B.K. Exploring microbial diversity for biotechnology: the way forward. Trends Biotechnol., **2010**, vol. 28, no 3, pp. 111–116.

Singh, M., and Chowdhury, S. Recent developments in solvent-free multicomponent reactions: a perfect synergy for eco-compatible organic synthesis. RSC Adv., **2012**, vol. 2, no 11, pp. 4547–4592.

Solatia, Z., Manevskia, K., Jørgensena, U., Labouriaub, R., Shahbazi, S., and P.E., L. Crude protein yield and theoretical extractable true protein of potential biorefinery feedstocks. Ind. Crops Prod., **2018**, vol. 115, pp. 214–226.

Šoštarič, M., Klinar, D., Bricelj, M., Golob, J., Berovič, M., and Likozar, B. Growth, lipid extraction and thermal degradation of the microalga *Chlorella vulgaris*. New Biotechnol., **2012**, vol. 29, no 3, pp. 325–331.

Speda, J., Jonson, B.-H., Carlsson, U., and Karlsson, M. Metaproteomics-guided selection of targeted enzymes for bioprospecting of mixed microbial communities. Biotechnol. Biofuels, **2017**, vol. 10, no 1, p. 128.

Spickermann, D., Kara, S., Barackov, I., Hollmanns, F., Schwaneberg, U., Duenkelmanns, P., and Leggewie, C. Alcohol dehydrogenase stabilization by additives under industrially relevant reaction conditions. J. Mol. Catal. B Enzym., **2014**, vol. 103, pp. 24–28.

Sukul, P., Schakermann, S., Bandow, J.E., Kusnezowa, A., Nowrousian, M., Leichert, and L.I. Simple discovery of bacterial biocatalysts from environmental samples through functional metaproteomics. Microbiome, **2017**, vol. 5, no 1, p. 28.

Szőllősi, G. Asymmetric one-pot reactions using heterogeneous chemical catalysis: recent steps towards sustainable processes. Catal. Sci. Technol., **2018**, vol. 8, no 2, pp. 389–422.

Tabah, B., Pulidindi, I.N., Chitturi, V.R., L.M.R., A., Varvak, A., Foran, E., and Gedanken, A. Solar-energy-driven conversion of biomass to bioethanol: a sustainable approach. J. Mater. Chem. A, **2017**, vol. 5, no 30, pp. 15486–15506.

Tarento, T.D., McClure, D.D., Talbot, A.M., Regtop, H.L., Biffin, J.R., Valtchev, P., and Kavanagh, J. M. A potential biotechnological process for the sustainable production of vitamin K1. Crit. Reviews Biotechnol., **2019**, vol. 39, no 1, pp. 1–19.

Tiwari, R., Nain, L., Labrou, N.E., and Shukla, P. Bioprospecting of functional cellulases from metagenome for second generation biofuel production: a review. Crit. Rev. Microbiol., **2018**, vol. 44, no 2, pp. 244–257.

Turner, P., Gashaw, M., and Karlsson, E.N. Potential and utilization of thermophiles and thermostable enzymes in biorefining. Microb. Cell Fact., **2007**, vol. 6, no 1, p. 9.

Turner, T.R., Ramakrishnan, K., Walshaw, J., Heavens, D., Alston, M., Swarbreck, D., Osbourn, A., Grant, A., and Poole, P.S. Comparative metatranscriptomics reveals kingdom level changes in the rhizosphere microbiome of plants. ISME J., **2013**, vol. 7, no 12, pp. 2248–2258.

Valadez-Carmona, L., Ortiz-Moreno, A., Ceballos-Reyes, G., Mendiola, J.A., and Ibáñez, E. Valorization of cacao pod husk through supercritical fluid extraction of phenolic compounds. J. Supercrit. Fluids, **2018**, vol. 131, pp. 99–105.

Van den Bosch, S., Schutyser, W., Vanholme, R., Driessen, T., Koelewijn, S.-F., Renders, T., Sels, B. Reductive lignocellulose fractionation into soluble lignin-derived phenolic monomers and dimers and processable carbohydrate pulps. Energy Environ. Sci., **2015**, vol. 8, no 6, pp. 1748–1763.

Venskutonis, P.R., and Jonušaite, K. High pressure biorefinery of essential oil yielding plants into valuable ingredients. In: XXIX International Horticultural Congress on

Horticulture: Sustaining Lives, Livelihoods and Landscapes (IHC2014): V World 1125. **2014**. pp. 399–406.

Verein Deutscher Ingenieure (VDI). Klassifikation und Gütekriterien von Bioraffinerien (Classification and Quality Criteria of Biorefineries, VDI 6310. Blatt1:2016-01). Beuth Verlag, Berlin. **2016**.

Verma, D., Kawarabayasi, Y., Miyazaki, K., and Satyanarayanam T. Cloning, expression and characteristics of a novel alkali stable and thermostable xylanase encoding gene (Mxyl) retrieved from compost-soil metagenome. PLoS One, **2013**, vol. 8, no 1.

Vernès, L., Li, Y., Chemat, F., and Abert-Vian, M. Biorefinery concept as a key for sustainable future to green chemistry—the case of microalgae. In: Plant Based "Green Chemistry 2.0". Springer, Singapore, **2019**. pp. 15–50.

Wahlström, R., and Suurnäkki, A. Enzymatic hydrolysis of lignocellulosic polysaccharides in the presence of ionic liquids. Green Chem., **2015**, vol. 17, no 2, pp. 694–714.

Wang, H., Hart, D.J., and An, Y. Functional metagenomic technologies for the discovery of novel enzymes for biomass degradation and biofuel production. BioEnergy Res., **2019**, vol. 12, no 3, pp. 457–470.

Wang, J., Sun, Z., Zhou, Y., Wang, Q., Ye, J.A., Chen, Z., and Liu, J. Screening of a xylanase clone from a fosmid library of rumen microbiota in Hu sheep. Anim. Biotechnol., **2012**, vol. 23, no 3, pp. 156–173.

Warnecke, F., Luginbühl, P., Ivanova, N., Ghassemian, M., Richardson, T.H., Stege, J.T., and Sorek, R. Metagenomic and functional analysis of hindgut microbiota of a wood-feeding higher termite. Nature, **2007**, vol. 450, no 7169, pp. 560–565.

WCED, S.W.S. World Commission on Environment and Development. Our Common Future. **1987**.

Wei, F., Gao, G.Z., Wang, X.F., Dong, X.Y., Li, P.P., Hua, W., Chen, H. Quantitative determination of oil content in small quantity of oilseed rape by ultrasound-assisted extraction combined with gas chromatography. Ultrason. Sonochem., **2008**, vol. 15, no 6, pp. 938–942.

Xu, J., Liu, B., Hou, H., and Hu, J. Pretreatment of eucalyptus with recycled ionic liquids for low-cost biorefinery. Bioresour. Technol., **2017**, vol. 234, pp. 406–414.

Yun, J., and Ryu, S. Screening for novel enzymes from metagenome and SIGEX, as a way to improve it. Microb. Cell Fact., **2005**, vol. 4, no 1, p. 8.

Zhu, N., Yang, J., Ji, L., Liu, J., Yang, Y., and Yuan, H. Metagenomic and metaproteomic analyses of a corn stover-adapted microbial consortium EMSD5 reveal its taxonomic and enzymatic basis for degrading lignocellulose. Biotechnol. Biofuels, **2016**, vol. 9, no 1, p. 243.

Zuhse, R., Leggewie, C., Hollmann, F., and Kara, S. Scaling-up of "smart cosubstrate" 1,4-butanediol promoted asymmetric reduction of ethyl-4,4,4-trifluoroacetoacetate in organic media. Org. Process Res. Dev., **2015**, vol. 19, no 2, pp. 369–372.

CHAPTER 2

Design of Green Chemical Processes

LORENA FARÍAS-CEPEDA, LUCERO ROSALES MARINES,
KARINA REYES ACOSTA, ADOLFO ROMERO GALARZA, and
ANILÚ RUBIO RÍOS*

*Chemical Engineering Department, Faculty of Chemical Sciences,
Autonomous University of Coahuila, Saltillo, Coahuila, Mexico*

*Corresponding author. E-mail: a.rubio@uadec.edu.mx

ABSTRACT

One of the priority issues in green chemistry is to devise of greener and sustainable chemical strategies and processes to reduce or eliminated the use of hazardous substances. The development of bio-based materials has been constantly increasing, and there is a global effort in agreement with the demands of sustainable management for reducing reliance of fossil resources. In this sense there are green polymerization systems improved to reach increased yields and produce polymers with high molar mass and with specific microstructure characteristics. In the same way, biocatalysis in green chemistry reactions are performer under mild conditions of temperature, pressure and pH. However, along with new materials, chemical processes and reactors must be redesign and developed to guarantee the requirements and meet the demand for green chemical products. Substantial differences with traditional chemical process are based on the synthesis of materials changing the solvents selections and the implementation of continuous reactors, as well as in evaluating the effectiveness of the process through factors such as atom economy, environmental factor, and process mass intensity to implement an industrial green process. This chapter is a review of the processes that have been design and developed to achieve to maximize efficiency and reduce waste.

2.1 GREEN CHEMISTRY PROCESSES

Since the beginning of its existence, human being has made efforts to develop and create products that allow them to have more comfort, and meet his needs, many of which are the result of the application of chemical engineering. However, as these developments are essential to human life, they are also contributing to the degradation of the environment. In this sense, chemical engineering has directed its actions to ensure that technological developments do not have such a harmful effect on the environment, and thus achieve a balance, which is considered the basis of green chemistry.

Green chemistry, or sustainable chemistry, is defined as the "design of chemical products and processes to reduce or eliminate the use and generation of hazardous substances" (Anastas, 2007). When talking about green chemistry, an aspect of utmost importance is designed, which includes not only systematic planning and conception but also the 12 principles that govern, considered in fact as design rules (Anastas and Eghbali, 2010). These 12 principles were introduced by Paul Anastas and John Warner in 1998, and they are based on the minimization or nonuse of toxic solvents in chemical processes, as well as the non-generation or residues from these processes. In this chapter, these principles are not detailed but are mentioned: (1) prevention, (2) atom economy, (3) less hazardous chemical synthesis, (4) designing safer chemicals, (5) safer solvents and auxiliaries, (6) design for energy efficiency, (7) use of renewable feedstocks, (8) reduce derivatives, (9) catalysis, (10) design for degradation, (11) real-time analysis for pollution prevention, (12) inherently safer chemistry for accident prevention. All the principles of green chemistry are important, and in all of them there is a lot of research, however, safer solvents and auxiliaries are one of the most addressed issues in terms of environmental issues. Therefore there is still much to study in this regard.

Since the concept of green chemistry began to have notoriety, many researchers have been dedicated to developing more sustainable products and processes, covering areas, such as pharmaceutical, medicine, food, agronomy, fuels and biorefinery, metallurgy, materials, and many others (de Marco et al., 2019; Lupette and Maréchal, 2018; Clark et al., 2012; Ilyas et al., 2019). de Marco et al. (2019) affirmed that green chemistry affects not only pharmaceutical analyses but also environment, population, analyst, and even company. In the case of chemical and pharmaceutical industries and laboratories, these should include green chemistry in all possible aspects, such as the analysis, including chosen method, reagents, accessories, personnel qualification and, time to evaluate the quality of a product. Although the work of de Marco et al. (2019), Lupette and Maréchal (2018), Clark et al. (2012), Ilyas et al. (2019), and de Marco et al. (2019) was focused on the pharmaceutical industry, the areas of impact that they mention should apply to any field of research, regardless of whether it is the pharmaceutical, food, paper, or materials industry.

As was mentioned, chemical engineering is responsible for achieving a balance between the development of chemical products and the sustainability of the processes and analysis through which those products are obtained. However, specifically, the process systems engineering is what allows to achieve such sustainable development, taking into account not only aspects such as metrics, product design, process design, process dynamics, and control but also economic, environmental and societal aspects of processes, products, and their life cycles (Bakshi, 2019). In this sense, whether chemical engineering would have to be directly related to green chemistry, which will allow achieving the sustainability to continue having the necessary scientific and technological development. That is why, process system engineering, which for a long time focused only on the economic feasibility, has evolved, taking into account other areas and aspects that allow processes and the products obtained from those processes to become more sustainable. Thus in the last decades, process systems engineering has covered areas as the process, enterprise, and supply chain, without neglecting the economic feasibility (Bakshi, 2019). On the other hand, in order to achieve sustainability, to establish a clear understanding of the processes and products obtained, it is necessary for process systems engineering to consider not only macroscopic scales but also microscopic scales such as atoms, molecules, and systems of particles, giving rise to methods for molecular design (Charpentier, 2016; Bakshi, 2019).

Also, the process systems engineering becomes an optimization problem, where the objective function and the restrictions that govern it are no longer only economic aspects but also aspects related to health, environment, and safety. In several cases, the main objective of the processes is the inclusion of the aspects above has led chemical engineering to a development that allows it to achieve sustainability. Bakshi poses the stages of evolution of chemical engineering toward sustainability, and the formulation of the corresponding optimization problems, among which the following can be mentioned (Bakshi, 2019):

1. Ignore the environment. Maximize profit.

2. Environment as a constraint. Maximize profit. Satisfy environmental constraints.
3. Local impact as an objective. Maximize profit. Minimize local environmental impact.
4. Life cycle impact as an objective. Maximize profit. Minimize the life cycle environmental impact.
5. Respecting nature's capacity as an objective. Maximize profit. Minimize the overshoot of ecosystem services.
6. Accounting for markets and human behavior. Maximize private and social profits. Minimize overshoot.

The previous stages show how the environmental and social impact of the processes has been playing a fundamental role in the process systems engineering development, and it is necessary to highlight that this trend continues to increase, since it is necessary to develop more sustainable products and processes.

On the other hand, Bakshi, in his review about the role of process systems engineering on sustainable chemical engineering, emphasizes the importance of four areas of research, which ones could ensure sustainable chemical engineering. These areas are (Bakshi, 2019) as follows:

1. Metrics. Used to quantify sustainability. In addition, they need to be well defined, easy to calculate and use for supporting the decision, and capture the requirements for claiming sustainability. This sustainability is commonly understood in terms of economy, society, and, environment. Metrics could be classified as one, two, and three dimensional, depending on whether the metric covers the previously cited aspects.
2. Design of chemical processes and their supply chains. In this case, the problems related to the processes design and supply chains may be formulated as optimization problems. Some of them are pinch analysis combines with process optimization, the waste reduction algorithm, and others.
3. Design of single and multimolecular products and their formulations. Recently, two types of products are identified as products designed by chemical engineers (Zhang et al, 2016). Single-species products include small and large molecules. On the other hand, multispecies products may be formulated or functional. There are many approaches to product design, however, in the case of design of single-species molecules, the approach of reaction network flux analysis has been extended to include economic and environmental aspects, while for the design of more complicated products, and efforts have been

directed toward satisfying the economic objectives.

4. Dynamics and control of chemical processes. Methods for process control are significant because a good understanding of the dynamic nature of sustainable systems, and the control of the process and societal systems would ensure the process of sustainable engineering.

Taking into account these mentioned aspects; it is possible to say that chemical engineering has evolved so that the activities concerning this field have reduced the environmental and social impact, making the processes and chemical products obtained from these processes more sustainable. In this way, chemical engineering involves the synthesis of nano- and microstructured materials, design, scale-up operation, control, and optimization of processes. Even though, the materials obtained from these chemical processes require the development of various technologies, including microtechnology (Charpentier, 2016).

Green chemistry and green process engineering involve the development of process design and operation from nano- and microsystems scales to industrial scale. Thus the chemical supply chain in a process begins with chemicals, synthesized and characterized at the molecular level (Charpentier, 2016). That is why the innovation of the processes is fundamentally based on two aspects, the multiscale and the multidisciplinary. When talking about nanometric scales, it is challenging to achieve a clear understanding of the properties and behavior of the system through experimental studies, which ones are done on a macroscopic scale. Therefore using molecular simulation techniques facilitates the work done, allowing them to study microscale systems from models and equations that describe their behavior. Elseways, there are also working conditions that are difficult to achieve, for example, high pressures or temperatures, being necessary in these cases, the use of macroscale simulation techniques that allow analyzing these scenarios. Thus modern computers allow modeling and simulating systems at different scales. Also, multidisciplinary is necessary for the development and innovation of sustainable processes.

Although many areas of scientific and technological development have shown a growing concern to develop more sustainable processes and products, one of the most interested in aspects related to such sustainability and green chemistry has been the polymer area. Development in polymer green chemistry covers several issues. These issues, as well as some examples of them, are listed in Table 2.1

There have been numerous developments in the area of polymer chemistry; however, biocatalysts and bio-based materials have been of attention to many researchers

TABLE 2.1 Efforts in Green Polymer Chemistry (Cheng, 2015; Hernández et al., 2015; Lerici et al., 2015; Li and Trost, 2018; Patel et al., 2015; Sheldon and Woodley, 2017; Prat et al., 2014)

	Examples
Biological catalysts	Enzymes, whole cells, catalytic antibodies
Biobased materials	Biobased building blocks
	Block copolymers from vegetable oils.
	Lignocellulosic feedstock
	Furan based structures
	Cyanate ester monomers derived from renewable resources.
	Natural fillers in composites.
	CO_2 as a monomer.
Degradable polymers	Agri-based natural renewable materials.
	Some polyesters and polyamides.
	Renewable thermoplastic polyacetal
Recycling of polymer products and catalysts	Catalytic degradation of polyolefin plastics.
	Catalytic degradation of polystyrene.
	catalyst degradation of a mixture of postconsumer polymer waste (PE/PP/PS
	Cracking of low-density polyethylene.
	Thermal catalytic transformation of HDPE
	Immobilized enzymes reuse.
Green Energy	Production and use of biodiesel or biofuels.
	Energy saving processes (e.g., active extrusion, jet cooking, microwave).
Molecular design and activity	Polymer with designed structures or functions.
	Improved biocatalysts.
	Protein engineering
	Organic synthesis
	Generic/metabolic engineering.
Design and usage of green solvents	Water, EtOH, i-PrOH, nBuOH, EtOAc, i-PrOAc, n-BuOAc.
	Solvent-free media
Green processing	Aim to minimize energy and chemical waste, to increase yield, and to decrease safety and toxicity hazards.
	Atom economy
	Environmental factor
	Process mass intensity
	Decreased byproducts

in recent years (Cheng and Gross, 2010). Bio-based materials are products that mainly consist of a substance or substances derived from biomass, or that have been obtained from processes that use biomass (Curran, 2010). These bio-based materials include polypeptides and proteins, polysaccharides, lipids and triglycerides, speciality polymeric materials, biomaterials, and others (Cheng and Gross, 2010).

On the other hand, catalysts play a fundamental role in the processes of biotransformation, so the development of biocatalysts that allow following the principles of green chemistry is of great importance. Some of these biocatalysts are enzymes and whole-cell approaches (Cheng and Gross, 2010).

It can be said that chemical engineering is the principal responsible for the development of sustainable processes and products, and that it involves many areas and aspects of the study. In conclusion, the green polymer processes and biocatalysis are two of the most important and most studied areas in green chemistry.

2.2 GREEN CHEMISTRY REACTORS

Chemical reaction engineering concerns to the all chemical or biological transformations of starting materials, derived from nonrenewable and renewable resources, in products like fuels, simple and speciality chemicals, fibers for clothing, materials for construction and communications, fertilizers, pharmaceuticals, and a variety of consumer products that depend to support their lifestyle. For economical-, environmental-friendly, and energy-efficient processes selecting the right chemical transformation, the catalyst, the reactor type, and be able to scale up these transformations to commercial production are crucial. For the reduction of global pollution, it is necessary to increase all measurable process efficiency variables like atoms, mass, and energy applying the knowledge of process intensification and scale-up (Dudukovic, 2009, 2010; Charpentier, 2016). In the context of process intensification, the reactors have received more consideration than other unit operations as the heart of chemical processes. In this part of the short-chapter tutorial on reactor theory and diverse intensified and greener reactors are presented.

2.2.1 REACTOR ENGINEERING THEORY

The engineering of a chemical reactor involves the reaction kinetic and the reactor design (mass and heat balance). The mass balance in the simple form is as follows:

In Out + Generation = Accumulation

For a simple reaction, $A-B$ the term generation is the rate at which a species is formed or lost by the reaction, and equals to rAV, where rA is the rate of loss of species, A, and V is the volume of the reactor. The kinetics of the reaction depends only upon the concentrations of the reactant (CA) and rate constant, for a first-order reaction $-rA=kCA$, where k is the rate constant that depends to the temperature by Arrhenius equation. This equation is difficult to solve when mass transfer limitations are present in the reactor because it is solved using the false supposition that a reaction will proceed at its inherent rate. When the reaction is not proceeding at its inherent rate, for example, the larger stirred tank, the mixing within the vessel restricts it; in other words, the reactants are not perfectly mixed together. The process intensification is used to eliminate such constraint in other to the reaction reach its inherent rate ensuring that the mixing and heat/mass transfer rates will be relatively fast compared to the fundamental process kinetics (Fogler, 2008).

For heterogeneous (multiple phases) reactions are more complicated, and this assumption does not necessarily apply. Many heterogeneous reactions are diffusion dependent more than kinetic dependent. For homogeneous reaction, if the reaction mixture is well-mixed and the phases are fully miscible, the process kinetics is determined by the rate constant that depends on the temperature defined by Arrhenius form expressions. In homogeneous reactions, if the mixing is slow, only a partial volume in the reactor is mixed, so the space–time yield of the system is low and are compromised the selectivity and byproducts.

Faster mixing, before a significant reaction time, produces an intimately mixed of reactive with each other, then reaction proceeds throughout the existing volume and controlled over stoichiometric proportion is reached; the reaction ceases to be "mixing controlled" to be "reaction controlled" (Reay et al., 2008).

Another issue in reactor design is the resident time distribution (RTD), which is the time that the molecules flow through a given geometry but not all the molecules stay in equal time, so it is a distribution. It is often easier to visualize an RTD as the answer to a perfect, infinitely narrow pulsed input ("Dirac delta" input). Therefore the ideal reactor, two extremes of the behavior of RTD are presented, the continuously stirred tank (CSTR) and plug flow reactor (PFR). An ideal CSTR is supposed to be fully back mixed or perfectly mixed, and it has a broad residence time distribution, on the other hand, ideal PFR's output is same as the input, and axial dispersion through the reactor are present, because of its flat velocity profile, so RTD is narrow. Real reactors RTDs lie between these two extremes, these

RTD are determined by the macro- and micromixing. This is necessary for greener processes that all the molecules of the fluid spend the same time in each atmosphere in the reactor, this reduces byproducts, or simply the molecules are less exposed to higher temperatures in order to achieve the desired conversion (Fogler, 2008; Reay et al., 2008).

As a result, some intensified process reactor has been designed, like cavitation reactor, oscillatory baffled reactors, spinning disc reactors, membrane bioreactors, microreactor, and others (Boodhoo, 2013; Elvira et al., 2013; Cao et al., 2015; Macedonio and Drioli, 2017; Powell, 2017; Zhang et al., 2018).

Scalability of mixing is the main problem in reactor design. For example, to ensure a perfect mix at laboratory scale batch stirred vessel, it is enough to apply a high agitation rate. In contrast, the mixing would not be perfect in larger containers, which can cause reactions to be limited since the costs to achieve homogeneous mixing at industrial levels are excessively expensive and technically difficult to achieve because the power number for the mixture is proportional to the diameter of the inverse power 5. In the case of CSTR, whose usage is common in the industry, there are problems in the mixing capacity depending on the dimensions of the reactor. If the scale-up is based on a constant impeller tip speed, and then the average circulation speed in the vortices is proportional to the tip speed selected, the rotation time is proportional to the vessel diameter. Therefore the turnover time of the vessel contents rises on a larger scale and a decrease in macromixing performance is presumed (Reay et al., 2008).

Heat transfer is critical to design and operate a reactor and that is why temperature must be controlled to the required accuracy. It is the vital importance that the reactor designer determines whether thermal runaway is possible, so the designer must use an accurate control system that can remove the heat produced by the reaction. If the reactor reaches the thermal runaway, the positive feedback must reach some other limit, for example, a safety valve or control system comes into play.

The process intensification improvements include smaller reactor size, enhanced safety, less waste generation, and higher product quality also energy and cost savings, because that improve energy and process efficiency through enhancing mixing, mass, and heat transfer as well as driving forces (Holkar et al., 2019). Some reactors used in are cavitation or sonochemical reactor, membrane reactors, field-enhanced reactor, microreactor, spinning disc reactors (SDR), and others, in this part of the section more detailed information about some of these reactors are offered.

2.2.2 CAVITATION REACTOR

Cavitation is the physicochemical phenomena of consecutive generation, development and, breakdown of a great number of microscopic cavities in the liquid medium. These breakdowns of cavities produce a large amount of energy from 1 to 1018 kW/m^3 over an insignificant period. The energy is produced in the form of high temperature and high pressure. Acoustic, hydrodynamic, optic, and particle are the form that cavitation is categorized; the most common use of these methods is hydrodynamic and acoustic cavitation. With cavitation free radicals, local hot spots, and microturbulence can be generated that result in an intensification of homogenous and heterogeneous reactions (Holkar et al., 2019).

When ultrasonic sound waves (16 kHz–100 MHz) pass through a fluid, pressure variations in the liquid propagate through it that consist of compression and rarefaction phase; this is called acoustic cavitation. In the rarefaction cycles, the negative acoustic pressure pulls fluid molecules away from each other, generating a void in the fluid after exceeding a critical molecular distance that causes the formation of cavities. As well, in the compression cycle, positive acoustic pressure impulses the molecules together and compresses the cavities which breakdown a fraction of time below the adiabatic condition that produces high local temperature and pressure condition for a small interval of time (millisecond to microsecond). The physicochemical transformations of cavitational breakdown affect depending upon the systems in which it occurs. In homogeneous liquid reactions, two effects occur, the first one is that a volume of the vapor from the liquid medium is enclosed by the cavity. In the breakdown, the extreme conditions of pressures and temperatures of the enclosed vapor is exposed that results in dissociation and generation of highly reactive radical species. Second, by the breakdown of the bubble also shear forces are produced and generate a shock wave in the surrounding bulk liquid, facilitating the mass transport due to the boundary layer. For heterogeneous systems, the asymmetric breakdown of the bubbles produces mechanical and structural defects; when bubbles breakdown close a solid surface high-pressure/high-velocity liquid jets are produced that clean the surface, activate the solid catalyst, and increase the mass transfer. In the solid phase, the asymmetric breakdown generates fragmentation and surface roughening, because of that ultrasound can increase the surface, micro/molecular level mixing and hence mass transport. Because of this properties, cavitation reactors can be effectively employed for the various physico-chemical processes, such as chemical synthesis, wastewater treatment, biotechnology, extraction, textile processing, polymer degradation, in petrochemical industries, emulsification, and crystallization. In this way, Chatel (2018) presents a very detailed analysis of the publications where the words sonochemistry and green chemistry are involved and are summarized in Table 2.2.

Cavitation reactors are most used in laboratory scale than industrial one because of the hydrodynamic behavior of different sonochemical reactor configurations and the relationships of mixing time, mass transfer, and flow patterns with operating parameters are not well understood. Cavitation reactors are very sensible to the operation conditions. Some of the most used sonochemical reactors are ultrasonic horn, ultrasonic bath, dual-frequency flow cell, triple-frequency flow cell, and longitudinally vibrating reactor, the efficiency of cavitation reactors are subject to geometric and operational parameters. For that Asgharzadehahmadi

TABLE 2.2 Biocatalytic Processes and Characteristics (Truppo, 2017; Lin and Tao, 2017; Sheldon and Woodley, 2018)

Biocatalysis Classifies	Characteristics
Resting whole-cell biocatalysis	The growth of the biocatalyst and the substrate to product conversion can be separate while still operating in whole-cell format.
	In comparison with chemical processes presents high selectivity and catalytic efficiency.
	Multistep reactions in single strain with cofactor regeneration.
	Recycling is sometimes possible.
	Milder operational conditions.
	Environmental friendly.
Isolated enzyme biocatalysis	The enzyme is used outside the cell in which was produced, overcoming the diffusional limitations of substrates into cells.
	Avoids the added cost of purification.
	The isolated enzyme can be easily separate from the product stream.
Immobilized enzyme biocatalysis	The enzyme is immobilized to facilitate removal from the product stream.
	Enables the use of biocatalysts in organic solvents systems.
	Facilitates reaction telescoping, enzyme recovery and reuse.
	Confers stability to the enzyme immobilizing it in a more stable conformation.

et al. (2016) and Holkar et al. (2019) present very detailed reviews of process intensification with sono-chemistry, the most important operation parameters (Asgharzadehahmadi et al., 2016; Holkar et al., 2019).

2.2.3 SPINNING DISC REACTOR

Spinning disc technology has practical importance in chemical engineering operations. This technology uses the flow of tinny liquid films on plane surfaces due to gravity, the surface wave formation induces mixing within the film, so increases heat and mass transfer rates (Boodhoo, 2013). A representation picture of a characteristic spinning disc is shown in Figure 2.1.

SDR can operate horizontally or vertically and are mounted on a rotating axle. The liquid is feed near the center and flows through the surface of the spinning disc under centrifugal force, which stretches and spreads the film. This thin liquid film permits high rates of mass transfer helping unit operations such as stripping, absorption, mixing, and reactions.

The viscosity of the film material, the speed of rotation, and surface geometry of the disk are some factors that determine the form of the film. Also, the surface geometry of the disc, fluid physical properties, radial location of the fluid are factors that affect the residence time and film thickness. 0.1–3 s is the range of the residence time of the fluid on the

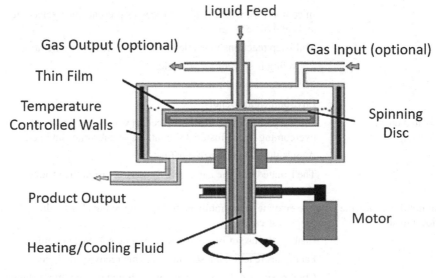

FIGURE 2.1   Schematic picture of a typical spinning disc.

disc. On exiting the periphery (edge) of the disc, the liquid is thrown onto the enclosing wall and then drains away. According to process requirements wall of SDR is heated or cooled. 0.01–0.03 cm/s area typical mass transfer coefficient values to perform mass transfer limited processes in fractions of a second for low-viscosity liquids (Doble and Kruthiventi, 2007; Pask et al., 2012). The working disc of SDR generally are between10 cm and 1 m of diameter and can be made with different materials, often with a base of copper (excellent thermal conductivity) with a tinny chrome plating for chemical resistance, or other material according to reaction conditions. In need of the application, the disc can be smooth, grooved, or meshed and the rotational speeds of an SDR can vary from 100 to 6000 rpm. The characteristics of the SDR are (Doble and Kruthiventi, 2007; Pask et al., 2012; Boodhoo, 2013) as follows:

1. Strong shearing and mixing in a tinny liquid film
2. Short liquid residence time
3. High solid/liquid heat/mass transfer
4. High liquid/vapor heat/mass transfer

The reaction investigated in SDR is photopolymerization of *n*-butyl acrylate, free radical, and cationic polymerization of styrene, Equilibrium-controlled condensation polymerization, phase transfer-catalyzed Darzens reaction, crystallization, preparation of nanoparticles, competitive organic reaction, food processing as pasteurization, concentrated sugar solutions, ice-cream making process, among others (Doble and Kruthiventi, 2007; Boodhoo, 2013).

2.2.4 MICROREACTOR

Microfluidics represents an innovative solution in terms of green chemistry; by the use of small amounts of reagents and solvents, producing less waste, the reaction conditions are easier to control, integrated, perform multiphase reactions and reliable scale-up a microfluid segment is defined as a minimum unit having micro properties that can be used to improve some unit operations and reactions in microspace (Yao et al., 2015; Chatel, 2018).

Applications of microreactor presented a rapid development above the last few decades because the advantage of microreactors comparted to batch processes and large-size continuous reactors.

Microreactors are the "micro" arrangement of a large-size reactor and can be used to improve reactions and some unit operations in microspace. Microreactors have shown higher heat and mass transfer rates (Figure 2.2), and the contact time, size, and shape of the interface between fluids can be easily and precisely controlled. This kind of chemical reactors offers controllable and high-throughput methods for the synthesis of chemicals with high yield, selectivity, stability, improved sample consistency, less energy consumption, low reaction volume, and homogeneity. These characteristics make microreactors ideal for fast reactions, highly exothermic reactions, and even explosive reactions (Yao et al., 2015; Suryawanshi et al., 2018).

Microreactors are fabricated in different materials, like ceramics, polymers, metals, stainless steel, glass, and silicon, instead of large pipes and vessels, these devices have channels with dimensions in the 0.1–0.3 mm range. The material whose microreactor is fabricated depends on various factors, operating conditions like pressure and temperature; physical properties of the reaction mixture like pH, phase, viscosity and reactivity, cost, the ability of mass production, and ease of fabrication. Microreactors are classified into two groups (Lerou et al., 2010; Suryawanshi et al., 2018):

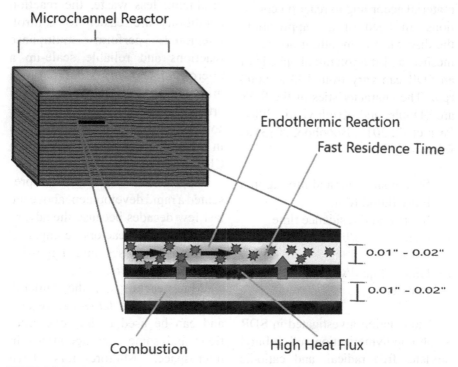

FIGURE 2.2   Depiction of process intensification facilitating improved heat and mass.

1. Microcapillary reactors are fabricated from appropriate pipes of desired length and material, which are principally used for chemical reactions to reach high yields and conversions and offer better control on heat and mass transfer within the reaction.

2. Chip microreactors are fabricated with glass, plastics, or silicon and can be made by micromachining with laser or microdrilling, dry or wet etching, injection molding, soft lithography techniques, or adaptations from the microelectronics industry like Lithographie Garbanoforming Abforming. Chip microreactors offer several advantages including easy control of microfluidics and integration of many processes into one reaction device. They can be operated at high pressure and temperature (Yao et al., 2015).

Lerou et al. (2010), Yao et al. (2015), and Suryawanshi et al. (2018) present excellent reviews about the fabrication and use of microreactors for the synthesis of inorganic nanoparticles, the preparation of metal nanoparticles, and the control of monodisperse emulsions, as well as for reactions of gas-liquid phase, liquid-phase liquid, gas-liquid systems-solid phases, polymerization reactions, and biosynthesis.

2.3 BIOCATALYSIS IN GREEN CHEMISTRY

Biocatalysis is understood as the use of natural substances such as enzymes from biological sources, whole cells, and catalytic antibodies (Li and Trost, 2008). These catalysts act as part of chemical reactions leading to high reaction rates and selectivity. The use of these natural substances, known as biodegradable catalysts, implies a reduction in energy use, avoid the use of organochlorine compounds and reduce the use of water or have a lower production of wastewater (Ivanković et al., 2017; Li and Trost, 2008). Biocatalysis as an ecological and sustainable technology includes the use of process design in order to improve the efficiency and sustainability of biocatalytic reactions (Sheldon and Woodley, 2018).

The use of biocatalysts is fully compatible with the green chemistry principles stipulated by Anastas and Wrangler in 1998 since biocatalysts significantly reduce waste, and they are obtained from renewable resources, renewable, biodegradable, and essentially nonhazardous and nontoxic. The use of these compounds is possible in environmental-friendly solvents, even in water, and their conditions of use make it possible to implement a more efficient use of energy in the process, so conditions in general are safe (Anastas and Wrangler, 1998; Sheldon and Woodley, 2018).

Sheldon and Woodley (2018) organized biocatalytic processes in the agreement of biocatalyst used for the conversion. Table 2.3 shows this classification, including the main characteristics of each one and the advantage that their use represents.

A key factor in the use of biocatalysts, since an appropriate catalyst, is to reduce residence time requirements and be more selective in the product to obtain. The use of enzymes as biocatalysts is focused on the decrease of the activation energy of an individual reaction, which causes a considerable acceleration

TABLE 2.3 Review of Recommended Green Solvents by Pharmaceuticals Solvents Selection Guides

Class		Conclusion	Company	Reference
Alcohols	Methanol	Recommended	Pfizer, Sanofi	Byrne et al.(2016), Prat et al. (2013)
		Some issues	GSK	Byrne et al. (2016)
	Ethanol	Recommended	Pfizer, Sanofi, Astra Zeneca	Byrne et al. (2016), Prat et al. (2014)
		Some issues	GSK	Byrne et al. (2016)
	1-Propanol	Recommended	Pfizer, Sanofi	Byrne et al. (2016), Prat et al. (2013)
		Some issues	GSK	Byrne et al. (2016)
	n-Propanol	Recommended	Sanofi	Prat et al. (2013)
	i-Propanol	Recommended	Pfizer, Sanofi, Astra Zeneca	Byrne et al. (2016), Prat et al. (2014)
		Some issues	GSK	Byrne et al. (2016)
	n-Butanol	Recommended	Pfizer, Sanofi, Astra Zeneca	Prat et al. (2014)
	1-Butanol	Recommended	Pfizer, Sanofi	Byrne et al. (2016), Prat et al. (2013)
		Few issues	GSK	Byrne et al. (2016)
	2-Butanol	Recommended	Sanofi	Byrne et al. (2016), Prat et al. (2013)
		Few issues	GSK	Byrne et al. (2016)
	t-Butanol	Recommended	Pfizer	Byrne et al. (2016)
		some issues	GSK	Byrne et al. (2016)
		Substitution advisable	Sanofi	Byrne et al. (2016)

TABLE 2.3 *(Continued)*

Class		Conclusion	Company	Reference
Ketones	Acetone	Recommended	Pfizer, Sanofi	Byrne et al. (2016), Prat et al. (2013)
	Methylethyl ketone	Recommended	Pfizer, Sanofi	Byrne et al. (2016), Prat et al. (2013)
Esters	Ethyl acetate	Recommended	Pfizer, Sanofi, Astra Zeneca	Byrne et al. (2016), Prat et al. (2013)
	n-Propyl acetate	Few issues	GSK	Byrne et al. (2016)
		Recommended	Sanofi	Prat et al. (2013)
	i-Propyl acetate	Recommended	Pfizer, Sanofi, Astra Zeneca	Byrne et al. (2016), Prat et al. (2014)
		Few issues	GSK	Byrne et al. (2016)
	n-Butyl acetate	Recommended	Sanofi	Prat et al. (2013)
Ethers	Methyl-THF	Recommended	Sanofi	Prat et al. (2013)
	Anisole	Recommended	GSK, Sanofi, Astra Zeneca	Prat et al. (2014)
Aprotic polar	Acetonitrile	Recommended	Sanofi	Prat et al. (2013)
		Usable	Ptfizer	Byrne et al. (2016)
Water	Water	Recommended	Pfizer, Sanofi	Byrne et al. (2016), Prat et al. (2013)
		Few issues	GSK	Byrne et al. (2016)

in the speed of the reaction, the enzyme remaining unchanged until the end of the reaction and without affecting the relative energy between reagents and products (Ivanković et al., 2017). Although enzymes stand out for their high stereochemical specificity, chemical selectivity, and specificity, they have deficiencies in the lack of thermal sensitivity and stability (Ivanković et al., 2017). Therefore, these parameters should be considered when designing the reaction process.

Undoubtedly, catalysis is a fundamental point in the development of green chemistry, and the use of appropriate biocatalysts has advantages that make processes highly competitive in terms of process utilization, selectivity, energy reduction and use of alternate reaction media. Development and use of biocatalysts have allowed implementing new strategies to achieve green chemistry practices. Methods to improve protein activity, specificity, and stability in biocatalysts are developing. Many inorganic materials, reagents in organic synthesis have been replaced by more atom and step-economical catalytic alternatives since biocatalytic procedures are environmentally attractive, more cost-effective, and more sustainable.

2.4 SOLVENT SELECTION

It is vitally important to consider the use of solvents in green chemistry procedures since these are present in greater quantity within a process. Green chemistry processes must be designed to require and generate substances with low or no toxicity, so it is necessary to incorporate the use of substances that ensure safe and clean industrial processes. Both the solvents used in the process, and the use of other supporting substances, should be avoided whenever possible, and when their use is elementary, it should be sought as a harmless material (Ivanković et al., 2017).

Green chemistry focuses on eliminating unwanted solvents, either eliminating them from the process or replacing them with environmentally, health, and safety materials (Byrne et al., 2016). The toxicity, flammability, and corrosion that most organic solvents have left them out of a classification of green chemistry. Even considering that energy-efficient distillation is carried out for post-process disposal, the economic and energy cost is high, so it is necessary to develop and seek to increase the use of environmentally friendly solvents inside the processes (Ivanković et al., 2017).

A concept that has gained a place within the processes of green chemistry are green solvents. Green solvents refer to those materials that allow carried out new synthetic routes of low environmental impact; eliminate emissions of volatile organic compounds; they are nontoxic and inflammable materials; and their use reduces the impact on the environment and human exposure to hazardous chemicals.

Utilization of solvents in the chemical processes is determining for the process itself; solvents play a fundamental function in chemical manufacture and synthesis. Solvents within the process fulfill functions of means of dispersion, dissolving reactants, extracting and washing products, separating mixtures, they act as a heat sink, temperature regulator, low mixture viscosity, and improve mass transfer, conversions, and selectivity (Byrne et al., 2019; Li and Trost, 2008).

The search for solvents that represent a green alternative has led to the generation of solvents selection guides for some process system. Solvent replacement within the processes in the pharmaceutical chemistry industry has been supported by the use of these guides.

The selection of suitable solvents is an important subject in a wide range of chemical processes. The chemical industry, in search of defining which solvents can enter into the classification of green solvents, has worked on their classification. Recommended solvents by pharmaceutical companies derive from worker and process safety, avoiding health and environmental hazards, the sustainability of the process and should be chemically and physically stable, low volatile, and easy to recycle (Ivanković et al., 2017; Byrne et al., 2016). It is essential to think about alternatives that downscale the impact caused by the use of organic solvents. Considering the recommendation for use or the degree of danger of solvents, allow from the beginning of the design to implement measures to avoid or limit the use of some of them.

Table 2.4 presents a review for the general classification of solvents recommended by pharmaceutical companies, label the solvents used in the chemical industry according to their recommendation for use. This classification is based on efforts to improve green chemistry processes, so they include a relevant number of recently evaluated solvents (Byrne, 2016; Prat et al., 2013, 2014).

2.5 EVALUATING GREEN CHEMISTRY PROCESS

The evaluation of green chemistry processes must be focused on compliance with the parameters that ensure the greenness of processes and products, covering disciplines of resources, materials, manufacture, maintenance, life-cycle assessment, and others.

Green chemistry is focused on seeking high efficiency in the use of raw materials, plus to reduce the use and generation of waste and hazardous substances (Song and He, 2018). It is also important to consider the selection of compounds that participate in green chemistry processes. Each reaction design involved in the process should consider the use of green components as solvent and reagents, catalyst and energy consumption must be controlled so that the process has zero or very low environmental impact (Ivanković et al., 2017). These conditions can be obtained by developing reactions that increase the production of the compound of interest and minimize by-products (Li and Trost, 2008).

Green chemical processes must be evaluated under the concepts of green chemistry metrics. The description of the reaction processes, their conversion, stages, and efficiency, in combination with the yield and selectivity are important factors for assessing the specificity in obtaining the compound of interest (Song and He, 2018). The use of chemical principles and ecological engineering has led to the design and implementation of an efficient production process. Metrics like atom economy, E factor, and process mass intensity have contributed to the realization of these objectives. Optimizing

TABLE 2.4 Overall Ranking of General Solvents Selection Guide

Family	Recommended	Recommended or Problematic	Problematic or Hazardous	Hazardous	Highly Hazardous
Water	Water	–	–	–	–
Alcohols	Ethanol i-Propanol n-Butanol	Methanol t-Butanol Benzyl Alcohol Ethylene glycol	–	–	–
Ketones	–	Acetone MEK MIBK Cyclohexanone	–	–	–
Esters	Ethyl acetate Isopropyl Acetate n-Butyl Acetate	Methyl Acetate	–	–	–
Ethers	Anisole	Me-THF	MTBE THF	Diisopropyl ether 1,4-dioxane DME	Diethyl ether
Hydrocarbons	–	Heptane Me-cyclohexane Toluene Xylenes	Cyclohexane	Pentane Hexane	Benzene
Halogenated	–	Chlorobenzene	DCM	–	Chloroform CCl4 DCE
Aprotic polar	Sulfolane	Acetonitrile DMPU DMSO	–	DMF DMAc NMP	Nitromethane
Miscellaneous acids	–	Acetic anhydride	Formic acid	Methoxy-ethanol	–
Amines	–	–	Pyridine	TEA	–

resources, improving environmental sustainability, reducing or eliminating the production of unwanted or harmful, and promoting the use of tools with a common measurement system, implies that the processes be designed under the vision of green chemistry.

2.5.1 ATOM ECONOMY

Efficiency evaluation of chemical processes from clearly defined principles can be done by determining the atom economy (Li and Trost, 2008). In order to measure the efficiency of a reaction, the concept of the atom economy can be used. One of the principles of green chemistry processes is the percent atom economy (%AE). It is defined as the relation of the molecular weight of the requested product with the molecular weight of all atoms in the processes, including solvents, separation and drying agents, but do not participate in the chemical reaction (Patel et al., 2013). The Atom economy relates the conversion efficiency of a chemical process considering the atoms involved in the process. It describes the relationship between the number of atoms of reagents and the number of atoms of generated products, and their maximum value is reached, in an ideal PFR there is no mixing of the medium along the x-axis, the feed

occurs while the product is obtained, presenting an axial dispersion through the reactor because its flat velocity profile, so RTD is narrow. A green chemistry driveway is to minimize as much of the substances involved in the process as possible, and if they cannot be eliminated, reuse or recycling is sought (Patel et al., 2013).

Achieve highly effective processes in function of the atom economy requires the design of chemical processes more efficient in the use of reactions in which the quantity of reactants than end up in the final desire product is the maximum possible. Considering that all reactants used in the process like solvents, separation agents, and drying agent do not participate directly of the chemical reaction, the atom economy should improve upon by a conscientious selection of starting material and catalyst system. The atom economy principle is linked to the principle of not generating waste, since synthetic methods must be designed to maximize the inclusion of all the materials used in the process into the final product (Ivanković et al., 2017).

Green chemistry contemplates that processes must maximize efficiency and reduce waste, including all materials used in the procedure into the final product. This means that all of the atoms from the starting materials should be incorporated into the product, avoiding the generation of by-products. Bu fulfilling this condition, an atom economy equivalent to 100% would be achieved, according to the equation proposed by Trost (1991).

In a chemical process, it is difficult to evaluate only through the economy of the atom if the principle of green chemistry is met, since the process may implicate the use of reagents that generate waste beyond the reaction. When there is an interest in defining the quantity of the desired product and the selectivity about the initial reagent the most common way is to estimate the reaction yield that is the quantity of a product generated by a chemical reaction from a given reactant.

2.5.2   ENVIRONMENTAL FACTOR

Due to the variability of the processes, the determination of the environmental factor (E factor) is suggested as an evaluation method. This method considers both the quantity and property of the by-product or waste. The factor E is defined as the ratio of waste on the product, and its calculation may or may not consider the water in the process. It is common to calculate the E factor without water; however, as there are cases in which the percentage of water involved in the process is high, it is important to determine the method to use.

The metric of the factor E is based on the mass of the waste generated by a given process. Based on the reaction path in the environment, it has been proposed that the E factor is a measure to assess the environmental footprint of chemical processes (Song and He, 2018). The calculation of E factor is determined by the amount of the mass of waste (kg) per unit of product, considering the total amount of waste generated in the process, including the losses of solvents, acids and bases, and the energy waste. Any material that does not correspond to the desired product, such as process products that do not have a subsequent use, reagents and solvents used during the process but that are not recycled or reused, can be considered as waste (Ivanković et al., 2017; Song and He, 2018).

2.5.3   PROCESS MASS INTENSITY (PMI)

Process mass intensity is described as the total mass of the materials used in the process about the mass of the desired product. This quantification method takes into account the quantity of reagents that are no reusable, the auxiliary reagents are the solvents used in the procedure, making this quantification system effective to evaluate the fulfillment of the objectives of a green chemistry process, such as sustainability and resource optimization (Madabhushi et al., 2018).

Process design, based on process mass intensity (PMI) data, can determine the degree of contribution of unit operations to the general process, which allows eliminating those that are not efficient or replacing them with innovative technologies them to achieve better results. This metric can be considered as a way of determining the impact of an operation about to other process routes, as well as determining which process is better in terms of efficiency (Budzinski et al., 2019; Madabhushi et al., 2018). In the case that only the manufacturing process is considered, the scale of the operation does not affect the PMI, so it could also be used to evaluate the efficiency of a process without having to take it to large-scale manufacturing operations (Madabhushi et al., 2018).

Although this metric is commonly applied to small-molecule manufacturing processes, it is applied to biological products also, providing a standard method for collecting mass data on the amount of water, raw material, and consumables, in addition to induce the search for higher efficiency in pharmaceutical syntheses (Budzinski et al., 2019; Jimenez-Gonzalez, 2011). Unlike the E factor, the PMI measures the mass of materials used to make 1 kg of product, although it can also

be determined with or without the inclusion of water. An advantage of PMI is that it is easier to calculate from the inputs in a reaction than the measurements of the waste generated (Dunn, 2012).

Metrics in green chemistry allow quantifying the improvements obtained by implementing sustainable methods, in order to evaluate the sustainability of the process. Although several alternative metrics have been proposed, each method involves different reference points, so that its use falls on the specific needs and conditions of the process to be evaluated. Even the development of new metrics for the evaluation of a process can be performed, which can start from the measurement of reagents, products, and energy to the generation of waste, measurement of the environmental footprint, or improvements in the process economy.

Green chemistry is already a reality in industrial processes; however, its approach to product planning and its synthesis process through environmental-friendly routes have opened new areas of opportunity orienting to new developments and processes that enter into this concept. One of the key challenges it is to achieve through new and innovative processes, products, and operations that are highly efficient in terms of sustainability and that reduce environmental impact.

KEYWORDS

- green chemistry
- process
- green reactors
- polymerization systems

REFERENCES

Anastas, P. Introduction green chemistry. Chem. Rev. 2007, 107, 6, 2167–2168.

Anastas, P.; Eghbali, N. Green chemistry: principles and practice. Chem. Soc. Rev. 2010, 39, 1, 301–312.

Asgharzadehahmadi, S.; Raman, A.; Parthasarathy, R.; Sajjadi, B. Sonochemical reactors: review on features, advantages and limitations. Renew. Sustain. Energy Rev. 2016, 63, 302–314.

Bakshi, B.R. Toward sustainable chemical engineering: the role of process systems engineering. Annu. Rev. Chem. Biomol. Eng. 2019, 10, 265–288.

Boodhoo, K. Process intensification for green chemistry: engineering solutions for sustainable chemical processing. In: Spinning Disc Reactor for Green Processing and Synthesis. Harvey A John Wiley & Sons. 2013, 59–90.

Budzinski, K.; Blewis, M.; Dahlin, P.; D'Aquila, D.; Esparza, J.; Gavin, J.; Ho, SV.; Hutchens, C-; Kahn, D.; Koenig, AG.; Kottmeier, R.; Millard, J.; Snyder, M.; Stanard, B.; Sun, L. Introduction of a process mass intensity metric for biologics. New Biotechnol. 2019, 49, 37–42.

Byrne, F.P.; Jin, S.; Paggiola, G.; Petchey, T.H.M.; Clark, J.H.; Farmer, T.J.; Hunt, A.J.; McElroy, C.R.; Sherwood, J. Tools and techniques for solvent selection: green solvent selection guides. Sustain. Chem. process. 2016, 4:7, 1–24.

Cao, C.; Zhang, N.; Chen, X.; Cheng, Y. A comparative study of Rh and Ni coated microchannel reactor for steam methane reforming using CFD with detailed chemistry. Chem. Eng. Sci. 2015, 137, 276–286.

Charpentier, J-C. What kind of modern "green" chemical engineering is required for the design of the "factory of future"? Proc. Eng. 2016, 138, 445–458.

Chatel, G. How sonochemistry contributes to green chemistry? Ultrason. Sonochem. 2018, 40, 117–122.

Cheng, H.N.; Gross, R.A. Green Polymer Chemistry: Biocatalysis and Biomaterials. ACS Symposium Series. American Chemical Society: Washington, DC, 2010.

Cheng, H.N.; Gross, R.A.; Smith, P.B. Green Polymer Chemistry: Biobased Materials and Biocatalysis. ACS Symposium Series 1192. American Chemical Society: Washington, DC, 2015.

Cheng, H.N.; Smith, P.B.; Gross, R.A. Green Polymer Chemistry: A Brief Review. ACS Symposium Series. American Chemical Society: Washington, DC, 2013.

Clark, J.H.; Luque, R.; Matharu, A.S. Green chemistry, biofuels, and biorefinery. Annu. Rev. Chem. Biomol. Eng. 2012, 3, 1, 183–207.

Curran, M.A. Biobased Materials. Kirk-Othmer Encyclopedia of Chemical Technology. John Wiley & Sons, Inc.: Hoboken, NJ, 2010, 1–19.

Doble, M.; Kruthiventi, A.K. Process and operations. In: Green Chemistry and Engineering. Elsevier, 2007, 105–170.

Dudukovic, M.P. Frontiers in reactor engineering. Science. 2009, 325, 698–701.

Dudukovic, M.P. Reaction engineering: status and future challenges. Chem. Eng. Sci. 2013, 65, 3–11.

Dunn, P.J. The importance of green chemistry in process research and development. Chem. Soc. Rev. 2012, 41, 1452–1461.

Elvira, K.S.; Casadevall I.S.; Wootton R.C.R.; deMello A.J. The past, present and potential for microfluidic reactor technology in chemical synthesis. Nat. Chem. 2013, 5, 905–915.

Fogler, H.S. Elementos de ingeniería de las reacciones químicas, 4th ed. Person Education: México, 2008.

Holkar, C.R.; Jadhav, A.J.; Pinjari, D.V.; Pandit, A.B. Cavitationally driven transformations: a technique of process intensification. Ind. Eng. Chem. Res. 2019, 58, 5797–5819.

Ilyas, S.; Farhan, M.; Bhatti, H.N. Role of green and integrated chemistry in sustainable metallurgy. In: Integrating Green Chemistry and Sustainable Engineering. Shahid, U-I. Scrivener Publishing LLC, 2019, 325–342.

Ivanković, A.; Dronjic, A.; Bevanda, A.M.; Talić S. Review of 12 principles of green chemistry in practice. Int. J. Sustain. Green Energy 2017, 6, 3, 39–48.

Jimenez-Gonzalez, C.; Ponder, C.S.; Broxterman, Q.B.; Manley, J.B. Using the right green yardstick: why process mass intensity is used in the pharmaceutical industry to drive more sustainable processes. Org. Process Res. Dev. 2011, 15, 912–917.

Lerou, J.J.; Tonkovich, A.L.; Silva, L.; Perry, S.; McDanie,l J. Microchannel reactor architecture enables greener processes. Chem. Eng. Sci. 2010, 65, 380–385.

Li, C.J.; Trost, B.M. Green chemistry for chemical synthesis. Proc. Natl. Acad. Sci. USA, 2008, 105, 36, 13197–13202.

Lin, B.; Tao, Y. Whole-cell biocatalysts by design. Microb. Cell Fact. 2017, 16, 106.

Lupette, J.; Maréchal, E. Phytoplankton glycerolipids: challenging but promising Prospects from Biomedicine to Green Chemistry and Biofuels. Blue Biotechnology, First ed. Wiley-VCH, 2018, 1, 191–215.

Macedonio, F.; Drioli, E. Membrane engineering for green process engineering. Engineering 2017, 3, 290–298.

Madabhushi, S.R.; Gavin, J.; Xu, S.; Cutler, C.; Chmielowski, R.; Rayfield, W.; Tugcu, N.; Chen, H. Quantitative assessment of environmental impact of biologics manufacturing using process mass intensity analysis. Biotechnol. Prog. 2018, 34, 6, 1566–1573.

de Marco, B.A.; Rechelo, B.S.; Tótoli, E.G.; Kogawa, A.C.; Salgado, H.R.N. Evolution of green chemistry and its multidimensional impacts: a review. Saudi Pharm. J. 2019, 27, 1–8.

Pask, S.D.; Nuyken, O.; Cai, Z. The spinning disk reactor: an example of a process intensification technology for polymers and particles. Polym. Chem. 2012, 3, 2698–2707.

Patel, K.R.; Sen, D.J.; Jatakiya, V.P. Atom economy in drug synthesis is a playground of functional groups. Am. J. Adv. Drug Deliv. 2013, 1, 073–083.

Powell, J.B. Application of multiphase reaction engineering and process intensification to the challenges of sustainable future energy and chemicals. Chem. Eng. Sci. 2016, 157, 15–25.

Prat, D.; Hayler, J.; Wells, A. A survey of solvent selection guides. Green Chem. 2014, 16, 4546–4551.

Prat, D.; Pardigon, O.; Flemming, H.W.; Letestu, S.; Ducandas, V.; Isnard, P.; Guntrum, E.; Senac, T.; Ruisseau, S.; Cruciani, P.; Hosek, P. Sanofi's solvent selection guide: a step toward more sustainable processes. Org. Process Res. Dev. 2013, 17, 1517–1525.

Reay, D.; Ramshaw, C.; Harvey, A. Reactors. In: Process Intensification. Elsevier, 2008, 103–186.

Sheldon, R.A.; Woodley, J.M. Role of biocatalysis in sustainable chemistry. Chem. Rev. 2018, 118, 801–838.

Song, Q.W.; He, L.N. Atom Economy. Encyclopedia of sustainability science and technology. Springer: New York, NY, 2018.

Suryawanshi, P.L.; Gumfekar, S.P.; Bhanvase, B.A.; Sonawane, S.H.; Pimplapure,

M.S. A review on microreactors: reactor fabrication, design, and cutting-edge applications. Chem. Eng. Sci. 2018, 189, 431–448.

Trost, B.M. The atom economy—a search for synthetic efficiency. Science. 1991, 254, 1471–1477.

Truppo, M.D. Biocatalysis in the pharmaceutical industry: the need for speed. ACS Med. Chem. Lett. 2017, 8, 5, 476–480.

Yao, X.; Zhang, Y.; Du, L.; Liu, J.; Yao, J. Review of the applications of microreactors. Renew. Sustain. Energy Rev. 2015, 47, 519–539.

Zhang, L.; Babi, D.K.; Gani, R. New vistas in chemical product and process design. Annu. Rev. Chem. Biomol. Eng. 2016, 7, 557–582.

Zhang, G.; Jin, W.; Xu, N. Design and fabrication of ceramic catalytic membrane reactors for green chemical engineering applications. Engineering. 2018, 4, 848–860.

CHAPTER 3

Catalytic Routes in Biomass Conversion: Synthesis of Furfural and HMF

DENIS A. CABRERA-MUNGUIA[1], HORACIO GONZÁLEZ[2],
ADOLFO ROMERO-GALARZA[3*], L. FARÍAS-CEPEDA[3],
LUCERO ROSALES-MARINES[3], and SANDRA L. CASTAÑÓN-ALONSO[4]

[1]*Advanced Materials, School of Chemistry, Autonomous University of Coahuila, México*

[2]*Faculty of Chemical Engineering, Universidad Michoacana de San Nicolás de Hidalgo, México*

[3]*Department of Chemical Engineering, School of Chemistry, Autonomous University of Coahuila, México*

[4]*Chemistry Department, Universidad Autónoma Metropolitana-Iztapalapa, Ciudad de México, México*

Corresponding author. E-mail: a_romero@uadec.edu.mx

ABSTRACT

Furfural and 5-(hydroxymethyl) furfural (HMF) are known as building block molecules for the production of value-added chemicals and oxygenated biofuels with better physicochemical properties than bioethanol in terms of its use as a fuel. The obtention of these furanic compounds is through the conversion of monosaccharides (pentoses and hexoses), available in lignocellulosic biomass, in the presence of acid catalysts. Given the industrial implementation of solid catalysts, many systems have been analyzed such as zeolites, mesoporous silicates, heteropoly acids, and ionic liquids. However, many aspects like catalytic deactivation and regeneration, economic cost and mechanical stability should be considered. Then, the purpose of this chapter is to give a glance to the reader about the mechanisms, reaction conditions, and catalytic systems that have been studied in the conversion of monosaccharides to obtain furfural and HMF, approaching on the most promising materials for the production of these furanic compounds.

3.1 INTRODUCTION

Nowadays, scientific researchers and politicians are working together to sum up efforts to find new alternatives resources of energy, reducing at the same time the dependence upon fossil fuels and therefore the greenhouse-gas emissions. Among the possible available feedstocks to synthesize biofuels and value-added chemical products, lignocellulose is the most abundant biomass resource, which is found in woody biomass and agricultural residues that consists of 35%–50% of cellulose, 20%–35% of hemicellulose and 10%–25% of lignin (Ma et al., 2019).

Hemicellulose is an amorphous heteropolymer made of D-glucose and pentoses such as xylose, rhamnose, and arabinose. Hemicellulose is hydrolyzed effortlessly in an acid or basic medium (Figure 3.1a), obtaining pentoses that under acid conditions are dehydrated giving rise to furfural. Whereas, cellulose is a polymer of D-glucose with an intramolecular hydrogen bonding interaction, obtaining a robust crystalline structure. Cellulose is hydrolyzed under acid or basic conditions leading to glucose (Figure 3.1b),

then glucose requires a step of isomerization to produce fructose before carrying out its dehydration that leads to 5-(hydroxymethyl)furfural (HMF).

Furfural and HMF are platform molecules obtained from the conversion of monosaccharides, with application in the synthesis of biofuels, fuel additives, biopolymers, and value-added chemicals, which are used in the pharmaceutical, agrochemical, and plastic industries (Zhang et al., 2018).

The production of furfural is by means of the dehydration of xylose using Brönsted acid catalysts, whereas glucose requires the first stage of isomerization (using Lewis acid catalysts) to produce fructose, followed by dehydration on Brönsted type acid catalyst to obtain HMF. Along with isomerization and dehydration of monosaccharides (pentoses and hexoses), several secondary reactions like rehydration and polymerization are carried out, leading to a loss of furfural and HMF yield.

Furfural was initially applied to the foundry industry. Later, its use was proposed as a furan-based chemical to produce furfuryl alcohol (FA), tetrahydrofurfuryl alcohol (THFA), 2-methyltetrahydrofuran (MTHF) and 2-furoic acid, which are useful as monomers and solvents in the industry. The current furfural producers at industrial scale are China, South Africa, and the

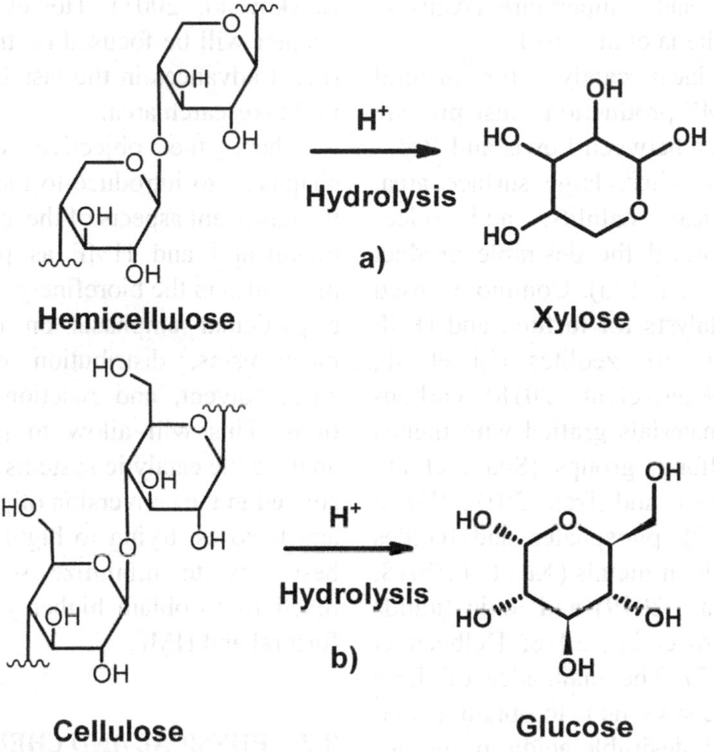

FIGURE 3.1 Hydrolysis under acid conditions of (a) hemicellulose and (b) cellulose.

Dominican Republic with a production estimated of 300,000 tons/year. The process involves the modification of the original batch production by Quaker Oats Technology to a one-step process consisting of hydrolysis of biomass, and dehydration of pentoses. However, all of them still need acidic streams to hydrolyze biomass (Dashtban et al., 2012). HMF is a furanic molecule from which can be derived ethers, esters, FA, formic acid, and levulinic acid that can be used as biofuels, resins, and pharmaceuticals. Nonetheless, the actual production of HMF at industrial scale is minimum, and most of it is sold as chemicals for laboratory and research purposes (Portillo Perez et al., 2019).

In order to enhance the production of furfural and HMF and at the same time to reduce the use of mineral acids, two strategies are proposed: (1) the use of tailor-made heterogeneous acid catalysts and (2) the improvement of reaction systems regarding reactor configuration, solvent, and temperature (Agirrez-abal-Telleria et al., 2014).

An ideal catalyst for furfural and HMF production must provide a balance between Lewis and Brönsted acid sites, large surface area, mechanical stability, and selectivity toward the desirable product (Yu et al., 2017a). Commonly used acid catalysts for furfural and HMF synthesis are zeolites (Li et al., 2018; Wang et al., 2018), carbon-based materials grafted with metals and sulfonic groups (Shaik et al., 2018; Tran and Tran, 2019; Wu et al., 2019), phosphates and halides of transition metals (Xu et al., 2018; Liu et al., 2019) and ionic liquids (Peleteiro et al., 2016; Delbecq et al., 2017). The main idea of these catalytic systems is to obtain a catalyst with desirable acidic properties to catalyze the reaction pathways of furfural and HMF synthesis (Jia et al., 2019). While boiling point, partition coefficient, and thermal stability are properties that must be analyzed to determine the performance and recyclability of the reaction solvent (Yu et al., 2017b).

Thus a deep understanding of the type and distribution of acid sites in the catalyst is necessary and also the role that Lewis and Brönsted acid sites play in every catalytic step during the production of furfural and HMF. In the literature, there are relevant reviews related to this field (Antal Jr, 1990; Kuster, 1990;

Lewkowski, 2001). However, this chapter will be focused on the most recent advances in the last 10 years in this research area.

Then, the objective of this chapter is to introduce to the reader to significant aspects of the synthesis of furfural and HMF as platform molecules in the biorefinery. Making a particular emphasis on reaction mechanisms, distribution of acid sites, solvent, and reaction conditions. This will allow to properly analyze the catalytic systems already studied in the conversion of pentoses and hexoses, trying to highlight the best way to minimize secondary reactions to obtain higher yields of furfural and HMF.

3.2 PHYSICAL AND CHEMICAL PROPERTIES OF FURFURAL AND 5-(HYDROXYMETHYL) FURFURAL

Furfural is an aromatic aldehyde with molecular formula $C_5H_4O_2$ and a molecular weight of 96.08. Physically is an oily and colorless liquid with an odor of almonds. Furfural is used as (1) a selective extractant for the removal of aromatics in lubricating oils, diesel fuels, and cross-linked polymers, (2) an effective fungicide especially in inhibiting the growing of wheat smut, and (3) intermediary in plastics production (Yan et al., 2014).

Furfural possesses as functional groups an aldehyde (–CHO) and a conjugated system of double bonds in the aromatic ring (C=C–C=C) that make it a reactive molecule and then a platform molecule for value-added products. Thus furfural can carry out the typical reactions of aldehydes such as reduction to alcohols, reductive amination to amines, and oxidation to carboxylic acids. The conjugated system in the ring can carry out alkylation, hydrogenation, oxidation, halogenation, ring-opening, and nitration reactions (Yan et al., 2014).

On the other hand, HMF is a white solid possessing a low melting point (30–34 °C), this compound comes from the dehydration of fructose produced from glucose isomerization, being soluble in water and organic solvents. Thus the hydrolysis of cellulose contained in lignocellulosic biomass by mineral acids or by an enzymatic method leads to its chemical transformation into hexoses. However, HMF can also be produced from food containing carbohydrates (e.g., juice, wine, bread, and syrup) when it is subjected to thermal treatment (cooking or drying) (Yu et al., 2017b). Indeed, HMF is produced at industrial scale by syrups extraction from energy crops.

Chemically, HMF possesses a furan ring bonded to a hydroxyl and aldehyde group at the exocyclic carbon atoms. The alcohol groups can undergo esterification, dehydration, oxidation, and halogenation reactions; whereas the aldehyde group can carry out reduction, decarboxylation, and reductive amination. Moreover, the ring structure can also react through halogenation, sulphonation, Friedel–Crafts alkylation or acylation, and Diels-Alder cycloaddition (Portillo Pérez et al., 2019). The general physicochemical properties of HMF and furfural are exhibited in Table 3.1.

TABLE 3.1 Physicochemical Properties of Furfural and HMF (Yan et al., 2014)

Compound	Furfural	HMF
Molecular weight, g/mol	96	126
Boiling point, °C	161.7	291
Density at 25°C	1.16	1.29
Solubility in water, wt.%	8.3	Soluble
Critical Pressure, MPa	5.5	5.64
Critical temperature, °C	397	515
Heat of combustion, kJ/mol	234.4	159
Heat of vaporization, kJ/mol	42.8	54.3
Flash point, °C	61.7	79.4

3.3 VALUE-ADDED CHEMICAL PRODUCTS FROM FURFURAL

Hemicellulose is a polymer present in biomass that is hydrolyzed to xylose, a C_5 sugar which is the precursor of chemicals, such as furfural and tetrahydrofuran (THF). Commonly, the synthesis of furfural is through dehydration of xylose in the presence of sulfuric acid. However, the degradation of furfural is a secondary reaction caused by the acid medium, decreasing the yield of this process. To overcome this issue is used a solvent system made of water and an organic solvent, where the furfural produced is extracted (Sato et al., 2019). The best way to transform furfural is by catalytic hydrogenation. In the next section, some of the main products of furfural are presented.

3.3.1 FURFURYL ALCOHOL

FA is a relevant chemical product that is commonly prepared by hydrogenation of furfural in the liquid or gas phase. The main applications of FA are the production of resins, polymers, and coatings, which have shown high resistance to acids and alkalis. Also, it serves as a solvent for resins and poorly insoluble pigments. Furthermore, FA is used as a precursor of THFA and 2,3-dihydropyran, an intermediary in the synthesis of lysines, ascorbic acid, lubricants, and plasticizers.

Synthesis of FA by liquid-phase hydrogenation has been carried out using solid catalysts based on Pd, Pt, Ru, and bimetals catalysts of Pd–Ni, Pd–Ir, Pd–Ru, Pt–Sn. However, a gas phase hydrogenation to produce FA is typically used with the application of a Cu–Cr catalyst at 130–200 °C and a hydrogen pressure of 30 bars (Figure 3.2a). Indeed, at 175 °C yields of FA around 96%–99% were obtained. However, when the reaction temperature increases to 250 °C, many other chemicals are produced, such as 2-methylfuran (36%), *n*-pentanol (35%), 1,5-pentanediol (15%), and 1,2-pentanediol (14%). Nonetheless, as Cr is a hazardous metal, there is still a need to design catalysts without Cr (Yan et al., 2014; Du et al., 2019; Salnikova et al., 2019).

3.3.2 TETRAHYDROFURFURYL ALCOHOL

THFA is a transparent, high-boiling liquid that is completely miscible with water. This compound is considered a green solvent as it is originated from furfural, which is obtained from a two-step process: (1) an acid hydrolysis of hemicellulose (lignocellulosic biomass) to obtain pentoses (xylose), and (2) acid-catalyzed dehydration of pentoses to yield furfural. It serves as a solvent in agricultural applications, printing inks, industrial and electronic cleaners. Also, it is applied as a solvent for fats and resins, and a building block in the synthesis of polyols. Traditionally it is produced by a two-step catalytic hydrogenation of furfural: (1) furfural hydrogenation to FA using Cu–Cr (Figure 3.2a), and (2) FA hydrogenation to THFA with noble metal catalysts, whereas in the industrial practice Ni catalyst is preferred (Figure 3.2b). Long reaction times, high energy consumption, and toxicity of the Cu–Cr catalytic systems are the main disadvantages, then it is of great importance to develop a single-step method for furfural hydrogenation to produce THFA (Yan et al., 2014; Feng et al., 2018; Li et al., 2018).

3.3.3 2-METHYL FURAN AND 2-METHYL TETRAHYDROFURAN

They are colorless liquids that act as powerful solvents. Also, 2-methyl furan (MF) and 2-methyl tetrahydrofuran (MTHF) are biofuel components that can be mixed with gasoline, and present physical properties comparable to bioethanol such as boiling point (ethanol: 78 °C, MF: 64 °C and MTHF: 80.3 °C) and flash point (ethanol: 13 °C, MF: −22 °C and MTHF: −12 °C). Additionally, MF is used as feedstock for the production of antimalarial drugs. Meanwhile, MTFH is applied as a substitute of THF and as the electrolyte of lithium electrodes.

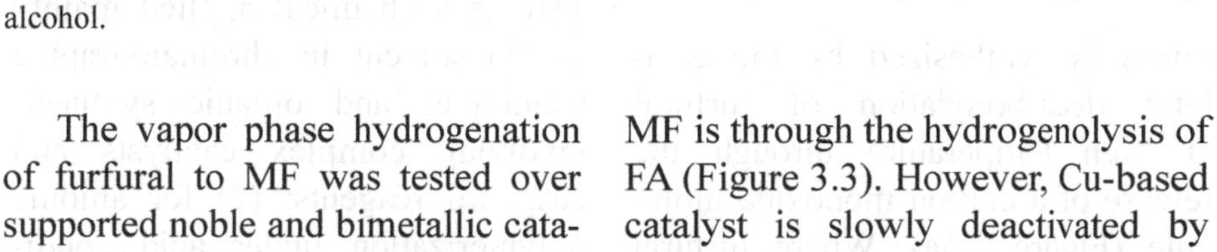

FIGURE 3.2 Reaction pathways of Furfural into a) Furfuryl alcohol, and b) Tetrahydrofurfuryl alcohol.

The vapor phase hydrogenation of furfural to MF was tested over supported noble and bimetallic catalysts. However, the selective conversion of furfural to MF has been obtained using Cu-based catalysts at high temperature and low pressure. Also, It is demonstrated that furfural conversion to MF proceeds via the formation of FA as an intermediary, where the production of MF is through the hydrogenolysis of FA (Figure 3.3). However, Cu-based catalyst is slowly deactivated by coke formation at low temperature (130 °C), which can be regenerated by coke burn off at 400 °C (Yan et al., 2014).

For the production of MTHF, two routes are identified (Figure 3.4): (1) hydrogenation through levulinic acid raw material and (2) hydrogenation

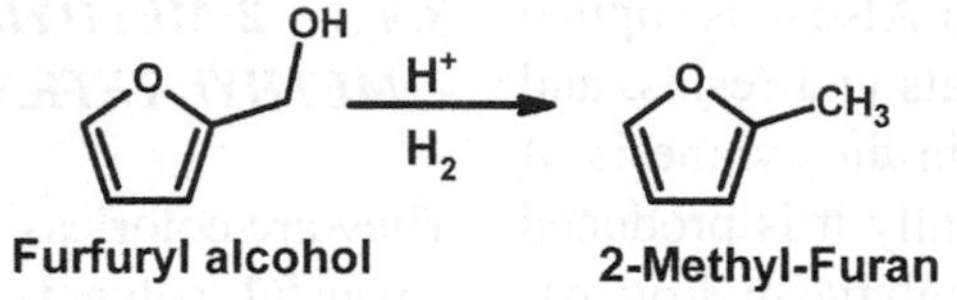

FIGURE 3.3   Hydrogenolysis of furfuryl alcohol into 2-methyl-furan.

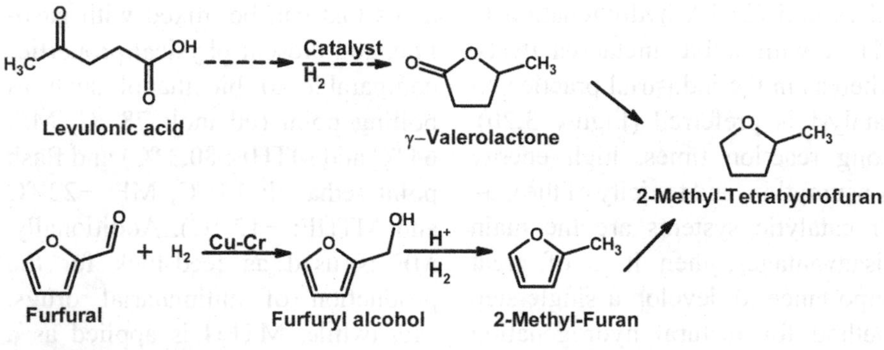

FIGURE 3.4   Main reaction pathways for MTHF production.

of MF. Homogeneous Ru-derived catalysts, Cu composites, and noble metal catalysts were used in this process (Yan et al., 2014)

3.3.4   FURAN

Furan is synthesized by the catalytic decarbonylation of furfural at high temperature through the release of a carbon monoxide molecule (Figure 3.5a). Where furfural goes through Cannizaro reaction to produce furoic acid, following the decarboxylation that produces furan. In this case, a noble metal catalyst is a preferred material for the decarbonylation reaction. Furan is produced by heating furfural to 158 °C over 5%

Pd catalyst supported on microporous carbon with potassium carbonate as promotor (Yan et al., 2014).

3.3.5   TETRAHYDROFURAN

THF is a chemical applied mainly as (1) solvent in chromatographic techniques and organic synthesis involving complex catalysts and Grignard reagents; (2) for anionic polymerization under acid conditions; and (3) adhesives. THF is synthesized from furan or furfural (Figure 3.5b) using noble metal catalysts, and Ni derived catalysts, being the main issue to solve the coke produced during the process (Yan et al., 2014).

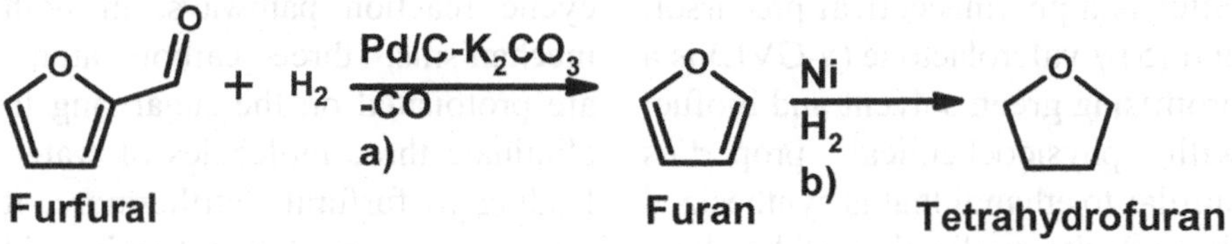

Furfural **Furan** **Tetrahydrofuran**

FIGURE 3.5 Reaction pathways of furfural into (a) furan, and (b) tetrahydrofuran.

3.4 VALUE-ADDED CHEMICAL PRODUCTS FROM 5-(HYDROXYMETHYL) FURFURAL

HMF is a platform molecule produced from cellulosic hexoses. The possible value-added chemicals synthesized from HMF are (Figure 3.6) (1) 2,5-furandicarboxylic acid from HMF oxidation that serves as a monomer for manufacturing of plastics (Lewkowski Jaroslaw, 2001); (2) 2,5-dimethylfuran a liquid with a higher octane number (119) than gasoline, and higher boiling point (92–94 °C) than ethanol; physicochemical properties that make it attractive as an alternative energy source, with better physicochemical properties than bioethanol in terms of energy (Nishimura et al., 2014); (3) 2,5-bis(hydroxymethyl) furan is a solid that comes from the reduction of HMF over Cu–Cr catalyst (Lewkowski Jaroslaw, 2001); (4) formic and levulinic acids are byproducts of HMF production, this

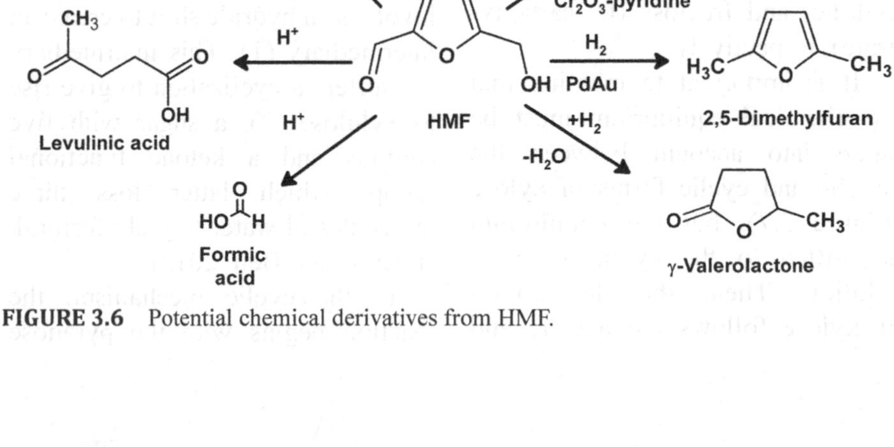

FIGURE 3.6 Potential chemical derivatives from HMF.

latter is a pharmaceutical precursor, and (5) γ-valerolactone (γ-GVL) is a promising green solvent and biofuel with physicochemical properties similar to ethanol that is synthesized through the cyclization of levulinic acid under acid conditions, and then hydrogenated to γ-GVL (Yan et al., 2014). The central idea is that these chemical compounds could replace their corresponding petroleum-derived compounds which are used in polymer production, as solvents, biofuels, fungicides, resins, and pharmaceuticals (Dou et al., 2018; Portillo Perez et al., 2019).

3.5 MECHANISM FOR FURFURAL SYNTHESIS

As mentioned before, the conversion of biomass to furfural or HMF in only one step is a difficult task. However, hemicellulose and cellulose can be hydrolyzed by dilution with H_2SO_4 to extract pentoses and hexoses (Jia et al., 2019). Hence, furfural and HMF are obtained from the dehydration of xylose (C_5 carbohydrate) and fructose (C_6 carbohydrate), respectively.

It is important to mention that the chemical equilibrium must be taken into account between the acyclic and cyclic forms of xylose (Figure 3.7), but the equilibrium is shifted to the cyclic form in solution. Then, the dehydration of xylose follows the acyclic and

cyclic reaction pathways, in both mechanisms, three carbon atoms are protonated on the sugar ring to eliminate three molecules of water, leading to furfural. Furthermore, it has been assessed that Lewis acid sites assist the isomerization of the carbohydrate, and then Brönsted acid sites promote the direct dehydration of the carbohydrate (Agirrezabal-Telleria et al., 2014; Yan et al., 2014).

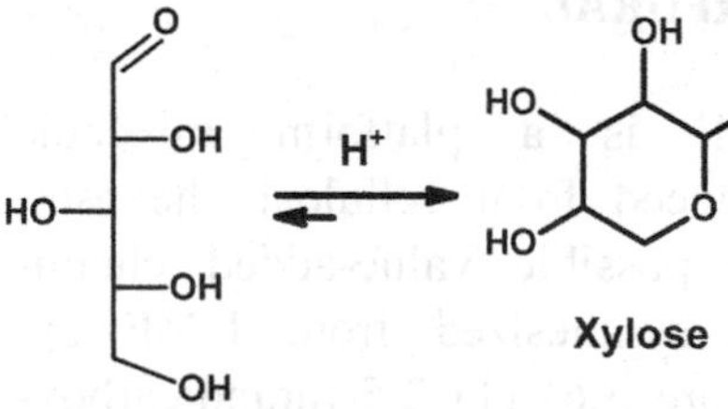

FIGURE 3.7   Chemical equilibrium of the acyclic and cyclic xylose forms.

3.5.1   *CYCLIC AND ACYCLIC DEHYDRATION OF PENTOSES (XYLOSE)*

The acyclic dehydration mechanism of pentoses (Figure 3.8) starts with the acyclic form of xylose and its subsequent enolization, which involves an hydride shift to create an intermediary (1). This intermediary (1) suffers a cyclization to give rise to xylulose (2), a sugar with five carbons and a ketone functional group, which latter loss three molecules of water to yield furfural. (Enslow and Bell, 2012).

In the cyclic mechanism, the reaction begins with the pyranose

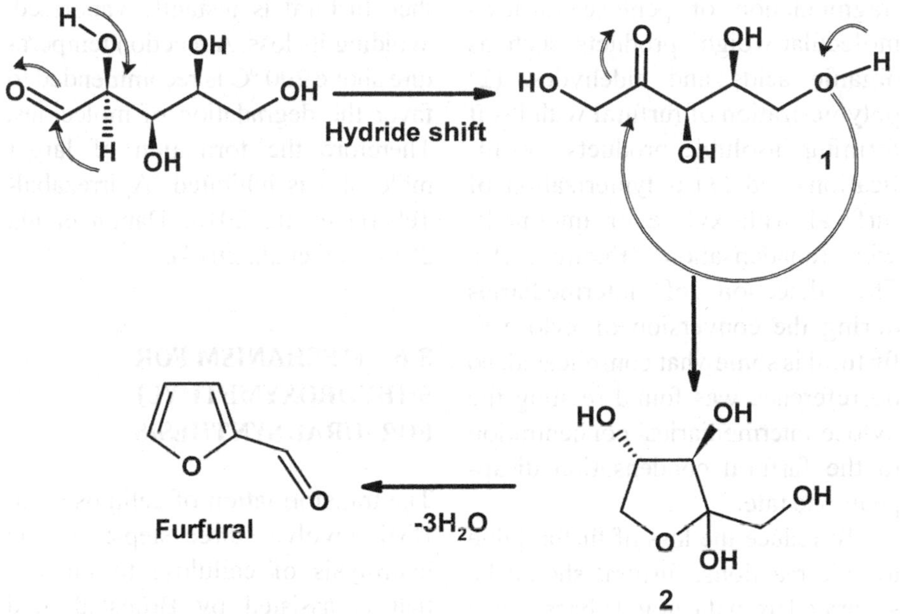

FIGURE 3.8 Acyclic mechanism of dehydration of xylose.

form of xylose, when a proton hydrogen (H+) attacks a nonbonding electron pair of a hydroxyl oxygen bond to a carbon atom, creates a positive charge on the oxygen atom (Figure 3.9). The positive charge is stable when is changed to the neighboring carbon, producing a cleavage of the C–O bond, the release of a water molecule (1,2 elimination) and the furanose form of pentose (1). The furanose intermediate is then dehydrated by means of a 1,2 elimination

(2), this unsaturated furanose form is one again dehydrated leading to furfural (Danon et al., 2014).

3.5.2 SIDE REACTIONS

When furfural is formed in a liquid medium and under acid conditions, side reactions occur. These yield-loss reactions include (Agirrezabal-Telleria et al., 2014; Danon et al., 2014; Yan et al., 2014) (1)

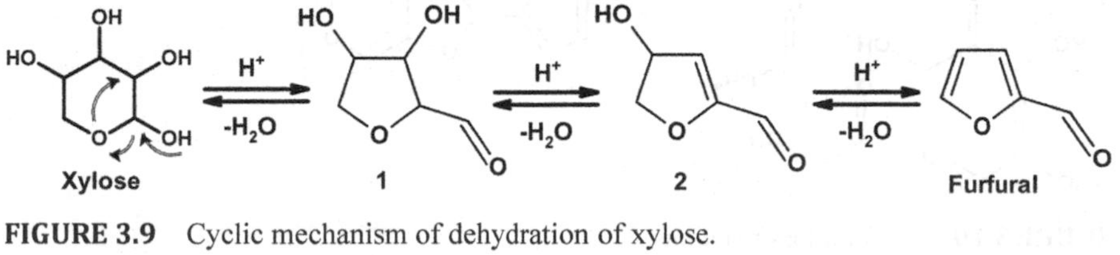

FIGURE 3.9 Cyclic mechanism of dehydration of xylose.

fragmentation of pentose to low-molecular-weight products such as organic acids and aldehydes, (2) polymerization of furfural with itself forming insoluble products (resinification) and (3) polymerization of furfural with xylose or intermediaries (condensation) (Figure 3.10). The detection of intermediaries during the conversion of xylose to furfural is somewhat complicated, so no reference was found relating the xylose intermediaries' concentration to the furfural condensation disappearance rate.

To reduce the loss of furfural due to side reactions, furfural should be separated from the liquid phase as fast as it is produced by steam stripping,

then furfural is instantly vaporized, avoiding its loss. A reaction temperature above 200 °C is recommended to favor the degradation of molecules. Therefore the formation of larger molecules is inhibited (Agirrezabal-Telleria et al., 2014; Danon et al., 2014; Yan et al., 2014).

3.6 MECHANISM FOR 5-(HYDROXYMETHYL) FURFURAL SYNTHESIS

The transformation of cellulose into HMF involves three steps: (1) the hydrolysis of cellulose to glucose that is assisted by Brönsted acid sites, (2) isomerization of glucose

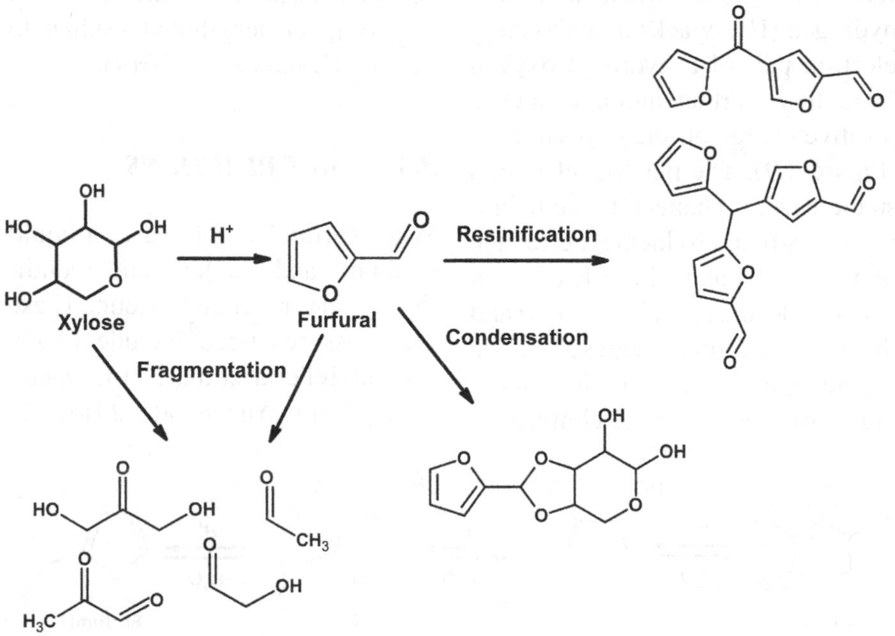

FIGURE 3.10 Furfural loss reactions.

to fructose, where Lewis acid sites are necessary; and (3) dehydration of fructose into HMF, promoted by Brönsted acid sites (Popa and Volf., 2018, Yu et al., 2017b). The chemical equilibrium between the acyclic and cyclic forms of fructose is illustrated in Figure 3.11, but the cyclic form is favored by the equilibrium, being predominant in solution.

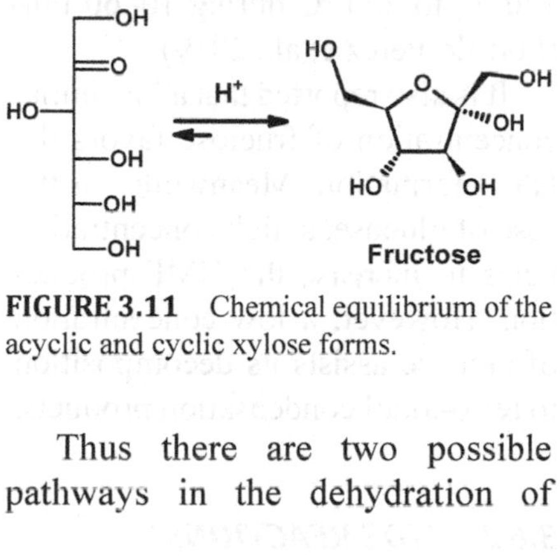

FIGURE 3.11 Chemical equilibrium of the acyclic and cyclic xylose forms.

Thus there are two possible pathways in the dehydration of

fructose to HMF: chain dehydration (acyclic) and annular dehydration (cyclic). The acyclic dehydration (Figure 3.12) involves the protonation of a hydroxyl group that under acid conditions follows to the loss of a water molecule (1,2 elimination); the intermediate suffers two consecutive (1,2 elimination) of water to obtain HMF. In the annular dehydration mechanism (Figure 3.13), a fructose molecule is dehydrated to form an enol-type intermediary (1); the intermediary (2) is generated by dehydration of the intermediary (1). Finally, a third water molecule is removed from the intermediary (2) to form HMF (Wang et al., 2019).

The literature suggests that Brönsted acid sites can assist the direct dehydration of glucose to synthesize HMF, which avoids the glucose–fructose isomerization; and it is proposed

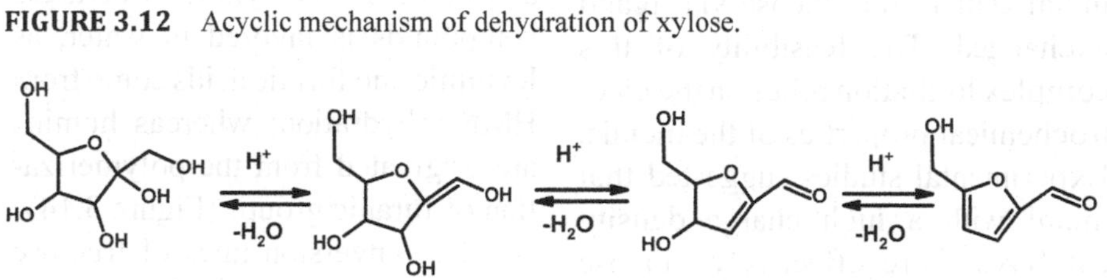

FIGURE 3.12 Acyclic mechanism of dehydration of xylose.

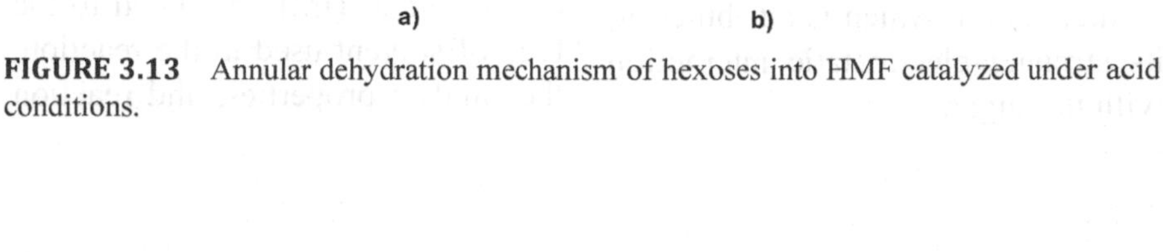

FIGURE 3.13 Annular dehydration mechanism of hexoses into HMF catalyzed under acid conditions.

a cyclic mechanism for that reaction pathway. Nonetheless, the activation energy for the direct glucose dehydration is 36.4 kcal mol^{-1} which is relatively high compared to the value of 29.4 kcal mol^{-1} for fructose dehydration (Yu et al., 2017b). Hence, only when Brönsted acidity is significant in the catalyst, the direct dehydration of glucose could take place.

Meanwhile, Lewis acid catalysts such as trivalent metal chlorides (CrCl$_3$, AlCl$_3$) facilitates the ring-opening of the glucose, which forms a complex with the acyclic glucose at its hydroxyl oxygen at C1 and C2. This configuration helps the Lewis acid site to polarize the carbonyl group (C1 glucose), promoting the transfer of a hydrogen atom from C2 to C1 and finally forming fructose (Yu et al., 2017b).

Among metal chlorides, chromium and aluminum chlorides have given the best yield of HMF during glucose conversion using dimethyl sulfoxide (DMSO) as a solvent. X-ray absorption spectroscopy and DFT calculation studies showed the formation of a complex between the metal center and glucose via ligand exchanged. The feasibility of this complex formation relies on the electrochemical properties of the metals. Experimental studies suggested that metal with a high charge density catalyzes more effectively glucose isomerization, which is attributed to its stronger electrostatic interaction with the sugar.

3.6.1 REACTION CONDITIONS

The temperature level determines the reaction pathways followed during the conversion of glucose or fructose. At 175–400 °C dehydration, retro-aldol condensation, or condensation can be presented to form humins, and their extent will depend on the temperature choice, being a possible, desirable temperature range from 170 °C to 220 °C during 10–60 min (Portillo Pérez et al., 2019).

It is also reported that a low initial concentration of fructose favors the HMF formation. Meanwhile, in the case of glucose, a high concentration helps to increase the HMF production. However, a low concentration of glucose assists its decomposition to retro-aldol condensation products.

3.6.2 SIDE REACTIONS

The straightforward route to produce HMF is the dehydration of fructose under acid conditions. Nonetheless, by-products are formed too, generating levulinic acid, formic acid, and humins. The synthesis of these compounds is favored in water, as levulinic and formic acids come from HMF rehydration; whereas humins are originated from the polymerization of furanic groups (Figure 3.14).

The conversion rates of fructose or glucose to HMF is related to the type of solvent used in the reaction, the catalyst properties, and reaction

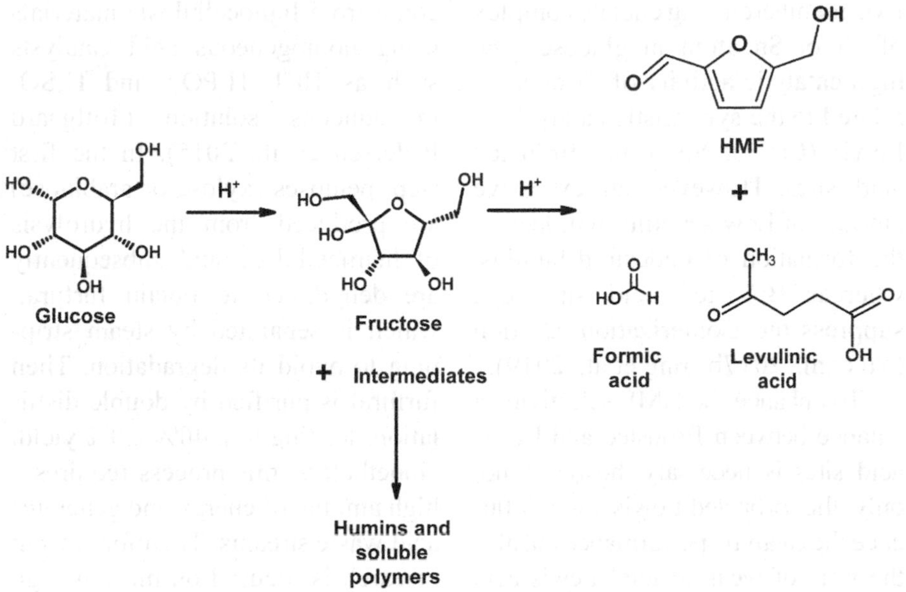

FIGURE 3.14 Side products during HMF synthesis.

temperature. To increase the HMF selectivity and reduced the yield of these side reactions, researchers have analyzed several alternatives such as the combined use of water and organic solvents (such as DMSO), the addition of halides salts, ionic liquids, deep eutectic solvents (DES), and also the water removal under reduced pressure (Lewkowski Jaroslaw, 2001; Agirrezabal-Telleria et al., 2014; Antonetti et al., 2017; Duo et al., 2018).

3.6.3 ROLE OF LEWIS AND BRÖNSTED ACID SITES

It is known that Brönsted acid sites assist the sugar hydrolysis as well as the dehydration of fructose and glucose. Meanwhile, Lewis acid sites catalyze the isomerization of glucose to fructose, which is a critical step, since glucose is a stable six-membered pyranose structure, whereas fructose is more reactive for dehydration reactions (Yu et al., 2017b).

Glucose conversion to HMF is a complex process since, as mentioned before, it requires an initial stage of isomerization of glucose to fructose. From the literature, it found that the use of metal chlorides ($CrCl_2$, $CrCl_3$, and $SnCl_4$) using as solvent ionic liquids enhanced the selectivity of HMF obtained from glucose. In this case, Cr or Sn is attributed to being responsible for the isomerization of glucose to fructose via a five-membered-ring chelate complex of Cr or Sn atom in glucose. The high catalytic activity of Cr or Sn is related to the synergistic catalysis of Lewis (Cr^{3+} or Sn^{+4}) and Brönsted acid sites. However, an excessive amount of Lewis acidity will lead to the formation of undesired humins, whereas Brönsted acid sites can suppress the isomerization reaction (Yu et al., 2017b; Jing et al., 2019).

To enhance the HMF selectivity, a balance between Brönsted and Lewis acid sites is necessary, however, not only the Brönsted/Lewis ratio influence the catalytic performance but also the ratio of weak to total Lewis acid sites are primordial in the synthesis of HMF (Jing et al., 2019). Also, it is suggested that desirable reaction pathways are less sensitive to the acid strength of the catalyst in comparison to the side reactions (Yu et al., 2017b).

It is reported that an excessive concentration of Brönsted acid sites could reduce the efficacy of Lewis acid sites on glucose isomerization. Therefore it is still necessary to further investigate to find the optimal acid strength distribution and the Lewis/Brönsted ratio of the catalysts.

3.7 SOLID CATALYST FOR FURFURAL AND 5-(HYDROXYMETHYL) FURFURAL PRODUCTION

In industrial processes, furfural is obtained from sugar pentoses, which come from lignocellulosic materials using homogeneous acid catalysts such as HCl, H_3PO_4, and H_2SO_4 in aqueous solution (Toftgaard Pedersen et al., 2015). In the first step, pentoses (xylose or arabinose) are produced from the hydrolysis of hemicellulose and subsequently are dehydrated to obtain furfural, which is separated by steam stripping to avoid its degradation. Then furfural is purified by double distillation, leading to a 40%–50% yield. Nonetheless, this process requires a high amount of energy and generates acid waste streams. Therefore recent research is focused on more benign furfural production strategies, being an attractive option the use of solid catalysts to transform pentoses carbohydrates to furfural (Kaiprommarat et al., 2016; Luo et al., 2019).

Concerning HMF synthesis, homogeneous catalysts such as HCl, H_2SO_4, H_3PO_4, HNO_3, and trifluoroacetic acid have also been widely used as Brönsted acids, to favor the tandem conversion of glucose to HMF. The main idea is to facilitate the hydrolysis and dehydration, due to its low cost and high HMF yields which are close to 85%–91% (Agirrezabal-Telleria et al., 2014; Agarwal et al., 2018; Portillo Perez et al., 2019). However, their industrial use is limited due to their high corrosion properties. Hence, heterogeneous Brönsted acid catalysts are studied in order to replace mineral acids.

Solid Brönsted acid catalysts containing sulfonate groups have shown high catalytic activity in the conversion of model compounds. Some examples are sulfonated resins, such as Dowex-50 and Amberlyst-38, which are capable of catalyzing the conversion of fructose to HMF with a yield of 83% and 88%, respectively. Other examples are sulfonated carbon structures and sulfonated polymers (polydivinyl-benzene-SO_3H). In such materials, the hydrophobicity of their catalytic surface plays a significant role in preventing the side reactions induced by water (Agirrezabal-Telleria et al., 2014; Agarwal et al., 2018; Portillo Perez et al., 2019).

In the case of Lewis acid catalysts, tin, aluminum, and chromium chlorides have performed isomerization of glucose to fructose, obtaining high HMF yields. Some other efficient Lewis catalysts have been metal-containing zeolites, such as Sn-beta, where tetrahedral tin sites act as Lewis acid sites to assist the isomerization reaction. Different types of Lewis acid sites have also been distinguished, the closed acid sites, and the open acid sites, the last one showing hydroxyl groups nearby the metal center (Agirrez-abal-Telleria et al., 2014; Agarwal et al., 2018; Portillo Perez et al., 2019).

An analogous catalytic system is mesoporous silicates whose acid functionality is given by the incorporation of –$PrSO_3H$ groups and metal loadings (Sn, Al, Ti, and Zr) into their framework, leading to bifunctional catalyst with Brönsted and Lewis acid sites, respectively (Agirrezabal-Telleria et al., 2014; De et al., 2016; Agarwal et al., 2018; Portillo Pérez et al., 2019). In addition, hydro-talcites are bifunctional materials known by their acid-basic Lewis actives sites (M-O^-) that have been tested in the conversion of sugars to furfural, obtaining a maximum xylose conversion of 78% and 50% of HMF yield at 170 °C during 4 h using Hf-Al hydrotalcite as solid catalyst (Delbecq et al., 2018).

Furthermore, Lewis acid can indirectly promote fructose dehy-dration, since protons (Brönsted acid sites) are a product of the metal (Lewis acid site) hydrolysis in water. However, this catalytic function may change according to the selection of the reactive solvents. As the metal may coordinate and polarize a carbonium intermediary to facilitate the limiting step-an internal hydride transfer. Thus a high electronega-tivity of the metal catalyst may result in more polar bonds in the interme-diary favoring the internal transfer of the hydride (Yu et al., 2017b).

Therefore a bifunctional catalyst, with both Lewis and Brönsted acid sites, is desirable to assist the one-pot conversion of glucose and polymeric carbohydrates (starch, cellulose) to furfural and HMF. In the following paragraphs, some of the principal catalytic systems applied in the conversion of sugars to furfural and HMF will be analyzed.

3.7.1 ZEOLITES: MICROPOROUS SILICATES

Zeolites are microporous materials, chemically composed of aluminosilicates with crosslinked tetrahedral of AlO_4 and SiO_4 moieties through an O-atom. Zeolites stand out by their shape selectivity and large open-channel like structure. Also, zeolites have shown high thermal stability, large surface area, and relevant acid properties that are tailored according to their Si/Al molar ratio. The use of zeolites in pentoses and hexoses conversion to furfural and HMF has been restricted mainly due to mass transfer limitations. Since glucose cannot be transported freely in the cavities of the zeolites, and the HMF formed (molecular diameter of 0.82 nm) is trapped in the catalytic system, to produce formic and levulinic acids in consecutive reactions (Agarwal et al., 2018, Portillo Perez et al., 2019).

Yoshida et al. (2017) studied furfural synthesis using bamboo powder as a feedstock and a water-toluene mixture as a solvent at 170 °C. The solid catalyst was a chabazite-type (CHA) zeolite prepared via an interzeolite conversion method from a faujasite-type zeolite. The CHA zeolite was capable for the hydrolysis of hemicellulose fraction in ball-milled bamboo powder to xylose and selective dehydration of xylose to furfural yielding a 60% of furfural.

Li et al. (2018) used H-ZSM-5 as heterogeneous acid catalyst for dehydration of corn cob using a green solvent γ-GVL, obtaining a maximum furfural yield of 71.7% at 190 °C during 1 h. Brönsted acid sites of H-ZSM-5 were related to this remarkable result.

Wang et al. (2018) used beta zeolite as solid catalyst for the conversion of fructose into HMF, in the presence of different solvents, where γ-butyrolactone was used as a solvent, obtaining an HMF yield of 50.25% and selectivity of 83.3% when the reactor is pressurized at 2MPa with N_2, and the reaction was carried out at 150 °C during 1 h. The values of HMF yield and selectivity are attributed to the presence of aluminum in tetrahedral and octahedral coordination in the framework of the beta zeolite.

Zhang et al. (2017) studied the use of silicoaluminophosphates (SAPOs), which are a class of molecular sieves with unique structure, high surface area, tunable acidity, pore size dimensions, and high thermal stability. SAPO-34 catalyst afforded an HMF yield of 93.6% from glucose at 170 °C in water/γ-GVL as a solvent media. Also, the SAPO-34 catalyst can be reused for five reaction cycles without significant loss of catalytic activity.

3.7.2 MESOPOROUS SILICATES

Among mesoporous silicate materials, acid functionalities supported on a hexagonal arrangement like MCM-41 and SBA-15 have shown high versatility. These materials possess limited acidity due to the surface silanol groups (Si-OH). However, their surface can be modified by incorporation of 3-mercaptopropyl-trimethoxysilane as the $-SO_3H$ precursor or even metals such as Al, Ti, Zr, and Sn, to provide Brönsted and Lewis acid sites. In this case, the cocondensation method has proven to be an alternative option to incorporate different functional groups to silicate materials (Agirrezabal-Telleria et al., 2014; De et al., 2016; Agarwal et al., 2018; Portillo Pérez et al., 2019).

MCM-41- and SBA-15-based materials generally show high ordered mesoporous structure along with large surface area and thermal stability. However, the catalytic activity shown by these materials is mainly attributed to their acid properties and a lesser extent to their textural properties (Portillo Pérez et al., 2019).

The main problems to overcome with these types of catalysts are the low thermal stability of their sulfonic acid sites ($-PrSO_3H$). One way to solve this problem is the enhancement of the silica-sulfonic interaction through the use of fluorocarbonate copolymers, where the samples suffered a remarkable reduction of their porosity. These samples, aged at 180 °C, showed high renewability through calcination and high furfural yields after repetitive reactions (Agarwal et al., 2018; Portillo Pérez et al., 2019). Also, to scale up the synthesis of these mesoporous materials to the industrial level, it is mandatory the use of templates or silicon precursors less expensive.

Kaiprommarat et al. (2016) reported the use of SO_3H-MCM-41 as an efficient catalyst for xylose dehydration, obtaining a furfural selectivity of 93% at 155 °C, which was related to its wide pore diameter in the range of 3–6 nm. Shirai et al. (2017) investigated the one-pot conversion of microcrystalline cellulose using various acid catalysts. According to their results, Al-SBA-15 stood out for its high HMF selectivity (26.9%) at 220 °C for 5 min and a hydrogen pressure of 1 MPa. Al-SBA-15 was also tested, leading to an HMF yield of 13%.

3.7.3 MIXED TRANSITION OXIDES

Significant studies have demonstrated the successful application of low-cost metal oxides, especially early transition metals such as titanium, zirconium, niobium, and tin; this obeys to their high specific surface area and large pore size since the reactants can reach the active sites inside the metal oxide pores (De et al., 2016).

Interestingly, the combination of TiO_2 and ZrO_2 can effectively suppress the rehydration of HMF into levulinic acid and formic acid. In the case of zirconia, an amphoteric compound, is attractive for hexose conversion to HMF due to its bifunctional character. The enhancement of its surface is made by coating with sulfonic groups and oxide metals of molybdenum, tungsten, iron, titanium, and tin to obtain superacid catalysts (Agarwal et al., 2018, Delbecq et al., 2018).

Also, only some particular phases of these metal oxides are catalytically active for the conversion of glucose and fructose. For example, anatase (TiO_2) due to its amphoteric character can carry out glucose conversion with a high HMF yield, meanwhile rutile phase is inactive. In the case of ZrO_2, tetragonal and cubic ZrO_2 species enhance the catalytic activity of fructose conversion to HMF. Nonetheless, in the case of glucose conversion anatase TiO_2 offers a better HMF yield than monoclinic/cubic-ZrO_2. Thus conversion of glucose to fructose is performed by the basic sites of the catalysts, whereas the synthesis of HMF is carried out by the acid sites of catalysts (De et al., 2016).

Kumar Mishra et al. (2019) used sonochemically synthesized Zn-doped CuO nanoparticles for furfural production from xylose. Its large surface area makes possible the complete conversion of xylose to furfural at 150 °C for 12 h, with a furfural yield of 86%, compared to 45% of ZnO and CuO nanoparticles. Also, niobium composites have been used as promising catalysts for furfural production. Candu et al. (2019) investigated the furfural production using niobium supported on beta zeolites, and in a solvent system of water/methyl isobutyl ketone (MIBK) in the presence of NaCl. The acid properties came from Al, extra-framework isolate Nb(V), and Nb_2O_5 pore-encapsulated clusters, where Nb(V)-OH exhibit Brönsted acidity of moderate strength. A furfural selectivity of 84.3% and a glucose conversion of 97.45% are achieved with this catalyst at 180 °C.

Sn-based catalysts have shown high activity in isomerization, retro-aldol conversion, and dehydration of glucose or its isomers. Marianou et al. (2018) studied the conversion of glucose in a water/DMSO two-phase system, using $SnCl_2$ and $SnCl_4$ as homogeneous catalysts, and SnO_2 supported on γ-alumina, ZSM-5 and beta zeolites or mesoporous aluminosilicates (Al-MCM-41). The results indicate that Lewis acid sites from tin oxides promote the retro-aldol reaction toward lactic acid, whereas Lewis acid sites of γ-alumina enhance the synthesis of both HMF and lactic acid. The highest HMF yield (27.5%) was achieved using Sn20/γ-alumina, accompanied by a lactic acid yield of 16.5%.

Mayer et al. (2019) obtained by ion-exchange $K_xSb_xTe_{(2-x)}O_6$ oxides (acid catalysts with layered structure-pyrochlore framework) that were used for the conversion of fructose to 5-hydroxymethylfurfural (HMF) in H_2O/MIBK; obtaining a fructose conversion of 99% and an HMF yield of 59% after 120 min.

Morales-Leal et al. (2019) grafted γ-alumina with thiol and organic sulfonic groups to improve the inherent acid properties of alumina to promote the tautomerization of fructose to its furanose form. This catalyst was packed into a continuous reactor-system to obtain HMF from fructose using a solvent system of THF/H_2O (4:1 w/w) yielding 95% of fructose conversion and 73% HMF selectivity.

3.7.4 METAL PHOSPHATES

Phosphates and pyrophosphates based on metals such as titanium, zirconium, niobium, vanadium, and chromium possess excellent acidic strengths, necessary for sugars conversion to Furfural and HMF production (Portillo Pérez et al., 2018). These materials are layered polymers, where each layer is made of a plane of tetravalent metal atoms surrounded by plane of phosphates species (Agarwal et al., 2018).

Metal phosphates have shown better catalytic activity than metal oxides due to their bifunctional acidity. Then, the ratio metal/phosphate in their framework regulates the amount of Lewis and Brönsted acid sites. Thus it is necessary a high Brönsted/Lewis acid site ratio to enhance the selectivity toward furfural and HMF (Delbecq et al., 2018). An effective metal phosphate catalyst requires an optimum Brönsted/Lewis ratio, since it was observed that an excessive amount of Brønsted acid sites had a detrimental effect on the isomerization process, whereas an excessive amount of Lewis acid sites led to the formation of by-products, decreasing the selectivity to HMF (De et al., 2016). During the conversion of sugars over metal phosphates using water as solvent, zirconium, and titanium phosphates gave the best results of activity and selectivity (Agarwal et al., 2018).

Xu et al. (2018) investigated furfural production from xyloses and wheat straw over chromium phosphate ($CrPO_4$) containing Lewis and Brönsted acid sites, using solvent system of water/THF. Under the best reaction conditions, a furfural yield of 88% was achieved at 160 °C after four reaction cycles. Furthermore, Chatterjee et al. (2018) synthesized metal-organic frameworks, materials possessing ultrahigh porosity, size, and shape-selective pores. In this case, researches used MIL-101 (Cr)-SO_3H as a catalyst, which presented Lewis (Cr) and Brönsted acid sites (SO₃H); and biomass conversion was conducted at 80–120 °C. The best result in this study gave rise to a furfural yield of 62.6% when MIL-101 is coated with *n*-octadecyltrichlorosilane.

Bifunctional catalysts with Lewis and Brönsted acid sites can enhance the selectivity of HMF. Thus Imteyaz Alam et al. (2014) prepared a titanium hydrogen phosphate material which contains Lewis and Brönsted acid sites, which was used as a catalyst during the dehydration of fructose, giving rise to an HMF yield of 55%, and 35% of yield when glucose was the feedstock.

Liu et al. (2019) studied a heterogeneous chromium-exchanged zirconium phosphate catalyst (ZrP-Cr) with a layered structure, with high thermal stability and high water tolerance. The optimized conditions of fructose dehydration lead to 94.55% of HMF and 43.2% of yield when glucose was the raw material. This catalyst was reused up to 6 reaction cycles at 120 °C without significant loss of catalytic activity. The idea of this catalyst is that zirconium phosphate with a layered structure has shown extreme insolubility in water, high thermal stability, and easy sedimentation; but also, the H⁺of the P-OH can be exchanged by divalent or trivalent cations, such as chromium whose salts have shown good results in the synthesis of furfural and HMF (Liu et al., 2019).

3.7.5 ION EXCHANGE RESINS

Sulfonic ion exchange resins are widely used in the conversion of pentoses and hexoses to furfural and HMF. These materials are known due to their large pore size and also they can be synthesized with fluorine and sulfonic groups, which increases their acid properties, and thus their catalytic activity (Agirrezabal-TelleriA et al., 2014; Agarwal et al., 2018). The most common type of these polymeric resins is styrene-based sulfonic acids (Amberlyst and Dow type resins) and perfluorinated ion exchange resins (Nafion) (Agarwal et al., 2018; Portillo Pérez et al., 2019).

Low HMF yields were obtained when the reaction was carried out using water as solvent, which is due to the Amberlyst resin is also active in HMF rehydration to levulinic and formic acids. Then, it has been demonstrated that the use of DMSO and the solvent systems: acetone-DMSO and acetone-water allow obtaining a better HMF yield (Agarwal et al., 2018). However, the application of the most relevant ion exchange resins: Amberlyst, Dowex, and Nafion in furfural and HMF production is jeopardized under elevated reaction conditions (> 130 °C) due to their poor thermal stability (De et al., 2016; Agarwal et al., 2018; Portillo Pérez et al., 2019).

Sato et al. (2019) studied the production of furfural by xylose degradation at 170 °C over ion exchange resins (Amberlyst 70). They found that toluene was an efficient solvent to assist the furfural production, leading to a maximum of 52.3%. In a similar work, Zhang et al. (2018) proposed polydivinylbenzene (PDVB-SO$_3$H) as a catalyst for the conversion of *Camellia oleifera* to furfural. The maximum furfural yield was 61.3% in a solvent system butyrolactone/water at 170 °C.

Recently, immobilized acid resins such as perfluorosulfonic acid resin and perfluorosulfonic acid have been receiving attention as solid acid catalysts because they are reusable, cost-efficient and environmental friendly compared with homogeneous catalyst such as H$_2$SO$_4$. Hence, in the following paragraphs, the most recent advances in the use of acid resins for dehydration of glucose or fructose to produce HMF are presented (Dou et al., 2018).

A porous organic hyper-cross-linked polymer (HCP) derived from benzene for the dehydration of carbohydrate to HMF was synthesized. The results obtained by Dong et al. (2018) indicate that HCP materials were very active in the dehydration of fructose, which gave rise to an HMF yield of 96.7% and almost 100% of fructose conversion in 30 min.

In a related work, Zhang et al. (2019b) modified sulfonated phenol-formaldehyde *p*-hydroxybenzene-sulfonic acid resin catalyst with Al^{+3} leading to the Al-SPFR material. This catalyst was evaluated in the dehydration of fructose, obtaining an HMF yield of 47.4% at 170 °C for 2 h using a water/γ-GVL solvent system.

Antonetti et al. (2017) used an Amberlyst-70 resin in the dehydration of fructose and inulin to produce HMF. This catalyst showed higher thermal stability (up to 190 °C) compared with other styrene-divinylbenzene resins, and higher acid concentration than perfluorinated sulfonic resins as Nafion. Under microwave heating, Amberlyst-70 reported an HMF yield up to 46% at 170 °C during 2 h, and using a water/γ-GVL as a solvent system.

Finally, Dou et al. (2018) reported the use of a mesoporous silica grafted with perfluorosulfonic acid resin (Aquivion@silica) as a catalyst. Amphiphilic Aquivion PFS is used as both acidic and template agent for the sol–gel synthesis of amorphous mesoporous silica; this latter was appropriated to immobilize Aquivion resin by the impregnation method, obtaining a catalyst with a large surface area, and highly accessible sulfonic acid groups. The Aquivion@silica allowed obtaining an HMF yield of 85% at low temperature (90 °C) using DMSO as a solvent. This material was catalytically stable during fourth reaction cycles, but the catalyst could be regenerated through a simple ion-exchange method.

3.7.6 CARBON-BASED CATALYSTS

Carbon-based materials are typically used as solid acid catalysts due to their high thermal stability, high chemical activity, and low production costs. They can be synthesized using as raw material biomass sources, obtaining cheaper, sustainable, and low toxicity wastes. Carbon-based materials are generally amorphous, and depending on the carbonization temperature (250 °C) they possess $-SO_3H$, carboxylic acids, and phenols groups, which are responsible for their remarkable catalytic activity (De et al., 2016, Delbecq et al., 2018; Portillo Pérez et al., 2019).

Currently, H_2SO_4 is used for the functionalization of these carbonaceous materials with sulfonic groups, which requires large amounts of water and other solvents to remove free acid and organic compounds. However, this kind of functionalization is not necessary as carbonaceous materials naturally contained carboxylic and phenol groups, as is the case of graphene oxide (GO) (Delbecq et al., 2018; Portillo Pérez et al., 2019). The presence of carboxylic and phenol groups enhance the surface interaction between the carbon-based material and cellulose, which simplify the mass transport of cellulose through the active sites (De et al., 2016).

Sulfonic groups, carboxylic acids, and phenol act as Brönsted acid sites improving the hydrolysis rate of cellulose. Generally, carbon supports are amphoteric and exhibit an acid–base character due to the presence of carboxylic and phenol groups, which enhances metal adsorption and catalyst dispersion. Meanwhile, Lewis acid sites can be added to supporting metals, generating a phenomenon called "hydrogen spillover," which is caused by the electrostatic interaction of the inherent oxygens in carbon-based materials and metal cations. The spillover originates the adsorption and dissociation of hydrogen, where a hydrogen atom migrates from a metal particle to the carbon surface (De et al., 2016).

Recently research on GO grafted with sulfonic groups has received attention for its Brönsted acid nature and its use in the dehydration of xylose to furfural. The most outstanding findings of Upare et al. (2019) allowed them to conclude that this material was superior leading to selective production of furfural of 86% using water as a solvent and a minimal loss of activity after three subsequent reaction cycles.

Another study was carried out by Li et al. (2017) using a robust acid catalyst (SC-CACt-700) with a matrix of carbon and functionalized with sulfonic groups using a greener approach. Sulfanilic acid was used instead of H_2SO_4 and γ-GVL was applied as a green solvent; leading to a furfural yield of 51.5% at 200 °C. Zhang et al. (2019) designed a solid catalyst that contains Cl⁻, pH-OH, and COOH⁻ groups that could act as bifunctional catalysts for the direct conversion of lignocellulosic material to furfural. Then, they were synthesized by hydrothermal carbonization of sucralose and subsequent sulfonation of the formed amorphous carbon (HSC-SO$_3$H). This catalyst was tested in the furfural synthesis using as a raw materials cellulose and corncob, leading to a furfural yield of 90.8% at 175 °C for 30 min in a γ-GVL/water system.

A free metal catalyst like GO presents high thermal conductivity, high surface area, and active sites such as carboxyl, epoxy, and hydroxyl groups that enhance the surface reactivity. Shaik et al. (2018) used GO as a solid catalyst for the dehydration of fructose to HMF under solvent-free conditions at 100 °C, obtaining a fructose conversion of 90% and an HMF selectivity of 87%, being stable under three reaction cycles. Furthermore, sulfonated amorphous carbon silica has been combined with DES to separate easily HMF in biomass processing. This strategic couple has led to an HMF yield of 67% and 100% of selectivity at 110 °C for 4 h (Tran and Tran, 2019).

Also, Wu et al. (2019) improved the HMF production from cellulose using a Ni-doped biomass-based carbon catalyst (Nin/CS) under hydrothermal degradation conditions. The characterization results found a large amount of hydroxyl and carboxylic groups on biomass-based carbon spheres (CSs). Also, the addition of a Ni source increased the inherent Lewis and Brönsted acid sites. Then, it was obtained an HMF yield of 85% when cellulose reacted with H$_2$ (6 MPa) at 200 °C for 60 min using Ni$_{2.0}$/CS as a catalyst.

3.7.7 IONIC LIQUIDS

Chemically, an ionic liquid is a salt of ions poorly coordinated, which makes it a liquid fluid at temperatures below 100 °C. Commonly, one component is organic, and the other ion presents a delocalized charge, avoiding the formation of a stable crystal. These physicochemical properties have allowed ionic liquids to be used as catalysts and solvent for furfural synthesis. However, the use of ionic liquids is limited from the commercial development of HMF production due to its high cost and complicated recovery (Agarwal et al., 2018).

Peleteiro et al. (2016) investigated the furfural synthesis from hemicellulose obtained by hydrothermal processing of *Eucalyptus globulus* wood, using a water solvent media catalyzed by 1-butyl-3-methylimidazolium hydrogen sulfate. In the optimized reaction conditions, the conversion of xylose and hemicellulose to furfural was 71.2% and 61.6%, respectively, with limited loss of catalytic activity under eight reaction cycles.

3.8 SOLVENT AND NACL EFFECT

The functions of solvents during the conversion of xylose or fructose to furfural and HMF are the dissolution of reactants and catalyst in the reaction medium, to serve as a medium to stabilize the intermediaries and products, and also, it is found that the solvent act as a cocatalyst improving the reaction kinetics. In the next paragraphs, the primary solvents used in sugars conversion to platform molecules are discussed.

3.8.1 WATER

Water is a polar protic solvent which offers a high polarity. Since the dehydration reaction is performed at high temperature (>175 °C), obtaining subcritical water (100 °C < T < 374 °C) in the process, that offers protons and hydroxyl ions. However, a low HMF yield resulted from cellulose even at high temperature (270 °C), extended reaction time, and also with the addition of Brönsted acids. According to the literature, this obeys to the performance of the

product rehydration yielding to levulinic and formic acids.

3.8.2 POLAR APROTIC SOLVENTS

3.8.2.1 DIMETHYL SULFOXIDE

The dehydration of fructose in DMSO is thermodynamically better than in other solvents. In this case, DMSO promotes the formation of a reactive tautomer of fructose and xylose, enhancing the dehydration reaction. Moreover, computational studies indicated that DMSO molecules are located nearby hydroxyl groups of pentoses and hexoses, protecting these reactive sites that lead to side reactions. Since fructose in DMSO is maintained in the furanose form (a five-membered ring) being available for dehydration, this furanose form is preserved by the strong interaction of fructose with the sulfonyl oxygen of DMSO (Yu et al., 2017a, 2017b).

Moreover, DMSO is coordinated to the carbonyl group of HMF, stabilizing the HMF molecules, and then preventing rehydration due to its anhydrous nature and polymerization reactions (Yu et al., 2017a, 2017b). A computational study remarks on the reduced susceptibility of HMF to the nucleophilic attack, and side reactions upon solvation in DMSO.

Furthermore, theoretical studies indicated that the dehydration of fructose to HMF displays a high selectivity over Brönsted acid sites, especially in DMSO solvent; as a Brönsted acid site (H⁺) prefers to interact with DMSO to form the active species [DMSOH]⁺, which assist the removal of the three water molecules from fructose. Nonetheless, the separation of HMF from the DMSO solvent remains a challenge due to the high boiling point of DMSO (Jiang et al., 2019).

3.8.2.2 TETRAHYDROFURAN

Opposite to DMSO, THF is a polar aprotic solvent miscible in water, that has been used in a binary solvent system with water in HMF production using as Lewis catalyst $AlCl_3$, giving an HMF yield of 52%. Also, the addition of salts as NaCl decreases the miscibility between water and THF, then, the HMF synthesis occurs in the aqueous phase, while the HMF extraction is performed in the organic phase (Yu et al., 2017b). However, one of the principal disadvantages of THF is its degradation at temperatures higher than 180 °C.

On the other hand, the incorporation of salt into the system enhances the reaction kinetics leading to an HMF yield of 61%. Sodium chloride (NaCl) is commonly added to the aqueous phase to increase the partition coefficient of HMF in solvent systems. Recent works have

indicated that Cl atoms can interact with hydrogen atoms in hydroxyl groups of glucose. Thus halide ligand (Cl⁻, Br⁻) is believed to facilitate the selective formation of HMF via dehydration, and to improve the cellulose hydrolysis (Jiang et al., 2019).

3.8.2.3 GREEN SOLVENTS

The solvents called "green" are derived from biomass. Some examples are MTHF that is similar in structure to THF, but with low solubility in water, high stability, and high boiling point. The principal factor is the MTHF/water ratio to control the distribution of acid catalysts (proton) between the layers of the aqueous and organic phases.

γ-GVL is another green solvent that could reach 60% of HMF using biomass directly as a raw material. This outstanding performance obeys to the low activation energy of cellobiose hydrolysis in the GVL/water solvent system. Also, another study indicates that the solvation of protons in GVL is higher than in water, accelerating the hydrolysis of cellobiose to glucose (Yu et al., 2017b).

3.8.2.4 IONIC LIQUIDS

Ionic liquids are another type of solvents that have been applied to convert glucose and fructose into

HMF. Ionic liquids are considered green solvents due to their low vapor pressure, flammability, and toxicity. Also, cellulose and other carbohydrates can be dissolved in a high concentration of ionic liquids (Portillo Perez et al., 2019). Delbecq et al. (2017) obtained an HMF yield of 82%, 55%, 54%, and 47% from glucose, water-soluble starch, and cellulose microcrystalline powder, respectively; when a solvent system of water/MIBK was used.

Ionic liquids disrupt the polysaccharide structure and dissolve it, to follow the hydrolysis of the polysaccharide. Hence, a solvent that presents a high hydrogen-bonding ability facilitates the cellulose conversion.

A similar candidate to hydrolyze cellulose are DESs that are synthesized from a quaternary ammonium salt (commonly choline chloride) and a hydrogen bond donor or metal halides. DESs possess similar physicochemical properties to ionic liquids but with a synthesis of low cost (Portillo Perez et al., 2019; Tran and Tran, 2019).

3.8.2.5 SOLVENT SYSTEMS

A solvent system possesses an organic and inorganic solvent. In this system furfural or HMF is produced in water, which contains the acid catalysts, and as soon as they are produced, they are transferred to the organic solvent to minimize the side reactions. To obtain a better performance for biomass conversion, a solvent system must (1) decrease the miscibility between the layers (aqueous and organic phase), (2) increase the partition coefficient of HMF and furfural into the extractive phase (organic phase), (3) accelerate the diffusion rate of HMF and furfural when is increased the volume of the organic phase (Dashtban et al., 2012; Yu et al., 2017b).

Also, HMF is obtained from sugars derived of food waste, Yu et al. (2017a) analyzed the synthesis of HMF using as a raw material starch from bread waste, $SnCl_4$ as a catalyst and the following solvent systems: acetone/water, acetonitrile/water, DMSO/water and THF/water. According to the study, solvent systems with acetone, acetonitrile and DMSO presented similar HMF yield and comparable yield of the side reactions (polymerization, rehydration), whereas THF showed the lowest HMF yield due to its low dipole moment.

The use of new green solvents such as γ-GVL, γ-butyrolactone, ionic liquids, and DESs have enhanced the selectivity to furfural and HMF production. Nonetheless, the molecular interaction between this new solvent, and the reactive systems, is still unknown. Thus it is required further research to have a better understanding of the synergistic role of catalysts and solvents for HMF and furfural production.

3.9 CONSOLIDATED PROCESSES FOR FURFURAL AND 5-(HYDROXYMETHYL) FURFURAL PRODUCTION

Nowadays, furfural is produced from lignocellulosic biomass by a modified process of Quaker Oats Technology (Dashtban et al., 2012), with an overall production of 300,000 tons/years. However, HMF synthesis is limited to a low scale by syrups extraction from energy crops (Yu et al., 2017b), which is sold as a chemical for laboratory and research purpose (Portillo Perez et al., 2019).

Then, a consolidated process is proposed for furfural and HMF production using as a raw material lignocellulosic biomass (Figure 3.15). To avoid the environmental problems related to the use of acidic streams for hydrolysis of biomass, a possible option is the application of cellulase enzymes to perform the depolymerization of cellulose and hemicellulose, which will yield to hexoses and pentoses, respectively. However, a pretreatment of the feedstock is necessary for removal of lignin and reduction of cellulose crystallinity (particle size). With this in mind, in a typical process the lignocellulosic biomass is grind in small particles that later suffer a physical treatment called steam explosion (autohydrolysis), where lignin is separated from hemicellulose and cellulose. Also, steam explosion causes the degradation of hemicellulose into pentoses and increase the potential hydrolysis of cellulose by cellulase enzymes, obtaining a 90% of enzymatic hydrolysis in 24 h (Sun and Cheng, 2012).

The pentoses obtained from hemicellulose degradation will serve as a raw material for furfural production, whereas the hexoses produced from the hydrolysis of cellulose will lead to HMF production through the dehydration reaction in a solvent system of water/γ-GVL, and a solid acid catalyst. The organic phase will avoid the yield reduction of furfural and HMF by side reactions. At the end, furfural, HMF, and possible by-products (formic acid and levulinic acid) will be separated, and the solvent should be recovered and reutilized.

3.10 CURRENT CHALLENGES

Until now, just a few works deal with the conversion of lignocellulosic biomass using acid catalyst. Research is focused on the synthesis of bifunctional catalyst, able to transform xylose to furfural with the maximum yield and selectivity. Nonetheless, it is also important to find out the most profitable solvent system, to decrease the possible side reactions when furfural is already on the reaction medium. Among the most successful applied solvents stand out DMSO, toluene, THF, γ-GVL, ionic liquids, and DES;

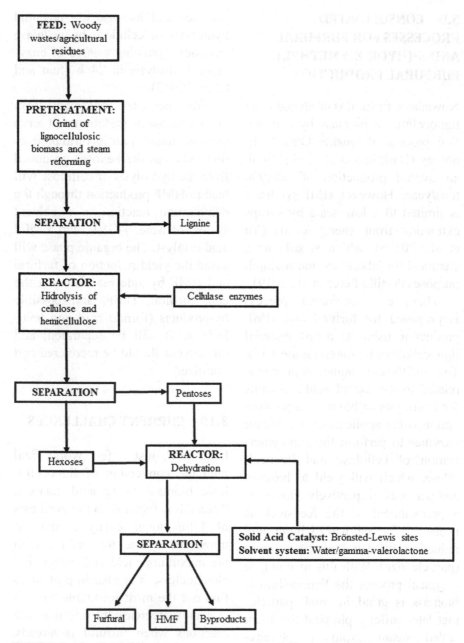

FIGURE 3.15 Consolidated process propose for furfural and HMF synthesis using lignocellulosic biomass as a feedstock.

which combined with halides salts (NaCl) improve the furfural and HMF yield.

However, there are still challenges such as (1) carrying out the one-pot conversion of cellulose to furfural or HMF, (2) the synthesis of bifunctional catalyst with the appropriate proportion of Lewis and Brönsted acid sites, and (3) the solvent system to create an adequate interaction between reactive and product molecules, in which the organic solvent should be greener and with lower cost than the toxic DMSO.

3.11 CONCLUSION

The synthesis of furfural and HMF furanic compounds through the conversion of xylose and fructose follows similar reaction pathways, where a bifunctional Lewis-Brönsted acid catalyst is necessary in order to catalyze isomerization and dehydration reaction paths. The distribution and strength of the acid sites in the catalysts is a crucial parameter to produce furfural and HMF. Therefore the distribution of acid sites in the catalyst must be appropriately modulated so that the reaction pathways of isomerization and dehydration that lead to the formation of the desired products (furfural and HMF) could proceed, avoiding at the same time degradation, retro-aldo condensation, and condensation of the reaction intermediaries.

Another parameter to consider during the synthesis of furfural and HMF is the solvent, both the catalytic system and solvent must function in a synergistically way. The solvent should promote the tautomerization of the feedstocks, facilitating the isomerization and dehydration reaction paths, it should also facilitate the isolation of the reaction products (furfural and HMF) to avoid degradation and polymerization reactions.

The best yields for furfural and HMF are obtained with the application of carbon-based catalysts (GO, waste carbon materials, sugars carbonization) with metal loadings (Sn, Ti, Zr) or functionalized with sulfonic groups. As an example, an amorphous carbon functionalized with sulfonic groups (HSC-SO$_3$H) led to a furfural yield of 90.8% at 175 °C for 30 min in a γ-GVL/water system using as a feedstock cellulose and corncob. Another catalytic system with outstanding results is a mesoporous silica grafted with perfluorosulfonic acid resin that leads an HMF yield of 85% at 90 °C using DMSO as a solvent. Interestingly, this catalyst can be regenerated by a simple ion-exchange method.

Nowadays, research suggests the trend to the use of green solvents that also come from biomass such as γ-GVL, a solvent that facilitates the hydrolysis of cellulose and hemicellulose without the need of a pretreatment step. Also, acid catalysts based on carbon represents a green catalyst that can be originated from biomass (woody waste, agricultural residues, sugar carbonization) leading to a sustainable process. Nonetheless, a deep understanding of the molecular interactions between these new green solvents, the reaction and product molecules, and the active sites of bifunctional catalysts is still necessary.

KEYWORDS

- biomass
- dehydration
- furfural
- HMF
- acid catalysts

REFERENCES

Agarwal, B.; Kailasam, K., Singh Sangwan, R., Elumalai, S. Traversing the history of solid catalyst for heterogeneous synthesis of 5-hydroxymethylfurfural from carbohydrate sugars: A review. *Renew. Sustain. Energy Rev.* **2018**, 82, 2408–2425.

Agirrezabal-Telleria, I.; Gandarias, I.; Arias, L. Heterogeneous acid-catalysts for the production of furan-derived compounds (furfural and hydroxymethylfurfural) from renewable carbohydrates: a review. *Catal. Today* **2014**, 234, 42–58.

Antal Jr, M.J., Mok, W.S.L., Richards, G.N. Mechanism of formation of 5-(hydroxymethyl)-2-furaldehyde from D-fructose and sucrose. *Carbohydr. Res.* **1990**, 199(1), 91–109.

Antonetti, C.; Raspolli, A.M.; Fulignati, G.S.; Licursi, D. Amberlyst A-70: A surprisingly active catalyst for the MW-assisted dehydration of fructose and inulin to HMF in water. *Catal. Commun.* **2017**, 97, 146–150.

Candu, N.; El Fergani, M.; Verziu, M.; Cojocaru, B.; Jurca, B.; Apostol, N.; Teodorescu, C.; Parvulescu, V.I.; Coman, S.M. Efficient glucose dehydration to HMF onto Nb-BEA catalysts. *Catal. Today* **2019**, 325, 109–116.

Chatterjee, A.; Hu, X.; Lam, F. L-Y. Towards a recyclable MOF catalyst for efficient production of furfural. *Catal. Today* **2018**, 314, 129–136.

Danon, B.; Marcotullio, G.; de Jong, W. Mechanistic and kinetic aspects of pentose dehydration towards furfural in aqueous media employing homogeneous catalysis. *Green Chem.* **2014**, 16, 39–54.

Dashtban, M.; Allan, G.; Fatehi, P. Production of furfural: overview and challenges. *J. FOR* **2012**, 2, 44–53.

De, S., Dutta, S., Saha, B. Critical design of heterogeneous catalysts for biomass valorization: current thrust and emerging prospects. *Catal. Sci. Technol.* **2016**, 6, 7364–7385.

Delbecq, F.; Wang, Y.T.; Leu, C. Various carbohydrate precursors dehydration to 5-HMF in an acidic biphasic system under microwave heating using betaine as co-catalyst. *Mol. Catal.* **2017**, 434, 80–85.

Delbecq, F.; Wang, Y., Muralidhara, A., El Ouardi, K., Marlair, G., Len, C. Hydrolysis of hemicellulose and derivatives-a review of recent advances in the production of furfural. *Front. Chem.* **2018**, 6, 1–29.

Dong, K.; Zhang, J.; Luo, W.; Su, L.; Huang, Z. Catalytic conversion of carbohydrates into 5-hydroxymethyl furfural over sulfonated hyper-cross-linked polymer in DMSO. *Chem. Eng. J.* **2018**, 334, 1055–1064.

Dou, Y.; Zhou, S.; Oldani, C.; Fang, W.; Cao, Q. 5-Hydroxymethylfurfural production from dehydration of fructose catalyzed by

Aquivion@silica solid acid. *Fuel* **2018**, 214, 45–54.

Du, H.; Ma, X.; Yan, P.; Jiang, M.; Zhao, Z.; Conrad Zhang, Z. Catalytic furfural hydrogenation to furfuryl alcohol over Cu/SiO$_2$ catalysts: a comparative study of the preparation methods. *Fuel Process. Technol.* **2019**, 193, 221–231.

Enslow, K. R.; Bell, A. The kinetics of Brönted acid-catalyzed hydrolysis of hemicellulose dissolved in 1-ethyl-3-methylimidazolium chloride. *RSC Adv.* **2012**, 2, 10028–10036.

Feng, S., Nagao, A., Aihara, T., Miura, H., Shishido, T. Selective hydrogenolysis of tetrahydrofurfuryl alcohol on Pt/WO$_3$/ZrO$_2$ catalysts: effect of WO$_3$ loading amount on activity. *Catal. Today* **2018**, 303, 207–212.

Imteyaz Alam, Md.; De, S.; Singh, B.; Saha, B.; Abu-Omar, M.M. Titanium hydrogenphosphate: An efficient dual acidic catalyst for 5-hydroxylmethylfurfural (HMF) production. *Appl. Catal. A-Gen.* **2014**, 486, 42–48.

Jing, Y.; Guo, Y.; Xia, Q.; Liu, X.; Wang, Y. Catalytic production of value added chemicals and liquid fuels from lignocellulosic biomass. *Chem* **2019**, 5(10), 2520–2546.

Kaiprommarat, S.; Kongparakul, S.; Reubroycharoen, P.; Guan, G.; Samart, C. Highly efficient sulfonic MCM-41 catalyst for furfural production: Furan-based biofuel agent. *Fuel* **2016**, 174, 189–196.

Kumar Mishra, R.; Bhooshan Kumar, V., Victor, A.; NeelPulidindi, I.; Gedanken, A. Selective production of furfural from the dehydration of xylose using Zn doped CuO catalyst. *Ultrason. Sonochem.* **2019**, 56, 55–62.

Kuster, B.F.M. A review focusing on its manufacture. *Starch* **1990**, 42(8), 314–321.

Lewkowski, J. (2001). Synthesis, chemistry and applications of 5-hydroxymethylfurfural and its derivatives. Arch. Org. Chem. 17–54.

Li, X.; Liu, Q.; Si, C.; Lu, L.; Luo, C.; Gu, X.; Liu, W. Green and efficient production of furfural from corn cob over H-ZSM-5 using γ-valerolactone as solvent. *Ind. Crop Prod.* **2018** 120, 343–350.

Li, W.; Zhu, Y.; Lu, Y.; Liu, Q.; Guan, S.; Chang, H.; Jameel, H.; Ma, L. Enhanced furfural production from raw corn stover employing a novel heterogeneous acid catalyst. *Bioresour. Technol.* **2017**, 245(A), 258–265.

Liu, B.; Ba, C.; Jin, M.; Zhang, Z. Effective conversion of carbohydrates into biofuel precursor 5-hydroxymethylfurfural (HMF) over Cr-incorporated mesoporous zirconium phosphate. *Ind. Crop. Prod.* **2015**, 76, 781–786.

Luo, Y.; Li, Z.; Li, X.; Liu, X.; Fan, J., Clark, J.L.; Hu, C. The production of furfural directly from hemicellulose in lignocellulosic biomass: a review. *Catal. Today.* **2019**, 319, 14–24.

Ma, J.; Shi, S.; Jia, X.; Xia, F.; Ma, H.; Gao, J.; Xu, J. Advances in catalytic conversion of lignocellulose to chemicals and liquid fuels. *J. Energy Chem.* **2019**, 36, 74–86.

Marianou, A.A.; Michailof, C.M.; Pineda, A.; Iliopoulou, E.F.; Triantafyllidis, K.S., Lappas, A.A. Effect of Lewis and Bronsted acidity on glucose conversion to 5-HMF and lactic acid in aqueous and organic media. *Appl. Catal. A-Gen.* **2018**, 555, 75–87.

Mayer, S.F.; Falcón, H.; Dipaola, R.; Ribota, P.; Moyano, L.; Morales-de la Rosa, S.; Mariscal, R.; Campos-Martin, J.M.; Alonso, J.A.; Fierro, J.L.G. Dehydration of fructose to HMF in presence of (H$_3$O) $_x$Sb$_x$Te$_{2x}$O$_6$(X=1,1.1,1.25) in H$_2$O-MIBK. *Mol. Catal.* **2019**. https://doi.org/10.1016/j.mcat.2018.12.025.

Morales-Leal, F.J.; Rivera de la Rosa, J.; Lucio-Ortiz, C.J.; De Haro-Del Río, D.A.; Solis-Maldonado, C.; Wi, S.; Casabianca, L.B.; García, C.D. Dehydration of fructose over thiol- and sulfonic-modified alumina in a continuous reactor for 5-HMF production: Study of catalyst by NMR. *Appl. Catal. B-Environ.* **2019**, 244, 250–261.

Nishimura, S.; Ikeda, N.; Ebitani, K. Selective hydrogenation of biomass-derived 5-hydroxymethylfurfural (HMF) to 2, 5-dimethylfuran (DMF) under atmospheric hydrogen pressure over carbon supported PdAu bimetallic catalyst. *Catal. Today.* **2014**, 232, 89–98.

Peleteiro, S.; Santos, V.; Parajó, J.C. Furfural production in biphasic media using an acidic ionic liquid as a catalyst. *Carbohydr. Polym.* **2016**, 153, 421–428.

Popa, V.; Volf, I.. Chapter 7, catalytic approaches to the production of furfural and levulinates from lignocelluloses. In: Biomass as Renewable Raw Material to Obtain bioproducts of high-tech value; Shen, Y.; Sun, J.; Wang, B.; Xu, F.; Sun, R.; Ed. Elsevier, **2018**, p. 239.

Portillo Perez, G.; Mukherjee, A.; Dumont, M-J. Insights into HMF catalysis. *J. Ind. Eng. Chem.* **2019**, 70, 1–34.

Salnikova, K.E., Matveeva, V.G., Larichev, Y.V., Bykov, A.V., Demidenko, G.N., Shkileva, I.P., Sulman, M.G. The liquid phase catalytic hydrogenation of furfural to furfuryl alcohol. *Catal. Today.* **2019**, 329, 142–148.

Sato, O.; Mimura, N.; Masuda, Y.; Shirai, M.; Yamaguchi, A. Effect of extraction on furfural production by solid acid-catalyzed dehydration in water. *J. Supercrit. Fluid* **2019**, 144, 14–18.

Shaikh, M.; Singh, S.K., Khilari, S.; Sahu, M.; Ranganath, K.V.S. Graphene oxide as a sustainable metal and solvent free catalyst for dehydration of fructose to 5-HMF: a new and green protocol. *Catal. Commun.* **2018**, 106, 64-67.

Shirai, H.; Ikeda, S.; Qian, E.W. One-pot production of 5-hydroxymethylfurfural from cellulose using solid acid catalyst. *Fuel Process. Technol.* **2017**, 159, 280–286.

Sun, Y.; Cheng, J. Hydrolysis of lignocellulosic materials for ethanol production: a review. *Bioresour. Technol.* **2002**, 83(1), 1–11.

Toftgaard Pedersen, A.; Ringborg, R.; Grotkjær, T.; Pedersen, S.; Woodley, J.M. Synthesis of 5-hydroxymethylfurfural (HMF) by acid catalyzed dehydration of glucose-fructose mixtures. *Chem. Eng. J.* **2015**, 273, 455–464.

Tran, P.H.; Tran, P.v. A highly selective and efficient method for the production of 5-hydroxymethylfurfural from dehydration of fructose using SACS/DES catalytic system. *Fuel.* **2019**, 246, 18–23.

Upare, P.P.; Hong, D-Y.; Kwak, J.; Lee, M.; Chitale, S.K.; Chang, J-S.; Hwang, D.W.; Hwang, Y.K. Direct chemical conversion of xylan into furfural over sulfonated graphene oxide. *Catal. Today* **2019**, 324, 66–72.

Wang, Y.; Ding, G.; Yang, X.; Zheng, H.; Zhu, Y.; Li, Y. Selectively convert fructose to furfural or hydroxymethylfurfural on beta zeolite: the manipulation of solvent effects. *Appl. Catal. B-Environ.* **2018**, 235, 150–157.

Wang, H.; Zhu, C.; Li, D.; Liu, Q.; Tan, J.; Wang, C.;Cai, C.; Ma, L. Recent advances in catalytic conversion of biomass to 5-hydroxymethylfurfural and 2, 5-dimethylfuran. *Renew. Sustain. Energy Rev.* **2019**, 103, 227–247.

Wu, Q.; Zhang, G.; Gao, M.; Cao, S.; Li, L.; Liu, S.; Xie, C.; Huang, L.; Yu, S.; Ragauskas, A.J. *J. Clean. Prod.* **2019**, 213, 1096–1102.

Xu, S.; Pan, D.; Wu, Y.; Song, X.; Gao, L.; Li, W.; Das, L.; Xiao, G. Efficient production of furfural from xylose and wheat straw by bifunctional chromium phosphate catalyst in biphasic systems. *Fuel Process. Technol.* **2018**, 175, 90–96.

Yan, L.; Ma, R.; Wei, H.; Li, L.; Zou, B.; Xu, Y. Ruthenium trichloride catalyzed conversion of cellulose into 5-hydroxymethylfurfural in biphasic system. *Bioresour. Technol.* **2019**, 279, 84–91.

Yan, K.; Wu, G.; Lafleur, T.; Jarvis, C. Production, properties and catalytic hydrogenation of furfural to fuel additives and value-added chemicals. *Renew. Sustain. Energy Rev.* **2014**, 38, 663-676.

Yoshida, K.; Nanao, H.; Kiyozumi, Y.; Sato, K.; Sato, O.; Yamaguchi, A.; Shirai, M. Furfural production from xylose and bamboo powder over chabazite-type zeolite prepared by interzeolite conversión method. *J. Taiwan Inst. Chem. E.* **2017**, 79, 55–59.

Yu, I.K.M.; Tsang, D.C.W.; Chen, S.S.; Wang, L.; Hunt, A.J.; Sherwood, J.; De Oliveira Vigier, K.; Jérôme, F.; Sik Ok, Y.; Sun Poon, C. Polar aprotic solvent-water mixture as the medium for catalytic production of hydroxymethylfurfural (HMF) from bread waste. *Bioresour. Technol.* **2017a**, 245(A), 456–462.

Yu, I.K.M.; Tsang, D.C.W. Conversion of biomass to hydroxymethylfurfural: A review of catalytic systems and underlying mechanisms. *Bioresour. Technol.* **2017b**, 238, 716–732.

Zhang, L.; He, Y.; Zhu, Y.; Liu, Y.; Wang, X. Camellia Olleifera shell as an alternative feedstock for furfural production using a high surface acidity solid acid catalyst. *Bioresour. Technol.* **2018**, 249, 536–541.

Zhang, L.; Tian, L.; Sun, R.; Liu, C.; Kou, Q.; Zuo, H. Transformation of corncob into furfural by a bifunctional solid acid catalyst. *Bioresour. Technol.* **2019a**, 276, 60–64.

Zhang, T.; Li, W.; Xin, H.; Jin, L.; Liu, Q. Production of HMF from glucose using an Al^{3+} promoted acidic phenol formaldehyde. *Catal. Commun.* **2019b**, 124, 56–61.

Zhang, L.; Xi, G.; Chen, Z.;Qi, Z.; Wang, X. Enhanced formation of 5-HMF from glucose using a highly selective and stable SAPO-34 catalyst. *Chem. Eng. J.* **2017**, 307, 877–883.

CHAPTER 4

Conversion of Biomass to Liquid Transportation Fuels Through Fischer–Tropsch Synthesis: A Global Perspective

C. LEYVA[1], A. ROMERO-GALARZA[2*], S. L. CASTAÑÓN-ALONSO[3],
L. FARÍAS-CEPEDA[2], L. ROSALES-MARINES[2], and A. RUBIO RÍOS[2]

[1]*Centro de Investigación en Ciencia Aplicada y Tecnología Avanzada-IPN-Unidad Legaria, Ciudad de México, México*

[2]*Department of Chemical Engineering, School of Chemistry, Autonomous University of Coahuila, México*

[3]*Chemistry Department, Universidad Autónoma Metropolitana-Iztapalapa, Ciudad de México, México*

Corresponding author. E-mail: a_romero@uadec.edu.mx

ABSTRACT

It has been reported that the temperature of the planet has increased in the last decades, as a result of the rising of greenhouse gases concentration in the atmosphere, this effect is known as a global warming (GW). The main cause of the GW is the generation of energy through the combustion of fossil fuel. In addition, the planet is facing a great energy crisis due to the depletion of the oil reserves, and the increasing demand for liquid fuels for the transport sector. This leads to produce clean and sustainable transport fuel, compelling the use of renewable energy source, such as biomass, to produce liquid fuel [biomass to liquid (BTL)]. The great advantage of BTL is that it can be sustainable, providing a carbon-neutral process. This chapter gives a global perspective of the current status in biomass to liquid processes through the well-established Fischer–Tropsch (FT) process. Aspects of BTL system are analyzed, and these include types of gasifiers and FT reactors, operation conditions, feedstock characteristics, FT catalysts, and FT reaction mechanism.

4.1 INTRODUCTION

The exhaustive use of fossil fuels is one of the prime reasons for global warming leading to climate change. Liquid fuels are expected to continue dominating the transportation sector despite raising prices. The current use of fossil fuels in different sectors continues to threaten global stability and sustainability. Sustainable energy sources are required due to the limited availability of fossil fuel reserves. The imminent world energy crisis and the increasing environmental concerns over global climate change is a top priority around the world in the present century. The increasing energy demand and environmental impact due to the use of fossil fuels have led to the use of other sources of energy (e.g., wind, solar, and biomass). Among these, biomass and biofuels are promising candidates able to mitigate greenhouse gas emission and to substitute fossil fuels (Demirbas, 2008; Nigam and Singh, 2011; Schneider et al., 2016). Nowadays, there is a great interest in the use of biomass as a source of renewable energy, since it does not generate polluting gases like CO_2, and secondly, because of its high availability throughout the world.

Several processes (e.g., physical, thermal, chemical, and biological) allow generating energy from biomass. Thus biomass can be converted into heat, electricity, and solid fuel

(coal), liquids (biodiesel, methanol, and ethanol), and gas (hydrogen and synthesis gas). Furthermore, it can also be obtained from a variety of sources such as wood, aquatic plant, crops, grass (Robbins et al., 2012; López-Bellido et al., 2014) agricultural and forestry wastes, and also municipal solid wastes are considered as a source.

In general, biomass consists of an organic, inorganic part, and water. The main chemical elements present in biomass are carbon (C), hydrogen (H), oxygen (O), nitrogen (N), sulfur (S), and chlorine (Cl). The H_2/CO ratio obtained from the gasification process depends strongly on the type of biomass used. In addition, the inorganic part of the biomass influences the combustion and forms the ashes and solid residues that remain after combustion. For these reasons, it is important to know the composition of the biomass (RWEDP, 2002).

So the characteristics of the biomass feedstock have a significant effect on the performance of the gasifier, especially the following features:

1. Moisture content: A high moisture content reduces the temperature achieved in the oxidation zone, resulting in the incomplete cracking of the hydrocarbons released from the pyrolysis zone.
2. Ash content: The oxidation temperature is often above the

melting point of the biomass ash, leading to clinkering/slagging problems in the hearth and subsequent feed blockage.

3. Volatile compounds: The gasifier must be designed to destruct tars and the heavy hydrocarbons released during the pyrolysis stage of the gasification process.

4. Particle size: The particle size of the feedstock material depends on the hearth dimensions but is typically 10%–20% of the hearth diameter. Large particles can form bridges which prevent the feed moving down, whereas smaller particles leading to a high-pressure drop and the subsequent shutdown of the gasifier.

An attractive way of using biomass is its thermochemical conversion to conventional fuel, where pyrolysis, gasification, reforming, and combustion are some conventional thermochemical process. Biomass pyrolysis without the use of catalysts is the simplest and cheapest method to generate energy. This technique is also a previous step to other thermal processes, such as combustion and gasification. Gases such as hydrogen and carbon monoxide can also be obtained by pyrolysis. These gases are useful for chemical synthesis and high-efficiency combustion systems. On the other hand, gasification can convert organic or carbonaceous materials into useful synthesis gas (syngas) composed of hydrogen, carbon monoxide, carbon dioxide, and methane.

Many synthetic fuels can be produced from biomass, such as hydrogen, substitute natural gas, dimethyl ether, and Fischer–Tropsch (FT) liquid fuel. This latter obtained through FT reaction, which is a heterogeneous catalytic process for the synthesis of transportation fuels and chemicals (Ail and Dasappa, 2016; Im-orb and Arpornwichanop, 2016). Fischer–Tropsch synthesis (FTS) is a very complex reaction process with the coproduction of paraffins, olefins, alcohols, and other products. Commonly, FT fuels are produced from coal (coal to liquid [CTL]), biomass (biomass to liquid [BTL]), and natural gas (gas to liquid [GTL]). Albeit BTL is reported to be roughly carbon neutral, which means that the amount of CO_2 released on burning biomass equals the amount taken from the atmosphere during the growth of the biomass (Liu et al., 2011), however, there are significant challenges associated with using biomass as a feedstock. This chapter briefly summarizes the types of gasifiers and FT reactors, FT catalysts, and factors affecting the production in BTL system.

4.2 BIOMASS GASIFICATION

Gasification is a thermochemical process with a change in the chemical structure of the biomass at 500–900 °C in the presence of a gasifying agent (i.e., air, oxygen, water steam, carbon dioxide, or mixtures of the components), which allows obtaining a gas rich in hydrogen. Gasification is an interesting process since (1) the process is fast and is efficient, and (2) biomass is environmental-friendly and renewable (Jin et al.,2010; Zhang et al. 2011). The gasification agent can be either air, oxygen, oxygen-enriched air, or steam. The type of gasification agent influences the H_2/CO molar ratio. The higher amount of hydrogen is obtained when steam is used as a gasification agent ($H_2/CO \sim 2$). When air or oxygen is used the H_2/CO molar ratio obtained is around 0.7 (Sánchez, 2014).

Overall the reactions taking place in the gasifier are the following: partial oxidation (4.1), complete oxidation (4.2), water gas reaction (4.3), and methane formation (4.4).

$$C + 0.5O_2 \leftrightarrow CO \ \Delta H = -268 \text{ MJ/kmol} \qquad (4.1)$$

$$C + O_2 \leftrightarrow CO_2 \ \Delta H = -406 \text{ MJ/kmol} \qquad (4.2)$$

$$C + H_2O \leftrightarrow CO + H_2$$
$$\Delta H = + 118 \text{ MJ/kmol} \qquad (4.3)$$

$$CO + 3H_2 \leftrightarrow CH_4 + H_2O$$
$$\Delta H = -88 \text{ MJ/kmol} \qquad (4.4)$$

The product gas from gasification includes hydrogen, carbon monoxide, carbon dioxide, methane, and water vapor. The main gasifying agent is usually air but oxygen/steam gasification are also used. The three types of product gas have different calorific values (CV). Low CV gas is used directly in combustion or as an engine fuel, like when air is used, whereas medium/high CV gases are obtained when steam or oxygen are used. These gases can be utilized as feedstock for subsequent conversion into liquid fuel via FTS.

Air is normally used for processes up to about 50 MW_{th}, since oxygen for gasification is expensive. The disadvantage is that the nitrogen introduced with the air dilutes the product gas, giving gas with CV lower than 6 MJ/m^3. Gasification with oxygen or with steam gives a gas with a net CV of 10–20 MJ/m^3. It can be seen that while a range of product gas qualities can be produced, economics factors are a primary consideration. Unlike the reaction with air/oxygen, the reaction of carbon with steam (the water gas reaction) is endothermic, requiring heat to be transferred at temperatures around 700°C, which is difficult to achieve. Gasifiers self-sufficient in heat are named autothermal and if they require heat, allothermal. Autothermal processes are the most common.

4.2.1 SYNGAS

The applicability of a biomass gasification technology strongly depends on the syngas quality and the H_2/CO_2 ratio (Abu El-Rub et al. 2004). Therefore the syngas to be used in an FTS has to be free from impurities like tar, char, and catalyst poisons (e.g., H_2S, COS, thiophenes). Furthermore, the total amount of inert gases (CO_2, N_2, and CH_4) is preferably low, in order to increase the partial pressure of hydrogen and carbon monoxide. One of the most important parameters in FTS is the H_2/CO ratio. FTS requires molar ratio between H_2 and CO near to 2: lower values reduce the yield reaction and cause risk of carbon deposition while, if hydrogen is in excess, methane and small chain hydrocarbons become the main product. Unfortunately, the syngas obtained through biomass gasification has a low hydrogen yield (ratio H_2/CO close to one): the composition and the level of contaminant depend on feedstock, reactor type, and operating parameters. Therefore, it can be

noted that the syngas produced from biomass involves a challenge in the BTL process. An option to overcome the problem, the gas composition has to be adjusted to have a H_2/CO ratio closer to 2 by means of a water–gas shift (WGS) reactor.

The syngas from biomass can be used directly as a fuel or as raw materials for producing higher value-added products through a subsequent synthesis process. A typical gasification scheme is integrated by the gasifier, gas cleaning system, and final application (direct burners, turbines, FT synthesis). An overall schematic of the Fischer Tropsch process is shown in Figure 4.1.

4.2.2 TYPES OF GASIFIERS

There are three main types of gasifiers: fixed or moving bed, entrained bed, and fluidized bed.

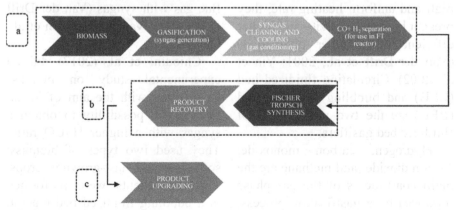

FIGURE 4.1 Full integrated system for biomass conversion: (a) syngas generation, (b) FT process, and (c) separation and refining of FT yield.

Fixed or moving bed gasifier has a simple design: the gasification medium is passed through a bed of solid fuel particles. Depending upon the flow direction of the gasifying medium, different configurations exist; the downdraft gasifiers are too popular. Both fixed and moving bed gasifiers produce syngas with a large amount of by-product due to the low and nonuniform heat and mass transfer between the solid biomass and the gasifying agent (Wang et al., 2008). Entrained flow systems gasify pulverized fuel particles suspended in a stream of oxygen or air and steam. Thus the use of biomass for gasification is limited due to its pulverization to obtain a fine powder with a size particle of around 80–100 mm. In the case of a fluidized bed gasifier, the fuel is gasified in a hotbed of small particles made of inert or active material fluidized by a gasification medium such as air or steam. Since they can achieve a high and uniform heating rate, they present high productivity and higher efficiency than other alternatives (Van der Drift et al., 2001; Yin et al. 2002). Circulating fluidized bed (CFB) and bubbling fluidized bed (BFB) are the two major types of fluidized bed gasifiers.

Hydrogen, carbon monoxide, carbon dioxide, and methane are the main constituents of the gas phase produced by a gasification process; other impurities, liquid (tar) and solid (char) (often less than 1%) remain in suspension in the gaseous stream. The MILENA gasifier (Van der Meijden, 2010) is very promising equipment consisting of two coupled fluidized bed reactors, one of them is a CFB and one BFB within the same apparatus and the bed materials are continuously recirculated between them. The CFB is located in the inner part of the equipment, allowing the physical separation of the phases of pyrolysis/gasification; whereas the combustion occurs in the outermost BFB avoiding the mixing of the flue gas with syngas. This configuration ensures the complete combustion of any fuel that is not converted into syngas, by increasing the total efficiency of the plant and minimizing purification, waste, and energy problems. The indirect circulating fluidized bed (MILENA gasification technology) is a different approach, which aims to separate the phases of char combustion and biomass gasification, to avoid the mixing of the flue gas with syngas (Van der Drift et al., 2005; Van der Meijden et al., 2007).

Chiodini et al. (2017) did an experimental study on biomass gasification with the aim of investigating the possibility to obtain a syngas with a higher H_2/CO ratio. They used two types of biomass: softwood and fast-growing crops. The experiments were performed in a bubbling fluidized bed reactor, and in an indirect gasifier with an internally circulating fluidized bed.

They obtained H_2/CO molar ratio close to 2 when a direct gasifier and a catalytic bed was used, even at low temperature and low steam/biomass ratio, whereas this is not the case in the indirect one. On the other hand, MILENA configuration, with the gasification taking place in the external BFB, could represent an interesting approach that allows the possibility of changing the residence time in the gasification zone, thus obtain syngas with a H_2/CO ratio of 2. Unfortunately, the use of MILENA gasifier operating in its standard configuration (internal CFB) did not allow achieving the expected results, a H_2/CO ratio remained below the unity.

4.2.3 BIOMASS TO LIQUID SYSTEM

In the BTL conversion system, the desired final product is a liquid. Generally, there are two types of BTL methods: direct liquefaction and indirect synthesis (Swain et al., 2011). The latter process involves the integration of three processes: gasification, syngas treatment, and FTS. In general, the H_2/CO ratio and amount of contaminants in the syngas are the most important parameters for the product yield and durability of a specific FT catalyst (Zhou et al., 2006; Lualdi et al., 2013). Table 4.1 shows BTL worldwide systems, even though the

demonstration plants of CUTEC, TRI, TUV, and Velocys showed successful results, they decided not to proceed with their plants to build a scale-up plant because of the cost issue. Air Liquide & CEA, KIT, Fulcrum Bioenergy Inc., and British Airways have ongoing projects to develop BTL fully integrated system (Kiener, 2008; Yeh, 2010; Viguié. et al., 2013).

Although there are different technologies of full integrated system (gasification gas, gas cleaning, and FT) is difficult to find the details of the fully integrated system, since most commercial approaches are confidential. Thus the details of fully integrated BTL results in a pilot-scale plant will be useful for the understanding and further optimization of the technologies as well as the development of a commercial scale BTL process. Below are cited some BTL process works carried out at pilot scale.

Kim et al. (2016) developed an integrated pilot-scale BTL process, which consists of a gasifier, syngas cleaning, and FT reactor. They reached an H_2/CO ratio of 1.67 through the indirect steam gasification of biomass. The syngas was purified in two gas cleaning stage; first cleaning process (solid impurities and tar) and second cleaning process (acid gas and CO_2). After the second cleaning process, the CO_2 content was less than 10%, and the COS and H_2S contents were less

TABLE 4.1 Biomass to Liquid (BTL) Worldwide Systems

				Research Status of Biofuel Production From BTL Process		
Institute	Year	Scale	Product	Gasification (scale, agent, fuel)	FT Rxn (scale, catalyst)	References
Technical University of Vienna (TUV-Austria)	###	Demo/Lab	Biodiesel, SNG	Dual fluidized bed (8 MWth, steamblown, wood chip)	Tubular slurry (2.5–5 kg/d product, Co or Fe catalyst)	Yeh (2010)
CHOREN (Germany)	###	Commercial	Biodiesel	Carbo-V (45 MWth, oxygen-blown, waste wood/wood chip)	Fixed bed (43 ton/d product, Co catalyst)	Kiener (2008)
Micro Energy co. (Japan)	###	Pilot	Biodiesel	Rotating reactor with electrical heating (2.5 TPD, steam, chipped wood)	Not provided	Mizushima and Takada (2014)
Velocys (United States)	2010	Demo/Pilot	Biodiesel	Dual fluidized bed (8 MWth, steamblown, wood chip)	Microchannel (115 kg/d product, Co catalyst)	Yeh (2010)
Clausthal Institute of Environmental Technology (CUTEC)	2010	Pilot/Lab	Biodiesel	Circulating fluidized bed (400 kWth, steam/oxygen-blown, mainly straw)	Fixed bed (150 mL/d product, Co catalyst)	Yeh (2010) and Clauben et al. (2005)
Korea Institute of Industrial Technology/ Korea Institute of Energy Research/Korea Research Institute of Chemical Technology (Korea)	2011	Bench	Biodiesel	Dual fluidized bed (20 kWth, steamblown, wood pellet)	Fixed bed (2.8 L/d product, Fe catalyst)	Kim et al. (2013)

TABLE 4.1 *(Continued)*

	Year	Status	Products	Process/Reactor	Additional	Reference
ThermoChem Recovery International (TRI-USA)	2011	Pilot	Biodiesel	Bubbling fluidized bed (5 TPD, steam-blown, black liquor)	Fixed bed (80 L/d product, Co catalyst)	Yeh (2010; 31) and Bain (2011)
French partner & Uhde (France)	2012	Demo	Biodiesel, Bio kerosene	Entrained flow (PDQ: PRENFLOTM gasification process with direct quench)	Axen GaselTM tech. (3200 L/d product, Co catalyst)	Viguié (2013); www.etipbioenergy.eu/ search?q=Biomass+to+ Liquids)
Neste Oil, Stora Enso (Finland)	2012	Commercial	Biodiesel	(12 MW)	Not provided	(www.etipbioenergy.eu/ search?q=Biomass+to+ Liquids)
Air liquide, Centre d'Etude Atomique (CEA); The French Alternative Energies and Atomic Energy Commission.	2013	Pilot/Demo	Biodiesel, Bio kerosene, bio naphtha	SYNDIÈSE-BtS project (1 ton/h, 10 ton/h, wood/ straw/green waste)	Not provided	(www.etipbioenergy.eu/ search?q=Biomass+to+ Liquids)
Vapo (Finland)	2014	Commercial	Biodiesel, Bio naphtha	(320 MW)	Not provided	(www.etipbioenergy.eu/ search?q=Biomass+to+ Liquids)
Karlsruhe Institute of Technology (KIT-Germany)	2015	Pilot	DME, Bio gasoline	Entrained flow, (5 MW, O2/steam, bioliqSyncrude from residual biomass)	Add DME synthesis prior to FT (2400 L/d biosynfuel)	(www.etipbioenergy.eu/ search?q=Biomass+to+ Liquids)
Fulcrum Bioenergy Inc. (United States)	2015	Commercial	Biodiesel, Bio jet	Fixed bed with plasma (500 TPD, steam, MSW)	Fixed bed (100,000 L/d, Co catalyst), by TRI tech	(www.etipbioenergy.eu/ search?q=Biomass+to+ Liquids)
British Airways (United Kingdom)	2015	Commercial	Bio jet fuel	Fixed bed with plasma (SPG: Solena's Plasma Gasification, 300 MW, O_2, MSW)	Microchannel (280 ton/d product, Co catalyst), by Velocys tech.	(www.etipbioenergy.eu/ search?q=Biomass+to+ Liquids)

than 1 ppm. Satisfying the minimum requirements of the FT process. FTS was conducted in two fixed-bed reactors using a $FeCu/Al_2O_3$ catalyst, and the integrated BTL process was operated for over 500 h.

Hanaoka et al. (2010) conducted a bench-scale hydrocarbon liquid fuel production experiment. They used a downdraft fix bed gasifier with oxygen-enriched air as a gasifying agent. Through FTS, 7.8 L of hydrocarbon liquid was produced. CUTEC (Yeh, 2010) developed a pilot-scale integrated system. A circulating fluidized bed gasifier and a fixed bed FT reactor were used for producing 150 mL/d FT diesel. The gasifier was operated for nearly 2500 h and the FT process was tested for nearly 900 h. Mizushima and Takada (2014) conducted a real vehicle test using BTL fuel from MicroEnergy, who developed a two-stage rotating gasifier with electrical heating, and 450 L/d of liquid fuel was produced by the system along with the generation of 125 kW of electricity.

Although the studies mentioned above obtained H_2/CO molar ratio above to 1 from biomass gasification, which is optimal for the production of liquid hydrocarbons via FTS, it is necessary to continue studying fully integrated BTL processes (gasification, gas cleaning, and reaction condition in the FT reactor) on a pilot scale and thus to optimize the operating conditions and costs, to be able to carry out a fully integrated BLT process at a commercial scale.

4.3 FISCHER–TROPSCH SYNTHESIS

The synthesis of FT is a chemical process that allows the catalytic synthesis of liquid hydrocarbons from the so-called synthesis syngas (a mixture of H_2 and CO). FTS consists of a series of polymerization reactions in which hydrocarbon chains are formed from carbon atoms derived from monoxide, under the influence of excess hydrogen and in the presence of a metal catalyst. The distribution of the products in the FTS depends on the reaction temperature, H_2/CO ratio, and the type of catalyst. The FT reaction is catalyzed by both iron and cobalt at pressures ranging from 10 to 60 bar and temperature ranging from 200 to 300 °C. Depending on the biomass used as feedstock, different molar ratios of H_2/CO are obtained. Generally, the H_2/CO ratio of iron catalysts should be around 1, whereas cobalt needs the H_2/CO ratio between 1 and 2. The polymerization reaction where the monomer $-CH_2-$ is formed on the catalyst surface from carbon monoxide and hydrogen, as well as secondary reactions, is showed in the following equations:

$$\text{Paraffins } 2(n + 1)H_2 + nCO$$
$$\rightarrow C_nH_{2n+2} + nH_2O \qquad (4.5)$$

Olefins $2nH_2 + nCO$

$\rightarrow C_nH_{2n} + nH_2O$ (4.6)

n-alcohols $2nH_2 + nCO$

$\rightarrow C_nH_{2n+1} + (n-1)H_2O$ (4.7)

Water gas shift (WGS)

$CO + H_2O \rightarrow CO_2 + H_2$ (4.8)

Boudouard reaction

$2CO \rightarrow C + CO_2$ (4.9)

4.3.1 REACTION MECHANISM

Based on the polymerization mechanism, the FT products follow the classical Anderson-Schulz-Flory (ASF) distribution (Abello and Montane, 2011) being its mathematical expression as follows:

$$\ln \frac{W_n}{n} = n\ln(\alpha) + \ln\left[\frac{(1-\alpha)^2}{\alpha}\right]$$

where α is the chain growth probability factor, and W_n represents the mass fraction for the hydrocarbon with the carbon number of n. According to the ASF model, product selectivity depends on the chain growth probability (α), and the maximum selectivity toward the target products is severely limited by the ASF product distribution. For example, the maximum selectivity toward C2-C4 hydrocarbons of about 60% is achieved with an α value between 0.4 and 0.5. Thus one of the most desirable challenging goals in the field of FTS is

to develop an efficient catalyst that can break the ASF distribution, and tailor the product selectivity. FT reaction mechanism is expected to obey the ASF distribution, although may be required to take into account the nature of the catalyst particles (Overett et al., 2000).

4.3.1.1 ALKYL MECHANISM

Alkyl mechanism is the most widely accepted reaction mechanism for chain growth in FTS with Fe, Co, and Ru catalysts, the schematic representation is showed in Figure 4.2. The first step of the reaction is the dissociation of CO on the surface of the catalyst, followed by the reaction between C and adsorbed hydrogen yielding in a consecutive reaction CH_2 and CH_3 surface species (Fontenelle and Fernandes, 2011). Chain growth takes place by successive incorporation of the CH_2. Chain growth can be interrupted by adding or removing hydrogen causing the corresponding paraffin or olefin (Brady and Pettit, 1981; Wang and Ekerdt, 1984). However, this mechanism does not explain the formation of oxygenated compounds (alcohols and aldehydes) that are also products of FT reaction.

4.3.1.2 ENOL MECHANISM

Enol mechanism is showed in Figure 4.2. This reaction mechanism suggests the formation of branched oxygenated compounds (CHROH) as a reaction intermediate (Huff and Satterfield, 1984; Davis, 2001). These surface species are formed by partial hydrogenation of chemisorbed CO. Chain growth occurs through a combination of two reactions; condensation reaction between enol species and by the elimination of water (Claeys and Van Steen, 2004).

4.3.1.3 ALKENYL MECHANISM

This mechanism is shown in Figure 4.2. Reaction initiation occurs by the formation of vinyl species ($-CH=CH_2$), which is generated by the reaction of methane and methylene species (Maitlis, 2004). Chain growth advances via the reaction between vinyl specie and monomer unit ($=CH_2$), generating an allyl species ($-CH_2CH=CH_2$). Chain termination results via reaction of surface hydrogen and the surface alkenyl species to produce α-olefins (Ndlovu and Phala, 2002).

4.3.1.4 CO-INSERTION MECHANISM

According to this mechanism surface methyl species is considered to be the chain initiator. The chain growth takes place by inserting a carbonyl intermediate into a metal-alkyl chain

bond results in the formation of a surface acyl species. Oxygen removal from the surface specie results in the generation of enlarged alkyl species. Subsequently, these species undergo various reactions could lead the termination steps to form of aldehydes and alcohols (Claeys and Van Steen, 2004). This mechanism is shown in Figure 4.2.

As it is observed, the alkyl mechanism is inadequate to account of branched hydrocarbons and oxygenates, the enol mechanism is unable to explain the formation of n-paraffins, the alkenyl mechanism only explains the formation of α-olefins as primary products, and the CO-insertion mechanism has a primary pathway for the formation of oxygenated compounds.

4.4 FISCHER–TROPSCH CATALYSTS

The metals used in the FTS are commonly Fe, Ru, Co, and Ni. The choice of the transition metal will contribute to the selectivity, for example, nickel is especially suitable for the formation of methane; iron, for C_5–C_{11} hydrocarbons and alcohols, cobalt for yielding C_9–C_{25} hydrocarbons, and ruthenium produces high molecular weight paraffinic waxes (C_{+20}). Ruthenium has the highest activity for the FT reaction but is an expensive catalyst compared to iron and cobalt. Hence,

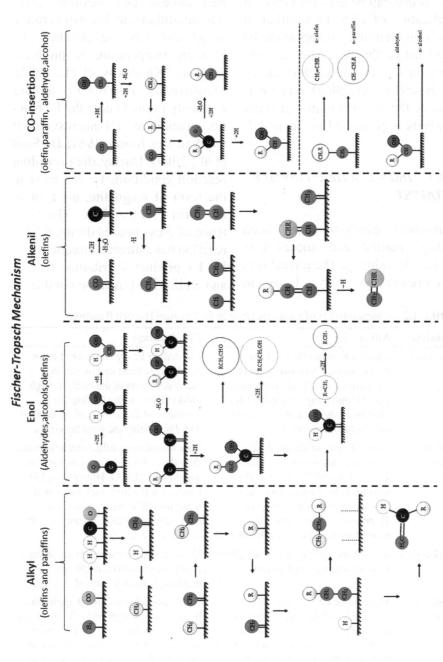

FIGURE 4.2 Fischer–Tropsch mechanism: alkyl; enol; alkenyl; CO-insertion.

Fe and Co are the only two ideal metals that can be used for practical application of FTS. In addition, it has been seen that Co appears to be more active than Fe. The turnover rates on Co are larger than on Fe catalysts (Li et al., 2002). A comparison of the characteristics of these catalysts is presented in Table 4.2.

4.4.1 IRON FISCHER–TROPSCH CATALYST

Iron-based catalysts have been widely studied and successfully applied in industry. The advantages that present the catalyst based on Fe are its low cost, high activity, easy access, and flexible operation conditions, at low-temperature diesel and wax are obtained and at high temperature it produces gasoline (Khodakov et al., 2007). Compared to Co or Ru, Fe-carbides are easily formed under the FT reaction condition (Niemantsverdriet and Van Der Kraan, 1981; De Smit et al., 2009). During the prevailing reaction condition, Fe can exist in the form of magnetite, α-Fe, or in the form of Fe-carbide. The existence of these phases during the FT reaction has a direct consequence on the FT product distribution (Dictor and Bell, 1986). A large number of

TABLE 4.2   Characteristics of Catalysts Used in the Fischer–Tropsch Reaction

Catalyst	Advantage	Disadvantage
Iron (Fe)	Minimum production of light hydrocarbons. A broad range of fractions of H_2/CO in synthesizes gas. At high temperature (613 K) is ideal to produce clear olefins with low methane selectivity.	Limited for heavy waxes production. Tendency to form carbon causing the deactivation of the catalyst. High production of water leading the inhibition of the activity of the catalyst and diminishing the growing chain.
Cobalt (Co)	More lifetime of the catalyst. Low tendency to form carbides at 473–573 K and 2.5–4 MPa. High selectivity to the wax formation. Low operative cost in the process. To prevent catalyst deactivation promoters are added (Ru, Re or Pt).	Low tolerance to sulfur and ammoniac than the iron catalyst. A low range of fractions of H_2/CO. High price (230 times than the iron one), because it is supported in metal oxides, that maximize the selectivity and activity of the catalyst.
Nickel (Ni)	It has more activity than pure cobalt. Low tendency to carbon production.	Volatile metal – carbonyls are formed easily. In industrial conditions, methane is mainly formed
Ruthenium (Ru)	More active FT catalysts. High molecular weight waxes are obtained at lower temperatures like 423 K. Activated in its metallic form without promoters for more stable activity.	Are exclude for industrial applications because of his high price (3×10^5 times more expensive than iron). Its use is also limited in academic research for the difficulty to control the reaction.

evidences suggest that Fe-carbides serve as the active phase (Jones et al., 1986; Shroff et al., 1995; Li et al., 2002; De Smit et al., 2008; De Smit et al., 2010). However, due to the coexistence of various carbide phases, such as Fe_2C, ε'- $Fe_{2.2}C$, Fe_7C_3, χ-Fe_5C_2 and θ-Fe_3C, during the FT reaction, it is difficult to differentiate the effect of each iron carbide phase. So, the actual active phase of iron is still under debate.

Gnanamani et al. (2016) researched the activity and selectivity of Hägg (χ-Fe_5C_2) and cementite (θ-Fe_3C). They found that both of the unpromoted Fe-carbides could be oxidized to Fe_3O_4 during FT reaction, and θ-Fe_3C displayed lower methane selectivity than χ-Fe_5C_2 at similar CO conversion, but χ-Fe_5C_2 initially displayed both higher overall, and WGS activities compared to θ-Fe_3C.

θ-Fe_3C is recently regarded as the active phase for C_{2-4} lower olefins. Liu et al. (2015) elaborated a microsphere catalyst of Mn-modified Fe_3O_4. They found that the addition of Mn promoter changed the electronic state of surface carbonaceous species and led to the formation of θ-Fe_3C as the dominant phase, by using XAFS and Mössbauer spectroscopy. The 6%Mn/Fe_3O_4 catalyst displayed the best selectivity to C_{2-4} lower olefins of around 60.1% and methane yield was as low as 9.7%. They suggested the enhanced selectivity to lower olefins was due to θ-Fe_3C.

Yang et al. (2012) reported that χ-Fe_5C_2 displayed higher intrinsic catalytic activity than the H_2-reduced hematite catalyst. Regarding the selectivity, the χ-Fe_5C_2 presented a good yielding of the C_{+5} hydrocarbon and the formation of long-chain olefins.

Promoters are added to Fe catalyst for increased activity and improved stability. It has been found that the addition of potassium on Fe catalyst enhances CO chemisorption and reduces H_2 chemisorption due to electron donation of potassium to iron, and CO readily accepted an electron from iron. Furthermore, hydrogen donates electrons to iron, and the existence of electron-donating alkali weakens the iron–hydrogen bond, boosting the Fe–C bond, and reducing the strength of the C–O bond (Bukur et al., 1990; Jeon et al., 2004). Tao et al. (2007) used manganese as a promoter for their Fe catalysts, they found that the initial FT activity was reduced when is increased the Mn content due to the weak carburization. However, Mn was able to restrict the reoxidation of Fe-carbides to Fe_3O_4, maintaining stability during the FT reaction. At the same time, Mn promoter restricted CH_4 formation and increased light olefin selectivity.

4.4.2 COBALT FISCHER-TROPSCH CATALYST

Cobalt catalysts mainly produce linear chain hydrocarbons (paraffins) with a few olefins and alcohols. Furthermore, it presents a low activity for the WGS reaction (Borg et al., 2007; Xiong et al., 2008). A disadvantage of cobalt catalyst is the difficulty in the regeneration of the deactivated catalyst, which requires sequential hydrogen and steam treatments or a solvent wash (Iglesia et al., 1993). Under FT conditions, cobalt mainly exists in the metallic state, which is the active phase for CO hydrogenation. Metallic Co may exist in two crystallographic phases, the face-centered cubic (FCC) and hexagonal close-packed (HCP). It has been observed that both crystallographic phases of metallic cobalt (FCC and HCP) exhibit different FT activity. The proportion of both phases affect the performance of the catalyst, for example, HCP Co was beneficial to CO conversion, whereas increasing the proportion of FCC Co led to a negative effect on catalyst activity (Ducreux et al., 2008; Fischer et al., 2011; Karaca et al., 2011; Eschemann et al., 2015). Gnanamani et al. (2013) prepared HCP Co and FCC Co respectively on SiO_2. They found that HCP Co presents higher FT activity and more C_{5+} compared with FCC Co.

Studying the effect of the particle size in the performance on the Co catalyst, Herranz et al. (2009) designed a gas flow reaction cell to make possible the acquisition of XAS spectra data at atmospheric pressure and at a temperature around to 360 °C, under this condition it was found that the rate of H_2 dissociation on smaller Co particles (size < 10 nm) was lower than on large Co particles, leading to the decrease of the mass-specific activity in the hydrogen-deuterium (H_2 dissociation) production (Yang et al. 2010). Tuxen et al. (2013) followed the relative concentration of dissociated CO by using in situ soft XAS technology on 4, 10, and 15 nm Co nanoparticles after exposure to CO/He at different temperatures, and found that the CO dissociation was much more effective on the large nanoparticles (15 nm) than on the smaller ones (4 nm). Bezemer et al. (2006) prepared a series of Co catalyst supported on carbon nanofibers with different particle sizes between 2.4 and 27 nm. It was observed that the catalytic activity decreased rapidly when the particle size of cobalt was below 6 nm. On the other hand, FT turnover rates are unaffected by the choice of support and cobalt dispersion within the range of 0.01-0.02 (Iglesia, 1997).

The great differences in activity and selectivity for HPC Co and FCC Co were attributed to their different surface structure and exposed facets. Through the Wulff construction, Liu et al. (2013a, b, 2016) found that FCC Co was predicted to have an octahedron-like shape with eight close-packed (111) facet, that conforms for about 80% of the total surface area, and another three exposed facets were (100), (110), and (310), whereas, HCP Co had a dihedral-like shape exposing six different facets, there are two close-packed (0001) facets, and the other four facets (1121), (1011), (1012), and (1120) take almost 70% of the overall surface area. From theoretical predictions, Huo et al. (2008), Liu et al. (2013a, b, 2016), Chen et al. (2016) and Zha et al. (2017) have shown that HCP Co is more active than FCC Co, and the most active four facets of the HPC Co are (1121), (1011), (1012), and (1120).

4.5 OPERATIONAL DEACTIVATION IN THE FISCHER–TROPSCH CATALYST

The catalyst of choice for the BTL process should have a reasonable lifetime, and be cost-efficient in terms of lifetime activity, that is, the activity per unit catalyst mass times the catalyst lifetime is the determining factor.

Figure 4.1 shows a schematic diagram of the fully integrated system for biomass conversion. The overall process involves the gasifier, gas cleaning system, and final application (direct burners, turbines, FT synthesis). In the product gas (syngas), various types of contaminants were formed during the gasification process, making it necessary to pass the product gas through several purifications (syngas cleaning) stages before the syngas compressor. The syngas cleaning process involves equipped for removing dust, tar, moisture from syngas, and absorption towers to remove impurities, such as CO_2, CO, H_2S, nitrogen, and halogenated compounds. The presence of poisons, such as sulfur, chloride, and nitrogen-containing compounds, in the syngas may lead to kinetic inhibition of the FT reaction or permanent deactivation of the catalyst. Permanent deactivation through the introduction of the poison in the syngas leads some additional transformation in the catalyst, for example, sintering, irreversible phase change, or enhanced carbon deposition.

Sulfur-containing compounds are known poison for FT catalysts (Dry, 2004). Espinoza et al. (1999) state that iron catalyst can deal with syngas containing more sulfur than cobalt catalysts. Furthermore, it has been inquiring that low level of sulfur poisoning may lead to enhanced carbon deposition (Visconti et al., 2007).

Halogenated compounds, such as HCl may interact with the irreducible oxides within the catalysts, as shown for the interaction of chlorine with TiO_2 (Shastri et al., 1985). This may result in creep of the support material over the catalytically active material resulting in a loss of catalytic activity.

Carbonyls may be found in the synthesis gas as a consequence of contacting syngas at high pressure and relatively low temperatures to steel pipelines. Deactivation of a cobalt-based catalyst has been reported due to iron carbonyl decomposition during FTS (De Jong et al., 1998).

4.6 FISCHER–TROPSCH REACTORS

The most common commercial reactors include a multitubular fixed-bed reactor, the slurry reactor, fluidized bed reactor, and a circulating fluidized bed reactor (Sie and Krishna, 1999). Each reactor has its advantage and disadvantage, depending on the type of product to be obtained, the operating conditions and the characteristics of the reactor (Guettel et al., 2008).

Depending on the operation temperature the FT reactors are classified on high-temperature Fischer–Tropsch (HTFT) reactor and low-temperature Fischer–Trospsch (LTFT) reactor (Figure 4.3). The key distinguishing characteristics between the HTFT and LTFT reactors is the fact that there is no liquid phase present outside the catalysts particles in the HTFT reactors. Iron catalysts are employed in HTFT reactors for the synthesis of gasoline and linear low molecular olefins. Two types of high-temperature fluidized bed systems are in commercial use; fixed fluidized bed reactors and CFB reactors. The HTFT reactors operate in the temperature range 320 °C–350 °C and up to 25 bar of pressure. Due to the high velocity in the narrowest part of the CFB reactor, the catalyst is dragged to the reaction zone in the heat exchangers, causing heat loss generated in the reaction. A new design made by Sasol for its first plant was a CFB reactor at high temperatures that was complex to operate. The feed gas enters at the bottom of the reactor about 200 °C and 25 bar, taking a catalyst stream (at a temperature of around 300 °C) and flows through the vertical column through the slide valve. Due to the high velocity in the narrowest part, the catalyst is quickly dragged into the reaction zone in the heat exchangers, largely eliminating the heat generated by the reaction. In the wide part where the catalyst is located (hopper), the catalyst and the gas are separated due to a speed drop and the gas leaves by a path of cyclones, which are dragged until returning to the hopper.

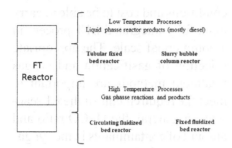

FIGURE 4.3 Fischer–Tropsch reactors scheme.

Fluidized bed reactors are diphasic (solid and gas), HTFT, and triphasic (solid, liquid and gas), LTFT, systems. When the main objective is the production of long-chain waxes the LTFT process is used and either multitubular fixed bed or triphasic bed slurry reactors can be considered. Either precipitated iron catalysts or supported cobalt catalysts may be used in LTFT reactors.

Limitation of the LTFT fixed bed reactor includes the pressure drop constraint, since the catalysts in fixed bed reactors have a diameter greater than about 1 mm. Intraparticle diffusion plays a limiting factor for the overall reaction rate for catalysts with sized greater than 1 mm. Thus intraparticle diffusion is an important factor that needs to be account for while choosing catalyst particle size and shape for a fixed bed FT process (Krishna and Sie, 2000). On the other hand, in the HTFT fluidized bed reactors the formation of a liquid phase will lead to serious problems due to particle agglomeration and loss of fluidization.

Both types of HTFT reactor are also currently in commercial operation. The CFB reactors are in use at the world's largest GTL plant in Mossel Bay, South Africa. The Sasol Advanced Synthol reactors are used at the world's largest synthetic hydrocarbon plant which is based on coal-derived synthesis gas in Secunda, South Africa.

The multitubular bed reactor consists of 2050 tubes that are filled with catalyst. The tubes must be surrounded by cooling water. The reactor operates at 225°C and up to 26 bar. Advantages of this reactor are that it has an effective heat elimination and there is not a problem of separation between the product and the catalyst. However, the reactor has many disadvantages, such as a high cost, and the mechanical difficulties that can present since the sheets of the tubes are very heavy. Also, the formation of carbon causes a rupture of the catalyst, and consequently the obstruction of the tubes, lost in conversion, and efficiency, and the need to replace the catalyst (Jager and Espinoza, 1995).

The slurry reactor operates with temperatures higher than the fixed bed reactor, eliminating the threat of catalyst deactivation, carbon formation, and catalyst dissolution. Another great advantage is the easy manufacture (lower cost) with respect to the fixed bed reactor.

In the last decade, several new plants with FT reactors have been built. In the design and construction of FT reactors, it is important to take theoretical aspects, such as kinetics and mechanisms. Basu (2007) summarized features of some initially developed FT reactors.

The most common type of fixed bed reactor is a multitubular reactor. The advantage of this reactor is the absence of catalyst–wax separator since the heavy wax products trickle-down the bed and get collected in the receiver pot. The catalyst in fixed bed reactors have a diameter greater than 1 mm, hence intraparticle diffusion is an important parameter that needs to be taken into account. The second type of commercial FT reactors is the slurry phase bubble column reactors use catalyst powder with a dimension of 10–200 μm. So, the effect of internal mass transfer resistances is negligible resulting in optimal activity and selectivity. However, solid catalyst separation represents an issue toward the commercialization of slurry bubble columns for FT process.

4.7 CONCLUSION

Technology providers such as CHOREN, CUTEC, TRI, TUV, and Velocys have developed integrated BTL system. However, it is necessary to continue studying fully integrated BTL processes on a pilot scale, and thus optimize the operating conditions and cost to be able to carry out a fully integrated BLT process in a commercial scale. The appropriate selection of gasification and syngas treatment methods are important to meet the requirements of the FT reaction. In general, the H_2/CO ratio and amount of contaminants in the syngas are the most important parameters for the product yield and durability of a specific FT catalyst. The most widely used catalysts in FT synthesis are Fe and Co. Generally, the H_2/CO ratio of Fe catalysts should be under 1, meanwhile Co needs the H_2/CO ratio between 1 and 2.

For Fe-based catalyst, it has been demonstrated that the χ-Fe_5C_2 phase constitutes the main active phase for the FT reaction with paraffins and olefins as the main products. Whereas for Co-based catalyst many studies indicated that the HCP Co phase is more active than the FCC Co phase. Size effect of Co suggests that FTS is a structure sensitive reaction. The TOF for CO conversion and C_{5+} selectivity increase with increasing particle size up to 6–10 nm, and the remained almost unchanged.

FT process is a well-established process that can be coupled to a biomass gasification. In general, syngas produced from biomass using air gasification process is deficient in H_2. The use of steam and oxygen as gasification agent enhances the H_2 content in the syngas. The steam to biomass ratio can be adjusted to vary the H_2/CO ratio in the syngas for compatibility with the FTS.

KEYWORDS

- **gasification**
- **biomass to liquid (BTL)**
- **H_2/CO ratio**
- **syngas**
- **active phase**
- **FT reaction mechanism**

REFERENCES

Abello, S; Montane, D. Exploring Iron-based multifunctional catalysts for Fischer-Tropsch synthesis: a review. *Chem. Sus. Chem.* **2011**, 4, 1538–1556.

Abu El-Rub, Z.; Bramer, E.A.; Brem, G. Review of Catalysts for Tar Elimination in Biomass Gasification Processes. *Ind. Eng. Chem. Res.* **2004**, 43(22), 6911–6919.

Ail, S.S.; Dasappa, S. Biomass to liquid transportation fuel via Fischer Tropsch synthesis technology review and current scenario. *Renew. Sustain. Energy Rev.* **2016**, 58, 267–286.

Bain, R. United States Country Report IEA Bioenergy, Task 33. NREL (National Renewable Energy Laboratory) Report. **2011**.

Basu, S. Design and development of Fischer Tropsch reactor and catalysts and their interrelationships. *Bull. Catal. Soc. India.* **2007**, 6(1), 1–22.

Bezemer, G.L.; Bitter, J.H.; Kuipers, H.P.C.E.; Oosterbeek, H.; Holewijn, J.E.; Xu, X.; Kapteijn, F.; van Dillen, A.J.; de Jong, K.P. Cobalt particle size effects in the Fischer–Tropsch reaction studied with carbon nanofiber supported catalysts. *J. Am. Chem. Soc.* **2006**, 128(12), 3956–3964.

Biomass to Liquids (BTL). Retrieved from www.etipbioenergy.eu/search?q=Biomass +to+Liquids. (accessed Jun 4, 2015).

Borg, O.; Eri, S.; Blekkan, E.; Storsater, S.; Wigum, H.; Rytter, E.; Holmen, A. Fischer–Tropsch synthesis over γ-alumina-supported cobalt catalysts: effect of support variables. *J. Catal.* **2007**, 248(1), 89–100.

Brady, R.C.; Pettit, R. Mechanism of the Fischer-Tropsch reaction. The chain propagation step. *J. Am. Chem. Soc.* **1981**, 103(5), 1287–1289.

Bukur, D.B.; Mukesh, D.; Patel, S.A. Promoter effects on precipitated iron catalysts for Fischer-Tropsch synthesis. Ind. Eng. Chem. Res. 1990, 29(2), 194–204.

Chen, C.; Wang, Q.; Wang, G.; Hou, B.; Jia, L.; Li, D. Mechanistic insight into the C2 hydrocarbons formation from syngas on fcc-Co(111) surface: a DFT study. *J. Phys. Chem. C* **2016**, 120(17), 9132–9147.

Chiodini, A.; Bua, L.; Carnelli, L.; Zwart, R.; Vreugdenhil, B.; Vocciante, M. Enhancements in Biomass-to-liquid processes: gasification aiming at high hydrogen/carbon monoxide ratios for direct Fischer-Tropsch synthesis applications. *Biomass Bioenergy* **2017**, 106, 104–114.

Claeys, M.; van Steen, E. Basic studies. *Studies in Surface Science and Catalysis.* **2004**, 601–680.

Clauben, M; Schindler, M; Vodegel, S; Carlowitz, O. The Biomass-to-Liquid Process at CUTEC: Optimization of the Fischer–Tropsch Synthesis with Carbon Dioxiderich Synthesis Gas. CUTEC Report. **2005**.

Davis, B. H. Fischer–Tropsch synthesis: current mechanism and futuristic needs. *Fuel Process. Technol.* **2001**, 71(1–3), 157–166.

De Jong, K.P.; Post, M.F.M.; Knoester, A. Deposition of iron from iron-carbonyl onto a working Co-based Fischer-Tropsch catalyst: the serendipitous discovery of a direct probe for diffusion limitation. *Stud. Surf. Sci. Catal.* **1998**, 119, 119–142.

De Klerk, A; Furimsky, E. Catalysis in the refining of Fischer-Tropsch syncrude. *Platinum Metals Rev.* **2011**, 55, 263–267.

De Smit, E.; Beale, A.M.; Nikitenko, S.; Weckhuysen, B.M. Local and long range order in promoted iron-based Fischer–Tropsch catalysts: a combined in situ X-ray absorption spectroscopy/wide angle X-ray scattering study. *J. Catal.* **2009**, 262(2), 244–256.

De Smit, E.; Cinquini, F.; Beale, A.M.; Safonova, O.V.; van Beek, W.; Sautet, P.; Weckhuysen, B.M. Stability and reactivity of ε–χ–θ iron carbide catalyst phases in Fischer–Tropsch synthesis: controlling μC. *J. Am. Chem. Soc.* **2010**, 132(42), 14928–14941.

De Smit, E.; Swart, I.; Creemer, J.F.; Hoveling, G.H.; Gilles, M.K.; Tyliszczak, T.; Kooyman, P.J.; Zandbergen, H.W.; Morin, C.; Weckhuysen, B.M.; de Groot, F.M.F. Nanoscale chemical imaging of a working catalyst by scanning transmission X-ray microscopy. *Nature.* **2008**, 456(7219), 222–225.

Demirbas, A. Biofuels sources, biofuel policy, biofuel economy and global biofuel projections. *Energy Convers. Manage.* **2008**, 49(8), 2106–2116.

Dictor, R.A.; Bell, A.T. Fischer-Tropsch synthesis over reduced and unreduced iron oxide catalysts. *J. Catal.* **1986**, 97(1), 121–136.

Dry, M.D. FT catalysts. Studies in Surface Science and Catalysis. 2004, 152, 533–600.

Ducreux, O.; Rebours, B.; Lynch, J.; Roy-Auberger, M.; Bazin, D. Microstructure of supported cobalt Fischer-Tropsch catalysts. *Oil Gas Sci. Technol Revue de l'IFP.* **2008**, 64(1), 49–62.

Eschemann, T.O.; Lamme, W.S.; Manchester, R.L.; Parmentier, T.E.; Cognigni, A.; Rønning, M.; de Jong, K.P. Effect of support surface treatment on the synthesis, structure, and performance of Co/CNT Fischer–Tropsch catalysts. *J. Catal.* **2015**, 328, 130–138.

Espinoza, R.L.; Steynberg, A.P.; Jager, B.; Vosloo, A.C. Low temperature Fischer-Tropsch synthesis from a Sasol perspective. *Appl. Catal. A: Gen.* **1999**, 186, 13–26.

Fischer, N.; van Steen, E.; Claeys, M. Preparation of supported nano-sized cobalt oxide and fcc cobalt crystallites. *Catal. Today* **2011**, 171(1), 174–179.

Fontenelle, A.; Fernandes, F.A. Comprehensive polymerization model for Fischer-Tropsch synthesis. *Chem. Eng. Technol.* **2011**, 6, 963–971.

Gnanamani, M.; Hamdeh, H.; Jacobs, G.; Shafer, W.; Sparks, D.; Davis, B. Fischer-Tropsch synthesis: activity and Selectivity of χ-Fe5C2 and θ-Fe3C carbides. Fischer-Tropsch Synthesis, Catalysts, and Catalysis. **2016**, 15–30.

Gnanamani, M.K.; Jacobs, G.; Shafer, W.D.; Davis, B.H. Fischer–Tropsch synthesis: Activity of metallic phases of cobalt supported on silica. *Catal. Today* **2013**, 215, 13–17.

Guettel, R.; Kunz, U.; Turek, T. Reactors for Fischer-Tropsch synthesis. *Chem. Eng. Technol.* **2008**, 31(5), 746–754.

Hanaoka, T.; Liu, Y.; Matsunaga, K.; Miyazawa, T.; Hirata, S.; Sakanishi, K. Bench-scale production of liquid fuel from woody biomass via gasification. *Fuel Process. Technol.* **2010**, 91(8), 859–865.

Herranz, T.; Deng, X.; Cabot, A.; Guo, J.; Salmeron, M. Influence of the cobalt particle size in the CO hydrogenation reaction studied by in situ X-ray absorption spectroscopy. J. Phys. Chem. B **2009**, 113(31), 10721–10727.

Huff, G.A.; Satterfield, C.N. Intrinsic kinetics of the Fischer-Tropsch synthesis on a reduced fused-magnetite catalyst. *Ind. Eng.*

Chem. Process Design Dev. **1984**, 23(4), 696–705.

Huo, C.-F.; Li, Y.-W.; Wang, J.; Jiao, H. Adsorption and Dissociation of CO as well as CHxCoupling and hydrogenation on the clean and oxygen pre-covered Co(0001) surfaces. *J. Phys. Chem. C* **2008**, 112(10), 3840–3848.

Iglesia, E. Design, synthesis, and use of cobalt-based Fischer-Tropsch synthesis catalysts. *Appl. Catal. A: Gen.* **1997**, 161 (1–2), 59–78.

Iglesia, E.; Soled, S.L.; Fiato, R.A.; Via, G.H. Bimetallic synergy in cobalt ruthenium Fischer-Tropsch synthesis catalysts. *J. Catal.* **1993**, 143(2), 345–368.

Im-orb, K.; Arpornwichanop, A. Techno-environmental analysis of the biomass gasification and Fischer-Tropsch integrated process for the co-production of bio-fuel and power. *Energy.* **2016**, 112, 121–132.

Jager, B.; Espinoza, R. Advances in low temperature Fischer-Tropsch synthesis. *Catal. Today* **1995**, 23, 17–28.

Jeon, J.-K.; Kim, C.-J.; Park, Y.-K.; Ihm, S.-K. Catalytic properties of potassium- or lanthanum-promoted Co/γ-Al₂O₃ catalysts in carbon monoxide hydrogenation. *Korean J. Chem. Eng.* **2004**, 21(2), 365–369.

Jin, H.; Lu, Y.; Guo, L.; Cao, C.; Zhang, X. Hydrogen production by partial oxidative gasification of biomass and its model compounds in supercritical water. *Int. J. Hydrogen Energy.* **2010**, 35(7), 3001–3010.

Jones, V.K.; Neubauer, L.R.; Bartholomew, C.H. Effects of crystallite size and support on the carbon monoxide hydrogenation activity/selectivity properties of iron/carbon. *J. Phys. Chem.* **1986**, 90(20), 4832–4839.

Karaca, H.; Safonova, O.V.; Chambrey, S.; Fongarland, P.; Roussel, P.; Griboval-Constant, A.; Lacroix, M.; Khodakov, A.Y. Structure and catalytic performance of Pt-promoted alumina-supported cobalt

catalysts under realistic conditions of Fischer–Tropsch synthesis. *J. Catal.* **2011**, 277(1), 14–26.

Khodakov, A.Y.; Chu, W.; Fongarland, P. Advances in the development of novel cobalt Fischer–Tropsch catalysts for synthesis of long-chain hydrocarbons and clean fuels. *Chem. Rev.* **2007**, 107(5), 1692–1744.

Kiener, C. Start-up of the first commercial BTL production facility the betaplant Freiberg. In: 16th European Biomass Conference & Exhibition. **2008**.

Kim, K.; Kim, Y.; Yang, C.; Moon, J.; Kim, B.; Lee, J.; Lee, U.; Lee, S.; Kim, J.; Eom, W.; Lee, S.; Kang, M.; Lee, Y. Long-term operation of biomass-to-liquid systems coupled to gasification and Fischer–Tropsch processes for biofuel production. *Bioresour. Technol.* **2013**, 127, 391–399.

Kim, Y.-D.; Yang, C.-W.; Kim, B.-J.; Moon, J.-H.; Jeong, J.-Y.; Jeong, S.-H.; Lee, S.-H.; Kim, J.-H.; Seo, M.-W.; Lee, S.-B.; Kim, J.-K.; Lee, U.-D. Fischer–Tropsch diesel production and evaluation as alternative automotive fuel in pilot-scale integrated biomass-to-liquid process. *Appl. Energy* **2016**, 180, 301–312.

Krishna, R.; Sie, S.T. Design and scale-up of the Fischer-Tropsch bubble column slurry reactor. *Fuel Process. Technol.* **2000**, 64, 73–105.

Li, S.; Ding, W.; Meitzner, G.D.; Iglesia, E. Spectroscopic and transient kinetic studies of site requirements in iron-catalyzed Fischer–Tropsch synthesis. *J. Phys. Chem B.* **2002**, 106(1), 85–91.

Li, S.; Krishnamoorthy, S.; Li, A.; Meitzner, G.D.; Iglesia, E. Promoted iron-based catalysts for the Fischer–Tropsch synthesis: design, synthesis, site densities, and catalytic properties. *J. Catal.* **2002**, 206(2), 202–217.

Liu, Y.; Chen, J.-F.; Bao, J.; Zhang, Y. Manganese-modified Fe3O4 microsphere catalyst with effective active phase of forming light olefins from syngas. *ACS Catal.* **2015**, 5(6), 3905–3909.

Liu, G.; Larson, E.D.; Williams, R.H.; Kreutz, T.G.; Guo, X. Making Fischer–Tropsch fuels and electricity from coal and biomass: performance and cost analysis. *Energy Fuels* **2011**, 25(1), 415–437.

Liu, J.-X.; Li, W.-X. Theoretical study of crystal phase effect in heterogeneous catalysis. *Wiley Interdiscip. Rev.: Comput. Mol. Sci.* **2016**, 6(5), 571–583.

Liu, S.; Li, Y.-W.; Wang, J.; Jiao, H. Mechanisms of H- and OH-assisted CO activation as well as C–C coupling on the flat Co (0001) surface-revisited. *Catal. Sci. Technol.* **2016**, 6(23), 8336–8343.

Liu, J.-X.; Su, H.-Y.; Li, W.-X. Structure sensitivity of CO methanation on Co (0001), and surfaces: density functional theory calculations. *Catal. Today* **2013a**, 215, 36–42.

Liu, J.-X.; Su, H.-Y.; Sun, D.-P.; Zhang, B.-Y.; Li, W.-X. Crystallographic dependence of CO activation on cobalt catalysts: HCP versus FCC. *J. Am. Chem. Soc.* **2013b**, 135(44), 16284–16287.

López-Bellido, L.; Wery, J.; López-Bellido, R.J. Energy crops: Prospects in the context of sustainable agriculture. *Eur. J. Agron.* **2014**, 60, 1–12.

Lualdi, M.; Lögdberg, S.; Boutonnet, M.; Järås, S. On the effect of water on the Fischer–Tropsch rate over a Co-based catalyst: the influence of the H_2/CO ratio. *Catal. Today* **2013**, 214, 25–29.

Maitlis, P.M. Fischer–Tropsch, organometallics, and other friends. *J. Organometal. Chem.* **2004**, 689(24), 4366–4374.

Mizushima N, Takada Y. Annex 38: Evaluation of Environmental Impact of Biodiesel Vehicles in Real Traffic Conditions. IEA-AMF (Advanced Motor Fuels) Final Report. **2014**.

Ndlovu, S.B.; Phala, N.S.; Hearshaw-Timme, M.; Beagly, P.; Moss, J.R.; Claeys, M.; van Steen, E. Some evidence refuting the alkenyl mechanism for chain growth in iron-based Fischer–Tropsch synthesis. *Catal. Today.* **2002**, 71(3–4), 343–349.

Niemantsverdriet, J.W.; Van Der Kraan, A.M. On the time-dependent behavior of iron catalysts in Fischer-Tropsch synthesis. *J. Catal.* **1981**, 72, 385–388.

Nigam, P.S.; Singh, A. Production of liquid biofuels from renewable resources. *Prog. Energy Combust. Sci.* **2011**, 37(1), 52–68.

Overett, M. J.; Hill, R.O.; Moss, J. R. Organometallic chemistry and surface science: mechanistic models for the Fischer–Tropsch synthesis. *Coord. Chem. Rev.* **2000**, 206-207, 581–605.

Robbins, M.P.; Evans, G.; Valentine, J.; Donnison, I.S.; Allison, G.G. New opportunities for the exploitation of energy crops by thermochemical conversion in Northern Europe and the UK. *Prog. Energy Combust. Sci.* **2012**, 38(2), 138–155.

RWEDP (2002). Wood energy basics. Regional wood energy development. http://www. rwedp.org (accessed July 14, 2014).

Sánchez, N. Obtención de gas de síntesis a partir de biomasa utilizando catalizadores de níquel. Bachelor dissertation, Valladolid University, 2014.

Schneider, J.; Grube, C.; Herrmann, A.; Rönsch, S. Atmospheric entrained-flow gasification of biomass and lignite for decentralized applications. *Fuel Process. Technol.* **2016**, 152, 72–82.

Shastri, A.G.; Dayte, A.K.; Schwank, J. Influence of chlorine on the surface area and morphology of TiO2. *Appl. Catal.* **1985**, 14, 119–131.

Shroff, M.D.; Kalakkad, D.S.; Coulter, K.E.; Kohler, S.D.; Harrington, M.S.; Jackson, N.B.; Sault, A.G.; Datye, A.K. Activation of precipitated iron Fischer-Tropsch synthesis catalysts. *J. Catal.* **1995**, 156(2), 185–207.

Sie, S. T.; Krishna, R. Fundamentals and selection of advanced Fischer–Tropsch reactors. *Appl. Catal. A: Gen.* **1999**, 186(1–2), 55–70.

Swain, P.K.; Das, L.M.; Naik, S.N. Biomass to liquid: a prospective challenge to research and development in 21st century. *Renew. Sustain. Energy Rev.* **2011**, 15(9), 4917–4933.

Tao, Z.; Yang, Y.; Zhang, C.; Li, T.; Ding, M.; Xiang, H.; Li, Y. Study of manganese promoter on a precipitated iron-based catalyst for Fischer-Tropsch synthesis. *J. Nat. Gas Chem.* **2007**, 16(3), 278–285.

Tuxen, A., Carenco, S., Chintapalli, M., Chuang, C.-H., Escudero, C., Pach, E., Jiang, P.; Borondics, F.; Beberwyck, B.; Alivisatos, A.P.; Thornton, G.; Pong, W.-F.; Guo, J.; Perez, R.; Besenbacher, F.; Salmeron, M. Size-dependent dissociation of carbon monoxide on cobalt nanoparticles. *J. Am. Chem. Soc.* **2013**, 135(6), 2273–2278.

Van der Drift, A.; van Doorn, J.; Vermeulen, J. Ten residual biomass fuels for circulating fluidized-bed gasification. *Biomass Bioenergy* **2001**, 20(1), 45–56.

Van der Drift, A; van der Meijden, C.M.; Boerrigter, H. MILENA gasification technology for high efficient SNG production from biomass. In: Proceedings of the 14th European Biomass Conference & Exhibition, Paris, France, **2005**, 628–630.

Van der Meijden, C.M. Development of the MILENA Gasification Technology for the Production of Bio-SNG, Eindhoven University of Technology. **2010**.

Van der Meijden, C.M.; van der Drift, A.; Vreugdenhil, B.J. Experimental results from the allothermal biomass gasifier MILENA. In: Proceedings of the 15th European Biomass Conference & Exhibition, Berlin, Germany, **2007**, 868–873.

Viguié, JC; Ullrich, N; Porot, P; Bournay, L; Hecquet, M; Rousseau, J. BioTfueL project: targeting the development of second-generation biodiesel and biojet fuels. OGST Revue d'IFP Energies Nouvelles; **2013**, 68, 935–946.

Visconti, C.G.; Lietti, L.; Forzatti, P.; Zennaro, R. Fischer-Tropsch synthesis on sulphur poisoned Co/Al2O$_3$ catalyst. *Appl. Catal. A Gen.* **2007**, 330, 49–56.

Wang, C.J.; Ekerdt, J.G. Evidence for alkyl intermediates during Fischer-Tropsch synthesis and their relation to hydrocarbon products. *J. Catal.* **1984**, 86(2), 239–244.

Wang, L.; Weller, C.L.; Jones, D.D.; Hanna, M.A. Contemporary issues in thermal gasification of biomass and its application to electricity and fuel production. *Biomass Bioenergy.* **2008**, 32(7), 573–581.

Xiong, H.; Zhang, Y.; Liew, K.; Li, J. Fischer–Tropsch synthesis: The role of pore size for Co/SBA-15 catalysts. J. Mol. Catal. A Chem. 2008, 295(1-2), 68–76.

Yang, J.; Tveten, E.Z.; Chen, D.; Holmen, A. Understanding the effect of cobalt particle size on Fischer–Tropsch synthesis: surface species and mechanistic studies by SSITKA and kinetic isotope effect. *Langmuir* **2010**, 26(21), 16558–16567.

Yang, C.; Zhao, H.; Hou, Y.; Ma, D. Fe5C2 nanoparticles: a facile bromide-induced synthesis and as an active phase for Fischer–Tropsch synthesis. *J. Am. Chem. Soc.* **2012**, 134(38), 15814–15821.

Yeh, B. Independent Assessment of Technology Characterizations to Support the Biomass Program Annual State-of-Technology Assessments. NREL (National Renewable Energy Laboratory) Report. 2010.

Yin, X.L.; Wu, C.Z.; Zheng, S.P.; Chen, Y. Design and operation of a CFB gasification and power generation system for rice husk. *Biomass Bioenergy* **2002**, 23(3), 181–187.

Zhai, P.; Chen, P.-P.; Xie, J.; Liu, J.-X.; Zhao, H.; Lin, L; Zhao, B.; Su, H.; Zhu, Q.; Li, W.; Ma, D. Carbon induced selective regulation of cobalt-based Fischer–Tropsch catalysts by ethylene treatment. *Far. Discuss.* **2017**, 197, 207–224.

Zhang, Y.; Li, B.; Li, H.; Liu, H. Thermodynamic evaluation of biomass gasification with air in autothermal gasifiers. *Thermochim. Acta.* **2011**, 519(1–2), 65–71.

Zhou, W.; Chen, J.-G.; Fang, K.-G.; Sun, Y.-H. The deactivation of Co/SiO$_2$ catalyst for Fischer–Tropsch synthesis at different ratios of H$_2$ to CO. *Fuel Process. Technol.* **2006**, 87(7), 609–616.

CHAPTER 5

Tailoring the Suitable Solid Catalyst for Biodiesel Production Using Second-Generation Feedstocks

D. A. CABRERA-MUNGUIA[1], A. ROMERO-GALARZA[2], H. GONZÁLEZ[3]*,
L. FARÍAS-CEPEDA[2], K. REYES-ACOSTA[2], and L. E. SERRATO-VILLEGAS[2]

[1]Advanced Materials, School of Chemistry, Autonomous University of Coahuila, México

[2]Department of Chemical Engineering, School of Chemistry, Autonomous University of Coahuila, México

[3]Chemical Engineering School, Universidad Michoacana de San Nicolás de Hidalgo, México

*Corresponding author. E-mail: hogoro@umich.mx

ABSTRACT

The high amount of free fatty acids (FFAs) contained in second-generation feedstocks, hinders the adequate performance of base catalysts, specially designed for trans-esterification reactions in biodiesel synthesis. Therefore simultaneous or sequential esterification-transesterification reactions of FFAs employing acid catalysts are necessary for the processing of second-generation raw materials. Then, this chapter focuses on explaining the main progress in the synthesis of solid acid and basic catalysts, particularly when second-generation feedstocks are processed. Also, special attention is paid to the recent advances in heterogeneous catalysts that have shown resistance to deactivation and could meet with the EN14214 regulation for biodiesel synthesis.

5.1 INTRODUCTION

The depletion of petroleum reserves around the world has focused the attention of scientific researches in biofuels, which not only could supply a portion of the energetic demand, but also with lower emissions of pollutants such as SO_x and particulate matter than petroleum derivatives (Hasan and Rahman, 2017). Thus biodiesel is a biofuel with the potential to replace petroleum diesel without further modification of the diesel engine. It can also be used as a mixture with petroleum diesel in a maximum diesel/biodiesel mass ratio of 4:1 (B20) (Hasan and Rahman, 2017; Kanaveli et al., 2017).

Biodiesel is synthesized generally through the transesterification reaction of vegetable oils. This reaction involves three subsequent reversible reactions where triglycerides are transformed into fatty acid methyl ester (FAME) or biodiesel, and glycerol (G) as a subproduct (Yang et al., 2018). Such reactions reduce the viscosity of oils and minimize the problems associated with obstruction of the injectors.

As the cost of the raw material directly affects the final price of biodiesel, new and economic feedstocks have been studied to replace the use of vegetable oils that are expensive and represent a problem of food security (Naylor et al., 2018). Hence, second-generation feedstocks like nonedible oils (NEOs), animal fat (AF), and waste cooking oils (WCOs) seem to be an appealing option because of its relatively low price. However, when homogeneous basic catalysts are used, the high content of free fatty acids (FFAs) and water in second-generation feedstocks limits their chemical conversion to biodiesel, forming soaps and gels that avoids the suitable purification and separation of biodiesel from glycerol (Atadashi et al., 2012).

Commonly two strategies have been used to process this kind of raw materials: (1) the simultaneous esterification–transesterification reaction in just one-step, and (2) the consecutive esterification–transesterification reactions in two steps. The first one is employed when the FFAs content does not exceed 2 wt.% in the raw material and it is used a strong base or acid catalyst. In the second method, the FFAs content is so high that an esterification pretreatment using acid catalysts is necessary, followed by water removal and then transesterification employing a base catalyst.

Therefore it is mandatory to make an overview of the different chemical compositions of acid and basic catalysts tested for esterification and transesterification reaction and emphasize the new advances in the design of solid catalysts for biodiesel production. Hence, this chapter serves to provide recent and accurate information about the new trends in the design of heterogeneous catalysts for the processing of second-generation feedstocks. First, it is defined the term of second-generation feedstock as well as their typical chemical composition, then is analyzed the processing of raw materials by simultaneous or consecutive esterification–transesterification reactions, after that are discussed the new advances in the synthesis and catalytic performance of base and acid catalysts. Finally, it is presented a discussion about recent trends in synthesis and tailoring of new heterogeneous catalysts for biodiesel production.

5.2 RAW MATERIAL: SECOND-GENERATION FEEDSTOCKS

Theoretically, any feedstock that contains high amounts of triglycerides (TGs) and/or FFAs can be transformed into biodiesel (Mandolesi de Araújo et al., 2013). However, biodiesel feedstocks must have large-scale production and low cost (Chakraborty et al., 2014; Bhuiya et al., 2016). To select the appropriate biodiesel raw material, biodiesel producers should take into account geographical location, regional climate, soil conditions, and agricultural practices of the country (Bhuiya et al., 2016; Kumar et al., 2015). All of them are relevant factors that affect the raw material price and thus, the final cost (75%) of biodiesel (Gülsen et al., 2014; Sajjadi et al., 2016; Haijari et al., 2017).

The raw materials available for biodiesel production are classified as follows (Chakraborty et al., 2014; Kumar et al., 2015; Haijari et al., 2017):

- *First generation:* Edible oils such as rapeseed, soybean, peanut, sunflower, palm, and coconut oil.
- *Second generation:* NEOs (Jatropha, Karanja, Mahua, Neem oil), AFs (tallow, lard, chicken fat) and WCOs (yellow and brown grease).
- *Third generation:* Microalga oil, fungi oil.
- *Fourth generation:* Genetically engineered oil crops.

Among these categories, edible vegetable oils represent the primary source of biodiesel production (95%) (Sajjadi et al., 2016). However, its use as a raw material involves a competition with the food market (Banković-Ilić et al., 2012; Wan Ghazali et al., 2015; Ambat et al., 2018), which raise the oil price. In the case of NEOs and microalgae, this is not a problem, since both can grow on marginal land and require less water and possesses CO_2 sequester capability (Kumar et al., 2015). In the case of AFs and WCOs feedstocks, the main advantage is that they are generally waste products and their cost is lower concerning edible vegetable oils (Banković-Ilić et al., 2012, Ambat et al., 2018).

Since second-generation feedstocks do not require a large investment cost (which is necessary for land cultivation, oil extraction, and residual management) as microalgae do (Ambat et al., 2018), this chapter focus on the advantages and disadvantages of second-generation feedstocks, their chemical composition and how to obtain biodiesel in a profitable manner.

5.2.1 NONEDIBLE OILS

When edible oil production in a country is not enough to satisfy the population consumption, the use of NEOs as a raw material for biodiesel synthesis represents an interesting option to reduce the manufacturing costs. The most often used nonedible plants for biodiesel are Jatropha, Karanja, Mahua, and Castor oil (Banković-Ilić et al., 2012). NEOs are economical in comparison with edible vegetable oils, can grow on lands that are not suitable for agriculture, and the establishment of these plants reduces the concentration of CO_2 in the atmosphere. Nevertheless, they possess a higher viscosity and are less volatile than edible vegetable oils (Bhuiya et al., 2016).

5.2.2 ANIMAL FAT

AFs, such as tallow, lard or white grease, yellow grease (<15% FFAs) and brown grease (>15% FFAs), are residues and economic raw materials when compared with edible vegetable oils (Banković-Ilić et al., 2012, Wan Ghazali et al., 2015). The biodiesel obtained using animal fats has a high cetane number, which is an indicator of the combustion speed of biodiesel. However, its large amount of triglycerides with saturated carbon chains that become solid wax at room temperature limits its use (Banković-Ilić et al., 2012), with the additional high amount of FFAs that hinders the common basic transesterification (Wan Ghazali et al., 2015).

5.2.3 WASTE COOKING OILS

The WCOs are significant economical raw materials for biodiesel production, until three times cheaper than edible vegetable oils (Banković-Ilić et al., 2012; Talebian-Kiakalaieh et al., 2013). Its utilization can also overcome the cost of water treatment and removal of this waste product, as well as the use of farmland (Talebian-Kiakalaieh et al., 2013). However, the main drawbacks are the high content of impurities (e.g., FFAs, water, polymers) and the suitable logistic for their collection from the households, restaurants, and food processing plants (Banković-Ilić et al., 2012; Wan Ghazali et al., 2015; Talebian-Kiakalaieh et al., 2013).

5.2.4 CHEMICAL COMPOSITION AND PROPERTIES

As mentioned before, the main components of second-generation feedstocks are TGs with impurities such as FFAs, polymers, and even water. These impurities interfere with the base transesterification reaction (see Figure 5.1a) employed to synthesize biodiesel, since the raw material should preferably be free of water and do not exceed a 2% w/w of FFAs (Atadashi et al., 2013), which affects the synthesis and quality of the biodiesel produced.

The presence of water in AF and WCOs benefits the hydrolysis of TGs (see Figure 5.1b), which generates additional amounts of FFAs. When these FFAs are in contact with basic homogeneous catalysts (NaOH, KOH, CH_3ONa), FFAs are transformed into soaps (see Figure 5.1c), which during biodiesel washing form foams limiting the separation of glycerol. Alternatively, the active sites in a solid base catalyst are poisoned by adsorption of these FFAs.

However, these FFAs are converted into more FAMEs or biodiesel through acid esterification reactions (see Figure 5.1d) (Banković-Ilić et al., 2012; Talebian-Kiakalaieh et al., 2013). In the next section, the treatment of feedstocks with different percentages of FFAs belonging to second-generation feedstocks is discussed.

5.3 CONSOLIDATED PROCESSES FOR BIODIESEL PRODUCTION USING SECOND-GENERATION FEEDSTOCKS

Biodiesel is produced on an industrial scale using refined oils with a low FFA content (<2% by weight) as raw material. The following paragraphs briefly describe a consolidated process (Figure 5.2) for the production of biodiesel using refined oils with low FFA content as raw material.

The reactor (batch or continuous) is usually fed with the oil, alcohol, and homogeneous catalyst (NaOH, KOH); it operates at ~60 °C and atmospheric pressure. The boiling point of methanol restricts these operating conditions since transesterification reactions occur in the liquid phase. In order to enhance the biodiesel yield, Tanaka et al. (1981) patented a process that includes the use of a dual system consisting of two consecutive reactors.

In the first batch reactor, only 80% of the alcohol and catalyst are used to perform the transesterification reactions. Then a 50% of the glycerol is removed from the first crude transesterification product and an additional transesterification step of the first crude product is carried out in the second batch reactor using the remaining catalyst and alcohol. This configuration leads to a FAME yield close to 100%, the general idea

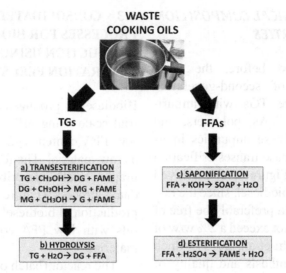

FIGURE 5.1 Main reaction pathways of TGs and FFAs during processing of WCOs for biodiesel synthesis.

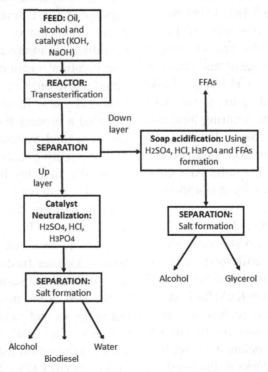

FIGURE 5.2 Industrial biodiesel production using refined oils.

of this process is the reduction of the amount of catalyst and alcohol employed, when compared with the use of a single batch reactor (Jeong and Park, 2005; Santori et al., 2012).

Glycerol separates from biodiesel due to its higher density. Normally a setting tank or a centrifuge is used for this purpose. The glycerol stream contains 50% glycerol together with an excess of methanol, catalyst and soap formed during the reaction of the homogeneous catalyst with FFAs. Then, glycerol is refined by the addition of mineral acids (HCl, H_2SO_4, H_3PO_4) to transform soaps into salts and FFAs, the latter are insoluble in glycerol and, therefore, are easily removed. After that, methanol is removed from glycerol by an instantaneous vacuum process that produces a glycerol purity of 85% (Van Gerpen, 2005; Santori et al., 2012).

Excess methanol can also be separated from crude biodiesel using a vacuum flash process. Then, the remaining catalyst in crude biodiesel is neutralized with a mineral acid, which also serves to separate any amount of soap formed in the transesterification reaction. The salts formed in the previous step are eliminated during the washing step with a large amount of water that is subsequently removed by a vacuum flash process. In addition, recovered methanol generally contains water, so before reuse, methanol goes through a process with molecular sieves

to remove remaining water (Van Gerpen, 2005; Santori et al., 2012).

From the previous discussion, it can be seen that a consolidated process is required for the production of biodiesel on an industrial scale using second-generation raw materials and heterogeneous catalysts. For this purpose, in the following sections several consolidated processes are discussed taking into account the content of FFAs in the raw material. It is expected that this information will guide the reader to select the best process for the production of biodiesel from a particular second-generation feedstock.

5.3.1 SIMULTANEOUS ESTERIFICATION–TRANSESTERIFICATION REACTIONS: ONE-STEP

When the FFAs content in raw materials does not exceed a 2 wt.%, as in edible vegetable oils, homogeneous or heterogeneous base catalysts are used, without a significant decreased in the FAME yield. However, the high content of FFAs in second-generation feedstocks makes necessary the use of a material capable of catalyzing both transesterification (conversion of TGs to FAMEs) and esterification (conversion of FFAs to FAMEs) reactions. In this sense, solid acid catalysts can be used for both esterification and transesterification reactions to obtain biodiesel.

Regarding base catalysts, the limiting step in the transesterification of TGs is the transformation of monoglycerides to glycerol (see Figure 5.1a). Li et al. (2014) confirm this result, they obtained the following values of activation energy for each one of the three consecutive reactions: 50.57 kJ/mol, 50.8 kJ/mol, and 86.67 kJ/mol.

Nonetheless, transesterification reactions can also be carried out using acid catalysis; its catalytic activity is lower compared to basic catalysis, which can be overcome by operating at higher reaction temperature (Islam et al., 2014). Even more, when the reaction proceeds through acid catalysis, the determinant step of transesterification reactions is the conversion of TGs to diglycerides (DGs), and the water formed during esterification can poison the active sites of heterogeneous acid catalysts, rendering in a low FAME yield (Cabrera-Munguia et al., 2017).

Thus bifunctional catalysts are a promising alternative since these solid catalysts contain acid and base catalytic sites, such as Lewis–Brönsted acid sites or acid–base sites or metal-containing acid–base sites, being able to carry out simultaneously the esterification of FFAs and transesterification of TGs (Semwal et al., 2011; Mardiah et al., 2017). In general, bifunctional solid catalysts are divided into two groups: inorganic and organically functionalized,

this latter contains both Brönsted and Lewis acid sites compared to inorganically functionalized catalysts. In the case of inorganic catalysts, the most representative involves a material containing a basic oxide like La_2O_3 with an acidic oxide as Bi_2O_3, ZnO, and Al_2O_3 or even CaO in this case as the basic oxide. Another example is a mixed oxide made of transition metals incorporated into alkaline metals oxides as in the framework of hydrotalcite (Mardiah et al., 2017). Whereas in organic bifunctional catalysts stands out SBA-15 with the incorporation of metal ions and $-SO_3H$ groups, waste carbon-based materials with $-OH$, $-COOH$, and $-SO_3H$ groups, and finally ionic liquids that are chemically composed by an anion and a cation, having then Lewis and Brönsted acid sites, respectively, which can be supported in a basic material. Another example is silica functionalized with various loadings of organophosphorous acid (Islam et al., 2014).

Regarding the production of biodiesel at the industrial scale with heterogeneous catalyst. In 2005, the French Institute of Petroleum developed a process (Esterfip-HTM) using a high reaction temperature (200 °C) and pressure (60 bar), where the catalyst is composed of a spinel of mixed oxides of aluminum and zinc without loss of the catalyst (Bournay et al., 2005). Correspondingly, in Figure 5.3a flow diagram is proposed

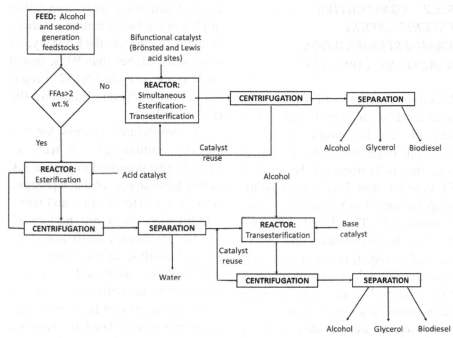

FIGURE 5.3 Consolidated processes for biodiesel production using second-generation feedstocks.

which shows a possible consolidated process for biodiesel synthesis using raw materials with < 2 wt.% of FFAs. In this feasible consolidated process esterification and transesterification are performed simultaneously using a bifunctional catalyst (Brönsted and Lewis acid sites) with a simple regeneration step, being the catalyst chemically stable to deactivation by water, TGs and FFAs adsorption, which is finally removed by centrifugation. The biodiesel and glycerol obtained can be split either by a setting tank or centrifugation. Then, the methanol in excess is removed from biodiesel and glycerol by a vacuum flash process. In this case, biodiesel and glycerol are pure and it is not necessary the use of additional step-processes such as acidification, neutralization, and washing.

To summarize this section, the literature has been reported that acid or bifunctional catalysts with high density of active sites can rapidly perform esterification and transesterification reactions. Also, hydrophobicity, specific surface area, and wide pore diameter are additional parameters that should be taken into account in the design of solid catalyst for biodiesel production from second-generation feedstocks.

136   *Handbook of Research on Bioenergy and Biomaterials*

5.3.2 CONSECUTIVE ESTERIFICATION–TRANSESTERIFICATION REACTIONS: TWO-STEP

When the content of FFAs is greater than 2 wt.%., it is preferred a two-step process for biodiesel production. The objective of the esterification step is to decrease the level of FFAs to less than 2 wt.% to avoid the soap formation in transesterification (second-step). The idea of a two-step reaction is to minimize the use of alcohol, catalyst, reaction time, and also the yield loss provoked when FFAs formed soap with homogeneous base catalysts or to avoid the poisoning of active sites in basic solid catalysts (Thoai et al., 2019).

Hence, simultaneous esterification–transesterification reactions showed a lower biodiesel yield than the consecutive one (Banković-Ilić et al., 2012) when using raw materials with a high content of FFAs (>2 wt.%). In the literature a two-step process is preferred, combining homogeneous acid catalyst and a solid base heterogeneous catalysis such as the use of CaO prepared from waste chicken eggshells by calcination in a two-step biodiesel production from Mahua and Karanja oil, obtaining a FAME yield of 95%. Another configuration is a two-step process combining a solid acid catalyst (sulfated zirconia) followed by a homogeneous base

catalyst achieving 94% conversion of FFA in the esterification reaction. With this approach, the FAME yield obtained is higher than 94%, which could meet with the EN14214 regulation for biodiesel (Banković-Ilić et al., 2012).

A consolidated process for raw materials containing > 2 wt.% of FFAs is also proposed in Figure 5.3. In this feasible consolidated process, consecutive esterification and transesterification steps are performed. The first step is the esterification with an acid catalyst to transform FFAs into biodiesel, afterward catalyst is removed by centrifugation, and the water produced can be removed by molecular sieves. Then, the reaction stream flows to the second reactor, where the basic catalyst and additional alcohol is added to carry out transesterification. Once again, the base catalyst is split by centrifugation; biodiesel and glycerol are separated by a setting tank or centrifugation. After that, methanol in excess is removed from biodiesel and glycerol by a vacuum flash process; leading to pure biodiesel and glycerol with not additional step-processes.

Among the advantages of this two-step system are a no acidic waste treatment, high efficiency, low equipment cost, simple recovery of the solid catalyst, a less time-consuming process and thus a lower energy consumption that generates a more economical product

(Talebian-Kiakalaieh et al., 2013). Therefore a consecutive esterification–transesterification system represents a convenient way to process raw materials with high FFAs content as second-generation feedstocks, which involves the esterification of the FFAs with an acid catalyst, an intermediary step to remove the water produced in esterification, followed by transesterification of TGs with a superacid or a base catalyst.

5.4 TAILORING HETEROGENEOUS CATALYST FOR BIODIESEL SYNTHESIS

This section has as a purpose to give a glance to the reader about the general physicochemical properties of heterogeneous catalysts commonly used for biodiesel synthesis (see Figure 5.4). Also, a classification of heterogeneous catalysts is made according to their chemical nature (basic or acid) and are described as the key factors taken into account during synthesis and activation of catalysts.

5.4.1 PHYSICOCHEMICAL CHARACTERISTICS OF SOLID CATALYSTS

Both esterification and transesterification reactions proceed through a sequence of steps, involving transport, adsorption, reaction, and desorption

of bulky and viscous molecules such as FFAs and TGs (2–5 nm).

5.4.1.1 POROSITY

Heterogeneous catalysts for biodiesel synthesis should possess a system of large pores (>2 nm) to avoid mass transport limitations (internal) (Chouhan and Sarma, 2011). For example, the use of zeolites for esterification and transesterification reactions is limited due to their microporous structure (Sasidharan and Kumar, 2004; Ramos et al., 2008; Martínez et al., 2011). Meanwhile, mesoporous materials such as SBA-15 (Chen et al., 2014; Melero et al., 2015; Cabrera-Munguia et al., 2018a), mesoporous–macroporous hydrotalcites (Tzompantzi et al., 2013; Liu et al., 2014; Cabrera-Munguia et al., 2018b), polymeric resins (Ma et al., 2017; Trombettoni et al., 2018; Zhang et al., 2018), and carbonaceous materials (Konwar et al., 2014a; Nata et al., 2017; Dhawane et al., 2018) have shown high catalytic activity in esterification and/or transesterification reactions, which may be associated with their porous structure that contains large pores. Furthermore, a system of interconnected pores in the catalyst is also desirable to enhance the transport of bulky molecules in liquid phase reactions such as esterification and transesterification. The synthesis of this kind of material with the bimodal

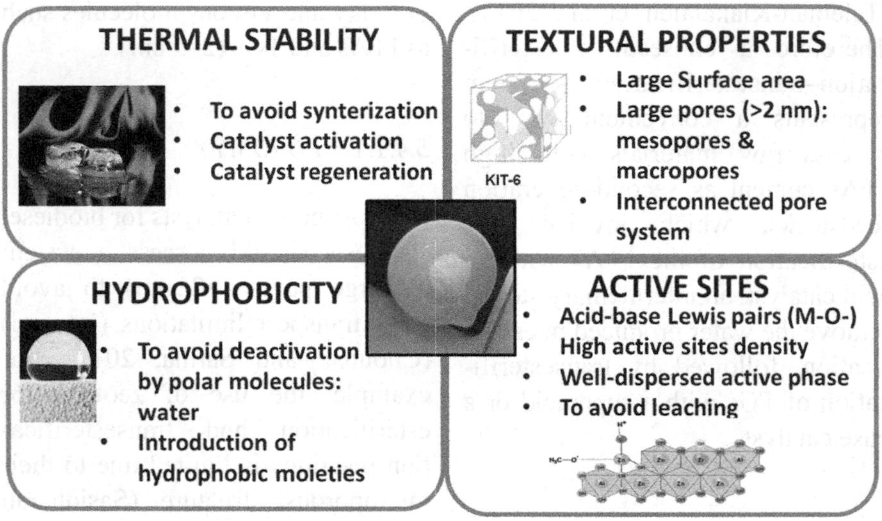

THERMAL STABILITY	TEXTURAL PROPERTIES
• To avoid synterization • Catalyst activation • Catalyst regeneration	• Large Surface area • Large pores (>2 nm): mesopores & macropores • Interconnected pore system
HYDROPHOBICITY	ACTIVE SITES
• To avoid deactivation by polar molecules: water • Introduction of hydrophobic moieties	• Acid-base Lewis pairs (M-O-) • High active sites density • Well-dispersed active phase • To avoid leaching

FIGURE 5.4   Desirable physicochemical properties of solid catalysts for biodiesel synthesis

meso- and macroporous network has been reported previously involving liquid crystalline and physical templating methods. For example, highly organized macro- and mesoporous SBA-15 catalysts have been developed, the macro- and mesopore diameters are tuned over the range of 200–500 nm and 5–20 nm, respectively (Wilson and Lee, 2012; Lee et al., 2014).

5.4.1.2   THERMAL STABILITY

Heterogeneous catalysts for biodiesel synthesis must be thermally stable in order to avoid the sintering phenomenon, due to the growing or agglomeration of microcrystals of the active phase, then reducing the area of active sites, and thus the catalytic

activity. This point must be attended as the activation and regeneration process generally involves the use of high temperature to desorb moieties, templates, or organic molecules. One of the main limitations of resins such as Amberlyst 15 for industrial applications obeys to its poor thermal stability, especially when it needs to be heated to eliminate water, a subproduct of the esterification reaction (Boz et al., 2015). The activation temperature is crucial for the generation of catalytic activity. Especially important for hydrotalcite materials whose activation temperature tunes its basic properties, and thus the transesterification yield (Liu et al., 2014; Cabrera-Munguia et al., 2018b). Furthermore, the activation temperature affects the structural properties of catalysts; for example, the use of tetragonal crystalline ZrO_2 phase gives rise to a better catalytic yield than the monoclinic phase (Islam et al., 2013).

5.4.1.3 REUSABILITY

The resistance to the dissolution of the active phase in the reaction medium is an important property of heterogeneous catalyst for biodiesel synthesis; this is one of the main drawbacks that promising basic and acid catalysts have shown, limiting their application in large scale production of biodiesel (Islam et al., 2013). For example, alkaline metal oxides (Na_2O, K_2O) and alkaline earth metal oxides (Na_2O, K_2O, MgO, SrO) heterogeneous base catalysts, and heteropolyacids corresponding to acid catalysts have shown leaching of their active phase, even though they presented high catalytic activities in transesterification and esterification reactions, respectively.

The particular characteristics that basic and acid heterogeneous catalysts should have are discussed below according to their type of active sites, hydrophobicity, and thermal activation. Figure 5.5 shows a general classification of the different types of solid catalysts used in the biodiesel production process.

5.4.2 ADVANCES IN THE DESIGN OF SOLID BASIC CATALYSTS

Heterogeneous base catalysts are divided into two groups: one with inherent active sites, and the other with basic supported species on nonbasic structures (Bing and Wei, 2019).

The type, concentration, and strength of surface basic active sites of a material are determined by its

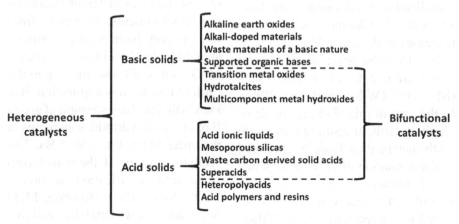

FIGURE 5.5 Solid catalysts for biodiesel production.

particle size, morphology, specific surface area, chemical composition, and electron transfer. Characterization techniques such as TPD, XPS, DRIFTS, EXAFS, calorimetry methods, and monitoring of acid probe molecules have been developed to determine the basic structure of active sites and its correlation with its catalytic performance (Bing and Wei, 2019).

Other classification of solid basic catalysts includes alkali or alkaline earth oxides (Dossin et al., 2006; Liu et al., 2008; Galadima and Muraza, 2014), mixed oxides containing elements of the groups IA, IIA and transitional metals such as Zr and Ti (Sun et al., 2010; Madhuvilakku et al., 2013; Hernández-Hipolito et al., 2014), and hydrotalcites (Tzompantzi et al., 2013; Liu et al., 2014; Cabrera-Munguia et al., 2018b).

The origin of basicity in alkaline earth oxides depends on the presence of M^{+2}–O^{-2} pairs in different coordination environments; the base strength of alkaline earth metals increases in the order $Mg < Ca < Sr < Ba$. Thus the basic sites of metal oxides are acid–base Lewis pairs sites (M^{+X}–O^{-2}). (Wilson and Lee, 2012). In the case of mixed oxides made of alkaline earth, transitional and rare earth metals, their basicity is related to the formation of O^- centers when M^{+x} is replaced by M^{+x-1}, obtaining an imbalance charge which is associated to a defect generation (Wilson and Lee, 2012). Meanwhile, the

formation of active sites in supported catalysts is explained by the insertion of the metal ion of the catalyst in a vacant site of the support, helping with the dispersion and decomposition of the active phase during the activation process (Islam et al., 2013).

Then, this section will make emphasis in classification and new advances on base solid catalysts, which are generally divided into (see Figure 5.6) alkaline earth oxides, alkaline doped materials, transitional metal oxides, and hydrotalcites. However, according to new research works, it will be adding three new group called waste materials of basic nature, multicomponent metal hydroxides, and also supported organic bases (see Figure 5.5).

5.4.2.1 ALKALINE EARTH OXIDES

Their basicity is attributed to M^{+2}–O^{-2} ion pairs in different coordination environments. Materials from this category have been extensively applied for biodiesel synthesis, being MgO and CaO the most popular earth oxides for this application. It is known that the base strength of group II oxides and hydroxides increases in the order $Mg < Ca < Sr < Ba$. The catalytic activity of these materials in the transesterification reaction is $CaO < SrO < BaO$. However, MgO does not show catalytic activity, strontium is dissolved in the reaction

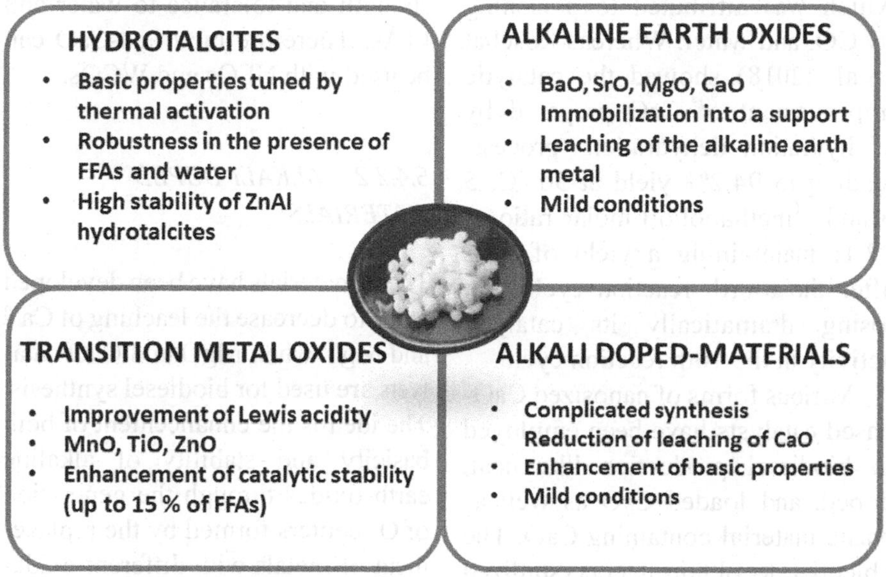

HYDROTALCITES	ALKALINE EARTH OXIDES
• Basic properties tuned by thermal activation • Robustness in the presence of FFAs and water • High stability of ZnAl hydrotalcites	• BaO, SrO, MgO, CaO • Immobilization into a support • Leaching of the alkaline earth metal • Mild conditions
TRANSITION METAL OXIDES	ALKALI DOPED-MATERIALS
• Improvement of Lewis acidity • MnO, TiO, ZnO • Enhancement of catalytic stability (up to 15 % of FFAs)	• Complicated synthesis • Reduction of leaching of CaO • Enhancement of basic properties • Mild conditions

FIGURE 5.6 Characteristics of basic heterogeneous catalysts for biodiesel synthesis

medium, and the leaching of barium in the ester phase is too high. Then, CaO is the most used due to its low price, low solubility in methanol, minor toxicity, and high availability from both natural and waste sources, easy preparation, mild reaction condition, and high ester yield (Marinković et al., 2016).

Calcium oxide is obtained by thermal decomposition of calcite or wastes such as mineral limestone or sea and eggshells which contain calcium carbonate ($CaCO_3$) (Marinković et al., 2016). The suitable activation temperature depends on the type of raw material, but it has been found that a calcination temperature of 900 °C generates a material with pure CaO without carbonates or hydroxyl groups

(Chouhan and Sarma, 2011; Lee et al., 2014), obtaining a 97% oil conversion at 75 °C.

The activity of CaO in transesterification reaction is due to the presence of the acid–basic Lewis pairs Ca^{+2}–O^{-2}, as calcium possess a small electronegativity (1.00), calcium is a weak acid, then its conjugated oxygen anion displays a robust basic property, which abstracts a proton from methanol, initializing the base-catalyzed reaction. Harsha Hebbar et al. (2018) synthesized calcium oxide nanoparticles and used it in the transesterification of *Bombax ceiba* oil, obtaining a 96.2% of yield at 65 °C, a methanol/oil molar ratio of 10.3:1 and 70 min for reaction time; nonetheless, in the sixth reaction cycle the FAME yield was reduced to 70%, which was attributed to poisoning by CO_2 and water. Whereas Roschat et al. (2018) showed the catalytic improvement of CaO prepared by a hydration–dehydration process, leading to 94.2% yield at 30 °C, 5 h and a methanol/oil molar ratio of 12:1; maintaining a yield of 90% after the fourth reaction cycle, but losing dramatically its catalytic activity in the 10th reaction cycle.

Various forms of nanosized CaO-based catalysts have been employed in biodiesel production like neat, doped, and loaded CaO as well as waste material containing CaO. The objective to obtain nanocrystallized CaO is made an efficient catalyst for transesterification reactions since its small crystallite size and low number of defects are associated with a large surface area and thus a better dispersion of the active sites. Thus it is required the assessment of the hydrothermal growth of flower-like CaO nanoparticles (Marinković et al., 2016).

Nonetheless, the loss of catalytic activity of CaO under subsequent cycles is attributed to the dissolution of Ca by methanol by the formation of calcium diglyceroxide (Chouhan and Sarma, 2011), and also to the deposition of organic materials in its catalytic surface (FFAs) (Chouhan and Sarma, 2011; Mardiah et al., 2017). To overcome these problems, CaO has been modified with La and Ce oxides, increasing its base strength and tolerance to water and FFAs. Therefore modified-CaO can be used with NEOs and WCOs.

5.4.2.2 ALKALI-DOPED MATERIALS

These materials have been developed to try to decrease the leaching of Ca^{+2} and Mg^{+2} when MgO and CaO catalysts are used for biodiesel synthesis. The idea is the enhancement of both basicity and stability of alkaline earth oxides through the generation of O^- centers formed by the replacement of metals with different oxidation numbers, creating an imbalance charge, and thus a superficial defect. Examples are CaO doped by Li and Cs cations (Lee et al., 2014), as well as CaMgO and CaZnO which are used as a way to stabilize the active sites of CaO (Mardiah et al., 2017).

Zhang et al. (2016) used CaO $(Sr_2Fe_2O_5–Fe_2O_3)$ magnetic catalyst that showed better catalytic performance than the pure one in the transesterification of soybean oil, finding that it was stable during five reaction cycles, maintaining its initial conversion (89%). Sudsakorn et al. (2017) synthesized CaO/MgO doped with Sr by the coprecipitation method, the catalytic results indicate that Sr^{+2} dopant enhanced the catalytic performance of CaO and MgO, reducing its particle size and providing catalytic stability during four reaction cycles.

5.4.2.3 TRANSITION METAL OXIDES

In this category alkaline earth elements are incorporated into metal oxides. The catalytic activity of these materials is very similar to CaO, but with higher stability, avoiding its dissolution (Lee et al., 2014). Some examples are MnO, ZnO, and TiO_2 with varying Lewis base character with good stability to biodiesel using feedstock with an acid content up to 15 wt.%. Also, sodium zirconates and sodium titanates have shown a high basicity, and thus a high catalytic activity, even though, its high sodium content will probably leach to the reaction medium decreasing their catalytic stability (Lee et al., 2014). Jamila et al. (2018) presented a heterogeneous metallic oxide catalyst (Mn–MgO–ZrO_2) which gave 96.4% of biodiesel yield at 90°C. These experiments were performed using 3% of catalyst, a methanol/oil molar ratio of 15:1, and a reaction time of 4 h. However, after the sixth reaction cycle, the biodiesel yield was reduced to 80%.

5.4.2.4 HYDROTALCITES

Hydrotalcites are double layered hydroxides with formula

$$[M^{2+}_{1-x}M^{3+}_x(OH)_2](A^n)_{x/n} \cdot H_2O$$

which are generally composed of Mg^{+2} and Al^{+3}. Being preferably the substitution of Mg by Zn to avoid the dissolution of the catalyst when Mg reacts with glycerol (Liu et al., 2014; Cabrera-Munguia et al., 2018b). The catalytic behavior of hydrotalcite is determined by the structure of its active sites on the surface, which can be controlled by its chemical composition and structural architecture. Also, the type and strength of its basic sites can be modulated by calcination, and rehydration treatment (memory effect of hydrotalcite) (Bing and Wei, 2019).

These materials are promising due to their basic strength that it is tuned according to its thermal stability. Thus at temperatures lower than 150 °C the interlamellar molecules of water are expelled generating mainly Brönsted basic sites (^-OH); between 300 and 500 °C mixed oxides of the corresponding metals are formed due to the process of dehydroxylation and loss of CO_2 leading to Lewis acid-basic pairs (M–O^-) of medium strength. At temperatures higher than 500 °C the structure collapses giving rise to the segregation of metal oxides and the formation of the spinel, forming terminal oxygens (O^{-2}) on the surface with very high basicity (Cavani et al., 1991).

Hence, ZnAl catalysts (Jiang et al., 2010; Moraes et al., 2014) are more robust catalysts in the presence of water, but with lower basic properties, however, these properties can be improved by the incorporation of metal cations such as La^{+3}, Ce^{+3}, and Zr^{+4} to the ZnAl hydrotalcite framework (Tzompantzi et al., 2013; Soares Dias et al., 2012).

In a recent study (Liu et al., 2014; Cabrera-Munguia et al., 2018; Pamatz et al., 2018], it was found that using a low activation temperature (200 °C) for ZnAl and ZnAl–Zr catalysts is just enough to catalyze transesterification reaction with a FAME yield of 82% at 200 °C during 2 h, and with a methanol/oil molar ratio of 30:1. Furthermore, the reusability studies indicated negligible deactivation by leaching of the active phase.

5.4.2.5 MULTICOMPONENT METAL HYDROXIDES

Sandesh et al. (2016) reported a mixed metal hydroxide as heterogeneous base catalysts for the transesterification of edible oil and NEO. $CaSn(OH)_6$ catalyst is a crystalline double perovskite hydroxide with alternating cations, which are coordinated with six oxygen atoms to form $Ca(OH)_6$ and $Sn(OH)_6$ polyhedra. This material shows an outstanding catalytic performance with a biodiesel yield of 94% for sunflower oil at 65 °C, using an activation temperature of 150 °C, with little deactivation after five subsequent reaction cycles and low hygroscopicity. Its high catalytic performance is attributed to its relatively abundant content of hydroxyl groups.

Da Silva et al. (2018) proposed stannosilicate of sodium as heterogeneous catalysts, based on mixed octahedral–pentahedral–tetrahedral microporous siliceous frameworks, which claims not only that is useful with edible oil and NEO but also can simultaneously perform both esterification and transesterification reactions. However, according to the analysis of adsorbed ammonia, this stannosilicate shows low acidity and its catalytic activity is mainly associated with the sodium species present in the material. Then, the esterification reaction can be related to the high reaction temperature employed in this research (150 °C). Unfortunately, it was observed a decrease of the FAME yield to 50% in the sixth reaction cycle.

5.4.2.6 WASTE MATERIAL OF A BASIC NATURE

These materials are classified into two groups: industrial wastes and biological wastes. The common characteristic of these materials is the presence of a high amount of $CaCO_3$ that with a thermal treatment could give rise to CaO. Among the industrial wastes are lime mud (a mixture of $CaCO_3$ and traces $MgCO_3$), limestone obtained from constructional sites, red mud (oxides of Ca, Fe, Si and Al) a waste produced from the refining of bauxite, and slag a waste matter from metals separation or reduction from their respective ores (metal, metal oxides, metal sulfides, SiO_2 (Marwaha et al., 2018).

Natural $CaCO_3$ and $MgCO_3$ are found in dolomites and biological wastes such as eggshells, snails, animal bones, and also plants (Marwaha et al; 2018). The objective of this research is the biodiesel production from waste material as feedstocks, to reduce the production cost and the environmental harm that chemical compounds cause with catalyst preparation.

5.4.2.7 SUPPORTED ORGANIC BASES

Organic nitrogenous compounds such as amine and guanidine exhibited comparable activities in reactions such as transesterification, without the formation of emulsions or soaps. In a recent research, tetraalkylammonium hydroxide-functionalized SBA-15 materials (SBA-15- pr-NR_3OH) have been prepared by anchoring dimethyloctadecyl [3-(trimethoxysilyl)propyl]ammonium hydroxides onto the surface of mesoporous SBA-15 silica, and then the catalytic activity was tested in the transesterification of soybean oil with methanol (Xie et al., 2014). This catalyst provides a FAMEs yield of 99.4% using a methanol/oil molar ratio of 12:1, catalyst loading 2.5 wt.%, and reaction time of 30 min. The obtained SBA-15-pr-NR_3OH catalyst could be recovered and reused for several recycle runs with a negligible loss of activity (Xie et al., 2014).

Moreover, prepared aminografted multiwalled carbon nanotubes (N-MWCNTs) were also used as an efficient base catalyst in the transesterification of TGs (glyceryl tributyrate). In these studies, nanotubes grafted with tertiary amines (Et_3N-CNTs) were identified as the most basic catalysts and thus the best catalytic performance in the transesterification reaction. The nanotubes grafted with tertiary amines exhibited the highest stability up to 130 °C, whereas secondary amine grafted amine was found to be the least stable. Reusability tests showed deactivation of the catalyst due to the adsorption of the triglycerides on the catalyst surface, but it was not observed a significant leaching of amino groups (Konwar et al., 2014a).

5.4.2.8 STATUS AND OPPORTUNITIES

Recent progress is made on the solid base catalyst for transesterification reaction, the current trend in this area leads to the use of industrial or biological wastes as heterogeneous catalysts for biodiesel production. Also, the enhancement of the hygroscopic and catalytic properties of CaO has been made with the incorporation of metal rare-earth elements (La, Ce) and alkali metals (Li, Cs), respectively. In the case of hydrotalcites their particular framework offers the possibility to modulate their basicity with the chemical composition and thermal activation, making these materials a promising option. Recent studies have shown negligible deactivation through several subsequent reaction cycles. Furthermore, the performance of materials like stannosilicates of sodium or $CaSn(OH)_6$ for transesterification reaction has been outstanding since these materials use low reaction temperature and in the case of $CaSn(OH)_6$ have shown little deactivation in subsequent reaction cycles. Also, supported organic bases such as ternary amines have presented a high FAME yield. However, the reusability tests have shown that these materials are deactivated by a high adsorption of TGs. Then, recent research in the design of solid base catalyst indicates that in order to obtain a more robust material, the basic species have to be involved in a consolidated framework. Then, scientists are building more elaborated crystalline structures to avoid the leaching of the

active species generating at the same time a hydrophobic surface.

However, the design of any catalyst involves much experimental work with the exploration and synthesis of new catalysts. Also, the characterization techniques for the identification of active sites are quite complicated, due to the complex heterogeneous reaction and deactivation processes. Then, the combination of advanced characterization techniques with DFT calculation is necessary to elucidate the chemical and electron environment of active sites that will give us some guidance about the catalytic performance of certain materials (Bing et al., 2019).

5.4.3   ADVANCES IN THE DESIGN OF SOLID ACID CATALYSTS

It is well known that solid acid catalysts can also transform feedstocks with a high content of FFAs (>2 wt.%) through simultaneous esterification and transesterification reactions. To circumvent the poisoning of acid active sites by strong adsorption of water or glycerol, it is necessary a hydrophobic surface with affinity to the oily species (FFAs, TGs). The hydrophobicity eludes the solvation of actives sites for the action of water (Park et al., 2010), which minimizes the catalyst deactivation. Thus the ideal characteristics of solid acid catalysts are a high concentration of active sites, moderate acid strength, hydrophobicity, and large pores to reduce mass transport limitation (Avhad and Marchetti, 2015).

Depending on the concentration of active sites and the hydrophobic nature of the surface catalysts, there are three possible scenarios of the approaching of lipophilic molecules as TGs or FFAs. In the first scenario, a hydrophobic surface with a low concentration of Brönsted acid sites would adsorb parallel the lipophilic tail of the FFA. In the second possibility, a hydrophobic surface with a moderate concentration of active sites would adsorb perpendicularly the FFAs molecules, with the tails forming a local hydrophobic environment. In the last case, an hydrophilic material with a high concentration of active sites would adsorb the water produced in esterification causing the catalyst deactivation.

However, the use of acid catalysts for biodiesel production requires severe reaction conditions to satisfactorily transform oil to biodiesel (Avhad and Marchetti., 2015), and also an excess of alcohol is necessary to prevent or minimize the catalyst deactivation by water. Hence, the ideal solid acid catalyst should be designed for a wide variety of feedstocks, reduce the reaction temperature and the amount of alcohol required (Trombetoni et al., 2018).

Since superacid catalyst naturally possesses hydrophilic active sites such as $-PrSO_3H$ or $-SO_3H$, then some hydrophobic moieties can be introduced in its catalytic surface to generate a more hydrophobic surface (León et al., 2015).

Solid acid catalysts are classified according to their respective acid sites: Lewis and Bronsted. Lewis acid sites are able to accept at least an electron pair free. Whereas Bronsted acid sites can donate or partially transfer a proton. Examples of solid acid with a high content of Bronsted acidity are sulfonated materials such as resins, heteropolyacids, and doped SBA-15 with $-PrSO_3H$ groups (Mbaraka and Shanks, 2005; Shah et al., 2014), whereas Lewis solid acid catalysts are WO_2/ZrO_2 or doped SBA-15 with metals such as Al^{+3}, Ti^{+4}, or Zr^{+4} (Kuzminska et al., 2014, Li et al., 2010).

Since acid catalysis proceeds at high reaction temperatures and long reaction times, only some works have tried to explain the role played by different types of acid sites during the reactions that produce biodiesel. For example, in the simultaneous esterification and transesterification of a model mixture of tricaprylin and palmitic acid, it was found that even though a high concentration of Bronsted acid sites gives rise to high catalytic activity, these are poisoning by water adsorption. Whereas Lewis acid sites are transformed into Bronsted acid sites when are in contact with water. Also, according to the GC-MS analysis of the main reaction products, it was assessed that the conversion of TGs to DGs was the determining step for transesterification (Cabrera-Munguia et al., 2017).

According to their chemical composition, heterogeneous acid catalysts are divided into (1) heteropolyacids, (2) acidic polymers and

HETEROPOLYACIDS	ACID POLYMERS AND RESINS	WASTE CARBON DERIVED SOLID ACIDS
• High Brönsted acid strength • Immobilization into a support • Leaching of HPA • Limited water tolerance	• High Brönsted acid strength • Hydrophobic surface • Large pore diameter and volume • Pore swelling a key factor • Low thermal stability	• Sulfur content and hydrophilicity/hydrophobicity ratio are key factor • Low Surface área • Deactivation by sulfur leaching and metanol adsorption
SUPERACIDS	MESOPOROUS SILICAS	ACID IONIC LIQUIDS
• Lewis and Brönsted acid sites • Hydrophobic surface • High thermal stability • Deactivation by sulfur leaching and WO₃ oxidation • High calcination temperature and long reaction times	• Lewis and Brönsted acid sites • Enhancement of hydrophobic surface by grafting with moieties of non-polar molecules • Tailoring of a interconnected system of large pores	• Lewis and Brönsted acid sites • Thermal stability • High cost and viscosity • High hydrophobicity • Immobilization into a support by solubilization in polar solvents

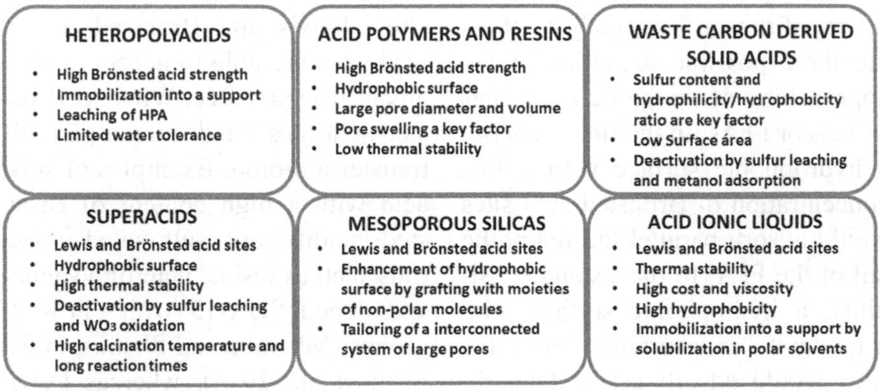

FIGURE 5.7　Classification of solid acid catalysts for biodiesel synthesis

resins, (3) waste carbon-derived solid acids, (4) superacids, (5) mesoporous silica, and (6) acid ionic liquids (see Figure 5.7).

5.4.3.1 HETEROPOLYACIDS

Heteropolyacids (HPAs) and polyoxometalates made of QO_4 central tetrahedron surrounded by octahedral metal-oxygen. The general formula of Keggin heteroatom anion is $QM_{12}O_{40}$, where Q represents the central atom-like P^{+5} or Si^{+4}. The Keggin heteropolyacids are $H_3PM_{12}O_{40}$ and $H_4SiM_{12}O_{40}$, where M denotes the metal ions (Mo^{+6} or W^{+6}) that can be substituted by other metal ions

such as V^{+5}, Co^{+2}, and Zn^{+2} during synthesis (Mansir et al., 2017). HPAs offers proton mobility which guarantees high catalytic performance, and its ionic structure displays a Brönsted acid strength similar to that of superacids catalysts. Nonetheless, HPAs are thermally unstable and are characterized by poor stabilization of molecule intermediates, which makes HPAs not appropriate for biodiesel production in their original form, as a result of their high solubility in polar medium, being the major drawback the leaching of the catalysts in methanol (Lee et al., 2014; Mansi et al., 2017).

In order to provide a large number of accessible acid sites, in the literature, it has been suggested to disperse HPAs onto the catalyst support with a large surface area. For example, tungstophosphoric acid (HPW) supported in Nb_2O_5, ZrO_2, and SBA-15 has been reported as catalysts with large surface area and a small amount of leaching.

Tropecelo et al. (2010) immobilized HPW in SBA-15 with a 7.3 wt.% HPA load yielding a conversion of 88% in the esterification of palmitic acid, where the reaction conditions were a methanol/palmitic acid molar ratio of 6:1, at 60 °C and 5 h of reaction. Similarly, a conversion of a 90% was obtained in the esterification of palmitic acid, when Alcañiz et al. (2018) studied the catalytic performance of HPW supported on ZrO_2 at 60 °C but with a 30 wt.% HPA load.

Cs salts of tungstophosphoric acid $Cs_xH_{(3-x)}PW_{12}O_{40}$ and $Cs_yH_{(4-y)}SiW_{12}O_{40}$ are additional examples of catalysts which are practically insoluble in water, whose catalytic activity requires a balance between the retention of acidic protons and generation of stable mesopores to facilitate molecular diffusion (Lee et al., 2014).

However, the preparation method of HPAs needs further improvement, and investigation of a promoter is required to increase the interaction between the support and HPA. It is also necessary to analyze how the reaction conditions affect their performance to be suitable for biodiesel production (Lee et al., 2014; Mansir et al., 2017). The principal drawback of HPA for its application in esterification and transesterification reactions is their limited water tolerance, which can only be enhanced by the dispersion of HPA on the support of the large surface area, modifying at the same time the acid strength and density of acid sites (Lee et al., 2014).

5.4.3.2 ACID POLYMERS AND RESINS

Acidic polymers and resins such as sulfonated mesoporous polymers or copolymers of styrene, divinylbenzene (DVB) are heterogeneous acid catalyst characterized by a hydrophobic surface with high Brönsted acid site density (sulfonic acid group attached to polymer chains) and large pore volume. The amount of cross-linking component (DVB) mainly determines the surface area and pore size distribution of porous acid resins. The gel-type resins are also synthesized from styrene in the presence of DVB (typically <5 mol%) with no other component (Trombettoni et al., 2018). Also, perfluorinated acidic ion exchangers, based on sulfonic acid groups, are nonporous solid acid material with acidity close to H_2SO_4; in this category, the performance of Nafion resin stands out, which after hydrolysis of the sulfonyl fluoride yields to the strongly acidic terminal $-CF_2CF_2SO_3H$ groups.

Materials such as Amberlyst-15 and Nafion NR50 are good esterification candidates. However, they suffer a rapid deactivation after 2 and 4.5 h. The fast deactivation of porous resins may be due to solvation (poisoning) of the active sites with water produced during esterification as well as water adsorbed on the inner/outer surface, which hinders the accessibility of the hydrophobic FFAs molecules. Moreover, the high acid strength of resins provokes the faster deactivation by cation impurities (Na^+, Ca^{2+}, Mg^{2+}) by forming salts (Trombettoni et al. 2018).

Studies regarding the influence of the morphology of acid resins on the catalytic performance during esterification of FFAs have demonstrated that the activity of catalyst mainly depends on its swelling capacity, being effective the gel-type resins than the porous Amberlyst-15. As porous resins possess a permanent porous structure, they adsorb more easily water. Furthermore, their pore diameter and hydrophobic nature allow the efficient diffusion of the substrate through the pore network (Lee et al., 2014); attracting TGs, FFAs, and tails of alcohol by the polymer support. Moreover, mass transfer limitations are reduced as a result of the pore-swelling of the resins in organic media, which determine the activity of these catalysts (Trombettoni et al., 2018).

For example, using a multi-SO_3H functionalized polymeric acid was obtained a yield of 96% of methyl oleate at 100 °C, when using a methanol/oleic acid molar ratio of 30:1 and 4 h of reaction time and maintaining its catalytic activity under four subsequent reaction cycles, however, reactivation by contact with H_2SO_4 resulted necessary (Pan et al., 2017a)

To summarize, the main disadvantages of acid resins are their low thermal stability and deactivation with polar molecules and metal ions. An excess of water in the raw material can dissociate the actives sites and hinder the accessibility to nonpolar molecules as FFAs and TGs. Furthermore, their active sites are regenerated by heating up to 200 °C. However, its thermal stability could be compromised (Trombettoni et al., 2018; Mansir et al., 2017).

5.4.3.3 WASTE CARBON DERIVED SOLID ACIDS

The main characteristic of carbon-based materials is that they come from waste materials such as biomass, alga residues, and even glycerol. Since plant biomass is mostly composed of carbon structures, they are transformed into carbon-based materials through direct carbonization. Indeed, biomass-based catalysts are known to be nontoxic, biodegradable possessing higher surface area than conventional catalysts such as Amberlyst-15 (Tang et al., 2018).

The methods to functionalize waste carbon materials are direct sulfonation and sulfonation via reductive alkylation/arylation. Among these methods, direct sulfonation is the most extensively studied, being sugars, biochar, starch, and lignin the most used raw materials. The parameters to study in this method are sulfonating agent, sulfonation time, and carbon precursor on the activity of such catalysts.

A similar method for preparing such materials is the incomplete carbonization of sulfonated polycyclic aromatic compounds in concentrated H_2SO_4. However, the objective is to obtain a rigid graphite-like network consisting of small polycyclic aromatic carbon sheets in a three-dimensional sp3-bonded structure (a structurally disorganized form of graphite). These graphite-like materials can thereafter be functionalized/modified similarly to graphite. Sulfonation of graphite-like materials yields to a highly stable solid with a high density of −SO3H groups, the robustness obtained avoid the leaching of sulfonic groups and thus obtaining a remarkable catalytic performance in the esterification of FFAs.

These studies indicate that direct sulfonation of carbon waste materials leads to catalytic surfaces containing Ph–OH, −COOH and −SO$_3$H groups. Meanwhile, the

incomplete carbonization of polycyclic aromatic compounds forms graphene carboxylate groups that stabilize the SO$_3$ amorphous carbon bearing −SO$_3$H groups. Their presence enhances catalytic performance by increasing overall acid density and by acting as sites for the attachment of substrates like TGs and FFAs. Further characterization also indicates that carbon-based sulfonated material possesses Lewis and Brönsted acid sites originated by −COOH and −SO$_3$H groups, respectively (Konwar et al., 2014b). Sulfonated carbonaceous materials are a promising material for FFA esterification and TGs transesterification. Its catalytic activity is proportional to the sulfur loading and the balance of hydrophobic/hydrophilic sites on the carbon; whose proportion is tuned by the experimental conditions of carbonization and sulfonation (Dhawane et al., 2018).

Thus short times of carbonization lead to smaller sheets with higher densities of −SO$_3$H groups, increasing their catalytic activity together with the sulfur leaching and hence their deactivation (Dhawane et al., 2018; Lee et al., 2014). For example, waste palm oil (5.2% of FFA) was used as a raw material in the simultaneous esterification–transesterification over sulfonated carbon catalyst from coconut residues yielding to a 92.7% of FAMEs at 65 °C during 8 h; however, the utilization of the catalyst during four reaction cycles leads to a reduction of the FAME yield up to 80% as a result of sulfur leaching (Thushari and Babel., 2018). Lathiya et al. (2018) synthesized sulfonated carbon catalysts obtained from the waste orange peel, and employed for esterification of corn acid oil obtaining a conversion of 91.68% at 65 °C and 4 h.

The outstanding catalytic performance of this material is related to its high acid site density and hydrophobicity, which prevents the hydration of active sites (Islam et al., 2014).

Unfortunately, these catalysts are prone to deactivation by adsorbed methanol, requiring the regeneration of the active phase by sulfuric acid. Moreover, waste carbon-derived solid acids possess a low surface area even though the surface activation of carbon is established; the method is harsh due to the low reactivity of carbon (Nata et al., 2017).

5.4.3.4   SUPERACIDS

These materials are solid catalyst formed basically by mixed metal oxides providing both Lewis (anion) and Brönsted acid sites (cations) necessaries for esterification and transesterification processes. These mixed metal oxides are generally composed of transition metals such as zirconium, zinc, titanium, iron, tungsten, and tin. Among these metal oxides, zirconia is the most popular catalyst with strong acid sites, high

thermal stability, and large pores. At about 600 °C, zirconia changes its crystal growth from the tetragonal phase to monoclinic improving its acid properties and surface area (Mansir et al., 2017).

Zirconia impregnated with sulfuric acid increases its acidity, and thus its catalytic performance.

After sulfate dispersion, zirconia catalysts were used to assist the esterification reaction between oleic acid and methanol, obtaining 90% of biodiesel yield after 12 h of esterification reaction at 60 °C using a methanol-to-oil molar ratio of 40:1 and 0.5 g of catalyst. Also, it is reported that the interaction between the sulfate species and zirconia is weak, reducing the catalytic activity during the reusability tests (Avhad and Marchetti, 2015). Therefore the use of chlorosulfonic acid instead of sulfuric acid as impregnation agent can enhance the concentration of sulfate species up to 4 times, obtaining 100% of biodiesel yield even after 5 catalysts recycle test for esterification, at 100 °C for 12 h using methanol-to-oil molar ratio of 8:1 and 3 wt.% of catalyst (Avhad and Marchetti, 2015).

Preparation of sulfated zirconia by solvent-free method was found to be more active in esterification and transesterification than sulfated zirconia prepared by standard precipitation conditions. This preparation method leads to an amorphous material, whereas sulfated zirconia by standard precipitation conditions is crystalline (tetragonal y monoclinic phases) (Semwal et al., 2011).

The main concern regarding the potential use of zirconia is its reduction in porosity and surface area after the thermal treatment. Nonetheless, the doping of the zirconia surface with hetero species could preserve its morphology at high calcination temperature. A recent study presented the doping of zirconia with tungsten oxide species, the resulting catalysts lead to 82% of biodiesel yield at 200 °C for 150 min, using a methanol-to-oil molar ratio of 12:1 and 1.5 g of catalyst amount. The results indicate that this material was found very active and hydrophobic until three reaction cycles and was resistant to 5% of the content of water and 9.1% of FFAs (Avhad and Marchetti, 2015).

A mixed oxide consisting of alumina with zirconia (Al_2O_3–ZrO_2) modified with tungsten oxide (WO_3) not only provides high mechanical strength but also enhances its acidity. The addition of alumina stabilizes the tetragonal phase of zirconia and prevents the growth of WO_3 particles. Comparing the catalytic performance between SO_4^{-2}/ZrO_2 and WO_3/ZrO_2, it is found that WO_3/ZrO_2 possesses higher stability than SO_4^{-2}/ZrO_2, eluding the leaching of acid sites into the reaction media. Nonetheless, further studies indicated the deactivation of the catalyst by the oxidation of WO_3 after long

term exposure to FFAs (Chouhan and Sarma, 2011).

Guo-liang et al. (2017) used tetragonal sulfated zirconia calcined at 500 °C as a catalyst in biodiesel production leading to 84.6% of FAME yield at 150 °C during 6 h and methanol/oil molar ratio of 20:1. Whereas Guldhe et al. (2017) used tungstate zirconia that showed a biodiesel yield of 94.58% at 100 °C during 3 h and with a methanol/oil molar ratio of 12:1.

Thus the main drawbacks of superacid catalysts are the high calcination temperatures (500–800 °C) required to meet the higher Brönsted acid sites needed. Also, a long reaction time is necessary to obtain a higher biodiesel yield (Chouhan and Sarma, 2011; Mansir et al., 2017; Mardiah et al., 2017).

5.4.3.5 MESOPOROUS SILICAS

Mesoporous silica (SBA-15, KIT-6, MCM-41) has been studied extensively as support of heterogeneous acid catalyst due to its mesostructured and hydrophobic or hydrophilic nature. These properties could be tuned via functionalization (sol–gel or grafting) with sulfonic acid groups (phenyl, propyl, and perfluoro sulfonic groups) and transitional metals (such as zirconium, aluminum, titanium, tin and recently niobium) (Lee et al., 2014; Avhad and Marchetti, 2015). Silica functionalized with phenyl sulfonic acid groups is reported to be more active than their corresponding propyl or perfluoro analogs, apart from the moiety dispersion, this obeys to direct incorporation of thermally stable covalent Si–C anchoring bonds during condensation method; in contrast, the Si–O–C bonds of perfluoro-SO_3H acid sites were thermally less stable and got released into the reaction medium at high reaction temperatures and methanol amount (Avhad and Marchetti, 2015).

Nonetheless, sulfonic acid functionalized silica possesses pore diameter in the order of 6 nm that is increased by swelling of Pluronic P123 micelles with trimethyl benzene, triethyl benzene, and triisopropyl benzene obtaining mesoporous silicas with a pore diameter of 5–30 nm. When these catalysts are used to perform esterification and transesterification under mild conditions, a high catalytic activity was observed, which is attributed to superior mass-transport of the bulky free fatty acid and triglycerides within the expanded $PrSO_3$–SBA-15.

For example, Cabrera-Munguia et al. (2018b) evaluated the effect of the insertion of Al^{+3}, Ti^{+4}, and – $PrSO_3H$ groups on the acid properties of SBA-15, observing the formation of Lewis and mainly Brönsted acid sites when metal ions are inserted. An opposite trend was observed when –$PrSO_3H$ are inserted in the SBA-15 framework, in this case, a higher proportion of Lewis acid sites along with its remarkable hydrophobicity were found to be responsible for a 90.75% of conversion of oleic acid at 100 °C, employing a methanol/oleic acid methanol molar ratio of 15:1, 5 wt.% of catalyst loading respect to the oleic acid mass and 5 h of reaction time.

However, the two-dimensional channels characteristic of the SBA-15 framework is hampered by molecular exchange with the bulk reaction media. Thus a three-dimensional interconnected channel such as KIT-6 mesoporous silica could enhance the in-pore accessibility of sulfonic acid sites. Hence, the tailoring of the mesoporous silica framework and surface functionality has produced solid acid catalysts suitable for esterification and transesterification with methanol under mild conditions. Nevertheless, there is still a challenge to extend the dimensions and types of pore interconnectivity in the silica frameworks, and also its hydrophobicity nature (Lee et al., 2014).

5.4.3.6 ACID IONIC LIQUIDS

A new line of catalysts named ionic liquids is chemically organic salt in the liquid state at room temperature, characterized by good thermal stability, high acidity, low vapor pressure, and poor solubility in the biodiesel phase (Pan et al., 2017b; Ullah et al., 2018].

The most common ionic liquids (ILs) are nitrogen-containing (such as alkylammonium, *N,N*-dialkylimidazolium, *N*-alkylpyridinium and *N,N* pyrrolidinium) or phosphorous-containing (such as alkyl phosphonium). The general choice of anions includes BF_4^-, PF_6^-, CH_3COO^-, CF_3COO^-, NO_3^- Tf_2N^-, $[(CF_3SO_2)_2N]^-$, $[RSO_4]^-$, and $[R_2PO_4]^-$, where Tf and R stand for bis (trifluoromethylsulfonyl) imide anions and alkyl group, respectively (Troter et al., 2016).

Acid ionic liquids have been tested successfully in esterification and transesterification reactions. In esterification, *N*-methyl-2-pyrrolidoniummethylsulfonate,[NMP] $[CH_3SO_3]$, is the best catalyst under mild reaction conditions, being able to convert a wide range of FFAs (stearic, myristic and palmitic) into alkyl esters, with conversions of 93.6%–95.3%, which is attributed to its longer carbon chain that facilitates the mass transfer in the reaction system. In the case of transesterification reaction, both single- and multi-SO_3H functionalized Brönsted acidic ILs are used as catalysts, being the most acidic and thus the best catalyst, 1-(4-sulfonic acid) butyl pyridinium hydrogen sulfate, $[(CH_2)_4SO_3HPy]$ $[HSO_4]$, obtaining a higher catalytic activity than H_2SO_4. Also, it is reported that the saturation degree of the oil has an inverse effect on the conversion since the oil with a higher saturation degree has a smaller chance to approach the active site of the catalyst. (Troter et al., 2016).

Ionic liquids are combined with metal chlorides and bentonite, and others are used as supports or supported on materials like Fe-SBA-15, Fe_3O_4–SiO_2, sulfhydryl- group-modified SiO_2 or polymers such as divinylbenzene. In ionic liquids supported on Fe-SBA-15, the catalytic activity on esterification reaction was attributed to the Lewis acid sites brought by Fe and the Brönsted acid sites that ionic liquids incorporate (Troter et al., 2016).

Basic ionic liquids are not as frequently as the acidic one, among the most used are imidazolium-based, such as imidazolium hydroxides, imidazoline; choline-based ionic liquids such as choline hydroxide, choline methoxide, and choline imidazolium. The catalytic activity of this type of ionic liquids decreases with the increase of the number of carbon atoms in alkyl-chain connected with the cation. Among the tested ionic liquids, bis-(3-methyl-1-imidazolium-)-ethylene dihydroxide (IMC2OH) is the most efficient. (Troter et al., 2016). Ding et al. (2018) analyzed the catalytic performance of acidic imidazolium ionic liquids in the biodiesel production from palm oil using microwave irradiation; obtaining a maximum yield of 98.93% with a methanol/oil molar ratio of 11:1 and 6.43 h, nonetheless its yield was reduced up to 84.76% after six cycles.

To summarize, the principal disadvantages of ionic liquids are its high cost and poor reusability. Then, the immobilization of ionic liquids onto support represents an alternative to overcome its solubility issue. With this objective, new acidic ionic liquid polymers and acidic ionic liquids-functionalized mesoporous resins are synthesized, obtaining solid catalyst with high acidity, hydrophobicity, large surface area, and thermal stability (Wu et al., 2015).

5.4.3.7 STATUS AND OPPORTUNITIES

Research on the heterogeneous acid catalyst for biodiesel production indicates that it is desirable that they have a high surface area, with hydrophobic properties and relevant acidity of both Lewis and Brönsted. Also, Brönsted acid sites help to develop esterification of FFAs whereas Lewis acid sites speed up transesterification of TGs. Then, bifunctional acid catalysts with Brönsted and Lewis acid type catalyst act synergistically in the production of biodiesel employing raw materials of low-quality (Pan et al., 2017b). Among the catalysts exhibit here waste carbon derived solid acids, mesoporous silica, and ionic liquids represent examples of these bifunctional catalysts which have

presented high catalytic yield in both esterification and transesterification reactions.

An extensive study of the above-mentioned materials is necessary in order to understand the role of the acid active sites (Brönsted and Lewis acid sites) during the process of biodiesel synthesis. The reusability of these materials is another key issue that must be further studied, acid catalysts must be designed to avoid dissolution of the active phase in the reaction medium.

Then, the concentration and strength of acid active sites and the hydrophobic nature of the surface should be modulated. One way to overcome these problems is the incorporation of sulfonic groups that came from organic acids such as chlorosulfonic acid or aryl sulfonic acid instead of sulfuric acid, to obtain a more hydrophobic sulfonic groups and also to augment its acid strength. Moreover, ionic liquids have these sulfonic groups bonded to a large chain of carbon molecules. Meanwhile, the hydrophobicity and pore structure in mesoporous silica are enhanced by the addition of molecules attached to the benzene ring.

Hence, the general idea in the synthesis of solid acid and basic catalyst is the obtention of a hydrophobic material to decrease the poisoning of the active sites by adsorption of water, FFAs, and TGs. Furthermore, the concentration and strength of the active sites should be enough to carry out esterification and transesterification with a high catalytic yield, but without the adsorption of FFAs or TGs. Thus the recent literature of base catalysts shows the synthesis of more complex structures where active metal oxides are present to circumvent the leaching of the active species. In the case of acid heterogeneous catalyst, nowadays the research is focused on the enhancement of the surface hydrophobicity with the addition of organic molecules by in situ methods and is also necessary in the incorporation of sulfonic groups bonded to large organic molecules with the purpose of giving robustness to the solid catalyst and reduce the leaching of the active species.

5.4 CONCLUSIONS

The synthesis of biodiesel using as a raw material nonedible vegetable oils, WCOs, and animal fats results complicated as a result of their high content of FFAs and water. The content of FFAs could be reduced through acid esterification leading to a better yield of biodiesel.

Heterogeneous acid catalysts should possess a hydrophobic surface to avoid deactivation of their acid sites by solvation by water, which hinders the accessibility to nonpolar molecules (FFAs and TGs). It is also desirable an interconnecting system of large mesopores to reduce mass transport limitations, together with the tailoring of acid catalysts that possess both Lewis and Brönsted acid sites to perform esterification and transesterification simultaneously. Being remarkable the performance of materials such as waste carbon derived solid acids, mesoporous silicas, and acidic ionic liquids due to its content of Lewis and Brönsted acid sites.

Regarding base solid catalyst, the main target to accomplish is to avoid the leaching of the active phase either for solubilization with methanol or by reaction with glycerol. Activation temperature also plays a crucial role in the tuning of their active sites and hence, their catalytic performance. Thus base materials such as transition metal oxides, hydrotalcites, and multicomponent metal hydroxides are promising catalysts due to their high catalytic stability and low thermal temperature activation.

KEYWORDS

- biodiesel
- transesterification
- esterification
- basicity
- acidity

REFERENCES

Alcañiz-Monge, J.; El Bakkali, B.; Trautwein, G.; Reinoso, S. Zirconia supported tungstophosphoric heteropolyacid as heterogeneous acid catalyst for biodiesel production. *Appl. Catal. B. Environ.* **2018**, 224, 194–203.

Ambat, I.; Srivastava, V.; Silanpää, M. Recent advancement in biodiesel production methodologies using various feedstock: a review. Renew. *Sust. Energy Rev.* **2018**, 90, 356–369.

Atadashi, I.M.; Aroua, M.K.; Abdul Aziz, A.R.; Sulaiman, N.M.N. Production of biodiesel using high free fatty acid feedstocks. *Renew. Sust. Energy Rev.* **2012**, 16, 3275–3285.

Atadashi, I.M.; Aroua, M.K.; Abdul Aziz, A.R.; Sulaiman, N.M.N. The effects of catalysts in biodiesel production: a review. *J. Ind. Eng. Chem.* **2013**, 19(1), 14–26.

Avhad, M.R.; Marchetti, J.M. A review on recent advancement in catalytic materials for biodiesel production. *Renew. Sust. Energy Rev.* **2015**, 50, 696–718.

Banković-Ilić, I.; Stamenković, O.S.; Veljković, V.B. Biodiesel production from non-edible plant oils. *Renew. Sust. Energy Rev.* **2012**, 16, 3621–3647.

Bhuiya, M.M.K.; Rasul, M.G.; Khan, M.M.K.; Ashwath, N.; Azad, A.K. Prospects of 2nd generation biodiesel as a sustainable fuel-Part: 1 selection of feedstocks, oil extraction techniques and conversion technologies. *Renew. Sust. Energy Rev.* **2016**, 55, 1109–1128.

Bing, W.; Wei, M. Recent advances for solid basic catalysts: structure design and catalytic performance. *J. Solid State Chem.* **2019**, 269, 184–194.

Boz, N.; Degirmenbasi, N.; Kalyon, D.M. Esterification and transesterification of waste cooking oil over Amberlyst 15 and modified Amberlyst 15 catalysts. *Appl. Catal. B Environ.* **2015**, 165, 723–730.

Bournay, L.; Casanave, D.; Delfort, B.; Hillion, G.; Chodorge, J.A.New heterogeneous process for biodiesel production: a way to improve the quality and the value of the crude glycerin produced by biodiesel plants. *Catal. Today.* **2005**, 106(1–4), 190–192.

Cabrera-Munguía, D.A.; González, H.; Gutiérrez-Alejandre, A.; Rico, J.L.; Huirache-Acuña, R.; Maya-Yescas, R.; del Río, R.E. Heterogeneous acid conversion of a tricaprylin-palmitic acid mixture over Al-SBA-15 catalysts: reaction study for biodiesel synthesis. *Catal. Today.* **2017**, 282, 195–203.

Cabrera-Munguia, D.A.; Tzompantzi, F.; Gutiérrez-Alejandre, A.; Rico, J.L.; González, H. New insights on the basicity of ZnAl-Zr hydrotalcites activated at low temperature and their application in transesterification of soybean oil. *J. Mater. Res.* **2018**, 33(21) 142, 3614–3624.

Cabrera-Munguia, D.A.; Tututi-Ríos, E.; Gutiérrez-Alejandre, A.; Rico, J.L.; González, H. Acid properties of M-SBA-15 and M-SBA-15-SO3H (M=Al, Ti) material san their role on esterification of oleic acid. *J. Mater. Res.* **2018**, 33(21), 3634–3645.

Cavani, F.; Trifiró, F.; Vaccari, A. Hydrotalcite-type anionic clays: preparation, properties and applications. *Catal. Today.* **1991**, 11, 173–301.

Chakraborty, R.; Gupta, A.K.; Chowdhury, R. Conversion of slaughterhouse and poultry farm animal fats and wastes to biodiesel: parametric sensitivity and fuel quality assessment. *Renew. Sust. Energy Rev.* **2014**, 29, 120–134.

Chen, S-Y.; Mochizuki, T.; Abe, Y.; Toba, M.; Yoshimura, Y. Ti-incorporated SBA-15 mesoporous silica as an efficient and robust Lewis solid acid catalyst for the production of high-quality biodiesel fuels. *Appl. Catal. B Environ.* **2014**, 148–149, 344–356.

Chouhan, A.P.S.; Sarma, A.K. Modern heterogeneous catalysts for biodiesel production:

a comprehensive review. *Renew. Sust. Energy* Rev. **2011**, 15, 4378–4399.

Da Silva, D.A.; Santisteban, O.A.N.; de Vasconcellos, A.; Silva Paula, A.; Aranda, D.A.G.; Giotto, M.V.; Jaeger, C.; Nery, J.G. Metallo-stannosilicate heterogeneous catalyst for biodiesel production using edible, non-edible and waste oils as feedstock. *J. Environ. Chem. Eng.* **2018**, 6(4), 5488–5497.

Dhawane, S.H.; Kumar, T.; Halder, G. Recent advancement and prospective of heterogeneous carbonaceous catalysts in chemical and enzymatic transformation of biodiesel. *Energy Convers. Manage.* **2018**, 167, 176–202.

Ding, H.; Ye, W.; Wang, Y.; Wang, X.; Li, L.; Liu, D.; Gui, J.; Song, C.; Ji, N. Process intensification of transesterification for biodiesel production from palm oil: microwave irradiation on transesterification reaction catalyzed by acidic imidazolium ionic liquids. *Energy.* **2018**, 144, 957–s967.

Dossin, T.F.; Reyniers, M-F.; Marin, G.B. Kinetics of heterogeneously MgO-catalyzed transesterification. *Appl. Catal. B Environ.* **2006**, 61, 35–45.

Galadima, A.; Muraza, O. Biodiesel production from algae by using heterogeneous catalysts: a critical review. *Energy.* **2014**, 78, 72–83.

Guldhe, A.; Singh, P.; Ansari, F.A.; Singh, B.; Bux, F. Biodiesel synthesis from microalgal lipids using tungstated zirconia as a heterogeneous acid catalyst and its comparison with homogeneous acid and enzyme catalysts. *Fuel.* **2017**, 187, 180–188.

Gülsen, E.; Olivetti, E.; Freire, F.; Dias, L.; Kirchain, R. Impact of feedstock diversification on the cost-effectiveness of biodiesel. *Appl. Energy.* **2014**, 126, 281–296.

Guo-liang, S.; Feng, Y.; Xiao-liang, Y.; Ruifeng, L. Synthesis of tetragonal sulfated zirconia via a nouvel route for biodiesel production. *J. Fuel Chem. Technol.* **2017**, 45(3), 311–316.

Haijari, M.; Tabatabaei, M.; Aghbashlo, M.; Ghanavati, H. A review on the prospects of sustainable biodiesel production: a globañ scenario with an emphasis on waste-oil biodiesel utilization. *Renew. Sust. Energy Rev.* **2017**, 72, 445–464.

Harsha Hebbar H.R.; Math, M.C.; Yatish, K.V. Optimization and kinetic study of CaO nano-particles catalyzed biodiesel production from *Bombax ceiba* oil. *Energy.* **2018**, 143, 25–34.

Hasan, M.M.; Rahman, M.M. Performance and emission characteristics of biodiesel-diesel blend and environmental and economic impacts of biodiesel production: a review. *Renew. Sust. Energy Rev.* **2017**, 74, 938–948.

Hernández-Hipólito, P.; García-Castillejos, M.; Martínez-Klimova, E.; Juárez-Flores, N.; Gómez-Cortes, A.; Klimova, T. Biodiesel production with nanotubular sodium titanate as a catalysts. *Catal. Today.* **2014**, 220–222, 4-11.

Islam, A.; Taufiq-Yap, Y.H.; Chan, E-S.; Mon-iruzzaman, M.; Islam, S.; Nurum Nabi, M.D. Advances in solid-catalytic and non-catalytic technologies for biodiesel production. *Energy Convers. Manage.* **2014**, 88, 1200–1218.

Islam, A.; Taufiq-Yap, Y.H.; Chu, C-M.; Chand, E-S.; Ravindra, P. Studies on design of heterogeneous catalysts for biodiesel production. *Process Saf. Environ.* **2013**, 91, 131–144.

Jamila, F.; Al-Muhtaeba, A.; Zar Myintb, M.T.Z.; Al-Hinaic, M.; Al-Hacj Baawaind, M.; Al-abria, M.; Kumare, G.; Atabanif, A.E. Energy conversion and Manangment Biodiesel production by valorizing waste *Phoenix dactylifera* L. Kernel oil in the presence of synthesized heterogeneous metallic oxide catalyst ($Mn@ MgO-ZrO_2$). *Energy Convers. Manage.* **2018**, 155, 128–137.

Jeong, G.T.; Park, D.H. Batch (one- and two-stage) production of biodiesel fuel from rapeseed oil. *Appl. Biochem. Biotechnol.* **2006**, 129–132, 668–679.

Jiang, W.; Lu, H.F.; Qi, T.; Yan, S.L.; Liang, B. Preparation, application, and optimization of Zn/Al complex oxides for biodiesel production under sub-critical conditions. *Biotechnol. Adv.* **2010**, 28. 620–627.

Kanaveli, I-P.; Atzemi, M.; Lois, E. Predicting the viscosity of diesel/biodiesel blends. *Fuel.* **2017**, 199, 248–263.

Konwar, L.J.; Boro, J.; Deka, D. Review on latest developments in biodiesel production using carbón-based catalysts. *Renew. Sust. Energy Rev.* **2014a**, 29, 546–564.

Konwar, L.J.; Das, R.; Thakur, A.J.; Salminen, E.; Arvela, P.M.; Kumar, N.; Mikkola, J-P.; Deka, D. Biodiesel production from acid oils using sulfonated carbon catalyst derived from oil-cake waste. *J. Mol. Catal. A Chem.* **2014b**, 167–176.

Kumar, M.; Sharma, M.P. Assessment of potential of oils for biodiesel production. *Renew. Sust. Energ. Rev.* **2015**, 44, 814–823.

Kuzminska, M.; Kovalchuk, T.V.; Backov, R.; Gaigneaux, E.M. Immobilizing heteropolyacids on zirconia-modified silica as catalysts for oleochemistry transesterification and esterification reactions. *J. Catal.* **2014**, 320, 1–8.

Lathiya, D.R.; Bhatt, D.V.; Maheria, K.C. Synthesis of sulfonated carbon catalyst from waste orange peel for cost effective biodiesel production. *Bioresour. Technol. Rep.* **2018**, 2, 69–76.

Lee, A.F.; Bennett, J.A.; Manayil, J.C.; Wilson, K. Heterogeneous catalysis for sustainable biodiesel production via esterification and transesterification. *Chem. Soc. Rev.* **2014**; 43, 7887–7916.

Léon, C.I.S.; Song, D.; Su, F.; An, S.; Liu, H.; Gao, J.; Guo, Y.; Leng, J. Propylsulfonic acid and methyl bifunctionalized Ti-SBA-15 silica as an efficient heterogeneous acid catalyst for esterification and transesterification. *Micropor. Mesopor. Mat.* **2015**, 204, 218–225.

Li, K.; Bai, L.; Yang, Y.; Jia, X. Kinetics of ionic liquid-heteropolyanion salts catalyzed transesterification of oleic acid methyl ester: a study by sequential method. *Catal. Today.* **2014**, 233, 155–161.

Li, W.; Xu, K.; Xu, L.; Hu, J.; Ma, F.; Guo, Y. Preparation of highly ordered mesoporous AlSBA-15-SO₃H hybrid material for the catalytic synthesis of chalcone under solvent-free condition. *Appl. Surf. Sci.* **2010**, 256, 3183–3190.

Liu, X.; He, H.; Wang, Y.; Zhu, S.; Piao, X. Transesterification of soybean oil to biodiesel using CaO as a solid base catalyst. *Fuel.* **2008**, 87, 216–221.

Liu, Q.; Wang, C.; Qu, W.; Wang, B.; Tian, Z.; Ma, H.; Xu, R. The application of Zr incorporated Zn-Al dehydrated hydrotalcites as solid base in transesterification. *Catal. Today.* **2014**, 234, 161–166.

Ma, Y.; Wang, Q.; Sun, X.; Wu, C.; Gao, Z. Kinetics studies of biodiesel production from waste cooking oil using FeCl₃-modified resin as heterogeneous catalyst. *Renew. Energy.* **2017**, 107, 522–530.

Mandolesi de Araújo, C.D.; de Andrade, C.C.; de Souza Silva, E.; Dupas, F.A. Biodiesel production from used cooking oil: a review. *Renew. Sust. Energ. Rev.* **2013**, 27, 445–452.

Mansir, N.; Taufiq-Yap, Y.H.; Rashid, U.; Lokman, I.M. Investigation of heterogeneous solid acid catalyst performance on low grade feedstocks for biodiesel production: a review. *Energy Convers. Manage.* **2017**, 141, 171–182.

Mardiah, H.H.; Ong, H.C.; Masjuki, H.H.; Lim, S.; Lee, H.V. A review on latest developments and future prospects of heterogeneous catalyst in biodiesel production from non-edible oils. *Renew. Sust. Energy Rev.* **2017**, 67, 1225–1236.

Martínez, S.L.; Romero, R.; López, J.C.; Romero, A.; Sánchez Mendieta, V.; Natividad, R. Preparation and characterization of CaO nanoparticles/NaX zeolite catalysts for the transesterification of sunflower oil. *Ind. Eng. Chem. Res.* **2011**, 50(5), 2665–2670.

Madhuvilakku, R.; Piraman, S. Biodiesel synthesis by TiO_2-ZnO mixed oxide nanocatalyst catalyzed palm oil transesterification process. *Bioresour. Technol.* **2013**, 150, 55–59.

Marinković, D.M.; Stanković, M.V.; Velićković, A.V.; Avramović, J.M.; Miladinović, M-J.; Stamenković, O.O.; Velijković, V.; Jovanović, D.M. Calcium oxide as promising heterogeneous catalyst for biodiesel production: current state and perspectives. *Renew. Sust. Energy Rev.* **2016**, 56, 1387–1408.

Marwaha, A.; Rosha, P.; Mohapatra, S.K.; Mahla, S.K.; Dhir, A. Waste materials as potential catalyst for biodiesel production: current state. *Fuel Process. Technol.* **2018**, 181, 175–186.

Mbaraka, I.K.; Shanks, B.H. Design of multifunctionalized mesoporous silicas for esterification of fatty acid. *J. Catal.* **2005**, 229, 365–373.

Melero, J.A.; Bautista, L.F.; Morales, G.; Iglesias, J.; Sánchez-Vázquez, R. Acid-catalyzed production of biodiesel over arenesulfonic SBA-15: insights into the role of water in the reaction network. *Renew. Energy.* **2015**, 75, 425–432.

Moraes, P.; Severino, A.; de Figueiredo, M.; de Oliveira, C.; Asumpção, C. Zn, Al-catalysts for heterogeneous biodiesel production: Basicity and process optimization. *Energy.* **2014**, 75, 453–462.

Nata, I.F.; Putra, M.D.; Irawan, C.; Lee, C-K. Catalytic performance of sulfonated carbon-based solid acid catalyst on esterification of waste cooking oil for biodiesel production. *J. Environ. Chem. Eng.* **2017**, 5, 2171–2175.

Naylor, R.L.; Higgins, M.M. The rise in global biodiesel production: implications for food security. *Glob. Food Sec.* **2018**, 16, 75–84.

Pan, H.; Liu, X.; Zhang, H.; Yang, K.; Huang, S.; Yang, S. Multi-SO3H functionalized mesoporous polymeric acid catalyst for biodiesel production and fructose to biodiesel additive conversion. *Renew. Energy.* **2017**, 107, 245–252.

Pan, H.; Zhang, H.; Yang, S. Production of biodiesel via simultaneous esterification and transesterification. In: Fang Z, Smith Jr RL, Li H, editors. Production of Biofuels and Chemicals with Bifunctional Catalysts. Singapore: Springer; **2017**. p. 311–314.

Pamatz-Bolaños, T.; Cabrera-Munguia, D.A.; Gonzalez, H.; del Rio, R.E.; Rodriguez-García, G.; Gutiérrez-Alejandre, A.; Tzompantzi, F.; Gómez-Hurtado, M.A. Transesterification of Caesalpinea eriostachys seed oil using heterogeneous and homogeneous basic catalysts. *Int. J. Green Energy.* **2018**, 15(8), 465–472.

Park, J-Y.; Kim, D-K.; Lee, J-S. Esterification of free fatty acids using water-tolerable Amberlyst as a heterogeneous catalyst. *Bioresource Technol.* **2010**, 101, 562–565.

Ramos, M.J.; Casas, A.; Rodríguez, L.; Romero, R.; Pérez, A. Transesterification of sunflower oil over zeolites using different metal loading: a case of leaching and agglomeration studies. *Appl. Catal. A Gen.* **2008**, 346, 79–85.

Roschat, W.; Phewphong, S.; Thangthong, A.; Moonsind, P.; Yoosuke, B.; Kaewpuang, T.; Promarak, V. Catalytic performance enhancement of CaO by hydration-dehydration process for biodiesel production at room temperature. *Energy Convers. Manage.* **2018**, 165, 1–7.

Sajjadi, B.; Abdul Raman, A.A.; Arandiyan, H. A comprehensive review on properties of edible and non-edible vegetable oil-based biodiesel: composition, specifications and prediction models. *Renew. Sust. Energy Rev.* **2016**, 63, 62–92.

Sandesh, S.; Kristachar, P.K.R.; Manjunathan, P.; Halgeri, A.B.; Shanbhag, G.V. Synthesis of biodiesel and acetins by transesterification reactions using novel

CaSn(OH)$_6$ heterogeneous base catalyst. *Appl. Catal. A Gen.* **2016**, 523, 1–11.

Santori, G.; Di Nicola, G., Moglie, M.; Polonora, F. A review analyzing the industrial biodiesel production practice starting from vegetable oil refining. *Appl. Energy.* **2012**, 92, 109–132.

Sasidharan, M.; Kumar, R. Transesterification over various zeolites under liquid-phase conditions. *J. Mol. Catal. A Chem.* **2004**, 210, 93–98.

Semwal, S.; Arora, A.K.; Badoni, R.P.; Tuli, D.K. Biodiesel production using heterogeneous catalysts. *Bioresour. Technol.* **2011**, 102, 2151–2161.

Shah, K.A.; Parikh, J.K.; Maheria, K.C. Biodiesel synthesis from acid oil over large pore sulfonic acid-modified mesostructured SBA-15: process optimization and reaction kinetics. *Catal. Today.* **2014**, 237, 29–37.

Soares Dias, A.P.; Bernardo, J.; Felizardo, P.; Neiva Correia, M.J. Biodiesel production over thermal activated cerium modified Mg-Al hydrotalcites. *Energy.* **2012**, 41, 344–353.

Sudsakorn, K.; Saiwuttikul, S.; Palitsakun, S.; Seubsai, A.; Limtrakul, J.; Biodiesel production from *Jatropha Curcas* oil using strontium-doped CaO/MgO catalyst. *J. Environ. Chem. Eng.* **2017**, 5, 2845–2852.

Sun, H.; Ding, Y.; Duan, J.; Zhang, Q.; Wang, Z.; Lou, H.; Zheng, X. Transesterification of sunflower oil to biodiesel on ZrO$_2$ supported La$_2$O$_3$ catalyst. *Bioresour. Technol.* **2010**, 101, 953–958.

Talebian-Kiakalaieh, A.; Saidina Amin, N.A.; Mazaheri, H. A review in novel processes of biodiesel production from waste cooking oil. *Appl. Energy* **2013**, 104, 683–710.

Tanaka, Y.; Okabe, A.; Ando, S. Method for the preparation of a lower alkyl ester of fatty acids. US Patent 4303590. 1981.

Tang, Z-E.; Lima, S.; Pang, Y-L.; Ong, H-C.; Lee, K-T. Synthesis of biomass as heterogenous catalyst for application in biodiesel. *Renew. Sust. Energy Rev.* **2018**, 92, 235–253.

Thoai, D.N.; Tongurai, C.; Prasertsit, K.; Kumar, A. Review on biodiesel production by two-step catalytic conversion. *Biocatal. Agric. Biotechnol.* **2019**, 18, 101023.

Thushari, I.; Babel, S. Sustainable utilization of waste palm oil and sulfonated carbon catalyst derived from coconut meal residue for biodiesel production. *Bioresour. Technol.* **2018**, 248, 199–203.

Trombettoni, V.; Lanari, D.; Prinsen, P.; Luque, R.; Marrocchi, A.; Vaccaro, L. Recent advance in sulfonated resin catalysts for efficient biodiesel and bio-derived additives production. *Prog. Energy Combust. Sci.* **2018**, 65, 136–162.

Tropecelo, A.I.; Casimiro, M.H.; Fonseca, I.M.; Ramos, A.M. Vital, J.; Castanheiro, J.E. Esterification of free fatty acids to biodiesel over heteropolyacids immobilized on mesoporous silica. *Appl. Catal. A Gen.* **2010**, 390, 183–189.

Troter, D.Z.; Todorović, Z.B.; Đokić-Stojanović, D.R.; Stamenković, O.S.; Veljković, V.B. Application of ionic liquids and deep eutectic solvents in biodiesel production: a review. *Renew. Sust. Energy Rev.* **2016**, 61, 473–500.

Tzompantzi, F.; Carrera, Y.; Morales-Mendoza, G.; Valverde-Aguilar, G.; Mantilla, A. ZnO-Al$_2$O$_3$-La$_2$O$_3$ layered double hydroxides as catalyst precursors for the esterification of oleic acid fatty grass at low temperature. *Catal. Today.* **2013**, 212, 164–168.

Ullah, Z.; Khan, A.S.; Muhammada, N.; Ullah, R.; Alqahtani, A.S.; Shah, S.N.; Ghanem, O.B.; Bustam, M.A.; Man, Z. A review on ionic liquids as perspective catalysts in transesterification of different feedstock oil into biodiesel. *J. Mol. Liq.* **2018**, 266, 673–686.

Van Gerpen, J. Biodiesel processing and production. *Fuel Process. Technol.* **2005**, 86, 1097–1107.

Wan Ghazali, W.N.M.; Mamat, R.; Masjuki, H.H.; Nafaji, G. Effects of biodiesel from different feedstocks on engine performance and emissions: a review. *Renew. Sust. Energy Rev.* **2015**, 51, 585–602.

Wilson, K.; Lee, A.F. Rational design of heterogeneous catalysts for biodiesel synthesis. Rational design of heterogeneous catalysts for biodiesel synthesis. *Catal. Sci. Technol.* **2012**, 2, 884–897.

Wu, J.; Gao, Y.; Zhang, W.; Tang, A.; Tan, Y.; Men, Y.; Tang, B. New imidazole-type acidic ionic liquid polymer for biodiesel synthesis from vegetable oil. *Chem. Eng. Process.* **2015**, 93, 61–65.

Xie, W.; Fan, M. Biodiesel production by transesterification using tetraalkylammonium hydroxides immobilized onto SBA-15 as a solid catalyst. *Chem. Eng. J.* **2014**, 239, 60–67.

Yang, X-X.; Wang, Y-T.; Yang, Y-T.; Feng, E-Z.; Luo, J.; Zhang, F.; Yang, W-J.; Bao, G-R. Catalytic transesterification to biodiesel at room temperature over several solid bases. *Energy Convers. Manage.* **2018**, 164, 112–121.

Zhang, H.; Li, H.; Pan, H.; Wang, A.; Souzanchi, S.; Xu, C.; Yang, S. Magnetically recyclable acidic polymeric ionic liquids decorated with hydrophobic regulator as highly efficient and stable catalysts for biodiesel production. *Appl. Energy.* **2018**, 223, 416–429.

Zhang, P.; Shi, M.; Liu, Y.; Fan, M.; Jiang, P.; Dong, Y. Sr doping magnetic CaO parcel ferrite improving catalytic activity on the synthesis of biodiesel by transesterification. *Fuel.* **2016**, 186, 787–791.

CHAPTER 6

Anaerobic Digestion as Consolidated Process Platform for Gaseous Biofuels Production and Other Value-Added Products

SALVADOR CARLOS HERNÁNDEZ*, DIANA SOFÍA SEGOVIA ARÉVALO, and LOURDES DÍAZ JIMÉNEZ

Centro de Investigación y de Estudios Avanzados del Instituto Politécnico Nacional Unidad Saltillo, México

Corresponding author. E-mail: salvador.carlos@cinvestav.edu.mx

ABSTRACT

Processes consolidation search to optimize the transformation of raw materials. The fundamental principle is to reduce the number of transformation steps obtaining maximal benefits. Consolidation theory is considered as an alternative to lignocellulose biomass valorization since it allows integrating the production of bioethanol and value-added products in a single step; this trend has been known as consolidated bioprocessing (CBP). However, consolidation application to bioprocesses other than bioethanol technology has been few addressed, even if several alternative pathways to transform biomass have been identified. Among these alternatives, anaerobic digestion (AD) has been detected as an efficient bioprocess to produce biofuels, such as biogas and hydrogen, and bioproducts such as organic acids and biosolids, from a series of raw materials. Then, in this chapter AD is studied under a consolidated bioprocessing perspective. The objective is to identify the advantages, drawbacks, and challenges imposed by the biomass revalorization through CBP by AD (CBP-AD). First, the fundamentals of consolidated theory are presented; after that, the biomass transformation through AD is analyzed. Once introduced to the principles of both topics, a configuration of CBP-AD to obtain gaseous, liquid, and solid products is proposed. Finally, a mathematical model is presented, and numerical simulations are performed to illustrate the CBP-AD.

6.1 INTRODUCTION

The English Oxford Dictionary includes two definitions for the term consolidation: (1) the action or process of making something stronger or more solid and (2) the action or process of combining several things into a single whole, which is more effective or coherent. The first concept is the basis of the application of consolidation in soil and material sciences. In soil mechanics, consolidation refers to the phenomena by which soils become compact; this occurs from the elimination of water, pores, and some other characteristics (Craig, 2004; Terzaghi et al., 1960); consolidation theory aim to explain and to improve soil behavior regarding liquid flows, permeability, elasticity, sedimentation, and other phenomena (García-Ros et al., 2019; Tamayo-Mas et al., 2018). Another important application of consolidation theory is in construction since the kind of soil can affect the stability of the building (Craig, 2004; Radhika et al., 2017). Also, this concept was used to study the sludge compaction phenomena in wastewater treatment plants

(Abusam and Keesman, 2009) and the formation of concrete (Josserand et al., 2006); in this context, consolidation theory helps to understand better and, then, to enhance management of materials.

The second definition is useful in manufacturing and biotechnology. Regarding unit manufacturing, consolidation processes concern the assembly of smaller objects into a single product to achieve a desired geometry, structure, or property. In this field, the effect of mechanical, chemical, and thermal energy on the interaction between materials to promote objects bonding is studied; powder metallurgy, ceramic molding, and polymer-matrix composite pressing are examples of consolidation processes (Finnie et al., 1995). In biotechnology, consolidation search for reducing the number of steps in the processing of raw materials and obtaining bioproducts in a single reactor (Lynd et al., 1996). The concept is relevant since it is expected that the reduction of steps also reduces energy requirements and improves economic benefits. It has been applied to lignocellulose biomass valorization as an alternative to integrate the production of bioethanol and value-added products in a single step (Carrillo-Nieves et al., 2019; Lynd et al., 2005; Mbaneme-Smith and Chinn, 2015); this trend has been known as consolidated bioprocessing (CBP). However, consolidation application to bioprocesses other than bio-ethanol technology has been few addressed, even if several alternative pathways to transform biomass have been identified. Among these alternatives, anaerobic digestion (AD) is identified as an efficient bioprocess to obtain biofuels (biogas, biohydrogen) and bioproducts (organic acids, biosolids) from a series of raw materials. Then, in this chapter, AD is studied under a consolidated bioprocessing perspective. The objective is to identify the advantages, drawbacks, and challenges of biomass transformation through CBP by AD (CBP-AD). In the first place, the basic concepts of consolidated processing are introduced; after that, the biomass conversion employing AD in liquid and solid medium is analyzed. Then, a configuration of CBP-AD is proposed in order to obtain gaseous, liquid, and solid products. A mathematical model is presented, and numerical simulations are performed to illustrate the CBP-AD configuration.

6.2 CONSOLIDATED PROCESSING: PRINCIPLES AND APPLICATIONS

6.2.1 SYNTHESIS OF MATERIALS

The synthesis of materials through consolidated processes relies on the combination of components promoting an optimal integration of them to produce a new material that achieves efficient behavior in a specific application. The integration can be influenced by the controlled application of chemical, thermal, and mechanical energy (Finnie et al., 1995). A group of consolidated processes is identified in this field. Some of them are powder processing, sintering, densification, shaping, composite production, welding, and joining; the application of laser beams, ultrasonic waves, and computer-aided design represents an evolution of this topic. A general representation of this concept is shown in Figure 6.1.

Some examples of consolidated synthesis of products are the formation of concretes, ceramics, and thermal insulators for buildings, among others.

It has been developed a chemical-based methodology to promote consolidation for recycling waste and demolished concretes; the objective is to use these materials instead of natural aggregates. The surface of these waste materials holds a layer of residual cement mortar introducing undesirable properties such as higher water absorption, lower crushing strength, and lower abrasion resistance in comparison with natural aggregates. Then, a diammonium hydrogen phosphate solution is employed to produce a chemical reaction with calcium-rich hydration products in the waste and demolished concretes; hydroxyapatite is

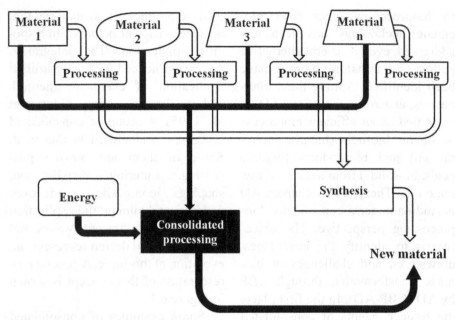

FIGURE 6.1   Schematic diagram of a consolidated process.

produced and precipitated filling pores and seals cracks. Then, the addition of hydroxyapatite improves the microstructure of the recycled materials; it was found that the concrete made with the recycled materials are comparable with the ones made with the natural aggregates (Wang et al., 2019). Besides, a self-consolidated concrete was synthesized and evaluated in the elimination of mechanical vibration, which is required in the process of filling formwork and encompass steel reinforcement in the fresh state under its weight. The properties of this kind of concrete formation, such as the strain at the strength and ultimate strain, elastic modulus, ductility, drift ration, and energy adsorption, are adequate to

construct bridge columns and other structures in high seismic regions (Ghadban et al., 2018).

Besides, sintering is defined as thermal treatment for bonding particles into a coherent solid structure; it is achieved through mass transport events that commonly occur on the atomic scale. Sintering requires high energy, which is usually provided by thermal energy equipment. Several improvements to the process have been implemented, such as microwaves (Singhal et al., 2018) and flash sintering (Biesuz and Sglavo, 2019). Microwave sintering improves mechanical and physical properties of consolidated materials, induces a better grain distribution and higher density, reduces processing time and energy requirements. Flash sintering allows ceramic densification in the order of seconds to a few minutes; this is its main advantage. A direct electric field is applied to the green specimen, and the current is forced to flow within the ceramic body. At a specific onset combination of electric field/furnace temperature, the material densifies in a very short time.

On the other side, the adobe has archaeological and architectural value since it has proven to be an efficient material to regulate the temperature in buildings. However, it is susceptible to erosion caused by wind and water, degrading efficiently. The circulation of water in the adobe pores induces several problems such as washouts, freeze-thaw cycles, and swelling-shrinkage cycles of the clay fraction; these phenomena lead to a reduction of grains cohesion, loss of mechanical properties, and even crumbling. In this sense, different studies focus on improving the consolidation of materials composing adobe bricks. The use of nanoparticles seems to be an alternative to the issues of adobe, as reported by Camerini et al. (2019). They proposed a ternary system (based on SiO_2, hydroxypropyl cellulose, and lime) to improve the consolidation of adobe; it was determined that these materials have full physicochemical compatibility with those of adobe. The use of nanoparticles permitted the preparation of stable and concentrated dispersions in water-solvent blends, which avoids the use of surfactants as stabilizers. The proposed system was applied to restore real adobe samples with high porosity and surface powdering caused by natural aging. The authors found that treated adobe mechanical properties were improved (resistance to peeling, abrasion, and wet–dry cycles). It was concluded that consolidation was due to the formation of calcium silicate hydrate from the ternary system.

6.2.2 CONSOLIDATED PROCESSING

Concerning the transformation of raw materials to obtain products, consolidated processing is based on the reduction of unitary processes minimizing time, energy requirements, and costs. This concept is illustrated in Figure 6.2. Instead of enchained steps to obtain a product, a single stage is implemented, integrating independent unitary processes, and optimizing the operating conditions and supplies. The strategies for step reduction depend on the kind of raw material and the target product; this is described in the next examples.

Additive manufacturing is an alternative to subtractive and formative manufacturing.

Subtractive manufacturing concerns the formation of a component

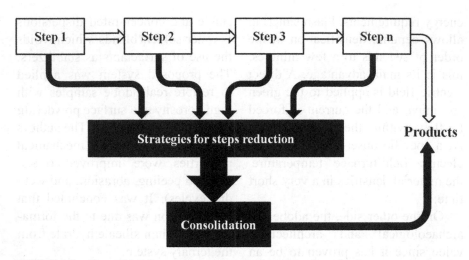

FIGURE 6.2   Consolidated processing as a reduction of number of unitary processes.

from a block of material by milling, turning, and other processes; this kind of manufacturing produces several wastes since the block is devastated to obtain a specific form. On the other side, formative manufacturing search to force materials to form a predefined geometry based on a die or a mold; examples of additive processes are forging and casting; frequently, the component requires additional treatment to obtain the final functionality. The additive manufacturing principle is the selective layer by layer deposition of material to create a 3D geometry of the finished component without the need for additional tooling and fixturing. This kind of manufacturing is an aided computer design approach; the required form is created in a computer before the physical transformation (Friel, 2015).

A laser consolidation process used to produce functional components using industrial materials such as Ni alloys, Co alloy, Ti alloy, Al alloy, and stainless steel has been reported recently. From an aided computer design, a layer-by-layer powder material deposition on a substrate is performed by a laser beam to build the target component. The component fabrication is developed in a single step reducing time, materials, and costs; it has been proven that the parts built are metallurgically sound, free of porosity or cracks (Xue, 2018).

Besides, consolidation of parts searches for the redesign of an assembled component with fewer, but more complex parts. Additive manufacturing is useful to facilitate parts consolidation since it allows consolidated component weight reduction and rapid fabrication. In this context, a hypothesis is that part consolidation allows simplifying the product structure, promote modularity, eliminate fasteners, joints, and connectors, and reduce assembly difficulties and cost (Mognol et al., 2006; Yang and Zhao, 2018). In this topic, three scenarios can be identified: (1) to consolidate multiple parts into one or more complex part eliminating unnecessary connectors and minimize required assembly times; (2) to eliminate a part and force other parts to accomplish the corresponding tasks; and (3) reconceptualization of the whole system and reallocate functions to other subsystems (Yang et al., 2017). A part consolidation method was proposed by Yang et al. (2015), considering function integration and structure optimization. The method involves two main modules: (1) functionality module, which is supposed to achieve part functionality through surface-level function integration and sequential part-level function integration; the modules use a based computer-aided design; (2) performance module, which is charged with the performance through the optimization of heterogeneous structures according to specific requirements. It was found that the proposed method aims to reach the perspective of functionality achievement and performance improvement. In the method validation, the results show a reduction of parts from 19 to 7 with less weight by 20% and demonstrating better performance. On the other side, a methodology to evaluate the environmental life cycle impact of parts consolidation was proposed (Yang et al., 2017), considering the production stages and using as indicators environmental impacts, energy consumption, and health impact. A floor attachment component on an underground train is selected as a case of study to evaluate the proposed methodology. The consolidated design shows a significant reduction in energy consumption and environmental impact on 20%, but it increases health toxicity. This study shows the importance of evaluating the environmental performance of consolidated parts.

On the other side, an analysis of costs related to maintenance of consolidation spare parts was performed (Knofius et al., 2019). The authors concluded that consolidation with additive manufacturing represents higher total costs than traditional manufacturing, which is due to the loss of flexibility: if a failure takes place, consolidated components require to be replaced entirely. The recommendation for industries is to include a careful cost analysis to assess the implementation of additive spare parts consolidation to avoid unforeseen effects (Knofius et al., 2019).

6.2.3 CONSOLIDATED BIOPROCESSING

CBP concept has been developed from the direct microbial conversion: a single microorganism performs cellulose and ethanol synthesis in one unit operation (Lynd et al., 1996). At present, CBP is defined as a sequential transformation of biomass (production of cellulolytic enzymes, hydrolysis of biomass, and sugars fermentation) to obtain desired bioproducts in a single process step via cellulolytic microorganisms (Carrillo-Nieves et al., 2019; Lynd et al., 2002). This concept implies the integration of independent steps into a single one, as shown in the schematic representation of the different pathways to transform lignocellulosic materials processing (Figure 6.3).

CBP concept applies to any fermentation product (Lynd et al., 2005). In this context, other authors concluded that the study of the transformation of lignocellulosic biomass performed by ruminants could help to develop optimal CBP schemes (Weimer et al., 2009). These animals produce milk, meat, wool, hides, biogas, and other products from lignocellulosic biomass. The rumen is a crucial element for microbial fermentation where volatile fatty acids (VFA), methane, carbon dioxide, and microbial cells are synthesized from pretreated feedstuffs. Pretreatment of the taken forage is performed by chewing it to reduce particle size near to 2 mm. In the rumen, particles stay around 72 h where they are hydrolyzed and fermented by ruminal microorganisms; this retention time is allowed

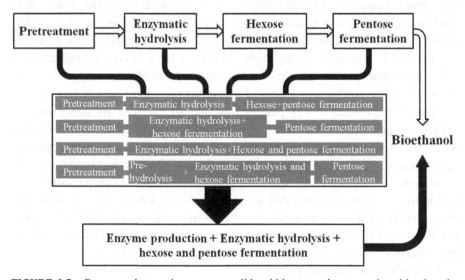

FIGURE 6.3   Processes integration on a consolidated bioprocessing to produce bioethanol.

by the omasum since its laminates trap large particles and return them to be reprocessed. Also, stable operating conditions (39 °C, neutral pH, the redox potential of 0.4 V, intermittent mixing force, and anaerobic conditions) enhance the fermentation process. Then, the fermentation in the rumen can be modeled as an anaerobic semicontinuous feedback stirred process. Figure 6.4 represents this degradation of lignocellulosic biomass as a CBP scheme (Weimer et al., 2009).

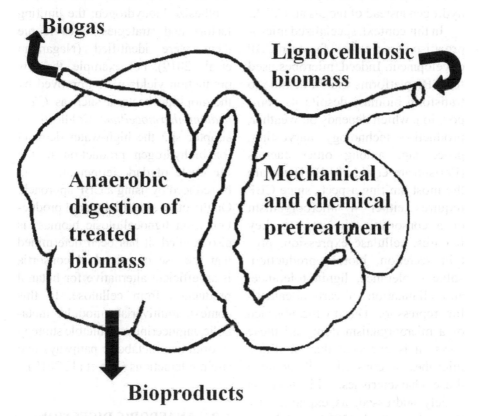

FIGURE 6.4 Representation of a natural consolidated processing in ruminants.

From the analysis of Weimer et al. (2009), it is possible to identify at least four strategies/challenges to replicate ruminal fermentation as a CBP platform: (1) the maximization of ethanol production, which implies to promote the growth of ruminal microorganisms producing this compound such as *Ruminococcus albus*, *Butyrivibrio fibrisolvens*, *Ruminobacter amylophilus*, and *Streptococcus bovis*; (2) the transformation of VFA into useful biofuels or bioproducts, this conversion can be done by chemical or biological pathways; (3) the maximization of biogas production and bioproducts, this requires to promote methanogenic bacteria growing after the VFA synthesis; (5) the boosting of the biological pathway to produce hydrogen instead of biogas and VFA.

In this context, specialized microorganisms are a critical factor in CBP development. Indeed, microbes used for CBP platforms should be able to transform biomass despite its composition, which depends on weather, production technology harvesting, processing, among other factors (Parisutham et al., 2014). These are the most limiting aspects since CBP requires either a microorganism or a consortium showing six key features: cellulase expression, protein secretion, biofuel production, solvent tolerance, lignin tolerance, and elimination of carbon catabolite repressors. Due to the absence of a microorganism doing all these tasks, it is expected that modified microbes or consortia will achieve these characteristics. This topic is widely addressed, as explained by Parisutham et al. (2014) and Mbaneme-Smith and Chinn (2015).

Most of CBP studies have been focused on ethanol production because there exists infrastructure allowing using ethanol as liquid biofuel (Weimer et al., 2009). However, other biofuels have been studied from the CBP perspective.

For example, since hydrogen has been identified as a possible renewable energy source, many research works have been addressed related to biohydrogen production, including the CBP approach. Dark fermentation of lignocellulosic biomass was identified as a feasible alternative to synthesize biohydrogen; the limiting factors and strategies to overcome them were identified (Nagarajan et al., 2019). For example, the low production yields were improved by thermophilic bacteria such as *Clostridium thermocellum*, Caldicellulosiruptor sp.; the high-water demand for biohydrogen production in the life cycle of dark fermentation can be reduced by using a CBP approach. On the other side, biobutanol production from lignocellulosic biomass is also studied; it has been determined that the use of microbial consortia is an efficient alternative for butanol production from cellulose. In this context, multivariate modular metabolic engineering is a suitable strategy to optimize metabolic pathways and strain interactions (Xin et al., 2019).

6.3 ANAEROBIC DIGESTION OF BIOMASS

6.3.1 FUNDAMENTALS OF ANAEROBIC DIGESTION

AD is a natural bioprocess devoted the progressive degradation of complex organic molecules, by the action of anaerobic bacteria, into a gaseous mixture (biogas) and a simple solid material (digestate), as reported in several works (Angelidaki et al., 1999; Meegoda et al., 2018; Möller and Müller, 2012). This process has been widely applied to transform organic wastes produced form agriculture, food industry, and municipalities into valuable products such as biogas and biofertilizers. All types of biomass can be used as a substrate for anaerobic processes since these materials are mainly composed of carbohydrates, proteins, fats, cellulose, and hemicelluloses (Möller and Müller, 2012; Vasco-Correa et al., 2018; Xie et al., 2016). Only highly lignified organic substances, such as wood, are not suitable for anaerobic transformation due to their slow decomposition; therefore a pretreatment stage should be implemented to allow cellulose and hemicellulose to be available for anaerobic transformation (Weiland, 2010).

The degradation of organic molecules is performed in sequential steps by bacteria populations developing complementing tasks. Based on several studies around the world, it is accepted that this bioprocess is developed in four sequential stages: hydrolysis, acidogenesis, acetogenesis, and methanogenesis (Batstone et al., 2002; Mosey, 1983; Xie et al., 2016). A schematic representation, adapted from Murphy and Thamsiriroj (2013) and Manchala et

al. (2017) of anaerobic digestion is introduced in Figure 6.5.

The action of enzymatic hydrolysis transforms the complex molecules composing biomass into soluble compounds: amino acids and long-chain fatty acids. The extracellular enzymes performing this degradation are produced by hydrolytic bacteria. This is the first step of AD, and it is known as hydrolysis. Hydrolytic bacteria grow fast in comparison with the other bacteria involved in AD, and they are less sensitive to parameters such as pH and temperature. However, this stage is affected by the substrate particle size and the enzymes performances in the raw material degradation; for this reason, hydrolysis has been identified as a limiting step, especially for lignin enriched substrates, in the biogas production from AD (Mao et al., 2015; Venkiteshwaran et al., 2015).

The second step is called acidogenesis or fermentation; acidogenic bacteria (involving obligatory and facultative anaerobic microorganisms) transform the hydrolysis products into VFA such as propionate, butyrate, valerate, and even acetate. Some alcohols, lactate, formate, carbon dioxide, and hydrogen are also produced in this stage (Manchala et al., 2017; Murphy and Thamsiriroj, 2013). Acidogenesis is achieved in a short time, introducing a risk regarding the accumulation of VFAs. A pH drop can be experienced when

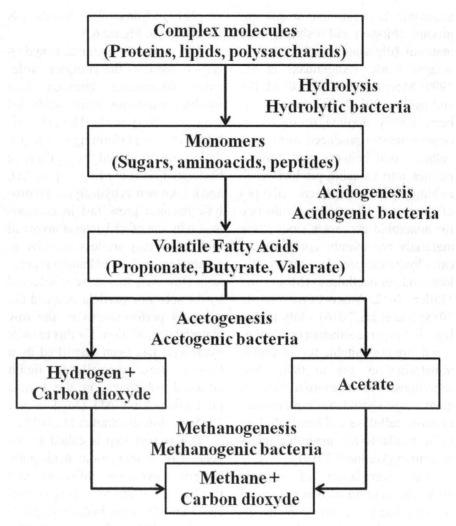

FIGURE 6.5 Stages involved in anaerobic digestion.

the transformation of acids is inhibited or performed slowly because of organic overload, toxic compounds, or rapid temperature changes. This situation should be taken into consideration in monitoring systems since large pH drops can inhibit or stop methanogenesis.

Acetogenesis is the third stage in AD; acetogenic bacteria convert VFA and alcohols in acetate, hydrogen gas, and carbon dioxide. Frequently this stage is considered as a single stage together with acidogenesis; this is because acetogenic bacteria are obligate to act simultaneously with acidogenic bacteria to transform the products of acidogenesis (Venkiteshwaran et al., 2015).

Finally, the last step is methanogenesis, where methane is synthesized, mainly by two pathways. Almost 70% of methane is produced from acetate, and the other 30% from hydrogen and CO_2 Methanogenesis is considered the slowest stage imposing the global dynamic of AD; also, this stage is the most sensitive to operating conditions; for this reason, it is considered as a limiting step (Batstone et al., 2002; Murphy and Thamsiriroj, 2013).

Since this process is developed by microorganism populations having different dynamics, several parameters could affect the development of the AD process (Meegoda et al., 2018). Among these parameters, temperature and pH are directly related to microorganisms. Meanwhile, carbon/ nitrogen ratio (C/N), organic loading rate (OLR), and retention time depend on operating conditions of the anaerobic system. The adequate pH for biological reactions, specifically for methanogenesis, is ranged from 6.5 to 8.5, where the optimal is 7. Below pH 6.5, methane synthesis is strongly inhibited; however, the biological activity remains even if slightly. If pH drops below 4, the anaerobic process could be stopped due to bacteria death. On the other side, above pH 8 the activity of the microorganisms is inhibited, and if the value returns to the operational range, the bacterial

growth can be reactivated (Mao et al., 2015).

Besides, three temperature regimes have been identified in the anaerobic process: thermophilic, mesophilic, and environmental (Kim et al., 2006). The thermophilic AD occurs at 50–70 °C; it allows obtaining high productivity since fast reaction rates and high load capacity are observed. However, it is more sensitive to operating conditions; acidification is frequently induced by this temperature range leading to lower biogas production. Moreover, the energy required to keep the thermophilic temperature affects the net energy balance. On the other side, mesophilic conditions consider 37 °C. Mesophilic systems are more stable and require less additional energy to reach and maintain the temperature; but, methane content in biogas is usually low; they are affected by poor biomass biodegradability and nutrient imbalance. Finally, environmental or seasonal systems operate at room temperature; then, the main advantage is that no additional energy is required. However, the lowest biogas yields are obtained with this temperature regime. Some authors recommend combining temperature regimes to achieve better performances (Mao et al., 2015).

On the other hand, the C/N is a nutrient content index in feedstock; the optimal reported is ranged from 20 to 35. Low C/N ratio due to high nitrogen concentration produces an excess of ammonia, which in turn leads to an increase of pH and inhibitory effects (Ren et al., 2018). High C/N causes low protein solubilization, which implies low total ammonia nitrogen and fatty acids concentrations; however, excessive-high C/N will cause insufficient nitrogen to bacteria inducing low biogas production (Puñal et al., 2000).

As the C/N ratio, the OLR and the retention time are directly related to the raw material feed into the anaerobic reactors. In this context, the number of organic compounds and the time they remain in contact with anaerobic bacteria influence the transformation efficiency and biogas production (Bi et al., 2019).

Different strategies have been developed in order to enhance AD performances; for example, bacteria immobilization, use of additives, raw material pretreatment, codigestion, and biorefinery schemes (Romero-Güiza et al., 2016; Vasco-Correa et al., 2018; Xie et al., 2016).

Bacteria immobilization allows concentrating the active microorganisms over solid support, improving the mass transfer. Then, immobilization induces the capacity to increase the OLR and to reduce retention time (Carlos-Hernández et al., 2014, 2018); in consequence, the transformation of organic wastes is also enhanced. Several materials have been evaluated as supports for anaerobic bacteria; for example, synthetic polymers, zeolites, sand, seashell, charcoal, ceramics, sintered glass, fire bricks, limestone, gravel, pumice, clay, and rock aggregates. In most of the cases, the immobilization allows the anaerobic systems to achieve high methane production, high chemical oxygen demand (COD) removal, and low retention time (Karadag et al., 2015; Romero-Güiza et al., 2016). Moreover, nanoparticles are also considered as solid support for the AD (Bania-merian et al., 2019) because nano-materials can be synthesized with metals and inorganic compounds, they are also considered as additive. Additives are a series of materials inducing other benefits such as the supply of nutrients for anaerobic bacteria (P, N, S, Fe, Ni, Mo, Co, and others) and the ability to mitigate ammonia inhibition (Romero-Güiza et al., 2016).

Besides, pretreatment of biomass is an early step to allow anaerobic systems to transform complex raw materials (André et al., 2018); that means, the hydrolysis stage (limiting step) is promoted or enhanced. Biological treatment through enzymes, fungi, or bacteria has proven to be a feasible alternative to improve hydrolysis stage; the application of electric fields, microwaves, and thermal energy has also been evaluated as pretreatment of several biomasses; also, acids and hydroxides are used to degrade raw materials before to transform them by anaerobic systems. All these treatments lead anaerobic systems to high conversion yields and high biogas production (Li et al., 2019).

Furthermore, the combination of two or more feedstock, known as codigestion, is studied as an alternative to enhance the properties, especially the C/N ratio, of the substrate for anaerobic systems and then to improve biogas production (Xie et al., 2016; Siddique and Wahid, 2018).

6.3.2 ANAEROBIC DIGESTION IN SOLID-STATE

AD in the solid-state (SS-AD) takes place when the level of total solids in the substrate is higher than 15%. Then, SS-AD is well situated to transform lignocellulosic biomass such as agricultural by-products, green waste, energy crops, and household wastes, among others. In this context, SS-AD is an alternative to revalorize solid organic wastes. Some of the advantages of SS-AD are higher volumetric methane productivity since there is more raw material to be transformed; reduced energy requirements for heating, because the biological activity promotes the required heat; less wastewater generation due to the higher solid contents, and a low-moisture digestate that is easier to handle and rich in nutrients which

requires few treatments before to be used as biofertilizer (Cavallo et al., 2018; Ge et al., 2016). On the other side, the main inconvenience of SS-AD is the low reaction rate, which is due to the slow hydrolysis and the low mass transfer rate. Since the solids content is high, the release of soluble compounds required by anaerobic bacteria is retarded in the hydrolysis stage. Moreover, the high solid content avoids a homogeneous agitation in the reaction medium reducing the microbial accessibility to the substrates and promoting the accumulation and even the dispersion of inhibitors (Xu et al., 2015). Furthermore, SS-AD requires higher organic loads that cause inhibition due to the accumulation of volatile short-chain fatty acids, causing a decrease in pH and biogas produced (Cavallo et al., 2018).

According to André et al. (2018), the most relevant issues to overcome regarding SS-AD are related to the substrate characterization [biochemical methane potential (BMP), codigestion, inhibition, pretreatment], the mass transfer, the bacteria populations involved in the biomass transformation (inoculum, dynamics), the development of monitoring systems in order to supervise the process evolution better, and the design and implementation of adequate technology. These issues are summarized as follows:

Substrate characterization: This is an important topic because the characteristic of the biomass imposes the transformation route. In this context, the BMP requires a standardized protocol to estimate with low error the biomethane production from heterogeneous substrates (Filer et al., 2019). Once the BMP of several substrates is known, it is feasible to determine efficient combinations for codigestion; this could also help to detect eventual inhibitors in the SS-AD. Also, from the characterization step, the necessity of substrate pretreatment and the technique to perform it can be identified. Even if there are several methods for pretreatment, frequently, the energy and economic requirements are high leading the whole system to a negative balance; for this reason, pretreatment is an important aspect of SS-AD.

Mass transfer: As said before, the high solid content limits the mass transfer leading to a nonhomogeneous reaction medium, which affects the substrate transformation yields. The diffusion of materials in the reactors is few studied, and then the parameters involved in the interaction mechanisms are also few known. These are some of the topics which should be addressed regarding the hydrodynamics of SS-AD (Sawatdeenarunat et al., 2015).

Bacteria populations: The bacteria dynamics is few known regarding ADSS; for example, the acclimation time and aging on the solid substrate should be useful to evaluate the transformation performances. The presence of solids can induce biofilm formation; even if this is desired in traditional AD, in SS biofilms could affect the biological reaction; then, this topic should be more in-depth studied. Another topic to be developed is the relation substrate/inoculums because this affects the hydrolysis and the methanogenesis, which are the limiting steps (Jiang et al., 2019).

Monitoring systems: The supervision of SS-AD is required to know the state of the process at any time. The methods used in liquid AD should be adapted to SS, for example, COD and oxygen measurements. Currently, biogas production is the most useful index to determine the process evolution; however, some other parameters are required to determine the state of substrate and bacteria populations better. The main drawback concerns the lack of online tools for monitoring purposes. In this context, the development of hard and soft sensors is a current requirement for SS-AD (André et al., 2018).

An alternative to deal with the lack of knowledge of SS-AD is the study by numerical simulations. This approach helps to understand the reaction mechanisms, predict the performance of the system, and facilitate the design of the process. The Anaerobic Digestion Model No. 1 (ADM1) was developed to study the AD in a liquid state, and several efforts have been addressed to adapt this model to SS-AD, but at present, there is not an accepted and extended model for SS. Currently, the models developed for SS-AD are classified as theoretical, empirical, and statistical; a comprehensive explanation of these models is presented in Xu et al. (2015) and is briefly described here.

Theoretical models provide information on the complex mechanisms involved in the system. For this reason, they are highly complicated for general applications because they require a series of inputs, outputs, and state variables; besides, they consider the biological, biochemical, and hydrodynamic phenomena taking place in anaerobic reactors. However, these kinds of models allow studying deeper the different mechanisms, for example, the bacteria grow, the intermediate products, the particle diffusion, the final products, and others. Some of the theoretical models are: ADM1-based, which adapt the ADM1 to the SS case; particles model, this approach considers that the reaction medium is composed by seeds (inoculums) and waste (substrates), based on the interaction of these particles a two particles model and a front reaction model are proposed; distributed parameters model, which considers column reactors where the biological activity depends on space and time (Abbassi-Guendouz et al., 2012).

On the other hand, the empirical models are based on experimental studies, and they search to explain specific behavior. Experimental data and general assumptions are stated as the basis of these kinds of models. The more representatives are the logistic model and the kinetic model. The first one is based on the population logistic growth model, which supposes bacteria growth is limited by the available resources; in this case, the substrate. The kinetic model is based on the simplified biochemical reaction rate law; it is presented as a relationship between the total solids and the bacteria population (Pastor-Poquet et al., 2018).

The statistical models do not require a deep understanding of the biological or physicochemical processes; they are based on a regression algorithm to fit a series of data by a simple mathematical expression. Simple linear regression, multiple linear regression, and even artificial neural networks are examples of those kinds of models (Xu et al., 2015).

In general, since mathematical models require a priori knowledge of SS-AD and experimental data, the real-life validation is one of the biggest challenges. The heterogeneous nature of the raw material and the reaction medium in the process make it difficult to collect the required data through experimentation. Assumptions are the basis for the development of a model for SS-AD, then more effective methods should be developed to verify these assumptions, which are used to simplify the model. This topic is useful to be addressed regarding a better understanding of SS-AD (Xu et al., 2015).

Besides biogas, the other final product of SS-AD is the remaining solids known as digestate; it is a nutrient-rich substance, a large number of microorganisms, and other organic substances, such as amino acids or proteins (Cavallo et al., 2018; Liu et al., 2019). Table 6.1 shows a general characterization of the digestate.

Since it is an abundant by-product, it has been studied for different applications, which are limited by the properties of the digestate, such as its fibrous nature, high oxygen content, and low calorific value. Therefore pretreatment is required to improve both the physical and chemical properties of the by-product. Some of the most recognized methods for this purpose are wet or liquid roasting since they can improve the qualities

TABLE 6.1 Physicochemical Characterization of Digestate

Item	Value
pH	7.6–8.3
Total solids (g/kg fresh matter)	32.2–78.8
Volatile solids (g/kg fresh matter)	23.9–63.7
Soluble COD (g/kg fresh matter)	7.3–18.5
COD (g/kg fresh matter)	77.1–100.3
Volatile fatty acids (g/kg fresh matter)	1.1–4.1
C/N	3.1–6.1
Kjeldahl total nitrogen (g/kg fresh matter)	4.5–8.7
NH_4–N (g/kg fresh matter)	1.7–4.5
NO_3–N (g/kg fresh matter)	0.011–0.013
PO_4–P (g/kg fresh matter)	0.14–0.27
Pb (mg/kg TS)	2.1–5.6
Ni (mg/kg TS)	16.6–42.4
Hg (mg/kg TS)	0.1–0.2
Cd (mg/kg TS)	0.1–0.3
As (mg/kg TS)	0.4–1
Cu (mg/kg TS)	21.7–25.6
Cr (mg/kg TS)	7.5–11.9
Zn (mg/kg TS)	94.6–175

COD, chemical oxygen demand; C/N, carbon/nitrogen ratio.

of the final product, such as homogeneity, carbon content, and calorific value (Zhang et al., 2019).

Some uses of digestate are briefly described in the next lines. It can be used as a recirculation medium for pretreatment in AD processes; this increases the conversion rate of the substrate reduces the amount of digestate generated at the end of the AD and promotes the efficiency of biogas production. Another application of this residue is in agriculture as a biofertilizer; it can be used directly after its generation or with a subsequent treatment (Liu et al., 2019; Timonen et al., 2019). Besides, it can be used to mitigate greenhouse gas emissions through the recycling of materials, the limitation of mineral fertilizers, and the improvement of soil properties. However, adequate techniques are essential for the management, processing, and distribution of digestate to avoid possible effects of acidification and eutrophication due to the increase in nutrient leaching, which depends on the quality of the soil, the weather conditions, and the characteristics of the digestate. One of the effects that digestate has on the soil is the increase in carbon balance, which leads to improved microbial processes and enzymatic activity, and in turn, increases the release of long-term nutrients in soils (Tampio et al., 2016). On the other side, digestate can also be used as raw material for thermochemical processes such as combustion, gasification, and pyrolysis; this kind of process allows to produce thermal energy and other valuable products such as synthesis gas, bio-oils, ash, and biochar (Zhang et al., 2019).

The generation and application of digestate also bring the potential risk of environmental contamination, due to the vast quantities of this by-product. Also, it contains a significant amount of methane formation potential and can, therefore, contribute to pollution and climate change. The added methane to the emissions of ammonia, carbon dioxide, and nitrous oxide from the digestate causes global pollution (Zeshan and Visvanathan, 2014). However, in general, the digestate can be seen as another relevant product of SS-AD, and it should be more in-depth studied.

6.3.3 VALUE-ADDED PRODUCTS FROM ANAEROBIC DIGESTION OF BIOMASS

The most recurrent application of AD is to convert organic wastes into biogas. Obtaining other products is a few addressed, and only in recent years is attracting attention. The development of biorefinery platforms is an excellent opportunity to take advantage of all the characteristics of AD. Since the raw materials are extensive, several biorefinery schemes based on AD could be implemented. For example, a revalorization scheme for food waste and lignocellulosic materials was proposed by Ren et al. (2018). First, this scheme includes a route for biodiesel production considering the separation of oils and grease from food waste; it is supposed that the remaining material is rich in fermentable components, which can be used to produce ethanol, organic acids, and microbial oil. After that, an anaerobic process is used to transform the effluents from previous processes, as well as some other organic wastes, into biogas, reusable liquid, and valuable solids. A pathway to use each one of these three components is presented. Biogas can be used for either heat and power production, or to obtain methane and carbon dioxide, both products are usable independently. Liquid in a combination of CO_2 produced from biogas is presented as a supply for algae raising, which is used to produce microbial oils. Finally, three alternatives for solids utilization are identified: composting for biofertilizer production; pyrolysis for bio-oil and biochar production; and thermal reforming for heat and power generation.

Another structure for biomass biorefining was introduced by Sawatdeenarunat et al. (2015). That structure considers the first transformation by AD of biomass; as for the previous case, biogas, liquid, and solids are obtained. More substantial processing of the liquid and solids is proposed in this scheme. The liquid is used for algae production, which is transformed in biodiesel by transesterification. Furthermore, the digestate is separated in solid and liquid fractions; the former is used together with the primary effluent to algae production and a series of value-added products; the solid fraction is processed by saccharification and thermal treatments to produce organic acids, alcohols, energy, and biochar.

An aspect few considered in the proposal of biorefining is the wastes produced in the different transformation processes; sludge and wastewater are commonly generated in the production of ethanol and biodiesel. Biodiesel produces another by-product (crude glycerol), which usually requires an additional treatment to be used. In this context, AD should be included as an element to treat all the wastes produced in biorefining. An example of this strategy is presented in Figure 6.6. The idea is to include biodiesel and bioethanol transformation in an integrated scheme, where AD is employed to treat the sludge produced by separation and fermentation processes. The crude glycerol obtained from transesterification is also included as a substrate for codigestion. The wastewater is considered to dilute sludge and crude glycerol and to increase the feasibility of AD transformation. The liquid is proposed

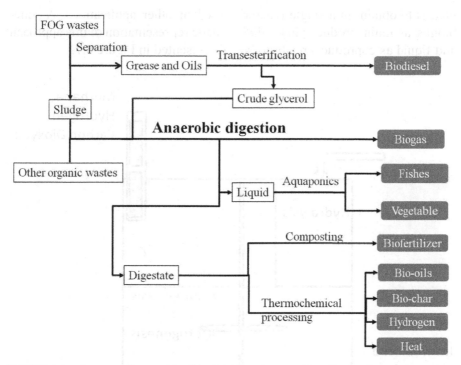

FIGURE 6.6 Strategy of anaerobic digestion inclusion on a biorefinery of biomass.

to be used in an aquaponic system, which allows producing fishes and vegetables. Similar pathways, as presented for the previous schemes, for using the digestate are also considered here.

6.4 CONSOLIDATED BIOPROCESSING BY ANAEROBIC DIGESTION

As said before, fundamentals of consolidation bioprocessing search for reducing the number of steps in the transformation of raw materials to obtain bioproducts. It is supposed that this principle can be applied to produce gaseous biofuels and solid bioproducts from the AD of dry or wet biomass. Then, based on the value-added products that can be obtained from the biorefinery scheme presented in the previous section, a processing scheme for CBP-AD is proposed. Then, the goal of this CBP-AD configuration is to obtain methane, hydrogen, and CO_2 in gaseous form, sludge for composting, and liquid for crop irrigation.

The proposal concerns a CBP-AD structure for transforming waste biomass into different products; the idea is to obtain, in a single reactor, biogas as main product plus solids and liquid as coproducts ready to be used in other applications. A schematic representation of this approach is presented in Figure 6.7.

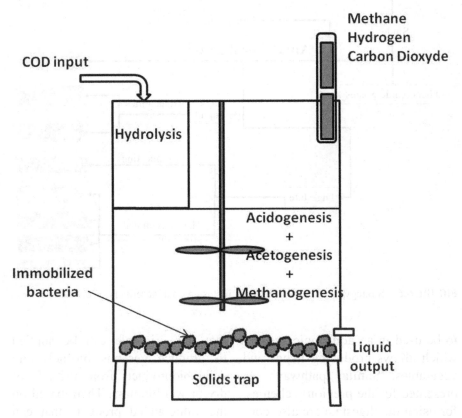

FIGURE 6.7   Approach of a consolidated bioprocessing by anaerobic digestion for biofuels production.

The organic matter is represented by the COD. It is injected into the reactor by a pump at a determined flow rate. In the first time, the complex molecules are transformed by hydrolytic bacteria into soluble monomers; after that, the acidogenesis, acetogenesis, and methanogenesis reactions, described in Section 6.3.1, are performed by the respective bacteria populations. These reactions lead to the production of a gaseous mixture that is composed of methane, hydrogen, and carbon dioxide. The biogas passes through a filter where the components are separated; then, this filter allows recovering each component that can be used in further applications. On the other side, the bioreactor includes a solids trap to recover the inactive biomass and the nontransformed solids. The immobilization of bacteria in solid support promotes the concentration of active biomass over the support; for this reason, only the inactive bacteria are released to the solids trap. Since this separation is performed all along with the biological reaction, it is expected that the recovered solids will require only small conditioning to be used. Finally, in a batch operation mode, the liquid is separated from the solids at the end of the biological reaction.

In continuous operation mode, the liquid is recovered all along with the reaction.

Numerical simulations of the CBP-AD are now introduced to illustrate the previous CBP-AD concept. The functional diagram in Figure 6.8 represents the process; this scheme is based on some reports (Angelidaki et al., 1999; Batstone et al., 2002; Costello et al., 1991; Dochain and Vanrolleghem, 2001). It is important to remark that in this scheme, some components and reactions in the different stages of AD are grouped to simplify the process modeling and better visualize the consolidating

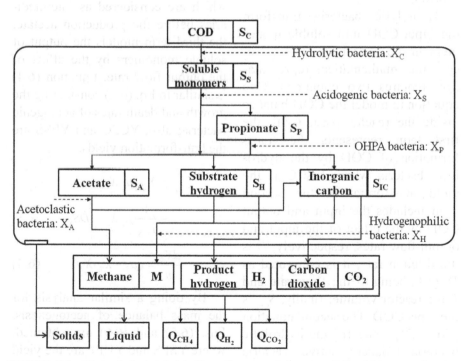

FIGURE 6.8 Numerical simulation flow diagram of the consolidated bioprocessing by anaerobic digestion for biofuels production.

bioprocessing approach. For example, all the organic matter is modeled by COD, a single substrate, named SS represent the different soluble monomers produced in hydrolysis; also, the intermediaries between acidogenesis and acetogenesis are grouped as propionate (SP).

From this representation, a mathematical model is developed by following the recommendations of Dochain and Vanrolleghem (2001) for a four population model and Batstone et al. (2002) for the ADM1. The obtained model uses the notation presented in Figure 6.8 for substrates and bacteria, and it is composed as follows.

Hydrolytic bacteria transform the input COD into soluble monomers in the hydrolysis. The corresponding mathematical representation requires two equations. One equation to model the COD balance inside the reactor [(Eq. 6.1)], the first term represents the transformation of COD by the hydrolytic bacteria where YSCC is the yield transformation; the second term includes the input and output COD by the effect of the input and output flow rates, respectively; D is the dilution rate, it is computed as $D=Q/V$, being Q the flow rate and V the reactor volume; finally, S_{Cin} is the input COD. The second equation [Eq. (6.2)], concerns the hydrolytic bacteria balance: active bacteria represented by the specific growth

rate (μ_c) minus inactive bacteria represented by the death rate (μ_c):

$$\frac{dS_C}{dt} = -\frac{1}{Y_{SCC}}\mu_C X_C + D$$
$$(S_{Cin} - S_C) \qquad (6.1)$$

$$\frac{dX_C}{dt} = (\mu_C - \kappa_C)X_C \qquad (6.2)$$

The acidogenesis mass balance considers the degradation of soluble monomers [(Eq. (6.3)] and the growth of acidogenic bacteria [Eq. (8.4)]. The first term in Eq. (8.3) represents the soluble monomers produced in the hydrolysis; the second term refers to the consumption of soluble monomers into long-chain volatile acids, which are considered as intermediaries before the production acetate; the third term models the output of soluble monomers by the effect of the output flow rate. Equation (6.4) is similar to Eq. (6.2) considering the growth and death rates of acidogenic bacteria; also, YCSC and YPSS are the transformation yields.

$$\frac{dS_S}{dt} = Y_{CSC}\mu_C X_C$$

$$-\frac{1}{Y_{PSS}}\mu_S X_S - DS_S \qquad (6.3)$$

$$\frac{dX_S}{dt} = (\mu_S - \kappa_S)X_S \qquad (6.4)$$

By doing a similar analysis for the mass balance of acetogenesis, Eqs. (6.5) and (6.6) are obtained, where YSPS and YHPP are the yield coefficients.

$$\frac{dS_P}{dt} = Y_{SPS}\mu_S X_S$$

$$-\frac{1}{Y_{HPP}}\mu_P X_P - DS_P \quad (6.5)$$

$$\frac{dX_P}{dt} = (\mu_P - \kappa_P)X_P \quad (6.6)$$

As said before, methane is produced by acetoclastic methanogenic and hydrogenotrophic methanogenic bacteria. It implies that the methanogenesis stage requires four equations to be represented. Equations (6.7) and (6.8) represent the mass balance for the first pathway, and the second one is modeled by Eqs. (6.9) and (6.10).

$$\frac{dS_A}{dt} = Y_{SAS}\mu_S X_S + Y_{PAP}\mu_P X_P$$

$$-\frac{1}{Y_{MAA}}\mu_A X_A - DS_A \quad (6.7)$$

$$\frac{dX_A}{dt} = (\mu_A - \kappa_A)X_A \quad (6.8)$$

$$\frac{dS_H}{dt} = Y_{SHS}\mu_S X_S + Y_{PHP}\mu_P X_P$$

$$-\frac{1}{Y_{MHH}}\mu_H X_H - DS_H \quad (6.9)$$

$$\frac{dX_H}{dt} = (\mu_H - \kappa_H)X_H \quad (6.10)$$

An additional Eq. (6.11) is introduced to represent the balance of inorganic carbon in the anaerobic reactor. This component is produced by acidogenic, acetogenic, and methanogenic acetoclastic bacteria; S_{IC} is used by methanogenic hydrogenophilic bacteria to produce methane,

and the dissolved carbon goes out from the reactor by the action of the output flow rate.

$$\frac{dS_{IC}}{dt} = Y_{SIS}\mu_S X_S + Y_{PIP}\mu_P X_P$$

$$+Y_{AIA}\mu_A X_A - \frac{1}{Y_{HIH}}\mu_H X_H - DS_{IC} \quad (6.11)$$

The biogas is a product of the biological activity inside the reactor. Methane and carbon dioxide are the main components of the produced gaseous mixture. Besides, the fraction of hydrogen which is not transformed into methane could also be released in the biogas; even if this fraction is small, the presence of hydrogen in biogas is relevant due to its high energy value. Therefore in this CBP-AD approach, it is proposed that it is feasible to promote the activity of OHPA bacteria to produce higher hydrogen volume. Then, the biogas equation [Eq. (6.12)] includes three components: methane produced from acetate by acetoclastic bacteria and hydrogen plus carbon dioxide by hydrogenotrophic bacteria; a fraction of hydrogen produced by OHPA bacteria; and gaseous carbon dioxide released from inorganic carbon.

$$Q_{Biogas} = Y_{ACH_4}\mu_A X_A + Y_{HCH_4}\mu_H X_H$$
$$+Y_{HH_2}\mu_S X_S + Y_{PH_2}\mu_P X_P + Y_{ICO_2}S_{IC} \quad (6.12)$$

According to CBP, the structure of the bioreactor should consider a separation of the components before being released. With this idea, the three components could be recovered separately to be used for further applications, and they are modeled by Eqs. (6.13)–(6.15). Since this is a first assay to conceptualize consolidated bioprocessing by AD, it is supposed that the biogas separation could be performed by adsorbent materials such as natural and synthetic zeolites, organic polymers, and even nanomaterials. At this time, a zero-order dynamics is proposed to model the separation of each component:

$$CH_4 = K_{CH_4} \cdot Q_{Biogas} \qquad (6.13)$$

$$CO_2 = K_{CO_2} \cdot Q_{Biogas} \qquad (6.14)$$

$$H_2 = K_{H_2} \cdot Q_{Biogas} \qquad (6.15)$$

On the other side, the separation of solids from liquids is performed by decantation. A fraction of solids, including most of the deadly bacteria which are not attached to the solid support, leaves the reactor by the effect of the output flow rate. The other fraction is recovered in the trap solids at the bottom of the reactor. It is supposed that the decantation of solids can be represented as a zero-order dynamics as follows:

$$Q_{Solids} = K_{SC}S_C + K_{SS}S_S + K_{SP}S_P$$
$$+ K_{SA}S_A + K_{SH}S_H + K_{SIC}S_{IC} \qquad (6.16)$$

This topic on AD is a few addressed regarding the obtaining value-added products (Sawatdeena-runat et al., 2015). The previous equation is a first approximation of quantification of solids, and more profound studies are required on this topic.

Finally, the liquid is directly determined from the dilution rate, which is computed as a function of the reactor volume and the output flow rate:

$$\frac{dL}{dt} = D \cdot V \qquad (6.17)$$

The model previously described has been implemented in the computational environment Matlab-Simulink. The parameters were adapted from the reported by Angelidaki et al. (1999), Batstone et al. (2002), Bernard et al. (2001), Carlos-Hernandez et al. (2009), Chorukova and Simeonov (2019), Rosen and Jeppsson (2006), and Schneider (2015). The initial conditions were computed by considering the derivative equal to zero in Eqs. (6.1)–(6.11); the input COD was set as 5 g/L. Also, it was supposed a flow rate of 10 L/h and a reactor of 1000 L. A series of simulations was performed to illustrate the CBP-AD model, and the obtained results are presented in the next figures.

Some of the results of the first simulation are presented in Figures 6.9 and 6.10, it considers a variation in the input COD from 5 to 20 g/L after to reach the steady-state; the flow rate is kept in the initial condition Q=10 L/h during all the simulation. It is observed at the beginning that the system has reached its steady-state, and it remains there until the variation on COD is induced at t=50 h (Figure 6.9a). After that, the substrate (in this case, acetate dynamics is presented) increases due to the biological reactions in the different stages of the process (Figure 6.9c); but, at the same time, bacteria grow (Figure 6.9d) since there are adequate conditions. As can be seen, bacteria lead the system to another equilibrium point, where they can transform the excedent substrate; this transformation is represented by the decreasing behavior of acetate after ~100 h. Also, due to biological activity, biogas production is increased (Figure 6.10a and b). This trend is also reproduced in the

biogas separation system, where the proposed model allows recovering around 90% of each component. On the other side, the solids production changes as a direct function of the increase on the input substrate since it is directly related to the bacteria adaptation to transform the substrate excedent. Finally, the cumulated liquid is presented in Figure 6.10d; since the input and output flow rate are equal and constant to keep a constant reaction volume inside the reactor, the obtained liquid is easy to compute,

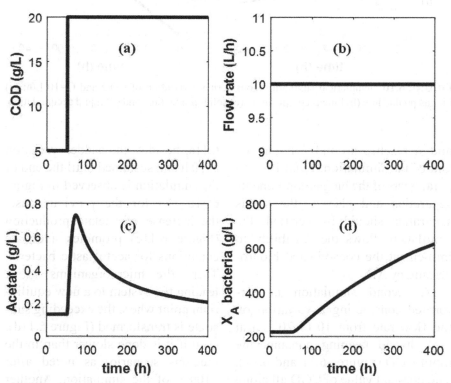

FIGURE 6.9 Numerical simulation considering a variation of COD and Q=10 L/h: (a) COD variation, (b) flow rate, (c) acetate production, and (d) bacterial grow.

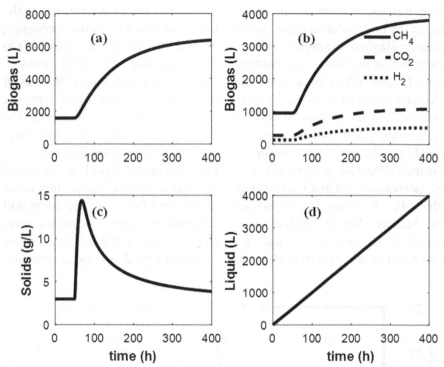

FIGURE 6.10 Numerical simulation considering a variation of COD and Q=10 L/h: (a) biogas production, (b) biogas composition, (c) solids production, and (d) liquid accumulation.

and it reaches around 4 m³ at the end of the simulation. Even if some parameters of the biogas components separation and also of the solids separation should be verified, the simulation shows the feasibility to implement the consolidated bioprocessing by AD.

A second simulation is performed considering a variation on the flow rate from 10 to 20 L/h at t=50 h and keeping constant the input COD (Figures 6.11 and 6.12). The constant value of COD all along the simulation is shown in Figure 6.11a; besides, the sudden variation at 50 h and sustained until the end of the simulation is observed in Figure 6.11b. As for the previous case, the increase of acetate production (Figure 6.11c) promotes adequate conditions for acetoclastic bacteria. Then, the microorganisms grow, leading the system to a new equilibrium point where the exceeding substrate is transformed (Figure 6.11d), even if it is done slower than in the previous scenario, as noted after ~100 h of the simulation. Another difference concerning the previous simulation is the transient state for biogas production; the increase in the input flow rate affects the biogas formation, which decreases before to return to the production on the initial equilibrium point. This difference is probably due to the adaptation time required by the different bacteria populations to transform the exceeding substrate; also, this could be linked to the residence time, since the output flow rate increases, it is possible that the substrate is not completely treated. This situation is also reflected in the separation of methane, hydrogen, and carbon dioxide, as observed in Figure 6.12b. On the other side, the solids recovery follows a similar dynamic as for the previous case (Figure 6.12c); the amount of recovered solids is lower since the concentration of COD is also lower than in the previous case. Moreover, since the output flow rate is increased, the cumulate liquid is also more significant, reaching a volume of ~7 m³.

An additional simulation considering variations in both COD and flow rate is performed. The

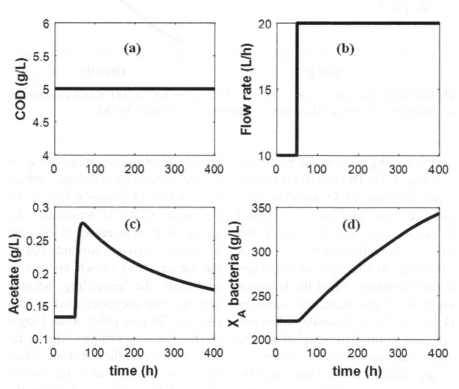

FIGURE 6.11 Numerical simulation of: (a) COD variation, (b) flow rate, (c) acetate production, and (d) bacterial grow (constant input of COD).

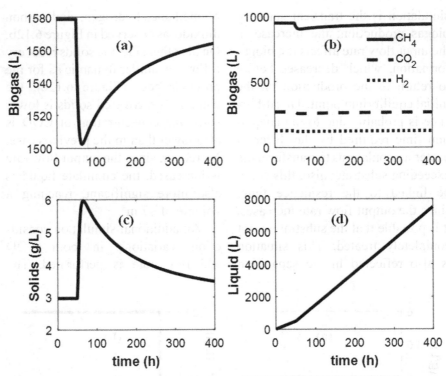

FIGURE 6.12 Numerical simulation of: (a) biogas production, (b) biogas composition, (c) solids production, and (d) liquid accumulation (constant input of COD).

corresponding results are presented in Figure 6.13. The variations in both input variables, COD, and flow rate, are presented in Figure 6.13a and b, respectively. The increase in the input COD produces a more relevant influence on the process behavior; it can be observed that the behavior on most of the presented variables (Figure 6.13c–g) is similar to the one obtained in the first simulation. The input variations lead the system to an equilibrium point where bacteria grow (Figure 6.13d), as in the previous scenarios. This bacterial growth implies the transformation of the exceeding substrate, which is observed in Figure 6.13c: when the input COD is increased, the acetate is also increased due to the biological activity. Around 100 h, the bacteria has grown enough to transform the exceeding acetate, and this one decreases tending to a new equilibrium point. Also, biogas production is according to the bacterial growth and the substrate transformation; it starts at a steady-state and leads to a new point where the production is higher (Figure 6.13e and f). In this scenario, the amount of recovered solids is more substantial than the recovery in previous cases due to the variations on the input variables and because of the bacteria dynamics (Figure 6.13g). Finally, the recovered liquid follows the dynamics of the second simulation (Figure 6.13h); this is evident since the output liquid depends only on the input flow rate, and this one is equivalent to the second scenario.

The numerical simulations show the feasibility of studying AD as a consolidated bioprocessing system. However, further studies are required to complement the proposed model, especially to represent better the separation of biogas components since the filtering materials could impose several technical and scientific challenges. Also, it is required to better quantify the solids recovery all along the AD stages. Besides, it is advisable to consider the most precise model to represent the hydrolysis stage; this can lead to the possibility of study SS-AD.

6.5 CHALLENGES AND OPPORTUNITIES

AD has been studied from several years ago; much of knowledge concerning the different issues of this process has been generated. Nevertheless, with the advances in technologies and materials, new opportunities are also identified. Regarding the CBP approach, AD is a promising alternative to be explored. Different strategies have been evaluated searching to overcome specific issues and consequently to improve the AD efficiency; therefore the main challenges are related to the integration of the developed knowledge in a single reactor, which requires more in-depth studies to ensure the whole process works as well as in independent experiments.

Among these challenges, it can be mentioned the improvement of hydrolysis to enhance the decomposition of complex molecules into compounds easy to transform by acidogenic bacteria. The different achievements relating to hydrolysis should be applied in a separate chamber of the CBP-AD reactor to avoid the separation of this step; this could be considered as an integrated pretreatment of raw materials.

Another situation to consider is the controlled supply of bioadditives to promote the production of specific gaseous compounds. For example, if hydrogen is desired as a priority, then bacteria producing this element could be added at a specific moment to allow a higher concentration of hydrogen producers and to induce inhibition of methanogenic bacteria; this action requires a microbiology approach to ensure that the global biologic activity is not affected. Also, automatic control

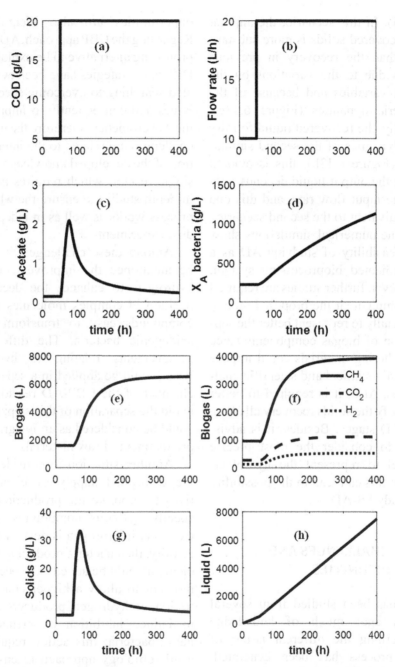

FIGURE 6.13 Numerical simulation considering variation in COD and flow rate of: (a) COD, (b) flow rate, (c) acetate production, (d) bacterial grow, (e) biogas production, f) biogas composition, (g) solids production, and (h) liquid accumulation.

algorithms should be implemented in order to compute the amounts and the way the additives must be applied.

The bacteria immobilization is a strategy to ensure inactive bacteria separation; death microorganisms are detached from the solid support, and they can precipitate together with suspended solids that are not transformed by biological action. Biosolids are studied only as a byproduct of AD and not as a principal result; nonetheless, there is enough information about digestates, which can be used as a basis to develop and value-added products at the same time. In this context, the solids trap design is an opportunity; for example, it is required to evaluate the possibility to integrate a posttreatment of solids in the same reactor.

Finally, the design of biogas components separation is also an opportunity from a technical and scientific viewpoint. Even if there exist several works related to biogas cleaning and upgrading, the integration in the CBP-AD reactor of a filter efficient and easy to operate is an important challenge. Filters based on the adsorption of gases are attractive candidates since the research on this topic is extensive; materials able to allow adsorption/desorption of methane, CO_2, and even hydrogen are identified. The challenge concerns in doing this at a medium large scale and in a short time.

KEYWORDS

- biogas
- waste revalorization
- anaerobic digestion
- bioproducts

REFERENCES

Abbassi-Guendouz, A., Brockmann, D., Trably, E., Dumas, C., Delgenès, J.P., Steyer, J.P., Escudié, R., 2012. Total solids content drives high solid anaerobic digestion via mass transfer limitation. Bioresour. Technol. 111, 55–61. https://doi.org/10.1016/j.biortech.2012.01.174

Abusam, A., Keesman, K.J., 2009. Dynamic modeling of sludge compaction and consolidation processes in wastewater secondary settling tanks. Water Environ. Res. 81, 51–56. https://doi.org/10.2175/106143008x304613

André, L., Pauss, A., Ribeiro, T., 2018. Solid anaerobic digestion: state-of-art, scientific and technological hurdles. Bioresour. Technol. 247, 1027–1037. https://doi.org/10.1016/j.biortech.2017.09.003

Angelidaki, I., Ellegaard, L., Ahring, B.K., 1999. A comprehensive model of anaerobic bioconversion of complex substrates to biogas. Biotechnol. Bioeng. 63, 363–372. https://doi.org/10.1002/(SICI)1097-0290(19990505)63:3<363::AID-BIT13>3.0.CO;2-Z

Baniamerian, H., Isfahani, P.G., Tsapekos, P., Alvarado-Morales, M., Shahrokhi, M., Vossoughi, M., Angelidaki, I., 2019.

Application of nano-structured materials in anaerobic digestion: current status and perspectives. Chemosphere 229, 188–199. https://doi.org/10.1016/j.chemosphere.2019.04.193

Batstone, D.J., Keller, J., Angelidaki, I., Kalyuzhnyi, S.V., Pavlostathis, S.G., Rozzi, A., Sandres, W.T.M., Siegrist, H., Vavilin, V.A., 2002. The IWA Anaerobic Digestion Model No 1 (ADM1). Water Sci. Technol. 45, 65–73. https://doi.org/10.2166/wst.2002.0292

Bernard, O., Hadj-Sadok, Z., Dochain, D., Genovesi, A., Steyer, J.-P., 2001. Dynamical model development and parameter identification for an anaerobic wastewater treatment process. Biotechnol. Bioeng. 75, 424–438.

Bi, S., Qiao, W., Xiong, L., Ricci, M., Adani, F., Dong, R., 2019. Effects of organic loading rate on anaerobic digestion of chicken manure under mesophilic and thermophilic conditions. Renew. Energy 139, 242–250. https://doi.org/10.1016/j.renene.2019.02.083

Biesuz, M., Sglavo, V.M., 2019. Flash sintering of ceramics. J. Eur. Ceram. Soc. 39, 115–143. https://doi.org/10.1016/j.jeurceramsoc.2018.08.048

Camerini, R., Chelazzi, D., Giorgi, R., Baglioni, P., 2019. Hybrid nano-composites for the consolidation of earthen masonry. J. Colloid Interface Sci. 539, 504–515. https://doi.org/10.1016/j.jcis.2018.12.082

Carlos-Hernandez, S., Sanchez, E.N., Béteau, J.F., 2009. Fuzzy observers for anaerobic WWTP: development and implementation. Control Eng. Pract. 17, 690–702. https://doi.org/10.1016/j.conengprac.2008.11.008

Carlos-Hernández, S., Sanchez, E.N., Béteau, J.F., Jiménez, L.D., 2014. Análisis de un proceso de tratamiento de efluentes para producción de metano. RIAI—Rev. Iberoam. Autom. e Inform. Ind. 11, 236–246. https://doi.org/10.1016/j.riai.2014.02.006

Carlos Hernández, S., Día Jiménez, L., Bueno García, A., 2018. Potential of energy production from slaughterhouse wastewater. Interciencia 43, 558–565.

Carrillo-Nieves, D., Rostro Alanís, M.J., de la Cruz Quiroz, R., Ruiz, H.A., Iqbal, H.M.N., Parra-Saldívar, R., 2019. Current status and future trends of bioethanol production from agro-industrial wastes in Mexico. Renew. Sustain. Energy Rev. 102, 63–74. https://doi.org/10.1016/j.rser.2018.11.031

Cavallo, O., de la Rosa, J.M., González-Pérez, J.A., Knicker, H., Pezzolla, D., Gigliotti, G., Provenzano, M.R., 2018. Molecular characterization of digestates from solid-state anaerobic digestion of pig slurry and straw using analytical pyrolysis. J. Anal. Appl. Pyrolysis 134, 73–82. https://doi.org/10.1016/j.jaap.2018.05.012

Chorukova, E., Simeonov, I., 2019. Mathematical modeling of the anaerobic digestion in two-stage system with production of hydrogen and methane including three intermediate products. Int. J. Hydrogen Energy, 1–9. https://doi.org/10.1016/j.ijhydene. z2019.01.228

Costello, D.J., Greenfield, P.F., Lee, P.L., 1991. Dynamic modelling of a single-stage high-rate anaerobic reactor-I. Model derivation. Water Res. 25, 847–858. https://doi.org/10.1016/0043-1354(91)90166-N

Craig, R.F., 2004. Craig's Soil Mechanics, Seventh. ed. Spon Press. Taylor & Francis Group, London. https://doi.org/10.4324/9780203494103

Dochain, D., Vanrolleghem, P.A., 2001. Dynamical Modeling and Estimation in Wastewater Treatment Processes. IWA Publishing, Cornwall.

Filer, J., Ding, H.H., Chang, S., 2019. Biochemical methane potential (BMP) assay method for anaerobic digestion research. Water 11, 1–29.

Finnie, I., Altan, T., Dornfeld, D.A., Posco, T.W.E., German, R.M., Jones, M.G., Kegg, R.L., Kuhn, H.A., Lindsay, R.P., Meyers, C.W., Pehlke, R.D., Ramaligam,

S., Richmond, O., Wang, K.K., Voelcker, H.B., Wright, P.K., 1995. Unit Manufacturing Processes: Issues and Opportunities in Research. National Academy Press, Washington, D.C.

Friel, R.J., 2015. Power ultrasonics for additive manufacturing and consolidating of materials, Power Ultrasonics: Applications of High-Intensity Ultrasound. Elsevier Ltd. https://doi.org/10.1016/B978-1-78242-028-6.00013-2

García-Ros, G., Alhama, I., Morales, J.L., 2019. Numerical simulation of nonlinear consolidation problems by models based on the network method. Appl. Math. Model., 69, 604–620. https://doi.org/10.1016/j.apm.2019.01.003

Ge, X., Xu, F., Li, Y., 2016. Solid-state anaerobic digestion of lignocellulosic biomass: recent progress and perspectives. Bioresour. Technol. 205, 239–249. https://doi.org/10.1016/j.biortech.2016.01.050

Ghadban, A.A., Wehbe, N.I., Pauly, T., 2018. Seismic performance of self-consolidating concrete bridge columns. Eng. Struct. 160, 461–472. https://doi.org/10.1016/j.engstruct.2018.01.065

Jiang, Y., Dennehy, C., Lawlor, P.G., Hu, Z., McCabe, M., Cormican, P., Zhan, X., Gardiner, G.E., 2019. Exploring the roles of and interactions among microbes in dry co-digestion of food waste and pig manure using high-throughput 16S rRNA gene amplicon sequencing. Biotechnol. Biofuels, 12, 1–16. https://doi.org/10.1186/s13068-018-1344-0

Josserand, L., Coussy, O., de Larrard, F., 2006. Bleeding of concrete as an ageing consolidation process. Cem. Concr. Res, 36, 1603–1608. https://doi.org/10.1016/j.cemconres.2004.10.006

Karadag, D., Köroılu, O.E., Ozkaya, B., Cakmakci, M., 2015. A review on anaerobic biofilm reactors for the treatment of dairy industry wastewater. Process Biochem. 50, 262–271. https://doi.org/10.1016/j.procbio.2014.11.005

Kim, J.K., Oh, B.R., Chun, Y.N., Kim, S.W., 2006. Effects of temperature and hydraulic retention time on anaerobic digestion of food waste. J. Biosci. Bioeng. 102, 328–332. https://doi.org/10.1263/jbb.102.328

Knofius, N., van der Heijden, M.C., Zijm, W.H.M., 2019. Consolidating spare parts for asset maintenance with additive manufacturing. Int. J. Prod. Econ. 208, 269–280. https://doi.org/10.1016/j.ijpe.2018.11.007

Li, Y., Chen, Y., Wu, J., 2019. Enhancement of methane production in anaerobic digestion process: a review. Appl. Energy 240, 120–137. https://doi.org/10.1016/j.apenergy.2019.01.243

Liu, T., Zhou, X., Li, Z., Wang, X., Sun, J., 2019. Effects of liquid digestate pretreatment on biogas production for anaerobic digestion of wheat straw. Bioresour. Technol. 280, 345–351. https://doi.org/10.1016/j.biortech.2019.01.147

Lynd, L.R., Elander, R.T., Wyman, C.E., 1996. Likely feature of cost of mature biomass ethanol technology. Appl. Biochem. Biotechnol. 57, 741–761.

Lynd, L.R., Van Zyl, W.H., McBride, J.E., Laser, M., 2005. Consolidated bioprocessing of cellulosic biomass: an update. Curr. Opin. Biotechnol. 16, 577–583. https://doi.org/10.1016/j.copbio.2005.08.009

Lynd, L.R., Weimer, P.J., van Zyl, W.H., Pretorius, I.S., 2002. Microbial cellulose utilization: fundamentals and biotechnology. Microbiol. Mol. Biol. Rev. 66, 506–577. https://doi.org/10.1128/MMBR.66.3.506-577.2002

Manchala, K.R., Sun, Y., Zhang, D., Wang, Z.-W., 2017. Anaerobic Digestion Modelling, Advances in Bioenergy. Elsevier Ltd. https://doi.org/10.1016/bs.aibe.2017.01.001

Mao, C., Feng, Y., Wang, X., Ren, G., 2015. Review on research achievements of biogas from anaerobic digestion. Renew. Sustain. Energy Rev. 45, 540–555. https://doi.org/10.1016/j.rser.2015.02.032

Mbaneme-Smith, V., Chinn, M.S., 2015. Consolidated bioprocessing for biofuel production: recent advances. Energy Emiss. Control Technol. 3, 23–44. https://doi.org/10.2147/eect.s63000

Meegoda, J.N., Li, B., Patel, K., Wang, L.B., 2018. A review of the processes, parameters, and optimization of anaerobic digestion. Int. J. Environ. Res. Public Health 15. https://doi.org/10.3390/ijerph15102224

Mognol, P., Lepicart, D., Perry, N., 2006. Rapid prototyping: energy and environment in the spotlight. Rapid Prototyp. J. 12, 26–34. https://doi.org/10.1108/13552540610637246

Möller, K., Müller, T., 2012. Effects of anaerobic digestion on digestate nutrient availability and crop growth: a review. Eng. Life Sci. 12, 242–257. https://doi.org/10.1002/elsc.201100085

Mosey, F.E., 1983. Mathematical modelling of the anaerobic digestion process: regulatory mechanism for the formation of short-chain volatile acids from glucose. Water Sci. Technol. 15, 209–232.

Murphy, J.D., Thamsiriroj, T., 2013. Fundamental science and engineering of the anaerobic digestion process for biogas production. The Biogas Handbook, 104–130. https://doi.org/10.1533/9780857097415.1.104

Nagarajan, D., Lee, D.J., Chang, J.S., 2019. Recent insights into consolidated bioprocessing for lignocellulosic biohydrogen production. Int. J. Hydrogen Energy 44, 14362–14379. https://doi.org/10.1016/j.ijhydene.2019.03.066

Parisutham, V., Kim, T.H., Lee, S.K., 2014. Feasibilities of consolidated bioprocessing microbes: From pretreatment to biofuel production. Bioresour. Technol. 161, 431–440. https://doi.org/10.1016/j.biortech.2014.03.114

Pastor-Poquet, V., Papirio, S., Steyer, J.P., Trably, E., Escudié, R., Esposito, G., 2018. High-solids anaerobic digestion model for homogenized reactors. Water Res. 142, 501–511. https://doi.org/10.1016/j.watres.2018.06.016

Puñal, A., Trevisan, M., Rozzi, A., Lema, J.M., 2000. Influence of C:N ratio on the start-up of up-flow anaerobic filter reactors. Water Res. 34, 2614–2619. https://doi.org/10.1016/S0043-1354(00)00161-5

Radhika, B.P., Krishnamoorthy, A., Rao, A.U., 2017. A review on consolidation theories and its application. Int. J. Geotech. Eng. 6362, 1–7. https://doi.org/10.1080/19386362.2017.1390899

Ren, Y., Yu, M., Wu, C., Wang, Q., Gao, M., Huang, Q., Liu, Y., 2018. A comprehensive review on food waste anaerobic digestion: research updates and tendencies. Bioresour. Technol. 247, 1069–1076. https://doi.org/10.1016/j.biortech.2017.09.109

Romero-Güiza, M.S., Vila, J., Mata-Alvarez, J., Chimenos, J.M., Astals, S., 2016. The role of additives on anaerobic digestion: a review. Renew. Sustain. Energy Rev. 58, 1486–1499. https://doi.org/10.1016/j.rser.2015.12.094

Rosen, C., Jeppsson, U., 2006. Aspects on ADM1 implementation within the BSM2 Framework, Technical Report.

Sawatdeenarunat, C., Nguyen, D., Surendra, K.C., Shrestha, S., Rajendran, K., Oechsner, H., Xie, L., Khanal, S.K., 2015. Anaerobic biorefinery: Current status, challenges, and opportunities. Bioresour. Technol. 215, 304–313. https://doi.org/10.1016/j.biortech.2016.03.074

Schneider, A., 2015. Dynamic Modeling and Simulation of Biogas Production Based on Anaerobic Digestion of Gelatine, Sucrose and Rapeseed Oil. Jacobs University.

Siddique, M.N.I., Wahid, Z.A., 2018. Achievements and perspectives of anaerobic co-digestion: a review. J. Clean. Prod. 194, 359–371. https://doi.org/10.1016/j.jclepro.2018.05.155

Singhal, C., Murtaza, Q., Parvej., 2018. Microwave sintering of advanced composites materials: a review. Mater. Today Proc.

5, 24287–24298. https://doi.org/10.1016/j.matpr.2018.10.224

Tamayo-Mas, E., Harrington, J.F., Graham, C.C., 2018. On modelling of consolidation processes in geological materials. Int. J. Eng. Sci. 131, 61–79. https://doi.org/10.1016/j.ijengsci.2018.05.008

Tampio, E., Salo, T., Rintala, J., 2016. Agronomic characteristics of five different urban waste digestates. J. Environ. Manage. 169, 293–302. https://doi.org/10.1016/j.jenvman.2016.01.001

Terzaghi, K., Bjerrum, L., Casagrande, A., Skempton, A.W., 1960. From Theory to Practice in Soil Mechanics. Selections from the writings of Karl Terzaghi. John Wiley & Sons, New York.

Timonen, K., Sinkko, T., Luostarinen, S., Tampio, E., Joensuu, K., 2019. LCA of anaerobic digestion: Emission allocation for energy and digestate. J. Clean. Prod. 235, 1567–1579. https://doi.org/10.1016/j.jclepro.2019.06.085

Vasco-Correa, J., Khanal, S., Manandhar, A., Shah, A., 2018. Bioresource Technology Anaerobic digestion for bioenergy production: global status, environmental and techno-economic implications, and government policies. Bioresour. Technol. 247, 1015–1026. https://doi.org/10.1016/j.biortech.2017.09.004

Venkiteshwaran, K., Bocher, B., Maki, J., Zitomer, D., 2015. Relating anaerobic digestion microbial community and process function: supplementary issue: water microbiology. Microbiol. Insights 8(S2), 37–44. https://doi.org/10.4137/mbi.s33593

Wang, L., Wang, J., Xu, Y., Cui, L., Qian, X., Chen, P., Fang, Y., 2019. Consolidating recycled concrete aggregates using phosphate solution. Constr. Build. Mater. 200, 703–712. https://doi.org/10.1016/j.conbuildmat.2018.12.129

Weiland, P., 2010. Biogas production: current state and perspectives. Appl. Microbiol. Biotechnol. 85, 849–860. https://doi.org/10.1007/s00253-009-2246-7

Weimer, P.J., Russell, J.B., Muck, R.E., 2009. Lessons from the cow: What the ruminant animal can teach us about consolidated bioprocessing of cellulosic biomass. Bioresour. Technol. 100, 5323–5331. https://doi.org/10.1016/j.biortech.2009.04.075

Xie, S., Hai, F.I., Zhan, X., Guo, W., Ngo, H.H., Price, W.E., Nghiem, L.D., 2016. Anaerobic co-digestion: A critical review of mathematical modelling for performance optimization. Bioresour. Technol. 222, 498–512. https://doi.org/10.1016/j.biortech.2016.10.015

Xin, F., Dong, W., Zhang, W., Ma, J., Jiang, M., 2019. Biobutanol production from crystalline cellulose through consolidated bioprocessing. Trends Biotechnol. 37, 167–180. https://doi.org/10.1016/j.tibtech.2018.08.007

Xu, F., Li, Y., Wang, Z.W. Mathematical modeling of solid-state anaerobic digestion. Prog. Energy Combust. Sci. 51, 49–66. https://doi.org/10.1016/j.pecs.2015.09.001

Xue, L., 2018. Laser consolidation: A rapid manufacturing process for making net-shape functional components, Second ed., Advances in Laser Materials Processing: Technology, Research and Application. Elsevier Ltd. https://doi.org/10.1533/9781845699819.6.492

Yang, S., Talekar, T., Sulthan, M.A., Zhao, Y.F., 2017. A Generic sustainability assessment model towards consolidated parts fabricated by additive manufacturing process. Proc. Manuf. 10, 831–844. https://doi.org/10.1016/j.promfg.2017.07.086

Yang, S., Tang, Y., Zhao, Y.F., 2015. A new part consolidation method to embrace the design freedom of additive manufacturing. J. Manuf. Process. 20, 444–449. https://doi.org/10.1016/j.jmapro.2015.06.024

Yang, S., Zhao, Y.F., 2018. Additive manufacturing-enabled part count reduction: a lifecycle perspective. J. Mech. Des. 140, 031702. https://doi.org/10.1115/1.4038922

Zeshan, Visvanathan, C., 2014. Evaluation of anaerobic digestate for greenhouse gas emissions at various stages of its management. Int. Biodeterior. Biodegrad. 95, 167–175. https://doi.org/10.1016/j.ibiod.2014.06.020

Zhang, D., Wang, F., Zhang, A., Yi, W., Li, Z., Shen, X., 2019. Effect of pretreatment on

chemical characteristic and thermal degradation behavior of corn stalk digestate: comparison of dry and wet torrefaction. Bioresour. Technol. 275, 239–246. https://doi.org/10.1016/j.biortech.2018.12.044

CHAPTER 7

Microbial Butanol Production From Lignocellulosic Biomass: Consolidated Bioprocessing (CBP)

JOSÉ A. RODRÍGUEZ-DE LA GARZA[1], THELMA K. MORALES-MARTÍNEZ[2], MIGUEL A. MEDINA-MORALES[1], ADOLFO ROMERO GALARZA[3], MAYELA MORENO-DÁVILA[1], CRISTÓBAL N. AGUILAR[4], and LEOPOLDO J. RÍOS-GONZÁLEZ[1*]

[1]Department of Biotechnology, School of Chemistry, Autonomous University of Coahuila, Mexico

[2]Bioprocess and Microbial Biochemistry Group, School of Chemistry, Autonomous University of Coahuila, Mexico

[3]Department of Chemical Engineering, School of Chemistry, Autonomous University of Coahuila, Mexico

[4]Food Research Department, School of Chemistry, Autonomous University of Coahuila, Mexico

*Corresponding author. E-mail: leopoldo.rios@uadec.edu.mx

ABSTRACT

Butanol is a major chemical precursor for the manufacture of paints, polymers, and plastics. Recently, it also has been considered as a possible alternative fuel for transport that can replace gasoline. Due to its physicochemical properties is considered the biofuel with greater similarity to gasoline compared to other biofuels (hydrogen, biodiesel, and ethanol). Compared to ethanol, butanol can be burned directly in current gasoline engines, generating 25% more energy and lower gas emissions to the environment. Currently, butanol is mainly produced from crude oil through a petrochemical reaction. However, its production costs are linked to the propylene production market, which is strongly related to the price of crude oil. Therefore another alternative independent to petroleum derivatives is the biotechnological pathway, known as ABE (acetone, butanol, ethanol). Simple sugars such as glucose are fermented by solventogenic strains (mainly of the genus *Clostridium*) producing ABE at a 3:6:1 ratio, respectively. Few countries, such as China, have maintained the production of butanol by fermentation from corn and sugarcane for decades. Sugarcane and corn starch are the raw materials commonly reported in several studies as the preferred substrate to produce butanol. However, the additional demand for these raw materials to produce biofuels can cause food security issues. In the butanol production process via fermentation, 75% of the final cost is directly related to feedstock prices. Due to its decisive role in the final product cost, the search for low-cost substrates is a persistent pursuit in the bioenergy industry. The abundance of lignocellulosic biomass and its low cost have outlined it as a potential alternative. However, its recalcitrant nature and the high costs of hydrolyzing cellulose have meant that there are no more plants on an industrial scale. This chapter describes recent advances in butanol production from lignocellulosic biomass with special emphasis on the processes known as consolidated bioprocessing.

7.1 INTRODUCTION

As time passes, oil reserves one day will be completely depleted and may be advisable to look and promote the search for new energy alternatives from renewable sources (Li et al., 2019). Before getting to the importance of fuel alternatives, we may comment that if the dependence of petroleum for fuels is decreased, the same resource can be redirected to establish the amount used for other high added-value products from petroleum (Volk and George, 2019). This way, the proportion of fuels would change in such a way that it may reflect positively in production costs of the previously mentioned petroleum derivates. In recent years, biotechnology is advancing to the point where it is having an impact in many industrial areas, offering environmental-friendly options for several other products. Even though butanol and ethanol production has been known for years, it has gained a renewed interest recently because both solvents can be used as car engine fuel. In the case of ethanol, it generates enough energy and can be used in current internal combustion engines but requires engine modifications to avoid corrosion (Yu et al., 2018). As useful as ethanol can be, it still represents a challenge, because it still has to be produced from cheap feedstocks to be able to compete with gasoline. The same can be stated about butanol, but it has advantages over ethanol such as higher energy value and boiling point, it's less hygroscopic and corrosive, has a higher-octane number, and also, it can be used directly in car engines without modifications (Buakhiaw and Sanguanchaipaiwong, 2017; Roberto and Gonçalves, 2017; Sarangi and Nanda, 2018). Since the 19th century, Pasteur discovered butanol production by bacteria and in the 20th century, butanol was already becoming an industrial level product, but its continued generation was hindered and later stopped because of the appearance of more competitive, cost-effective, and high yielding petrochemical processes in the 1950s (Oliva-Rodríguez et al., 2019). Taking into consideration the fact that most of the current research and many industrial processes are avoiding the use of direct food sources as raw material, interest in using residual materials has been on the rise. For instance, if a direct food source is the main material for biofuels production, it is named first-generation biofuels (Ding et al., 2019). The processes that employ residual materials, biomass, agro-industrial wastes, and/or by-products are named second-generation biofuels (Diaz et al., 2018). Third-generation biofuels are produced from algae, which its energy storage polysaccharides are the resource exploited for biofuels production

(Gao et al., 2016). Now in context, second-generation biofuels may be the most attractive type of biofuel, mainly because it can be considered cheaper than the others and there are many sources to choose from. Nowadays, for butanol production, many of the materials mentioned contain substrate that can be converted to butanol by microorganisms. Since its first isolation by Chaim Weizmann in 1915, solventogenic bacteria have been used to produce solvents via its ABE fermentation pathway which yields acetone, butanol, and ethanol (Niglio et al., 2019). As residues are being considered to be the main source for butanol production, it is extremely important to establish an adequate pretreatment process that can allow the release of more sugars for its fermentation and subsequent production of the compound of interest. The main components of biomass and agroindustrial wastes is lignocellulose. This polymer is formed by cellulose, hemicellulose, and lignin. The first is a homopolysaccharide of glucose monomers linked by β-1,4 glycosidic bonds. The second is a heteropolysaccharide of multiple sugars where xylose and arabinose are the main components. The last structure is a polymer of phenolic units which are cross-linked among itself (Parisutham et al., 2017). To produce butanol from this type of polymers, these last must be pretreated and enzymatically degraded to release sugars. The last part of the process is ABE fermentation, which is commonly performed by a bacterium of the genus *Clostridium*, which may include *Clostridium acetobutylicum, Clostridium beijerinckii, Clostridium saccharoperbutylicum*, or *Clostridium saccharoperbutylacetonicum* (Patakova et al., 2018). There are still many aspects of biotechnological production of butanol that require research and development of genetic engineering tools that can improve bacterial strains, so that bacteria could be able to tolerate higher product accumulation, improvement in pentose degradation, direct butanol production from cellulose, among others. Also, pretreatments are of utmost importance because it must be customized depending on the material to yield degradable cellulose and hemicellulose, removal of lignin, and limited formation of fermentation inhibitory compounds. Since all these processes are complicated to some extent, a techno-economic feasibility analysis must be carried out to assess the monetary impact at an industrial level (Jang and Choi, 2018). In this regard, there must be adjustments to develop a feasible methodology that can provide a competitive economic process that can render higher butanol concentration and yields (Dalle Ave and Adams, 2018). A strategy that helps to achieve this purpose is the metabolic engineering

of solventogenic strains to produce higher quantities of butanol (Krivoruchko et al., 2013). In this case, researchers are working in genetic manipulation for product accumulation resistance and/or the ability for the bacteria to produce enzymes that are not expressed naturally, such as cellulases and β-glucosidases, among others (Rahnama et al., 2014). There is an extensive array of aspects that can be manipulated to improve butanol production and the techno-economic factors that must be taken into consideration for future industrial applications (Dumitrescu et al., 2018). It is worth noting that renewable sources of biomass and or agroindustrial wastes can serve as a substrate, and the rational exploitation of these resources can prove feasible and profitable in medium to long term.

7.2 MICROBIAL PRODUCTION OF BUTANOL

As previously mentioned, butanol production was observed in *Clostridia* since research studies began more than a hundred years ago (Sauer, 2016). These bacteria are able to produce the solvent of interest from different carbon sources such as glucose, xylose, glycerol, and proteins. In particular, the metabolic pathway of glucose degradation produces and accumulates solvents such as acetone, butanol, and ethanol, thus named ABE fermentation (Charubin et al., 2018). The fermentation process occurs in two distinct phases: the acidogenic phase, where cell accumulation takes place, as well as the organic acid production, which includes acetic and butyric acids. After this, the solventogenic phase takes hold and the acids produced are reintegrated to the metabolism and acetone, butanol, and ethanol are formed (Liu et al., 2019). Microbial production of solvents is mainly carried out by obligated anaerobic bacteria of the genus *Clostridium* (Nanda et al., 2017). Clostridia are spore-forming bacteria, which are natural ABE fermentation microorganisms. It has been observed that in the solventogenic phase growth is not occurring due to the sporulation of the cells. It is in this time frame that solventogenesis is taking place where acetone, butanol, and ethanol are being produced in an approximate ratio of 3:6:1, respectively (Zhang et al., 2018). It has been implied that, while in this phase, there is no cell growth, the previously accumulated organic acids serve as substrates and precursors of butanol and the other solvents. The main *Clostridium* species known to perform the ABE fermentation are, *C. acetobutylicum, C. beijerinckii, C. butylicum, C. saccharoperbutylicum, Clostridium pasteurianum, C. saccharoperbutylacetonicum,* and *Clostridium aurantibutyricum*

(Patakova et al., 2018). The ABE metabolic pathway has been studied primarily in *C. acetobutylicum*; in this case, with glucose as a substrate, the common glycolysis pathway is followed to produce pyruvate. From this molecule, acetyl-CoA is formed, to subsequently form acetoacetyl-CoA. From this, 3-hydroxybutyryl-CoA is synthesized, followed by crotonyl-CoA and butyryl-CoA. Having reached this molecule, butyraldehyde is produced to finally form butanol. Also, in the metabolic pathway, from acetyl-CoA and acetoacetyl-CoA, it is transformed into acetaldehyde and acetoacetate, and finally being reduced to ethanol and acetone, respectively (Ndaba et al., 2015). Although, as previously mentioned, butanol has a greater proportion of the produced solvents (Zhang et al., 2018). In the case of butanol produced from glycerol (Krasňan et al., 2018), it is first converted into dihydroxyacetone, to be further phosphorylated to dihydroxyacetone phosphate, to produce pyruvate. From pyruvate, the metabolic route will be carried out as previously mentioned, leading to solventogenesis (Sarchami et al., 2016; Wen et al., 2014). As mentioned before, *Clostridium* is able to use several substrates for butanol production, in the following pages, the use of more complex biomolecules from lignocellulosic wastes will be addressed for its biotechnological processing.

7.3 BIOTECHNOLOGICAL BUTANOL FROM BIORESOURCES

Since the initial stages of butanol production at industrial levels in the 20th century, the substrate of choice was starch. *Clostridium* has the ability to produce the enzymes required to degrade starch, such as amylases, pullulanases, α-amylases, α-glycosidases, glucoamylases, and amylopullulanases (Yang et al., 2017). Because starchy materials come from primary food sources, it is necessary to seek other types of sources to widen the substrate possibilities for butanol production (Ujor et al., 2014). However, it is important to consider that there are many starch sources that do not generate food security issues or minimal effects at least. Starch has been used for bioethanol production, and this being a similar process, it can be also used as a substrate for butanol production. Butanol production from starch should not represent a problem because *Clostridium* is able to produce enzymes for its degradation and starch itself is easier to be converted into fermentable sugars than other polysaccharides (Yang et al., 2017). The more challenging way to produce butanol is to bioprocess lignocellulosic materials from wastes from agroindustry and/or biomass. It is well known that lignocellulose is the most abundant polymer on our planet and if used as

a resource, it can be renewable. As it is formed by cellulose, hemicellulose, and lignin, from the first two, which are polysaccharides, several sugars can be obtained, such as glucose, xylose, galactose, mannose, arabinose, and cellobiose and can be converted to butanol using solventogenic strains (Seifollahi and Amiri, 2019). Evidently, these substrates can be obtained from several materials (rice straw, corncob and waste of corn processing, sugarcane bagasse, brown seaweed, grass, and agave bagasse, among others), from which there has been progressing in the process of butanol production (Oliva-Rodríguez et al., 2019). Regarding lignocellulose processing for butanol production, to release fermentable sugars, it is necessary to resort to pretreatments to favor cellulose breakdown. Hemicellulose represents a lesser challenge as its structure is easier to degrade (Ibrahim et al., 2017). Enzymatic hydrolysis is required in order to achieve the depolymerization of cellulose and hemicellulose. All of these steps must be integrated to promote sugar accumulation for them to be fermented by *Clostridia* for butanol production (Amiri and Karimi, 2018). Another source of fermentable molecules is macro- and microalgae. These are considered raw materials for third-generation biofuel production, where nonarable land is involved. In biofuel production, microalgae have been recently cultivated, primarily to produce lipids for biodiesel. From here, a residue containing proteins and polysaccharides remains, and it can be used as feedstock for butanol production. Depending on the species, the content of degradable polysaccharides in the algae cell varies, and in particular, the species of *Chlorella* presents a high content of polysaccharides (Yan et al., 2014). In this case, algae cultures require CO_2 and light energy, among other aspects, and can manifest autotrophic, heterotrophic, and mixotrophic metabolism. Since *Clostridia* is able to produce amylases, the starch contained in algae can be converted into butanol via ABE fermentation (Adeniyi et al., 2018).

7.4 CONSOLIDATED BIOPROCESSING

The butanol production process from lignocellulosic biomass in one step is considered consolidated bioprocessing (CBP) (Vivek et al., 2019). In principle, it can apply a strategy of CBP to produce a wide range of high-value chemicals from biomass (Wang et al., 2018; Kylilis et al., 2018; Sgobba et al., 2018; Shahab et al., 2018; Feng et al., 2018; Liu et al., 2018). This requires recalcitrant biomass substrates to be degraded into soluble sugars and a metabolic intervention to direct the metabolic flux toward the desired products

in a process that can render high yield and concentration (Charubin et al., 2018; Majidian et al., 2018; Mazzoli, 2012).

The conventional process of production of butanol from lignocellulose (2G) is divided into three main steps: (1) physical, chemical, or biological pretreatment of biomass; (2) enzymatic hydrolysis (saccharification), and (3) fermentation ABE → acetone–butanol–ethanol (Gottumukkala et al., 2017). In a CBP strategy to produce butanol, the enzymatic hydrolysis and fermentation are carried out in one step using microorganisms capable of degrading cellulose and carried out solventogenesis from pretreated biomass (Ibrahim et al., 2018; Olson et al., 2012; Xin et al., 2018; Yan and Fong, 2017; Mbaneme and Chinn, 2015).

Lignocellulosic biomass can contain up to 75% of complex carbohydrate and is recalcitrant (Zabed et al., 2017); therefore an initial pretreatment step is essential for breaking the complex structure of the lignocellulosic material and provides access to hydrolytic enzymes to release sugars (dos Santos et al., 2019). The hydrolysis of cellulose by microorganisms requires a set of multiple enzymes to hydrolyze cellulose into soluble sugar monomers that can be fermented to produce butanol. At least three types of enzymes are required to hydrolyze cellulose: (1) endoglucanases; (2) exoglucanases, and (3) β-glucosidases (Gusakov, 2011). Endoglucanases randomly hydrolyze β-glucoside internal amorphous regions of the cellulose to produce oligosaccharides of various degrees of polymerization and generate new chain ends. The exoglucanases continue the action of endoglucanases, hydrolyzing the reducing and nonreducing ends cellulose chains to generate glucose or cellobiose as the main products. The exoglucanases can also hydrolyze microcrystalline cellulose, possibly separating the cellulose chains of the microcrystalline structure. β-glucosidase breaks down the soluble cellodextrins into glucose and cellobiose (Figure 7.1). The correct combination of the activities and the level of production of each enzyme is essential for the efficient hydrolysis of lignocellulosic biomass (Qian et al., 2017).

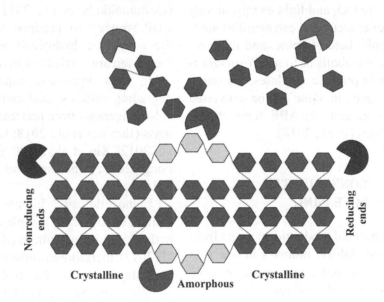

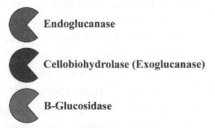

FIGURE 7.1   Schematic representation of the hydrolysis of amorphous and microcrystalline cellulose by noncomplexed cellulase systems.

Fungi are microorganisms that have shown the higher levels of production of these enzymes and with a great variety. Fungi belonging to the genus *Trichoderma* have been the subject of extensive research due to its high levels of cellulase production. Mainly, the *Trichoderma reesei* cellulase system has been studied for over 50 years (Hasunuma et al., 2013). Although *T. reesei* excretes β-glucosidases, the level of production of this enzyme is significantly lower compared to other fungi strains, such as *Aspergillus* species. These cellulolytic microorganisms are capable of producing ethanol or hydrogen gas, but only at very low concentrations. Furthermore, these microbes are obligate aerobes, making genetic modifications or reconfiguration process to enhance the butanol production more complicated (Parisutham et al., 2014).

Recent research suggests that some *Clostridium* strains are feasible to be genetically modified to produce butanol directly from cellulose. *Clostridium* cells present one of the broadest and more flexible systems that allow the use of a vast range of substrates (Linger and Darzins, 2012). Its theoretical ability to use almost all simple and complex carbohydrates (including cellulose, xylans, and many oligo- and polysaccharides) gives these microorganisms a unique advantage that almost no other microorganisms present, mainly because the cost of raw materials represent more 32% of the operating costs of production of butanol from biomass (Gaida et al., 2016).

Anaerobic bacteria that degrade vegetative biomass, such as *Clostridium cellulolyticum*, *Clostridium cellulovorans*, *Clostridium thermocellum*, and *Clostridium celevecrescens*, have multienzyme complexes known as cellulosomes (Lin et al., 2015; Ou et al., 2017; Bao et al., 2019; Qin et al., 2018; Singh et al., 2018; Jiang et al., 2017). In these systems, the different types of cellulose-degrading enzymes are assembled in the structural scaffolding subunits through strong protein noncovalent interactions in the coupling (dockerin) and complementary (cohesins) modules. The scaffolding subunits also contain a carbohydrate-binding module that connects all enzyme complex to the cellulose surface (Figure 7.2) (Hasunuma et al., 2013). Due to that, scaffolding subunits are joint together covalently to the cell wall of the microorganism by its anchoring proteins, the microorganisms that have cellulosome system can use cellulose as carbon and energy source. Efficient synergistic degradation of vegetative biomass is the result of the combination of the orientation of the substrate and enzyme complex and its spatial proximity of the different types of cellulases to each other (Xin et al., 2018).

The cellulosomes of *Clostridium* species, particularly *C. thermocellum*, contain a wide variety of hydrolytic

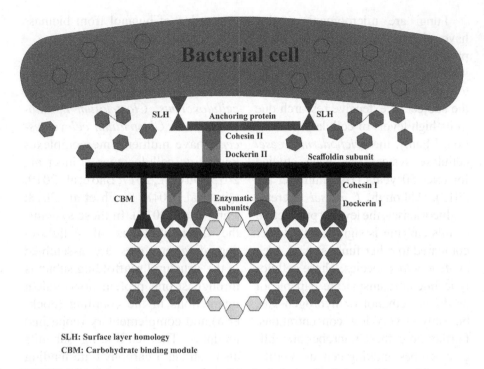

SLH: Surface layer homology
CBM: Carbohydrate binding module

FIGURE 7.2   Schematic representation of the hydrolysis of cellulose with a multienzyme complex (cellulosome) by cellulolytic *Clostridium* species.

enzymes (including endoglucanases, exoglucanases, and β-glucosidase) and show high enzyme activity to degrade vegetative biomass. However, its cellulase production is hundred times less than aerobic microorganisms as *T. reesei* (Qian et al., 2017).

Another disadvantage of the cellulosome system is that the cellulase activity in *C. thermocellum* and bacteria such as *C. cellulolyticum* is strongly inhibited by the accumulation of cellobiose, due to that in the cellulosome, the β-glucosidases can only be partially involved in the hydrolysis of cellobiose.

Although external supplementation of β-glucosidase can help counteract the inhibitory effect by the accumulation of cellobiose, this entails an extra expense (Gaida et al., 2016). The integration of β-glucosidase in the cellulosome not only would eliminate the effect of inhibition of cellobiose but also the rate of degradation of cellulose would be enhanced (Charubin et al., 2018).

The concept of CBP for butanol production can be carried out in the three different configurations: (1) genetic modification of solventogenic microorganisms with cellulolytic pathways engineering or vice versa; (2) coculturing of cellulolytic microorganisms and solventogenic; and (3) a two-stage process with cellulolytic and solventogenic microorganisms (Shanmugam et al., 2019).

7.4.1 GENETIC MODIFICATION TO ACHIEVE SIMULTANEOUS SACCHARIFICATION FOR SOLVENTOGENESIS

The successful production of butanol from lignocellulosic materials through a CBP strategy by strain genetic modification will depend largely on the ability of the microorganism to degrade cellulose and its efficiency in the solventogenesis process (Charubin et al., 2018). For this purpose, recently it has been an increase of attention to developing strategies with the use of genetic engineering tools primarily to modified natural cellulolytic microbes such as *Clostridium* to produce butanol or butanol-producing microbes with the ability to degrade lignocellulosic biomass (Li and He, 2016).

Knowledge associated with the metabolic pathway for the production of butanol is of great importance to carry out strategies for the enhancement of butanol production. In this regard, in 2001, the genomic sequence of *C. acetobutylicum* was characterized and currently remains as the model microorganism for

ABE fermentation. This analysis may provide more information on the primary metabolism and, in turn, help to elucidate the functional characterization of various enzymes that favor the production of butanol (Nölling et al., 2001).

Butanol concentrations can be improved by the removal of the byproducts, including acids and ethanol, using a combined approach of proteomic analysis of specific metabolites. However, due to the complexity of the multistep enzymatic conversion, the engineering of *C. acetobutylicum* to increase butanol production possesses a challenge for genetic engineers (Zhao et al., 2014). This strategy is complicated because all the genes involved in the solventogenic pathway of this microorganism are present in pSOL1 megaplasmid, and there is a possibility that the strain can lose the plasmid resulting in a strain without solventogenic capability. Instead of modifying the pSOL1 plasmid, strategies of gene deletion and overexpression have been carried out (Xin et al., 2018). For example, it was discovered that suppressing the genes butyrate kinase (*buk*), acetate kinase (*ack*), and phosphotransacetylase (*pta*) of *C. acetobutylicum* of the acidogenic pathway increases the butanol production (Table 7.1) (Jang et al., 2012).

Due to that, the *adhE* enzyme can convert acetyl-CoA and butyryl-CoA

TABLE 7.1 Metabolic Strategies for Butanol Production Enhancement Using Different Strains of *Clostridium* and Glucose as Substrate

Strain	Metabolic Construction	Butanol (g/L)	Butanol Yield (g/g)	Reference
Clostridium tyrobutyricum ATCC 25755	Knockout of CoA transferase (*coat*); upregulation of aldehyde/alcohol dehydrogenase (*adhE2*)	26.2	0.31	Zhang et al. (2018)
Clostridium acetobutylicum ATCC 824	Knockout of phosphotransacetylase (*pta*) and acetate kinase (*ack*); upregulation of aldehyde/alcohol dehydrogenase (*adhE2*)	18.9	0.29	Jang et al. (2012)
Clostridium tyrobutyricum ATCC 25755	Knockout of acetate kinase (*ack*); upregulation of aldehyde/alcohol dehydrogenase (*adhE2*)	10.0	0.27	Yu et al. (2011)
Clostridium acetobutylicum ATCC 824	Knockout of butyrate kinase (*buk*)	16.7	–	Harris et al. (2000)

to ethanol and butanol, respectively, the inclusion of a specific butanol dehydrogenase gene (*bdh*) can eliminate the production of ethanol and increase the production of butanol (Ji et al., 2013). Another strategy is to reduce the hydrogenase enzyme responsible for the production of hydrogen that could further increase the production of butanol (Biswas et al., 2015).

C. *cellulolyticum* was the first metabolically engineered strain for the direct production of butanol from cellulose (Gaida et al., 2016). Although no native microorganism can produce a significant amount of butanol from cellulose directly, various cellulosic *Clostridium*, such as C. *thermocellum* and *cellulolyticum*, were engineered to produce isobutanol (5.4 g/L) (Lin et al., 2015)

and *n*-butanol (0.12 g/L) (Gaida et al., 2016) after inserting the keto-acid and CoA-dependent pathways, respectively (Table 7.2). In addition, recently the ability to biosynthesize *n*-butanol was also added into C. *cellulovorans* overexpressing the aldehyde and alcohol dehydrogenase enzymes (*adhE1* and/or *adhE2*) of *Clostridium acetobutylicum*, which allowed to reach concentrations of butanol up to 4.0 g/L (Ou et al., 2017).

Metabolic engineering of microorganisms as CBP platforms has received enormous interest in the last decade. However, progress in the development of these strains (nonmodel) is slow, mainly due to the lack of genetic engineering tools available and to the short knowledge of its systems.

TABLE 7.2 Consolidated Bioprocessing (CBP) for Butanol Production Through Genetic Modification of *Clostridium* Cellulolytic Strains

Strains	Substrate	Metabolic Construction	Product (g/L)	Yield (g/g)	Productivity (g/L/h)	Reference
Clostridium cellulovorans	Cellulose	Overexpression of *bdhB*, *adhE1* and *adhE2* from *Clostridium acetobutylicum*	Butanol 4.0	0.22	0.0128	Bao et al. (2019)
C. cellulovorans	Corncob	Introduction of *adhE1* and *ctfA-ctfB-adc* genes from *Clostridium acetobutylicum* ATCC 824	Butanol 3.4	–	–	Wen et al. (2019)
C. cellulovorans	Corncob	Overexpression of *adhE2*	Butanol 3.3	0.14	0.046	Ou et al. (2017)
Clostridium cellulolyticum	Cellulose	Introduction of the CoA-dependent pathway	Butanol 0.12	0.01	0.00025	Gaida et al. (2016)
Clostridium thermocellum	Cellulose	Introduction of heterologous ketoisovalerate decarboxylase (KIVS)	Isobutanol 5.4	0.068	0.072	Lin et al. (2015)

7.4.2 COCULTURE OF CELLULOLYTIC AND SOLVENTOGENIC MICROORGANISMS

To date, there is no native strain available that can use cellulose as a sole carbon source to produce butanol. Genetically modified strains, either by overexpression heterologous or homologous, had unsatisfactory performances of butanol concentrations (Jiang et al., 2018b). Another strategy is the use of two microorganisms in synergy capable of carrying out the task of converting lignocellulosic biomass into fermentable sugars for subsequent conversion to butanol. In nature, 99% of microorganisms live in association with other species in some microbial consortia (Jiang et al., 2018a). Microbial consortia or artificial microbial cocultures could perform more complicated tasks and support more challenging environments compared with pure strains. These communities have several superior features compared with monocultures, such as (1) greater diversification of products, (2) robust metabolic capabilities, and (3) ability to use a broader spectrum of substrates.

Recent research has evaluated the design and creation of synthetic microbial consortia, including bacteria–bacteria, yeast–yeast, and even fungus–bacteria in the production of biofuels and high value-added bioproducts (Figure 7.3) (Jiang 2018b).

Currently, the most capable cellulose-degrading microorganism remains to be the aerobic filamentous fungus, *T. reesei*, which is considered to be a hyper-cellulase producer. However, due to that *T. reesei* growth conditions (aerobic) are very different from the solventogenic *Clostridium* species (strictly anaerobic), few studies of butanol production using bacteria–fungal consortia have been reported. In this regard, the study of coculture of *T. reesei* with *Escherichia coli* engineered via ABE, using pretreated corncobs as substrate, resulted in low butanol concentration (1.8 g/L) (Vivek et al., 2019).

As already mentioned, the cellulolytic *Clostridium* species (*C. cellulolyticum, C. cellulovorans*, and *C. thermocellum*) can grow in complex polysaccharides such as cellulose that can be further digested to produce fermentable sugars and subsequently lactate, acetate, H_2, and CO_2. Although the coculture of *C. thermocellum* and *C. saccharoperbutylacetonicum* was able to produce 8.3 g/L butanol from xylan (Table 7.3) (Jiang et al., 2018a). Another inconvenience is that the optimum temperatures for both microorganisms are different (60 °C and 30 °C). This drawback would imply an additional cooling system that can significantly influence the process costs. Another coculture system used primarily in starch-based raw materials

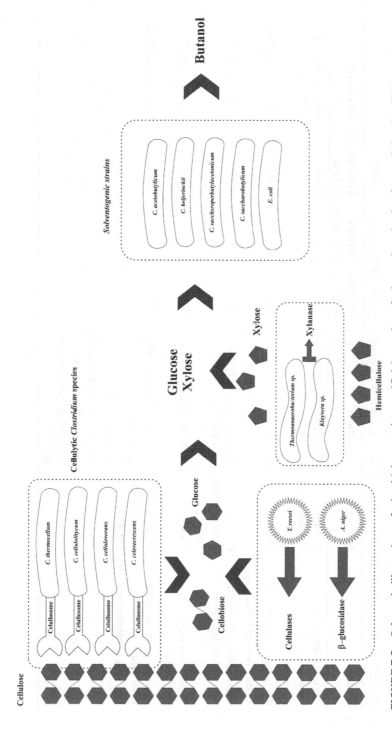

FIGURE 7.3 Schematic illustrations of microbial consortia or artificial co-culture for butanol production from cellulose by CBP.

TABLE 7.3 Butanol Production From Lignocellulose by Coculture Through Consolidated Bioprocessing (CBP)

Strains	Substrate	Conditions	Butanol (g/L)	Butanol Yield (g/g)	Productivity (g/L/h)	Reference
Microbial consortia, Clostridium sp.	Corncob pretreated with alkali	37 °C, 150 rpm, 2% of Clostridium sp.	10.8	0.54	0.064	Shanmugam et al. (2019)
Thermoanaerobacterium, Clostridium acetobutylicum	Xylan	-	8.3	–	0.043	Jiang et al. (2018a)
Clostridium cellulovorans, Clostridium beijerinckii	Corncob pretreated with alkali	pH controlled at ≈7; 150 rpm and of 37 °C	11.5	–	0.095	Wen et al. (2017)
Bacillus cereus, C. beijerinckii	Corn mash	37 °C in static conditions	6.78	–	0.070	Mai et al. (2017)
Clostridium thermocellum, Clostridium saccharoperbutylacetonicum	Rice straw pretreated with NaOH	First step: C. thermocellum at 60 °C for 24 h Second step: C. Saccharoperbutylacetonicum was inoculated and the temperature was decreased at 30 °C	5.5	–	0.016	Kiyoshi et al. (2015)

is *Bacillus* sp.–*Clostridium* sp. The first can efficiently hydrolyze the starch to glucose due to efficient excretion of amylase and consume the oxygen, which promotes bacterial growth of *Clostridium* sp., which grows under strictly anaerobic conditions. Hydrolyzed glucose could then be consumed simultaneously by *Clostridium* sp. for ABE production (Table 7.3) (Mai et al., 2017). An interesting approach is to cultivate anaerobic consortia with hydrolytic capabilities, with some bioaugmented solventogenic *Clostridium* species. This strategy can allow achieving higher concentration and yield of butanol (10.8 g/L and 0.54 g/g) using corncob pretreated with alkali (Shanmugam et al., 2019).

Production strategy by coculture shows great potential to diversify the type of raw material that can be used. Compared to genetic modification, the configuration of the microbial coculture saves time and is less susceptible to contamination. However, concentrations and yields are still low, so it is necessary to continue developing research that promotes technical and economic feasibility.

7.5 SEPARATION TECHNIQUES IN THE ACETONE–BUTANOL–ETHANOL FERMENTATION

The production of butanol via ABE fermentation is facing significant challenges due to the toxicity of butanol to microorganism, low solvents concentration (<3 wt.% butanol), and the enormous amount of water required for fermentation broth. So, the cost of product removal from the dilute fermentation broth still hinders the industrial-scale production of bio-butanol. To address the obstacles of ABE fermentation and improve the economic feasibility, one of the most effective ways is the development of more efficient processes for butanol recovery. Techniques for ABE recovery such as gas stripping, pervaporation, and distillation are discussed.

7.5.1 GAS STRIPPING

The process consists of bubbling a gas stream through the ABE fermentation broth to strip away acetone, butanol, and ethanol (Qureshi and Ezeji, 2008). The stripping gas is passed through a condenser where the vapors are partly condensed, while the depleted gas stream is recycled to the fermenter for another extraction cycle. It has been demonstrated that circulating a large amount of gas into the fermenter does not cause any cell damage but leads to foam formation (Abdehagh et al., 2014). Moreover, the performance of the gas stripping technique depends on the gas flow, antifoam, the size of the bubble, and the presence of other components in the fermentation broth (Ezeji et al., 2005; Xue et al., 2014). Park et al. (1991) observed that an increase in gas flow and temperature led to an increase in the conversion of glucose and on the separation of butanol (Liao et al., 2014). Additionally, it has been mentioned that an efficient butanol recovery and a high superficial velocity are recommended. However, gas stripping leads to a generation of a large amount of foam, thus the necessity of usage of larger amounts of an antifoaming agent, which can be toxic to the bacteria, and consequently reduces the production of the solvents. The main advantages of gas stripping in comparison with other recovery methods are that it is harmless to the culture and its known simplicity. The main concerns are the selectivity (butanol recovered) and the energy requirements. In this technique, the energy demand is highly dependent on the energy consumed by the heat exchanger and condenser (Abdehagh et al., 2014).

7.5.2 PERVAPORATION

Pervaporation is an unconventional separation technique where one or more components are separated from a liquid mixture. For butanol, acetone, and ethanol separation, the liquid feed mixture is in direct contact with one side of a hydrophobic membrane in which the permeate is removed as vapor (Chapman et al., 2008). The permeating vapor can then be recovered by condensation (Fontalvo et al., 2005).

The mass transfer in butanol pervaporation includes three steps: (1) sorption of the liquid feed (permeate) to the upstream side of the membrane; (2) diffusion of the permeate through the membrane, and (3) desorption and evaporation of the permeate at the downstream side of the membrane (Ezeji et al., 2010). Thus for a membrane to be efficient, it has to present high diffusivity, as well as the product selectivity to be removed (Zheng et al., 2009). It has been observed that the membrane thickness affects the flux and the selectivity (Qureshi et al., 1999) as well as the porosity of the membrane (Peters et al., 2006). Membranes based on silicates have shown high selectivity for all alcohols, and this selectivity increases as the length of the chain increase from C1 to C5 (Ezeji et al., 2010; Huang, 2001).

In recent years, a two-stage hybrid separation process has been developed in order to achieve a high ABE concentration with fewer energy requirements (Wen et al., 2018). However, there is a disadvantage of this process compared with the gas stripping pervaporation process, due to relatively low efficiency in the second-stage separation. Nevertheless, the two-stage hybrid separation process is a promising alternative in terms of easy operation, energy-saving, and high efficiency.

7.5.3 DISTILLATION

Distillation is one of the most technologically developed separation processes in which separation occurs to the difference of the boiling points of the separated components. The mixture of substances to be separated are relatively volatile so that the composition of the vapors released will be different from the content of solvents in the boiling liquid. Despite progress in the ABE fermentation, the extraction process is the bottleneck for biobutanol production at an industrial scale (Kraemer et al., 2011). It is challenging to separate butanol from water acetone–butanol–ethanol mixture using simple distillation methods (Jee and Lee, 2014) since azeotropic mixture can be formed during the process. The use of distillation for butanol recovery is considered too demanding in terms of energy requirements, due to the large water content in the broth, since the process must evaporate butanol from water the in the distillation column (Bîldea et al., 2016), which it has a boiling point higher than water, using up to 220% of the energy content of butanol. However,

replacing two distillation columns with a DWC unit leads to about 15% energy saving, whereas the heat integration manages to add another 10% saving, leading to an overall total of 25% energy savings (Patraşcu et al., 2017).

In order to improve butanol separation techniques, it will be necessary to carry out more research to develop integrated processes that can allow combining one or more techniques to enhance butanol selectivity and lower costs.

7.6 CONCLUSION

Successful utilization of agroindustrial and other lignocellulosic biomass to produce butanol is a challenging task, in which pretreatment and hydrolysis of lignocelluloses and strain selection play crucial roles. Understanding the genomics and physiology of clostridial strains and integrating consolidated bioprocessing can result in efficient biomass hydrolysis and ABE fermentation in a single step. Therefore optimization and improvement of existing genetic engineering tools, life-cycle assessment, and more studies focused on economic feasibility analysis to identify promising feedstock technology can lead to a sustainable butanol production that can be cost-effective in the nearby future.

KEYWORDS

- **butanol**
- **biofuel**
- **fermentation**
- **lignocellulosic biomass**
- **consolidated bioprocessing**

REFERENCES

Abdehagh, N.; Tezel, F.H.; Thibault, J. Separation techniques in butanol production: Challenges and developments. *Biomass Bioenergy.* **2014**, 60, 222–246. doi:10.1016/j.biombioe.2013.10.003

Adeniyi, O.M.; Azimov, U.; Burluka, A. Algae biofuel: Current status and future applications. *Renew. Sustain. Energy Rev.* **2018**, 90, 316–335. doi:10.1016/j.rser.2018.03.067

Amiri, H.; Karimi, K. Pretreatment and hydrolysis of lignocellulosic wastes for butanol production: challenges and perspectives. *Bioresour. Technol.* **2018**, 270, 702–721. doi:10.1016/j.biortech.2018.08.117

Bao, T.; Zhao, J.; Li, J.; Liu, X.; Yang, S.T. n-Butanol and ethanol production from cellulose by *Clostridium cellulovorans* overexpressing heterologous aldehyde/alcohol dehydrogenases. *Bioresour. Technol.* **2019**, 285, 121316. doi.org/10.1016/j.biortech.2019.121316

Bîldea, C.S.; Patraşcu, I.; Segovia Hernandez, J.G.; Kiss, A.A. Enhanced down-stream processing of biobutanol in the ABE fermentation process. *Comput. Aided Chem. Eng.* **2016**, 38, 979–984. doi:10.1016/b978-0-444-63428-3.50168-5

Biswas, R.; Zheng, T.; Olson, D.G.; Lynd, L.R.; Guss, A.M. Elimination of hydrogenase active site assembly blocks H_2 production and increases ethanol yield in *Clostridium thermocellum*. *Biotechnol. Biofuels*. **2015**, 8, 1–8. doi.org/10.1186/s13068-015-0204-4.

Buakhiaw, B.; Sanguanchaipaiwong, V. Effect of media on acetone-butanol-ethanol fermentation by isolated *Clostridium* spp. *Energy Proc.* **2017**, 138, 864–869. doi:10.1016/j.egypro.2017.10.104.

Chapman, P.D.; Oliveira, T.; Livingston, A.G.; Li, K. Membranes for the dehydration of solvents by pervaporation. *J. Membr. Sci.* **2008**, 318, 5–37. doi:10.1016/j.memsci.2008.02.061

Charubin, K.; Bennett, R.K.; Fast, A.G.; Papoutsakis, E.T. Engineering *Clostridium* organisms as microbial cell-factories: challenges & opportunities. *Metab. Eng.* **2018**, 50, 173–191. doi:10.1016/j.ymben.2018.07.012

Dalle Ave, G.; Adams, T.A. Techno-economic comparison of acetone-butanol-ethanol fermentation using various extractants. *Energy Convers. Manage.* **2018**, 156, 288–300. doi:10.1016/j.enconman.2017.11.020

Diaz, A.B.; Blandino, A.; Caro, I. Value added products from fermentation of sugars derived from agro-food residues. *Trends Food Sci. Technol.* **2018**, 71, 52–64. doi:10.1016/j.tifs.2017.10.016

Ding, J.; Xu, M.; Xie, F.; Chen, C.; Shi, Z. Efficient butanol production using cornstarch and waste *Pichia pastoris* semi-solid mixture as the substrate. *Biochem. Eng. J.* **2019**, 143, 41–47. doi:10.1016/j.bej.2018.12.017

dos Santos Vieira, C.F.; Maugeri Filho, F.; Maciel Filho, R.; Pinto Mariano, A. Acetone-free biobutanol production: past and recent advances in the isopropanol-butanol-ethanol (IBE) fermentation. *Bioresour. Technol.* **2019**, 121425. doi.org/10.1016/j.biortech.2019.121425

Dumitrescu, A.M.; Banu, I.; Bumbac, G. Process modeling and simulation for butanol removing from fermentation broth by extraction with biodiesel. *Renew. Energy.* **2018**, 131, 137–143. doi:10.1016/j. renene.2018.07.040

Ezeji, T.C.; Karcher, P.M.; Qureshi, N.; Blaschek, H.P. Improving performance of a gas stripping-based recovery system to remove butanol from *Clostridium beijerinckii* fermentation. *Bioprocess Biosyst. Eng.* **2005**, 27, 207–214. doi:0.1007/s00449-005-0403-7

Ezeji, T.C.; Milne, C.; Price, N.D.; Blaschek, H.P. Achievements and perspectives to overcome the poor solvent resistance in acetone and butanol-producing microorganisms. *Appl. Microbiol. Biotech.* **2010**, 85, 1797–1712. doi:10.1007/s00253-009-2390-0

Feng, Y.; Zhao, Y.; Guo, Y.; Liu, S. Microbial transcript and metabolome analysis uncover discrepant metabolic pathways in autotrophic and mixotrophic anammox consortia. *Water Res.* **2018**, 128, 402–411. doi.org/10.1016/j.watres.2017.10.069

Fontalvo, J.; Cuellar, P.; Timmer, J.M.K.; Vorstman, M.A.G.; Wijers, J.G.; Keurentjes, J.T.F. Comparing pervaporation and vapor permeation hybrid distillation processes. *Ind. Eng. Chem.* **2005**, 44, 5259–5266. doi:10.1021/ie049225z

Gaida, S.M.; Liedtke, A.; Jentges, A.H.W.; Engels, B.; Jennewein, S. Metabolic engineering of *Clostridium cellulolyticum* for the production of n-butanol from crystalline cellulose. *Microb. Cell. Fact.* **2016**, 15, 1–11. doi.org/10.1186/s12934-015-0406-2

Gao, K.; Orr, V.; Rehmann, L. Butanol fermentation from microalgae-derived carbohydrates after ionic liquid extraction. *Bioresour. Technol.* **2016**, 206, 77–85. doi:10.1016/j.biortech.2016.01.036

Gottumukkala, L.D.; Haigh, K.; Görgens, J. Trends and advances in conversion of lignocellulosic biomass to biobutanol: microbes, bioprocesses and industrial viability. *Renew. Sustain. Energy Rev.* **2017**, 76, 963–973. doi.org/10.1016/j.rser.2017.03.030

Gusakov, A.V. Alternatives to *Trichoderma reesei* in biofuel production. *Trends*

Biotechnol. **2011**, 29, 419–425. doi. org/10.1016/j.tibtech.2011.04.004

Harris, L.M.; Desai, R.P.; Welker, N.E.; Papoutsakis, E.T. Characterization of recombinant strains of the *Clostridium acetobutylicum* butyrate kinase inactivation mutant: need for new phenomenological models for solventogenesis and butanol inhibition. *Biotechnol. Bioeng.* **2000**, 67, 1–11. doi.org/10.1002/(SICI)1097-0290(20000105)67:1<1::AID-BIT1>3.0.CO;2-G

Hasunuma, T.; Okazaki, F.; Okai, N. Hara, K.Y.; Ishii, J.; Kondo, A. A review of enzymes and microbes for lignocellulosic biorefinery and the possibility of their application to consolidated bioprocessing technology. *Bioresour. Technol.* **2013**, 135, 513–522. doi.org/10.1016/j. biortech.2012.10.047

Huang, J. Pervaporative recovery of n-butanol from aqueous solutions and ABE fermentation broth using thin-film silicalite-filled silicone composite membranes. *J. Membr. Sci.* **2001**, 192, 231–242. doi:10.1016/s0376-7388(01)00507-5

Ibrahim, M.F.; Ramli, N.; Bahrin, E.K.; Abd-Aziz, S. Cellulosic biobutanol by *Clostridia*: challenges and improvements. *Renew. Sust. Energy* Rev. **2017**, 79, 1241–1254. doi:10.1016/j.rser.2017.05.184

Jang, M.O.; Choi, G. Techno-economic analysis of butanol production from lignocellulosic biomass by concentrated acid pretreatment and hydrolysis plus continuous fermentation. *Biochem. Eng. J.* **2018**, 134, 30–43. doi:10.1016/j.bej.2018.03.002

Jang, Y.-S.; Lee, J.Y.; Lee, J.; Park, J.H.; Im, J.A.; Eom, M.-H.; Lee, J.; Lee, S.-H.; Song, H.; Cho, J.-H.; Seung, D.Y.; Lee, S.Y. Enhanced butanol production obtained by reinforcing the direct butanol-forming route in *Clostridium acetobutylicum*. *MBio.* **2012**, 3, 1–9. doi.org/10.1128/mbio.00314-12

Jee, K.Y.; Lee, Y.T. Preparation and characterization of siloxane composite membranes for n-butanol concentration from ABE solution by pervaporation. *J. Membr. Sci.* **2014**, 456, 1–10. doi:10.1016/j.memsci.2013.12.061

Ji, Y.; Mao, G.; Wang, Y.; Bartlam, M. Crystallization and preliminary X-ray characterization of an NAD(P)-dependent butanol dehydrogenase A from *Geobacillus thermodenitrificans* NG80-2. *Acta Crystallogr. F Struct. Biol. Cryst. Commun.* **2013**, 69, 184–187. doi.org/10.1107/S1744309113000766

Jiang, Y.; Guo, D.; Lu, J.; Dürre, P.; Dong, W.; Yan, W.; Zhang, W.; Ma, J.; Jiang, M.; Xin, F. Consolidated bioprocessing of butanol production from xylan by a thermophilic and butanologenic *Thermoanaerobacterium* sp. M5. *Biotechnol. Biofuels.* **2018a**, 11, 1–14. doi.org/10.1186/s13068-018-1092-1

Jiang, Y.; Liu, J.; Dong, W.; Zhang, W.; Fang, Y.; Ma, J.; Jiang, M.; Xin, F. The draft genome sequence of thermophilic *Thermoanaerobacterium thermosaccharolyticum* M5 capable of directly producing butanol from hemicellulose. *Curr. Microbiol.* **2018b**, 75, 620–623. doi.org/10.1007/s00284-017-1425-5

Jiang, Y.; Zhang, T.; Lu, J.; Dürre, P.; Zhang, W.; Dong, W.; Zhou, J.; Jiang, M.; Xin, F. Microbial co-culturing systems: butanol production from organic wastes through consolidated bioprocessing. *Appl. Microbiol. Biotechnol.* **2018**, 102, 5419–5425. doi.org/10.1007/s00253-018-8970-0

Jiang, L.L.; Zhou, J.J.; Quan, C.S.; Xiu, Z.L. Advances in industrial microbiome based on microbial consortium for biorefinery. *Bioresour. Bioprocess.* **2017**, 4, 1–10. doi.org/10.1186/s40643-017-0141-0

Kiyoshi, K.; Furukawa, M.; Seyama, T.; Kadokura, T.; Nakazato, A.; Nakayama, S. Butanol production from alkali-pretreated rice straw by co-culture of *Clostridium thermocellum* and *Clostridium saccharoperbutylacetonicum*. *Bioresour. Technol.* **2015**, 186, 325–328. doi.org/10.1016/j.biortech.2015.03.061

Kraemer, K.; Harwardt, A.; Bronneberg, R.; Marquardt, W. Separation of butanol from acetone–butanol–ethanol fermentation by a hybrid extraction–distillation process. *Comput. Chem. Eng.* **2011**, 35, 949–963. doi:10.1016/j.compchemeng.2011.01.028

Krasňan, V.; Plž, M.; Kras, V.; Marr, A.C.; Markošová, K.; Rosenberg, M.; Rebroš, M. Intensified crude glycerol conversion to butanol by immobilized *Clostridium pasteurianum*. *Biochem. Eng. J.* **2018**, 134, 114–119. doi:10.1016/j.bej.2018.03.005

Krivoruchko, A.; Serrano-Amatriain, C.; Chen, Y.; Siewers, V.; Nielsen, J. Improving biobutanol production in engineered *Saccharomyces cerevisiae* by manipulation of acetyl-CoA metabolism. *J. Ind. Microbiol. Biotechnol.* **2013**, 40, 1051–1056. doi:10.1007/s10295-013-1296-0

Kylilis, N.; Tuza, Z.A.; Stan, G.B.; Polizzi, K.M. Tools for engineering coordinated system behaviour in synthetic microbial consortia. *Nat. Commun.* **2018**, 9, 2677. doi.org/10.1038/s41467-018-05046-2

Li, T.; He, J. Simultaneous saccharification and fermentation of hemicellulose to butanol by a non-sporulating *Clostridium* species. *Bioresour. Technol.* **2016**, 219, 430–438. doi.org/10.1016/j.biortech.2016.07.138

Li, Y.; Tang, W.; Chen, Y.; Liu, J.; Lee, C.F. Potential of acetone-butanol-ethanol (ABE) as a biofuel. *Fuel.* **2019**, 242, 673–686. doi:10.1016/j.fuel.2019.01.063

Liao, Y.C.; Lu, K.M.; Li, S.Y. Process parameters for operating 1-butanol gas stripping in a fermentor. *J. Biosci. Bioeng.* **2014**, 118, 558–564. doi:10.1016/j.jbiosc.2014.04.020

Lin, P.P.; Mi, L.; Morioka, A.H.; Yoshino, K.M.; Konishi, S.; Xu, S.C.; Papanek, B.A.; Riley, L.A.; Guss, A.M.; Liao, J.C. Consolidated bioprocessing of cellulose to isobutanol using *Clostridium thermocellum*. *Metab. Eng.* **2015**, 31, 44–52. doi.org/10.1016/j.ymben.2015.07.001

Linger, J.G.; Darzins, A. Consolidated bioprocessing. In: Lee J. (Ed.) *Advanced*

Biofuels and Bioproducts. Springer, New York, NY. **2012**, 267–280. doi.org/10.1007/978-1-4614-3348-4

Liu, X.; Li, X.B.; Jiang, J.; Liu, Z.N.; Qiao, B.; Li, F.F.; Cheng, J.S.; Sun, X.; Yuan, Y.J.; Qiao, J.; Zhao, G.R. Convergent engineering of syntrophic *Escherichia coli* coculture for efficient production of glycosides. *Metab. Eng.* **2018**, 47, 243–253.

Liu, J.; Zhou, W.; Fan, S.; Qiu, B.; Wang, Y.; Xiao, Z.; Tang, X.; Wang, W. Coproduction of hydrogen and butanol by *Clostridium acetobutylicum* with the biofilm immobilized on porous particulate carriers. *Int. J. Hydrogen Energy*. **2019**, 44, 11617–11624. doi:10.1016/j.ijhydene.2019.03.099 doi.org/10.1016/j.ymben.2018.03.016

Mai, S.; Wang, G.; Wu, P.; Gu, C.; Liu, H.; Zhang, J.; Wang, G. Interactions between *Bacillus cereus* CGMCC 1.895 and *Clostridium beijerinckii* NCIMB 8052 in co-culture for butanol production under non-anaerobic conditions. *Biotechnol. Appl. Biochem.* **2017**, 64, 719–726.doi. org/10.1002/bab.1522

Majidian, P.; Tabatabaei, M.; Zeinolabedini, M.; Naghshbandi, M.P.; Chisti, Y. Metabolic engineering of microorganisms for biofuel production. *Renew. Sustain. Energy Rev.* **2018**, 82, 3863–3885. doi. org/10.1016/j.rser.2017.10.085

Mazzoli, R. Development of Microorganisms for cellulose-biofuel consolidated bioprocessings: metabolic engineers' tricks. *Comput. Struct. Biotechnol. J.* **2012**, 3, e201210007.doi.org/10.5936/ csbj.201210007

Mbaneme, V.; Chinn, S.S.M. Consolidated bioprocessing for biofuel production: recent advances. *Energy Emiss. Control Technol.* **2015**, 3, 23–44. doi.org/10.2147/EECT. S63000

Nanda, S.; Golemi-Kotra, D.; McDermott, J.C., Dalai, A.K.; Gökalp, I.; Kozinski, J.A. Fermentative production of butanol: Perspectives on synthetic biology. *N. Biotechnol.* **2017**, 37, 210–221. doi:10.1016/j.nbt.2017.02.006

Ndaba, B.; Chiyanzu, I.; Marx, S. N-Butanol derived from biochemical and chemical routes: a review. *Biotechnol. Rep.* **2015**, 8, 1–9. doi:10.1016/j.btre.2015.08.001

Niglio, S.; Marzocchella, A.; Rehmann, L. Clostridial conversion of corn syrup to acetone-butanol- ethanol (ABE) via batch and fed-batch fermentation. *Heliyon*. **2019**, e01401. doi:10.1016/j.heliyon.2019. e01401

Nölling, J.; Breton, G.; Omelchenko, M.V.; Makarova, K.S.; Zeng, Q.; Gibson, R.; Lee, H.M.; Dubois, J.; Qiu, D.; Hitti, J.; Wolf, Y.I.; Tatusov, R.L.; Sabathe, F.; Doucette-Stamm, L.; Soucaille, P.; Daly, M.J.; Bennett, G.N.; Koonin, E.V.; Smith, D.R. Genome sequence and comparative analysis of the solvent-producing bacterium *Clostridium acetobutylicum*. *J. Bacteriol.* **2001**, 183, 4823–4838. doi. org/10.1128/JB.183.16.4823-4838.2001

Oliva-Rodríguez, A.G.; Quintero, J.; Medina-Morales, M.A.; Morales-Martínez, T.K.; Rodríguez-De la Garza, J.A.; Moreno-Dávila, M.; Aroca, G.; Ríos-González, L.J. *Clostridium* strain selection for co-culture with *Bacillus subtilis* for butanol production from agave hydrolysates. *Bioresour. Technol.* **2019**, 275, 410–451. doi:10.1016/j.biortech.2018.12.085

Olson, D.G.; McBride, J.E.; Joe Shaw, A.; Lynd, L.R. Recent progress in consolidated bioprocessing. *Curr. Opin. Biotechnol.* **2012**, 23, 396–405. doi.org/10.1016/j.copbio.2011.11.026

Ou, J.; Xu, N.; Ernst, P.; Ma, C.; Bush, M.; Goh, K.Y.; Zhao, J.; Zhou, L.; Yang, S.T.; Liu, X. Process engineering of cellulosic *n*-butanol production from corn-based biomass using *Clostridium cellulovorans*. *Process Biochem.* **2017**, 62, 144–150. doi. org/10.1016/j.procbio.2017.07.009

Parisutham, V.; Chandran, S.P.; Mukhopadhyay, A.; Lee, S.K.; Keasling, J.D. Intracellular cellobiose metabolism and its applications in lignocellulose-based biorefineries. *Bioresour. Technol.* **2017**, 239, 496–506. doi:10.1016/j.biortech.2017.05.001

Parisutham, V.; Kim, T.H.; Lee, S.K. Feasibilities of consolidated bioprocessing microbes: from pretreatment to biofuel production. *Bioresour. Technol.* **2014**, 161, 431–440. doi.org/10.1016/j.biortech.2014.03.114

Park, C.H.; Okos, M.R.; Wankat, P.C. Acetone-butanol-ethanol (ABE) fermentation and simultaneous separation in a trickle bed reactor. *Biotechnol. Prog.* **1991**, 7, 185–194. doi:10.1021/bp00008a014

Patakova, P.; Kolek, J.; Sedlar, K.; Koscova, P.; Branska, B.; Kupkova, K.; Paulova, L.; Provaznik, I. Comparative analysis of high butanol tolerance and production in Clostridia. *Biotechnol. Adv.* **2018**, 36, 721–738. doi:10.1016/j.biotechadv.2017.12.004

Patraşcu, I.; Bîldea, C.S.; Kiss, A.A. Eco-efficient butanol separation in the ABE fermentation process. *Sep. Purif. Technol.* **2017**, 177, 49–61. doi:10.1016/j.seppur.2016.12.008

Peters, T.A.; Poeth, C.H.S.; Benes, N.E.; Bujis, H.C.W.M.; Vercauteren, F.F.; Keurentjes, J.T.F. Ceramic-supported thin PVA pervaporation membranes combining high flux and high selectivity; contradicting the flux-selectivity paradigm. *J. Membr. Sci.* **2006**, 276, 42–50. doi:10.1016/j.memsci.2005.06.066

Qian, Y.; Zhong, L.; Gao, J.; Sun, N.; Wang, Y.; Sun, G.; Qu, Y.; Zhong, Y. Production of highly efficient cellulase mixtures by genetically exploiting the potentials of *Trichoderma reesei* endogenous cellulases for hydrolysis of corncob residues. *Microb. Cell Fact.* **2017**, 16, 1–16. doi.org/10.1186/s12934-017-0825-3

Qin, Z.; Duns, G.J.; Pan, T.; Xin, F. Consolidated processing of biobutanol production from food wastes by solventogenic *Clostridium* sp. strain HN4. *Bioresour. Technol.* **2018**, 264, 148–153. doi.org/10.1016/j.biortech.2018.05.076

Qureshi, N.; Blaschek, H.P. Production of acetone, butanol and ethanol (ABE) by a hyper-producing mutant strain of *Clostridium beijerinckii* BA101 and recovery by pervaporation. *Biotechnol. Prog.* **1999**, 15, 594–602. doi: 10.1021/bp990080e

Qureshi, N.; Ezeji, T.C. Butanol, "a superior biofuel" production from agricultural residues (renewable biomass): recent progress in technology. *Biofuel. Bioprod. Biorefin.* **2008**, 2, 319–330. doi:10.1002/bbb.85

Rahnama, N.; Foo, H.L.; Aini, N.; Rahman, A.; Ariff, A. Saccharification of rice straw by cellulase from a local *Trichoderma harzianum* SNRS3 for biobutanol production. *BMC Biotechnol.* **2014**, 14, 103. doi: 10.1186/s12896-014-0103-y

Roberto, W.; Gonçalves, R. Review on the characteristics of butanol, its production and use as fuel in internal combustion engines. *Renew. Sustain. Energy Rev.* **2017**, 69, 642–651. doi:10.1016/j.rser.2016.11.213

Sarangi, P.K.; Nanda, S. Recent developments and challenges of acetone-butanol-ethanol fermentation. In: Sarangi, P.K., Nanda, S., Mohanty, P. (Eds.) Recent Advancements in Biofuels and Bioenergy Utilization. 11– Springer Nature: Singapore. **2018**, pp. 111–123. doi.org/ 10.1007/978-981-13-1307-3_5

Sarchami, T.; Munch, G.; Johnson, E.; Kießlich, S.; Rehmann, L. A review of process-design challenges for industrial fermentation of butanol from crude glycerol by non-biphasic *Clostridium pasteurianum*. *Fermentation*. **2016**, 2, 13. doi:10.3390/fermentation2020013

Sauer, M. Industrial production of acetone and butanol by fermentation—100 years later. *FEMS Microbiol. Lett.* **2016**, 363, 1525–1534. doi:10.1093/femsle/fnw134

Seifollahi, M.; Amiri, H. Enzymatic post-hydrolysis of water-soluble cellulose oligomers released by chemical hydrolysis for cellulosic butanol production. *Cellulose*. **2019**, 26, 4479–4494. doi:10.1007/s10570-019-02397-x

Sgobba, E.; Stumpf, A.K.; Vortmann, M.; Jagmann, N.; Krehenbrink, M.; Dirks-Hofmeister, M.E.; Moerschbacher, B.; Philipp, B.; Wendisch, V.F. Synthetic *Escherichia coli-Corynebacterium glutamicum* consortia for L-lysine production from starch and sucrose. *Bioresour. Technol.* **2018**, 260, 302–310. doi.org/10.1016/j.biortech.2018.03.113

Shahab, R.L.; Luterbacher, J.S.; Brethauer, S.; Studer, M.H. Consolidated bioprocessing of lignocellulosic biomass to lactic acid by a synthetic fungal-bacterial consortium. *Biotechnol. Bioeng.* **2018**, 115, 1207–1215. doi.org/10.1002/bit.26541

Shanmugam, S.; Sun, C.; Chen, Z.; Wu, Y.R. Enhanced bioconversion of hemicellulosic biomass by microbial consortium for biobutanol production with bioaugmentation strategy. *Bioresour. Technol.* **2019**, 279, 149–155. doi.org/10.1016/j.biortech.2019.01.121

Singh, N.; Mathur, A.S.; Gupta, R.P.; Barrow, C.J.; Tuli, D.; Puri, M. Enhanced cellulosic ethanol production via consolidated bioprocessing by *Clostridium thermocellum* ATCC 31924. *Bioresour. Technol.* **2018**, 250, 860–867. doi.org/10.1016/j.biortech.2017.11.048

Ujor, V.; Bharathidasan, A.K.; Cornish, K.; Ezeji, T.C. Feasibility of producing butanol from industrial starchy food wastes. *Appl. Energy.* **2014**, 136, 590–598. doi:10.1016/j.apenergy.2014.09.040

Vivek, N.; Nair, L.M.; Mohan, B.; Nair, S.C.; Sindhu, R.; Pandey, A.; Shurpali, N.; Binod, P. Bio-butanol production from rice straw—recent trends, possibilities, and challenges. *Bioresour. Technol. Rep.* **2019**, 100224. doi.org/10.1016/j.biteb.2019.100224

Volk, H.; George, S.C. Organic Geochemistry Using petroleum inclusions to trace petroleum systems – a review. *Org. Geochem.* **2019**, 129, 99–123. doi:10.1016/j.orggeochem.2019.01.012

Wang, J.; Lu, X.; Ying, H.; Ma, W.; Xu, S.; Wang, X.; Chen, K.; Ouyang, P. A novel process for cadaverine bio-production using a consortium of two engineered *Escherichia coli. Front. Microbiol.* **2018**, 9, 1–12. doi.org/10.3389/fmicb.2018.01312

Wen, H.; Gao, H.; Zhang, T.; Gong, P.; Li Zhuangzhuang, Chen, H.; Cai, D.; Qin, P.; Tan, T. Hybrid pervaporation and salting-out for effective acetone-butanol-ethanol separation from fermentation broth. *Bioresource Technol. Rep.* **2018**, 2, 45–52. doi: 10.1016/j.biteb.2018.04.005

Wen, Z.; Ledesma-Amaro, R.; Lin, J.; Jiang, Y.; Yang, S. Improved n-butanol production from *Clostridium cellulovorans* by integrated metabolic and evolutionary engineering. *Appl. Environ. Microbiol.* **2019**, 85, 2560–2618. doi.org/10.1128/AEM.02560-18

Wen, Z.; Minton, N.P.; Zhang, Y.; Li, Q.; Liu, J.; Jiang, Y.; Yang, S. Enhanced solvent production by metabolic engineering of a twin-clostridial consortium. *Metab. Eng.* **2017**, 39, 38-48. doi.org/10.1016/j.ymben.2016.10.013

Wen, Z.; Wu, M.; Lin, Y.; Yang, L.; Lin, J.; Cen, P. Artificial symbiosis for acetone-butanol-ethanol (ABE) fermentation from alkali extracted deshelled corn cobs by co-culture of *Clostridium beijerinckii* and *Clostridium cellulovorans. Microb. Cell Fact.* **2014**, 13, 92. doi:10.1186/s12934-014-0092-5

Xin, F.; Dong, W.; Zhang, W.; Ma, J.; Jiang, M. Biobutanol production from crystalline cellulose through consolidated bioprocessing. *Trends Biotechnol.* **2018**, 37, 167–180. doi.org/10.1016/j.tibtech.2018.08.007

Xue, C.; Du, G.Q.; Sun, J.X.; Chen, L.J.; Gao, S.S.; Yu, M.L.; Yang, S.T.; Bai, F.W. Characterization of gas stripping and its integration with acetone-butanol-ethanol fermentation for high-efficient butanol production and recovery. *Biochem. Eng. J.* **2014**, 83, 55–61. doi:10.1016/j.bej.2013.12.003

Yan, Q.; Fong, S.S. Challenges and advances for genetic engineering of non-model bacteria and uses in consolidated biopro-cessing. *Front. Microbiol.* **2017**, 8, 1–16. doi.org/10.3389/fmicb.2017.02060

Yan, Y.; Li, X.; Wang, G.; Gui, X.; Li, G.; Su, F.; Wang, X.; Liu, T. Biotechnological preparation of biodiesel and its high-valued derivatives: a review. *Appl. Energy.* **2014**, 113, 1614–1631. doi:10.1016/j.apenergy.2013.09.029

Yang, M.; Kuittinen, S.; Vepsäläinen, J.; Zhang, J.; Pappinen, A. Enhanced acetone-butanol-ethanol production from ligno-cellulosic hydrolysates by using starchy slurry as supplement. *Bioresour. Technol.* **2017**, 243, 126–134. doi:10.1016/j.biortech.2017.06.021

Yu, X.; Guo, Z.; He, L.; Dong, W.; Sun, P.; Shi, W.; Du, Y. Effect of gasoline/n-butanol blends on gaseous and particle emis-sions from an SI direct injection engine. *Fuel.* **2018**, 229, 1–10. doi:10.1016/j.fuel.2018.05.003

Yu, M.; Zhang, Y.; Tang, I.C.; Yang, S.T. Metabolic engineering of *Clostridium tyrobutyricum* for n-butanol production. *Metab. Eng.* **2011**, 13, 373–382. doi.org/10.1016/j.ymben.2011.04.002

Zabed, H.; Sahu, J.N.; Suely, A.; Boyce, A.N.; Faruq, G. Bioethanol production from renewable sources: Current perspec-tives and technological progress. *Renew.*

Sustain. Energy Rev. **2017**, 71, 475-501. doi.org/10.1016/j.rser.2016.12.076

Zhang, J.; Wang, P.; Wang, X.; Feng, J.; Sandhu, H.S.; Wang, Y. Enhancement of sucrose metabolism *in Clostridium saccha-roperbutylacetonicum* N1-4 through meta-bolic engineering for improved acetone–butanol–ethanol (ABE) fermentation. *Bioresour. Technol.* **2018**, 270, 430–438. doi:10.1016/j.biortech.2018.09.059

Zhang, J.; Zong, W.; Hong, W.; Zhang, Z.T.; Wang, Y. Exploiting endogenous CRISPR-Cas system for multiplex genome editing in *Clostridium tyrobutyricum* and engi-neer the strain for high-level butanol production. *Metab. Eng.* **2018**, 47, 49-59. doi.org/10.1016/j.ymben.2018.03.007

Zhao, L.; Cao, G.L.; Wang, A.J.; Ren, H.Y.; Zhang, K.; Ren, N.Q. Consolidated bioprocessing performance of *Thermo-anaerobacterium thermosaccharolyticum* M18 on fungal pretreated cornstalk for enhanced hydrogen production. *Biotechnol. Biofuels.* **2014**, 7, 1–10. doi.org/10.1186/s13068-014-0178-7

Zheng, Y.N.; Li, L.Z.; Xian, M.; Ma, Y.J.; Yang, J.M.; Xu, X.; He, D.Z. Problems with the microbial production of butanol. *J. Ind. Microbiol. Biotech.* **2009**, 36, 1127–1138. doi:10.1007/s10295-009-0609-9

CHAPTER 8

Consolidated Process for Bioenergy Production and Added Value Molecules from Microalgae

B. A. AYIL-GÚTIERREZ[1], Y. J. TAMAYO-ORDOÑEZ[2],
M. C. TAMAYO-ORDOÑEZ[3], A. V. CÓRDOVA-QUIROZ[4],
L. J. RIOS-GONZÁLEZ[5], and I. M. M. MORENO-DAVILA[5*]

[1]CONACYT—Centro de Biotecnología Genómica, Instituto Politécnico Nacional, Blvd. del Maestro, s/n, Esq. Elías Piña 88710. Reynosa, Tamaulipas, México

[2]Estancia Posdoctoral Nacional-CONACyT, Posgrado en Ciencia y Tecnología de Alimentos. Facultad de Ciencias Químicas. Universidad Autónoma de Coahuila, Ing J. Cardenas Valdez S/N, República, 25280 Saltillo, Coahuila, México

[3]Laboratorio de Ingeniería Genética, Departamento de Biotecnología. Facultad de Ciencias Químicas. Universidad Autónoma de Coahuila, Ing J. Cardenas Valdez S/N, República, 25280 Saltillo, Coahuila, México

[4]Facultad de Química, Dependencia Académica de Ciencias Química y Petrolera, Universidad Autónoma del Carmen, Carmen, México

[5]Laboratorio de Biotecnología Ambiental. Departamento de Biotecnología. Facultad de Ciencias Químicas. Universidad Autónoma de Coahuila, Ing J. Cardenas Valdez S/N, República, 25280 Saltillo, Coahuila, México

*Corresponding author. E-mail: mayela.morenodavila@uadec.edu.mx

ABSTRACT

Microalgae are a favorable source for the production of renewable bioenergy. There are extensive reviews of the use of microalgae for the production of biohydrogen, bioethanol, biofuels, variety of metabolites, and other products of commercial interest such as polysaccharides, triacylglycerols (TAGs), and pigments, among others. Nowadays there are two strategies implemented to increase the different products from microalgae. These strategies comprise culture medium optimization and genetic engineering. In this chapter, we summarize what are the best conditions in culture of microalgae and genetic engineering strategies that have been established in some genera of microalgae for the obtaining of different products.

8.1 INTRODUCTION

The search of alternative sources for producing of fuels instead of the fossil sources has gained relevance due to negative effects of fossil fuels, their delimitation (no renewables), and the rising energy demands (Dandu and Nanthagopal, 2019; Lee et al., 2018). The production of biofuels (fuels from organic sources) has been explored and divided into mainly two generations (Demirbas, 2011; Islam et al., 2010). The first generation is based on biofuel produced from food crops (e.g., sugarcane, corn, or starch) in cultivated and harvested biomass. The second generation corresponds to new pathways of energy production from sources that are not used for food supply. Furthermore, there exists a third generation that includes microalgae biomass that represents several advantages in comparison with the first and second generation (Dragone et al., 2017; Behera

et al., 2015). This third generation is mainly based on feedstocks used specifically for biofuel production and has lower land competition (Islam et al., 2010).

Microalgae fit in the third generation because they do not need an enormous amount of agricultural land as terrestrial plants, plus their high growth rate, the potential growth in many places, environmental tolerance, easy harvesting and extraction, a large diversity of species and high capacity to accumulate triacylglycerols (TAGs) or fatty acids, needed for biodiesel, have made to microalgae been a good alternative for biofuels production (Dandu and Nanthagopal, 2019; Gouvela and Oliveira, 2008; Hannon et al., 2010).

A broad range of biofuels can be obtained from microalgae biomass such as biodiesel, bioethanol, biobutanol, biogas, biohydrogen, biojet fuel, bio-oil, syngas, and biogasoline (Dandu and Nanthagopal, 2019; Behera et al., 2015; Hannon et al., 2010). All these biofuels can be obtained by different processes as biochemical conversions, which include anaerobic digestion (Biogas) and fermentation (bioethanol), thermochemical conversion, which include gasification (syngas) and liquefaction (bio-oil), and chemical conversion—transesterification, this one is specifically for biodiesel production (Behera et al., 2015; Gouvela and Oliveira, 2008; Galadima et al., 2014; Kraan, 2013).

On the other hand, obtaining algal biomass to obtain different products of commercial interest, such as alginates, carotenoids, sugars, and other metabolites of commercial interest is already a reality (Borowitzka, 2013; Suganya et al., 2016; Hu et al., 2017).

In this chapter, we describe the genetic strategies and culture optimization conditions used to obtain different products of economic interest or that are used for the generation of new sources of energy used microalgae.

8.2 ALGAL CULTURE STRATEGIES

The use of microalgae for producing biodiesel presents several advantages in comparison with plants used for the same porpoise (Brennan and Owende, 2010). The microalgae can store high lipids amounts (>72%), moreover, the lipid composition in microalgae oil is mainly formed by lauric acid, palmitic acid, myristic acid, and oleic acid, and these are desirable for biodiesel production (Chandra et al., 2019).

The microalgae growth rate is faster than plants and also they do not require large fertile grounds as much as plants. Plants used for biodiesel production reduce the portion of them destined to human food consumption, for its part with microalgae this does not happen.

The microalgae adapt themselves to diverse cultures medium, even with pretreated wastewater and grow in a wide temperatures range (from −2 °C to 50 °C) (Zhang et al., 2014; Patel et al., 2016). And finally, there exist several species of microalgae, it is estimated there are near of 1 million species of microalgae.

There are two strategies implemented to increase the amount of lipids in microalgae. The increasing of lipids in microalgae would allow a production costs reduction, in order to compete with petro diesel in economical themes. These strategies comprise culture medium optimization and genetic engineering and we describe the progress of the investigations regarding these two approaches.

8.2.1 MICROALGAE

The term "microalgae" covers a wide variety of eukaryotic microorganisms commonly called green algae (Chlorophyta), red algae (Rhodophyta), and diatoms (Bacillariophyta). Many prokaryotes, such as blue-green algae (Cyanophyceae), are included in this chapter (Brennan and Owende, 2010; Zhang et al., 2014; Patel et al., 2016). Microalgae can be found in different environments because some species thrive in freshwater and others in saline conditions or seawater; they are aquatic organisms that lack the complex cellular structures found in higher plants; most species are photoautotrophic, this means that they use solar energy to convert it into chemical forms through photosynthesis (Slade and Bauen, 2013). Each species is adapted to its environment and all have particular characteristics; for this reason, the native characteristics of the strains should be considered when designing the production and harvest process.

8.2.2 CULTIVATION OF MICROALGAE IN REACTORS FOR THE GENERATION OF BIOMASS AND APPLICATIONS

In general, four parameters can be listed as the most important for the operation of a microalgae cultivation system: the strain to be cultivated, the light source (solar or artificial), the source of nutrients in the environment, and the system temperature (Chew et al., 2018).

The cultivation of microalgae at large scales is mainly done in two ways, the most discussed in the literature are the modality in open tanks and photobioreactors that are closed systems (Brennan and Owende, 2010; Patel et al., 2016; Draaisma et al., 2013, Chiaramonti et al., 2013). Each modality can acquire different variants with different configurations depending on the purposes of the culture (Table 8.1).

8.2.3 CULTIVATION CONDITIONS FOR MICROALGAE WITH HIGH PRODUCTION OF FERMENTABLE BIOMASS

Strains of microalgae with high carbohydrate production have been reported. Among the most studied species are *Chlorella, Dunaliella, Chlamydomonas*, and *Scenedesmus* (Karatay et al., 2016).

The emphasis in the study of these strains are the advantages they have in addition to the production of fermentative carbohydrates, such as the generation of metabolites of great commercial value, in the case of *Dunaliella*, or for *Chlorella, Chlamydomonas* sp. and *Scendesmus* sp. its possible use in wastewater treatment with effective nutrient removal capacities (Cheah et al., 2016).

For example, in the case of chlorella, in particular, *Chlorella vulgaris* can accumulate a large amount of carbohydrates, from 37% to 55% of its dry biomass (Chen et al., 2016). For example, Ho et al. (2013) isolated the strain of *C. vulgaris* FSP-E, which had a biomass yield of 1,437 mg/L per day with a percentage of carbohydrates 51.3%, using a tubular photobioreactor reactor of 1 L with 2% CO_2 at an aeration rate of 0.2 vvm, illumination at an intensity of 450 µmol/m^2 s with a period of nitrogen starvation (2013). This same strain was later evaluated, performing a growth under mixotrophic conditions, adding

TABLE 8.1 Comparison Between Open System and Closed System Cultivation Models

Parameter	Open System	Closed System
Capital cost	Moderate	High
Commercial applications	Reported 8–10 000 tons of algal biomass per year.	Limited and only for high added value compounds that are used in food and cosmetics
Growth period	6–8 weeks	2–4 weeks
Harvesting cost	High, dependent on species	Low due to high biomass concentration
Light utilization	Low	High
Operating cost	Relatively cheap	Expensive
Product quality	Low	High
Production flexibility	Only few species possible	Switching possible as the growth parameter
Shear	Low due to gentle mixing	High due to turbulent flow and pumping through gas exchange
Water evaporation	High	Low

2 g/L of sodium acetate, which had a biomass productivity of 1022.3 mg/L and 498.5 mg/L of carbohydrates per day (Chen et al., 2016). Another study conducted with *C. vulgaris* showed good percentages of carbohydrates, in this case the microalga *C. vulgaris* KMMCC-9 UTEX 26, grown in basal medium cultivated during 14 days with a cycle 16–8 h light-darkness, temperatures of 20–22 °C, a light intensity condition of 150 μmol m^2/s and restriction of nitrogen or sulfur sources for 3 days, resulted in a total carbohydrate content of 22.4% of the total mass of the component (Kim et al., 2014).

C. vulgaris has not been the only one of the genus to have high concentrations of usable biomass for the production of bioethanol. *Chlorella minutissima* reached a production of 60.3% of carbohydrates using a mixotrophic growth in a channeling reactor with the addition of arabinose, calculating a theoretical ethanol production equal to 39.1 mL per 100 g of biomass (Freitas et al., 2013). On the other hand, *Chlorella variabilis* NC64A accumulated 37.8% starch from the dry weight of the cells cultured in N8 medium plus 0.05% yeast extract at 25 °C with a light/dark cycle of 16/8 h, illumination of 150 μmol m^2/s and continuous supply of 2% CO_2 in air at 0.5 vvm (Cheng et al., 2013).

Dunaliella is a microalga of high pharmaceutical and industrial value, due to its accumulation capacity of β-carotene (Morowvat and Ghasemi, 2016); however, it also has a high content of fatty acids, proteins, and carbohydrates. Pavón et al. evaluated the effect of temperatures on a *Dunaliella salina* DUS1 strain, obtaining a 25% carbohydrate content in photobioreactors at 20,000 K (2017). In another investigation, *Dunaliella tertiolecta* present a 40% (P/P) yield of reducing sugar in enzymatic saccharification using 1,4-α-D-glucanohydrolase (AMG 300L) and an ethanol production of 140 mg of ethanol/g of residual biomass by fermentation with *Saccharomyces cerevisiae* (Lee et al., 2013).

In the case of *Chlamydomonas reinhardtii* UTEX90 reached a production of 44% of starch in dry weight in a stirred photobioreactor of 2.5 L at 23 °C and 130 rpm at 450 μmol m²/s, its ethanol production was 235 mg/L of ethanol from of 1 g of algae biomass by a separate method of hydrolysis and fermentation (SHF) (Choi et al., 2010). On the other hand, *Chlamydomonas fasciata* Ettl 437 had a carbohydrate percentage of 43.5% and a theoretical ethanol yield corresponding to 79.5%, by using a stirred photobioreactor with a supply of CO_2 at an aeration rate of 4 vvm and a luminous intensity of 3000 lux (Asada et al., 2012).

For this part, the *Scenedesmus* microalgae have a percentage of carbohydrates more than 30%. Investigations with two different strains of *Scenedesmus obliquus* (SC and CNW-N) showed carbohydrate yields of 30% and 51%, respectively.

It was reached in Bristol medium in a column photobioreactor with bubbling air and a luminous intensity of 150 μmol m²/s; the second yield was obtained using a photobioreactor at 28 °C and an agitation speed of 300 rpm with a CO_2 feed of 2.5% (Miranda et al., 2012; Ho et al., 2013).

In addition to these microalgae, there are recombinant microalgae strains with good bioethanol theoretical yields. Such is the case of the cyanobacterium *Synechococus elongatus* PCC7942 that increased its photosynthetic rate and its carbohydrate content by coexpressing ictB, ecaA, and acsAB, achieving a glucose recovery of almost 90%, at a biomass concentration of 80 g/L with a theoretical yield of 91% in fermentation SHF with *Zymomonas mobilis* (Chow et al., 2015).

8.3 CHALLENGES TO FACE TO IMPROVE THE YIELDS OF BIOETHANOL IN MICROALGAE

Biofuels are a promising alternative for the replacement of fossil fuels. Among these, bioethanol stands out because it is an alternative to the use of gasoline to reduce the emission of atmospheric pollutants, the thanks goes to the presence of oxygen in its molecular form, which allows a relatively low temperature combustion with reduction of the emission of several toxic gases such as carbon monoxide (CO), nitric oxide (NO), and volatile organic compounds (Rodionova et al., 2016). However, the main challenge in the bioethanol industry is the need to discover a suitable raw material, together with a respectful approach to the environment and an economically viable production process (Rodionova et al., 2016; Jambo et al., 2016). Microalgae are a favorable source for renewable bioenergy production, although microalgae do not produce ethanol proper, they serve as a carbon source in the process. They have advantages over other sources of fermentable biomass as they are fast-growing organisms with high biomass productivity; some species accumulate a high content of carbohydrates with low levels of lignin and hemicellulose, which makes it much easier to become simple reducing sugar compared to lignocellulosic biomass, its biofuel production being 100 times greater than that of plants superiors (Rodionova et al et al., 2016; Simionato et al., 2013). In addition to avoiding competition for farmland with land crops (Bibi et al, 2016).

Although the production of bioethanol from microalgae is favorable for the bioenergy industry, the main restriction for its commercialization is the cost of production that later represents the price of the biofuel. The profitability can be maximized by the productivity of biomass (in terms of carbohydrates, proteins, lipids), this factor depends on the choice of strain and/or farming systems, which has led the efforts to search for strains with high biomass production and efficient crop system and even the use of genetic engineering, with the hope of commercial-scale production of microalgae biofuels in the near future (Rodionova et al., 2016; Chaudhary et al., 2014).

Although microalgae with a good production of fermentable biomass can improve the profitability of the bioethanol production process, stages that slow down the process on a commercial scale are the fermentative stages (hydrolysis and fermentation) of the biomass. In the case of hydrolysis, high yields have been obtained in the conversion of biomass to low reagent concentrations (acid and enzymatic hydrolysis) (Chaudhary et al., 2014).

In the case of enzymatic hydrolysis, it has been pointed out that although microalgae and cyanobacteria are diverse in terms of their cellular structure and that each species requires a specific enzyme to be effectively saccharified. In general, the use of amylase contributes significantly to the hydrolysis of cultivated microalgae under conditions of nutrient limitation (Môllers et al., 2014). On the other hand, acid hydrolysis (sulfuric, nitric or hydrochloric acid) at temperatures between 120 and 140 °C for 15–30 min results in percentages higher than 80% saccharification and fermentation (SSF). However, the use of high concentrations of catalyst (acids and alkalis) can inhibit the fermentation stage due to salt formation after liquor neutralization (Markou et al., 2013). It is important to mention that chemical hydrolysis methods can also facilitate solvent extraction of lipids, thus serving to recover both fermentable sugars and lipids in the same strain (Lorente et al., 2015).

The simple sugars released during hydrolysis can be easily converted to bioethanol with the help of some microorganisms. The most commonly used for ethanolic fermentation are saccharomyces yeasts or bacteria of the genus *Zymomonas* (Rodionova et al., 2016).

In advanced bioethanol production, the assimilation of new fermentation technologies has been progressing annually. Instead of a simple fermentation process, more viable steps are being taken to increase the production speed in an economically feasible way, such as the process of simultaneous SSF where the enzymes and the culture put together, so the sugars released by the hydrolysis are quickly converted to bioethanol, thus avoiding the inhibition produced in the conventional processing of hydrolysis and two-stage fermentation (SHF) (Chaudhary et al., 2014).

Although the production of bioethanol from microalgae is a promising process for the replacement of fossil fuels, several research workers agree that the greatest bottleneck for the industrial implementation of bioethanol from microalgae could fall on the fermentative processes (hydrolysis and fermentation); therefore, it is necessary to carry out detailed studies to optimize these processes toward a competitive production with respect to fossil fuels (Rodionova et al., 2016; Chaudhary et al., 2014).

8.4 ADVANCES IN LIPID YIELDS IN ALGAE

8.4.1 CULTURE CONDITIONS TO IMPROVE THE PRODUCTION OF FATTY ACIDS IN MICROALGAE

The lipid composition of microalgae can be changed by environmental stress, it is known what factors such as nutrients, light, and culture temperature affect both the growth of algae and their biomass composition (Cheah et al., 2016; Cheng et al., 2013).

Sulfur in limiting quantity inhibits cell division, whereas low concentration inhibits photosynthetic assimilation of carbon-rich compounds, such as polysaccharides (starch) (Markou et al., 2013), which would allow obtaining less biomass, for the extraction of lipids in algae. It has been described that sulfide depletion causes a decrease in the activity of photosystem II; therefore, antenna complexes (LHCII) are transferred to photosystem I (PSI). Then, there is a decrease in O_2, as a result of the decrease in PSII activity. So the residual activity of PSII is used for oxidative degradation of organic substrates such as starch (Hemschemeir et al., 2009; Zhang et al., 2002; Melis et al., 2000; WyKoff et al., 1998).

Nitrogen depletion is a strategy widely studied in a variety of microalgae, this can lead to a sharp increase in lipid or carbohydrate content, since this forces them to transform proteins or peptides into lipids or carbohydrates (Vitova et al., 2015), on the other hand, the CO_2 and the light intensity help to a greater efficiency in the absorption of carbon for the conversion of carbohydrates, under nitrogen-limiting conditions (Chen et al., 2016). Below we describe some research conducted in this regard.

In some studies, it has been documented that the cultivation of microalgae such as *C. reinhardtii*, *S. obliquus*, and *C. vulgaris* in a medium lacking nitrogen (N) and phosphorus (P) acts as an inducing medium resulting in a higher lipid concentration (Sharma et al., 2012). There are three types of culture: autotrophic, mixotrophic, and heterotrophic, the last two are the most popular for microalgae growing. The mixotrophic culture is characterized by the use of photosynthesis and

other micronutrients in the culture for feeding. According to Yeh and Chang, this culture favors the maximum concentration of lipids in *C. vulgaris* (from 67 to 144 mg/Ld) (2012). The heterotrophic culture depends only on the micronutrients in the culture because the light flow is minimal or nonexistent, therefore it is necessary to supplement this absence with nutrients such as carbon or glucose. Species such as *Chlorella saccharophila* in this culture has demonstrated increasing of its lipid content of 54% (Isleten-Hosoglu et al., 2012). The micronutrients in algal culture are, mainly, potassium, iron, manganese, zinc, and copper, also some cultures have vitamins such as biotin (Christenson et al., 2011).

Also the intensity and quality of light affect the composition of fatty acids. Khoeyi et al. demonstrated in *C. vulgaris* that the maximum percentage of saturated fatty acids (33.38%) was obtained at a light intensity of 100 μmol m²/s in 16:8 h (light:dark) photoperiods, whereas the monounsaturated acids and polyunsaturated acids reach their maximum percentage (15.93% and 27.40%, respectively) at 37.5 μmol m²/s and photoperiods in 8:16 h, that is, they decreased with increasing irradiance (2012). Yeesang and Cheirsilp in four species of *Botryococcus* showed that with a moderately high light flux (82.5 μE/m² s) and a high level of iron (0.74 mM), lipid accumulation was improved to 35.9%. In *Scenedesmus sp.* it was reported that this microalga had the highest lipid content (41.1%) and neutral lipid contents (32.9%) when it grew under a light intensity of 400 μmol m²/s (2011). Most of the fatty acids found were oleic acid (43%–52%), palmitic acid (24%–27%) and linoleic acid (7%–11%) (Liu et al., 2012). In addition to these reports in Table 8.2, we summarize advances in research where we described the major culture conditions for increases lipids production in microalgae species.

8.4.2 GENETIC ENGINEERING IN ALGAE TO INCREASE LIPID YIELDS

8.4.2.1 BIOSYNTHESIS OF LIPIDS PATH IN ALGAE

Previous works have described the metabolic route of lipids biosynthesis, in order to understand the whole lipids production process. This helps to design the pathways toward microalgae transformation to improve lipid content. Figure 8.1 illustrates the roles of each enzyme involved in biosynthesis of fatty acids. In general, production of lipids comes from two different carbon sources for fatty acid synthesis, one involving part of photosynthesis, other one implying a biochemical

pathway for converting polysaccharides into lipids. Both pathways are related in the process of lipid biosynthesis (Chen et al., 2012; Bellou et al., 2014; Tamayo-Ordóñez et al., 2017).

The first pathway takes place in the plastids. The CO_2 produced in photosynthesis is converted to glycerate-3-phosphate (G3P). G3P is converted to pyruvate and afterward to acetyl-CoA through a reaction catalyzed by the pyruvate dehydrogenase complex. In a later step, ATP dependent carboxylation of acetyl-CoA converts it to malonyl-CoA; a reaction catalyzed by acetyl-CoA carboxylase (ACC) that is considered as a limiting step of the process because it depends on the flux of acetyl-CoA toward the lipid biosynthesis pathway. Conversion to malonyl-CoA is followed by cycles of decarboxylation addition of malonyl-CoA to acyl units and ß-reduction, which are catalyzed by the fatty acid synthase system (FAS), until saturated molecules carbons 16 (16C) and 18 (18C) are produced.

Palmitic and oleic acids are the precursors of polyunsaturated molecules produced by aerobic desaturation and elongation mechanisms.

The second pathway is used by oleaginous heterotrophic microalgae during assimilation of sugar in conditions of nitrogen (N) or phosphate (P) starvation (Bellou et al., 2012, 2014; Tamayo-Ordóñez et al., 2017).

TABLE 8.2 Oil Content of Some Microalgaea

Microalgal Species	Culture Conditions	Lipid Content (% dry weight)	Biomass Productivity (g/L/day)	Lipid Productivity (mg/L/day)	Reference
Chlorophyta					
Chlorella ellipsoidea YSR03	Phototrophic	32 ± 5.9	0.07	22.38	Abou et al. (2011)
Chlorella protothecoides	Heterotrophic	49	1.2	586.8	Gao et al. (2010)
Chlorella sp.	Phototrophic	32.6–66.1	0.077–0.338	51–124	Hsieh et al. (2009)
Chlorella vulgaris	Phototrophic	20–42	0.21–0.346	44–147	Feng et al. (2011)
C. vulgaris	Phototrophic, mixotrophic, heterotrophic	21–38	0.01–0.254	4–54	Liang et al. (2009)
Dunaliella sp.	Phototrophic	12.0–30.12	1.3–3.0	360–390	Araujo et al. (2011)
Haematococcus pluvialis	Phototrophic	15.61–34.85	–	–	Damiani et al. (2010)
Neochloris oleoabundans	Phototrophic	7–40	0.31–0.63	38–133	Li et al. (2008)
N. oleabundans UTEX #1185	Phototrophic	19–56	0.03–0.15	10.67–38.78	Gouveia et al. (2010)
Pseudochlorococcum sp.	Phototrophic	24.6–52.1	0.234–0.76	53–350	Li et al. (2011)
Scenedesmus obliquus	Phototrophic	21–58	0.070–0.094	19.0–43.3	Abou et al. (2009)
Tetraselmis chui	Phototrophic	17.25–23.5	1.0–2.6	240–440	Araujo et al. (2011)
Tetraselmis sp.	Phototrophic	8.2–33.0	0.158–0.214	18.6–22.7	Huerlimann et al. (2010)
Bacillariophyceae					
Chaetoceros calcitrans CS 178	Phototrophic	39.8	0.04	17.6	Rodolfi et al. (2009)
Chaetoceros gracilis	Phototrophic	15.5–60.28	3.4–3.7	530–2210	Araujo et al. (2011)
Chaetoceros muelleri	Phototrophic	11.67–25.25	1.2–2.7	140–670	Araujo et al. (2011)

TABLE 8.2 *(Continued)*

Microalgal Species	Culture Conditions	Lipid Content (% dry weight)	Biomass Productivity (g/L/day)	Lipid Productivity (mg/L/day)	Reference
Nitzschia cf. pusilla YSR02	Phototrophic	48 ± 3.1	0065	31.4	Abou et al. (2011)
Phaeodactylum tricornutum F&M-M40	Phototrophic	18.7	0.24	44.8	Rodolfi et al. (2009)
Skeletonema sp. CS 252	Phototrophic	31.8	0.09	27.3	Rodolfi et al. (2009)
Thalassiosira pseudonana CS 173	Phototrophic	20.6	0.08	17.4	Rodolfi et al. (2009)
Others					
Crypthecodinium cohnii	Heterotrophic	19.9	2.236	444.9	Chen and Smith (2012)
Isochrysis sp.	Phototrophic	6.5–21.25	0.7–2.7	150–180	Arauo et al. (2011)
Isochrysis sp.	Phototrophic	22.0–34.1	0.029–0.090	6.44–21.1	Huerlimann et al. (2010)
Isochrysis zhangjiangensis	Phototrophic	29.8–40.9	0.667–3.1	66.2–140.9	Bellou et al., 2014
Nannochloropsis oculata	Phototrophic	22.75–23.0	2.4–3.4	550–790	Araujo et al. (2011)
N. oculata NCTU-3	Phototrophic	22.7–41.2	0.296–0.497	84–151	Tamayo et al. (2017)
Nannochloropsis sp.	Phototrophic	21.3–37.8	0.021–0.064	4.59–20.0	Huerlimann et al. (2010)
Pavlova salina CS 49	Phototrophic	30.9	0.16	49.4	Rodolfi et al. (2009)
Rhodomonas sp.	Phototrophic	9.5–20.5	0.018–0.064	2.06–6.04	Huerlimann et al. (2010)
Thalassiosira weissflogii	Phototrophic	6.25–13.21	0.5–1.5	20	Araujo et al. (2011)

[a] Reported by Liang and Jiang (2013).

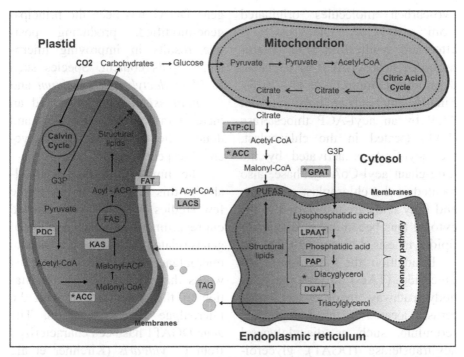

FIGURE 8.1 Lipid synthesis in algae. Lipid biosynthesis involves two different carbon sources for fatty acid synthesis: one being part of photosynthesis, and a second one, part of the biochemical pathway converting polysaccharides into lipids.

This pathway involves the metabolic processes of glycolysis and the citric acid cycle. Under conditions of N and P depletion, NAD^+ isocitrate dehydrogenase can be inhibited leading to the accumulation of citrate in the mitochondria, and afterward to its excretion to the cytosol. Citrate is converted to oxaloacetate and acetyl-CoA by cytosolic ATP-citrate lyase (Avidan and Pick, 2015; Avidan et al., 2015), and the produced acetyl-CoA is converted to malonyl-CoA by cytosolic ACC (Mühlroth et al., 2013). Malonyl-CoA is transferred to one of the subunits of the FAS, specifically to acyl carrier proteins (ACP) (Greenwell et al., 2010; Blatti et al., 2012). Afterward, malonyl-ACP enters the fatty acid synthesis cycle through 3-ketoacyl-ACP synthase (KAS); a reaction in which both pathways become related.

KAS catalyzes the condensation of an acetyl group with malonyl-ACP to form ketobutyryl-ACP, which through iterative sequential reduction and dehydration reactions is first converted to butyryl-ACP and after several cycle repetitions to palmitoyl-ACP. Palmitoyl-AVP is converted to stearoyl-ACP after the addition of two-carbon molecules originated from acetyl-CoA. Oleoyl-ACP is afterward synthesized after desaturation of stearoyl-ACP by a plastid desaturase (Mühlroth et al., 2013). Finally, fatty acids are unbound from ACP by an acyl-ACP thioesterase (FAT) located in the chloroplast, the acyl-CoA is activated by the long-chain acyl-CoA synthase, also located in the chloroplast envelope, and fatty acids are transferred to the cytosol thus becoming available for lipid synthesis (Lei et al., 2012).

In addition, the formation of triglycerides (TAG) through the Kennedy pathway involves acyltransferases located in the endoplasmic reticulum, such as diacylglycerol acyltransferase (DGAT), glycerol-3-phosphate acyltransferase, lysophosphatidic acid acyltransferase, and lysophosphatidylcholine acyltransferase (Chen et al., 2012; Liu and Benning, 2013), each of these transferases being in charge of the consequential transformation of lysophosphatidic acid to phosphatidic acid, diacylglycerol, and triacylglycerol.

8.4.2.2 ADVANCES IN GENETIC MANIPULATION OF THE ENZYMES INVOLVED IN THE PATHWAY OF LIPID BIOSYNTHESIS IN ALGAE

There are advances in genetic engineering in microalgae through the overexpression of the gene. The gene DGAT has been the principal gene-modified, producing positive results in improving microalgae lipids content. Species such as *Phaeodactylum tricornutum* and *C. minutissima* have presented an increase in lipid content of 35% and double the content of triacylglycerols, respectively.

The limited information available in this field may be due to the few studies that have been carried out regarding the characterization of genes related to lipid biosynthesis in microalgae. Nevertheless, there exist works that have achieved information of the genes related with the microalgae lipid production. The gene DGAT1 has been characterized from *C. vulgaris* (Kirchner et al., 2016), also the level expression of the gene ACL (codifies for the ATP-citrate lyase enzyme) was studied by Ikaran et al. (2005), concluding that under N limitation the expression level was the lowest at 120 h of the time of growth but the TAG's content increased. Wan et al. (2011) determined that the increment of the level expression of the *aacD* gene (codifies for the acetyl CoA carboxylase enzyme) increments the lipid content in microalgae.

C. reinhardtii is a microalgae species widely studied for genetic transformation (Pratheesh et al. 2014). Other species have been genetically transformed with aim of increment yields lipids are *Volvox carteri, Dunalliela salina, Chlorella sorokiniana, C. vulgaris, and Haematococcus pluviales* (Schiedlmeier et al., 1994; Dawson et al., 1997; Kumar et al., 2004; Kathiresan et al., 2009).

8.5 MICROALGAE AS A SOURCE OF VARIOUS PRODUCTS

The world production of biomass from microalgae is estimated at approximately 7000 tons per year, with a market value between 3800 million and 5400 million dollars (Brasil et al., 2017).

The use of microalgal biomass has great potential in multiple areas; however, traditional methods for the separation of compounds are focused on a single use at the same time damaging the other useful fractions of biomass. Given this scenario, the concept of biorefinery is being adopted at a higher level; a biorefinery is a facility (similar to petroleum refineries) whose raw material is biomass, this concept directs efforts toward a sustainable processing of biomass in a range of commercial products and energy, maximizes production, and improves the use of materials premiums, enhancing their value (Chew et al., 2017; Vanthoor-Koopmans et al., 2013).

The first group of products derived from microalgae are the food supplements based on dry biomass of *Spirulina* and *Chlorella*, these form the largest market of products based on microalgae (Sili et al., 2012).

On the other hand, the microalgae biomass has received a considerable interest as a potential raw material for the production of biofuels because depending on the species and the cultivation conditions they can produce large quantities, mainly, of polysaccharides (sugars) and triacylglycerides (lipids), which are the building blocks to produce bioethanol and biodiesel (Slade and Bauen, 2013; Moreno-Garcia et al., 2017). Other fuels such as methane and hydrogen can be obtained from the anaerobic digestion and fermentation of microalgal biomass. The use of biomass codigestion with substrates with a high C/N ratio has also been proposed to improve the efficiency of the process (Ward et al., 2014; Jankowska et al., 2016).

An option with great potential to improve the competitiveness of the production of fuels from fatty acids obtained from microalgae (biodiesel) is the genetic improvement of strains to increase the productivity of lipids and thus be able to design 'desirable' fatty acid profiles by engineering metabolic. An enzyme of interest is diacetylglycerol acyltransferase (DGAT) since in searches of sequenced microalgae genomes it was found that most microalgae have several isoforms for this enzyme (Chen and Smith, 2012).

Microalgae can accumulate between 30% and 50% of dry weight in lipids (depending on the culture conditions) (Chew et al., 2017). Due to this, in addition to the production of biodiesel, obtaining polyunsaturated fatty acids from this material is of great interest. Fatty acids are used for feeding in infant formulas and as food supplements because they have been proposed as nutraceutical compounds, mainly polyunsaturated fatty acids such as eicosapentaenoic acid (EPA) and docosahexaenoic acid (DHA); FDA reports in 2004 indicate foods that are rich in polyunsaturated fatty acids, particularly EPA and DHA, reduce the risk of coronary heart disease (da Silva Vaz et al., 2016; Borowitzka, 2013).

Carotenoids are a group of high-value molecules that can be found in the same way as products of microalgae metabolism. ß-carotene was the first high-value product produced commercially from microalgae cultures *(D. salina)*, whose production began in the 1980s by four producers: Koor Foods (Nature Beta Technology) in Israel, Western Biotechnology Ltd and Betatene Ltd in Australia, as well as Nutralite in the United States; these products have an estimated market of around 270 million dollars (Borowitzka, 2013). More examples of these high-value molecules are astaxanthin, which is also a pigment that can be obtained from *H. pluvialis*, the lutein from *Muriellopsis* sp., the canthaxanthin from *Chlorella zofigiensis*, to mention just a few (Suganya et al., 2016; Hu et al., 2017; Zhang et al., 2014). Lutein is known for its

protective function against macular degeneration of the eye, the importance of which is that it cannot be synthesized by humans, so diet is the only source that supplies this compound; on the other hand, astaxanthin is a known antioxidant, with protective cardioprotective, neuroprotective, anticancer, and antidiabetic properties (Hu et al., 2017; Zhang et al., 2014).

Phycobiliproteins are a particular group of pigments found only in algae, for example, Spirulina has been reported as an excellent source (Suganya et al., 2016; Sonani et al., 2016). Because philicoproteins have a bright color, a nontoxic nature and antioxidant properties, these characteristics open the door to their potential application in various industries as additives in food and cosmetics, they have also been reported to have pharmacological effects and also have various activities such as the antioxidant, anticancer, neuroprotective, anti-inflammatory, hepatoprotective, and hypocholesterolemic (Sonani et al., 2016) Unlike most of the compounds mentioned in this review, the phycocyanin, phycoerythrin, and allophycocyanin pigments can only be found in cyanobacteria such as Spirulina (Arthrospira), red algae such as Porphyridium, Rhodella, and Galdieria, Cryptophyta, and Glaucophyta. The market value for phycobiliprotein products exceeds 60 million dollars (Borowitzka, 2013).

Until now, fuels, pigments, and lipids have been mentioned; however, another major fraction in the composition of the microalgal biomass are the proteins, which can constitute between 50% and 70% of the biomass (Chew et al., 2017). They can be used as feed improvers for livestock or as food supplements for human use (Suganya et al., 2016). In this section Spirulina (55%–70%) and Chlorella (53%) are reported as models of microorganisms that may contain high levels of proteins in dry weight, although it is also the source of many other bioactive molecules (Chew et al., 2017; de Morais et al., 2015).

Bioactive molecules of protein nature can also be found among the metabolites of microalgae. Among the enzymes synthesized by some microalgae are cellulases, galactosidases, proteases, lipases, phytases, laccases, amylases, antioxidant enzymes, and enzymes associated with the accumulation of carbohydrates and carbon concentration (Brasil et al., 2017). Protein hydrolysates can also be part of the products obtained from microalgae, these hydrolysates (also known as biopeptides) can be used as food supplements for human use for the increase of protein digestibility; as well as for animal feed increasing the protein value of forages (da Silva Vaz et al., 2016).

Finally, to a lesser extent, microalgae cultures can be used as an alternative to chemical remediation methods, for example, for the sequestration (fixation) of carbon dioxide, reducing the emission of this pollutant to the environment. Jacob-Lopes and Franco (2013) present a study in which the possibility of removing CO_2 from a gas stream from an oil refinery is analyzed by measuring the release of oxygen and the growth kinetics of Aphanothece microscopica Nägeli (RSMan92) with obtaining up to 92% of volatile solid compounds and 3.64% of biomass formation, which could be coupled to a biogas production system. There are also reports in which microalgae cultures are used for the removal of nutrients from wastewater, helping to eliminate high phosphorus and nitrogen contents (Christenson and Sims, 2011; Fenton and ÓhÚallacháin, 2012). As well as the removal of heavy metals, since they have a great tolerance capacity, reaching promising results (Kumar et al., 2015).

8.6 CONCLUSIONS

The great capacity and flexibility that microalgae present for the synthesis and accumulation of metabolites of interest make them an attractive model for the study of obtaining these molecules at industrial levels; although it is necessary to carry out all the procedures from an approach like the one carried out in the biorefineries, achieving the maximum use of the products and the biomass.

Still continue to study processes such as the improvement of strains to achieve higher yields and lower costs, because even with all that is provided by current technological advances is difficult to achieve processes with a favorable cost-benefit in some cases. The knowledge in basic aspects of the regulation of metabolic pathways involved in the generation of a product of interest from microalgae, and the use of this knowledge to generate strains of genetically modified algae, is an alternative that remains a powerful tool for improving the yields of any product of interest in microalgae.

KEYWORDS

- **microalgae**
- **biofuels**
- **metabolites**
- **genetic engineering**

REFERENCES

Abou-Shanab, R.A.I., Hwang, J.H., Cho, Y., Min, B., & Jeon, B.H. Characterization of microalgal species isolated from fresh water bodies as a potential source for biodiesel production. *Appl. Energy.*, **2011**, 88, 3300–3306.

Araujo, G.S., Matos, L.J., Goncalves, L.R., Fernandes, F.A., & Farias, W.R. Bioprospecting for oil producing microalgal strains: evaluation of oil and biomass production for ten microalgal strains. *Bioresour. Technol.*, **2011**, 102, 5248–5250.

Asada, C., Doi, K., Sasaki, C., & Nakamura, Y. Efficient extraction of starch from microalgae using ultrasonic homogenizer and its conversion into ethanol by simultaneous saccharification and fermentation. *Nat. Resour.*, **2012**, 3 (04), 175.

Avidan, O. and Pick, U. Acetyl-CoA synthetase is activated as part of the PDH-bypass in the oleaginous green alga *Chlorella desiccata*. *J. Exp. Bot.*, **2015**, 66, 7287–7298.

Avidan, O., Brandis, A., Rogachev, I. and Pick, U. Enhanced acetyl-CoA production is associated with increased triglyceride accumulation in the green alga *Chlorella desiccata*. *J. Exp. Bot.*, **2015**, 66, 3725–3735.

Behera, S., Singh, R., Arora, R., Kumar Sharma, N., Shukla, M., & Kumar, S.Scope of algae as third generation biofuels. *Front. Bioeng. Biotechnol.*, **2015**, 2, 90.

Bellou, S., Baeshen, M. N., Elazzazy, A. M., Aggeli, D., Sayegh, F. & Aggelis, G.Microalgal lipids biochemistry and biotechnological perspectives. *Biotechnol. Adv.*, **2014**, 32, 1476–1493.

Bellou, S. and Aggelis, G.Biochemical activities in *Chlorella sp.* and *Nannochloropsis salina* during lipid and sugar synthesis in a lab-scale open pond simulating reactor. *J. Biotechnol.*, **2012**, 164, 318–329.

Bibi, R., Ahmad, Z., Imran, M., Hussain, S., Ditta, A., Mahmood, S., & Khalid, A. Algal bioethanol production technology: a trend towards sustainable development. *Renew. Sust. Energy Rev.*, **2016** 71, 976–985.

Blatti, J. L., Beld, J., Behnke, C. A., Mendez, M., Mayfield, S. P. and Burkart, M. D. Manipulating fatty acid biosynthesis in microalgae for biofuel through protein–protein interactions. *PLoS One*, 2012, 7, e42949.

Borowitzka, M. A. High-value products from microalgae—their development and commercialization. *J. Appl. Phycol.*, **2013**, 25(3), 743–756.

Brasil, B. D. S. A. F., de Siqueira, F. G., Salum, T. F. C., Zanette, C. M., & Spier, M. R. Microalgae and cyanobacteria as enzyme biofactories. *Algal Res.*, 2017,25, 76–89.

Brennan, L., & Owende, P. Biofuels from microalgae—a review of technologies for production, processing, and extractions of biofuels and co-products. *Renew. Sust. Energy Rev.*, 2010, 14(2), 557–577.

Chandra, R., Iqbal, H.M., Vishal, G., Lee, H.S., and Nagra, S. Agal biorefinery: a sustainable approach to valorize algal-based biomass towards multiple product recovery. *Bioresour. Technol.* 2019, 131, 105398.

Chaudhary, L., Pradhan, P., Soni, N., Singh, P., & Tiwari, A. Algae as a feedstock for bioethanol production: new entrance in biofuel world. *Int. J. Chem. Technol. Res.*, 2014, 6, 1381–1389.

Cheah, W. Y., Ling, T. C., Show, P. L., Juan, J. C., Chang, J. S., & Lee, D. J. Cultivation in wastewaters for energy: a microalgae platform. *Appl. Energy*, 2016, 179, 609–625.

Chen, C. Y., Chang, H. Y., & Chang, J. S. Producing carbohydrate-rich microalgal biomass grown under mixotrophic conditions as feedstock for biohydrogen production. *Int. J. Hydrogen Energy.*, 2016, 41(7), 4413–4420.

Chen, C. Y., Zhao, X. Q., Yen, H. W., Ho, S. H., Cheng, C. L., Lee, D. J., et al., (2013). Microalgae-based carbohydrates for biofuel production. *Biochem. Eng. J.*, 2013, 78, 1–10.

Chen, J. E. and Smith, A. G. A look at diacylglycerol acyltransferases (DGATs) in algae. *J. Biotechnol.*, 2012, 162, 28–39.

Cheng, Y. S., Zheng, Y., Labavitch, J. M., & VanderGheynst, J. S. Virus infection of *Chlorella variabilis* and enzymatic saccharification of algal biomass for bioethanol production. *Bioresour. Technol.*, 2013, 137, 326–331.

Chew, K. W., Yap, J. Y., Show, P. L., Suan, N. H., Juan, J. C., Ling, T. C., Lee d. j. &

Chang, J. S. Microalgae biorefinery: high value products perspectives. *Bioresour. Technol.*, 2017, 229, 53–62.

Chiaramonti, D., Prussi, M., Casini, D., Tredici, M. R., Rodolfi, L., Bassi, N., Zitelli G. C. & Bondioli, P.Review of energy balance in raceway ponds for microalgae cultivation: re-thinking a traditional system is possible. *Appl. Energy*, 2013, 102, 101–111.

Chiu, S.Y., Kao, C.Y., Tsai, M.T., Ong, S.C., Chen, C.H., & Lin, C.S. Lipid accumulation and CO_2 utilization of *Nannochloropsis oculata* in response to CO_2 aeration. *Bioresour. Technol.*, 2009, 100, 833–8.

Choi, S. P., Nguyen, M. T., & Sim, S. J. Enzymatic pretreatment of *Chlamydomonas reinhardtii* biomass for ethanol production. *Bioresour. Technol.*, 2010, 101(14), 5330–5336.

Chow, T. J., Su, H. Y., Tsai, T. Y., Chou, H. H., Lee, T. M., & Chang, J. S. Using recombinant cyanobacterium (*Synechococcus elongatus*) with increased carbohydrate productivity as feedstock for bioethanol production via separate hydrolysis and fermentation process. *Bioresour. Technol.*, 2015, 184, 33–41.

Christenson, L., & Ronald, S.Production and harvesting of microalgae for wastewater treatment, biofuels and bioproducts. *Biotechnol. Adv.*, 2011, 29 (6), 686–702.

Couto, R.M., Simoes, P.C., Reis, A., Da Silva, T.L., Martins, V.H., & Sanchez-Vicente, Y. Supercritical fluid extraction of lipids from the heterotrophic microalga *Crypthecodinium cohnii. Eng. Life Sci.*, 2010, 10, 158–64.

Christenson, L., & Sims, R. Production and harvesting of microalgae for wastewater treatment, biofuels, and bioproducts. *Biotechnol. Adv.*, 2011, 29(6), 686–702.

da Silva Vaz, B., Moreira, J. B., de Morais, M. G., & Costa, J. A. V. Microalgae as a new source of bioactive compounds in food supplements. *Curr Opin Food Sci.*, 2016, 7, 73–77.

Damiani, M.C., Popovich, C.A., Constenla, D., & Leonardi, P.I. Lipid analysis in *Haematococcus pluvialis* to assess its potential use as a biodiesel feedstock. *Bioresour. Technol.*, **2010**, 101, 3801–7.

Dandu, M.S.R. & Nanthagopal, K. Tribological aspects of biofuels—a review. *Fuel*, **2019**, 258, 116066.

Dawson, H., Burlingame, R., & Cannons, A. Stable transformation of Chlorella: rescue of nitrate reductase-deficient mutants with the nitrate reductase gene. *Curr. Microbiol.*, **1997**, 35(6), 356–362.

de Morais, M. G., Vaz, B. D. S., de Morais, E. G., & Costa, J. A. V. Biologically active metabolites synthesized by microalgae. *Biomed. Res. Int., 2015.*

Demirbas, M. F. Biofuels from algae for sustainable development. *Appl. Energy.*, **2011**, 88(10), 3473–3480.

Draaisma, R. B., Wijffels, R. H., Slegers, P. E., Brentner, L. B., Roy, A., & Barbosa, M. J. Food commodities from microalgae. *Curr. Opin Biotech.*, **2013**, 24(2), 169–177.

Dragone, G., Fernandes, B., Vicente, A. A., & Teixeira, J. A. Third generation biofuels from microalgae. En A. Mendez-Vilas, Current Research, Technology and Education Topics in Applied Microbiology and Microbial Biotechnology, **2010**, 1355–1366.

Feng, D.N., Chen, Z.A., Xue, S., & Zhang, W. Increased lipid production of the marine oleaginous microalgae *Isochrysis zhangjiangensis* (Chrysophyta) by nitrogen supplement. *Bioresour. Technol.*, **2011**, 102, 6710–6716.

Feng, Y.J., Li, C., & Zhang, D.W. Lipid production of *Chlorella vulgaris* cultured in artificial wastewater medium. *Bioresour. Technol.*, **2011**, 102, 101–105.

Fenton, O., & Ó hÚallacháin, D. Agricultural nutrient surpluses as potential input sources to grow third generation biomass (microalgae): a review. *Algal Res.*, **2012**, 1(1), 49–56.

Freitas, B. C. B., Cassuriaga, A. P. A., Morais, M. G., & Costa, J. A. V. Pentoses

and light intensity increase the growth and carbohydrate production and alter the protein profile of *Chlorella minutissima*. *Bioresour. Technol.*, **2017**, 238, 248–253.

Galadima, A., & Muraza, O.Biodiesel production from algae by using heterogeneous catalysts: a critical review. *Energy*, **2014**, 78, 72–83.

Gao, C.F., Zhai, Y., Ding, Y., & Wu, Q.Y. Application of sweet sorghum for biodiesel production by heterotrophic microalga *Chlorella protothecoides*. *Appl. Energy*, **2010**, 87, 756–761.

Gouveia, L., Marques, A.E., da Silva, T.L., & Reis, A. *Neochloris oleabundans* UTEX#1185: a suitable renewable lipid source for biofuel production. *J. Ind. Microbiol. Biotechnol.*, **2009**, 36, 821–826.

Gouvela, L., & Oliveira, A. C.Microalgae as raw material for biofuels production. *J. Ind. Microbiol. Biotechnol.*, **2008**, 36(2), 269–274.

Greenwell, H. C., Laurens, L. M., Shields, R. J., Lovitt, R. W. & Flynn, K. J. Placing microalgae on the biofuels priority list: a review of the technological challenges. *J. R. Soc. Interface,* **2010**, 7, 703–726.

Hannon, M., Gimpel, J., Tran, M., Rasala, B., & Mayfield, S. Biofuels from algae: challenges and potential. *Biofuels*, **2010**, 1(5), 763–784.

Hemschemeier A., Melis A., Happe T. Analytical approaches to photobiological hydrogen production in unicellular green algae. *Photosynth. Res.*, **2009**, 102, 523–540.

Ho, S. H., Huang, S. W., Chen, C. Y., Hasunuma, T., Kondo, A., & Chang, J. S. Characterization and optimization of carbohydrate production from an indigenous microalga *Chlorella vulgaris* FSP-E. *Bioresour. Technol.*, **2013**, 135, 157–165.

Ho, S. H., Kondo, A., Hasunuma, T., & Chang, J. S. Engineering strategies for improving the CO_2 fixation and carbohydrate productivity of *Scenedesmus obliquus* CNW-N used for bioethanol fermentation. *Bioresour. Technol.*, **2013**, 143, 163–171.

Hsieh, C.H., & Wu, W.T. Cultivation of microalgae for oil production with a cultivation strategy of urea limitation. *Bioresour. Technol.*, **2009**, 100, 3921–3926.

Hu, J., Nagarajan, D., Zhang, Q., Chang, J. S., & Lee, D. J. Heterotrophic cultivation of microalgae for pigment production: a review. *Biotechnol. Adv.* **2017**, 36(1), 54–67.

Huerlimann, R., de Nys, R., & Heimann, K. Growth, lipid content, productivity, and fatty acid composition of tropical microalgae for scale-up production. *Biotechnol. Bioeng.*, **2010**, 107, 245–57.

Ikaran, Z., Suárez Alvarez, S., & Castañón, S. The effect of nitrogen limitation on the physiology and metabolism of chlorella vulgaris var L3. *Algal Res.*, **2015**, 10, 134–144.

Islam, M. A., Heimann, K., & Brown, R. J. Microalgae biodiesel: current status and future needs for engine performance and emissions. *Renew. Sust. Energy Rev.*, **2017**, 79, 1160–1170.

Isleten-Hosoglu, M., Gultepe, I., & Elibol, M.Optimization of carbon and nitrogen sources for biomass and lipid production by *Chlorella saccharophila* under heterotrophic conditions and development of Nile red fluorescence based method for quantification of its neutral lipid content. *Biochem. Eng. J.*, **2012**, 61, 11–19.

Jacob-Lopes, E., & Franco, T. T. From oil refinery to microalgal biorefinery. *J. CO₂ Util.*, **2013**, 2, 1–7.

Jambo, S. A., Abdulla, R., Azhar, S. H. M., Marbawi, H., Gansau, J. A., & Ravindra, P. A review on third generation bioethanol feedstock. *Renew. Sust. Energy Rev.*, **2016**, 65, 756–769.

Jankowska, E., Sahu, A. K., & Oleskowicz-Popiel, P. Biogas from microalgae: review on microalgae's cultivation, harvesting and pretreatment for anaerobic digestion. *Renew. Sust. Energy Rev.*, **2016**, 75, 692–709.

Karatay, S. E., Erdoğan, M., Dönmez, S., & Dönmez, G. Experimental investigations on bioethanol production from halophilic microalgal biomass. *Ecol. Eng.*, **216**, 95, 266–270.

Kathiresan, S., Chandrashekar, A., Ravishankar, G., & sarada, R. *Agrobacterium*-mediated transformation in the green alga *Haematococcus pluvialis* (Chlorophyceae, Volvocales). *J. Phycol.*, **2009**, 45(3), 642–649.

Khoeyi, Z. A., Seyfabadi, J., & Ramezanpour, Z. Effect of light intensity and photoperiod on biomass and fatty acid composition of the microalgae, *Chlorella vulgaris*. *Aquacult. Int.*, **2012**, 20(1), 41–49.

Kim, K. H., Choi, I. S., Kim, H. M., Wi, S. G., & Bae, H. J. Bioethanol production from the nutrient stress-induced microalga *Chlorella vulgaris* by enzymatic hydrolysis and immobilized yeast fermentation. *Bioresour. Technol.*, **2014**, 153, 47–54.

Kirchner, L., Wirshing, A., Kurt, L., Reinard, T., Glick, J., Gram, E. J., et al.Identification, characterization, and expression of diacylgylcerol acyltransferase type-1 from *Chlorella vulgaris*. *Algal Res.*, **2016**, 13, 167–181.

Kraan, S. Mass-cultivation of carbohydrate rich macroalgae, a possible solution for sustainable biofuel production. *Mitig. Adapt. Strateg. Glob. Change.* **2013**, 18(1), 27–46.

Kumar, K. S., Dahms, H. U., Won, E. J., Lee, J. S., & Shin, K. H. Microalgae–a promising tool for heavy metal remediation. *Ecotox. Environ Safe*, **2015**, 113, 329–352.

Kumar, S., Misquitta, R., Reddy, V., Rao, B., & Rajam, M. Genetic transformation of the green alga-*Chlamydomonas reinhardtii* by *Agrobacterium tumefaciens*. *Plant Sci.*, **2004**, 731–738.

Lee, O. K., Kim, A. L., Seong, D. H., Lee, C. G., Jung, Y. T., Lee, J. W., & Lee, E. Y. Chemo-enzymatic saccharification and bioethanol fermentation of lipid-extracted residual biomass of the microalga,

250

Handbook of Research on Bioenergy and Biomaterials

Dunaliella tertiolecta. Bioresour. Technol.,
2013, 132, 197–201.

Lei, A., Chen, H., Shen, G., Hu, Z., Chen,
L. & Wang, J. Expression of fatty acid
synthesis genes and fatty acid accumulation in *Haematococcus pluvialis* under
different stressors. *Biotechnol. Biofuels.*,
2012, 5, 1–11.

Li, Y.T., Han, D.X., Sommerfeld, M., &
Hu, Q.A. Photosynthetic carbon partitioning and lipid production in the oleaginous microalga *Pseudochlorococcum* sp.
(Chlorophyceae) under nitrogen-limited
conditions. *Bioresour. Technol.*, **2011**, 102,
123–129.

Li, Y.Q., Horsman, M., Wang, B. & Lan,
C.Q. Effects of nitrogen sources on cell
growth and lipid accumulation of green
alga *Neochloris oleoabundans*. *Appl.
Microbiol. Biotechnol.*, **2008**, 81, 629–36.

Liang, M.H., & Jiang, J.G. Advancing oleaginous microorganisms to produce lipid via
metabolic engineering technology. *Prog.
Lipid Res.*, **2013**, 52(4), 395–408.

Liang, Y., Sarkany, N., & Cui, Y. Biomass
and lipid productivities of *Chlorella
vulgaris* under autotrophic, heterotrophic and mixotrophic growth conditions.
Biotechnol. Lett., **2009**, 31, 1043–1049.

Liu, B. & Benning, C. Lipid metabolism
in microalgae distinguishes itself. *Curr.
Opin. Biotechnol.*, **2013**, 24, 300–309.

Liu, J., Yuan, C., Hu, G., & Li, F. Effects
of light intensity on the growth and lipid
accumulation of microalga Scenedesmus
sp. 11-1 under nitrogen Limitation. *Appl.
Biochem. Biotechnol.*, **2012**, 166 (8),
2127–2137.

Lorente, E., Farriol, X., & Salvadó, J. Steam
explosion as a fractionation step in biofuel
production from microalgae. *Fuel Progress Technol.*, **2015**, 131, 93–98.

Markou, G., Angelidaki, I., Nerantzis, E., &
Georgakakis, D. Bioethanol production by
carbohydrate-enriched biomass of Arthrospira (Spirulina) platensis. *Energies*, 2013,
6(8), 3937–3950.

Melis, A., Zhang, L., Forestier, M., Ghirardi,
M. L., & Seibert, M. Sustained photobiological hydrogen gas production upon
reversible inactivation of oxygen evolution in the green alga *Chlamydomonas
reinhardtii*. *Plant Physiol.*, **2000**, 122(1),
127–136.

Miranda, J. R., Passarinho, P. C., &
Gouveia, L. Pre-treatment optimization
of *Scenedesmus obliquus* microalga for
bioethanol production. *Bioresour. Technol.*,
2012, 104, 342–348.

Möllers, K. B., Cannella, D., Jørgensen, H.,
& Frigaard, N. U. Cyanobacterial biomass
as carbohydrate and nutrient feedstock for
bioethanol production by yeast fermentation. *Biotechnol. Biofuels.*, **2014**, 7(1), 64.

Mühlroth, A., Li, K., Rokke, G., Winge,
P., Olsen, Y., Hohmann-Marriott, M. F.,
Vadstein, O. & Bones, A. M. Pathways
of lipid metabolism in marine algae,
co-expression network, bottlenecks and
candidate genes for enhanced production
of EPA and DHA in species of *Chromista*.
Mar. Drugs, **2013**, 11, 4662–4697.

Oh, Y.K., Hwang, K.R., Kim, C., J.R., & Lee,
J.S. Recent development and key barriers
to advanced biofuels: a short review.
Bioresour. Technol., **2018**, 257, 320–333.

Patel, A., Gami, B., Patel, P., & Patel, B.
Microalgae: antiquity to era of integrated
technology. *Renew. Sust. Energy Rev.*,
2016, 71, 535–547.

Pavón-Suriano, S. G., Ortega-Clemente, L. A.,
Curiel-Ramírez, S., Jiménez-García, M. I.,
Pérez-Legaspi, I. A., & Robledo-Narváez,
P. N. Evaluation of colour temperatures
in the cultivation of *Dunaliella salina* and
Nannochloropsis oculata in the production
of lipids and carbohydrates. *Environ. Sci.
Pollut. R.* **2017**, 1–9.

Morowvat, M. H., & Ghasemi, Y. Culture medium optimization for enhanced
β-carotene and biomass production by
Dunaliella salina in mixotrophic culture. *Biocatal Agric Biotechnol.*, **2016**, 7,
217–223.

Radakovits, R., Jinkerson, R., Darzins, A., & Posewitz, M. Genetic engineering of algae for enhanced biofuel production. *Eukaryot. Cell*, **2010**, 9 (4), 486–501.

Rodionova, M. V., Poudyal, R. S., Tiwari, I., Voloshin, R. A., Zharmukhamedov, S. K., Nam, H. G., & Allakhverdiev, S. I. Biofuel production: challenges and opportunities. *Int. J. Hydrogen Energy*, **2016**, 42(12), 8450–8461.

Rodolfi, L., Zittelli, G.C., Bassi, N., Padovani, G., Biondi, N., Bonini, G., et al. Microalgae for oil: strain selection, induction of lipid synthesis and outdoor mass cultivation in a low-cost photobioreactor. *Biotechnol. Bioeng.*, **2009**, 102, 100–112.

Schiedlmeier, B., Schmitt, R., Müller, W., Kirk, M., Gruber, H., Mages, W., et al. Nuclear transformation of *Volvox carteri*. *PNAS U.S.A.*, **1994**, 91(11), 5080–5084.

Sharma, K. K., Schuhmann, H., & Schenk, P. M. High Lipid Induction in Microalgae for Biodiesel Production. *Energies.*, **2012**, 5(5), 1532–1553.

Sili, C., Torzillo, G., & Vonshak, A. Arthrospira (Spirulina). In *Ecology of Cyanobacteria II*, **2012**, 677-705. Springer: The Netherlands.

Simionato, D., Basso, S., Giacometti, G. M., & Morosinotto, T. Optimization of light use efficiency for biofuel production in algae. *Biophys. Chem.*, **2013**, 182, 71–78.

Slade, R., & Bauen, A. Micro-algae cultivation for biofuels: cost, energy balance, environmental impacts and future prospects. *Biomass Bioenergy*, **2013**, 53, 29–38.

Sonani, R. R., Rastogi, R. P., Patel, R., & Madamwar, D. Recent advances in production, purification and applications of phycobiliproteins. *WJBC.*, **2016**, 7(1), 100.

Suganya, T., Varman, M., Masjuki, H. H., & Renganathan, S. Macroalgae and microalgae as a potential source for commercial applications along with biofuels production: a biorefinery approach. *Renew. Sust. Energy Rev.*, **2016**, 55, 909–941.

Tamayo-Ordóñez, Y. J., Ayil-Gutiérrez, B. A., Sánchez-Teyer, F. L., De la Cruz-Arguijo, E. A., Tamayo-Ordóñez, F. A., Córdova-Quiroz, A. V., & Tamayo-Ordóñez, M. C. Advances in culture and genetic modification approaches to lipid biosynthesis for biofuel production and in silico analysis of enzymatic dominions in proteins related to lipid biosynthesis in algae. *Phycol. Res.*, **2017**, 65(1), 14–28.

Vitova, M., Bisova, K., Kawano, S., & Zachleder, V. Accumulation of energy reserves in algae: from cell cycles to biotechnological applications. *Biotechnol. Adv.*, **2015**, 33(6), 1204–1218.

Wan, M., Liu, P., Xia, J., Rosenberg, J. N., Oyler, G. A., Betenbaugh, M. J., et al. The effect of mixotrophy on microalgal growth, lipid content and expression levels of three pathway genes in *Chlorella sorokiniana*. *Appl. Microbiol. Biotechnol.*, **2011**, 91(3), 835–844.

Ward, A. J., Lewis, D. M., & Green, F. B. Anaerobic digestion of algae biomass: a review. *Algal Res.*, **2014**, 5, 204–214.

Yeesang, C., & Cheirsilp, B. Effect of nitrogen, salt, and iron content in the growth medium and light intensity on lipid production by microalgae isolated from freshwater sources in Thailand. *Bioresour. Technol.*, **2011**, 102 (3), 3034–3040.

Yeh, K. L., & Chang, J. S. Effects of cultivation conditions and media composition on cell growth and lipid productivity of indigenous microalga *Chlorella vulgaris* ESP-31. *Bioresour. Technol.*, **2012**, 105, 120–127.

Zhang L., Happe T., Melis A. Biochemical and morphological characterization of sulfur deprived and H2-producing *Chlamydomonas reinhardtii* (green alga). *Planta* **2002**, 214, 552–561.

Zhang, J., Sun, Z., Sun, P., Chen, T., & Chen, F. Microalgal carotenoids: beneficial effects and potential in human health. *Food Funct.*, **2014**, 5(3), 413–425.

CHAPTER 9

Bioelectrochemical Systems for Wastewater Treatment and Energy Recovery

S. Y. MARTÍNEZ-AMADOR[1], L. J. RÍOS-GONZALEZ[2],
M. M. RODRÍGUEZ-GARZA[2], I. M. M MORENO-DÁVILA[2],
T. K. MORALES-MARTINEZ[3], M. A. MEDINA-MORALES[2], and
J. A. RODRÍGUEZ-DE LA GARZA[2*]

[1]Botany Department, Agronomic Division, Universidad Autónoma Agraria Antonio Narro, Mexico

[2]Department of Biotechnology, School of Chemistry, Universidad Autonoma de Coahuila, Mexico

[3]Bioprocess and Microbial Biochemistry Group, School of Chemistry, Universidad Autonoma de Coahuila, Mexico

*Corresponding author. E-mail: antonio.rodriguez@uadec.edu.mx

ABSTRACT

At present, civilization is facing several problems, being pollution and new energy sources at the top of the list. Billions of dollars are spent annually, to clean up the environment that has been polluted due to our diverse activities. Hence, the need to develop new technologies allows us to carry out this task with minimal energy consumption and high cleanup efficiency. In the case of wastewater, it can be treated sustainably with safety to minimize energy demand during the process, therefore maximize the recovery of valuable resources such as energy, water, and high value-added by-products. Bioelectrochemical system platform technology may carry out efficiently and at a lower cost the remediation process of different pollutants, compared to the traditional biological treatment. BES are capable of degrading diverse residuals or wastes, whereas the electrodes present in the BES serve as electron acceptors or donors and additionally the energy produced during the process may be used as an energy source for small devices.

9.1   INTRODUCTION

Energy demand has not stopped growing and the environmental problems caused are mainly due to a high percentage of this demand that is provided by fossil fuels. Additionally, clean water sources are also in decline rapidly due to population growth and industrialization. In order to overcome these modern problems, efficient wastewater treatment with low energy demand should be taken into consideration, in current and future infrastructure. Additionally, if during this treatment, energy recovery and other high value-added by-or-end products are a priority, a promising future can be guaranteed (Jung et al., 2020). Bioelectrochemical systems (BES) have received much attention over the past decade, since this technology is quite attractive, due to its feature of having the capacity to produce electricity during the wastewater treatment process (Tian et al., 2014; De Vrieze et al, 2018). BES are an emerging treatment technology based on microbial interaction (Zhang et al., 2014; Butti et al., 2016). Microorganisms convert the chemical energy stored in biodegradable materials to generate an electrical current and other high value-added by-products. BES offer a novel solution for integrated waste treatment and energy recovery (Haavisto et al., 2020). Almost all BES share a common principle within the anode chamber or compartment (Figure 9.1), in which biodegradable substrates, such as waste materials, are oxidized by microorganisms and generate an electrical current due to the electrons and protons released during the process (Zhuwei et al., 2007). The current can be captured directly for electricity generation or used to produce hydrogen and other value-added chemicals. Current production was first reported a century ago by Potter in 1911, but scientific interest in this concept has recently flourished (Wang and Ren, 2013). Unlike conventional wastewater treatment systems, ecological treatment technologies have the advantages of limiting or eliminating the use of chemicals, not generating annoying odors, are easy to operate, and also more economical (Chiranjeevi et al., 2013; Bajracharya et al., 2016; Meena et al., 2019). The electrons released by the microbial oxidation of organic matter, flow from the anode to the cathode through an external load (electric resistance), whereas the protons diffuse through the selective membrane (Zhuwei

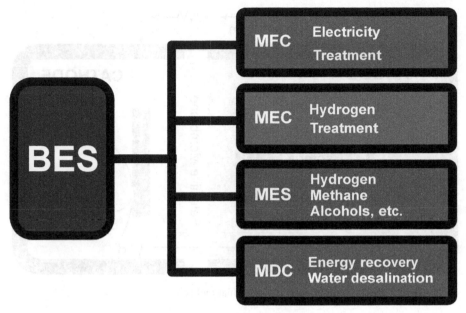

FIGURE 9.1 Different types of BES and main application.

et al., 2007; Gouveia et al., 2013). On the cathode, the electrons and protons react with oxygen forming water as an end product (Chiranjeevi et al., 2013; Venkata et al., 2014b). BES not only can be efficient in treating easily degradable substrates, such as domestic wastewater but also are considered to perform well-treating recalcitrant pollutants and toxic wastewaters (Ramirez-Vargas et al., 2018; Wang et al., 2020). This chapter provides an insight into the different types of BES and their applications, the basic principles involving electron transfer in the different types of BES, as well as the more recent and relevant developments of BES.

9.2 APPLICATIONS OF BIOELECTROCHEMICAL SYSTEMS

The different types of BES can be labeled as MXCs (e.g., microbial fuel cell; microbial electrosynthesis cell; microbial desalination cell; microbial electrolysis cell), in which the X simply represents the main function and the benefit of the cell (Jin and Fallgren, 2014). A microbial fuel cell (MFC) is the most characteristic type of BES, whose main function is the direct generation of electricity (Figure 9.2). When an external power supply is added to an MFC reactor to reduce the potential of the cathode, the system becomes a microbial

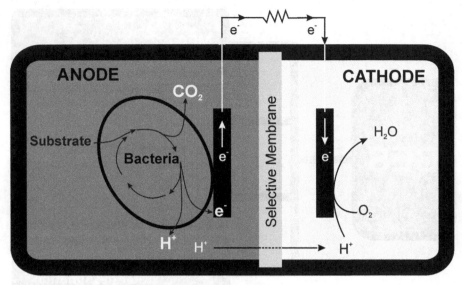

FIGURE 9.2   Basic configuration of a microbial fuel cell.

electrolysis cell (MEC), where hydrogen gas and other products are generated (Wang and Ren, 2013). More efficiently, hydrogen can be produced directly from biomass sources, using a bioelectrochemically assisted microbial cell (Pan et al., 2013). If the primary objective of the BES is to reduce oxidized contaminants, such as uranium, perchlorate, or chlorinated solvents, the cell can be named as a microbial remediation cell. If the main objective of the system is to synthesize high-value-added chemicals through cathodic reduction by microorganisms, the system can be named as a microbial electrosynthesis. Another system called microbial desalination cell includes an additional chamber between the anode and cathode and uses the internal potential to drive water desalination (Wang and Ren, 2013). Nutrients such as nitrogen with the uses of BES can be removed through bioelectrochemical denitrification or recovered through ammonium migration driven by electricity generation (Zhang et al., 2014; Nancharaiah et al., 2019).

9.3 MICROBIAL FUEL CELLS

MFC refers to a bioreactor system focused on the production of electricity from biodegradable materials, based on microbial catalytic reactions (Wang and Ren, 2013; Rashid et al., 2013a, b). The process implies the oxidation of organic carbon in the anode, under anaerobic conditions, producing electrons, protons, and products such as volatile fatty acids, biomass, and carbon dioxide (Velasquez-Orta et al., 2011). The process allows to treat wastewater with simultaneous recovery of electricity (Zhou et al., 2012; Luo et al., 2017; Kokko et al., 2018). Besides, the recovery of final high-value-added products confers an attractive biorefinery alternative (Venkata et al., 2014a; Nancharaiah et al., 2016; Jadhav et al., 2017). The typical MFC consists of an anodic chamber and a cathode chamber with electrodes, separated by a selective membrane that could only allow the exchange of anions or cations, the more common selective membranes are the proton exchange membrane (PEM) (Fu et al., 2010; Shentan et al., 2014). PEM accounts for most of the cost of MFC, the main function is as an insulator to maintain redox potential. This is achieved, because PEM separates different types and/or concentrations of electrolytes into two chambers, and restricts the exchange of specific ions. It is also the main internal resistance of MFC (Fu et al., 2009). Internal resistance is one of the key factors that affect the performance of fuel cells (Velasquez-Orta et al., 2011; Jafary et al., 2013;). Nafion has been considered one of the most popular PEMs, due to its highly selective proton permeability. However, many works have focused on developing less expensive and more durable substitutes (Zhuwei et al., 2007; Fu et al., 2010; Chiranjeevi et al., 2013; Rahimnejad et al., 2014; Yousefi et al., 2016). Microorganisms metabolize organic matter through biochemical reactions and transferring electrons and protons, these electrons and protons move through a series of redox components (NAD +, FAD, etc.) toward an available electron terminal acceptor, thus generating the proton motive force that facilitates the generation of rich phosphate bonds that microorganism use for growth and other metabolic activities (Venkata et al., 2014a). The electrons are collected in the anaerobic chamber, by an electrode at the anode, and passed through an external circuit (external load) and transferred to the cathode. At the same time, protons released from organic carbon oxidation, migrate to the cathode, through the selective membrane that limits the diffusion of oxygen in the anodic chamber (Velsquez-Orta et al., 2011). The following aspects limit the flow of electrons: (1) the activation overpotential required to transfer of electrons from the substrate to the anode, (2) the extracellular electron transfer mechanism and (3) internal resistance (Fu et al., 2009). The performance of MFC depends on several factors, such as cell configuration, nature of the anolyte, electrode material, electrode spacing, nature of microorganisms, electron transfer mechanism, pH, and temperature, among others (Zhuwei et al., 2007; Wang et al., 2017). An MFC inoculated with bacterial consortia has proven to be more stable and productive than MFC inoculated with pure strains, due to nutrient adaptability and resistance to stress factors (Jiang et al., 2013). The electricity produced in MFC through wastewater treatment has the potential to be used as an energy source of small or microrobots such as Gastrobots or could also be used for water electrolysis (Pant et al., 2011; Venkata et al., 2014a, b).

9.4 MICROBIAL ELECTROLYSIS CELLS

A MEC as previously described for the case of MFC uses the property of bacteria to convert chemical energy stored in the substrate to electrical energy, and allowing the electrolysis process of water (Liu et al., 2016). External power applied onto the electrical circuit of BES drives electrons from anode to the cathode and also supports the hydrogen generation at the cathode (Wang and Ren, 2013; Santoro et al., 2017). In contrast to MFC, the cathode of MEC works in anaerobic conditions to allow the production of hydrogen (Zou and He, 2018). However, the anaerobic conditions of the MEC's cathode can also promote the production of methane, once CO_2 and methanogens are available. A few of the methods to mitigate methane production include the aeration of the cathode chamber between batches, lowering the pH or adding methanogens inhibitors, short retention times, or thermal pretreatment of the inoculum to eliminate methanogens (Butti et al., 2016; Kokko et al., 2018). Higher electric currents are typically observed in MEC, when compared to MFC, due to the additional applied voltage that helps to overcome the cathode limitations in MEC. Another separate MFC can also provide the energy required for MEC operation as a power source. However, for the supply of constant voltage for hydrogen production in MEC a stack configuration of MFCs might be required. In a MEC with bioanode and biocathode, the expensive catalyst can be avoided, due to microorganisms present in the cathode decrease the start-up time and produces comparable current densities to those of bioanode (Haavisto et al., 2020). Furthermore, the hydrogen synthesized in MEC can also drive the biochemical production of other high-added-value products (Kumar et al., 2017; Santoro et al., 2017). The most common examples of reduction reactions at the cathode are protons reduction to hydrogen, oxygen reduction to hydrogen peroxide, and CO_2 reduction to methane and acetate (Bajracharya et al., 2017b).

9.5 MICROBIAL ELECTROSYNTHESIS

Microbial electrosynthesis (MES), also known as bioelectrosynthesis, is another variant of BES, which utilizes the reducing power generated from the anodic oxidation to produce high value-added products in the cathode.

The microorganisms present in the cathode that are commonly present as a microbial biofilm attached to the cathode's surface are responsible for reducing the available terminal electron acceptor to produce high value-added products, such as acetate, ethanol, and butyrate (Bajracharya et al., 2016; Zou and He, 2018). As described by Bajracharya et al. (2017a), the bioelectrosynthesis of diverse chemical compounds is driven by CO_2 reduction as well as reduction/oxidation of other organic feedstocks using microbes as biocatalysts. Several reports have described the reduction of CO_2 to acetate and other multicarbon extracellular products with hydrogen as the electron donor (Bajracharya et al., 2017a, b; Fazal et al., 2019). In addition, MES can have also been proposed as an alternative strategy to capture electrical energy in covalent chemical bonds of organic products (Bajracharya et al., 2016).

9.6 MICROBIAL DESALINATION CELLS

Compared to conventional BES such as MFC, MEC, and MES, microbial desalination cells (MDC) (see Figure 9.3) generally have an extra chamber located in the middle of the MDC, this chamber is in which the desalination is carried out (Yang et al., 2019). The chambers of the MDC are separated by two different selective membranes, an anion exchange membrane and a cation exchange membrane. As in the other variants of BES, the microorganism present in the anodic chamber oxidize the organic matter into CO_2 and generating electrons and protons in the process. These electrons as in travel from the anode to the cathode through an external circuit, and are then used to reduce electron acceptors in the cathode chamber. Anions and cations present in the desalination chamber migrate to the anode and cathode chambers passing through the selective membranes and maintain a balance in the electric charge of the MDC, allowing the desalination of water to be carried out without any additional energy input. Desalination efficiency will significantly depend on the electric generation by the MDC (Yang et al., 2019).

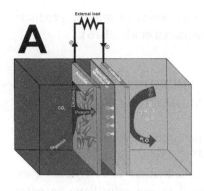
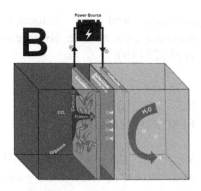
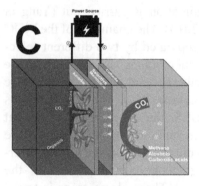
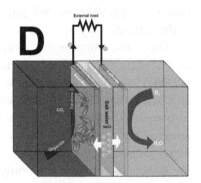

FIGURE 9.3 Different configurations of main bioelectrochemical systems: (A) microbial fuel cell (MFC); (B) microbial electrolysis cell (MEC); (C) microbial electrosynthesis (MES); (D) microbial desalination cell (MDC).

9.7 TRANSFER OF ELECTRONS IN THE ANODIC CHAMBER

BES uses microorganisms, to oxidize donors of organic and inorganic electrons, mainly from waste materials, and transfer electrons to the anode electrode. A group of microorganisms, called electrochemically active bacteria, of exoelectrogenic nature, and capable of converting stored chemical energy into organic or inorganic substrates, are used to produce electrical energy during anaerobic respiration (Wang and Ren, 2013). There exist two possible ways to transfer electrons: direct contact, and bacteria donating electrons directly to the electrode, through several "C" type cytochromes, present in their cell membrane (see Figures 9.4 and 9.5). The latter requires 5 specific proteins, namely, OmcA, MtrC, MtrA, MtrB, and CymA, belonging to the "C" type

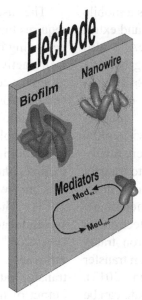

FIGURE 9.4 Different extracellular electron transfer to electrodes carried out by microorganisms in BES.

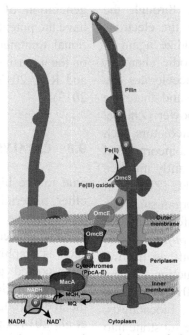

FIGURE 9.5 Extracellular electron transfer mechanism of *Geobacter sulfurreducens* as described by Lovley (2006).

cytochrome family. MtrA is a mobile protein that links the intra- and extracellular environment. OmcA is also known for its role in fixing bacteria to electrodes forming biofilms. A second way to transfer electrons is through electrically conductive appendages (nanowires/pilis) (Venkata et al., 2014b), where microorganisms have the ability to transfer electrons through protein structures attached to the cell membrane; Another way to transfer electrons is the use of a mobile electron transport, such as indirect electron transfer mediators (Wang and Ren, 2013). Mediators in an oxidized state can be easily reduced, by capturing electrons from the inside of the membrane. The mediators move through the membrane and release the electrons to the anode, and oxidize again in the solution in the anodic chamber. This cyclic process accelerates the electron transfer rate, and therefore increases the output power (Zhuwei et al., 2007). Electron mediators, such as Mn4 +, or neutral red incorporated into the anode, significantly improve the performance of MFC (Zhuwei et al., 2007). Each microorganism uses a different electron transfer system, for example, *Geobacter sulfurreducens* uses its conductive pili as nanowires for electron conduction (Logan, 2009; Venkata et al., 2014b). The use of microorganisms as biocatalysts, convert any degradable substrate into energy and chemical end products.

The use of complex materials requires the cooperation of polymer-degrading bacteria and electrochemically active bacteria, with the first group the complex polymers, such as cellulose or proteins that are broken down into simple organic matter, such as volatile fatty acids, alcohol, or ammonium, and the second group oxidizes these simple organic products in the anode (Wang and Ren, 2013). The degradability of the substrate and the anodic microbial ecosystem are the factors that most influence the microbial electron transfer rate, and the final current output of the MFC (Velásquez-Orta et al., 2011). Waste treatment in the anode chamber represents a new generation of technology, as they have the potential to transform traditional treatment processes, focused on the intensive use of energy (Wang and Ren, 2013; Nancharaiah et al., 2019).

9.8 CATALYSTS

Due to the high cost of platinum, other materials are often used as cost-effective alternatives at the expense of catalytic performance, such as graphite, graphite felt, platinum-coated graphite, and carbon cloth electrodes and other coated metal electrodes (Fu et al., 2009; Liu et al., 2014; Li et al., 2017; Santoro et al., 2017). Ferricianide has also been widely used in MFC systems, and though it is not that expensive, it requires replacement and is toxic to the environment (Gajda et al., 2013). The replacement of catalysts with photosynthetic microalgae species is a very promising approach to improve cathodic performance (Yufeng et al., 2014).

9.9 MICROALGAE IN THE CATHODE CHAMBER

Microalgae have received much attention in recent years, due to their potential application for industrial CO_2 extraction (see Figure 9.6)

and the production of many valuable metabolites (Yang et al., 2000; Azwar, et al., 2013; Elmekawy et al., 2018). The use of microalgae has been extended for wastewater treatment, biodiesel production, power generation, and as photosynthetic gas exchangers for space travel (Fu et al., 2009; Subashchandrabose et al., 2011; Subashchandrabose et al., 2011; Zhou et al., 2012; Cui et al., 2014; Elmekawy et al., 2018). Microalgae are included in the definition of plants as unicellular and microscopic eukaryotes, with size ranging from 1/1000 to 2 mm. These organisms have been isolated, mutated, and used in genetic engineering for the effective

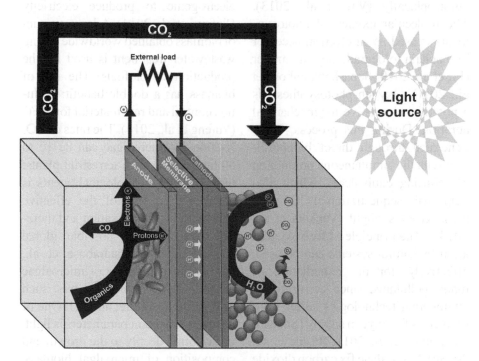

FIGURE 9.6 Photobioelectrochemical cell with microalgae in the cathode.

bioremediation of recalcitrant organic pollutants, achieving an improvement of the degradation rate, and guaranteeing a better survival and colonization in contaminated areas (Subashchadrabose et al., 2011). At present, the applications for expanding biofuels derived from microalgae are not yet economical (Yufeng et al., 2014). Most microalgae are commonly found in association with other microorganisms (Subashchadrabose et al., 2011). Microalgae can perform oxygenic photosynthesis, producing up to 75% of the Earth's oxygen (Gajda et al., 2013; Zhou et al., 2012) and simultaneously consume the greenhouse gas (carbon dioxide), to grow photoautotrophically (Wu et al., 2013). The molecular oxygen of photosynthesis is used as an electron acceptor by bacteria to degrade organic matter (Fu et al., 2010; Subashchadrabose et al., 2011). Oxygen photosynthesis is a potential alternative for mechanical aeration. During this process, O_2 is generated through direct biophotolysis with the simultaneous production of reducing equivalents, which also helps in the sequestration of CO_2 in the presence of sunlight (Venkata et al., 2014a). The principle of free oxygenation by natural systems can be used, effectively, for the remediation of many pollutants, since conventional engineering technologies suffer from high costs for oxygen supply (Subashchadrabose et al., 2011; Rashid et al., 2013a). Microalgae fix carbon dioxide through the Calvin cycle, using light as a source of energy and assimilate CO_2 as a source of carbon (Yang et al., 2000; Rashid et al., 2013b). Microalgae can absorb both nitrogen and phosphorus, with biomass generation, that may be used for energy production (Zhang et al., 2014). Due to their high efficiency of photosynthesis and lipid content, microalgae have the potential to produce new biofuel energy (Wu et al., 2013). The use of microalgae biomass has been reported as a substrate for bacteria in the anodic compartments (Gouveia et al., 2013), since it contains high levels of lipids (32%) and carbohydrates (51%) that are easily degradable by bacteria electrogenic to produce electricity (Yufeng et al., 2014). A large amount of biomass obtained worldwide during wastewater treatment is used for the production of biofuels. The use of biomass has a double benefit: pollution control and raw material for MFC (Yufeng et al., 2014). The rates of CO_2 fixation by microalgae can be 10 to 50 times faster than terrestrial plants; the use of these biological agents is considered as one of the effective approaches to CO_2 fixation and therefore mitigation of possible global warming (Subashchadrabose et al., 2011). The cultivation of microalgae depends on several parameters, such as light, temperature, and pH. One of the most important parameters is light, contributing deeply to the growth and composition of microalgal biomass,

(Gouveia et al., 2013; Wu et al., 2013), since this factor affects photosynthesis and the metabolic pathway of microalgae (Wu et al., 2013). The depletion of ammonium, nitrate, and phosphate due to the growth of microalgae, is fairly advantageous for the removal of nutrients from wastewater (Subashchadrabose et al., 2011). Under conditions of nutrient limitation, especially nitrogen, microalgae increase their production of lipids, carbohydrates and /or pigments (Gouveia et al., 2013). Microalgae release a variety of organic molecules, which include low molecular weight compounds and extrapolymeric substances composed of proteins, nucleic acids, mannitol and arabinose as the excretion products; acetate, propionate, lactate and ethanol as fermentation products. All these molecules serve as substrates for the growth of bacteria (Subashchadrabose et al., 2011). The microalgae cell walls are composed of polysaccharides and carbohydrates, which have a negative charge. Most metals bind to negatively charged groups of ligands, being the basis for the removal of metals from wastewater. However, heavy metals are potent inhibitors of photosynthesis, since they can replace or block the active sites of certain enzymes (Subashchadrabose et al., 2011). The studies are still in laboratory stages and the power generation yields are still relatively low (Wu et al., 2013).

9.9.1 MICROALGAE METABOLISM

Microalgae cells can use light as a source of energy, the light absorbed by pigments drives the photosynthetic transport of electrons, resulting in the reduction of NADP + and the production of ATP. It has been concluded that the number of ATP molecules formed from a pair of electrons that move through the photosynthetic electron transport chain is around 1.3. NADPH and ATP are formed by the action of light; CO_2 is reduced by a series of dark reactions called the Calvin cycle. The first step is the fixation of CO_2, catalyzed by ribulose 1,5-bisphosphate carboxylase. This enzyme is also an oxygenase, which can react with O_2 and give rise to a different pathway called photorespiration. Algae have the path of photorespiration and photosynthesis inhibited by a high concentration of O_2. Photosynthesis, including light reactions, the Calvin cycle and starch synthesis, is found in chloroplasts (Yang et al., 2000). Photosynthesis under conditions of light, carbon dioxide, and water produces oxygen as a by-product. Under dark conditions, the starch stored inside the microalgae will be consumed by oxidation (Fu et al., 2009). Glyceraldehyde phosphate (GAP) is exported to the cytoplasm for consumption. After the export of GAP, the carbon flux is divided into sugar synthesis or oxidation, through the glycolytic pathway to obtain pyruvate. Glucose can be converted directly to starch without the need for conversion before GAP. In mixotrophic metabolism, a part of the exogenous glucose is converted directly into starch, and the rest is oxidized through the glycolytic pathway. The pentose phosphate (PP) pathway has been reported to operate at the same time as the Calvin cycle in the cytoplasm. In the autotrophic and mixotrophic metabolic pathways, only the pentose phosphate pathway supplies PP for the synthesis of nucleic acids. Nitrate is the predominant form of nitrogen available to most plants. Nitrate is reduced to ammonia in two steps: nitrate is reduced by a cytoplasmic NADH nitrate reductase to nitrite, which is further reduced to ammonia. Glutamate dehydrogenase and glutamine synthetase are considered as important inputs of ammonia in organic form (Yang et al., 2000; Feijuan et al., 2012).

9.10　LIFE-CYCLE ASSESSMENT OF BIOELECTROCHEMICAL SYSTEMS

With each new technology, the environmental footprints must be clearly marked before it is applied on a large scale. It is even more important in the case of bioenergy systems because bioenergetic technologies use renewable biomass resources to produce a range of energy-related products, including electricity, chemical end products product, and other materials (Pant et al., 2011; Zhang et al., 2019). Research so far has been limited to laboratory-scale reactors, but in the coming years technology is expected to be possible on a large scale. To avoid the unwanted consequences of emerging technologies, a life-cycle assessment of bioelectrochemical systems should be carried out (Pant et al., 2011; Zhang et al., 2019). Life-cycle assessment has been the method of choice for bioenergy and carbon-sequestration technologies, it is an approach to determine the environmental consequences, impacts on human health, and the depletion of sources of a particular product, throughout its entire production cycle. This approach over time will reveal the true potential of the product evaluated, so that cautionary measures can be suggested, to reduce the negative environmental impact. It is a method that defines and reduces the environmental burdens of a product, process or activity, identifying and quantifying the use of energy and materials, waste discharges, and evaluation of opportunities for improvement throughout the life cycle. The potential of a bioelectrochemical system, in relation to greenhouse gas emissions, must be taken into consideration, when evaluating the life cycle of these systems (Pant et al., 2011).

9.11 CONCLUSION

MFC are a special type of bioelectrochemical systems, which use active biocatalysts to generate energy. They are one of the most promising approaches to bioremediation of wastewater and simultaneously produce electrical energy. Although these technologies are barely made on a laboratory scale, they have received much attention by researchers in the last decade, due to their novel approach in solving environmental pollution problems and the depletion of fossil fuels. This technology opens a new interdisciplinary field, which involves different areas for research and development, including microbiology and electrochemistry. Microalgae have been proposed as an alternative for the extraction of greenhouse gas (CO_2) during the wastewater treatment process and for obtaining valuable products such as ethanol and hydrogen.

KEYWORDS

- **bioelectrochemical systems**
- **electron transfer**
- **energy**
- **high value-added by-products**
- **wastewater treatment**

REFERENCES

Azwar, M.Y., Hussain, M.A., Abdul-Wahab, A.K. Development of biohydrogen production by photobiological, fermentation and electrochemical processes: a review. *Renew. Sust. Energy* Rev. **2013**, 31, 158–173.

Bajracharya, S., Sharma, M., Mohanakrishna, G., Dominguez-Benneton, X., Strik, D.P.B. T. B., Sarma, P.M., Pant, D. An overview on emerging bioelectrochemical systems (BESs): technology for sustainable electricity, waste remediation, resource recovery, chemical production and beyond. *Renew. Energy* **2016**, 98, 153–170.

Bajracharya, S., Srikanth, S., Mohanakrishna, G., Zacharia, R., Strik, D. P., Pant, D. Biotransformation of carbon dioxide in bioelectrochemical systems: state of the art and future prospects. *J. Power Sources* **2017a**, 356, 256–273.

Bajracharya, S., Vanbroekhoven, K., Buisman, C.J.N., Strik D.P.B.T.B., Pant D. Bioelectrochemical conversion of CO_2 to chemicals: CO_2 as a next generation feedstock for electricity-driven bioproduction in batch and continuous modes. *Faraday Discuss.* **2017b**, 202, 433–449.

Butti, S. K., Velvizhi, G., Sulonen, M. L. K., Haavisto, J. M., Oguz Koroglu, E., Yusuf Cetinkaya, A., Singh, S., Arya, D., Annie-Modestra, J., Vamsi-Krishna, K., Anil, V., Bestami, O., Aino-Maija, L., Jaakko, A.P., Venkata, M.S. Microbial electrochemical technologies with the perspective of harnessing bioenergy: maneuvering towards upscaling. *Renew. Sustain. Energy* Rev. **2016**, 53, 462–476.

Chiranjeevi, P., Chandra, R., Venkata, M.S. Ecologically engineered submerged and emergent macrophyte based system: An integrated eco-electrogenic design for harnessing power with simultaneous wastewater treatment. *Ecol. Eng.* **2013**, 51, 181–190.

Cui, M. H., Gao, L., Lee, H. S., Wang, A. J. Mixed dye wastewater treatment in a bioelectrochemical system-centered process. *Bioresour. Technol.* **2019**, 297, 122420. https://doi.org/10.1016/j.biortech.2019.122420

De Vrieze, J., Arends, J. B. A., Verbeeck, K., Gildemyn, S., Rabaey, K. Interfacing anaerobic digestion with (bio)electrochemical systems: potentials and challenges. *Water Res.* **2018**, 146, 244–255.

Elmekawy, A., Hegab, H. M., Vanbroekhoven, K., Pant, D. Techno-productive potential of photosynthetic microbial fuel cells through different configurations. *Renew. Sustain. Energy Rev.* **2014**, 39, 617–627.

Fazal, T., Saif ur Rehman, M., Mushtaq, A., Hafeez, A., Javed, F., Aslam, M., Fatima, M., Faisal, A., Iqbal, J., Rehman, F., Farooq, R. Simultaneous production of bioelectricity and biogas from chicken droppings and dairy industry wastewater employing bioelectrochemical system. *Fuel* **2019**, 256, 115902.

Feijuan, W., Zhu Cheng, Z. Effects of nitrogen and light intensity on tomato (*Lycopersicon esculentum* Mill) production under soil water control. *Afr. J. Agric. Res.* **2012**, 7, 4408–4415.

Fu, C.C., Hung, T.C., Wu, W.T., Wen, T.C., Su, C.H. Current and voltage responses in instant photosynthetic microbial cells with *Spirulina platensis. Biochem. Eng. J.* **2010**, 52, 175–180.

Fu, C.C., Su C.H., Hung, T.C., Hsieh, C.H., Suryani, D., Wu, W.T. Effects of biomass weight and light intensity on the performance of photosynthetic microbial fuel cells with *Spirulina platensis. Bioresour. Technol.* **2009**, 100, 4183–4186.

Gajda, I., Greenman, J., Melhuish, C., Ieropoulos, I. Photosynthetic cathodes for Microbial Fuel Cells. *Int. J. Hydrogen Energy* **2013**, 38, 11559–11564.

Gouveia Luisa, Neves Carole, Sebastiao Diogo, Nobre Beatriz P., Matos Cristina T. Effect of light on the production of bioelectricity and added-value microalgae biomass in a photosynthetic alga microbial

fuel cell. *Bioresour. Technol.* **2013**, 154, 171–177.

Haavisto, J. M., Kokko, M. E., Lakaniemi, A. M., Sulonen, M. L. K., Puhakka, J. A. The effect of start-up on energy recovery and compositional changes in brewery wastewater in bioelectrochemical systems. *Bioelectrochemistry.* **2020**, 132, 107402. Pre-Proof. https://doi.org/10.1016/j.bioelechem. 2019.107402

Jadhav, D. A., Ghosh Ray, S., Ghangrekar, M. M. Third generation in bio-electrochemical system research—a systematic review on mechanisms for recovery of valuable by-products from wastewater. *Renew. Sustain. Energy Rev.* **2017**, 76, 1022–1031.

Jafary, T., Rahimnejad, M., Ghoreyshi, A.A., Najafpour, G., Hghparast, F., Wan Daud, W.R. Assessment of bioelectricity production in microbial fuel cells through series and parallel connections. *Energy Convers. Manage.* **2013**, 75, 256–262.

Jiang, Y., Ulrich, A.C., Liu Y.. Coupling bioelectricity generation and oil sands tailings treatment using microbial fuel cells. *Bioresour. Technol.* **2013**, 139, 349–354.

Jin, S. Fallgren, P. H. Feasibility of using bioelectrochemical systems for bioremediation. *Microb. Biodegrad. Bioremediation.* **2014**, 389–405.

Jung, S., Lee, J., Park, Y. K., Kwon, E. E. Bioelectrochemical systems for a circular bioeconomy. *Bioresour. Technol.* **2020**, 300, 122748. https://doi.org/10.1016/j.biortech.2020.122748

Kokko, M., Epple, S., Gescher, J., Kerzenmacher, S. Effects of wastewater constituents and operational conditions on the composition and dynamics of anodic microbial communities in bioelectrochemical systems. *Bioresour. Technol.* **2018**, 258, 376–389.

Kumar, G., Saratale, R. G., Kadier, A., Sivagurunathan, P., Zhen, G., Kim, S. H., Saratale, G. D. A Review on bio-electrochemical systems (BESs) for the syngas and value added biochemicals production. *Chemosphere* **2017**, 177, 84–92.

Li, J., Hu, L., Zhang, L., Ye, D. ding, Zhu, X., Liao, Q. Uneven biofilm and current distribution in three-dimensional macroporous anodes of bio-electrochemical systems composed of graphite electrode arrays. *Bioresour. Technol.* **2017**, 228, 25–30.

Liu, W., Cai, W., Guo, Z., Wang, L., Yang, C., Varrone, C., Wang, A. Microbial electrolysis contribution to anaerobic digestion of waste activated sludge, leading to accelerated methane production. *Renew. Energy.* **2016**, 91, 334–339.

Liu, S., Song, H., Wei, S., Yang, F., Li, X. Biocathode materials evaluation and configuration optimization for power output of vertical subsurface flow constructed wetland-microbial fuel cell systems. *Bioresour. Technol.* **2014**, 166, 575–583.

Logan, B.E. Exoelectrogenic bacteria that power microbial fuel cells. *Nat. Rev. Microbiol.* **2009**, 7, 375–381.

Lovley, D. R. Bug Juice: Harvesting Electricity with Microorganisms. *Nat. Rev. Microbiol.* **2006**, 4, 497–508.

Luo, S., Wang, Z. W., He, Z. Mathematical modeling of the dynamic behavior of an integrated photo-bioelectrochemical system for simultaneous wastewater treatment and bioenergy recovery. *Energy.* **2017**, 124, 227–237.

Meena, R. A. A., Yukesh Kannah, R., Sindhu, J., Ragavi, J., Kumar, G., Gunasekaran, M.; Rajesh Banu, J. Trends and resource recovery in biological wastewater treatment system. *Bioresour. Technol. Rep.* **2019**, *7* (September), 100235.

Nancharaiah, Y. V., Mohan, S. V., Lens, P. N. L. *Removal and Recovery of Metals and Nutrients From Wastewater Using Bioelectrochemical Systems*, Elsevier BV, **2019**.

Nancharaiah, Y. V., Venkata Mohan, S., Lens, P. N. L. Recent advances in nutrient removal and recovery in biological and bioelectrochemical systems. *Bioresour. Technol.* **2016**, 215, 173–185.

Pant Deepak, Singh Anoop, Van Bogaert Gilbert, Alvarez Gallego Yolanda, Diels

Ludo, Vanbroekhoven Karolien. An introduction to the life cycle assessment (LCA) of bioelectrochemical systems (BES) for sustainable energy and product generation: relevance and key aspects. *Renew. Sust. Energy Rev.* **2011**, 15, 1305–1313.

Rahimnejad, M., Bakeri, G., Najafpour, G., Ghasemi, M., Oh S.E. A review on the effect of proton exchange membranes in microbial fuel cells. *Biofuel Res. J.* **2014**, 1, 7–15.

Ramírez-Vargas, C. A., Prado, A., Arias, C. A., Carvalho, P. N., Esteve-Núñez, A., Brix, H. Microbial electrochemical technologies for wastewater treatment: principles and evolution from microbial fuel cells to bioelectrochemical-based constructed wetlands. *Water* **2018**, 10, 1–29.

Rashid N., Cui, Y.F., Rehman, M.S.U., Han, J.I. Enhanced electricity generation by using algae biomass and activated sludge in microbial fuel cell. *Sci. Total Environ.* **2013a**, 456–457, 91–94.

Rashid, N., Rehman, M.S.U., Memon, S., Rahman Z.U., Lee K., Han J.I. Current status, barriers and developments in biohydrogen production by microalgae. *Renew. Sust. Energy Rev.* **2013b**, 22, 571–579.

Santoro, C., Arbizzani, C., Erable, B., Ieropoulos, I. Microbial fuel cells: from fundamentals to applications. A review. *J. Power Sources* **2017**, 356, 225–244.

Subashchandrabose, S.R., Ramakrishnan, B., Megharaj, M., Venkateswarlu, K., Naidu, R. Consortia of cyanobacteria/microalgae and bacteria: biotechnological potential. *Biotechnol. Adv.* **2011**, 29, 896–907.

Tian, Y., Ji, C., Wang, K., Le-Clech, P. Assessment of an anaerobic membrane bio-electrochemical reactor (AnMBER) for wastewater treatment and energy recovery. *J. Membr. Sci.* **2014**, 450, 242–248.

Velasquez-Orta S.B., Head I.M., Curtis T.P., Scott K. Factors affecting current production in microbial fuel cells using different industrial wastewaters. *Bioresour Technol.* **2011**, 102, 5105–5112.

Venkata M.S., Srikanth S., Chiranjeevi P., Arora Somya, Chandra Rashmi. Algal biocathode for in situ terminal electron acceptor (TEA) production: synergetic of bacteria-microalgae metabolism for the functioning of biofuel cell. *Bioresour. Technol.* **2014a**, 166, 566–574.

Venkata M.S., Velvizhi G., K. Vamshi K., Babu M.L. Microbial catalyzed electrochemical systems: a biofactory with multifacet applications. *Bioresour. Technol.* **2014b**, 165, 355–364.

Wang, H. C., Cheng, H. Y., Cui, D., Zhang, B., Wang, S. Sen, Han, J. L., Su, S. G., Chen, R., Wang, A. J. Corrugated stainless-steel mesh as a simple engineerable electrode module in bio-electrochemical system: hydrodynamics and the effects on decolorization performance. *J. Hazard. Mater.* **2017**, 338, 287–295.

Wang, Z., He, Z. Demystifying terms for understanding bioelectrochemical systems towards sustainable wastewater treatment. *Curr. Opin. Electrochem.* **2020**, 19, 14–19.

Wang, H., Ren, Z. J. A comprehensive review of microbial electrochemical systems as a platform technology. *Biotechnol. Adv.* **2013**, 31, 1796–1807.

Wu, Y.C., Wang, Z.J., Zheng, Y., Xiao, Y., Yang, Z.H., Zhao, F.. Light intensity affects the performance of photo microbial fuel cells with *Desmodesmus s.p.* A8 as cathodic microorganism. *Appl. Energy.* **2013**, 116, 86–90.

Yang, E., Chae, K. J., Choi, M. J., He, Z., Kim, I. S. Critical review of bioelectrochemical systems integrated with membrane-based technologies for desalination, energy self-sufficiency, and high-efficiency water and wastewater treatment. *Desalination* **2019**, 452, 40–67.

Yang Chen, Hua Qiang, Shimizu Kazuyuki. Energetics and carbon metabolism during growth of microalgal cells under photo-autotrophic, mixotrophic and cyclic light autotrophic/dark-heterotrophic conditions. *Biochem. Eng. J.* **2000**, 6, 87–102.

Yousefi, V., Mohebbi-Kalhori, D., Samimi, A., Salari, M. Effect of separator electrode assembly (SEA) design and mode of operation on the performance of continuous tubular microbial fuel cells (MFCs). *Int. J. Hydrogen Energy* **2016**, 41, 597–606.

Yufeng Cui, Naim Rashid, Naixu Hu, Muhammad Saif Ur Reham, Jong.In Han. Electricity generation and microalgae cultivation in microbial fuel cell using microalgae-enriched anode and biocathode. *Energy Convers. Manage.* **2014**, 674–680.

Zhang Fei, Li Jian, He Zhen. A new method for nutrients removal and recovery from wastewater using a bioelectrochemical system. *Bioresour. Technol.* **2014**, 166, 630–634.

Zhang, J., Yuan, H., Abu-Reesh, I. M., He, Z., Yuan, C. Life cycle environmental impact comparison of bioelectrochemical systems for wastewater treatment. *Procedia CIRP* **2019**, 80, 382–388.

Zhou Minghua, He Huanhuan, Jin Tao, Wang Hongyu. Power generation enhancement in novel microbial carbon capture cells with immobilized *Chlorella vulgaris*. *J. Power Sources* **2012**, 214, 216–219.

Zhuwei Du, Haoran Li, Tingyue Gu. A state of the art review on microbial fuel cells: a promising technology for wastewater treatment and bioenergy. *Biotechnol. Adv.* **2007**, 25, 464–482.

Zou, S. He, Z. Efficiently "pumping out" value-added resources from wastewater by bioelectrochemical systems: a review from energy perspectives. *Water Res.* **2018**, 131, 62–73.

CHAPTER 10

Valorization of Nonnative Aquatic Weeds Biomass Through Their Conversion to Biofuel

FERNANDO MÉNDEZ-GONZÁLEZ[1], ALEJANDRA PICHARDO-SÁNCHEZ[1], BEN HUR ESPINOSA-RAMÍREZ[2], NUBIA R. RODRÍGUEZ-DURÁN[3], GUADALUPE BUSTOS-VÁZQUEZ[3], and LUIS V. RODRÍGUEZ-DURÁN[3*]

[1]Departamento de Biotecnología, Universidad Autónoma Metropolitana-Iztapalapa, Ciudad de México, México

[2]División de Electromecánica Industrial, Universidad Tecnológica de Tecámac, Tecámac, Estado de México, México

[3]Departamento de Ingeniería Bioquímica, Universidad Autónoma de Tamaulipas, Unidad Académica Multidisciplinaria Mante, Blvd. Enrique Cárdenas González No. 1201 Pte. Col. Jardín, C.P. 89840 Ciudad Mante, Tamaulipas, México

*Corresponding author. E-mail: luis.duran@docentesuat.edu.mx

ABSTRACT

The climate change suffered by our planet has contributed to the mobility and proliferation of nonnative species, some of them causing serious ecological damage and economic problems. Among the nonnative species are organisms considered aquatic weeds such as plants and macroalgae, which have spread massively in various aquatic ecosystems. The strategies implemented for their removal have a high cost and are not capable of eradicating the plague. Aquatic weeds have exceptionally high reproduction rates and their cellulose and hemicellulose content makes them attractive feedstocks for biofuels production. Therefore this chapter presents some strategies implemented for the use and valorization of the biomass of these aquatic organisms through their conversion to biofuels.

10.1 INTRODUCTION

Worldwide there is a high energy demand; however, in recent years, a specific interest in fossil fuel replacement has been generated. Which causes economic and environmental damage (Silva et al., 2011). That is why, in recent years, the production and use of biofuels have gained importance as an alternative to the use of fossil fuels (Inderwildi and King, 2009). Among the most produced biofuels are bioethanol and biodiesel (Azadi et al., 2017). However, other types of fuels such as biogas, bio-oil, and syngas/H$_2$ have high prospects for generation and use. Biofuels come from three different sources: (1) Starch, sugar, and triglyceride; (2) Lignocellolosics; and (3) micro-algae (Azadi et al., 2017). The biological sources of the lignocel-lulosic group are the most used for the generation of bioethanol, bio-oil, biogas, and syngas/H$_2$. Within this group, we can locate the biomass from aquatic weeds that include aquatic plants (fresh-water weeds) and macroalgae (salt-water weeds). They are nonedible biomass and in different areas of the world causes various economic (Engelen and Santos, 2009; Cook et al., 2013) and environmental prob-lems (Gunnarsson and Petersen, 2007). However, their valorization through their conversion to second-generation biofuels such as ethanol

and biogas could contribute to the population control of these species in the infestation zones (Figure 10.1). Also, its high reproduction rate would reduce problems associ-ated with other raw materials such as deforestation and the increase in food prices (Mussatto et al., 2010; Borines et al., 2011). Therefore this chapter presents the charac-teristics of aquatic weeds in terms of their potential for conversion to biofuel, scientific and techno-logical advances aimed at the valo-rization of the biomass of aquatic weeds due to their conversion to biofuel (bioethanol and biogas) and the prospects related to the use of biomass of aquatic weeds for the production of biofuels.

10.2 POTENTIAL OF AQUATIC WEED BIOMASS FOR THE BIOFUELS PRODUCTION

The aquatic weeds (macroalgae and plants) are considered a potential plague that causes high costs for their removal. However, their biomass is a resource with high potential growth rates and high content of hydrolyz-able carbohydrates (Bruhn et al., 2011). Therefore they have great potential as a feedstock for biofuels production, which could allow to raise incomes, provide employment and rural development in prolifera-tion zones. In this section, some prop-erties of aquatic weeds (macroalgae

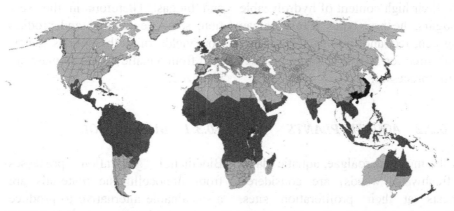

FIGURE 10.1 Areas with high aquatic weed infestation. The colors blue, brown, red, and green symbolize freshwater weeds, brown macroalgae, red macroalgae, and green macroalgae, respectively. Created at mapchart.net.

and plants) for their conversion to biofuel were analyzed.

10.2.1 MACROALGAE

Macroalgae are organisms with the ability to absorb nutrients in most of their surface area (John et al., 2011); This reduces the energy requirements associated with nutrient transport (Wi et al., 2009) and as a consequence, they have a high growth rate (compared to land crops) (Wi et al., 2009; Goh and Lee, 2010). Macroalgae composition has a high content of hydrolyzable carbohydrates (Nkemka and Murto, 2010) and therefore a viable energy source. The energy potential from the marine biomass was estimated at around 100 EJ/yr, which is higher than terrestrial biomass and municipal solid wastes (22 and 7 EJ/yr, respectively) (Chynoweth et al.,

2011). Macroalgae are divided into three groups: Chlorophyta (green algae), Rhodophyta (red algae), and Phaeophyta (brown algae) (Borines et al., 2011). Green algae contain a high content of starch, pectin, and cellulose (Ortiz et al., 2006; Lahaye and Robic, 2007). Red algae contain a low presence of cellulose, the compounds that predominate in its composition are sulfated galactan polymers, such as agar, carrageenan, and funoran (Saravanan et al., 2018). On the other hand, brown seaweed has alginic acid, cellulose, hemicellulose, laminarin, fucose, xylose, and mannitol (Bertagnolli et al., 2014; Davis et al., 2003; Marquez et al., 2014). By containing compounds of commercial interest, some types of macroalgae are used for extraction processes. However, the residues of these processes have the potential to be used for biofuels production. Due to their high content of hydrolyzable sugars, the three types of macroalgae (green, red, and brown) are a viable resource for feeding biofuel production processes.

10.2.2 AQUATIC PLANTS

Like most macroalgae, aquatic plants (freshwater weeds) are considered pests at their proliferation sites. However, aquatic plants such as water hyacinth (*Eichhornia crassipes*), water lettuce (*Pistia stratiotes* L.), and duckweed (*Landoltia punctata*) represent a promising feedstock source for biofuels production processes due to their high availability, resistance to extreme environmental conditions (water level, nutrient availability, pH, temperature, and toxic substances) (Ganguly et al., 2012), biomass yields (60-100 ton/ ha yr) (Mishima et al., 2008), and high content of cellulose and hemicellulose (Gressel, 2008). Because these plants proliferate in freshwater reservoirs their salt content is lower compared to macroalgae and, consequently, biomass pretreatment is less expensive.

10.3 BIOFUELS PRODUCTION FROM BIOMASS OF AQUATIC WEEDS

Among the most studied and produced biofuels generated from aquatic weeds biomass are bioethanol

and biogas. Therefore in this section, some generalities and production yields of both biofuels generated from aquatic weeds biomass are analyzed.

10.3.1 BIOETHANOL

Bioethanol generation processes from lignocellulosic materials are a sustainable alternative to produce energy-efficient sources. In those processes, to implement some type of pretreatment is necessary to increase the biomass availability for microbial and/or enzymatic action (Wi et al., 2009; Taherzadeh and Karimi, 2007), to minimize the loss of sugar, and to consume energy (Das et al., 2016). After the pretreatment, the hydrolyzed matter is fermented and, subsequently, the ethanol is recovered by distillation. The fermentation process can be done by microbial action (yeasts and bacteria alcoholic fermentation) or for the thermochemical process (gasification). The aquatic weeds contents high pentoses concentration (up to 45%) (Kumar et al., 2009); these sugars are not assimilable by *Saccharomyces cerevisiae*. Therefore studies are oriented toward the use of other microorganisms such as *Pichia stipitis* (Gunan-Yucel and Aksu, 2015) and *Escherichia coli* (recombinant strain) (Mishima et al., 2008), which have shown effectiveness in converting xylose to ethanol.

Macroalgae such as *Sargassum* sp., *Gracilaria* sp., *Prymnesium parvum*, *Gelsium amansii*, and *Laminaria* sp. are characterized by having a high carbohydrate content (Wi et al., 2009; Adams et al., 2009; Horn et al., 2000a); which can be hydrolyzed and converted to ethanol (Table 10.1). Biomass (regularly in brown algae) with a high content of mannitol makes it necessary to add an oxidation process to fructose for subsequent conversion; this process can be carried out by the action of the enzyme mannitol dehydrogenase. Yields of 0.38 g ethanol/g mannitol from *Laminaria hyporbea* have been achieved by fermentation with *Pichia angophorea* (Horn et al., 2000b). On the other hand, fresh-water plants such as *E. crassipes*, *P. stratiotes*, and *L. punctata* are used for ethanol production. Both feedstock sources (macroalgae and plants) have a similar ethanol production yield, which is from 13 to 30 g/L (Table 10.1). The ethanol

TABLE 10.1 Ethanol Production From Aquatic Weeds

Organism Type	Macroalgae	Fermentative Microorganism	Yield	Reference
Brown macroalgae	*Saccharina latissimi*	*Saccharomyces cerevisiae*	0.45 % v/v	Adams et al. (2009)
	Laminaria japonica	*S. cerevisiae*	23.3 g/L	Ge et al. (2011)
	Sargassum sp.	*S. cerevisiae*	19.9 ± 0.3 g/L	Saravanan et al. (2018)
	Sargassum sp.	*Hanseniaspora opuntiae*	18.37 ± .03 g/L	Saravanan et al. (2018)
Red macroalgae	*Gracilaria* sp.	*S. cerevisiae*	28.7 ± 0.4 g/L	Saravanan et al. (2018)
	Gracilaria sp.	*H. opuntiae*	27 ± 0.6 g/L	Saravanan et al. (2018)
Plant	*Eichhornia crassipes*	*S. cerevisiae and Zymmonas mobilis*	13.6 g/L	Das et al. (2016)
	E. crassipes	*Pichia stiptis*	23 g/L	Kumar et al. (2009)
	E. crassipes	*P. stiptis*	18 g/L	Nigam (2002)
	E. crassipes	*S. cerevisiae*	14.4 g/L	Mishima et al. (2008)
	E. crassipes	*Escherichia coli*	16.9 g/L	Mishima et al. (2008)
	Pistia stratiotes L.	*S. cerevisiae*	14.9 g/L	Mishima et al. (2008)
	P. stratiotes L.	*E. coli*	16.2 g/L	Mishima et al. (2008)
	Landoltia punctata	*S. cerevisiae*	30.80 ± 0.80 g/L	Chen et al. (2012)
	L. punctata	*S. cerevisiae*	24.06 g/L	Su et al. (2014)

production yields from aquatic weeds are similar to those obtained with other lignocellulosic materials ($\geq$ 40 g/L) (Vohra et al., 2014). The high availability of this resource and high convergence performance make aquatic weeds a sustainable option for ethanol production.

10.3.2 BIOGAS

Anaerobic digestion for biogas production is a widely studied and used method for the valorization of aquatic weed biomass (Milledge and Harvey, 2016). Theoretical production yields of up to 0.2 L_{CH_4} per g_{VS} (Volatile Solids) can be obtained from anaerobic digestion (Alvarado-Morales et al., 2013; Chen et al., 2015). Also, unlike other processes, anaerobic digestion does not require the use of dehydrated biomass (Alvarado-Morales et al., 2013; Milledge et al., 2014), which represents an advantage in areas where climatic conditions make drying difficult, since, because aquatic weeds have a high water content, their removal may require more energy than the content in biomass (Murphy et al., 2013). The digested biomass can have different applications as fertilizer, sorbent, and soil conditioner (Appels et al., 2011; Soto et al., 2015), which increases its valorization rate. Among the current challenges for the valorization of

aquatic weed biomass are low biodegradability, ammonia and sulfur inhibition (Jard et al., 2013; Ward et al., 2014), and potential toxicity caused by the presence of sodium (Hierholtzer and Akunna, 2012), heavy metals (Nkemka and Murto, 2010), tannins, furanic, and phenolic compounds (Monlau et al., 2014).

The biogas production from aquatic weeds biomass has been a widely studied process. Macroalgae and freshwater plants from genus such as *Sargassum, Turbinaria, Hydroclathrus, Caulerpa, Ulva, Eichhornia,* and *Salvina* have been used for methane production (Marquez et al., 2014; Soto et al., 2015). Some aquatic weeds species contain low C/N ratio (> 30) (Oliveira et al., 2015), which prevents the inhibition of methanogenesis by the formation of ammonia (Jard et al., 2013). However, for some sargassum species, C/N ratio values of 12–22 have been reported (Marquez et al., 2014). Recent studies have focused on increasing the C/N ratio by feeding glycerol and cooking oil waste (Oliveira et al., 2015); thus methane production has increased between 19% and 56%. Macroalgae and aquatic plants show yields of biogas production from 0.11 to 0.55 L_{CH_4}/g_{VS} (Table 10.2). Those yields are similar to those obtained from other lignocellulosic materials (0.33–0.45 L_{CH_4}/g_{VS}) (Lin et al., 2017).

TABLE 10.2 Biogas Production From Aquatic Weeds

Organism Type	Macroalgae	Yield (L_{CH_4}/g_{VS})	Reference
Brown algae	*Sargassum muticum*	0.11	Milledge and Harvey (2016)
	S. muticum	0.13	Jard et al. (2013)
	S. muticum	0.11–0.15	Soto et al. (2015)
	Macrocystis sp.	0.41	Chynoweth (2005)
	Laminaria sp.	0.28	Chynoweth (2005)
	Sargassum sp.	0.26–0.38	Chynoweth (2005)
Green algae	*Ulva lactuca*	0.15–0.27	Bruhn et al. (2011)
	Ulva sp.	0.19	Costa et al. (2012)
	Enteromorpha sp.	0.15	Costa et al. (2012)
Red algae	*Gracilaria* sp.	0.18	Costa et al. (2012)
Plant	*Eichhornia crassipes*	0.22	Mathew et al. (2015)
	Salvinia molesta	0.55	Mathew et al. (2015)

10.4 PROSPECTS FOR THE BIOFUELS PRODUCTION FROM AQUATIC WEEDS

The biomass of aquatic weeds has proven to be a viable resource for bioethanol and biogas production. However, for the operation of large-scale processes, some aspects involved in the biomass harvest, transport, preprocessing, and processing must be considered. Because the operation of large-scale plants for cellulosic bioethanol and biogas production still has several limitations, including high capital investment, technical knowledge, and high transportation cost of feedstock (Das et al., 2016).

In terms of harvest, most aquatic weeds have different compositions depending on the season (Lamare and Wing, 2001). Therefore this variable should be considered so as not to affect the feasibility of the large-scale process. Alternatives such as biomass silage could be implemented as a conservation method, without affecting the energy content (energy losses are low <8%) (Herrmann et al., 2015; Milledge and Harvey, 2016). The biomass transport is limited by its high-water content. Therefore it has been chosen to press the biomass to remove most of the moisture and/or adapt medium-scale production plants near the harvester site (Hronich et al., 2008).

Also, some problems associated with the presence of toxic and inhibitory compounds have limited the production of energy from aquatic weeds. However, strategies such as codigestion with other wastes (glycerol, cooking oil, paper, etc.) have shown efficacy in reducing the toxic and inhibitory effects of various compounds (Yen and Brune, 2007; Costa et al., 2012; Oliveira et al., 2015). Another opportunity for improvement for the optimal operation of anaerobic digesters could be to operate them continuously, guaranteeing their productivity. Therefore efforts aimed at generating technical and logistical knowledge could strengthen the biofuel production industry from aquatic weed biomass.

10.5 CONCLUDING REMARKS

The aquatic weeds biomass are a sustainable resource for the production of bioethanol and biogas due to their high availability, resistance to extreme environmental conditions, growth rate, hydrolyzable carbohydrates content. The biofuels production yields from aquatic weeds are similar to other cellulosic raw materials. The exploitation of this resource could allow us to control the aquatic weeds infestations, raise incomes, provide employment and rural development in proliferation zones. However, for the development of large-scale processes, diverse aspects involved in the process harvest, transport, preprocessing, and processing must be studied.

KEYWORDS

- **aquatic weeds**
- **macroalgae**
- **aquatic plants**
- **biomass valorization**
- **biofuels**

REFERENCES

Adams, J.; Gallagher, J.; Donnison, I. Fermentation study on *Saccharina latissima* for bioethanol production considering variable pre-treatments. *J. Appl. Phycol.* 2009; 21, 569–574.

Alvarado-Morales, M.; Boldrin, A.; Karakashev, D.B.; Holdt, S.L.; Angelidaki, I.; Astrup, T. Life cycle assessment of biofuel production from Brown seaweed in Nordic conditions. *Bioresour. Technol.* 2013, 129, 92–99.

Appels, L.; Lauwers, J.; Degrève, J.; Helsen, L.; Lievens, B.; Willems, K.; Van Impe, J.; Dewil, R. Anaerobic digestion in global bioenergy production: potential and research challenges. *Renew. Sust. Energy Rev.* 2011, 15, 4295–4301.

Azadi, P.; Malina, R.; Barrett, S.R.H.; Kraft M. The evolution of the biofuel science. *Renew. Sust. Energy Rev.* 2017, 76, 1479–1484.

Bertagnolli, C.; Espindola, A.P.D.M.; Kleinübing, S.J.; Tasic, L.; da Silva, M.G.C. *Sargassum filipendula* alginate from Brazil: seasonal influence and characteristics. *Carbohydr. Polym.* 2014, 111, 619–623.

Borines, M.G.; de Léon, R. L.; McHenry, M.P. Bioethanol production from farming non-food macroalgae in Pacific island nations: chemical constituents, bioetanol yields, and prospective species in the

Philippines. *Renew. Sust. Energy Rev.* 2011, 15, 4432–4435.

Bruhn, A.; Dahl, J.; Nielsen, H.B.; Niko-laisen, L.; Rasmussen, M.B.; Markager, S.; Olesen, B.; Arias, C.; Jensen, P.D. Bioenergy potential of *Ulva lactuca*: biomass yield, methane production and combustion. *Bioresour. Technol.* 2011, 102, 2595–2604.

Chen, H.; Zhou, D.; Luo, G.; Zhang, S.; Chen, J. Macroalgae for biofuels production: progress and perspectives. *Renew. Sust. Energy Rev.* 2015, 47, 427–437.

Chynoweth, D.P. Renewable biomethane from land and ocean energy crops and organic wastes. *HortScience* 2005, 40, 283–286.

Chynoweth, D.P.; Owens, J.M.; Legrand, R. Renewable methane from anaerobic digestion of biomass. *Renew. Energ.* 2011, 22, 1–8.

Cook, E.J.; Jenkins, S.; Maggs, C.; Minchin, D.; Mineur, F.; Nall, C.; Sewell, J. Impacts of climate change on non-native species. *MCCIP. Sci Rev.* 2013, 2013, 155–166.

Costa, J.C.; Gonçalves, P.R.; Nobre, A.; Alves, M.M. Biomethanation potential of macroalgae *Ulva* spp. and *Gracilaria* spp. and in co-digestion with waste activated sludge. *Bioresour. Technol.* 2012, 114, 320–326.

Das, A.; Ghosh, P.; Paul, T.; Ghosh, U.; Ranjan, B.P.; Chandra, M. K. Production of bioethanol as useful biofuel through the bioconversion of water hyacinth (*Eichhornia crassipes*). *3 Biotech.* 2016, 6, 70.

Davis, T.A.; Volesky, B.; Mucci, A. A review of the biochemistry of heavy metal biosorption by brown algae. *Water Res.* 2003, 37, 4311–4330.

Engelen, A.; Santos, R. Which demographic traits determine population growth in the invasive brown seaweed *Sargassum muticum*? *J. Ecol.* 2009, 97, 675–684.

Ganguly, A.; Chatterjee, P.K.; Dey, A. Studies on ethanol production from water hyacinth—a review. *Renew. Sust. Energy Rev.* 2012, 16, 966–972.

Ge, L.; Wang, P.; Mou, H. Study on saccharification techniques of seaweed wastes for the transformation of ethanol. *Renew. Energy* 2011, 36, 84–89.

Goh, C.S.; Lee, K.T. A visionary and conceptual macroalgae-based third generation bioethanol (TGB) biorefinery in Sabah, Malaysia as an underlay for renewable and sustainable development. *Renew. Sust. Energy Rev.* 2010, 14(2), 42–848.

Gressel, J. Transgenics are imperative for biofuel crops. *Plant Sci.* 2008, 174, 246–263.

Gunan-Yucel, H.; Aksu, Z. Ethanol fermentation characteristics of *Pichia stipitis* yeast from sugar beet pulp hydrolysate: use of new detoxification methods. *Fuel* 2015, 158, 793–799.

Gunnarsson, C.C.; Petersen, C. M. Water hyacinths as a resource in agriculture and energy production: a literature review. *Waste Manage.* 2007, 27, 117–129.

Herrmann, C.; FitzGerald, J.; O'Shea, R.; Xia, A.; O'Kiely, P.; Murphy, J.D. Ensiling of seaweed for a seaweed biofuel industry. *Bioresour. Technol.* 2015, 196, 301–313.

Hierholtzer, A.; Akunna, J.C. Modelling sodium inhibition on the anaerobic digestion process. *Water Sci. Technol.* 2012, 66, 1565–1573.

Horn, S.J.; Aasen, I.M.; Ostgaard, K. Ethanol production from seaweed extract. *J. Ind. Microbiol. Biotechnol.* 2000a, 25, 249–254.

Horn, S.J.; Aasen, I.M.; Ostgaard, K. Production of ethanol from mannitol by *Zymobacter palmae*. *J. Ind. Microbiol. Biotechnol.* 2000b, 24, 51–57.

Hronich, J.E.; Martin, L.; Plawsky, J.; Bungay, H.R. Potential of *Eichhornia crassipes* for biomass refining. *J. Ind. Microbiol. Biotechnol.* 2008, 35, 393–402.

Inderwildi, O.; King, D. Quo vadis biofuels?. *Energy Environ. Sci.* 2009, 2, 343–346.

Jard, G.; Marfaing, H.; Carrere, H.; Delgenes, J.P.; Steyer, J.P.; Dumas, C. French Brittany macroalgae screening: composition and methane Brittany macroalgae screening: composition and methane potential for potential alternative sources of energy and products. *Bioresour. Technol.* 2013, 144, 492–498.

John, R.P.; Anisha, G.S.; Nampoothiri, K.M.; Pandey, A. Micro and macroalgal biomass: a renewable source for bioethanol. *Bioresour. Technol.* 2011, 102, 186–193.

Kumar, A.; Singh, L.K.; Ghosh, S. Bioconversion of lignocellulosic fraction of water-hyacinth (*Eichhornia crassipes*) hemicellulose acid hydrolysate to ethanol by *Pichia stipitis*. *Bioresour. Technol.* 2009, 100, 3293–3297.

Lahaye, M.; Robic, A. Structure and functional properties of Ulvan, a polysaccharide from green seaweeds. *Biomacromolecules*. 2007, 8, 1765–1774.

Lamare, M.D.; Wing, S.R. Calorific content of New Zealand marine macrophytes. *N. Z. J. Mar. Freshwater Res.* 2001, 35, 335–341.

Lin, Y.; Liang, J.; Zeng, C.; Wang, D.; Lin, H. Anaerobic digestion of pulp and paper mill sludge pretreated by microbial consortium OEM1 with simultaneous degradation of lignocellulose and chlorophenols. *Renew. Energy* 2017, 108, 108–115.

Marquez, P.G.B.; Santiañez, W.J.E.; Trono Jr., G.C.; Montaño, M.N.E.; Araki, H.; Takeuchi, H.; Hasegawa, T. Seaweed biomass of the Philippines: sustainable feedstock for biogas production. *Renew. Sustain. Energy Rev.* 2014, 38, 1056–1068.

Mathew, A. K.; Bhui, I.; Banerjee, S. N.; et al. Biogas production from locally available aquatic weeds of Santiniketan through anaerobic digestion. *Clean Technol. Environ. Policy*. 2015, 17, 1681–1688.

Milledge, J.J.; Harvey, P.J. Ensilage and anaerobic digestion of *Sargassum muticum*. *J. Appl. Phycol*. 2016, 28, 3021–3030.

Milledge, J.J.; Heaven, S. Methods of energy extraction from microalgal biomass: a review. *Rev. Environ. Sci. Biotechnol*. 2014, 13, 301–320.

Mishima, D.; Kuniki, M.; Sei, K.; Soda, S.; Ike, M.; Fujita, M. Ethanol production from candidate energy crops: water hyacinth (*Eichhornia crassipes*) and water lettuce (*Pistia stratiotes* L.). *Bioresour. Technol*. 2008, 99, 2495–2500.

Monlau, F.; Sambusiti, C.; Barakat, A.; Quéméneur, M.; Trably, E.; Steyer, J.P.; Carrère, H. Do furanic and phenolic compounds of lignocellulosic and algae biomass hydrolyzate inhibit anaerobic mixed cultures? A comprehensive review. *Biotechnol. Adv*. 2014, 32 (5), 934–951.

Murphy, F.; Devlin, G.; Deverell, R.; McDonnell, K. Biofuel production in Ireland—an approach to 2020 targets with a focus on algal biomass. *Energies*. 2013, 6, 6391–6412.

Mussatto, S.I.; Dragone, G.; Guimaraes, P.M.R.; Silva, J.P.A.; Carneiro, L.M.; Roberto, I.C.; et al. Technological trends, global market, and challenges of bioethanol production. *Biotechnol. Adv*. 2010, 28, 817–830

Nigam, J.N. Bioconversion of water-hyacinth (*Eichhornia crassipes*) hemicellulose acid hydrolysate to motor fuel ethanol by xylose-fermenting yeast. *J. Biotechnol*. 2002, 97, 107–116.

Nkemka, V.N.; Murto, M. Evaluation of biogas production from seaweed in batch tests and in UASB reactors combined with the removal of heavy metals. *J. Environ. Manage*. 2010, 91, 1573–1579.

Oliveira, J.V.; Alves, M.M.; Costa, J.C. Optimization of biogas production from *Sargassum* sp. using a design of experiments to assess the co-digestion with glycerol and waste frying oil. *Bioresour. Technol*. 2015, 175, 480–485.

Ortiz, J.; Romero, N.; Robert, P.; Araya, J.; Lopez-Hernandez, J.; Bozzo, C.; Navarrete, E.; Osorio, A.; Rios, A. Dietary fiber, amino acid, fatty acid and tocopherol contents of the edible seaweeds *Ulva lactuca* and *Durvillaea antarctica*. *Food Chem*. 2006, 99, 98–104.

Saravanan, K.; Duraisamy, S.; Ramasamy, G.; Kumarasamy, A.; Balakrishnan, S. Evaluation of the saccharification and fermentation process of two different seaweeds for an ecofriendly bioethanol production. *Biocatal. Agric. Biotechnol.* 2018, 14, 444–449.

Silva, E.E.; Escobar, J.C.; Rocha, M.H.; Reno, M.L.; Venturini, O.J.; del Olmo, O.A. Issues to consider, existing tools and constraints in biofuels sustainability assessments. *Energy* 2011, 36, 2097–2110.

Soto, M.; Vazquez, M.A.; de Vega, A.; Vilarino, J.M.; Fernandez, G.; de Vicente, M.E. Methane potential and anaerobic treatment feasibility of *Sargassum muticum*. *Bioresour. Technol.* 2015, 189, 53–61.

Su, H.; Zhao, Y.; Jiang, J.; Lu, Q.; Li, Q.; Luo, Y; Wang, M. Use of duckweed (*Landoltia punctata*) as a fermentation substrate for the production of higher alcohols as biofuels. *Energy Fuels* 2014, 28(5), 3206–3216.

Taherzadeh, M.J.; Karimi, K. Enzyme-based hydrolysis processes for ethanol from lignocellulosic materials: a review. *Bioresources* 2007, 2, 707–738.

Vohra, M.; Manwar, J.; Manmode, R.; Padgilwar, S.; Patil, S. Bioethanol production: feedstock and current technologies. *J. Environ. Chem. Eng.* 2014, 2, 573–584.

Ward, A.J; Lewia, D.M.; Green, F.B. Anaerobic digestion of algae biomass: a review. *Algal Res.* 2014, 5, 204–214.

Wi, S.G.; Kim, H.J.; Mahadevan, S.A.; Yang, D.; Bae, H. The potential value of the seaweed Ceylon moss (*Gelidium amansii*) as an alternative bioenergy resource. *Bioresour. Technol.* 2009, 100, 6658–6660.

Yen, H.W.; Brune, D.E. Anaerobic co-digestion of algal sludge and waste paper to produce methane. *Bioresour. Technol.* 2007, 98, 130–134.

CHAPTER 11

Engineering Microorganisms for Chemicals and Biofuels Production

Y. J. TAMAYO-ORDOÑEZ[1], B. AYÍL-GÚTIERREZ[2], A. RUIZ-MARÍN[3],
V. M. INTERIÁN-KU[4], W. POOT-POOT[5], G. J. SOSA-SANTILLÁN[6],
V.H. RAMOS-GARCIA[7], and M. C. TAMAYO-ORDOÑEZ[8*]

[1]Estancia Posdoctoral Nacional-CONACyT. Posgrado en Ciencia y Tecnología de Alimentos, Facultad de Ciencias Químicas, Universidad Autónoma de Coahuila, Ing J. Cardenas Valdez S/N, República, 25280 Saltillo, Coahuila, México

[2]CONACYT- Centro de Biotecnología Genómica, Instituto Politécnico Nacional, Blvd. del Maestro, s/n, Esq. Elías Piña, 88710 Reynosa, México

[3]Facultad de Química, Dependencia Académica de Ciencias Química y Petrolera, Universidad Autónoma del Carmen, Carmen, México

[4]Instituto Tecnológico de la Zona Maya, carretera Chetumal Escárcega Km. 21.5, Ejido Juan Sarabia, C.P. 77960 Quintana Roo, México

[5]Laboratorio de Biotecnología, Facultad de Ingeniería y Ciencias, Universidad Autónoma de Tamaulipas, Centro Universitario, 87120 Ciudad Victoria, Tamaulipas, México

[6]Departamento de Biotecnología. Facultad de Ciencias Químicas, Universidad Autónoma de Coahuila, Ing J. Cardenas Valdez S/N, República, 25280 Saltillo, Coahuila, México

[7]Biotecnologia Vegetal. Centro de Biotecnología Genómica, Instituto Politécnico Nacional, Blvd. del Maestro, s/n, Esq. Elías Piña, 88710 Reynosa, México

[8]Laboratorio de Ingeniería Genética. Departamento de Biotecnología. Facultad de Ciencias Químicas, Universidad Autónoma de Coahuila, Ing J. Cardenas Valdez S/N, República, 25280 Saltillo, Coahuila, México

*Corresponding author. E-mail: mtamayo@uadec.edu.mx

ABSTRACT

Biofuels and chemicals are a promising and highly attractive alternative for minimizing the use of fossil fuels. Breakthrough in obtaining higher performance in triacylglycerols and obtaining different chemicals, such as alcohols, alkanes, carboxylic acids and fatty acids; commodity chemicals, such as monoalcohols, diols, carboxylic, and dicarboxylic acids; and biopolymers. Several microorganisms have been genetically manipulated to produce desirable molecules, for application in biofuels, biohydrogen, and chemicals. This chapter describes the advances made in the genetic manipulation for obtaining different biochemical and biofuels, in models of microorganisms. It also discusses the perspective of using the enzymes involved in the biosynthesis pathway of the different compounds in biocatalysis in order to make efficient the production of compounds of interest.

11.1 INTRODUCTION

Nowadays, the search of alternative sources of producing of fuels instead fossil sources has gained relevance due to the negative effects of fossil fuels, their delimitation (no renewables), and the rising energy demands (Kumar et al., 2015; Wan et al., 2011). Production of biofuels such as biodiesel is based on plant oils, animal fats, and algal oils (Tamayo-Ordoñez et al., 2017; Karmakar et al., 2010; Cunha et al., 2013) and more recently, an area that has drawn interest is single cell oil (SCO) production by heterotrophic oleaginous microorganisms (Liang and Jiang, 2013; Vicente et al., 2009).

Under adequate culture conditions, the yields of some natural microorganisms or genetically modified organisms can reach values between 40% and 70% of lipids in cellular composition (Liang and Jiang, 2013; Vicente et al., 2009), suggesting that obtaining biofuels and biochemicals from microbial strains is feasible (Nel and Cooper, 2009). Besides the obtaining of biofuels and biochemical, using microorganisms (bacteria, yeast and fungi) have many advantages, such as short life cycle, abundant and cheap raw materials, less labor required, crops are little affected by biotic, and abiotic factors (Donot et al., 2014).

According to Kosa and Ragauskas (2011), the use of oleaginous microorganisms presents great potential for obtaining biodiesel, under the scheme of (1) the sustainable use of biological products of low-cost raw materials, (2) high yields of SCOs, and (3) production in a reproducible, high quality, and sustainable fashion. For its part, Marella et al. (2018) suggested that obtaining oleochemicals and its application in biodiesel, detergents, soaps, personal care products, industrial lubricants, plastic enhancers, bioplastics, emulsifiers, coatings, food and feed additives is now a sustainable reality of low economic cost and with less damage to the environment.

In this chapter of the book, we summarize the advances obtained in genetic engineering reported in different oleaginous microorganisms such as bacteria, yeast, and fungi, for lipids production, microbial triacylglycerol (TAG) biosynthesis, and obtaining oleochemicals for their application in a variety of products.

11.2 BIOSYNTHETIC ROUTE FOR OBTAINING LIPIDS IN MICROORGANISMS

Numerous research works have focused on the increase of lipids accumulation via genetic manipulation of enzymes including acetyl-CoA carboxylase (ACC), glycerol-sn-3-phosphate acyl-transferase (GPAT), diacylglycerol acyltransferase (DGAT), acetyl-CoA synthetase, ATP citrate lyase (ACL; malicious enzyme), glycerol-3-phosphate dehydrogenase (G3PDH) (Figure 11.1). These enzymes were shown to be involved in lipid accumulation pathways in microorganisms (Liang and Jiang, 2013; Wang et al., 2011). Below we describe basic aspects of what is the function of these enzymes and why the importance of regulating their function genetically.

In prokaryotes, due to its structural simplicity, the TAG formation occurs in the cytoplasm. For its part in eukaryotes, TAG formation takes place in mitochondria and plastid located in the endoplasmic reticulum. The most important route to TAG biosynthesis is the glycerol-3-phosphate (G3P) or Kennedy pathway (Marella et al., 2018) and de novo formation of PA (Figure 11.1).

11.2.1   KENNEDY PATHWAY

The first step of TAG synthesis is the acylation of G3P with an acyl-CoA to form lysophosphatidate (LPA), which is catalyzed by acyl-CoA:GPAT (Yang et al., 2000; Cao et al., 2006). The LPA is then further condensed, catalyzed by lysophosphatidate acyl-transferase (LPAT), with another acyl-CoA to produce phosphatidate (PA) (Athenstaedt and Daum, 1999). Afterward, PA can be dephosphorylated by phosphatidic acid phosphatase to produce diacylglycerol (DAG). At last, synthesis of TAG is catalyzed by acyl-CoA:DGAT, which incorporates the third acyl-CoA into DAG.

11.2.2   DE NOVO FORMATION OF PHOSPHATIDATE

The former route of PA synthesis is present in bacteria and all types of

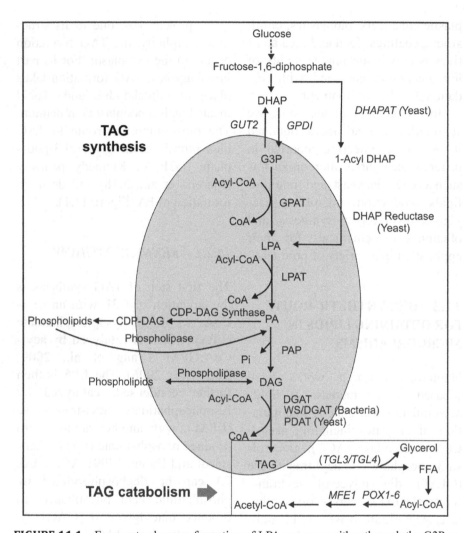

FIGURE 11.1 For yeasts, de novo formation of LPA can occur either through the G3P or DHAP pathways. In yeasts, the DGAT and PDAT catalyze TAG formation. In *Acinetobacter calcoaceticus* ADP1 (bacteria), WS/DGAT exhibits the DGAT activity. GAP: glyceraldehyde 3-phosphate; DHAP: dihydroxyacetone phosphate; PEP: phosphoenolpyruvate; ACP: acyl-carrier protein; FFA: free fatty acid; G3P: glycerol-3-phosphate; LPA: lysophosphatidate; PA: phosphatidate; DAG: diacylglycerol; CDPDAG: CDP-diacylglycerol; TAG: triacylglycerol; PDH: pyruvate dehydrogenase; GPAT: glycerol-3-phosphate acyltransferase; LPAT: lysophosphatidate acyl-transferase; PAP: phosphatidic acid phosphatase; DGAT: diacylglycerol acyl-transferase; WS/DGAT: wax ester synthase/acyl-CoA:diacylglycerol acyltransferase; PDAT: phospholipid:diacylglycerol acyltransferase; DHAPAT: DHAP acyltransferase; GPD1 and GUT2: encoding glycerol 3-phosphate dehydrogenase; TGL3 and TGL4: encoding triacylglycerol lipases; POX1–6: encoding the six acyl-CoA oxidases; MFE1: encoding multi-functional enzyme.

eukaryotes, the DHAP pathway is restricted to yeast and mammalian cells (Athenstaedt and Daum, 1999; Racenis et al., 1992; Minskoff et al., 1994). DHAP is acylated at the sn-1 position by DHAP acyltransferase, and the product 1-acyl-DHAP is reduced by 1-acyl-DHAP reductase to yield LPA, which is further acylated to PA by LPAT.

11.3 GENETIC ENGINEERING STRATEGIES TO ENHANCE BIODIESEL, BIOCHEMICAL, AND BIOHYDROGEN PRODUCTION

To obtain different products from microorganisms, different genetic strategies have been used, such as CRISPR (Wu et al., 2017), deletions, gene silencing, heterologous transformation, and homologous target genes of interest (Zhou et al., 2016; Choi et al., 2019). In general, to increase the production of lipids, biochemical (ethanol and isopropanol, among others) in microorganisms and their application in biodiesel, the following lines of research have been followed: (1) modification of precursors of acetyl-CoA, (2) overexpressing enzymes of the fatty acid biosynthesis pathway, (3) partially blocking competing pathways, and (4) the multigene transgenic approach.

To date there are several reports of bacteria, fungi, and yeast in which genes encoding enzymes related to the biosynthesis of fatty acids have already been genetically modified. In Table 11.1, we summarize some researches in which the most outstanding performances and productivity of lipids were obtained.

1. *Modification of Acetyl-CoA precursors.* One of the proposed strategies to increase fatty acids in microorganisms has been based on increasing the metabolic precursors of fatty acids such as acetyl-CoA, malonyl-CoA, and acyl-CoA. In *Escherichia coli*, it has been shown that if we limit these metabolic precursors to the fermentative pathways resulting in lactate, acetate, succinate, and ethanol, it is possible that higher yield of fatty acids will be obtained, since the main acid precursor fatty acid is malonyl-CoA. In *Saccharomyces cerevisiae*, studies of heterologous expression transforming the gene encoding the enzyme ACL from *Mus musculus* and overexpression of native mitochondrial citrate transporter and malate dehydrogenase showed an increase of 20% of free fatty acid in *S. cerevisiae* (Zhou et al., 2016). Also in *S. cerevisiae* using the ATP:citrate lyase from *Yarrowia lipolytica* showed an increase of 5% of free fatty acid. This strain showed that the low increase in fatty acids was caused by the flux of acetyl-CoA

TABLE 11.1 Researches About Lipid Synthesis by Overexpressing Genes or Knockout Genes

Genes/Enzymes	Source-species/ Receiver-species	Function	Reference
accA-D (ACC), tesA/thioesterase I	*Escherichia coli/E. coli*	Increases yields of six times more fatty acid synthesis	Davis et al. (2000)
ACC1 (ACC)/Acetyl-CoA carboxylase	*Mucor rouxii/Hansenula polymorpha*	Produces more than 40% fatty acid content	Ruenwai et al. (2009)
ACC (ACC)/Acetyl-CoA carboxylase	*E. coli/E. coli*	Increase lipid content three times more	Meng et al. (2011)
ACC1 (ACC)/Acetyl-CoA carboxylase	*Yarrowia lipolytica/Y. lipolytica*	Increase lipid content twice more	Tai and Stephanopoulos (2013)
ACC1 (ACC)/Acetyl-CoA carboxylase	*M. rouxii/H. polymorpha*	Increase total FA content once again	Ruenwai et al. (2009)
ACS (FAA3)/Acetyl-CoA synthase	*Saccharomyces cerevisiae* Dsnf2/*S. cerevisiae* Dsnf2	A total of 29.8% of lipids by dry weight is obtained	Kamisaka et al. (2007)
Acyl-ACP-thioesterase (FAT), fabD gene encoding malonyl CoA: ACP transacylase (MAT)	*E. coli, Streptomyces avermitilis* MA-4680, *Streptomyces coelicolor* A3/*E. coli*	Produces more than 11% of fatty acid content	Zhang et al. (2012)
DLRO1 (PDAT)/Phospholipid: diacylglycerol acyltransferase	*Y. lipolytica/Y. lipolytica*	Produces 40% of TAG content by dry weight	Beopoulos et al. (2009)
ΔDGUT2 (GPDH)/Glycerol 3-phosphate dehydrogenase	*Y. lipolytica/Y. lipolytica*	Incrementa más de tres veces el lipid content	Beopoulos et al. (2008)
GPD1 (GPDH)/Glycerol 3-phosphate dehydrogenase	*Y. lipolytica/Y. lipolytica*	Increase lipid content more than three times	Dulermo and Nicaud (2011)
ΔDGUT2 (GPDH)/Glycerol 3-phosphate dehydrogenase	*Y. lipolytica/Y. lipolytica*	Increase lipid content more than three times	Dulermo and Nicaud (2011)
GPD1, ΔDGUT2 (GPDH)/Glycerol 3-phosphate dehydrogenase	*Y. lipolytica/Y. lipolytica*	Increase five times more TAG content	Dulermo and Nicaud (2011)
(ACS)/Acetyl-CoA synthase	*E. coli/E. coli*	Increase ACS activity nine times more	Lin et al. (2006)

TABLE 11.1 *(Continued)*

Gene/Enzyme	Organism	Result	Reference
ME/Malic enzyme malA gene encoded for the isoforms III and IV of ME	E. coli K-1/E. coli BL21	Increase four times more lipid content; Increase lipid content twice more	Meng et al. (2011); Zhang et al. (2007)
ACL/ATP:citrate lyase	Aspergillus oryzae/Aspergillus oryzae	Increase the fatty acid content twice more	Tamano et al. (2013)
Δ6 FA desaturase = D6DM	Cunninghamella echinulata M1AN6/Lipomices kononenkoae	Produces 1.2% in total FA	Wang et al. (2011)
ΔTGL3, ΔTGL4/Triacylglycerol lipases	Y. lipolytica/Y. lipolytica	Greater lipid production is obtained	Dulermo and Nicaud (2011)
ACC/Acetyl-CoA carboxylase thioesterase; fadD/Acyl-CoA synthetase	E. coli/E. coli	Increase twenty times more fatty acid content	Lu et al. (2008)
ACC1/Acetyl-CoA carboxylase; DGAT1/Diacylglycerol acyl-transferase.	Y. lipolytica/Y. lipolytica	Increase five times more lipid content	Tai and Stephanopoulos(2013)
D6 and D12 desaturases	Mortiella alpina/Y. lipolytica	Produces 20.2% in total lipid content by dry weight	Chuang et al. (2010)
Exo-inulase (INU1)	Klyveromyces marxianus CBS 6556/Y. lipolytica ACA-DC SO104	Produces 50.6% in oil content by dry weight	Zhao et al. (2010)
POX1-6 gene/acyl-CoA oxidase-AOXs; MFE1/encodes the multifunctional enzyme of the β-oxidation pathway, GPDH isoform (GPD1), glycerol 3-phosphate dehydrogenase (DGUT2)	Y. lipolytica/Y. lipolytica	Produce more lipid accumulation	Dulermo and Nicaud (2011)

that was channeled into malate via malate synthase (Mls1p) and that additional carbon was lost to glycerol via glycerol-3-phosphate dehydrogenase (Gpd1p), so that the downregulation of MLS1 and deletion of GPD1 in addition to ACL overexpression, showed 70% increase in the genetically manipulated strain (Ghosh et al., 2016).

2. *Overexpressing enzymes of the fatty acid biosynthesis pathway.* In bacteria, yeasts, and fungi, enzymes involved in the synthesis of fatty acids have been genetically modified (Liang and Jiang, 2013; Kosa and Ragauskas, 2011). In Table 11.1, we summarize some of the heterologous genetic transformations reported in bacteria, fungi, and yeasts.

Kalscheuer et al. (2004) demonstrated the bifunctionality of wax ester synthase/acyl-coenzyme A: diacylglycerol acyltransferase WS/ DGAT from the bacterium *Acinetobacter calcoaceticus* ADP1 that was heterologously expressed in *S. cerevisiae*. The mutant strain H1246 (pESC-URA atfA) showed an increase in total lipids of 21% respect to the wild type. Also in *E. coli* the ACC (encoded by *accA*, *accB*, *accC*, accD) and thioesterase I (encoded by the *tesA* gene) have been shown to increase up to 6 times more the rate of fatty acid synthesis with 6.6 nmol of free

fatty acid (Davis et al., 2000). Related to this, the heterologous expression of the ACC1 from the oleaginous fungus *Mucor rouxii* in *Hansenula polymorpha* showed a 40% increase in the total fatty acid content (Ruenwai et al., 2009) and the regulation of ACC in *Y. lipolytica* showed an increase in the total lipid content of 2 times more compared to the wild strain (Tai, 2013).

On the other hand, heterologous transformation of genes encoding acyl-ACP thiosterase (FAT) and Malonyl CoA: ACP transacylase (MAT) from *Streptomyces* spp., into *E. coli*, was shown to increase up to 11% more free fatty acid (Zhang et al., 2012). Also, the transformation of *E. coli* with FAT gene from *Diploknema butyracea* showed production of free fatty acid greater than 0.2 g/L (Zhang et al., 2011).

In literature it has been described that the DGAT enzyme, presents isoforms, many investigations have been directed to know which of all the DGAT isoforms, is responsible for the accumulation of free fatty acid. The generation of mutants in ΔDGAT1 and ΔDGAT2 of *S. cerevisiae* has indicated that DGAT2 is possibly responsible for a 50% increase in the accumulation of free fatty acid (Zaremberg and McMaster, 2002).

Also knockout mutant analysis affecting the phospholipid: diacylglycerol acyltransferase of *Y. lipolytica* resulted in a 40% loss of TAG, demonstrating the importance of this enzyme for TA biosynthesis (Beopoulos et al., 2009).

Studies involving the generation of mutants affecting GPDH in *Y. lipolytica* showed a three-fold increase in lipids compared to the wild strain (Beopoulos et al., 2009).

3. *Partially blocking competing pathways.* Blocking lipid b-oxidation has proven to be a good alternative to induce fatty acid accumulation in Yeast. AOX2p (catalyze the limiting step of peroximal b-oxidation) expression in *Y. lipolytica* showed an increase in the size and number of lipid bodies, where fatty acids are accumulated (Mlickova et al., 2004). Also in this same model, it has been shown that an increase in NADPH production with upregulation of oxidative stress defense pathways can be achieved up to lipid yield of 82% and increased productivity by fivefold (Qiao et al., 2017). The regulation of the degradation of TAG, through the repression of lipases (TGL3 and TGL4), is also a metabolic pathway that has demonstrated high levels of lipids in *Y. lipolytica,*

4. *The multigene transgenic approach.* The manipulation of more than one target gene within an organism is estimated could result in better yields of the desired product (Tamayo-Ordoñez et al., 2016). However, for the transforming microorganism to display several of features, it is usually required to insert two or more genes (Liang and Jiang, 2013). As more genes are inserted, deleted, or regulated in a genome, complex genetic regulations can be caused by not reaching the desired product. In microorganisms, even though the generation of transgenic the multigene is viable, few studies have been successful.

For example, the co-overexpression of ACC1 and DGAT1, into oleaginous yeast *Y. lipolytica*, showed that the mutant strain was able to accumulate up to 62% of its cell dry weight, almost fivefold greater than the control (Tai, 2013). Also the inactivation of the genes POX1-6 (encoded the six AOXs) and GUT2 of *Y. lipolytica*, into strains where GPD1 was overexpressed, demonstrated to affect lipid accumulation in strains that are defective for b-oxidation (Dulermo et al., 2011). According to these investigations, multigene approach is shown a great prospect in lipid accumulation for biodiesel production, as long as the basic aspects of regulation and metabolism pathways are known where genes of interest to be genetically modified are involved.

11.4 OBTAINING DIFFERENT CHEMICALS BY MODIFYING CHAIN LENGTH OF FATTY ACIDS

The genetic modification to change the length of the carbon chains of fatty acids is relevant according to the application that will be given to the final product. For example, gasoline-like and diesel-like biofuels require short/medium-chain fatty acids, waxes require very-long-chain (VLC) fatty acids and alcohols (Marella et al., 2018). Additionally, lactone fragrances and polyunsaturated fatty acids require a precise chain length.

For the modification of the length of the carbon chains of fatty acids, attention has been focused on genetically modifying the Fungal type I fatty acid synthases (FAS I) and type II fatty acid synthases. Below we describe some research carried out regarding the genetic modification of these enzymes.

Some research reported in order to obtain fatty acids with carbon chains in length C14–C26 is that published by Xu et al. (2016) replaced the malonyl/palmitoyl transferase (MPT) domain in FAS1 of *Y. lipolytica* with thioesterases,

the resulting transformant demonstrated a threefold increase in C14 acid, contributing 29% of the total free fatty acids. Also, the heterologous transformation of type I FAS from mycobacteria in *S. cerevisiae* was shown to produce fatty acids of C16–C26 (Yu et al., 2017).

On the other hand, some genetic engineering studies, with the aim of obtaining fatty acids with carbon chains in length of C6–C8, is the one reported by Gajewski et al. (2017), these authors inserted point mutations located in the active sites of the b-ketoacyl synthase, acetyl-CoA: ACP transacetyltransferase, and MPT domains, demonstrated that the mutant strain produced mostly C6 and C8 chains (Gajewski et al., 2017). Also the creation of chimeras of fungal FAS I (*R. toruloides*) and thioesterases (*Acinetobacter baylyi*) into *S. cerevisiae* strains showed produced up to 1.7 mg/L of C6 and C8 (Zhu et al., 2017).

11.5 GENETIC ENGINEERING OF MICROORGANISM TO PRODUCE BIOHYDROGEN

Microorganisms produce hydrogen via two main pathways: photosynthesis and fermentation. Photosynthesis is a light-dependent process, including direct biophotolysis, indirect biophotolysis, and photo-fermentation, whereas, anaerobic fermentation, also known as dark fermentation, is a light-independent process (Jiang et al., 2010; Strong et al., 2016).

Fermentative hydrogen production is conducted by fermentative microorganisms, such as *Clostridium*, *Enterobacter*, *E. coli*, and *Citrobacter* species or mixed cultures.

At present, derived from the fact that we obtain hydrogen from the photosynthetic processes and fermentative process, genetic engineering has focused its attention on the enzymes that participate in both processes. To be able to efficiently produce hydrogen from both processes, researchers have always focused on target genes, which are regulated about the production of hydrogen in microorganisms.

Although hydrogen production yields are usually higher with photosynthetic processes, oxygen is evolved during photosynthesis that inhibits the hydrogenase enzyme that is responsible for H2 production. On the other hand, in fermentative microorganisms have rapid growth and are not affected by oxygen as much as the main process is anaerobic (any residual oxygen is rapidly consumed at the onset). Therefore fermentative hydrogen production is more advantageous than the photosynthetic hydrogen production and appears to have more potential for practical applications (Das and Veziroglu, 2008; Mathews and Wang, 2009).

In this subtopic, we will describe progress in enzymes involved in the processes of photosynthesis and fermentation, which have been investigated to increase the production of hydrogen. Some of the strategies used to generate recombinant strains that lead to higher levels of hydrogen production are (1) knocking out the genes encoding uptake, (2) overexpression of hydrogen-evolving hydrogenases, (3) shutting down metabolic pathways that compete for hydrogen production, and (4) overexpression of cellulases, hemicellulases, and lignases that can maximize glucose availability (Nath and Das, 2004).

In *E. coli*, the fermentation products are hydrogen, acetate, ethanol, lactate, formate, and some succinate. To enhance hydrogen production, recombinant strains of *E. coli* were developed having mutations in several genes, for example, in the large subunit of uptake *hyaB* and *hybC*, in lactate dehydrogenase (*ldhA*), in the formate hydrogenlyase system (FHL) repressor (*hycA*), in the FHL activator (*fhlA*), in fumarate reductase (*frdBC*), in the Tat system (*tatA–E*), in the alpha subunit of the formate dehydrogenase-*N* and -*O* (*fdnG* and *fdoG* respectively), in the alpha subunit of nitrate reductase A (*narG*), in pyruvate dehydrogenase (*aceE*), in pyruvate oxidase (*poxB*), and in proteins that transport formate (*focA* and *focB*) (Maeda et al., 2007a, 2007b; Vardar-Schara et al., 2008).

11.5.1 GENETIC MODIFICATION OF FHL REPRESSOR AND ACTIVATOR

In recombinant strain *E. coli* SR13, overexpression of FHL activator (fhlA), accompanied by inactivation of the FHL repressor (*hyc*A) was shown to increase hydrogen production up to 2.8 times (Yoshida et al., 2005). Also Penfold et al. (2003), performing analyses in *E. coli*, by inactivating the FHL repressor (*hyc*A), showed increase in hydrogen production two more times, compared to the wild strain. Moreover, Bisaillon et al. (2006), conducting double mutant experiments (*ldh*A and *fhl*A) in *E. coli*, showed a 47% increase in hydrogen production.

11.5.2 GENETIC MANIPULATION IN THE HYDROGENASE ENZYME

The genetic manipulation of hydrogenases in different microorganisms is an investigation that was initiated during the 1980s. Overexpression of the hydrogenase gene from *Clostridium butyricum* into *E. coli* strain HK16 was shown to increase the catalytic activity of this enzyme 3.5 times to produce H2 (Karube et al., 1983). Also, overexpression of *hyd*A gene encoding a [Fe]-hydrogenase

in *Clostridium paraputrificum* M-21 resulted in a 1.7-fold enhancement in hydrogen production from *N*-acetyl-glucosamine (GlcNAc) (Morimoto et al., 2005). In another study, where strains of *E. coli* that have lost *Hyd*-1 and *Hyd*-2 showed a 37% increase in hydrogen production rate compared with wild-type strain BW545 (Bisaillon et al., 2006).

11.5.3 GENETIC MODIFICATION OF THE TAT SYSTEM

The modification of the Tat system, in *E. coli*, was shown to increase hydrogen production twice more (4.4 mmol (mg protein), this increase resulted from the deletion of *tat*C or *tat*A–E (Penfold et al., 2003).

11.5.4 BLOCKING COMPETITIVE PATHWAYS TO PRODUCE COMPOUNDS OF INTEREST

Yoshida et al. (2006) increased hydrogen yields (1.08 mol of H2 per mole of glucose), when they blocked the competitive routes of lactate (deletion of *ldh*A) and succinate (deletion of *frd*BC) (Yoshida et al., 2005, 2006). Also, Maeda et al. generated sevenfold mutants of *E. coli*. In this mutated strain, the fumarate reductase (*frd*C) was deleted to prevent phosphoenol pyruvate from forming succinate, and to prevent pyruvate from forming anything but formate, the lactate dehydrogenase (*ldh*A), the pyruvate dehydrogenase (*ace*E), and the pyruvate oxidase (*pox*B) were deleted. The mutated strain showed a hydrogen yield of 1.3 mol of H2 per mole of glucose compared with 0.65 mol of H2 per mole of glucose with the wild-type BW25113 cells. Also in an investigation reported by Liu et al. (2006), in which the enzyme acetate kinase was inactivated in *C. tyrobutyricum*, was shown to increase 1.5-fold enhancement in hydrogen production from glucose (Liu et al., 2006).

11.5.5   THE MULTIGENE TRANSGENIC APPROACH

In *E. coli*, the quintuple mutant strain of BW25113 with *hya*B, *hyb*C, *hyc*A, and *fdo*G deleted and *fhl*A over-expressed was shown to increase its hydrogen production, up to 141-fold higher (113 ± 12 mmol (mg protein)$^{-1}$/h) compared with the wild-type strain (Maeda et al., 2007b).

11.6   MODIFICATION GENETIC OF METHANOTROPHS BACTERIA PARA PRODUCER DESIRABLE MOLECULES

The genetic modification of methanotrophs bacteria, so that they efficiently use CH$_4$ and/or MeOH as substrates or cosubstrates with sugars to produce desirable molecules, such as biofuels or precursors to biofuel molecules (such as alcohols, alkanes, carboxylic acids, and fatty acids), commodity chemicals (such as mono-alcohols, diols, carboxylic, and di-carboxylic acids), and biopolymers, is a near goal today (Jiang et al., 2010; Whitaker et al., 2015).

Aerobic methanotrophs are Gram-negative bacteria capable of utilizing methane as the sole carbon and energy source (Trotsenko *and* Murrell, 2008). These bacteria are present in several environments and are important for the oxidation of methane in the natural world.

The phylogenetic classification has been grouped into three types of aerobic methanotrophs (type I, type II, and type X). This classification was based on morphological differences, structures of intracytoplasmic membranes, pathways of carbon assimilation, abilities to fix nitrogen, and some other physiological characteristics. Thus two filamentous methane oxidizers, *Crenothrix polyspora* (Stoecker et al., 2006) and *Clonothrix fusca* (Vigliotta et al., 2007), and three extremely acidophilic bacteria, namely *Methylacidiphilum infernorum* V4T, *Methylacidiphilum fumariolicum* SOLV, and *Methylacidiphilum kamchatkense* Kam1 have been isolated and characterized (Dunfield et al., 2007; den Camp et al., 2009; Islam et al., 2008).

These bacteria have demonstrated the ability to use C1 substrates (CH$_4$, and/or MeOH and sometimes formaldehyde, HCHO) as a sole substrate (Hakemian and Rosenzweig, 2007; Anthony, 1982). Aerobic methylotrophs use either the serine cycle, the ribulose bisphosphate pathway (RuBP), or the ribulose monophosphate (RuMP) pathway to assimilate, in all of them it is produced as product pyruvate, which is usable in other metabolic pathways. Figure 11.2 illustrates the main metabolism of methanotrophs bacteria, and the key feature is the oxidation of methane via methanol to formaldehyde, which serves as an intermediate in catabolism and anabolism. Even though the C–H bond in methane is strong and notoriously stable, methane monooxygenase (MMO) can break it under ambient conditions (Dalton, 2000). This enzyme is thus an important model biocatalyst for C1 chemists. Experiments with methanotrophs bacteria have indicated that in cultures where the MMO substrate can be alkanes, alkenes, alicyclic hydrocarbons, halogenated aliphatics, and aromatic compounds, suggesting that these bacteria can use these substrates

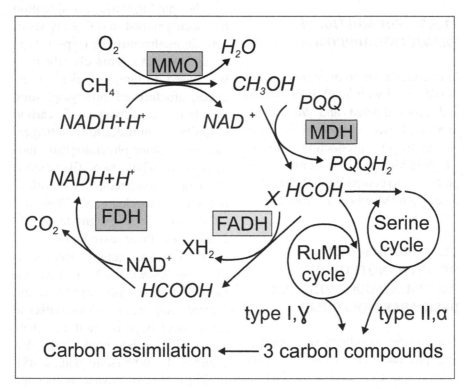

FIGURE 11.2 Metabolic pathway of methanotrophs. MDH, methanol dehydrogenase; FADH, formaldehyde dehydrogenase; FDH, formate dehydrogenase.

and generate molecules of commercial value or important molecules, such as pyruvate, which is used or redirected to metabolic pathways to produce molecules of interest.

The stoichiometry if bacteria use the serine cycle is CO_2 + 2HCHO+ 2NADH + 2ATP = pyruvate +2NAD$^+$ + 2ADP + FPH$_2$ In the case of RuBP the stoichiometry is 3HCHO + 7ATP+ NAD$^+$ = pyruvate + NADH + 7ADP, and the stoichiometry for RuMP-pathway transaldolase is 3HCHO + NAD+ + ADP = pyruvate + NADH + ATP.

Because of the importance of the mechanism used by these bacteria to use C1 substrates, it has brought new attention to genetically manipulate the enzymes involved in the metabolic process. The enzymes where genetic and biochemical characterization investigations have focused on are methanol dehydrogenase (MDH); formaldehyde dehydrogenase (FADH); formate dehydrogenase (FDH) (Figure 11.2).

11.6.1 METHANOL DEHYDROGENASE

MDHs are classified in PQQ-dependent, NAD-dependent, or oxygen-dependent. PQQ-dependent MDHs encoded by multigene clusters (Kalyuzhnaya et al., 2008). Expression analysis of the PQQ operon (pqqABCDEF) from *Klebsiella pneumoniae* in *E. coli* demonstrated

that the heterologous PQQ production is conditioned to the presence of oxygen (Puehringer et al., 2008; Velterop et al., 1995).

11.6.2 NAD-DEPENDENT METHANOL DEHYDROGENASES

The best enzyme characterized is the MDHs of *Bacillus stearothermophilus* and *Bacillus methanolicus*. Brautaset et al. (2013) demonstrated in vivo activity of five NAD MDHs isolated from *B. methanolicus* strains and overexpressed into *E. coli*. The genes *Mdh2* and *Mdh3* from *B. methanolicus* MGA3 demonstrated to express MDH activity in *E. coli*, in the future this strain could be used to increase the production of 1,4-butanediol (Burgard et al., 2014) and other chemicals from MeOH in *E. coli*.

11.6.3 3-HEXULOSE-6-PHOSPHATE SYNTHASE AND 6-PHOSPHO-3-HEXULOISOMERASE ENZYMES

Other enzymes that have deserved attention are the 3-hexulose-6-phosphate synthase (HPS), and fructose-6-phosphate by a 6-phospho-3-hexuloisomerase (PHI) enzymes, involved in the RuMP pathway. The genetic modification of both enzymes has been a complicated challenge to carry out because the *hps* and *phi* genes are differentially transcribed according to the type of methylotrophs bacteria. For example, in obligate methylotrophs the transcription of the *hps* and *phi* genes are transcribed separately as monocistronic messages, whereas in facultative methylotrophs typically transcribe the *hps* and *phi* as a polycistronic message in response to increasing HCHO concentrations (Yurimoto et al., 2009; Chistoserdova et al., 2007; Taylor et al., 2004).

Some advances in genetic engineering, for the manipulation of HPS and PHI enzymes, have indicated that the recombinant fusion of the HPS and PHI proteins from *Mycobacterium gastri*, expressed in *E. coli*, shown to exhibit approximately twofold higher catalytic efficiency compared to the single HPS (Orita et al., 2007). Also, the heterologous transformation of *hps* and *phi* genes from *Bacillus brevis* S1 in *Pseudomonas putida* S12 (Koopman et al., 2009) demonstrated that the mutant strain exhibited higher biomass yields when grown in the presence of glucose and HCHO versus glucose alone. Also, the overexpression of the HPS and PHI genes from *Methylomonas aminofaciens* in *Burkholderia cepacia* TM1 (Mitsui et al., 2003) demonstrated that the mutant strain grows on vanillin or vanillic acid, produces HCHO, and thus HCHO detoxification during growth on vanillin or vanillic acid becomes necessary.

Thus these genetic modifications have been shown to improve the metabolic processes of HCHO oxidation to CO_2, can be widely applied to obtain any product of interest or eliminate formaldehyde and obtain CO_2 that is widely applied in the processing of different polymers (Tomasko et al., 2003; Wang et al., 2011)

11.7 CONCLUSION

Metabolic engineering is clearly an important approach to improve different products of interest like enzymes, biohydrogen, fatty acids, biofuels, and alcohols, among others. The majority of research carried out in this regard has been carried out on bacteria and yeasts, this is due to the facilities (replication, cost, short lifetime) offered by these biological models. However, this recombinant technology is also being adopted by plant and animal systems. Although eukaryotic and prokaryotic systems are far from the degree of complexity in regulation, the specific study of enzymes involved in different metabolic pathways and the manipulation of specific genes in different biological systems offers a good biotechnology alternative to improve yields of some product or to be able to obtain some product with the biotechnological application. This positions the genetic engineering of microorganisms as an excellent tool to improve the chances of improving the quality of life of man, always minimizing risks to the environment.

KEYWORDS

- **biofuels**
- **genetic engineering**
- **biohydrogen**
- **enzyme**

REFERENCES

Anthony, C. *The Biochemistry of Methylotrophs*. Academic Press; **1982**.

Athenstaedt, K.; Daum, G. Phosphatidic acid, a key intermediate in lipid metabolism. *Eur. J. Biochem.* **1999**, 266, 1–16.

Beopoulos, A.; Chardot, T.; Nicaud, J.M. *Yarrowia lipolytica*: a model and a tool to understand the mechanisms implicated in lipid accumulation. *Biochimie.* **2009**, 91, 692–696.

Beopoulos, A.; Mrozova, Z.; Thevenieau, F.; Le Dall, M.T.; Hapala, I.; Papanikolaou, S. et al. Control of lipid accumulation in the yeast *Yarrowia lipolytica. Appl. Environ. Microbiol.* **2008**, 74, 7779–7789.

Bisaillon, A.; Turcot, J.; Hallenbeck, P.C. The effect of nutrient limitation on hydrogen production by batch cultures of *Escherichia coli. Int. J. Hydrogen Energy.* **2006**, 31, 1504–1508.

Brautaset, T.; Heggeset, T.M.B.; Krog, A;, Quax, W.J.;, Sibbald, M.J.J.B.; Vorholt, J.; Muller, J.; Kiefer, P.; Potthoff, E.; Wendisch, V.F. Novel methanol dehydrogenase enzymes from *Bacillus*. World patent publication WO/2013/110797 A1.

2013, Assignees: Sinvent As, Rijksuniversiteit Groningen, Eth Zurich.

Burgard, A.P.; Osterhout, R.E.; Van Dien, S.J.; Tracewell, C.A.; Pharkya, P.; Andrae, S. Microorganisms and methods for enhancing the availability of reducing equivalents in the presence of methanol, and for producing 1,4-butanediol related thereto. US Patent Publication 20140058056. **2014**, Assignee: Genomatica, Inc. (San Diego, CA, USA).

Cao, Z.; Gao, H.; Liu, M.; Jiao, P. Engineering the acetyl-CoA transportation system of *Candida tropicalis* enhances the production of dicarboxylic acid. *Biotechnol. J.* **2006**, 1, 68–74.

Chistoserdova, L.; Lapidus, A.; Han, C.; Goodwin, L.; Saunders, L.; Brettin, T.; Tapia, R.; Gilna, P.; Lucas, S.; Richardson, P.M. et al. Genome of *Methylobacillus flagellatus*, molecular basis for obligate methylotrophy, and polyphyletic origin of methylotrophy. *J. Bacteriol.* **2007**, 189, 4020–4027.

Choi, K.R.; Jang, W.D.; Yang, D.; Cho, J.S.; Park, D.; Lee, S.Y. Systems metabolic engineering strategies: integrating systems and synthetic biology with metabolic engineering. *Trends Biotechnol.* **2019**, 37, 817–837.

Chuang, L.T.; Chen, D.C.; Nicaud, J.M.; Madzak, C.; Chen, Y.H.; Huang, Y.S. Co-expression of heterologous desaturase genes in *Yarrowia lipolytica. New Biotechnol.* **2010**, 27, 277–282.

Cunha Jr, A.; Feddern, V.; Marina, C.; Higarashi, M.M.; de Abreu, P.G.; Coldebella, A. Synthesis and characterization of ethylic biodiesel from animal fat wastes. *Fuel.* **2013**, 105, 228–234.

Das, D.; Veziroglu, T.N. Advances in biological hydrogen production processes. *Int. J. Hydrogen Energy.* **2008**, 33, 6046–6057.

Davis, M.S.; Solbiati, J.: Cronan, Jr. J.E. Overproduction of Acetyl-CoA carboxylase activity increases the rate of fatty acid biosynthesis in *Escherichia coli. J. Biol. Chem.* **2000**, 275, 28593–28598.

den Camp, H.; Islam, T.; Stott, M.; Harhangi, H.; Hynes, A.; Schouten, S.; Jetten, M.; Birkeland, N.; Pol, A.; Dunfield, P. Environmental, genomic and taxonomic perspectives on methanotrophic *Verrucomicrobia*. *Environ. Microbiol. Rep.* **2009**, 1, 293–306.

Donot, F.; Fontana, A.; Baccou, J.C.; Strub, C.; Schorr-Galindo, S. Single cell oils (SCOs) from oleaginous yeasts and moulds: production and genetics. *Biomass Bioenergy.* **2014**, *68*, 135–150.

Dulermo, T.; Nicaud, J.M. Involvement of the G3P shuttle and beta-oxidation pathway in the control of TAG synthesis and lipid accumulation in *Yarrowia lipolytica. Metab. Eng.* **2011**, 13, 482–491.

Dunfield, P.F.; Yuryev, A.; Senin, P.; Smirnova, A.V.; Stott, M.B.; Hou, S.; Ly, B.; Saw, J.H.; Zhou, Z.; Ren, Y.; Wang J.; Mountain, B.W.; Crowe, M.A.; Weatherby, T.M.; Bodelier, P.L.; Liesack, W.; Feng, L.; Wang, L.; Alam, M. Methane oxidation by an extremely acidophilic bacterium of the phylum *Verrucomicrobia, Nature.* **2007**, 450, 879–882.

Gajewski, J.; Pavlovic, R.; Fischer, M.; Boles, E.; Grininger M. Engineering fungal de novo fatty acid synthesis for short chain fatty acid production. *Nat. Commun.* **2017**, 8, ncomms14650.

Ghosh, A.; Ando, D.; Gin, J.; Runguphan, W.; Denby, C., Wang, G.; García Martín, H. 13C metabolic flux analysis for systematic metabolic engineering of *S. cerevisiae* for overproduction of fatty acids. *Front. Bioeng. Biotechnol.* **2016**, 4, 76.

Dalton, H. The Leeuwenhoek Lecture 2000 the natural and unnatural history of methane-oxidizing bacteria. *Philos. Trans. R. Soc. Lond. B Biol. Sci.* **2005**, 360, 1207–1222.

Hakemian, A.S.; Rosenzweig, A.C. The biochemistry of methane oxidation. *Annu. Rev. Biochem.* **2007**, 76, 223–241.

Islam, T.; Jensen, S.; Reigstad, L.J.; Larsen, O.; Birkeland, N.K. Methane oxidation at 55 degrees C and pH2 by a thermoacidophilic bacterium belonging to the *Verrucomicrobia* phylum. *Proc. Natl. Acad. Sci. USA.* **2008**, 105, 300–304.

Jiang, H.; Chen, Y.; Jiang, P.; Zhang, C.; Smith, T.J.; Murrell, J.C.; Xing, X.H. Methanotrophs: multifunctional bacteria with promising applications in environmental bioengineering. *Biochem. Eng. J.* **2010**, 49, 277–288.

Kalscheuer, R.; Luftmann, H.; Steinbüchel, A. Synthesis of novel lipids in *Saccharomyces cerevisiae* by heterologous expression of an unspecific bacterial acyltransferase. *Appl. Environ. Microbiol.* **2004**, 70, 7119–7125.

Kalyuzhnaya, M.G.; Hristova, K.R.; Lidstrom, M.E., Chistoserdova, L. Characterization of a novel methanol dehydrogenase in representatives of Burkholderiales: implications for environmental detection of methylotrophy and evidence for convergent evolution. *J. Bacteriol.* **2008**, 190, 3817–3823.

Kamisaka, Y.; Tomita, N.; Kimura, K.; Kainou, K.; Uemura, H. DGA1 (diacylglycerol acyltransferase gene) overexpression and leucine biosynthesis significantly increase lipid accumulation in the Δ snf2 disruptant of *Saccharomyces cerevisiae. Biochem. J.* **2007**, 408, 61–68.

Karmakar, A.; Karmakar, S.; Mukherjee, S. Properties of various plants and animals feedstocks for biodiesel production. *Bioresour. Technol.* **2010**, 101, 7201–7210.

Karube, I.; Urano, N.; Yamada, T.; Hirochika, H.; Sakaguchi, K. Cloning and expression of the hydrogenase gene from *Clostridium butyricum* in *Escherichia coli. FEBS Lett.* **1983**, 158, 119–122.

Koopman, F.W.; de Winde, J.H.; Ruijssenaars, H.J. C(1) compounds as auxiliary substrate for engineered *Pseudomonas putida* S12. *Appl. Microbiol. Biotechnol.* **2009**, 83, 705–713.

Kosa, M.; Ragauskas, A.J. Lipids from heterotrophic microbes: advances in metabolism research. *Trends Biotechnol.* **2011**, 29, 53–61.

Kumar, R.R.; Rao, P.H.; Arumugam, M. Lipid extraction methods from microalgae: a comprehensive review. *Front. Energy Res.* **2015**, 2, 61–70.

Liang, M.H.; Jiang, J.G. Advancing oleaginous microorganisms to produce lipid via metabolic engineering technology. *Prog. Lipid Res.* **2013**, 52, 395–408.

Lin, H.; Castro, N.M.; Bennett, G.N.; San, K.Y. Acetyl-CoA synthetase overexpression in *Escherichia coli* demonstrates more efficient acetate assimilation and lower acetate accumulation: a potential tool in metabolic engineering. *Appl. Microbiol. Biotechnol.* **2006**, 71, 870–874.

Liu, X.; Zhu, Y.; Yang, S.-T. Construction and characterization of *ack* deleted mutant of *Clostridium tyrobutyricum* for enhanced butyric acid and hydrogen production. *Biotechnol. Prog.* **2006**, 22, 1265–1275.

Lu, X.; Vora, H.; Khosla, C. Overproduction of free fatty acids in *E. coli*: implications for biodiesel production. *Metab. Eng.* **2008**,10, 333–339.

Maeda, T.; Sanchez-Torres, V.; Wood, T.K. Enhanced hydrogen production from glucose by a metabolically-engineered *Escherichia coli*. *Appl. Environ. Microbiol.* **2007a**, 77, 879–890.

Maeda, T.; Sanchez-Torres, V.; Wood, T.K. *Escherichia coli* hydrogenase 3 is a reversible enzyme possessing hydrogen uptake and synthesis activities. *Appl. Microbiol. Biotechnol.* **2007b**, 76, 1035–1042.

Maeda, T.; Sanchez-Torres, V.; Wood, T.K. Metabolic engineering to enhance bacterial hydrogen production. *Microb. Biotechnol.* **2008**, 1, 30–39.

Marella, E.R.; Holkenbrink, C.; Siewers, V.; Borodina, I. Engineering microbial fatty acid metabolism for biofuels and biochemicals. *Curr. Opin. Biotechnol.* **2018**, 50, 39–46.

Mathews, J.; Wang, G. Metabolic pathway engineering for enhanced biohydrogen production. *Int. J. Hydrogen Energy.* **2009**, 34, 7404–7416.

Meng, X.; Yang, J.; Cao, Y.; Li, L.; Jiang, X.; Xu, X. et al. Increasing fatty acid production in *E. coli* by simulating the lipid accumulation of oleaginous microorganisms. *J. Ind. Microbiol. Biotechnol.* **2011**, 38, 919–925.

Minskoff, S.A.; Racenis, P.V.; Granger, J.; Larkins, L.; Hajra, A.K.; Greenberg, M.L. Regulation of phosphatidic acid biosynthetic enzymes in *Saccharomyces cerevisiae*. *J. Lipid. Res.* **1994**, 35, 2254–2262.

Mitsui, R.; Kusano, Y.; Yurimoto, H.; Sakai, Y.; Kato, N.; Tanaka, M. Formaldehyde fixation contributes to detoxification for growth of a nonmethylotroph, *Burkholderia cepacia* TM1, on vanillic acid. *Appl. Environ. Microbiol.* **2003**, 69, 6128–6132.

Mlickova, K.; Roux, E.; Athenstaedt, K.; d'Andrea, S.; Daum, G.; Chardot, T. et al. Lipid accumulation, lipid body formation, and acyl coenzyme A oxidases of the yeast *Yarrowia lipolytica*. *Appl. Environ. Microbiol.* **2004**, 70, 3918–3924.

Morimoto, K.; Kimura, T.; Sakka, K.; Ohmiya, K. Overexpression of a hydrogenase gene in *Clostridium paraputrificum* to enhance hydrogen gas production. *FEMS Microbiol. Lett.* **2005**, 246, 229–234.

Nel, W.P.; Cooper, C.J. Implications of fossil fuel constraints on economic growth and global warming. *Energy Policy.* **2009**, 37, 166–180.

Orita, I.; Sakamoto, N.; Kato, N.; Yurimoto, H.; Sakai, Y. Bifunctional enzyme fusion of 3-hexulose-6-phosphate synthase and 6-phospho-3-hexuloisomerase. *Appl. Microbiol. Biotechnol.* **2007**, 76, 439–445.

Penfold, D.W.; Forster, C.F.; Macaskie, L.E. Increased hydrogen production by *Escherichia coli* strain HD701 in comparison with the wild-type parent strain MC4100. *Enzyme Microb. Technol.* **2003**, 33, 185–189.

Puehringer, S.; Metlitzky, M.; Schwarzenbacher, R. The pyrroloquinoline quinone biosynthesis pathway revisited: a structural approach. *BMC Biochem.* **2008**, 9, 8.

Qiao, K.; Wasylenko, T.M.; Zhou, K.; Xu, P.;Stephanopoulos G. Lipid production in *Yarrowia lipolytica* is maximized by engineering cytosolic redox metabolism. *Nat. Biotechnol.* **2017**, 35, 173–177.

Racenis, P.V., Lai, J.L., Das, A.K., Mullick, P.C., Hajra, A.K., Greenberg, M.L. The acyl dihydroxyacetone phosphate pathway enzymes for glycerolipid biosynthesis are present in the yeast *Saccharomyces cerevisiae*. *J. Bacteriol.* **1992**, 174, 5702–5710.

Ruenwai, R.; Cheevadhanarak, S.; Laoteng, K. Overexpression of acetyl-CoA carboxylase gene of *Mucor rouxii* enhanced fatty acid content in *Hansenula polymorpha*. *Mol. Biotechnol.* **2009**, 42, 327–332.

Stoecker, K.; Bendinger, B.; Schoning, B.; Nielsen, P.H.; Nielsen, J.L.; Baranyi, C.; Toenshoff, E.R.; Daims, H.; Wagner, M.; *Cohn's Crenothrix* is a filamentous methane oxidizer with an unusual methane monooxygenase. *Proc. Natl. Acad. Sci. USA.* **2006**, 103, 2363–2367.

Strong, P.J.; Kalyuzhnaya, M.; Silverman, J.; Clarke, W.P. A methanotroph-based biorefinery: potential scenarios for generating multiple products from a single fermentation. *Bioresour. Technol.* **2016**, 215, 314–323.

Tai, M.; Stephanopoulos G. Engineering the push and pull of lipid biosynthesis in oleaginous yeast *Yarrowia lipolytica* for biofuel production. *Metab. Eng.* **2013**, 15, 1–9.

Tamano, K.; Bruno, K.S.; Karagiosis, S.A.; Culley, D.E.; Deng, S.; Collett J.R. Increased production of fatty acids and triglycerides in *Aspergillus oryzae* by enhancing expressions of fatty acid synthesis-related genes. *Appl. Microbiol. Biotechnol.* **2013**, 97, 269–281.

Tamayo-Ordóñez, M.C.; Espinosa-Barrera, L.A.;Tamayo-Ordóñez, Y.J.; Ayil-Gutiérrez, B.; Sánchez-Teyer, L.F. Advances and perspectives in the generation of polyploid plant species. *Euphytica.* **2016**, 209, 1–22.

Tamayo-Ordóñez, Y.J.; Ayil-Gutiérrez, B.A.; Sánchez-Teyer, F.L.; De la Cruz-Arguijo,

E. A.; Tamayo-Ordóñez, F.A.; Córdova-Quiroz, A.V., Tamayo-Ordóñez, M.C. Advances in culture and genetic modification approaches to lipid biosynthesis for biofuel production and in silico analysis of enzymatic dominions in proteins related to lipid biosynthesis in algae. *Phycol. Res.* **2017**, 65, 14–28.

Taylor, E.J.; Smith, N.L.; Colby, J.; Charnock, S.J.; Black, G.W. The gene encoding the ribulose monophosphate pathway enzyme, 3-hexulose-6-phosphate synthase, from *Aminomonas aminovorus* C2A1 is adjacent to coding sequences that exhibit similarity to histidine biosynthesis enzymes. *Antonie Van Leeuwenhoek.* **2004**, 86, 167–172.

Tomasko, D.L.; Li, H.; Liu, D.; Han, X.; Wingert, M.J.; Lee, L.J.; Koelling, K.W. A review of CO_2 applications in the processing of polymers. *Ind. Eng. Chem. Res.* **2003**, 42, 6431–6456.

Trotsenko, Y.A.; Murrell, J.C. Metabolic aspects of aerobic obligate methanotrophy. *Adv. Appl. Microbiol.* **2008**, 63, 183–229.

Vardar-Schara, G.; Maeda, T.; Wood, T.K. Metabolically engineered bacteria for producing hydrogen via fermentation. *Microb. Biotechnol.* **2008**,1, 107–125.

Velterop, J.S.; Sellink, E.; Meulenberg, J.J.M.; David, S.; Bulder, I.; Postma, P.W. Synthesis of pyrroloquinoline quinone in-vivo and in-vitro and detection of an intermediate in the biosynthetic-pathway. *J. Bacteriol.* **1995**, 177, 5088–5098.

Vicente, G.; Bautista, L.F.; Rodríguez, R.; Gutiérrez, F.J.; Sádaba, I.; Ruiz-Vázquez, R.M.; Garre, V. Biodiesel production from biomass of an oleaginous fungus. *Biochem. Eng. J.* **2009**, 48, 22–27.

Vigliotta, G.; Nutricati, E.; Carata, E.; Tredici, S.M.; De Stefano, M.; Pontieri, P.; Massardo, D.R.; Prati, M.V.; De Bellis, L.; Alifano, P. *Clonothrix fusca* Roze 1896, a filamentous, sheathed, methanotrophic gamma-proteobacterium, *Appl. Environ. Microbiol.* **2007**, 73, 3556–3565.

Wan, M.; Liu, P.; Xia, J.; Rosenberg, J.N.; Oyler, G.A.; Betenbaugh, M.J. et al. The effect of mixotrophy on microalgal growth, lipid content and expression levels of three pathway genes in *Chlorella sorokiniana*. *Appl. Microbiol. Biotechnol.* **2011**, 91, 835–844.

Wang, Q.; Luo, J.; Zhong, Z.; Borgna, A. CO_2 capture by solid adsorbents and their applications: current status and new trends. *Energy Environ. Sci.* **2011**, 4, 42–55.

Wang, P.; Wan, X.; Zhang, Y.; Jiang, M. Production of γ-linolenic acid using a novel heterologous expression system in the oleaginous yeast *Lipomyces kononenkoae*. *Biotechnol. Lett.* **2011b**, 33, 1993.

Wang, J.J.; Zhang, B.R.; Chen, S.L. Oleaginous yeast *Yarrowia lipolytica* mutants with a disrupted fatty acyl-CoA synthetase gene accumulate saturated fatty acid. *Process Biochem.* **2011a**, 46, 1436e41.

Whitaker, W.B.; Sandoval, N.R.; Bennett, R.K.; Fast, A.G.; Papoutsakis, E.T. Synthetic methylotrophy: engineering the production of biofuels and chemicals based on the biology of aerobic methanol utilization. *Curr. Opin. Biotechnol.* **2015**, 33, 165–175.

Wu, J.; Zhang, X.; Xia, X.; Dong, M. A systematic optimization of medium chain fatty acid biosynthesis via the reverse beta-oxidation cycle in *Escherichia coli*. *Metabolic Eng.*. **2017**, 41, 115–124.

Wynn, J.P.; Ratledge, C. Malic enzyme is a major source of NADPH for lipid accumulation by *Aspergillus nidulans*. *Microbiology*. **1997**, 143, 253–257.

Xu, P.; Qiao, K.; Ahn, W.S., Stephanopoulos, G. Engineering *Yarrowia lipolytica* as a platform for synthesis of drop-in transportation fuels and oleochemicals. *Proc. Natl. Acad. Sci.* **2016**,113, 10848–10853.

Yang, C.; Hua, Q.; Shimizu, K. Energetics and carbon metabolism during growth of microalgal cells under photoautotrophic, mixotrophic and cyclic light-autotrophic/dark-heterotrophic conditions. *Biochem. Eng. J.* **2000**, 6, 87–102.

Yoshida, A.; Nishimura, T.; Kawaguchi, H.; Inui, M.; Yukawa, H. Enhanced hydrogen production from formic acid by formate hydrogen lyase-overexpressing *Escherichia coli* strains. *Appl. Environ. Microbiol.* **2005**, 71, 6762–6768.

Yoshida, A.; Nishimura, T.; Kawaguchi, H.; Inui, M.; Yukawa, H. Enhanced hydrogen production from glucose using *ldh*- and *frd*-inactivated *Escherichia coli* strains. *Appl. Microbiol. Biotechnol.* **2006**,73, 67–72.

Yu, T.; Zhou, Y.J.; Wenning, L.; Liu, Q., Krivoruchko, A.; Siewers, V.; Nielsen, J.; David F. Metabolic engineering of *Saccharomyces cerevisiae* for production of very long chain fatty acid-derived chemicals. *Nat. Commun.* **2017**, 8, ncomms15587.

Yurimoto, H.; Kato, N.; Sakai, Y. Genomic organization and biochemistry of the ribulose monophosphate pathway and its application in biotechnology. *Appl. Microbiol. Biotechnol.* **2009**, 84, 407–416.

Zaremberg, V.; McMaster, C.R. Differential partitioning of lipids metabolized by separate yeast glycerol-3-phosphate acyltransferases reveals that phospholipase D generation of phosphatidic acid mediates sensitivity to choline-containing lysolipids and drugs. *J. Biol. Chem.* **2002**, 277, 39035–39044.

Zhang, X.; Agrawal, A.; San, K.Y. Improving fatty acid production in *Escherichia coli* through the overexpression of malonyl CoA-acyl carrier protein transacylase. *Biotechnol. Prog.* **2012**, 28, 60–65.

Zhang, X.; Li, M.; Agrawal, A.; San, K.Y. Efficient free fatty acid production in *Escherichia coli* using plant acyl-ACP thioesterases. *Metab. Eng.* **2011**,13, 713–722.

Zhang, Y.; Adams, I.P.; Ratledge, C. Malic enzyme: the controlling activity for lipid production? Overexpression of malic enzyme in *Mucor circinelloides* leads to a 2.5-fold increase in lipid accumulation. *Microbiol.* **2007**, 153, 2013–2025.

Zhao, C.H.; Cui, W.; Liu, X.Y.; Chi, Z.M.; Madzak, C. Expression of inulinase gene in the oleaginous yeast *Yarrowia lipolytica* and single cell oil production from inulin-containing materials. *Metab. Eng.* **2010**, 12, 510–517.

Zhou, Y.J.; Buijs, N.A.; Zhu, Z.; Qin, J.; Siewers, V.; Nielsen, J. Production of fatty acid-derived oleochemicals and biofuels by synthetic yeast cell factories. *Nat. Commun.* **2016**, 7, 11709.

Zhu, Z.; Zhou, Y.J.; Krivoruchko, A., Grininger, M.; Zhao, Z.K.; Nielsen J. Expanding the product portfolio of fungal type I fatty acid synthases. *Nat. Chem. Biol.* **2017**, 13, 360–362.

CHAPTER 12

Development of Useful Microbial Strains by Genetic Engineering

MARÍA TAMAYO-ORDOÑEZ[1], HUMBERTO MARTÍNEZ MONTOYA[2],
L. MARGARITA LÓPEZ-CASTILLO[3], PAOLA ANGULO-BEJARANO[4],
RODOLFO TORRES-DE LOS SANTOS[5], and ERIKA ACOSTA-CRUZ[1*]

[1]Departamento de Biotecnología. Facultad de Ciencias Químicas.
Universidad Autónoma de Coahuila, Blvd. Venustiano Carranza,
25280 Saltillo, Coahuila, México

[2]Laboratory of Genetics and Comparative Genomics, UAM Reynosa Aztlán—
Universidad Autónoma de Tamaulipas, Reynosa, Tamaulipas, México

[3]Centro de Biotecnología Femsa. Escuela de Ingeniería y Ciencias,
Tecnológico de Monterrey, Monterrey, Nuevo León, México

[4]Tecnologico de Monterrey, Centre of Bioengineering, School of
Engineering and Sciences, Campus Queretaro, Querétaro, Qro, Mexico

[5]Unidad académica multidisciplinaria Mante Centro. Universidad
Autónoma de Tamaulipas, Cd. Mante, Tamaulipas, México

*Corresponding author. E-mail: erika.acosta@uadec.edu.mx

ABSTRACT

Bacteria can be described as microscopical factories for primary and secondary metabolites, by the use of their enzyme machinery. For centuries, we have been using bacteria to meet our needs. Today, the use of bacteria has been extended, and genetic engineering plays a key role in the transformation of microbes into the desired cell factories with high efficiency for metabolite production. Here, we present an overview of recent advances in metabolic engineering for the modification required. Overexpression or deletion of the related enzymes for de novo synthesis of products of interest will be highlighted with relevant examples.

12.1 THE GENETIC ENGINEERING OF MICROBIAL STRAINS: IMPLICATIONS AND RISK

The history of mankind and economic growth have always partnered with a variety of foods, such as bread, beer, wine, meat, and dairy products; so as knowledge of the microorganisms involved in the processing and conservation of food has increased in the last 200 years and the rapid development of molecular biology techniques has allowed a deeper understanding of history genetic and specific microbial metabolic pathways (Choi et al., 2019; Donot et al., 2014; Wu et al., 2016).

Therefore the genetic improvement of the microbial strains involved in food processing may represent a viable solution for the globalization of the food and agriculture markets (Csutak and Sarbu, 2018; Choi and Lee, 2004). On the other hand, advances in molecular biology and genetics of microorganisms have led to the emergence of new experimental technology, so that the real possibility of designing strains of microorganisms with the desired properties has arisen (Liang and Jiang, 2013; Es et al., 2018). Therefore the production of biologically active compounds by microorganisms to be profitable, it is necessary to use strains of highly efficient production as regards, genetically modified microorganisms are increasingly used as platforms for the production of amino acids, vitamins, and enzymes for food use (Choi et al., 2019; Wu et al., 2016).

There is also an active development of the genetically modified organisms (GMO) for use in biofuels (Liang and Jiang, 2013; Es et al., 2018), health (Kärenlampi, 2016), and human and animal nutrition (Patel et al., 2017; Elena et al., 2018), as well as in the production of a wide range of compounds of interests (Padkina and Sambuk, 2016). Microorganisms have a high adaptive capacity to the environment where they develop, allowing them to adjust to adverse conditions and thus ensure their survival. These adaptive phenomena include resistance to antimicrobials and other toxic substances as disinfectants (Troncoso, 2017).

Small (2007) described the genetic engineering (GE) as "an application of biotechnology involving the manipulation of DNA and the transfer of gene components between species in order to encourage replication of desired traits." Genetic engineering (GE) includes sets of techniques and methods that are used to build molecules of DNA recombinant and then insert them into receptor molecules (Gaj et al., 2013). On the other hand, one of the microorganisms that were used as the first object of biotechnology was *Escherichia coli*, since, whose process of replication, transcription, translation, and regulation mechanisms of genetic activity were studied in detail to these genetic procedures (Padkina and Sambuk, 2016), they are specialized in decomposing substrates present in its habitat; this evolutionary process has resulted in the production of molecules with catalytic activity adapted to the substrate, molecular stability, efficient intracellular processing in a mechanism for transport out of the cell's extracellular enzymes (Choi and Lee, 2004; Bindal et al., 2018).

The use of these enzymes, obtained from GMO, has had an important application in biocatalysis. The term biocatalysis was established in the present century as the use of enzymes in chemical synthesis and refers to the use of cells or their isolated enzymes to catalyze reactions or transformations leading to the obtention of compounds of interest for the satisfaction of numerous human needs (Arroyo et al., 2014; Truppo, 2017). Therefore biocatalysts are seen as an important alternative because of their attractive advantages including their availability from renewable resources, their biodegradability, their capacity to act in mild conditions of pH and temperature, their high catalytic activity, and unmatched levels of chemocontrol, regiocontrol, and stereocontrol (Wallace and Balskus, 2014; Jemli et al., 2016).

So apparently, the improvement in the quality of life for man has gone hand in hand with engineered microorganisms; however, when speaking of GMOs, we always have to consider the risks and ethical concerns. Multiple genetic modification projects are focused on an intensive debate concerning their application, their evolutionary achievements, and ethical implications. In these terms, fields of food obtained from GMOs and their impact on human health have received more attention (Small, 2007; Lassen et al., 2002; Robinson, 1999).

GE, being a discipline that includes the management of a group of techniques, which aims to improve the processes of production of enzymes, compounds, or molecules with commercial value by editing of genes to produce strains of microorganisms with a property in particular, is one of the most active and least understood and most questioned the biotechnology subdisciplines. Thus when attempting to modify genetically a microorganism, we must always consider there is still controversy because some genetically modified microorganisms have shown unintended harm to other organisms, reduced effectiveness of pesticides, and gene transfer to nontarget species (Robinson, 1999; Verma et al., 2011; Barrows et al., 2014). Also has been discussed how to select genes target to edit in the organism, according to Gould and Cohen (2000), "since genes do not act alone but that there are multiple interactions between them," so if we perform a wrong selection of the genes to edit, it is possible that we would affect not only the metabolic pathway of the compound, enzyme, or molecule of interest but also others alternate routes of metabolism, which could compromise the survival of the species. As more genes are inserted, deleted, or regulated in a genome, the higher is the risk of unknown effects on human health (Tamayo-Ordoñez et al., 2016).

Also, another point of debate in the generation of GMOs has been the presence of cassettes of antibiotics, the presence of reporter genes, and horizontal and vertical gene transfer in microorganism nontarget. For this concern, the GE offers several alternatives, such as the transformation via chloroplast and vectors that allow the removal of cassettes of antibiotics as foreign DNA is inserted into the host (Day and Goldschmidt-Clermont, 2011). Additionally, gene editing techniques like the use of zinc-finger nucleases (ZFNs), transcription activator-like effector nucleases (TALEN), and the clustered regulatory interspaced short palindromic repeats (CRISPR) associated nuclease (Cas) have allowed site-directed modifications in the gene of interest (Gaj et al., 2013; Curtin et al., 2012; Feng et al., 2013; Upadhyay et al., 2013).

These techniques have shown that the methods for the edition of genomes through GE have progressed with the aim of minimizing the risk

of the creation of GMOs on human health and the environment. Interestingly, these advances in GE are the result of ethics debates and discussions arising when we observe, question, and analyze the impacts and risks of GMOs on the environment and human health. This suggests that it is possible to adopt the techniques of GE to obtain molecules of interest, but always evaluating the implications and the risks that could lead to the creation of GMOs.

Borlaug (1997) argues that "the development and production of new products through GE technology will contribute to economic growth and prosperity and that this technology will be necessary in order to feed the burgeoning world population by 2050."

The acceptance and inclusion of GMOs in each country, continent, and geographical region should be evaluated separately according to the needs of each population, focused on safeguarding native resources and strains of microorganisms.

12.2 GENETIC ENGINEERING FOR DEVELOPMENT OF MICROBIAL STRAINS. FROM CELL TRANSFORMATION TO CRISPR-*CAS*9

The development of techniques focused on genome manipulation in living cells has been a long-standing goal for researchers. Genetic edition research has grown exponentially in the last few years due to the numerous applications involving the use of genetically modified microorganisms in different fields such as gene therapy, bioremediation, industrial processes, synthetic biology, and others. Nowadays, the biotechnology industry relies on the efficiency of bacterial and fungal strains to produce useful products; the yields obtained may be improved, optimizing the culture medium and conditions. However, this approach is limited by the organism's ability to synthesize the desired product or metabolite; therefore the production of bacterial products may be improved significantly by manipulating the bacterial genome either by recombination or introducing punctual changes into the gene structure.

In previous years, the bacterial modification was achieved by selecting and isolating strains with desired natural variants, but late methods included selecting induced mutants and selecting recombinants. Natural selection of bacterial strains may improve industrial yields of desired products, but the risk of changes within the genome of bacterial strains over generations makes this technique unreliable on such improvements. On the other hand, induction of mutations in the bacterial genome may be achieved by different stress-induced chemical and physical methods, such as ultraviolet light, temperature, radiation, nutritional deprivation, or exposure to antibiotics (Foster 2007). However, mutagen exposition may lead to induce random changes across the genome and most mutations may result in deleterious to the organism of interest. Indeed, a small proportion could be more productive than their nonmutated ancestors. The process involving the introduction of foreign DNA into bacteria is called transformation, despite their stability, bacterial genomes are plastic in evolutionary terms, its structure is dynamic and shaped by horizontal gene transfer, rearrangements, and gain/loss events of mobile elements such as plasmids and genomic islands (Dobrindt et al., 2004) (Figure 12.1). Currently, genetic engineered strains are created to produce several primary and secondary metabolites such as amino acids, vitamins, organic acids, alcohols, carotenoids (Adrio and Demain, 2010). By definition, primary metabolites are those products synthesized during the exponential phase of growth and are part of the cellular growth process, whereas secondary metabolites are compounds commercially relevant for health, nutrition, and other industries such as antibiotics, pesticides, pigments, and toxins (Adrio and Demain, 2010).

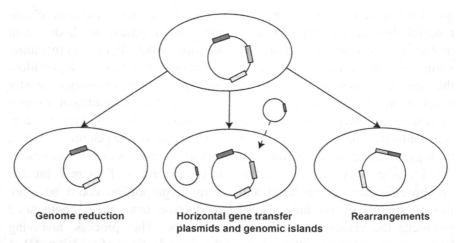

Genome reduction   Horizontal gene transfer   Rearrangements
                         plasmids and genomic islands

FIGURE 12.1   Acquisition and loss of genetic material events in bacteria.

Bacterial chromosomes can be manipulated as result of transposition or recombination events, transposable elements are also called "jumping genes" and they can move from one genomic position to another within the genome, inducing a chromosomal mutation that can result in the insertion of a gene of interest or silencing of genes when they interrupt a gene sequence (Kidwell 2005; Choi and Kim, 2009). Recombinant DNA approaches (homologous recombination and site-specific recombination) imply the addition, alteration, or inactivation of specific genes in order to modify or block specific pathways. In the homologous recombination process, usually result of the recombinase *Rec*A catalytic activity between two homologous sequences (Baitin et al., 2003; Choi and Kim, 2009), exogenous DNA is introduced in vivo to the bacteria through the cell membrane,

later DNA may be integrated enabling recipient bacteria to express novel genes and ultimately acquire new traits. Site-specific recombination moves specific nucleotide sequences between nonhomologous sites within a genome and is driven by two types of major families of enzymes, the resolvase-invertase and the Int family (FLP-*FRT*, Cre-*lox*P) (Choi and Kim, 2009). The artificial process of transformation can be engineered using PCR products and oligonucleotides; thus modifications allow gene knockouts, replacements, deletions, and point mutations (Thomason et al., 2014; Stanbury et al., 2016). In order to improve strains of commercial importance, transformation can be achieved in vivo using designed plasmids capable of accepting foreign sequences and carrying antibiotic resistance genes or lethal genes that may be expressed if the cell is not transformed (Figure 12.2). Bacterial transformation through the use of recombinant plasmids currently is widely used to express interest genes to produce functional proteins within the bacteria. These cells are cultured in bioreactors under controlled conditions and proteins purified from cells and other cellular components. To date, *E. coli* is the most widely and the first-choice bacteria to produce recombinant proteins due to its complete genetic characterization and its ability to proliferate on inexpensive substrates (Ferrer-Miralles and Villaverde, 2013; Rosano and Ceccarelli, 2014) but the use of other bacterial species are being considered as potential "cell factories" due to their metabolic diversity and biosynthetic potential under different environmental conditions (Ferrer-Miralles and Villaverde, 2013). Among the diverse bacterial phyla considered

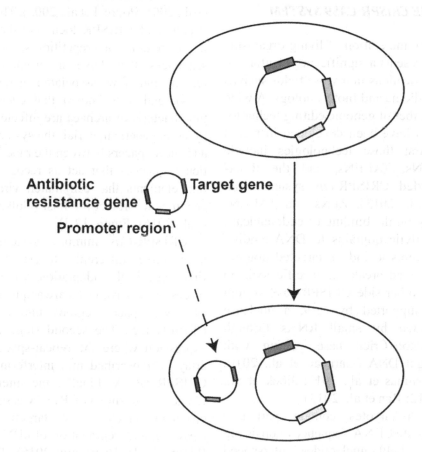

Antibiotic resistance gene

Target gene

Promoter region

FIGURE 12.2 Bacterial transformation by exogenous DNA carried by a plasmid.

useful for recombinant protein production are Proteobacteria (*Caulobacter, Rodhobacter, Pseudoalteromonas, Pseudomonas, Halomonas, Chromohalobacter*), Actinobacteria (*Streptomyces, Nocardia, Mycobacterium, Corynebacterium, Brevibacterium*), and Firmicutes (*Bacillus, Lactococcus*) (Ferrer-Miralles and Villaverde, 2013).

12.2.1 GENOME EDITION WITH THE CRISPR-CAS9 SYSTEM

Genome edition of living organisms represents a significant potential for applications in diverse fields such as medicine and biotechnology. A wide number of genome editing technologies have been developed in recent years; those technologies include ZFNs, TALENs, and the RNA-guided CRISPR-*cas* system (Ran et al., 2013). ZFNs and TALENs rely on the binding of endonuclease catalytic domains to DNA-binding proteins to induce targeted double-stranded breaks at specific loci, on the other side CRISPR-cas9 system is supported by cas9, a nuclease guided by small RNAs through Watson–Crick base pairing with target DNA (Garneau et al., 2010; Gasiunas et al., 2012; Jinek et al., 2012; Ran et al., 2013).

Prokaryotes contain particular repeated DNA sequences consisting of 24–40 nucleotides interspaced for similar size sequences; those sequences were named short regularly spaced repeats and renamed in 2002 as CRISPR (Jansen et al., 2002; Mojica et al., 2005). CRISPR locus is widespread among prokaryotes; archaea, Gram-negative and Gram-positive bacteria (including thermophilic) and they act as an adaptive immune system for prokaryotes against undesired DNA or RNA from viruses or mobile elements such as plasmids (Jansen et al., 2002; Mojica et al., 2005; Pourcel et al., 2005). The structure of CRISPR locus consists in a region of repetition-spacer sequences that have a promoter region and few associated genes (*Cas*), and it is known that every time bacteria or archaea are infected by virus genetic material, the system adds new spacers between the repetition sequences that act as recognition elements that can match viral genomes to protect the host (Golkar et al., 2016) (Figure 12.3).

CRISPR-*Cas* immune systems have three different stages; the first stage is the adaptation, where pieces of invasive DNA are acquired as novel spacer regions into the CRISPR loci. The second stage is expression where the repeat-spacer array is transcribed into interfering CRISPR rRNA. Finally, the interference stage directs CRISPR-associated endonucleases for targeting, cleavage, and degradation of cDNA (Figure 12.4) (Barrangou 2015). To

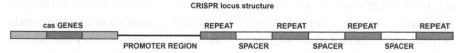

FIGURE 12.3 CRISPR locus structure.

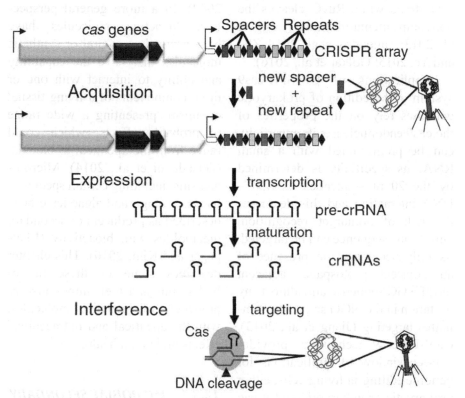

FIGURE 12.4 CRISPR-*Cas* bacterial immune system. Acquisition, expression and interference stages (Taken from Barrangou (2015)).

date, the CRISPR-*Cas* system has been classified into three categories, type I, II, and III. Types I and III are found in both bacteria and archaea and it is known that these types contain multiple *Cas* proteins that form complexes with crRNA, the CASCADE complex in type I and *Cmr* or *Csm* RAMP in type III.

Type II is exclusive of bacteria and is the most studied; this mechanism includes the family of *Cas*9 proteins that contains two domains HNH and *Ruv*C in order to perform the gene silencing (Haft et al., 2005; Makarova et al., 2011; Golkar et al., 2016). In the type II CRISPR mechanism, the HNH and RuvC nuclease cut both strands of the target DNA generating double-stranded breaks defined by a 20nt target sequence within associated crRNA transcript. HNH cleaves the complementary strand while RuvC cleaves the noncomplementary strand (Cong et al., 2013; Almendros et al., 2014; Zhu and Ye, 2015; Golkar et al., 2016).

Applications of CRISPR-cas9 system in the edition of prokaryotic genomes rely on the properties of the cas9 endonuclease: its specificity can be programmed with a small RNA, its specificity is determined by the 20 nt sequences of RNA–DNA interaction and this decreases the risk of nontarget recognition, almost any sequence can be targeted, its only necessary the presence of an adjacent protospacer adjacent motif NGG sequence and almost any mutation in the NGG sequence eliminates targeting (Jiang et al., 2013). CRISPR-cas9 technology provides a new window of applications for genome editing in living cells, either prokaryotic or eukaryotic and it can apply to microorganisms where plasmids can be introduced.

12.3 BIOACTIVE COMPOUNDS DERIVED FROM MICROBIAL SOURCES

Bioactive molecules have been defined considering diverse perspectives along time. In terms of nutrition, bioactive compounds are defined as essential or nonessential compounds that occur in nature, they are part of the food chain and can be shown to affect human health (Biesalski et al., 2009). In a more general perspective, bioactive molecules have been defined as natural or synthetic molecules that have the capability and ability to interact with one or more components of a living tissue/organism, presenting a wide range of probable effects, which could range from therapeutical to biocidal (Guaadaoui et al., 2014). Microorganisms, including certain species of bacteria, fungi, and algae have been described as producers of secondary metabolites with bioactivity (Bhatnagar and Kim, 2010). This chapter describes some of these microbial strains and their importance as producers of bioactive molecules, with therapeutical and nutraceutical effects on human health.

12.3.1 MICROBIAL SECONDARY METABOLITES AS BIOACTIVE COMPOUNDS

Since the discovery of penicillin by Fleming in 1928, microbes have been considered as an extensive source of bioactive compounds (Bérdy, 2005; Genilloud, 2014). In 2005, among 22,000–23,000 microbial bioactive compounds were identified, including antibiotics, enzyme inhibitors, anticancer, antimalarial, pesticides, and herbicides (Bérdy, 2005). In recent years, among 23,000 bacterial metabolites were identified as antibiotics (Manivasagan et al., 2014). Thus, microbes are considered nowadays as biofactories of bioactive molecules with broad applications, and some strains have been even exploited to commercial scale. In this section, we will highlight the importance of microbes as sources of bioactive molecules.

12.3.2 BIOACTIVE COMPOUNDS FROM ACTINOMYCETES

Actinomycetes are a highly diverse group of Gram-positive bacteria (Gomes et al., 2018). To date, at least 350 genera have been identified ubiquitously and could be found in terrestrial and marine ecosystems (Gomes et al., 2018; Lee et al., 2018). Since the discovery of the aminoglycoside streptomycin in 1944, from *Streptomyces griseus*, actinomycetes have been considered as an important source of valuable compounds, including antimicrobial molecules. Some historical examples with commercial impact include the chloramphenicol from *Streptomyces venezuelae*, discovered in 1947, or the tetracyclin, isolated from *Streptomyces rimosus* in 1948, or the erythromycin, obtained from *Saccharopolyspora erythraea*, in 1952 or the vancomycin from *Streptomyces*

orientales, reported in 1952 (Genilloud, 2014; Takahashi and Nakashima, 2018). In the decade of the 1960s, gentamycin was isolated from *Micromonas purpurea*. This discovery was crucial for antibiotics research, directing several efforts to the exploration of the so-called rare actinomycetes as sources of new bioactive compounds (Genilloud, 2014). Since then, approximately 10,000 compounds, identified as antimicrobials, derived from up to 8000 species of actinomycetes have been identified, representing up to 65% of the antibiotics used in medicine (Bérdy, 2005; Lee et al., 2018; Takahashi and Nakashima, 2018).

Even in recent years, molecules derived from actinomycetes have been identified as antibiotics, such as platensimycin, a selective FabF inhibitor isolated from *Streptomyces platensis* that demonstrated potent activity against a broad spectrum of Gram-positive bacteria, by selective inhibition of cellular lipid biosynthesis (Wang et al., 2006; Das, Sakha-Ghosh and Manna, 2016). Taromycin A displays another recent example, a lipopeptide discovered from a gene cluster of a marine isolate of *Saccharomonospora* sp. This lipopeptide displayed activity against methicillin-resistant *Staphylococcus aureus* and daptomycin-sensitive strains of *Enterococcus faecalis* and *Enterococcus faecium* (Yamanaka et al., 2014; Genilloud, 2017).

The isolation of actinomycin in the decade of 1940s from *Streptomyces antibioticus* demonstrated that antimicrobial compounds produced by actinomycetes could be multifunctional molecules. In the particular case of actinomycin, this molecule displayed both antimicrobial and antitumorigenic activity (Demain and Sanchez, 2009). It has been estimated that, in general, actinomycetes are the source of approximately 61% of all microorganism-derived bioactive molecules, with 16% of the total obtained from "rare actinomycetes," such as *Micromonosporaceae, Pseudocardinaceae, and Thermomonosporaceae* (Genilloud, 2014). The principal therapeutical bioactivities attributed to actinomycete-derivated molecules, besides their antimicrobial potential, are antifungal, anticancer, antitumor, cytotoxic, cytostatic, anti-inflammatory, antiparasitic, antimalarial, antiviral, antioxidant, and anti-angiogenesis (Bérdy, 2005; Manivasagan et al., 2014).

Various species of actinomycetes are soil microorganisms and have developed diverse survival strategies, including saprophytic habits or the synthesis of secondary metabolites that often result toxic for competitor organisms, such as fungi or algae (Gomes et al., 2018; Lee et al., 2018). For example, Chandranamycin A, an antibiotic isolated from *Actinomadura* sp. also displayed antifungal activity against *Mucor miehei* and antialgal activity against *Chlorella vulgaris* and *Chlorella solokirniana* (Maskey et al., 2004; Manivasagan et al., 2014).

Besides the antimicrobial potential of several compounds produced by actinomycetes, various molecules with important therapeutical uses have been discovered. For example, salinosporamide A, extracted from *Salinispora tropica*, is an atypical beta-lactone-gamma-lactam which displayed activity as a proteasome inhibitor and as apoptosis inductor in diverse types of myeloma cells. For both activities, this compound has been explored as a possible anticancer agent (Gulder and Moore, 2008; Manivasagan et al., 2014).

There are several examples of actinomycetes-derived anticancer compounds, including anthracyclines, which have demonstrated their effectivity as antineoplastic molecules and are used as therapeutical options against a wide range of cancers, including leukemias, lymphomas, breast, uterine, lung, and ovarian cancers (Hortobagyi, 1997; Demain and Sanchez, 2009). Another classic example is the bleomycin, isolated from *Streptomyces verticillium*. This nonribosomal glycopeptide was discovered in the decade of the 1960s and approved as a chemotherapeutic agent by the US Food and Drug Administration (FDA) in 1973 (Demain and Sanchez, 2009).

Drugs such as mitomycins, anthracenones, enediynes, and epothilones are approved products widely recognized as antitumorigenic (Demain and Sanchez, 2009; Manivasagan et al., 2014).

Other examples of anticancer molecules reported in the literature are resistoflavin, from *Streptomyces* sp. (Kock et al., 2005; Bhatnagar and Kim, 2010), streptanoate, from *Streptomyces* sp. DC3 (Noomnual et al., 2016), marinomycin A and D from *Marinispora sp.* (Kwon et al., 2006), daryamide A and C from *Streptomyces* strain CNQ-085 (Asolkar et al., 2006), daryamide D and F from *Streptomyces* strain SNE-011 (Fu et al., 2017) and carpatamide, from *Streptomyces* strain SNE-011 (Fu et al., 2017) are some examples of these anticancer compounds.

Metabolites with antimalarial activity, such as trioxacarcin A, B, and C, from *Streptomyces ochraceus* and *Streptomyces bottropensis* have also been identified as therapeutical compounds (Maskey et al., 2004). Nevertheless, not only therapeutical compounds have been identified in actinomycetes. For example, the herbicide Bialaphos is derived from *Streptomyces hygroscopicus* (Demain and Sanchez, 2009). Thus considering the enormous diversity of actinomycetes, their adaptation to diverse ecosystems and biosynthetic potential, these organisms will continue attracting researchers toward the identification of bioactive molecules (Manivasagan et al., 2014).

12.3.3 BIOACTIVE COMPOUNDS FROM NONACTINOMYCETE BACTERIA

Excluding the actinomycetes, bacteria, in general, are excellent sources of bioactive compounds. For example, around 93 bioactive compounds have been identified in *Vibronaceae*, including antibacterial, siderophores, and anticancer compounds (Mansson et al., 2011). In general, thousands of bacterial metabolites have been reported as bioactive compounds (Bérdy, 2005). Epothilones, for example, are macrolides isolated from the myxobacterium *Sorangium cellulosum*, which have been identified as antitumorigenic compounds (Demain and Sanchez, 2009; Cheng and Huang, 2018). Spinosyns, a family of macrolides produced by *Saccharopolyspora spinosa*, are used as insecticides (Demain and Sanchez, 2009). Prodigiosin, produced by *Serratia marcescens*, *Pseudomonas magneslorubra*, *Vibrio psychroerythrous*, *Serratia rubidaea*, *Vibrio gazogenes*, *Alteromonas rubra*, and *Rugamonas rubra* have demonstrated their effectiveness as a proapoptotic agent against diverse cancer cell lines (Darshan and Manonmani, 2015).

Carotenoids as astaxanthin and zeaxanthin have been identified as nutraceutical compounds with antioxidant properties. They are produced by species such as *Agrobacterium aurantiacum*, *Paracoccus carotinifaciens*, and *Xanthophyllomyces dendrorhous* (astaxanthin), *Flavobacterium* sp., *Paracoccus zeaxanthinifaciens* (zeaxanthin) (Malik et al., 2012; Darshan and Manonmani, 2015).

Bioactive compounds have also been isolated from anaerobes, such as the closthioamide from *Clostridium cellulolyticum*, a polythioamide antibiotic (Lincke et al., 2010; Challinor and Bode, 2015). Diverse bioactive compounds have been identified in *Burkholderiales* and symbiotic bacteria (Challinor and Bode, 2015). *Xenorhabdus* and *Photorhabdus* species are sources of anthraquinones and peptide metabolites derived from nonribosomal peptide synthase and polyketide synthase, which display antibiotic activity (Challinor and Bode, 2015; Tobias et al., 2017). Cyanobacteria are also sources of bioactive metabolites, including phenolic compounds, phycocyanin, polyunsaturated fatty acids, sulfate polysaccharides, and carotenoids, which display diverse bioactivities, such as antimicrobial, antioxidant and antibacterial, anticancer, antiviral, and in some cases as toxins (Singh et al., 2005; Vijayakumar and Menakha, 2015; Nowruzi et al., 2018). Some examples are given by the anticancer/antimicrobial borophycin, isolated from *Nostoc linckia* and *Nostoc spongiaeforme var. tenue* (Vijayakumar and Menakha, 2015; Nowruzi et al., 2018), or the cryptophycin, from *Nostoc* sp., which is a potent fungicide (Singh et al., 2005; Nowruzi et al., 2018).

In conclusion, considering the diversity and the ubiquity of bacteria, microbes should be considered as relevant sources for the obtaining of bioactive compounds, and the discovery of novel molecules. Once identified the molecules of interest, and their biosynthetic pathway, the usage of GE tools become crucial in the optimization of the microorganisms as biofactories, increasing the efficiency in metabolite production, and reducing production costs.

12.4 GENETIC ENGINEERING OF MICROORGANISMS FOR ENZYME PRODUCTION AND ITS APPLICATION IN THE FOOD INDUSTRY

12.4.1 MICROORGANISMS AND FOOD INDUSTRY: A GENERAL OVERVIEW

The development of human civilization is tightly linked to food biotechnology. The first evidence for the use of microorganisms in this sense might be traced back thousands of years (~10,000 BCE), specifically as a food preservation technology (fermentation) (Nair and Prajapati, 2003). However, since this was developed altogether around the globe, establishing a particular date, place or time might not be completely accurate.

Fermentation processes were the first ones to emerge and be applied in a plethora of different food sources: meat, beverages, cheese, bread, to mention a few. Nevertheless, the understanding of how the different microorganisms involved in all these food biotechnology processes is a rather recent event (~200 years) that coincides with the invention of the microscopes and the development of proper scientific techniques to understand the complex microscopic world (Soetaert and Vandamme, 2010; Csutak and Sarbu, 2018).

The use of microbiological enzymes in the food industry can be linked to the dawn of fermented food products like beers, bread, soy sauces, pickles, cheese making, wines, among others (Singh et al., 2019). However, routine applications in the food industry became common since the 20th century (Singh et al., 2019).

Enzymes are the quintessential nature chemical catalysts; among the most common ones applied for food processing, we can distinguish: amylases, proteases, pectinases, invertases, cellulases, and lactases, among others. The main

microbiological genera employed for this purpose include *Bacillus (Bacillus subtilis, Bacillus amyloliquifaciens, Bacillus licheniformis)* (Saengsanga et al., 2016; Kumar et al., 2017), *Aspergillus (Aspergillus niger, Aspergillus oryzae)* (Di Ghionno et al., 2017; Dias et al., 2017), *Saccharomyces (Saccharomyces cerevisiae)* (Gallone et al., 2018), *Lactobacillus (Lactobacillus casei, Lactobacillus acidophilus, Lactobacillus delbruekii)*, *Schizosaccharomyces pombe* (Mylona et al., 2016), and *Rhizopus (Rhizopus oryzae, Rhizopus oligosporus)* (Mohamed et al., 2015; Lee et al., 2016).

Food applications of these microorganisms and their derived enzymes in the industrial process and research require thorough analysis in microbial collections (Bourdichon et al., 2012). Even when the increase in genetic and genomic data nowadays might seem as an advantage for this analysis, the potential application of microbial-derived enzymes can be decreased due to limitations or misinterpretations of the Nagoya protocol (Börner et al., 2018).

On the other hand, since the advent of microbial GE in the late 1970s, (Itakura et al., 1977) the protocols for genetic modification are available for a high number of microorganisms and constitute a very important strategy to obtain enzymes with higher yields or new applications within the food industry (Börner et al., 2018).

12.4.2 GENETICALLY ENGINEERED MICROORGANISMS WITHIN THE FOOD INDUSTRY

As mentioned before, since the late 1970s scientists have been able to restructure the genetic constitution of microorganisms, with potential applications in diverse processes including food biotechnology. "GMOs" is a term applied to any organism which has been subjected to alterations in its genetic material employing technologies that are not exactly natural (mating or genetic recombination from parental lines). This technology is also regarded as *recombinant DNA* technology or *GE* (World Health Organization, 2017). In line with this idea, the term GMOs within the food industry is applied to the use of foods that have been produced by using *genetically modified microbes* (GMM) or *genetically modified plants* (GMP). GMMs are responsible for major changes in desirable food qualities or organoleptic characteristics such as changes in taste and texture as well as changes associated with productivity (yield, shelf life, maturation). Either GMMs or GMPs can both account for an increase in nutritional values (Mallikarjuna and Yellamma, 2019).

Nowadays, even though a significant amount of *genetically modified* *enzymes* (GME) with application in the food industry are available in the market, still the major source for enzymes comes from nongenetically modified enzymes that are obtained from diverse sources (animals, plants, or microorganisms). Nevertheless, these natural enzymes have important drawbacks regarding their capacity to meet expected demands in terms of food processing environments (high pressure, extreme pH conditions, salt levels, and solvents, among others). Thus the use of GME allows generating alternatives with more amiable qualities (yield, specificity, purity, stability, surface property, cost-effective, etc.) (Zhang et al., 2019).

Diverse strategies are utilized to obtain these GME and many of them are expressed in model microorganisms such as *E. coli* or *Pichia pastoris*. Among these strategies, we can highlight (1) gene sequence optimization; (2) gene truncation or fusion; (3) site-directed mutagenesis; (4) directed evolution; (5) semirational design (Zhang et al., 2019).

Depending on the final product, the scientist must decide which strategy is more suitable. In general terms, the expression of recombinant in a heterologous background requires to analyze codon preferences in the host cell; thus a gene sequence optimization pathway must be followed (Rosano and Ceccarelli, 2014). If, however, the idea is to enhance enzyme function, a gene truncation could be used, this technique allows to "shut down" other DNA sequences which are not important for the production of the enzyme, increasing the overall outcome or the activity of the enzyme (Dediu, 2015). Conversely, site-directed mutagenesis is a finer strategy, since it involves genetic alterations at a more precise level. For instance, in order to succeed, scientists must have a fair understanding of important details regarding the enzyme to be modified, such as structure, catalytic mechanism, active sites, among others. This information is then used for decision making in terms of which specific DNA sequences are to be modified (Hua et al., 2018).

On the other hand, strategies that not necessarily imply the "in-depth" understanding of enzyme structure or other important details are also utilized. Such is the case of directed evolution, which is done mainly by generating libraries that are used for profound screening and selection of new enzymatic activities, which are a result of random mutagenesis (Labrou 2010). Finally, induced mutations that require an understanding of DNA sequence, the enzyme structure, or in silico modeling approaches are all part of the pipeline for a semirational design for new enzyme generation (Özgün et al., 2016).

Emerging technologies such as genome editing by CRISPR-CAS9 are also starting to gain popularity for the use of new enzymes with application in the food industry, specifically in lactic acid bacteria (LAB) (Börner et al., 2018). Since the technology for genome editing is relatively recent, some optimization efforts have been reported in *Lactobacillus reuteri* (Oh and Van Pijkeren, 2014), *Lactobacillus plantarum* (Leenay et al., 2019), and *Lactobacillus lactis* (Van Der Els et al., 2018). Indeed, modifications for the CRISPR-Cas9 methodology applied in *L. lactis* will allow the use of single plasmid systems for inducing the expression of various proteins with potential applications in the dairy industry (Berlec et al., 2018). Other microorganisms with reported use of this technology include filamentous fungi from the *Aspergillus* genus (Nødvig et al., 2015). Fungi are known for the difficulty they pose for genetic manipulation; thus the development of genomic editing techniques opens the possibility for multiple applications within the food industry.

Important considerations that have an impact on genetic transformation strategies for the production of food enzymes from microbiological origin must be carefully analyzed. For instance, a routine step, such as choosing the correct plasmid, can be more delicate when the final product is included in a particular step of the food processing. In this sense, plasmid election must be based firstly on the purpose of the genetic modification, and also on the plasmid type: replicating, integrative, or cryptic (Landete, 2017). Care must be taken when choosing these kinds of plasmids since common selectable markers (antibiotic resistance) for other GMO technologies are not used in food-grade GE. Plasmid replication mechanisms impact directly in the results obtained since they regulate the host range, stability, and copy number (Shareck et al., 2004). Also, when more than one plasmid is required for the genetic modification strategy, the compatibility among vectors must be taken into consideration. If care is not taken in this sense, significant drawbacks might appear in the host cell leading to the loss of expression of the selected proteins (De Vos and Simons, 1994). Besides, the type of integration of the transformation vectors must also be revised carefully. According to Douglas et al. (2011), three strategies can be followed: transposition of IS elements, site-specific recombination (attP/integrase), and homologous recombination (Douglas et al., 2011). In line with this, food-grade vectors are designed with special care and considering regulations that will allow them to be labeled as *Generally Regarded As Safe* (*GRAS*). In addition, the use of the same DNA or the host or DNA coming from GRAS microorganisms is not classified as genetically modified and will not receive negative perceptions as compared to their nonfood grade counterparts (Lu et al., 2013). Thus GMM derived from these strategies can be recognized as GRAS by the US FDA and the Qualified Presumption of Safety in Europe, therefore allowing them to be used in downstream food industry applications (Landete, 2017).

Nevertheless, even when food vectors can be designed based on GRAS DNA, the use of selective agents that do not include antibiotic resistance is imperative since this will not allow the use of in food applications. Therefore for food-grade vectors selective agents are based on marked changes in microorganism strain phenotypes or if they restore impaired functions (Landete, 2017). Since the modulation of gene expression results of importance in the food industry, different mechanisms have been developed. For instance, the use of two-component signal transduction systems such as nisin-inducible vectors is available (Sørvig et al., 2003). Other strategies include the use of high-affinity ABC transporters that are only regulated by specific metals (Zn^{2+}) (Llull and Poquet 2004). Also, sugar transport mechanisms such as phosphotransferase system for lactose (inducible *lac* promoters), which at least for LAB can help as dominant or complementation markers and regulators of gene expression (Ma et al.,

2014). Some others include thermolabile promoters (D'Souza et al., 2012) and phage induced systems (O'Sullivan et al., 1996).

In general, a thorough understanding of plasmid design is imperative for the GE of microorganisms for enzyme production in the food industry. A plethora of potential applications come along once a stable and efficient transformation protocol is established in microorganisms. Thus many of the genetically modified enzymes will find application mainly in the food industry; however, other strategies involving food-grade vectors point out their potential application also in the production of probiotic cultures, protein production, metabolic engineering, and in health (Landete, 2017; Börner et al., 2018).

Since the public acceptance of GMO derived technologies and their application in the food industry is still divided among consumers, which nowadays exert their right to ask for correct labeling indicating a non-GMO origin of foods or associated derivatives, innovation in this area is of particular interest. Strategies such as genome editing as the ones described in this chapter hold a promising future in the biotechnological food sector. The potential use of GMM and GME is enormous and it is not only limited to the food sector but other important industrial processes.

12.5 CONTRIBUTION TO HIGHER RESISTANCE AND TOLERANCE TO ANTIBIOTICS AND OTHER COMPOUNDS USED FOR BIOLOGICAL CONTROL OF PATHOGENS

The emergence of bacteria resistant to multiple antimicrobial agents, including antibiotics, has made infectious diseases difficult to treat or intractable, making it a global health problem and one of the serious challenges facing humanity today. Alarmingly, the number of drug-resistant diseases increases rapidly, while the discovery of new drugs occurs slowly (Martínez, 2014; Courvalin, 2016; Ogawara, 2018). Consequently, there is an urgent need to understand the mechanisms of bacterial resistance and the development of alternatives to control it. The term antibiotic, coined by Selman Waksman in 1942, is derived from "anti," "opposite," and "biotikos," of living beings, and is defined as a substance secreted by a microorganism that has the ability to alter the growth of pathogenic microorganisms (Ogawara, 2018; Blair et al., 2015). The World Health Organization (World Health Organization, 2017) has made a list of the main microorganisms that have developed resistance to antibiotics worldwide; based on its clinical importance in humans and its increasing resistance to third-generation carbapenems, cephalosporins, fluoroquinolones, and aminoglycosides. This list includes *E. faecium, S. aureus, Klebsiella pneumoniae, Acinetobacter baumanii, Pseudomonas aeruginosa,* and *Enterobacter* spp., *Enterobacteriaceae* spp. (*E. coli, K. pneumoniae, Enterobacter* spp., *Serratia* spp., *Proteus* spp., *Providencia* spp.), *Helicobacter pylori, Salmonella* spp., *Neisseria gonorrhoeae, Streptococcus pneumoniae, Haemophilus influenza,* and *Shigella* spp. (Martínez, 2014).

Antibiotics can have multiple cell targets; most affect some of the cellular processes of translation, transcription, replication, and cell wall synthesis (Brown, 2016). For example, commonly used antibiotics, β-lactams (cephalosporins, carbapenems, and penicillins), inhibit cell wall synthesis. Others are inhibitors of protein synthesis (aminoglycosides, tetracyclines, chloramphenicol, macrolides oxazolidinones, and streptogramins); rifamycins constitute inhibitors of transcription, and quinolones inhibit DNA synthesis by binding to gyrase (Petchiappan and Chatterji, 2017). In bacteria, antibiotic resistance may be directly related to their genome (e.g., *Mycobacterium tuberculosis* is intrinsically resistant to β-lactam because it encodes a β-lactamase enzyme in its genome); or, they can acquire resistance through chromosomal mutations or by horizontal gene transfer from other bacteria (Blair et al., 2015).

In general, there are three mechanisms by which a bacterium can develop antibiotic resistance (Figure 12.5):

1. Increase in the expression of porin genes to reduce the cellular input of antibiotics (Fajardo-Lubián et al., 2019); or, over-expression of multidrug efflux pumps genes that are responsible for the active export of antibiotics to the cell outside. Therefore intracellular concentrations of the antibiotic are minimized as a result of poor penetration into the bacteria or the flow of antibiotics (Hinchliffe et al., 2013). In the case of *Enterobacteriaceae, Pseudomonas* spp. and *Acinetobacter* spp., reductions in expression porin contribute significantly to the resistance of drugs such as carbapenems and cephalosporins (Baroud et al., 2013).

2. Degradation or biochemical modification of the structure of antibiotics: β-lactamase enzymes can degrade β-lactam antibiotics (e.g., *K. pneumoniae, E. coli, P. aeruginosa,* etc.); There are several known enzymes that can transfer chemical groups such as phosphate, acyl, and nucleotidyl to antibiotic molecules (acetyltransferases, nucleotidyltransferases, as well as phosphotransferases) (Shen et al., 2013). The enzyme chloramphenicol-florfenicol

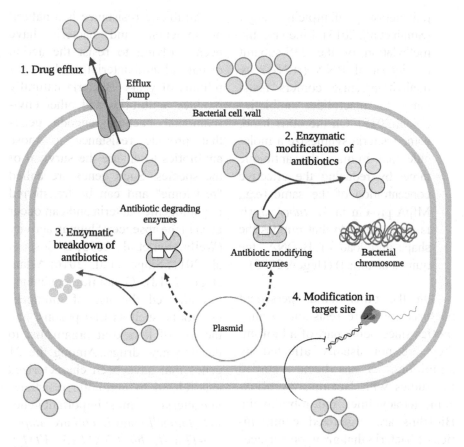

FIGURE 12.5 Mechanisms of antibiotic resistance. There are three main ways by which bacteria can increase their resistance: (a) varying the efflux of antibiotics; (b) modifying the antibiotics; or (c) modifying their targets (Adapted from Aslam, B., Wang. *Infection and drug resistance, 11,* 1645, 2018).

resistance (cfr) methyltransferase specifically methylates the A2503 site in the 23S rRNA, thereby conferring resistance to a wide range of drugs that have targets near this site, such as phenols, pleuromutilins, streptogramins, lincosamides, and oxazolidinones (including linezolid) (Blair et al., 2015).

3. Mutation, modification, or protection of targets through mutations, methylations or mimicry of the antibiotic binding points (Blair et al., 2015; Petchiappan and Chatterji, 2017). For example, the resistance of *M. tuberculosis* strains to rifampicin is due to the presence of mutations in the *rpoB* gene that encodes the β-subunit of RNA

polymerase, rifampicin target (Smith et al., 2013). Likewise, the methylation of the 23S-subunit of ribosomal RNA (rRNA) by methyltransferase confers resistance to macrolide antibiotics (Blair, 2015). On the other hand, some bacteria synthesize a molecule that mimics the antibiotic target by reducing the effective concentration of the same (e.g., MfpA protein in *M. tuberculosis* is a pentapeptide that mimics the shape and load of B-DNA, the quinolone target) (Hegde, 2005).

On the other hand, persistent bacteria are more resistant to antibiotics since being part of a biofilm; they are not usually affected by antibiotics, as has been recorded in studies with lactams, ciprofloxacin, tetracycline, and tobramycin. Biofilms are organized communities of bacteria that grow on surfaces or at liquid–air interfaces when microorganisms are integrated into an extracellular matrix of proteins, extracellular DNA, lipids, polysaccharides, and water, for example, *P. aeruginosa, K. pneumoniae, S. aureus,* and *M. tuberculosis.* Thus several nanoparticles (e.g., silver, selenium) have been reported to have an inhibitory effect against biofilms: also, the (p) ppGpp has been implicated in biofilm formation in multiple species like *P. aeruginosa, M. tuberculosis,* and so forth (Algburi et al., 2017).

Antibiotic resistance is a natural phenomenon, and bacteria have been evolving to resist the action of natural antibacterial products for billions of years. Bacteria naturally produce antibiotics and other environmental bacteria encode genes that provide resistance to those antibiotics to ensure the survival of the species. Such genes are called "resistome" and can be transferred to pathogenic bacteria and can occur even in the absence of human activity (Wellington et al., 2013; D'Costa et al., 2011; Singer et al., 2016; Aslam et al., 2018). Genomic sequencing has allowed the use of mutagenesis tools, such as transposons, and the use of knockout organisms, to develop new drugs. Among the 21 genes that have been characterized and related to antibiotic resistance in bacteria, the most important genes in *Salmonella* and *E. coli* are *AmpC, bla-TEM-1, bla-CTXM-15, VIM-1, NDM-1, floR, tetG,* and *mcr-1* gene, related to colistin resistance (Aslam et al., 2018). In this sense, the alterations of the essential genes of the cell have been explored, although the cellular phenotype can be determined by the alteration of this gene or by the environmental conditions (Domalaon et al., 2018).

CRISPR systems can be an alternative to disable antibiotic resistance genes in bacteria by eliminating crucial domains from their genome. CRISPR-Cas technology is part of the adaptive immune system in bathtubs and arches that acts against foreign nucleotide material. CRISPR-Cas systems are composed of a CRISPR RNA (crRNA) and Cas proteins and the crRNA is complementary to the target sequence and thus guides the Cas proteins for the sequence-specific recognition and cleavage (Yao et al., 2018), for example, the genetic modification of *S. aureus* through the basic edition of nCas9-APOBEC1 and CRISPR-Cas9 (Chen et al., 2017) and the dCas9 system in *S. aureus* (Dong et al., 2017). Subsequently, reprogramming of CRISPR-Cas9 directed to virulence genes in *S. aureus*, directed to the kanamycin resistance gene, *aph-3*, was performed, this resulted in a strong inhibition of *S. aureus* growth due to chromosomal excision and subsequent cell death (Bikard et al., 2014). The CRISPR-Cas9/Cas12a system in *Pseudomonas putida* and the CRISPRi has been applied for the repression of genes in *P. aeruginosa, Pseudomonas fluorescens*, and *P. putida* (Tan et al., 2018). In addition, CRISPR-Cas9 and its interference in the genome have been performed in *K. pneumoniae* (Loureiro and Da Silva, 2019).

12.6 FINAL REMARKS

In nature, we can find a diversity of environments and habitats, which has allowed the evolution of microorganisms. Derived from this process,

each strain of microorganisms has developed mechanisms that allow its adaptation to different substrates (Crespi et al., 2001; Elena and Lenski, 2003). It has been described that even isolates belonging to the same species can display new characteristics making them more efficient in the transformation of different substrates, production of different bioactive molecules, and production of different compounds of interest (Rabaey and Verstraete, 2005; Subramaniyan and Prema, 2002). Because of this, the identification and isolation of new species of microorganisms is an arduous task that still represents a biotechnological potential.

According to this book, the knowledge generated in microorganisms with specific properties and characteristics applied to the fields of medicine, food, energy generation, and restoration of environments (Suardana et al., 2012; Lee and Kim, 2015), and it has evolved as GE strategies advance. These advances aim to minimize the adverse effects of gene transfer on nontargeted microorganisms (Verma et al., 2011; Barrows et al., 2014).

One of the limitations in the knowledge of microorganisms with biotechnology applications has been that, in most cases, the culture requires conditions that are only generated in their habitats. It is estimated that at the present day, only 1% of existing microorganisms have been cultivated. This leads to continue working on two nearby objectives: the first to optimize, formulate, and improve the culture conditions to be able to isolate and identify more microorganisms; and as a second objective, improve the use of other molecular strategies such as metagenomics, barcodes, metaproteomics, and genomic sequencing of uncultured microorganisms from single cells. These technologies would allow us to identify microorganisms that are not cultivable to date, and help identify genes with future biotechnology application in microorganisms with distinctive characteristics (Giraffa and Neviani, 2001; DeLong, 2009; Maron et al., 2007).

ACKNOWLEDGMENT

To Professor Luis Antonio Muñiz Cervantes for grammar and spelling corrections.

KEYWORDS

- bacteria
- genetic engineering
- metabolite production
- bioactive compounds
- enzyme production

REFERENCES

Adrio, J.L.; Demain, A.L. Recombinant organisms for production of industrial products. *Bioeng. Bugs*. 2010, 1, 116–131.

Algburi, A.; Comito, N.; Kashtanov, D.; Dicks, L.; Chikindas, M. L. Control of biofilm formation: antibiotics and beyond. *Appl. Environ. Microbiol.* 2017, 83, e02508–e02516. doi:10.1128/AEM.02508-16.

Almendros, C.; Mojica, F.J.; Diez-Villasenor, C.; Guzman, N.M.; Garcia-Martinez, J. CRISPR-Cas functional module exchange in *Escherichia coli*. *MBio*. 2014, 5, e00767–e00713.

Arroyo, M.; Acebal, C; de la Mata, I. Biocatálisis y biotecnología. *Arbor*. 2014, 190 a156.

Aslam, B.; Wang, W.; Arshad, M.I.; Khurshid, M.; Muzammil, S.; Rasool, M. H.; Baloch, Z. Antibiotic resistance: a rundown of a global crisis. *Infect. Drug Resist.* 2018, 11, 1645–1658. doi:10.2147/IDR.S173867.

Asolkar R.N.; Jensen P.R.; Kauffman C.A.; Fenical, W. Daryamides A-C, weakly cytotoxic polyketides from a marine-derived actinomycete of the genus *Streptomyces* strain CNQ-085. *J. Nat. Prod.* 2006, 69, 1756–1759.

Baitin, D.M., Zaitsev, E.N.; Lanzov, V.A. Hyper-recombinogenic RecA protein from *Pseudomonas aeruginosa* with enhanced activity of its primary DNA binding site. *J. Mol. Biol.* 2003, 328, 1–7.

Baroud, M.; Dandachea, I.; Araj, G.F.; Wakim, R.; Kanj, S.; Kanafani, Z.; Hkairallah, M.; Sabrá, A.; Shehab, M.; Dbaibo, G.; Matar, G.M. Underlying mechanisms of carbapenem resistance in extended-spectrum β-lactamase-producing *Klebsiella pneumoniae* and *Escherichia coli* isolates at a tertiary care centre in Lebanon: role of OXA-48 and NDM-1 carbapenemases. *Int. J. Antimicrob. Agents*, 2013, 41, 75–79.

Barrangou, R. The roles of CRISPR–Cas systems in adaptive immunity and beyond. *Curr. Opin. Immunol.* 2015, 32, 36–41.

Barrows, G.; Sexton, S.; Zilberman, D. Agricultural biotechnology: the promise and prospects of genetically modified crops. *J. Econ. Perspect.* 2014, 28, 99–120.

Bérdy J. Bioactive microbial metabolites—a personal view. *J. Antibiot.* 2005, 58, 1–26.

Berlec, A.; Škrlec, K.; Kocjan, J.; Olenic, M.; Štrukelj, B. Single plasmid systems for inducible dual protein expression and for CRISPR-Cas9/CRISPRi gene regulation in lactic acid bacterium *Lactococcus lactis*. *Sci. Rep.* 2018, 8, 1:1009.

Bhatnagar I.; Kim S.K. Immense essence of excellence: marine microbial bioactive compounds. *Mar. Drugs.* 2010, 8, 2673–2701.

Biesalski H.K.; Dragsted L.O.; Elmadfa, I. Grossklaus, R.; Müller, M.; Schrenk, D.; Walter, P; Weber, P. Bioactive compounds: definition and assessment of activity. *Nutrition.* 2009, 25, 1202–1205.

Bikard, D.; Euler, C.W.; Jiang, W.; Nussenzweig, P.M.; Goldberg, G.W.; Duportet, X.; Fischetti, V.A.; Marraffini, L.A. Exploiting CRISPR-Cas nucleases to produce sequence-specific antimicrobials. *Nat. Biotechnol.* 2014, 32, 1146–1150.

Bindal, S.; Dagar, V.K.; Saini, M.; Khasa, Y.P.; Gupta, R. High level extracellular production of recombinant γ-glutamyl transpeptidase from *Bacillus licheniformis* in *Escherichia coli* fed-batch culture. *Enzyme Microb. Technol.* 2018, 116, 23–32.

Blair, J.M.; Webber, M.A.; Baylay, A.J.; Ogbolu, D.O.; Piddock, L.J. Molecular mechanisms of antibiotic resistance. *Nat. Rev. Microbiol.* 2015, 13, 42–51.

Borlaug, N.E. Feeding a world of 10 billion people: the miracle ahead. *Biotechnol. Biotec. Eq.* 1997, 11, 3–13.

Börner, R.A.; Kandasamy, V.; Axelsen, A.M.; Nielsen, A.T.; Bosma, E.F. Genome editing of lactic acid bacteria: opportunities for food, feed, pharma and biotech. *FEMS Microbiol. Lett.* 2018, 366, 1: fny291.

Bourdichon, F.; Casaregola, S.; Farrokh, C.; Frisvad, J.C.; Gerds, M.L.; Hammes, W.P.; Harnett, J.; Huys, G.; Laulund, S.; Ouwehand, A. Food fermentations: microorganisms with technological beneficial use. *Int. J. Food Microbiol.* 2012, 154, 87–97.

Brown, E.D.; Wright, G.D. Antibacterial drug discovery in the resistance era. *Nature*, 2016, 529, 336–343.

Challinor V.L.; Bode H.B. Bioactive natural products from novel microbial sources. *Ann. N. Y. Acad. Sci.* 2015, 1354, 82–97.

Chen, W.; Zhang, Y.; Yeo, W.S.; Bae, T.; Ji, Q. Rapid and efficient genome editing in *Staphylococcus aureus* by using an engineered CRISPR/Cas9 system. *J. Am. Chem. Soc.* 2017, 139, 3790–3795.

Cheng, H.; Huang, G. Synthesis & antitumor activity of epothilones B and D and their analogs. *Future Med. Chem.* 2018, 10, 1483–1496.

Choi, K.R.; Jang, W.D.; Yang, D.; Cho, J.S.; Park, D.; Lee, S.Y. Systems metabolic engineering strategies: integrating systems and synthetic biology with metabolic engineering. *Trends Biotechnol.* 2019, 1744, 1–21.

Choi, K.H.; Kim, K.J. Applications of transposon-based gene delivery system in bacteria. *J. Microbiol. Biotechnol.* 2009, 19, 217–228.

Choi, J.H.; Lee, S.Y. Secretory and extracellular production of recombinant proteins using *Escherichia coli*. *Appl. Microbiol. Biot.* 2004, 64, 625–635.

Cong, L., Ran, F.A., Cox, D.; Lin, S., Barretto, R.; Habib, N.; Hsu, P.D.; Wu, X.; Jiang, W.; Marraffini, L.A.; Zhang, F. Multiplex genome engineering using CRISPR/Cas systems. *Science.* 2013, 339, 819–823.

Courvalin, P. Why is antibiotic resistance a deadly emerging disease? *Clin. Microbiol. Infect.* 2016, 22, 405–407.

Crespi, B.J. The evolution of social behavior in microorganisms. *Trends Ecol. Evol.* 2001, 16, 178–183.

Csutak, O.; Sarbu, I. Genetically Modified Microorganisms: Harmful or Helpful? In:

Genetically Engineered Foods. Elsevier. 2018, 143–175.

Curtin, S.J.; Voytas, D.F.; Stupar, R.M. Genome engineering of crops with designer nucleases. *Plant Genome*. 2012, 5, 42–50.

Day, A.; Goldschmidt-Clermont, M. The chloroplast transformation toolbox: selectable markers and marker removal. *Plant Biotechnol. J.* 2011, 9, 540–553.

Darshan, N., Manonmani, H.K. Prodigiosin and its potential applications. *J. Food Sci. Technol.* 2015, 52, 5393–5407.

Das, M.; Sakha-Ghosh, P.; Manna, K. A review on Platensimycin: a selective FabF inhibitor. *Int. J. Med. Chem.* 2016, DOI: 10.1155/2016/9706753.

D'Costa, V.M.; King, C.E.; Kalan, L.; Morar, M.; Sung, W.W.L.; Schwarz, C.; Wright, G.D. (Antibiotic resistance is ancient. *Nature*, 2011, 477, 457–461.

DeLong, E.F. The microbial ocean from genomes to biomes. *Nature*. 2009, 459, 200.

De Vos, W.; Simons, G. Gene cloning and expression systems in *lactococci*. In: Genetics and Biotechnology of Lactic acid Bacteria. Springer. 1994, 52–105.

Dediu, D. *An Introduction to Genetics for Language Scientists*. Cambridge University Press. 2015.

Demain, A.L. Sanchez, S. Microbial drug discovery: 80 years of progress. *J. Antibiot.* 2009, 62, 5–16.

Dias, F.F.G.; Junior, S.B.; Hantao, L.W.; Augusto, F.; Sato, H.H. Acrylamide mitigation in French fries using native l-asparaginase from *Aspergillus oryzae* CCT 3940. *LWT–Food Sci. Technol.* 2017, 76, 222–229.

Domalaon, R.; Idowu, T.; Zhanel, G.G.; Schweizer, F. Antibiotic hybrids: the next generation of agents and adjuvants against Gram-negative pathogens?. *Clin. Microbial. Rev.* 2018, 31, e00077–17. doi:10.1128/CMR.00077-17.

Dong, X.; Jin, Y.; Ming, D.; Li, B.; Dong, H.; Wang, L.; Wang, T.; Wang, D. CRISPR/dCas9- mediated inhibition of gene expression in *Staphylococcus aureus*. *J. Microbiol. Methods*, 2017, 139, 79–86.

Di Ghionno, L.; Marconi, O.; Sileoni, V.; De Francesco, G.; Perretti, G. Brewing with prolyl endopeptidase from *Aspergillus niger*: the impact of enzymatic treatment on gluten levels, quality attributes and sensory profile. *Int. J. Food Sci. Technol.* 2017, 52, 1367–1374.

Douglas, G.L.; Goh, Y.J.; Klaenhammer, T.R. Integrative food grade expression system for lactic acid bacteria. In: *Strain Engineering*. Springer. 2011, 373–387.

Donot, F.; Fontana, A.; Baccou, J.C.; Strub, C.; Schorr-Galindo, S. Single cell oils (SCOs) from oleaginous yeasts and moulds: production and genetics. *Biomass Bioenerg.* 2014, 68, 35–150.

Dobrindt, U.; Hochhut, B.; Hentschel, U.; Hacker, J. Genomic islands in pathogenic and environmental microorganisms. *Nat. Rev. Microbiol.* 2004, 2, 414–424.

D'Souza, R.; Pandeya, D.R.; Hong, S.T. *Lactococcus lactis*: an efficient Gram positive cell factory for the production and secretion of recombinant protein. *Biomed. Res.* 2012, 23, 1–7.

Elena, G.M.; Ramona, B. E.; Holban, A.M. Approved genetically engineered foods: types, properties, and economic concerns. In: *Genetically Engineered Foods*. Academic Press. 2018, 85.

Elena, S.F.; Lenski, R.E. Microbial genetics: evolution experiments with microorganisms: the dynamics and genetic bases of adaptation. *Nature Rev. Genet.* 2003, 4, 457.

Eş, I.; Khaneghah, A.M.; Barba, F.J.; Saraiva, J.A.; Sant'Ana, A.S.; Hashemi, S.M.B. Recent advancements in lactic acid production—a review. *Food Res. Int.* 2018, 107, 763–770.

Fajardo-Lubián, A.; Ben Zakour, N.L.; Agyekum, A.; Qi, Q.; Iredell, J.R. Host adaptation and convergent evolution increases antibiotic resistance without loss of virulence in a major human pathogen.

PLoS Pathogens, 2019, 15, e1007218. doi:10.1371/journal.ppat.1007218.

Feng, Z.; Zhang, B.; Ding, W.; Liu, X.; Yang, D.L.; Wei, P.; Cao F.; Zhu S.; Zhang F.; Mao Y.; Zhu, J.K. Efficient genome editing in plants using a CRISPR/Cas system. *Cell Res.* 2013, 23, 1229.

Ferrer-Miralles, N.; Villaverde, A. Bacterial cell factories for recombinant protein production; expanding the catalogue. *Microb. Cell. Fact.* 2013, 12, 113.

Foster, P. L. Stress-induced mutagenesis in bacteria. Crit. Rev. Biochem. Mol. Biol. 2007, 42, 373–397.

Fu, P.; La, S.; MacMillan, J.B. Daryamide analogues from a marine-derived Streptomyces species. *J. Nat. Prod.* 2017, 80, 1096–1101.

Gaj, T.; Gersbach, C. A.; Barbas III, C.F. ZFN, TALEN, and CRISPR/Cas-based methods for genome engineering. *Trends Biotechnol.* 2013, 31, 397–405.

Gallone, B.; Mertens, S.; Gordon, J.L.; Maere, S.; Verstrepen, K.J.; Steensels, J. Origins, evolution, domestication and diversity of *Saccharomyces* beer yeasts. *Curr. Opin. Biotechnol.* 2018, 49, 148–155.

Garneau, J.E.; Dupuis, M.E.; Villion, M.; Romero, D.A.; Barrangou, R.; Boyaval, P.; Fremaux, C.; Horvath, P.; Magadan, A.H.; Moineau, S. The CRISPR/Cas bacterial immune system cleaves bacteriophage and plasmid DNA. *Nature.* 2010, 468, 67–71.

Gasiunas, G.; Barrangou, R.; Horvath, P.; Siksnys, V. Cas9-crRNA ribonucleoprotein complex mediates specific DNA cleavage for adaptive immunity in bacteria. *Proc. Natl. Acad. Sci. USA.* 2012, 109, E2579–E2586.

Genilloud, O. The re-emerging role of microbial natural products in antibiotic discovery. *Antonie van Leeuwenhoek, Int. J. Gen. Mol. Microbiol.* 2014, 106, 173–88.

Genilloud, O. Actinomycetes: still a source of novel antibiotics. *Nat. Prod. Rep.* 2017, 34, 1203–1232.

Giraffa, G.; Neviani, E. DNA-based, culture-independent strategies for evaluating microbial communities in food-associated ecosystems. *Int. J. Food Microbiol.* 2001, 67, 19–34.

Golkar, Z.; Rochelle, L.; Bagasra, O. Crisprs/Cas9 may provide new method for drug discovery and development. *J. Mol. Biomark. Diagn.* 2016, 7, 280.

Gomes, E. de B.; Dias, R.L.; de Cassia, M.; de Miranda, R. Actinomycetes bioactive compounds: Biological control of fungi and phytopathogenic insect. *Afr. J. Biotechnol.* 2018, 17, 552–559.

Gould, F.; Cohen, M.B. Sustainable use of genetically modified crops in developing countries. In Agricultural Biotechnology and the Poor: An International Conference. CGIAR: Washington, DC, USA. 2000. p. 139.

Guaadaoui, A.; Benaicha, S.; Elmajdoub, N.; Hamal, A.; Bellaoui, M. What is a bioactive compound? A combined definition for a preliminary consensus. *Int. J. Food. Sci. Nutr.* 2014, 3, 17–179.

Gulder, T.A.M.; Moore, B.S. Salinosporamide natural products: potent 20S proteasome inhibitors as promising cancer chemotherapeutics. *Bone.* 2008, 23, 1–7.

Haft, D.H.; Selengut, J.; Mongodin, E.F.; Nelson, K.E. A guild of 45 CRISPR-associated (Cas) protein families and multiple CRISPR/Cas subtypes exist in prokaryotic genomes. *PLoS Comput. Biol.* 2005, 1, e60.

Hegde, S.S.; Vetting, M.W.; Roderick, S.L.; Mitchenall, L.A.; Maxwell, A.; Takiff, H.E.; Blanchard, J.S.A. Fluoroquinolone resistance protein from *Mycobacterium tuberculosis* that mimics DNA. *Science,* 2005, 308, 1480–1483.

Hinchliffe, P.; Symmons, M.F.; Hughes, C.; Koronakis, V. Structure and operation of bacterial tripartite pumps. *Annu. Rev. Microbiol.* 2013, 67, 221–242.

Hua, W.; El Sheikha, A.F.; Xu, J. Molecular techniques for making recombinant enzymes used in food processing. In: Molecular Techniques in Food Biology: Safety, Biotechnology, Authenticity and Traceability. Wiley. 2018, 95–114.

Hortobagyi, G.N. Anthracyclines in the treatment of cancer. An overview. *Drugs.* 1997, 54, 1–7.

Itakura, K.; Hirose, T.; Crea, R.; Riggs, A.D.; Heyneker, H.L.; Bolivar, F.; Boyer, H.W. Expression in *Escherichia coli* of a chemically synthesized gene for the hormone somatostatin. *Science.* 1977, 198, 4321: 1056–1063.

Jansen, R.; Embden, J.D.; Gaastra, W.; Schouls, M.L. Identification of genes that are associated with DNA repeats in prokaryotes. *Mol. Microbiol.* 2002, 43, 1565–1575.

Jemli, S.; Ayadi-Zouari, D.; Hlima, H.B., Bejar, S. Biocatalysts: application and engineering for industrial purposes. *Crit. Rev. Biotechnol.* 2016, 36, 246–258.

Jiang, W.; Bikard, D.; Cox, D.; Zhang, F.; Marraffini, L.A. RNA-guided editing of bacterial genomes using CRISPR-Cas systems. *Nat. Biotechnol.* 2013, 31, 233–239.

Jinek, M.; Chylinski, K.; Fonfara, I.; Hauer, M.; Doudna, J.A.; Charpentier, E. A programmable dual-RNA-guided DNA endonuclease in adaptive bacterial immunity. 2012, 337, 816–821.

Kärenlampi, A. V. *Genetically Modified Microorganisms. Encyclopedia of Food and Health.* 2016. p. 211.

Kidwell, M.G. Transposable elements. In: T. R. Gregory, ed. *The Evolution of the Genome.* Elsevier: San Diego, 2005.

Kock, I.; Maskey, R.P.; Biabani, M.A.F. Helmke, E.; Laatsch, H. 1-Hydroxy-1-norresistomycin and resistoflavin methyl ether: New antibiotics from marine-derived streptomycetes. *J. Antibiot.* 2005, 58, 530–534.

Kumar, S.; Haq, I.; Prakash, J.; Raj, A. Improved enzyme properties upon glutaraldehyde cross-linking of alginate

entrapped xylanase from *Bacillus licheniformis. Int. J. Biol. Macromol.* 2017, 98, 24–33.

Kwon H.C.; Kauffman, C.A.; Jensen, P.R.; Fenical, W. Marinomycins A-D, anti-tumor–antibiotics of a new structure class from a marine actinomycete of the recently discovered genus "*Marinispora.*" *J. Am. Chem. Soc.* 2006, 128, 1622–1632.

Labrou, N.E. Random mutagenesis methods for in vitro directed enzyme evolution. *Curr. Protein Pept. Sci.* 2010, 11, 1:91–100.

Landete, J.M. A review of food-grade vectors in lactic acid bacteria: from the laboratory to their application. *Crit. Rev. Biotechnol.* 2017, 37, 3: 296–308.

Lasken, R. S. Genomic sequencing of uncultured microorganisms from single cells. *Nat. Rev. Microbiol.* 2012, 10, 631.

Lassen, J.; Madsen, K. H.; Sandøe, P. Ethics and genetic engineering–lessons to be learned from GM foods. *Bioproc. Biosyst. Eng.* 2002, 24, 263–271.

Lee, L.H.; Chan, K.G.; Stach, J.; Wellington, E.M.H.; Goh, B.H. Editorial: the search for biological active agent(s) from actinobacteria. *Front. Microbiol.* 2018, 9, 1–4.

Lee, L.W.; Cheong, M.W.; Curran, P.; Yu, B.; Liu, S.Q. Modulation of coffee aroma via the fermentation of green coffee beans with *Rhizopus oligosporus:* I. Green coffee. *Food Chem.* 2016, 211, 916–924.

Lee, S.Y.; Kim, H. U. Systems strategies for developing industrial microbial strains. *Nat. Biotechnol.* 2015, 33, 1061.

Leenay, R.T.; Vento, J.M.; Shah, M.; Martino, M.E.; Leulier, F.; Beisel, C.L. Genome Editing with CRISPR-Cas9 in *Lactobacillus plantarum* revealed that editing outcomes can vary across strains and between methods. *Biotechnol. J.* 2019, 14, 3: 1700583.

Liang, M.H.; Jiang, J.G. Advancing oleaginous microorganisms to produce lipid via metabolic engineering technology. *Prog. Lipid Res.* 2013, 52, 395–408.

Lincke, T.; Behnken, S.; Ishida, K.; Roth, M.; Hertweck, C. Closthioamide: an unprecedented polythioamide antibiotic from the strictly anaerobic bacterium *Clostridium cellulolyticum. Angew. Chem. Int. Ed.* 2010, 49, 2011–2013.

Llull, D.; Poquet, I. New expression system tightly controlled by zinc availability in *Lactococcus lactis. Appl. Environ. Microbiol.* 2004, 70, 9: 5398–5406.

Loureiro, A.; da Silva, G.J. CRISPR-Cas: converting a bacterial defence mechanism into a state-of-the-art genetic manipulation tool. *Antibiotics.* 2019, 8, 18. doi:10.3390/antibiotics8010018.

Lu, W.; Kong, J.; Kong, W. Construction and application of a food-grade expression system for *Lactococcus lactis. Mol. Biotechnol.* 2013, 54, 2: 170–176.

Ma, S.J.; Li, K.; Li, X.S.; Guo, X.Q.; Fu, P.F.; Yang, M.F.; Chen, H.Y. Expression of bioactive porcine interferon-alpha in *Lactobacillus casei. World J. Microbiol. Biotechnol.* 2014, 30, 9: 2379–2386.

Makarova, K. S., Haft, D.H.; Barrangou, R.; Brouns, S.J.; Charpentier, E.; Horvath, P.; Moineau, S.; Mojica, F.J.; Wolf, Y.I.; Yakunin, A.F.; Van der Oost, J.; Koonin., E.V. Evolution and classification of the CRISPR-Cas systems. *Nat. Rev. Microbiol.* 2011, 9, 467–477.

Malik, K.; Tokkas, J.; Goyal, S. Microbial pigments: a review. 2012, 361–365.

Mallikarjuna, N.; Yellamma, K. Genetic and Metabolic Engineering of Microorganisms for the Production of Various Food Products. In: *Recent Developments in Applied Microbiology and Biochemistry.* Elsevier. 2019, 167–182.

Manivasagan, P.; Venkatesan, J.; Sivakumar, K.; Kim, S.K. Pharmaceutically active secondary metabolites of marine actinobacteria. *Microbiol. Res.* 2014, 169, 262–278.

Mansson, M.; Gram, L.; Larsen, T.O. Production of bioactive secondary metabolites by marine Vibrionaceae. *Mar. Drugs.* 2011, 9, 1440–1468.

Maron, P.A.; Ranjard, L.; Mougel, C.; Lemanceau, P. Metaproteomics: a new approach for studying functional microbial ecology. *Microb. Ecol.* 2007, 53, 486–493.

Martinez, J.L. General principles of antibiotic resistance in bacteria. *Drug Discov. Today Technol.* 2014, 11, 33–39.

Maskey, P.; Helmke, E.; Kayser, O.; Fiebig, H.H.; Maier, A.; Busche, A.; Laatsch, H. Anti-cancer and antibacterial trioxacarcins with high anti-malaria activity from a marine Streptomycete and their absolute stereochemistry. *J. Antibiot.* 2004, 57, 771.

Mohamed, S.A.; Salah, H.A.; Moharam, M.E.; Foda, M.; Fahmy, A.S. Characterization of two thermostable inulinases from *Rhizopus oligosporus* NRRL 2710. *J. Genet. Eng. Biotechnol.* 2015, 13, 1: 65–69.

Mojica, F.J., Diez-Villasenor, C.; Garcia-Martinez, J.; Soria, E. Intervening sequences of regularly spaced prokaryotic repeats derive from foreign genetic elements. *J. Mol. Evol.* 2005, 60, 174–182.

Mylona, A.; Del Fresno, J.; Palomero, F.; Loira, I.; Bañuelos, M.; Morata, A.; Calderón, F.; Benito, S.; Suárez-Lepe, J.A. Use of *Schizosaccharomyces* strains for wine fermentation—effect on the wine composition and food safety. *Int. J. Food Microbiol.* 2016, 232, 63–72.

Nair, B.M.; Prajapati, J.B. The history of fermented foods. In: *Handbook of Fermented Functional Foods.* CRC Press. 2003, 17–42.

Nødvig, C.S.; Nielsen, J.B.; Kogle, M.E.; Mortensen, U.H. A CRISPR-Cas9 system for genetic engineering of filamentous fungi. *PLoS One.* 2015, 10, 7: e0133085.

Noomnual, S.; Thasana, N; Sungkeeree, P.; Mongkolsuk, S.; Loprasert, S. Streptanoate, a new anticancer butanoate from *Streptomyces* sp. DC3. *J. Antibiot.* 2016, 69, 124–127.

Nowruzi, B.; Haghighat, S.; Fahimi, H.; Mohammadi, E. *Nostoc* cyanobacteria species: a new and rich source of novel bioactive compounds with pharmaceutical potential. *J. Pharm. Heal. Serv. Res.* 2018, 9, 5–12.

Ogawara, H. Comparison of strategies to overcome drug resistance: learning from various kingdoms. *Molecules*, 2018, 23(6), 1476. doi:10.3390/molecules23061476

Oh, J.H.; Van Pijkeren, J.P. CRISPR–Cas9-assisted recombineering in *Lactobacillus reuteri*. *Nucleic Acids Res.* 2014, 42, 17: e131–e131.

O'Sullivan, D.J.; Walker, S.A.; West, S.G.; Klaenhammer, T.R. Development of an expression strategy using a lytic phage to trigger explosive plasmid amplification and gene expression. *Biotechnology.* 1996, 14, 1: 82.

Özgün, G.P.; Ordu, E.B.; Tütüncü, H.E.; Yelboğa, E.; Sessions, R.B.; Gül Karagüler, N. Site saturation mutagenesis applications on *Candida methylica* formate dehydrogenase. *Scientifica.* 2016. https://doi.org/10.1155/2016/4902450

Padkina, M.V.; Sambuk, E.V. Genetically modified microorganisms as producers of biologically active compounds. *Russ. J. Genet. Appl. Res.* 2016, 6, 669–683.

Patel, A.K.; Singhania, R.R.; Pandey, A. Production, purification, and application of microbial enzymes. In: *Biotechnology of Microbial Enzymes*. Academic Press. 2017. p. 13.

Petchiappan, A, Chatterji, D. Antibiotic resistance: current perspectives. *ACS Omega*, 2, 2017, 7400–7409.

Pourcel, C.; Salvignol, G.; Vergnaud, G. CRISPR elements in Yersinia pestis acquire new repeats by preferential uptake of bacteriophage DNA, and provide additional tools for evolutionary studies. *Microbiology.* 2005, 151, 653–663.

Rabaey, K.; Verstraete, W. Microbial fuel cells: novel biotechnology for energy generation. *Trends Biotechnol.* 2005, 23, 291–298.

Ran, F.A.; Hsu, P.D.; Wright, J.; Agarwala, V.; Scott, D.A.; Zhang, F. Genome engineering using the CRISPR-Cas9 system. *Nat. Protoc.* 2013, 8, 2281–2308.

Rosano, G.L.; Ceccarelli, E.A. Recombinant protein expression in *Escherichia coli*: advances and challenges. *Front. Microbiol.* 2014, 5, 172.

Rosano, G.L.; Ceccarelli, E.A. Recombinant protein expression in microbial systems. *Front. Microbiol.* 2014, 5, 341.

Stanbury, P.F.; Whitaker, A.; Hall, S. *Principles of Fermentation Technology*. Elsevier. 2016.

Robinson, J. Ethics and transgenic crops: a review. Electron. *J. Biotechnol.* 1999, 2, 5–6.

Saengsanga, T.; Siripornadulsil, W.; Siripornadulsil, S. Molecular and enzymatic characterization of alkaline lipase from *Bacillus amyloliquefaciens* E1PA isolated from lipid-rich food waste. *Enzyme Microb. Technol.* 2016, 82: 23–33.

Shareck, J.; Choi, Y.; Lee, B.; Miguez, C.B. Cloning vectors based on cryptic plasmids isolated from lactic acid bacteria: their characteristics and potential applications in biotechnology. *Crit. Rev. Biotechnol.* 2004, 24, 4: 155–208.

Shen, J.; Wang, Y.; Schwarz, S. Presence and dissemination of the multiresistance gene cfr in Grampositive and Gram-negative bacteria. *J. Antimicrob. Chemother.* 2013, 68, 1697–1706.

Singer, A.C.; Shaw, H.; Rhodes, V.; Hart, A. Review of antimicrobial resistance in the environment and its relevance to environmental regulators. *Front. Microbiol.* 2016, 7, 1728.

Singh, R.; Singh, A.; Sachan, S. Enzymes used in the food industry: friends or foes? In: *Enzymes in Food Biotechnology*, Elsevier. 2019, 827–843.

Singh, S.; Kate, B.N.; Banecjee, U.C. Bioactive compounds from cyanobacteria and microalgae: an overview. *Crit. Rev. Biotechnol.* 2005, 25, 73–95.

Small, B. Sustainable development and technology: genetic engineering, social sustainability and empirical ethics. *Int. J. Sust. Dev.*, 2007, 10, 402–435.

Smith, T.; Wolff, K.A.; Nguyen, L. Molecular biology of drug resistance in *Mycobacterium tuberculosis. Curr. Top. Microbiol. Immunol.*, 2013, 374, 53–80.

Soetaert, W.; Vandamme, E.J. *Industrial Biotechnology: Sustainable Growth and Economic Success.* John Wiley & Sons. 2010.

Sørvig, E.; Grönqvist, S.; Naterstad, K.; Mathiesen, G.; Eijsink, V.G.; Axelsson, L. Construction of vectors for inducible gene expression in *Lactobacillus sakei* and *L. plantarum. FEMS Microbiol. Lett.* 2003, 229, 1: 119–126.

Suardana, I.W.; Sujaya, I.N., Artama, W.T. Probe aplication to diagnostic programe of Shiga like toxin-2 (stx2) gen from *Escherichia coli* O157: H7. *Jurnal Veteriner.* 2012, 13, 434–439.

Subramaniyan, S.; Prema, P. Biotechnology of microbial xylanases: enzymology, molecular biology, and application. *Crit. Rev. Biotechnol.* 2002, 22, 33–64.

Takahashi, Y.; Nakashima, T. Actinomycetes, an inexhaustible source of naturally occurring antibiotics. *Antibiotics.* 2018, 7, 45.

Tamayo-Ordoñez, M.C.; Espinosa-Barrera, L.; Tamayo-Ordoñez, Y.J.; Ayil-Gutierrez, B.; A.; Sánchez-Teyer, L.F. Advances and perspectives in the generation of polyploid plant species. *Euphytica.* 2016, 209, 1–22.

Tan, S.Z.; Reisch, C.R.; Prather, K.L.J. A robust CRISPR interference gene repression system in *Pseudomonas. J. Bacteriol.* 2018, 200, 10.1128/JB.00575-17.

Thomason, L.C.; Sawitzke, J.A.; Li, X.; Costantino, N.; Court, D.L. 2014. Recombineering: genetic engineering in bacteria using homologous recombination. *Curr. Protoc. Mol. Biol.* 2014, 106, 11–39.

Tobias, NJ, Wolff, H, Djahanschiri, B.; Grundmann, F.; Kronenwerth, M.; Shi, Y.M.; Simonyi, S.; Grün, P.; Shapiro-Ilan, D.; Pidot, S.J.; Stinear, T.P.; Ebersberger, I.; Bode, H. B. Natural product diversity associated with the nematode symbionts *Photorhabdus* and *Xenorhabdus. Nat. Microbiol.* 2017, 2, 1676–1685.

Troncoso, C.; Pavez, M.; Santos, A.; Salazar, R; Barrientos, L. Implicancias estructurales y fisiológicas de la célula bacteriana en los mecanismos de resistencia antibiótica. *Int. J. Morphol.* 2017, 35, 1214–1223.

Truppo, M.D. Biocatalysis in the pharmaceutical industry: the need for speed. *ACS Med. Chem. Lett.* 2017, 8, 476–480.

Upadhyay, S.K.; Kumar, J.; Alok, A.; Tuli, R. RNA-guided genome editing for target gene mutations in wheat. *G3: Genes Genom. Genet.* 2013, 3, 2233–2238.

Van der Els, S.; James, J.K.; Kleerebezem, M.; Bron, P.A. Versatile Cas9-driven subpopulation selection toolbox for *Lactococcus lactis. Appl. Environ. Microbiol.* 2018, 84, 8: e02752–e02717.

Verma, C.; Nanda, S.K.; Singh, R.B.; Singh, R.; Mishra, S. A review on impacts of genetically modified food on human health. *Open Nutraceuticals J.* 2011, 4, 1–10.

Vijayakumar, S.; Menakha, M.; Pharmaceutical applications of cyanobacteria—a review. *J. Acute Med.* 2015, 5, 15–23.

Wallace, S.; Balskus, E.P. Opportunities for merging chemical and biological synthesis. *Curr. Opin. Biotech.* 2014, 30, 1–8.

Wang, J.; Soisson, S.M.; Young, K.; Shoop, W.; Kodali, S.; Galgoci, A.; Painter, R.; Parthasarathy, G.; Tang, Y. S.; Cummings, R.; Ha, S.; Dorso, K.; Motyl, M.; Jayasuriya, H.; Ondeyka, J.; Herath, K.; Zhang, C.; Hernandez, L.; Allocco, J.; Basilio, Á.; Tormo, J. R.; Genilloud, O.; Vicente, F.; Pelaez, F.; Colwell, L.; Lee, S. H.; Michael, B.; Felcetto, T.; Gill, C.; Silver, L. L.; Hermes, J. D.; Bartizal, K.; Barrett, J.; Schmatz, D.; Becker, J. W.; Cully, D.; Singh, S. B. Platensimycin is a selective FabF inhibitor with potent antibiotic properties. *Nature.* 2006, 441, 358–361.

Wellington E.M., Boxall A.B., Cross P., et al. The role of the natural environment in the emergence of antibiotic resistance in Gram-negative bacteria. *Lancet Infect. Dis.* 2013, 13, 155–165.

World Health Organization. Antibacterial Agents in Clinical Development: An Analysis of the Antibacterial Clinical Development Pipeline, Including Tuberculosis. WHO: Geneva. 2017, WHO/EMP/IAU/2017.12.

Wu, G.; Yan, Q.; Jones, J.A.; Tang, Y.J.; Fong, S.S.; Koffas, M.A. Metabolic burden: cornerstones in synthetic biology and metabolic engineering applications. *Trends Biotechnol*. 2016, 34, 652–664.

Yamanaka, K.; Reynolds, K.A.; Kersten, R.D.; Ryan, K.S.; Gonzalez, D.J.; Nizet, V.; Dorrestein, P.C.; Moore, B.S. Direct cloning and refactoring of a silent lipopeptide biosynthetic gene cluster yields the antibiotic taromycin A. *Proc. Natl. Acad. Sci*. 2014, 111, 1957–1962.

Yao, R.; Liu, D.; Jia, X.; Zheng, Y.; Liu, W.; Xiao, Y. CRISPR-Cas9/Cas12a biotechnology and application in bacteria. *Synth. Syst. Biotechnol*. 2018, 3, 135–149.

Zhang, Y.; Geary, T.; Simpson, B. K. Genetically modified food enzymes: a review. *Curr. Opin. Food Sci*. 2019, 25, 14–18.

Zhu, X.; Ye, K. Cmr4 is the slicer in the RNA-targeting Cmr CRISPR complex. *Nucleic Acids Res*. 2015, 43, 1257–1267.

PART II
Biomaterials and Biomolecules

CHAPTER 13

Biotechnological Production of Biomaterials and Their Applications

MAURICIO CARRILLO-TRIPP[1], RODOLFO TORRES-DE LOS SANTOS[2], MIGUEL A. MEDINA-MORALES[3], LEOPOLDO RÍOS-GONZÁLEZ[3], MARÍA ANTONIA CRUZ-HERNÁNDEZ[4], SALVADOR CASTELL-GONZÁLEZ[5], GERARDO DE JESÚS SOSA-SANTILLÁN[3], and ERIKA ACOSTA-CRUZ[3*]

[1]Laboratorio de la Diversidad Biomolecular, Centro de Investigación y de Estudios Avanzados del Instituto Politécnico Nacional Unidad Monterrey, Apodaca, Nuevo León, México

[2]Unidad académica multidisciplinaria Mante Centro, Universidad Autónoma de Tamaulipas, Cd. Mante, Tamaulipas, México

[3]Departamento de Biotecnología, Facultad de Ciencias Químicas, Universidad Autónoma de Coahuila, Blvd. Venustiano Carranza, 25280 Saltillo, Coahuila, México

[4]Laboratorio de Interacción Ambiente-Microorganismo, Centro de Biotecnología Genómica, Instituto Politécnico Nacional, Reynosa, Tamaulipas, México

[5]Colegio de Posgraduados en Ciencias Ambientales y Biotecnología del Sureste, A.C. Mérida, Yucatán, México

*Corresponding author. E-mail: erika.acosta@uadec.edu.mx.

ABSTRACT

Biotechnology is a diverse and thriving sector that overlaps with many industries, from energy to medicine, manufacturing to materials. There is a global interest in the development of advanced sustainable biomaterials as well as high-efficiency, low-cost manufacturing strategies. In the first part of this chapter, we review the bacteria bioproduction of polymers like polyhydroxyalkanoates (PHA), polylactates, cellulose, and amyloid aggregates, or compounds like 1-3 propanediol as a precursor of various materials. We also discuss applications in the emerging field of bioconcretes. Although viruses are not considered living microorganisms, the biopolymers they produce have been successfully used in the life sciences, bioelectronics, and nanomaterials. Hence, in the second part, we review recent biotechnological applications of the viral capsid because of its tremendous potential as a biocompatible material.

13.1 MICROBES AND ITS PRODUCTS FOR BIOMATERIALS, POLYMERS, AND BIOCONSTRUCTION

13.1.1 INTRODUCTION

In a strict sense, biotechnology refers to the use of biological systems or its parts to solve problems and make products. Modern biotechnology has a growing impact on society, as is one of the most important scientific and technological revolutions in the 21st century. One of the expanding fields during the last decades has been the development of novel material for diverse applications, including biomedicine, sustainable containers, and construction. There are classical concepts that have been

evolving along with the area, and for this chapter context, we would like to review some of them.

The concept of biomaterial has been under discussion for years. Since the prefix bio-implies life, it could broadly mean a substance derived from a living organism, or used for it. The European Society for Biomaterials defined in 1987 the term as "a nonviable material used in a medical device, intended to interact with biological systems" (Williams, 1987). This definition limited the word biomaterials to health care issues. Since then, there has been a vast evolution, and metallic components, polymers, biopolymers, ceramics, or composite materials were included as biomaterials, and this definition was a subject of controversy. In consequence, in later years, Williams made a very profound analysis about the nature of biomaterials, and concluded this definition: "A biomaterial is a substance that has been engineered to take a form which, alone or as a part of a complex system, is used to direct, by control of interactions with components of living systems, the course of any therapeutic or diagnostic procedure, in human or veterinary medicine" (Williams, 2009). Therefore, it applies also to substances and systems. Interestingly, some biopolymers are applied biomaterials.

Biopolymers are organic macromolecules produced by diverse living organisms, including microorganisms, composed of repeated monomers. In this case, the prefix bio- is used for the first acception mentioned above: derived from living organisms, or used for it. Homopolymers are such molecules that are composed of one type of monomer, while heteropolymers are composed of more than one type of monomer. Such polymers can be complex and often branched (Stal, 2011). Bacteria are especially prolific in the production of biopolymers, and they are used at the industrial level since their growth is relatively fast and simple.

The economic activities of modern society, and the massive use and demand of synthetic polymers (typically plastics), formed from petroleum-derived products in combination with synthetic additives have led to the contamination of large amounts of soil by xenobiotic compounds which also accumulate in water and air (Barbeau et al., 2015, Gerhardt et al., 2017). Several authors claim that the speed of accumulation is greater than the detoxification ability of the components of the ecosystems. Plastic products have very versatile qualities, including their durability and resistance to degradation, which is the main cause of their extended usage in the manufacture of a wide variety of products, which are essential for the current lifestyle. Global climate change and the depletion of oil resources used for the generation of these plastics are worrisome

(McGuinness and Dowling, 2009, Kuppusamy et al., 2017). Thus, the option of replacing petroleum-derived plastics with bioplastics and biopolymers has been explored in recent years as a sustainable and ecological alternative.

Some examples of biopolymers are polyesters, polyhydroxy acids, polysaccharides, polyalkylene dicarboxylates, polyether-esters, polyamides, polyesteramides, polyether amides, polyurethanes, aliphatic polycarbonates, extracellular polymeric substances (EPSs), among others (Nofar et al., 2018; Zia et al., 2015; Goonoo et al., 2016; Zhang et al., 2018; Singh et al., 2017; Xiao y Zheng, 2016; Kim y Rhee, 2003; Antonsczak et al., 2014; Chudecka-Glaz et al., 2016; Debbabi et al., 2017; Bajestani et al., 2017; Zhou et al., 2015; Tan et al., 2018; Kawaguchi et al., 2017; Sun et al., 2018). Some of these biopolymers are derived from a diverse set of polysaccharides, proteins, lipids, polyphenols, and specialty polymers produced by bacteria, fungi, plants, and animals, so they are considered renewable resources. Biopolymers of bacterial origin have very interesting and sustainable features, such as biodegradability and eco-friendly manufacturing processes, and have a vast range of applications, including medical science.

Also, there is a rich field of research that can be expanded by utilizing various novel approaches for the guidance of living organisms for "biomediated" material structuring purposes. The usage of biological systems, via microbial fabrication, biotechnology, or biomediated material structuring implies the second range of usage or abstraction of biological materials or systems, as elegantly discussed by Duerling et al. (2018). An example of this is bioconcretes, and we will address the subject in this chapter.

13.1.2 POLYLACTIC ACID

Nowadays, plastics are still produced mainly from oil, and most of these persist in the environment as nonbiodegradable waste (Singhvi et al., 2019). As a result, over the past decade, numerous research projects have been carried out for its production from renewable and environmentally friendly raw materials, such as some fermentation products (López-Gomez et al., 2019).

Lactic acid (LA) is one of six main compounds that serve as a building block of biopolymers and can be produced from sugars by anaerobic fermentation (Komesu et al., 2017). LA contains a chiral carbon and, therefore, can be found in two different isomeric forms: L-(+)-lactic acid (L-LA) and D-(-)-lactic acid (D-LA) (Cubas-Cano et al., 2018). Preferably the L-LA has been commercially used in the food

and pharmaceutical industry since it can be metabolized by humans due to the presence of the enzyme L-lactate dehydrogenase (Singhvi et al., 2019). On the contrary, the D-LA isomer is considered harmful to humans in high doses and may cause acidosis or decalcification (Alves de Oliveira et al., 2018).

The polymerization of LA generates PLA, which is one of the most used bioplastics due to its physical–mechanical properties and biodegradability (Figure 13.1). During their polymerization, racemic mixtures result in an amorphous and unstable polymer, while the use of pure LA enantiomers in appropriate proportions generates high crystallinity (Nofar et al., 2019). The LA produced by fermentation offers some advantages compared to the petrochemical process, such as (1) use of renewable and cheap substrates, (2) lower energy consumption, and (3) D-acid production or L-LA optically pure by selecting appropriate microorganisms (Singhvi et al., 2018).

Currently, about 90% of the LA used worldwide is produced by fermentation via LA bacteria (LAB), such as *Lactobacillus, Pediococcus, Enterococcus, Leuconostoc, Streptococcus,* and *Lactococcus,* using raw materials rich in sugar or starch (Mazzoli et al., 2014). However, the high costs of these materials and food security related issues are the main obstacles limiting their development

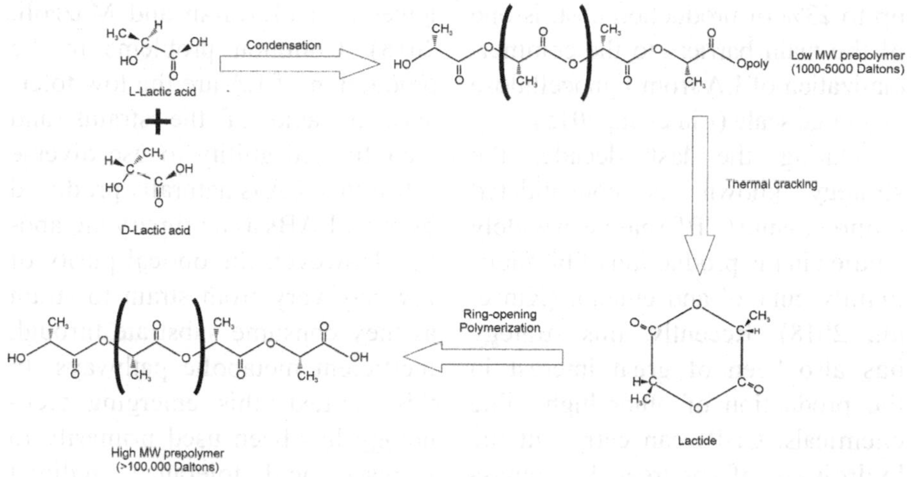

FIGURE 13.1 Polymerization of LA to produce PLA.

on a large scale as a renewable alternative to traditional oil-based plastics (Komesu et al., 2017).

Alternatively, lignocellulosic biomass, the most abundant form of fixed renewable carbon on the planet, is a promising raw material for the sustainable production of LA and other chemicals at low cost without needing more arable land or interfering with food production (Mazzoli et al., 2014). The world production of lignocellulosic biomass is approximately 2×10^{11} ton/year, where around 8×10^9–20×10^9 tons of the total can be used for the production of chemicals such as LA (Mazzoli et al., 2014). The main polysaccharides present in the lignocellulosic biomass are cellulose (38%–50%), hemicellulose (23%–32%), and lignin (12%–25%) (Ponnusamy et al., 2019). Cellulose is a long chain of D-glucose formed by glycosidic linkages β-1$\rightarrow$ 4 converting cellulose into its amorphous state due to the strong hydrogen bonding between the microfibrils. However, the presence of hemicellulose and mainly lignin confers rigidity to the cell wall but hinders the hydrolysis of cellulose (Liu et al., 2019).

Typically, lignocellulosic biomass processing through biotechnology strategies involves three main stages: (1) pretreatment of biomass to enhance digestibility by removing lignin and/or hemicellulose, (2) Enzymatic hydrolysis of pretreated biomass to depolymerize cellulose and hemicellulose, and (3) fermentation of enzymatic hydrolysates to obtain the products of interest (Rios-González et al., 2017). However, the high costs of commercial cellulase enzyme complexes, which contribute up to 23% of production cost, is one of the main barriers to the commercialization of LA from lignocellulose at a large scale (Liu et al., 2015).

During the last decade, the strategy known as consolidated bioprocessing (CBP) has been widely studied in the production of biofuels, mainly butanol and ethanol (Xin et al., 2018). Recently, this strategy has also been of great interest in the production of other high-value chemicals. CBPs can carry out the hydrolysis of pretreated biomass and its fermentation in a one-stage process, decreasing the costs of production (Shahab et al., 2017). The production of LA through CBP can be carried out by genetic modification of LAB so that they can have the capacity to hydrolyze cellulose or starch; using coculture between LAB and microorganisms that hydrolyze cellulose or starch; and through the use of microbial consortia (Tarraran and Mazzoli, 2018). Although LA production reports using CBP in literature are scarce, recent works on this area have shown the feasibility of this strategy (Eş et al., 2018).

13.1.2.1 *GENETIC AND METABOLIC ENGINEERING IN LA PRODUCTION*

Metabolic engineering in fermentation processes is an advanced tool for developing competitive strains that render high productivity at a lower cost (Tarraran and Mazzoli, 2018). Common problems in the production of LA are the low tolerance of acid of the strains and their limited ability to use diverse substrates. LA is naturally produced by most LABs as a primary metabolite. However, the optical purity of LA may vary from strain to strain as they consume substrate through inefficient metabolic pathways. In this context, this emerging technology has been used primarily to increase acid tolerance, redirect metabolic pathways to consume substrate more efficiently, and produce a purer LA (Eş et al., 2018).

Nowadays, due to the less recalcitrant nature of starch compared to lignocellulose, the most successful studies in the CBP have been those focused on the development of recombinant amylolytic LAB to hydrolyze starch (Okano et al., 2018). Relatively few natural LAB strains can degrade starch, and so far, none have been reported with cellulase activity (Hassan et al., 2019). Lignocellulose hydrolysis involves the synergistic interaction of different enzymes with affinity to different specific substrates (cellulases, xylanases, and other hemicellulases) with different catalytic mechanisms (endoglucanases, exoglucanases, and β-glucosidases). The filamentous fungi (e.g. *Trichoderma reesei* and *Aspergillus niger*) and actinomycetes are the most studied microorganisms due to their ability to produce "free" cellulases in amounts of 1–10 g/L (Mazzoli et al., 2014). However, also some anaerobic bacteria such as *Clostridium* spp. and *Ruminoccoccus* spp. can hydrolyze cellulose (although with a lower concentration of cellulases compared to fungi) by binding to the lignocellulosic surface using "complexed" multienzymatic systems known as cellulosomes (Kawaguchi et al., 2016).

Different genetic engineering strategies have been carried out to allow LABs the capacity to hydrolyze lignocellulosic biomass by heterologous cellulase expression (Mazzoli et al., 2012; Yamada et al., 2013). The strains of *Lactobacillus plantarum* and *Lactococcus lactis* have been commonly used as "chassis" microorganisms by heterologous expression of proteins from anaerobic cellulolytic bacteria. Currently, recombinant cellulosic microbial LAB is developed by heterologous expression of what is called minicellulosomes or through artificial cellulomes (Bayer et al., 2007).

The endoglucanase Cel8A from *C. thermocellum* was successfully expressed in *Lactobacillus gasseri* ATCC 33323, *Lactobacillus jonhsonii* NCK 88 (Cho et al., 2000), and *Lactobacillus plantarum* (Rossi et al., 2001; Okano et al., 2010). *Clostridium thermocellum* is one of the most studied anaerobic bacteria that can produce the cellulosome complex. The genome of *C.* *thermocellum* ATCC 27405 contains at least 79 cellulosomal genes, of which ~70 encode the type-I dockerin-containing proteins (Hirano et al., 2016). The endoglucanase Cel6A and the endoxylanase Xyn11A of *Termobifida fusca* were successfully introduced in *Lb. plantarum* WCF1 (Morais et al., 2013) with the potential to be used to bioaugment consortia that biodegrade biomass.

Recently, in a report by Stern et al. (2018), the endoglucanase GH10 and GH11 xylanase of *Clostridium papyrosolvens* introduced also in *Lb. plantarum* WCF1 promoted an improvement in the hydrolysis of wheat straw pretreated with hypochlorite. Rarely the presence of genes encoding enzymes involved in the depolymerization of xylooligosaccharides, arabinoxylans, or arabinans (i.e., β-xylosidases and arabinofuranosidases) has been detected in different LAB strains (Tarraran and Mazzoli, 2018). However, the strain of *Lc. lactis* is one of the LAB bacteria most used for the expression of heterologous proteins to hydrolyze xylooligosaccharides, although proteins with cellulolytic characteristics have also been expressed (Gandini et al., 2017). The xylanase produced by *Bacillus coagulans* and Neucallimastrix sp. were expressed in *Lc. lactis* MG1316 allowing improvement in the hydrolysis of xylans and CMC, respectively (Raha et al., 2006; Ozkose et al., 2009).

Due to its low protein secretion capacity of bacteria compared to eukaryotic cells, but with a higher specific activity than free cellulases, it seems to be the best enzymatic systems that can confer cellulolytic capacity on LABs. In addition, LABs are phylogenetically close, and their content of G + C is low and similar to *C. cellulovorans, C. thermocellum*, and *C. cellulolyticum* (Mazzoli et al., 2014). These similarities are essential for efficient biosynthesis of heterological proteins, with particular emphasis on efficient translation, which is often skewed by different use of codons in very distant organisms (Mazzoli et al., 2012). Modulation of the stability of the mRNA can be an alternative tool to optimize the heterological expression of cellulase in LAB (Okano et al., 2010b). Recent studies suggest that the mechanisms of protein secretion in Clostridia and LAB may be similar (Mingardon et al., 2011). Gene products encoding cellulosic components of cellulosic Clostridia, including their original signal peptide, could be efficiently secreted by *Lb. plantarum* (Mingardon et al., 2011; Morais et al., 2013; Okano et al., 2010b). This significantly reduces problems related to cell heterologous expression. As regards the secretion of other proteins in heterologous hosts, the expression of heterological cellulose may be hampered by the saturation of the

host's transmembrane transport mechanisms, which causes reduction/loss of cell viability (Mazzoli et al., 2012). Such limitations are still a crucial bottleneck of recombinant cellulolytic strategies (Tarraran and Mazzoli et al., 2018).

13.1.3 POLYHYDROXYAL-KANOATES

Currently, the use of bioplastics produced by fermentation, like the generically called PHA and particularly polyhydroxybutyrate (P3HB) and polyhydroxybutyrate-cohydroxyvalerate P (3HBco3HV), has gained much more attention from the industrial and research field since they are microbial polymers friendly to the environment and have a great potential in biomedical, agricultural, and industrial applications (Mozejko-Ciesielska and Kiewisz, 2016).

PHA are biopolyesters, polymers of hydroxyalkanoic acids that some bacteria, archaea, and microalgae accumulate intracellularly as reserve material, to later use it as a source of carbon and energy. Its characteristics of biocompatibility and biodegradability allow its use in a wide range of applications, such as the biomedical sector (Ali-Raza et al., 2018). These bioplastics represent a renewable and sustainable resource without being persistent or causing pollution.

Currently, numerous bacterial genera have been reported as PHA producers that can occur in aerobic and anaerobic conditions (Kim et al., 2007; González-García, 2013). Some of these microorganisms can store intracellular inorganic and organic inclusions that are surrounded by phospholipids. If the inclusion has a core of iron oxide, it is called inorganic inclusion such as magnetosomes, while if the inclusion has a polyester core, it is called organic inclusion as PHA (Poli et al., 2011; Rehm, 2009).

The bacteria that produce PHA are divided into two groups, in the first are the bacteria that require the limitation of phosphorus (*Rhodospirillum rubrum, Rhodobacter sphaeroides, Caulobacter crescentus, Pseudomonas oleovorans*), nitrogen (*Ralstonia cepacian, Ralstonia eutropha, Alcaligenes latus, Pseudomonas oleovorans*), oxygen (*Azotobacter vinelandii, Azotobacter beijerinckii*), or magnesium *(Pseudomonas* sp.) to accumulate PHA and do not accumulate PHA during their growth phase (Reddy et al., 2003; Nitschke et al., 2011).

However, some bacteria produce the polymer associated with growth and do not require any limitation of nutrients, although this case is less frequent. Recombinant *Escherichia coli* belongs to this second (Muhammadi et al., 2015).

13.1.3.1 MICROORGANISMS USED FOR THE PRODUCTION OF PHA

With respect to the diversity of the microorganisms with the capacity to produce PHA, as well as the carbon sources that they use and the accumulation percentages, to date more than 300 species have been reported, among which are the bacteria that they use substrates, such as carbohydrates, lipids, proteins, aromatic compounds, agroindustrial residues, and gases. Among the bacterial species that are used to accumulate large quantities of PHA and its copolymers are native and recombinant strains, such as *Ralstonia eutropha, Cupriavidus necator, Wautersia eutropha, Azotobacter* sp., *Bacillus* sp., *Pseudomonas* sp., *Burkholderia* sp., *Azospirillum brasilense,* Activated sludge, Cyanobacteria, *Chromobacterium* sp., *Erwinia* sp., recombinant *E. coli* (Gonzalez-Garcia et al., 2013; Poli et al., 2011, Tugarova et al., 2017; Ali-Raza et al., 2018).

13.1.3.2 APPLICATIONS OF PHA

PHAs exhibit relatively high molecular weights, thermoplastic or elastomeric characteristics, and other physical and mechanical properties that make them candidates for various applications in the packaging industry, medicine, pharmacy, agriculture and food, or as raw materials for the synthesis of chemicals enantiomerically pure and for the production of paints (Babel and Steinbüchel, 2001, Ruth et al., 2007).

It is also possible to use them in the form of aqueous latex for covering fibrous materials, such as paper or cardboard. Due to its water resistance, this cover protects against deterioration caused by moisture. Another application of interest is its use as a packaging material since PHA has a low diffusivity of oxygen (Babel and Steinbüchel, 2001). In addition, they can be used as carriers for the long-term release of herbicides or insecticides (Galego et al., 2000). Also, the high molecular weight of PHA can be useful to produce ultra-strong fibers for the fishing industry (Bugnicourt et al., 2014).

PHA offer a promising potential in the area of biomaterials in pharmacy, as mentioned in a study conducted by Chung et al. (2013), which reported the production of 3-hydroxyalkanoic acids (3HA) using the genetically modified bacterium *Pseudomonas entomophila* with precursors for the synthesis of value-added chemical products, including pharmaceuticals, antibiotics, food additives, fragrances, and vitamins. The homopolymer of length, 3-hydroxyoctanoic acid (3HO), showed potential

antimicrobial activities. Studies have also been conducted with *Burkholderia sacchari*, which has been reported as a potential producer of APHA. In this study, 30 different carbon sources were evaluated as cosubstrate to incorporate different monomers in the PHA in the culture of this species (Mendonca et al., 2014).

Studies in *Pseudomonas oleovorans* on the synthesis and intracellular degradation of P3HO showed that before the bacterium completes the utilization of the carbon source of the medium, the cell decreases the polymerase concentration (from 20 to 12%) while which increases the production of depolymerase (from 10 to 15%) (Lenz and Marchessault, 2005).

Biodegradability and biocompatibility are key characteristics of bioplastics. A property of PHA that distinguishes them from traditional petroleum-based plastics is their biodegradability. These biopolyesters are air-stable, inert, resistant to moisture, and insoluble in water (Ojumu et al., 2004). They can be completely degraded to water and carbon dioxide under aerobic conditions and to methane and carbon dioxide under anaerobic conditions by microorganisms in soil, sea, lake water, and wastewater (Khamma et al., 2005). PHAs can be biodegraded by a wide variety of ubiquitous microorganisms in many ecosystems to carbon dioxide or methane, both in aerobic and anaerobic conditions, without the formation of toxic products (Reddy et al., 2003).

13.1.3.3 PROSPECTIVES

The prospects for PHA are promising. Numerous research projects have proposed the evaluation of a wide range of carbon sources, bacterial strains, fermentation conditions, and recovery methods to obtain better performance and economic prospects. However, the cost plays an important role: if the price of these materials were reduced, they would have expansive applications. Currently, studies have been focused on producing PHA from plants, like corn and potatoes, as well as the search for cheaper technologies for its production and use in the manufacture of products for daily use. Also, recent advances in synthetic biology and genetic engineering have led to the production of PHA from strains that do not produce PHA in its wild type and that do not produce toxins. The progression in recovery techniques has improved the efficiency of the extraction of biomass with high purity. In this context, it is resorting to the search of microbial strains that on the one hand are capable of producing PHA, and on the other, they can use as substrates residues of the agricultural and livestock industries.

13.1.4 BACTERIAL CELLULOSE

Cellulose is the most abundant natural polymer on the planet obtained from renewable sources. Although wood is the primary natural resource for obtaining cellulose, other plants such as hemp, flax, or cotton contain substantial amounts. It consists of a linear homopolysaccharide composed of β-D-glucose chains linked by β-1,4-glycosidic bonds (Fernandes et al., 2017). In addition to these sources, cellulose can be produced by algae, fungi, and some species of bacteria, these being the most important. Bacterial strains that synthesize cellulose (BC) include *Aerobacter*, *Xanthococcus*, *Azotobacter*, *Acetobacter*, *Rhizobium*, *Pseudomonas*, and *Alcaligenes*. Among them, *Komagataeibacter xylinum* (formerly *Gluconacetobacter xylinum*) is the microorganism most commonly used to synthesize BC due to its high yield and stable production of BC biofilms (Gao et al., 2019). BC was first described in 1886 by AJ Brown, who observed the production of cellulose by *Acetobacter* in the presence of oxygen and glucose forming a biofilm of variable thickness in order to maintain high oxygenation of the colonies near the surface as a protective barrier against natural desiccation and radiation. In addition, BC is important, and in cell morphology, adhesion, and proliferation (Brown et al., 2000; Portela et al., 2019). The growing interest in the production of BC is due to its excellent physical and chemical properties, including high safety, low cost, high flexibility and adaptability, hydrophilicity, transparency, excellent biocompatibility, and biodegradability (Khalid et al., 2017). In this sense, bacterial cellulose is pure cellulose, microfibrils are smaller than the ones in plant produced cellulose, and therefore, are more porous and with greater water retention, while vegetable cellulose contains impurities such as hemicellulose and lignin and is less crystalline (Hickey et al., 2019).

BC biosynthesis is performed by fermentation of glucose to linear β-1,4-glucan chains that are secreted extracellularly and subsequently crystallized to form cellulose monofilaments. A certain number of these monofilaments are added to form gelatinous filamentous fibers on the surface of a liquid medium (Jacek et al., 2019). Bacterial nanocellulose is biosynthesized in two stages, glucose is first converted to β-1,4-glucan chains, and then cellulose crystallization occurs. The conversion of glucose into cellulose requires four enzymatic steps (Figure 13.2): the phosphorylation of glucose by glucokinase to glucose-6-phosphate (G6P); isomerization of G6P to glucose-1-phosphate (G1P) by the enzyme phosphoglucomutase (PGM); conversion of glucose-1-phosphate to uridine glucose diphosphate (UDP-glucose)

by UDP-glucose pyrophosphorylase; and finally, the synthesis of cellulose from UDPglucose by cellulose synthase (Bcs), a complex of four subunits, BcsA, BcsB, BcsC, and BcsD, which are encoded by three (*bcsAB, bcsC,* and *bcsD*) or four (*bcsA, bcsB, bcsC,* and *bcsD*) genes. It has been determined that UDP-glucose-pyrophosphorylase is the key enzyme in the biosynthesis of BC since its activity in cellulose producing bacteria is about one hundred times higher than in nonproducing bacteria (Jacek et al., 2019; Liu et al., 2018). BC biosynthesis is related to other metabolic pathways, such as the pentose phosphate route, the Embden-Meyerhof-Parnas route, the Krebs cycle, and gluconeogenesis. In addition to glucose, hexoses, glycerol, dihydroxyacetone, pyruvate, or dicarboxylic acids can also be converted into cellulose (Nagashima et al., 2016).

It has been determined that the genetic regulation of BC production is determined by the genetic complex of *BC synthase* and cyclic diguanylate (c-di-GMP) as the key activator of the BcsA-BcsB subunit of *bcs*. On the other hand, Liu et al. (2018) found that the quorum detection system (QS) demonstrated a possible mechanism for regulating the QS action in the production of BC in *G. xylinus* CGMCC 2955. In this sense, recently Augimeri and Strap (2015) showed that an increase in

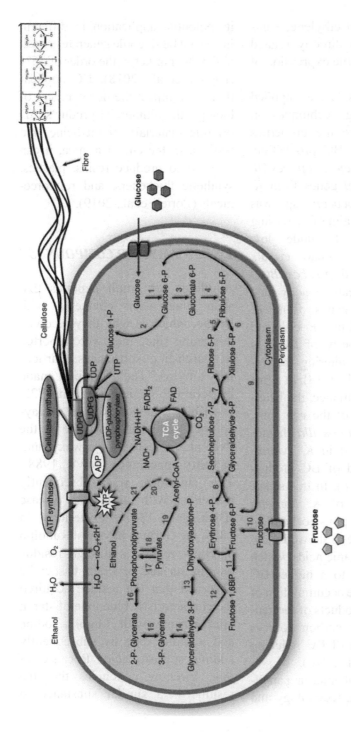

FIGURE 13.2 Pathways for the biosynthesis of BNC by *K. xyiinus* and assembly of cellulose molecules into nanofibrils: (1) Glucokinase-ATP, (2) PGM, (3) Glucose-6-phosphate dehydrogenase, (4) 6-phosphogluconate dehydrogenase, (5) Phosphorribulose isomerase, (6) Phosphorribulose epimeraase, (7) Transaketolase, (8) Transaldolase, (9) Phosphoglucoisomerase, (10) Fructokinase, (11) Fructokinase ATP, (12) Aldolase, (13) Triosephosphate isomerase, (14) Glyceraldehyde 3-phosphate dehydrogenase, (15) Phosphoglycerate mutase, (16) Enolase, (17) Pyruvate kinase (18) Pyruvate biphosphate kinase, (19) Pyruvate dehydrogenase, (20) Alcohol dehydrogenase, and (21) Aldehyde dehydrogenase. (Reprinted from Jacek et al. (2019). http://creativecommons.org/licenses/by/4.0/)

the concentration of ethylene, a gas phytohormone was directly related to the regulation of the expression of *bcsA* and *bcsB*.

Several studies have applied genetic engineering techniques in bacteria to optimize the properties and profitability of BC production. The transfer of the *cmc, ccp, cesAB, cesC, cesD,* and *bgl* genes from *K. xylinus* to *Synechococcus* sp. was performed, with the aim of increasing the production of BC under low salinity conditions (Zhao et al., 2015). In another study, the *bcsABCD* operon and upstream *cmcax* and *ccpAx* regulatory genes from *K. xylinus* were cloned into *E. coli,* and the crystalline property of BC was improved without altering performance (Buldum et al., 2018). In terms of improving performance, the heterologous expression of the *vgb* gene, which encodes *Vitreoscilla* hemoglobin, was obtained in *K. xylinus*, improving the yield of BC production (Liu et al., 2018). In this sense, the mutant strains of *K. hansenii* obtained through a physicochemical strategy of mutagenesis showed a low accumulation of organic acids, which are directly related to a higher BC production, since the accumulation of organic acids, by-products of fermentation, compete by carbon sources with the production of BC reducing its synthesis (Li et al., 2016).

It is predicted that emerging applications of BC-based technology and its potential application in different industrial fields could generate growth in the BC market in the order of 15% (Digital et al., 2018). BC applications can impact the development of biological, commercial products, 3D printing materials, biomedicine, and textiles. In the chemical area, it can be used to produce resins, plastics, synthetic thickeners, and reinforcements (Portela et al., 2019).

13.1.5 AMYLOID COMPOUNDS

Amyloids are small peptides that stain blue when contacted with Iodine, which is why they received this name since they were reported to starch-like substances, particularly amylose. The first report, made by Jeremiah Wainewright in 1722, described a deposition of a clay-colored, pituitous substance in the liver and spleen with a waxy and/ or fluffy appearance (Doyle, 1988). This substance was subsequently associated with Lardaceous disease (Dickinson, 1867). Posteriorly in 1854, the German physicist Rudolph Virchow reports a foreign residue on the surface of brain tissue: he observed that the residue acquired the characteristic blue tone of starch after rinsing with Lugol's iodine and a violet tone by subsequently adding sulfuric acid. Because of this, Virchow concludes that the residue is a similar substance to cellulose and reports it only as amyloid (Virchow, 1867).

Currently, amyloids have been extensively studied as they can naturally accumulate in tissues and organs, which interferes with normal interstitial functions and have been associated with chronic inflammatory and/or neurodegenerative diseases such as Parkinson's, Huntington's, Alzheimer's, and even diabetes (Bharadwaj et al., 2017; Gustot et al., 2015; Joachim et al., 1989; Lêdo et al., 2016; Roher et al., 1993). In some cases, these peptides are not only difficult to process but also have the ability to change the usual configuration of adjacent peptides, in some kind of self-replication process, acquiring cytotoxic properties, and becoming a "proteinaceous infectious particle" (Prion), generating degenerative damage to tissue as in spongiform encephalopathy (Eraña et al., 2017). All prions are amyloid compounds, but not all amyloids are prions (Jarosz-Griffiths et al., 2016).

Amyloid compounds are aggregated short-chain peptides that, under normal physiological conditions, are widely distributed recognition proteins with several functions like anchoring, cell recognition, and even immune response (Gras et al., 2008; Müller et al., 2017; Nhan et al., 2015). In the process of amyloids formation, it is possible to find intermediaries that are cytotoxic, but mature amyloids are not

(Westwell-Roper and Verchere, 2019; Zhao and Lukiw, 2015), which suggests that amyloids intermediaries could be prion-like peptides and force a configurational change of proteins with biological importance.

At the structural level, amyloids have a very interesting peculiarity: they have a negative beta-sheet configuration (left turn), which is a mirror image of the normal protein configuration that is the alfa sheet (right turn) (Knowles and Mezzenga, 2016; Tycko, 2015). An amyloid is a fiber formed of small protein particles (subamyloids) that have 36 to 43 amino acids residues and are added by affinity but do not form covalent bonds between the peptides (Figure 13.3).

Subamyloid makes hydrogen bonds, allowing protein fibers to slide between them, something like muscle tissue elongation, giving it the property to be elastic. The pulling force between subamyloid particles is determined by the electron affinity of the residues of each amino acid in the chain, although it has not been completely elucidated. These particles have an electrostatic affinity, and like any hybridization, the affinity force can be manipulated by astringency (Shammas et al., 2011). In order to alter the electrochemical attraction that exists between the subamyloids, the conditions of the chemical environment can be manipulated by salts, pH,

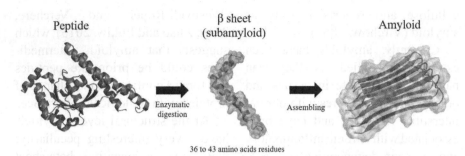

FIGURE 13.3   Amyloid formation and hierarchical length scales. The peptides are digested in 36–43 residues b-sheets with an unusual secondary configuration. These b-sheets or subamyloids assemble by affinity to form fibrils. The kind of amyloids and the affinity of the amino acids set up the physical properties of the fiber. This file is licensed under the Creative Commons Attribution-Share Alike 4.0 International.

electric current, among other factors, impacting the elasticity and strength (Hughson et al., 2016; Yoo et al., 2018). This tension by affinity is a property shared with nucleic acids (Nakano et al., 2014) and polysaccharides (Mendes et al., 2017).

This amyloid structure is very robust, can tolerate high temperatures, has no catalytic availability, so proteases neither degrade them nor are they sensible to detergents or denaturing agents, and even hold up strong mechanical forces without breaking (Marshall et al., 2011).

As the research on amyloid complex advances, it becomes more evident that they are more common than we suspected, and this configuration could naturally set if the needed denaturing conditions are presented (Hammarström et al., 2002). With the improvement of the methods of detection and elucidation of amyloid bodies, nonpathogenic natural uses have been observed in organisms (Maury, 2009), such as the silkworm that uses amyloid fibrous structures on the surface of the cocoon as a natural barrier against bacteria and spores, and favor moth survival (Iconomidou et al., 2000; Otzen, 2010).

13.1.5.1   PHYSICAL PROPERTIES OF AMYLOIDS

Amyloid fibers are one of the stiffest known biomaterials, with an elastic resistance of 3–20 GPa (1 GPa = 10,200 kg/cm^2). On the other hand, they also have a considerable fracture resistance of 0.6 GPa, similar to silk resistance (Schleeger et al., 2013). This is interesting since amyloid complexes resistance is related to their conformation, with a helical structure that favors the formation of noncovalent interactions toward the fiber core. These noncovalent interactions between the residues vary in strength according to the amino acids sequence, favoring the formation of hydrogen bonds and thus increasing the stability and strength of the fiber (Greenwald and Riek, 2010). A great advantage of amyloid is that they are present in all the life forms, so there are several alternatives for their industrial production (Hammarström et al., 2002).

These characteristics of being inert, insoluble, and resistant to physical and chemical stress, make it a highly desirable biomaterial. In the science of biomaterials, the self-assembling molecules are very valuable (Gazit, 2007; Lu et al., 2003), and if we consider that native proteins (such as lactoglobulin), highly known proteins as insulin, and de novo proteins or proteins overdesign can be used for the formation of this complexes (Gallardo et al., 2016), the amyloids could be a promising material for the development of new and innovated products.

13.1.5.2 BACTERIAL INCLUSION BODIES

The inclusion bodies (IBs) are amyloid complexes of beta-sheet protein–protein interactions (Speed et al., 1996; Gonzalez-Montalban et al., 2006), and the result of this aggregation is a high degree purity protein granulates (Morell et al., 2008). The IBs are common when transgenic bacteria produce recombinant proteins and store them in the cytoplasm. The IBs could be formed by almost any protein of interest if we stimulate the amyloid complex, and they could easily cross the membrane if the amyloid is inert (not pathogenic). This makes IBs a promising approach for amyloid application since they could be made by proteins with therapeutic interest. Initial experiments indicate than IBs could be used as vehicles for cell therapy (Liovic et al., 2012; Vazquez et al., 2012), something similar to cellular nanopills or bioscaffolds, since this complex trick the systems and slowly release the subamyloids for interaction with other proteins within the target cell (Table 13.1).

13.1.5.3 NANOWIRES

On the other hand, a metal can be attached to the peptide sequence and used it like conductive nanopolymers (Hauser et al., 2014). The first nanotechnology applications investigated were the design of nanowires and nanoelectronics materials and has been very successful. The electron transfers can be mediated by a metal, an enzyme, or even graphene (Hauser et al., 2014). Using peptides

TABLE 13.1   Therapeutic Proteins Delivered as Bacterial IBs (modified from Ferrer-Miralles et al., 2015)

Therapeutic Protein	Biological Effect	Delivery Route	Reference
Human dihydrofolate reductase	Recovery of cell viability in DHFR-deficient cells	Nanopills	Vazquez et al. (2012)
Human catalase	Recovery of cell viability during an oxidative stress	Nanopills	Vazquez et al. (2012)
Chaperone Hsp70	Inhibition of cell apoptosis	Nanopills and bioscaffolds	Seras-Franzoso et al. (2013) and Vazquez et al. (2012)
Leukemia inhibitory factor	Recovery of cell viability upon growth factor removal from the medium	Nanopills	Vazquez et al. (2012)
Keratin-14	Construction of cell filaments	Nanopills	Liovic et al. (2012)
Fibroblast growth factor	Cell proliferation in the absence of soluble growth factors	Nanopills and bioscaffolds	Seras-Franzoso et al. (2013)

attached to metals, or enzymes, we can build nanowire protein structures, and using the vapor deposition technique the peptides can be tuned by simply adjusting the amount of peptide building blocks in the gas phase. For example, this amyloid complex can be employed for the construction of nanocapacitors to store energy (Adler-Abramovich et al., 2009).

In general, amyloids are a potential new biomaterial with many uses in industry and medicine. The amyloid biomaterial science is promising since the application of the self-assembly capacity leads to a robust, reproducible, practical, and affordable way to produce nanometer-scale building blocks at an important scale.

13.1.6   1,3-PROPANEDIOL

For the production of polymeric materials such as polyurethanes, polyesters, and polyethers, 1,3-propanediol (1,3-PD) is an important precursor for such compounds (Figure 13.4), which has been recently transcended to cosmetic, drugs, and lubricants production (Vieira et al., 2015; Ruan et al., 2019).

FIGURE 13.4   Structure of a 1,3-propanediol molecule.

One of the most relevant uses of 1,3-PD is its role as a precursor of polypropylene or polymethylene terephthalate (Figure 13.5) (Lindl-bauer et al., 2017). As this molecule is not available or obtainable from petroleum oil, biological genera-tion is a viable option. By chemical synthesis, scientists have devel-oped conversions of fatty acids into polyols (Ruan et al., 2019). These can be used as building blocks for several materials such as polyure-thanes with many industrial appli-cations (Figure 13.6) (Kaur et al., 2012). Also, the products derived from 1,3-PD have better proper-ties than others produced from butanediols, 1,2-propanediols, and ethylene glycols. Thus, 1,3-PD has an increasing demand (Wischral et al., 2018). Several microorganisms can produce 1,3-PD, such as bacteria strains of the genres *Citrobacter*, *Lactobacillus, Klebsiella,* and *Clostridium* (Celińska, 2010). The advan-tage of producing this molecule by biotechnological means is that it's more environmental-friendlier and cost-effective compared to chemical synthesis.

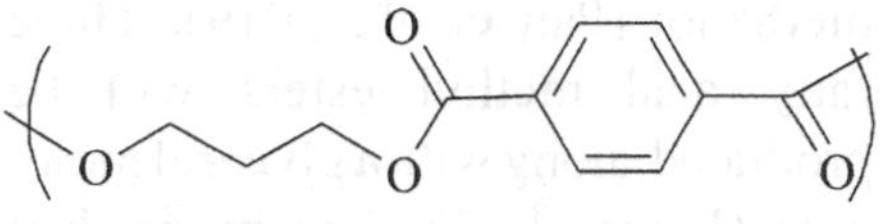

FIGURE 13.5 Structure of a polymethylene terephthalate, a 1,3-PD derivate.

13.1.6.1 PRODUCTION BY FERMENTATION

For 1,3-PD, the most common way of production involves the use of glycerol as it is the only known substrate for microbes. By oxidative and reduction reactions, a dismuta-tion occurs yielding 1,3-PD (Yun et al., 2018). The theoretical yield of 1,3-PD is 072 g/g glycerol, which is considered high and represents an attractive process for its production (Vivek et al., 2018). An advanta-geous aspect is that microbes 1,3-PD producers also produce other added value metabolites such as lactate, acetate, succinate, H_2, butanol, 2,3-butanediol. The most acces-sible strategy for 1,3-PD production is the use of glycerol as a substrate

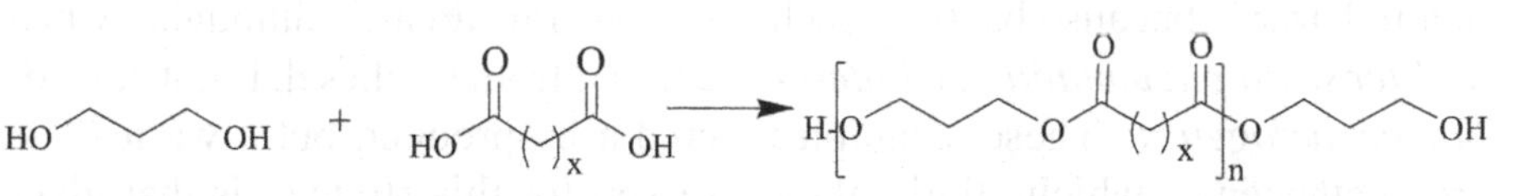

FIGURE 13.6 Synthetic route for the production of a polydiol from 1,3-PD.

(Xafenias et al., 2015). There are studies where glycerol coming from biodiesel production plants is used as a substrate (Lee et al., 2015).

Glycerol is a by-product of biodiesel production. For the biofuel to be generated, a triglyceride must react with alcohol, most commonly methanol (Pan et al., 2019). Three fatty acid methyl esters will be produced, along with a glycerol molecule (Figure 13.7). The by-product can be separated from the biodiesel

and it can serve as the substrate for 1,3-PD. In metabolism, glycerol is dehydrated for 3-hydroxypropionaldehyde to be formed and by a 1,3-propanediol oxidoreductase and

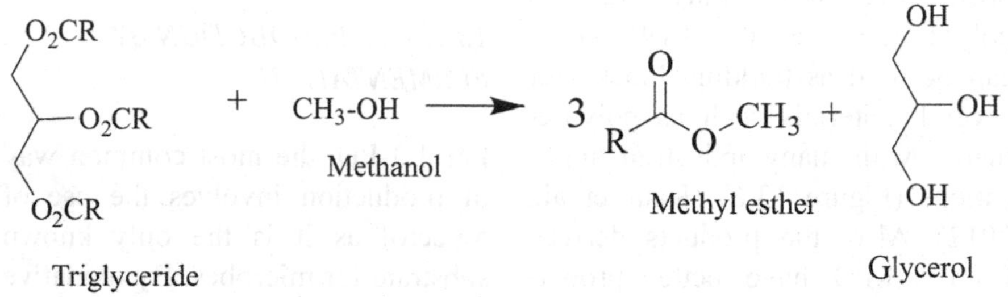

FIGURE 13.7   Biodiesel production yielding glycerol as a substrate for 1,3-propanediol biotechnological production.

$NADH_2$ to NAD conversion, 1,3-PD is formed (Figure 13.8). Along with the metabolite of interest production, mandatory biomass is formed, and by further metabolic processing from pyruvate, ethanol, acetic acid, citric acid, and butanol may be produced (Lee et al., 2015).

As previously mentioned, several microorganisms can produce 1,3-PD but some of them might represent health hazards because bacteria such as *Klebsiella pneumoniae* or *Enterobacter aerogenes*. These examples are pathogens, which limit their industrial potential. Other cases, such as *Citrobacter freundii, Lactobacillus brevis,* and *Clostridium pasteurianum* produce lower titers of 1,3-PD but are not pathogens (Metsoviti et al., 2013; Johnson and

Rehmann, 2016; Vivek et al., 2016). The production of 1,3-propanediol by biotechnological process requires glycerol, yielding the metabolite by using the more suitable microbe. An approach that scientists are taking is the production of 1,3-propanediol from glycerol derived from biodiesel (Figure 13.9). For a more specific explanation about the substrate, glycerol has great commercial value in several areas, although, when coming from biodiesel, is not considered a by-product, but a waste. The reason for this strategy is that glycerol has many pollutants such as methanol, soaps, and salts. If used in metabolite production, glycerol from biodiesel becomes feasible (Vivek et al., 2018; Dams et al., 2018).

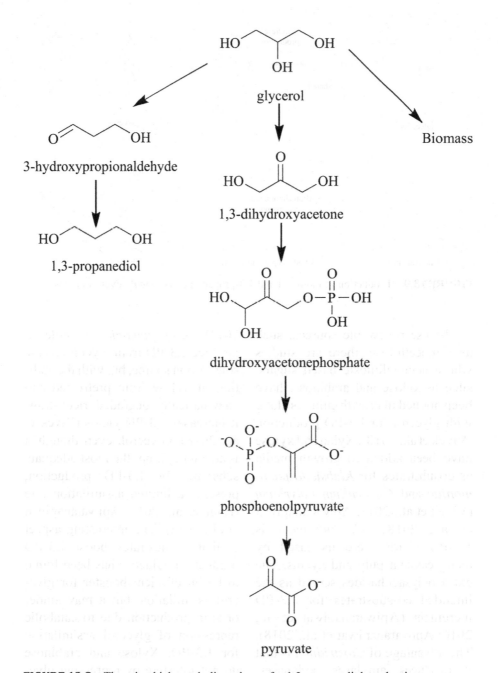

FIGURE 13.8 The microbial metabolic pathway for 1,3-propanediol production.

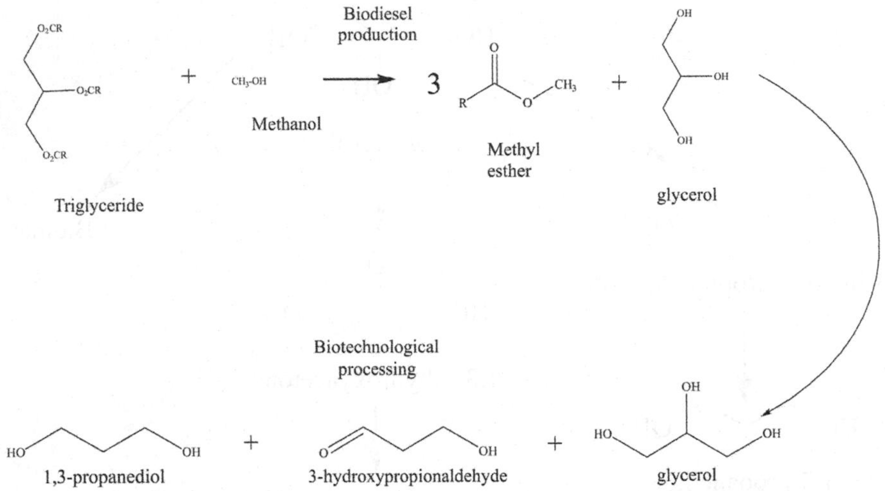

FIGURE 13.9   Production of biodiesel and 1,3-propanediol as a single overall process.

To use renewable sources, such as lignocellulose, there are studies where hemicellulosic based sugars, such as xylose and arabinose, have been applied in growth cultures along with glycerol for 1,3-PD production (Vivek et al., 2018). Xylan and xylose have been added to growth media as cosubstrates for *Klebsiella pneumoniae* and *Clostridium butyricum* (Vivek et al., 2018; Apiwatanapiwat et al., 2018). *C. butyricum* is 1,3-PD yields were increased by using cassava pulp and xylans. The extra polysaccharides served as the intended co-substrates for 1,3-PD increment (Apiwatanapiwat et al., 2016; Apiwatanapiwat et al., 2018). The advantage of *Clostridium* is that it produces amylases, xylanases, and cellulases, and certain strains are nonpathogenic (Thomas et al., 2014). *K. pneumoniae* is able to produce 1,3-PD from glycerol as the sole carbon source, but with the addition of xylose from pretreated and enzymatically degraded rice straw, it increases 1,3-PD yields (Vivek et al., 2018). Glycerol, even though it is adequate, if not the most adequate substrate for 1,3-PD production, presents a limited assimilation rate (Kaur et al., 2012; Apiwatanapiwat et al., 2018). The interesting aspect is that cosubstrates boost 1,3-PD production. Glucose has been found to be an efficient booster for glycerol assimilation, but it may hinder or stop production due to catabolic repression of glycerol assimilation for 1,3-PD. Xylose and arabinose do not interfere by repressing glycerol catabolism, increases biomass production, and 1,3-PD as well.

There is a different case with lactobacilli, such as *Lactobacillus reuteri* is unable to use glycerol as sole carbon source, thus, a co-substrate fermentation is needed to achieve 1,3-PD (Vieira et al., 2015). *Lactobacillus diolivorans* has also been efficient in 1,3-PD (Lindlbauer et al., 2017). The main advantage of using strains such as *Lactobacillus* or *Clostridium* is that the selected strains of *Clostridium* are not pathogenic, and *Lactobacillus* strains are considered safe (Vieira et al., 2015). This is very important to remark due to the comparison of *K. pneumoniae* with other strains. *K. pneumoniae* has produced more than 55 g/L of 1,3-PD, but its pathogenicity arises concerns for its use at an industrial level. It is because of this that strains as the already mentioned, along with *Clostridium pasteurianum, Clostridium acetobutylicum, Lactobacillus brevis*, or *Citrobacter freundii* are much safer to work with. *C. butyricum* has produced 94 g/L and *Clostridium acetobutylicum* 84 g/L (Pflügl et al., 2012; Vivek et al., 2017). There are also research works that include genetically modified strains of *E. coli* for 1,3-PD production. The reason for modifying *E. coli* metabolism is that, as previously mentioned, many of the 1,3-PD producers are pathogenic, require anaerobiosis or also produces 2,3-butanediol which makes its separation from 1,3-PD more difficult. This problem can be

addressed by using *E. coli* because of its relative safety, and it is well-studied for this type of applications (Yang et al., 2018).

13.1.7 BIOCONCRETES

Biomineralization is a process by which living beings can produce minerals. Organisms that can precipitate minerals are present in all major taxonomic groups, from bacteria to chordates. Bacteria are the second important group in this aspect, only after several genera of fungus (Jiménez-López et al., 2007).

Bacteria and other prokaryotes usually precipitate minerals by the production of nucleation sites generated by their metabolism, either by changing the physicochemical conditions of the surrounding extracellular environment or by the production of EPSs that, as a whole, they create the saturation conditions necessary for mineral production (Konhauser and Riding, 2012). Since these mechanisms are not regulated by genetic instructions but only by the presence of microorganisms, this type of biomineralization is known as biologically induced mineralization or as biologically influenced mineralization if it is caused by the presence of cellular products without the need for living cells. In both cases, the minerals produced by these mechanisms are known as organominerals (Dupraz et al., 2009).

One of the applications of these bacterial biomineralization processes is the production of bioconcrete: a self-healing construction material. The bacteria present in the concrete are reactivated by forming a crack and entering water or moisture through it. As a product of their metabolism, they produce calcium carbonate crystals that resurface surfaces, extending their useful life, and reducing maintenance costs (Jonkers et al., 2007; Seifan et al., 2016).

The treatments of fissures and pores in concrete are generally divided into passive and active. Passive treatments only repair exterior cracks while active can fix both exterior and interior cracks (Pacheco-Torgal and Labrincha, 2013). Active treatment techniques are known as self-hearing techniques and can operate independently of the conditions and position of the fissure. These techniques have the ability to activate once the fissure occurs, repairing it.

A concrete self-hearing mechanism can be established in three ways: (1) autogenous healing, (2) encapsulation of polymeric materials, and (3) microbial production of calcium carbonate (Wu et al., 2012). The ideal treatment should have quality, long lifespan, and the ability to repair cracks repeatedly on an unlimited number of occasions (Li and Herbert, 2012). Autogenous healing is the natural process of concrete repair that occurs in the presence of the mixture on aqueous solution. It repairs fissures through the hydration or dehydration of cement particles or carbonation of dissolved calcium hydroxide (Edvardsen, 1999). In the method of encapsulation of polymeric material, the repair of the fissures occurs by conversion of the healing agent to foam in the presence of mixture. Although the generated chemical fibers can heal cracks (Dry, 1994), these materials do not behave similarly to concrete and, in some cases, can even cause existing cracks to extend (Seifan et al., 2016).

Recently, biotechnological alternatives have become common. According to Seifan et al. (2016), biological healing processes are based on the production of calcium carbonate through biomineralization. The successful implementation of these innovative treatment methods will result in a longer lifespan of concrete structures, as well as a significant reduction in cement production and structural replacement.

As already mentioned, the biomineralization process involves the formation of calcium carbonate due to the metabolic activity of microorganisms where the enzyme urease plays an important role. (Kim et al., 2012; Achal et al., 2015). Urease catalyzes the hydrolysis of urea to CO_2 and ammonium, resulting in an increase in pH and carbonate concentration in the bacterial environment.

The application of ureolytic bacteria as healing agents in the concrete mix is an environmentally friendly technology and minimizes costs (De Muynck, 2013). It has also been reported that it improves compression strength, tension, elasticity, and other structural characteristics of concrete (Othman et al., 2016).

According to Castro Alonso et al. (2017), the characteristics of durability, strength, and low cost of concrete compared to other materials, make concrete a material widely used in the construction industry. However, concrete is susceptible to deterioration caused by various physicochemical and biological factors (Al-Salloum et al., 2016). Cracks in the internal structure of concrete derived from these factors cause irreversible damage which, according to Silva et al. (2015), entails a cost of repair and maintenance of around 230 euros per cubic meter of concrete.

In recent years, different bacteria have been used in research work related to the development of bioconcretes. According to Seifan et al. (2016), among the most used are: *Bacillus pseudofirmus*, *Bacillus cohnii*, *Bacillus alkalinitrilicus*, *Bacillus sphaericus*, *Sporosarcina pasteurii*, *Pseudomonas aeruginosa*, *Bacillus sphaericus*, *Bacillus cereus*, *Bacillus amyloliquefaciens*, *Sporosarcina soli*, *Bacillus massiliensis*, *Arthrobacter crystallopoietes*, *Lysinibacillus fusiformis*, and *Diaphorobacter nitroreducens*.

Microorganisms such as *Bacillus sphaericus* and *Bacillus pasteurii* can produce biominerals through metabolic reactions in the presence of a calcium source (Wang et al., 2012). These urease positive microorganisms are involved in the nitrogen cycle and can produce calcium carbonate through the hydrolysis of urea (Wiktor and Jonkers, 2011). This metabolic conversion promotes the precipitation of calcium carbonate, which acts as a barrier and blocks the entry of corrosive chemicals into cracks (Dick et al., 2006).

Dissimilatory nitrate reduction is another known pathway to produce minerals. Denitrification is defined as the respiratory process that results in the reduction of nitrate to nitrite, nitric oxide, nitrous oxide, and nitrogen gas. Minerals are precipitated through the oxidation of organic compounds by the reduction of nitrate via denitrifying bacteria. The most significant attribute of this route is its application in anaerobic zones. The microorganisms that are involved in denitrification processes are facultative anaerobes, mainly species of *Denitrobacillus*, *Thiobacillus*, *Alcaligenes*, *Pseudomonas*, *Spirillum*, *Achromobacter*, and *Micrococcus* (Seifan et al., 2016).

Usually, bacteria and nutrients are added directly into the concrete mix during preparation and emptying. In this process, the healing agents are dissolved in water and then the mixture is added to the cement and sand. Alkalophilic bacteria such as *Bacillus* species can tolerate the extreme environment of concrete and are therefore the most attractive species for self-hearing concretes. Several studies show that these spore-forming bacteria can survive without nutrients for several hundred years (Schlegel, 1993). Moreover, endospores can resist environmental or chemical changes, as well as ultraviolet radiation and mechanical stress (Van Tittelboom and De Belie, 2013).

To protect the bacteria from the harsh concrete preparation conditions, their encapsulation is preferred before being added to the mixture. A suitable carrier should act as a shell species that have no effect on carbonate precipitation and have no limiting effect on the concrete formation (Wang et al., 2015). Wang et al. (2014) developed self-healing systems based on bacteria using glass capillaries, porous powders, and bacteria encapsulated in hydrogels. They determined that hydrogels are the most suitable means to perform this type of process and that water is essential to ensure bacterial activity.

In this way, research has been carried out with promising results regarding the development of bioconcretes. Some examples illustrate this trend. Wang et al. (2015) explored the application of a *Bacillus sphaericus* strain, a calcium carbonate precipitator, to generate a self-healing concrete. They demonstrated in situ activity of the bacterial spores encapsulated in alginate hydrogels, concluding the potential that this type of system has for the generation of bioconcrete. Anneza et al. (2016) identified and studied *Enterococcus faecalis* (ureolytic bacteria) and *Bacillus* sp. (sulfate reduction bacteria) in the manufacture of concrete to determine their effect on the compression strength. Their results showed that the bacteria play a crucial role in the improvement of this strength. Irwan et al. (2016) studied the effect of the addition of calcium lactate on the compression strength and the penetration of water into a bioconcrete, using strains of *Enterococcus faecalis* and *Bacillus* sp. They determined that the addition of calcium lactate in the bioconcrete is promising in improving the properties and durability of concrete. Alonso et al. (2018) tested different bacterial strains isolated from soils of the Laguna Region in Mexico; their results proved the role of microbial induced calcium carbonate precipitation in the improvement of the physical–mechanical properties of the bioconcrete.

In summary, the enormous perspective that the development and use of bioconcrete can have in the short-term future, which would lead to true innovation in the construction industry, is unobjectionable. However, according to Seifan et al. (2016), for an industrial application in the near future, several critical challenges must be faced.

Despite recent advances in protocols to produce self-healing concrete, existing studies still suffer from the lack of numerical simulation to reduce experimental costs and time (Vaghari et al., 2015). In addition, the feasibility of using repairing agents during the mixing of the material and extending the activity of bacteria under difficult conditions within the concrete for long periods require further investigation.

13.1.8   FINAL REMARKS

Biotechnology, bioengineering, and materials science have advanced to the creation of new biomaterials that possess enhanced properties (Scott et al., 2018). Biological-made materials have a wide range of applications, from medical to agricultural, industrial, and even construction fields. Some of the metabolic pathways that naturally occur in living organisms to produce them are hard to replicate since some of them have not been elucidated yet.

Synthetic biology provides a new approach to reprogram natural biosynthesis, and it has enabled the design of novel biomaterials by combining desired domains in vivo (Keating and Young, 2019). This technology addresses toward the production of biomaterials as a sustainable alternative, with improved production rates and reduced costs. Also, it will make available a more extensive use of these materials, allowing feasible and effective means to new or improved materials design and synthesis for wider applications.

13.2   VIRAL CAPSIDS AS BIOMATERIALS

13.2.1   INTRODUCTION

The use of the viral capsid for the development of biomaterials has seen a significant increase in several fields of science and a diversity of applications. Capsids are highly ordered protein cage assemblies that have been used for the development of vaccines since the last century. However, capsids have recently caught the attention as useful biotemplates for nanomaterials with high translational potential as nanodevices for biological and medical applications such as molecular imaging biosensors, conducting wires for microelectronics, fuel cells, controlled drug delivery systems in anticancer therapy, biological vehicles in gene therapy, and tissue regeneration, due to their unique structure-function properties of self-assembly under physiological conditions, monodispersed nanoparticle size, biodegradability, low-to-none toxicity, and easy modulation through genetic and chemical engineering. In this review, we describe prominent advances made in the last decade in these existing multidisciplinary areas of research through a small selection of representative works.

To this date, there are a plethora of virus alternatives that had been used for biomaterial purposes. One accepted taxonomy classifies viruses according to their natural host. For example, bacteriophages (bacteria "eaters") are a type of virus that infects and replicates within bacteria. Regardless of their life cycle, lytic (virulent) or lysogenic (temperate), phages have caught the attention of researchers and biotechnologists (Salmond and Fineran, 2015; Harada et al., 2018). Phages are particularly attractive carriers for drug delivery and as a platform for vaccine materials. Among these, some common types that have been studied are T4 phage (lytic phage), T7 phage (lytic phage), P22 phage (lytic phage), l phage (temperate phage), M13 phage (nonlytic phage), fd phage (nonlytic phage), MS2 phage (lytic phage), and f29 phage (lytic phage).

Plant viruses are another major group that has been utilized for technology development (Scholthof et al., 2002; Sainsbury et al., 2010). They have been genetically engineered to display immunogenic epitopes on their surfaces to provide effective vaccination against several disease pathogens and for immunotherapy. In this case, we can mention viruses, such as bamboo mosaic virus, cardamom mosaic virus, johnsongrass mosaic virus, papaya

mosaic virus (PapMV), papaya ringspot virus, plum pox potyvirus, potato virus X (PVX), potato virus Y, tobacco etch virus, tobacco mosaic virus (TMV), zucchini yellow mosaic virus, tobacco mild green mosaic tobamovirus (TMGMV), artichoke mottled crinkle virus, cucumber mosaic virus (CMV), cowpea mosaic virus (CPMV), brome mosaic virus, carnation mottle virus, cowpea chlorotic mottle virus (CCMV), turnip yellow mosaic virus, and maize rayado fino virus.

Finally, human and animal viruses, like *Lentivirus* and *Adenovirus*, have also been the focus of attention in recent years for the development of novel biotechnology applications (Wen et al., 2016; Fuenmayor et al., 2017; Jeevanandam et al., 2019). In this category, we can mention viruses such as the human immunodeficiency virus (HIV), human enterovirus, hepatitis B virus (HBV), Rous sarcoma virus, rotavirus, simian virus 40 (SV40), and chicken anemia virus.

Each one of these viruses might have advantages and disadvantages over the rest at different phases of development. On the one hand, some phages have been used in their *virion* form, that is, including the original genetic material, maintaining infectivity. On the other hand, some viruses have been "emptied" and only the outer shell, known as the *capsid*, has been used, decreasing the particle stability in some cases. Either way, one thing in common among all viruses with novel applications in biotechnology is the *coat protein* (CP).

The CP is the fundamental unit that builds up the capsid. Hence, alterations to the CP renders capsids with new properties. Such alterations can be genetic or chemical modifications. In order to rationally plan for the right modification, molecular knowledge of the CP structure and how they interact with each other to spontaneously self-assemble to form a capsid has proved to be critical for further technological advancements.

13.2.2 CAPSID STRUCTURE

Viral capsids show two basic kinds of morphological geometries (symmetric quaternary structures); icosahedral (spherical, isotropic); or

helical (elongated rod-shaped or filamentous, anisotropic) (Figure 13.10). A combination of these two (complex kind) has also been seen, for example, phages, however, technological applications have not taken full advantage of it yet. Icosahedral capsids are adopted by more than half of the virus families that we currently know. In contrast, helical capsids are found in about 10% of virus families (Castón and Carrascosa, 2013).

A hexagonal lattice is the starting point to build and describe both helical and icosahedral capsids. We can imagine a flat hexagonal network built from many copies of one CP, where all proteins interact with their neighbors in identical environments. To produce a closed structure, the edges of a 2D lattice have to bind through bending. This process produces a tubular geometry, with

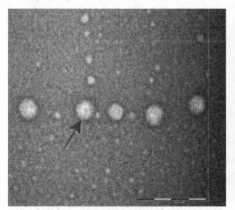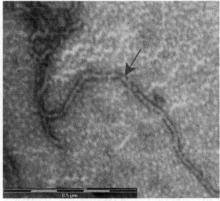

FIGURE 13.10 Morphological geometries of viral capsids. (Left) Icosahedral virus-like particles of CCMV. (Right) Helical filamentous virus-like particles of PapMV. Both types of particles (red arrows) were self-assembled in vitro using recombinant CPs. Images obtained by negative-stained transmission electron microscopy. Modified from Chávez-Calvillo (2011).

the CPs forming a helix or a stack of discs depending on the direction of the bending axis. The number of CPs in each turn determines the curvature and diameter of the tube. In helical capsids, their length is defined by the length of the nucleic acid it encloses. In this case, the total number of CPs per capsid can reach a few thousand. Filamentous phages are semiflexible bionanofibers with approximately 900 nm in length and 7 nm in width (Cao et al., 2018).

Accordingly, an icosahedral structure can be generated from a hexagonal 2D lattice by transforming some hexagonal nodes into pentagonal nodes by removing one CP. Pentamers introduce a convex curvature in the lattice. Enough number of pentamers render a closed structure. The simplest icosahedral capsid is formed by 12 pentagonal vertices comprising a total of 60 CPs. The introduction of hexagonal nodes in-between the 12 pentamers, increasing the number of CPs, produces icosahedral particles with larger diameters. An investigation of the type of viruses that exist in the world's oceans showed that the nontailed icosahedral morphotype dominates the globe for as much as 90% of all the viral particles observed (Brum et al., 2013). On average, icosahedral capsids are 50 nm in diameter, regardless of depth or oceanic region. One known extreme case in this category,

although still not reported as a platform for biotechnological applications, is amoeba-infecting viruses in the *Mimiviridae* family. A mimivirus has an icosahedral capsid 400 nm in diameter, comprising a total of 72,000 CPs.

More than 50 years ago, two seminal works laid out the foundations of modern structural virology. First, Watson and Crick proposed the basic principles for the construction of icosahedral viruses (Crick and Watson, 1956). A few years later, Caspar and Klug developed the quasi-equivalence theory (Caspar and Klug, 1962). Since then, significant advances in the understanding of the molecular aspects of capsid quaternary structure have been made. In most part, this was due to a large amount of structural data now readily available. For instance, density maps of virus particles have been determined by electron microscopy, available at the Electron Microscopy Data Bank (Patwardhan, 2017). Atomic coordinates and high-resolution structural models of virus particles determined by X-ray crystallography and NMR spectroscopy are available at the Protein Data Bank (Berman et al., 2003). Furthermore, the specialized VIPERdb Science Gateway (Carrillo-Tripp et al., 2009; Ho et al., 2018) has leveraged the abundant and diverse 3D structural information on viruses determined at near-atomic resolution with bioinformatic analysis (Carrillo-Tripp et al., 2008) and visualization (Carrillo-Tripp et al., 2015) tools through a friendly web interface for the study of icosahedral capsids. At the time of this writing, VIPERdb provides the curated and standardized X-ray and Cryo-EM atomic-level models of close to 600 unique capsid structures comprising more than 50 families and more than 100 genera.

Data-mining of this ever-increasing amount of virus structural information has provided a better understanding of the macromolecular structure–function relationship, always in the context of evolution. It is now known that, at least in the case of capsids, natural selection maintains primarily the structural fold of the CP rather than the sequence. Even for viruses that infect cells from different kingdoms, there seems to be a structural convergence when it comes to their capsid proteins. Not surprisingly, the number of structural folds currently observed in viral CPs is small (Abrescia et al., 2012). Nonetheless, it has been shown that, in general, the buried core residues in the CP are the most conserved, followed by the residues at the protein–protein interfaces. The solvent-exposed surface of the capsid shows more naturally occurring sequence variations (Montiel-García et al., 2016). Hence, the capsids' surface and the protein's

N and C termini are the regions that have been generally exploited for the development of novel bionanotechnology platforms.

13.2.3 BIOMATERIAL INNOVATIONS

One of the main advantages to work with capsid proteins is that in many cases, even in the lack of nucleic acid, they spontaneously self-assemble into empty shells, known as *virus-like particles* (VLPs), preserving the size and shape of the original virion's capsid. In the laboratory, the CP can be obtained in different ways. The virion can be purified from an infected organism, usually a plant, and then disassembled to filter out unwanted genetic material. Another option is to express the CP in a recombinant system, usually bacteria or yeast. The latter strategy allows for the genetic modification of the CP gene, which is cloned into a plasmid to transform the expression system (Díaz-Valle et al., 2015).

13.2.3.1 GENETIC CP MODIFICATIONS TO HARVEST LIGHT

One remarkable example of a rational genetic CP modification to create novel functions was done to the CP of the M13 bacteriophage

(Chen et al., 2013). The strategy allowed to produce particles with distinct substrate-specific peptides expressed on the capsid's surface (phage-display). This mutant was a generic template later used to build gold-binding capsids, assembling Au nanocrystals with 2D or 3D morphologies with interesting electronic and optical properties for different device applications, such as fuel cells, batteries, and photocatalysts. In particular, the photoanode of a dye-sensitized solar cell was improved by the integration of a 3D viral network, which can template titanium dioxide nanowires for efficient electron transport. At the same time, the viral network can also bind gold nanoparticles to improve light harvesting. Such devices need a solid scaffold, which in this case, it was a cross-linked virus hydrogel.

13.2.3.2 VIRUS HYDROGEL

Networks of hydrophilic polymer chains are called *hydrogels*. When held together by cross-linking, this matrix produces a solid 3D material that does not dissolve in water. Virus hydrogels can be built in different ways. One example is the strategy that combines DNA with the CP of CMV (Tao and Xu, 2014). The capacity of this CP to assemble with nonspecific DNA was first employed to synthesize linear nanotubes. Later, engineered

branched DNA molecules (Y-shaped and cross-shaped) were prepared and mixed with the CMV's CP in a 5bp:1cp ratio. One advantage of this method over peptide-based, DNA-polyacrylamide-based, or pure-DNA-based hydrogels is that the DNA-CP coassembly allows for the modification of the protein surface without affecting the gel's overall architecture provided by the DNA scaffold. Hence, DNA-CP can be the base for other hydrogel preparations, either by the design of tailored DNA sequences, or engineering the CP through functionalization.

13.2.3.3 METAL DETECTION AND MAGNETIC BIOCOMPOSITES BY FUNCTIONALIZATION

Functionalization refers to the process of providing new functions to a given system through structure modifications. For example, a dual functionalization on a TMV variant was produced by introducing lysine and cysteine residues at solvent accessible sites on the CP. This modification allows for pairwise immobilization of two different functional groups close to each other on the particle's surface by chemical conjugation protocols (Wege and Geiger, 2018). On a different work, TMV VLPs were obtained with randomly distributed functional groups, different functional groups arranged in striped but randomized arrangements, or distinct groups clustered in adjacent domains in the particle (Eiben, 2018).

A diversity of metal ions is critical for many bioprocesses and cell functions. However, they can also be a dangerous source of contamination and even deadly toxic in high concentrations. Hence, there is a need for methods and cheap devices to detect and quantify metals with a high degree of specificity. A potential biosensor for ferric (Fe^{3+}) and ferrous (Fe^{2+}) ions has been studied by the functionalization of the bacteriophage M13's CP P8 (Guo et al., 2015). A mutant phage was engineered to introduce a tyrosine residue on the N terminus of the CP through a YG peptide. When incubated with ferric or ferrous ions, the functionalized phage aggregated into amorphous structures that blocked their amplification after infecting bacteria, that is, the phage was much less infective. The metal detection limit achieved by this method was orders of magnitude lower than the sanitary security limit restricted by the World Health Organization for Fe^{3+}. Furthermore, the phage sensor was tested against several metal ions, namely Ni^{2+}, Pb^{2+}, Zn^{2+}, Mn^{2+}, Co^{2+}, Ca^{2+}, Cu^{2+}, Cr^{3+}, Ba^{2+}, and K^+, showing high specificity for Fe^{3+} and Fe^{2+} through their interaction with the phenol group of the inserted tyrosine residue in the CP.

Crystals of iron oxide in the form of magnetite (Fe_3O_4) or iron sulfide in the form of greigite (Fe_3S_4) are found in the interior of magnetotactic bacteria inside the invaginations of their inner membrane called *magnetosomes*. It is believed that magnetosomes function as a nanocompass allowing the cell to align and move along magnetic field lines. Because of their unique characteristics, namely high crystallinity, strong magnetization, uniform size and shape, and highly regulated biosynthesis, magnetosomes are very attractive for biotechnological applications. They can be functionalized by modifications in their naturally occurring membrane proteins or the addition of foreign functional structures. A magnetic biocomposite was achieved by combining a genetically functionalized magnetosome with TMV particles (Mickoleit et al., 2018). The use of nanobodies as highly specific binding domains, expressed in the surface of a magnetosome, allowed for the coupling of TMV previously functionalized with the corresponding complementary antigens. This strategy produced mesoscopic magnetic structures, and with the right functionalization, TMV-magnetosome particles with a 1:1 stoichiometry. This multifunctional magnetic platform is a technology with great potential in novel biomaterial development and biomedicine.

13.2.3.4 *PHOTOELECTROCHEM-ISTRY BY MINERALIZATION*

Given that VLPs are oligomeric protein hollow cages, they have been used to direct the synthesis of inorganic molecules in their interior. This general strategy has led the way toward the so-called green chemistry. For instance, CCMV VLPs have been used as containers to produce monodisperse titanium dioxide (TiO_2, a photoactive semiconductor) nanoparticles, employed in a photo-electrochemical cell (Jolley et al., 2011). The protein-TiO_2 composites obtained have photocatalytic activity and are similar to nanocrystalline anatase (one of the natural forms of TiO_2). It was shown that titanium crystals nearly fill the VLPs with cores of about 22 nm, allowing them to be integrated into nanodevices as photoactive materials. TiO_2 nanoparticles have also been immobilized on the outside surface of a virus-derived protein nanoassembly. The addition of a peptide with a specific sequence to the C terminus of the TMGMV CP allowed the bioconjugation between them (Song et al., 2015). It was speculated that such composites could be applied in the future in nanomedicine, energy conversion and nanoelectronic devices, quantum dots-based biotracking and bioimaging, and even bioremediation.

Another kind of mineralization used for biotechnological purposes is the one produced by silicon oxide in the form of silica (SiO_2). This approach has been employed to mineralize the outside surface of some plant viruses, both virions and VLPs. Most widely used in this case have been the CPMV (icosahedral 28 nm diameter), TMV (rod-shaped 300 × 18 nm), and PVX (filamentous 500 × 15 nm). Once silica biomineralized, the particles show modified mechanical properties. Several features can be controlled, such as silica shell thickness, surface quality, monodispersity, or integrity of the templates. Optimized protocols for site-selective silica mineralization to build spherical, elongated curved, or tubular nanostructures are readily available (Dickmeis et al., 2018).

12.2.3.5 *MASTERING ENCAPSULATION*

It is crucial to be able to control different aspects of the physical characteristics and self-assembly process of the VLPs. As mentioned earlier, the virus CP can be obtained in different ways. One is by infected-plant purification. Another is to use bacteria or yeast to express the recombinant protein. The yield of both strategies was combined, leveraging the advantages offered by each method. The TMV CP coassembly procedure allowed the production of hybrid nanotubes of uniform length and improved stability when genomic RNA was used as a scaffold (Zhou et al., 2016).

Elastin-like polypeptides (ELPs) are stimulus-responsive reversible switches that go from soluble to insoluble in water by changes in temperature, pH, or salt concentration. The integration of an ELP block, together with a hexahistidine sequence (H_6) able to chelate divalent metal ions, at the N terminal of the CCMV CP, allowed for the Ni^{2+} controlled assembly of VLPs with two distinct diameters (van Eldijk et al., 2016). An additional property of the H_6-ELP-CP hybrid is the ability to encapsulate H_6-tagged proteins selectively.

The HIV CP can be assembled in vitro to yield hollow tubes, cone-shaped VLPs, or hexagonal lattice monolayer sheets deposited on a solid substrate that has a negatively charged surface. The latter case was characterized and proved to maintain the architecture of the wildtype HIV capsid. In terms of physical properties and kinetics, it was shown that lattice response to disruptions was self-healing, malleable by chemical additives and that the assembly goes through a nucleation and growth process when the CP concentration is kept below some threshold value (Valbuena and Mateu, 2017).

Encapsulation of quantum dots or metal nanoparticles, although toxic for the human body, has been achieved with the use of different CPs. One example is the case of the SV40, a polyomavirus, which has an icosahedral cage-like structure 45 nm in diameter formed by its capsid proteins VP1, VP2, and VP3. SV40 VP1 can self-assemble into polymorphic VLPs in vitro. Like many other CPs, the self-assembly is controlled by pH, ionic strength, and in this case, calcium ions. Depending on the solution, SV40 VP1 can form icosahedral particles with different sizes (smaller than the virion) or tube-like assemblies (Zhang et al., 2018). Also, the encapsulation of active enzymes by VLPs derived from bacteriophage P22 has been used as a strategy to protect them from inactivation or degradation. These biocomposites can act as the base of new nanomaterials that incorporate the catalytic function of the sequestered molecules (Patterson, 2018).

More recently, the so-called nanostar colloids have been reported (Wege and Eber, 2018). They have a gold core and an adjustable number of nanotube "arms," available for further functionalization. In this case, the Au core acts as the solid support, whereas the arms are made by encapsulating RNA strands, previously complexed to the core, with TMV CPs. The number of functional arms is adjusted by RNA concentration during the complexing step. This type of biomaterial can be adapted to specific applications given the multivalent nature of TMV and the flexibility to design arms of different lengths.

13.2.3.6 VIRUSES HELPING HUMAN TISSUE REGENERATION

In general, a synthetic extracellular matrix has been used as solid support for stem cells during the process of differentiation in regenerative medicine applications. Such biomaterials need to have physical and chemical characteristics similar to that of living tissue, and most importantly, be biocompatible and biodegradable. It has been shown that native viruses arranged on 2D supports accelerate mesenchymal stem cell differentiation. To increase cell attachment to the supporting material and available adhesion area, an extensively used strategy has been to functionalize a highly porous hydrogel with a genetically modified human-safe bacteriophage (e.g., M13) or plant virus (e.g., TMV) to produce 3D biomaterials. Examples of this technique have been reported for neural cell growth (Chung et al., 2010), bone differentiation (Lee et al., 2012; Luckanagul et al., 2012; Luckanagul et al., 2014), and other tissues like nerves, cartilage, skin, and heart (Cao et al., 2018).

13.2.3.7 THERANOSTIC BIOMATERIALS

By now, it should be easy to imagine how the structural properties of capsids have been employed for bioimaging applications as a new generation of contrast agents in procedures like positron emission tomography by using them as carriers of [^{18}F] fluorodeoxyglucose, or iron oxide or Gd^{3+} for magnetic resonance imaging (MRI). However, some aspects like immunogenicity or loading efficiency still need to be addressed (Shukla and Steinmetz, 2015). The capsid of HBV forms spike-like structures on its surface. In total, 240 recombinant HBV CPs self-assemble in-vitro forming highly multivalent VLPs 36 nm in diameter. It has been shown that it is possible to genetically encode foreign proteins on the tip of each spike by modifying the HBV CP with different purposes (e.g., sensitive diagnostic probes). Following this strategy, red and green fluorescent nanoparticles were built with the HBV capsid as a scaffold by inserting naturally occurring DsRed or GFP fluorescent proteins into the CP (Yoo et al., 2012). The hybrid nanoparticles not only showed a 170-fold increase in fluorescence compared to the original DsRed and GFP but also high in-vivo stability, promising great potential as active noncytotoxic biomaterials for optical imaging.

In recent years, efforts have been made to develop platforms able to combine diagnostics with therapeutics within the same nanodevice. Rotavirus VLPs coated with a magnetic compound that seems to be a potential theranostic biomaterial because they should have valuable properties for both magnetic resonance or fluorescence cellular imaging and drug delivery. The recombinant CP VP4 of rotavirus was chemically conjugated with Fe_3O_4 nanoparticles in vitro, which then were loaded with doxorubicin (DOX), a typical anticancer drug (Chen et al., 2012). Improved cellular MRI sensitivity was obtained by the internalization of the nanoparticles into the studied cells (African green monkey kidney MA104, and the human liver carcinoma HepG2). Furthermore, cytotoxicity to cancer cells was observed.

Another identified theranostic biomarker/target is collagens. They are structural proteins found in the space between cells in various connective tissues, making it the most abundant protein in mammals. Collagens are over-expressed and then degraded in some cancers. M13 phages were genetically modified to integrate two functional peptides, namely CMP and HPQ, which selectively bind to disrupted collagen and streptavidin-conjugated molecules, respectively. While the CMP site targets the cancer tissue selectively, the HPQ functionalized site has two purposes. One is for imaging when a fluorescent dye binds to it. The second purpose is to provide directed therapeutic agents when the mutated phages bind specific drugs (Jin et al., 2014).

13.2.4 FINAL REMARKS

As we have seen, viral nanoparticles are now widely developed to produce nanostructured materials for several biomedical uses and nanotechnological applications. Here, we reviewed a small set of representative works to exemplify recent efforts invested in those fields. For further learning on the subject, the reader can see the work of Douglas (2006), Soto and Ratna (2010), Yildiz et al. (2011), Farr et al. (2014), Wen and Steinmetz (2016), and Krupovic et al. (2019). Without a doubt, the near future holds even more discoveries that will improve and extend the use of virus-based biomaterials.

FUNDING

Part of the work described in this chapter was funded by the Consejo Nacional de Ciencia y Tecnología México (grant number 132376) and Fondo Sectorial de Investigación para la Educación CONACYT México (grant number A1-S-17041) to M.C.-T.

KEYWORDS

- **biomaterials**
- **bacteria**
- **viral capsid**
- **biopolymers**

REFERENCES

Abrescia, N.G.; Bamford, D.H.; Grimes, J.M.; Stuart, D.I. Structure unifies the viral universe. Annu. Rev. Biochem. 2012,81, 795–822.

Achal, V.; Mukherjee, A.; Kumari, D.; Zhang, Q. Biomineralization for sustainable construction: a review of processes and applications. Earth-Sci. Rev. 2015,148, 1–17.

Adler-Abramovich, L., Aronov, D., Beker, P., Yevnin, M., Stempler, S., Buzhansky, L., Rosenman, G., Gazit, E. Self-assembled arrays of peptide nanotubes by vapour deposition. Nat. Nanotechnol. 2009,4(12), 849–854

Ali, Z.; Sharjeel, A.; Ibrahim, M. Polyhydroxyalkanoates: characteristics, production, recent developments and applications. Int. Biodeter. Biodegr. 2018,126, 45–56.

Al-Salloum, Y.; Abbas, H.; Sheikh, Q.I.; Hadi, S.; Alsayed, S.; Almusallam, T. Effect of some biotic factors on microbially induced calcite precipitation in cement mortar. Saudi. J. Biol. Sci. 2016,24, 286–294.

Alves, R.; Oliveira, D.; Komesu, A.; Eduardo, C.; Rossell, V. Challenges and opportunities in lactic acid bioprocess design— from economic to production aspects. Biochem. Eng. J. 2018,133, 219–239. doi.org/10.1016/j.bej.2018.03.003.

Anneza, L.H.; Irwan, J.M.; Othman, N.; Faisal, A. Identification of bacteria and the effect on compressive strength of concrete. MATEC Web of Conference. 2016,47, 01008.

Antonsczak, M.; Popiel, K.; Stefaińska, J.; Wietrzyk, J.; Maj, E.; Janczak, J.; Michalska, G.; Brzezinski, B.; Hucziński, A. Synthesis, cytotoxicity and antibacterial activity of new esters of polyether antibiotic—salinomycin. Eur. Med. Chem. 2014,76, 435–444. doi 10.1016/j.ejmech.2014.02.031.

Apiwatanapiwat, W.; Vaithanomsat, P.; Tachaapaikoon, C.; Ratanakhanokchai, K.; Kosugi, A. Effect of cassava pulp supplement on 1,3-propanediol production by Clostridium butyricum. J. Biotechnol., 2016,230, 44–46.

Apiwatanapiwat, W.; Vaithanomsat, P.; Thanapase, W.; Ratanakhanokchai, K.; Kosugi, A. Xylan supplement improves 1,3-propanediol fermentation by Clostridium butyricum. J. Biosci. Bioeng. 2018,125, 662–668.

Augimeri, R.V.; Strap, J.L. The phytohormone ethylene enhances cellulose production, regulates CRP/FNRKx transcription and causes differential gene expression within the bacterial cellulose synthesis operon of Komagataeibacter (Gluconacetobacter) xylinus ATCC 53582. Front. Microbiol. 2015,6, 633–619.

Babel, W.; Steinbüchel, A. Biopolyesters (Advances in Biochemical Engineering/ Biotechnology). 1st ed. Springer-Verlag: Berlin, 2001, 342.

Bajestani, M.I.; Mousavi, S.M.; Jafari, A.; Shojausadati, S.A. Biosynthesis and physicochemical characterization of a bacterial polysaccharide/polyamide blend, applied for microfluidics study in porous media. Int. J. Biol. Macromol. 2017,96, 100–110. doi 10.1016/j.ijbiomac.2016.11.048.

Barbeau, D.; Lutier, S.; Bonneterre, V.; Persoons, R.; Marques, M.; Herve, C.; Maitre, A. Occupational exposure to polycyclic aromatic hydrocarbons: relations between atmospheric mixtures, urinary metabolites and sampling times. Int. Arch. Occup. Environ. Health 2015, 88, 19–29.

Bayer, E.A.; Lamed, R.; Himmel, M.E. The potential of cellulases and cellulosomes for cellulosic waste management. Curr. Opin. Biotechnol. 2007,18, 237–245. doi.org/10.1016/j.copbio.2007.04.004.

Berman, H.; Henrick, K.; Nakamura, H. Announcing the worldwide protein data bank. Nat. Struct. Mol. Biol. 2003,10, 980–980.

Bharadwaj, P.; Wijesekara, N.; Liyanapathirana, M.; Newsholme, P.; Ittner, L.; Fraser, P.; Verdile, G. The link between type 2 diabetes and neurodegeneration: ROLES for Amyloid-β, Amylin, and Tau proteins. J. Alzheimer's Dis., 2017,59(2), 421–432.

Bobik, T.A. Polyhedral organelles compartmenting bacterial metabolic processes. Appl. Microbiol. Biotechnol, 2006,70, 517–25.

Brown, R.M.; Saxena, I.M. Cellulose biosynthesis: a model for understanding the assembly of biopolymers. Plant Physiol. Biochem. 2000,38, 57–67.

Brum, J.R.; Schenck, R.O.; Sullivan, M.B. Global morphological analysis of marine viruses shows minimal regional variation and dominance of non-tailed viruses. ISME J. 2013,7, 1738–1751.

Bugnicourt, E.; Cinelli, P.; Lazzeri, A.; Alvarez, V. Polyhydroxyalkanoate (PHA): review of synthesis, characteristics, processing and potential applications in packaging. Express Polym. Lett. 2014,8, 791–808.

Buldum, G.; Bismarck, A.; Mantalaris, A. Recombinant biosynthesis of bacterial cellulose in genetically modified *Escherichia coli*. Bioproc. Biosyst. Eng. 2018,41, 265–279.

Cao, B.; Li, Y.; Yang, T.; Bao, Q.; Yang, M.; Mao, C. Bacteriophage-based biomaterials for tissue regeneration. Adv. Drug Deliv. Rev. 2018,145, 73–95. doi: 10.1016/j.addr.2018.11.004.

Carrillo-Tripp, M.; Brooks, C.L.; Reddy, V.S. A novel method to map and compare protein-protein interactions in spherical viral capsids. Proteins 2008,73, 644–655.

Carrillo-Tripp, M.; Montiel-García, D.J.; Brooks, C.L.; Reddy, V.S. Capsidmaps: protein-protein interaction pattern discovery platform for the structural analysis of virus capsids using google maps. J. Struct. Biol. 2015,190, 47–55.

Carrillo-Tripp, M.; Shepherd, C.M.; Borelli, I.A.; Venkataraman, S.; Lander, G.; Natarajan, P.; Johnson, J.E.; Brooks, C.L.; Reddy, V.S. VIPERdb2: an enhanced and web API enabled relational database for

structural virology. Nucleic Acids Res. 2009,37, D436–D442.

Caspar, D.L.D.; Klug, A. Physical principles in the construction of regular viruses. Cold Spring Harbor Symp. Quant. Biol. 1962,27, 1–24.

Castón, J.R.; Carrascosa, J.L. The basic architecture of viruses. In Structure and Physics of Viruses: An Integrated Textbook. Mateu, M.G., ed. Springer: Dordrecht, 2013, 53–75.

Castro, M.J.; López, C.E.; García, S.O.; Narayanasamy, R.; Fajardo, G.J.; Herrera, H.; Balagurusamy, N. Improved strength and durability of concrete through metabolic activity of ureolytic bacteria. Environ. Sci. Pollut. Res. 2018,25, 21451–21458.

Celińska, E. 2010. Debottlenecking the 1,3-propanediol pathway by metabolic engineering. Biotechnol. Adv. 2010,28, 519–530.

Chávez-Calvillo, G. La proteína de la cápside de los virus: un acercamiento a su estructura y ensamblaje. Masters Dissertation, Cinvestav, México, 2011.

Chen, P.Y.; Dang, X.; Klug, M.T.; Qi, J.; Courchesne, N.M.D.; Burpo, F.J.; Fang, N.; Hammond, P.T.; Belcher, A.M. Versatile three-dimensional virus-based template for dye-sensitized solar cells with improved electron transport and light harvesting. ACS Nano. 2013, 7, 6563–6574.

Chen, W.; Cao, Y.; Liu, M.; Zhao, Q.; Huang, J.; Zhang, H.; Deng, Z.; Dai, J.; Williams, D.F.; Zhang, Z. Rotavirus capsid surface protein VP4-coated Fe3O4 nanoparticles as a theranostic platform for cellular imaging and drug delivery. Biomaterials 2012, 33, 7895–7902.

Cho, H.; Choi, J.Y.; Hwang, M. S.; Kim, Y. J.; Lee, H.M.; Lee, H.S.; Lyoo, C. H. In vivo cortical spreading pattern of tau and amyloid in the Alzheimer disease spectrum. Ann. Neurol. 2016, 80(2), 247–258.

Cho, J.S.; Choi, Y.J.; Chung, D.K. Expression of *Clostridium thermocellum* endoglucanase gene in *Lactobacillus gasseri* and *Lactobacillus johnsonii* and characterization of the genetically modified probiotic *Lactobacilli*. Curr. Microbiol. 2000, 40, 257–263.

Chudecka-glaz, A.; Szczeblińska, J.; Cymbaluk-Ploska, A.; Kohn, J.; El Fray, M. New poly(ester-amide) copolymers modified with polyester (PEAE) for anticancer drug encapsulation. J. Microencapsul. 2016, 33, 702–711. doi 10.1080/02652048.2006.1228708.

Chung, A.L.; Zeng, G.D.; Jin, H.L.; Wu, Q.; Chen, J.C.; Chen, G.Q. Production of medium-chain-length 3-hydroxyalkanoic acids by β-oxidation and phaC operon deleted Pseudomonas entomophila harboring thioesterase gene. Metab. Eng. 2013, 17, 23–29.

Chung, W.J.; Merzlyak, A.; Yoo, S.Y.; Lee, S.W. Genetically engineered liquid-crystalline viral films for directing neural cell growth. Langmuir 2010, 26, 9885–9890.

Crick, F.H.C.; Watson, J.D. Structure of small viruses. Nature 1956, 177, 473–475.

Cubas-Cano, E.; González-Fernández, C.; Ballesteros, M.; Tomás-Pejó, E. Biotechnological advances in lactic acid production by lactic acid bacteria: lignocellulose as novel substrate. Biofuels, Bioprod. Bioref. 2018, 12, 290–303. doi. org/10.1002/bbb.1852.

Dams, R.I.; Viana, M.B.; Guilherme, A.A.; Silva, C.M.; André, B.; Angenent, L.T.; Santaella, S.; Leitão, R. Production of medium-chain carboxylic acids by anaerobic fermentation of glycerol using a bioaugmented open culture. Biomass Bioenerg. 2018, 118, 1–7.

Das, S.; Jacob, R.S.; Patel, K.; Singh, N.; Maji, S.K. Amyloid fibrils: versatile biomaterials for cell adhesion and tissue engineering applications. Biomacromolecules 2018, 19(6), 1826–1839.

Debbabi, F.; Gargoubi, S.; Hadj Ayed, M.A.; Abdessalem, S.B. Development and characterization of antibacterial braided polyamide suture coated with chitosan-citric acid biopolymer. J. Biomater. Appl. 2017, 32, 384–398. doi 10.1177/0885328217721868.

Deuerling, S.; Kluger, S.; Klotz, M.; Zollfrank, C.; Van Opdenbosch, D. A perspective on bio-mediated material structuring. Adv. Mater. 2018, 30, e1703656. doi 10.1002/adma.201703656.

Díaz-Valle, A.; García-Salcedo, Y.M.; Chávez-Calvillo, G.; Silva-Rosales, L.; Carrillo-Tripp, M. Highly efficient strategy for the heterologous expression and purification of soluble cowpea chlorotic mottle virus capsid protein and in vitro pH-dependent assembly of virus-like particles. J. Virol. Methods. 2015, 225, 23–29.

Dick, J.; De Windt, W.; Graef, B.; Saveyn, H.; Van Der Meeren, P.; De Belie, N.; Verstraete, W. Bio-deposition of a calcium carbonate layer on degraded limestone by Bacillus species. Biodegradation 2006, 17, 357–367.

Dickinson, W.H. On the nature of the waxy, lardaceous, or amyloid deposit. Med. Chir. Trans., 1867, 50(1), 39–55.

Dickmeis, C.; Altintoprak, K.; van Rijn, P.; Wege, C.; Commandeur, U. Bioinspired silica mineralization on viral templates. In Methods in Molecular Biology. Springer: New York, NY, 2018, 337–362.

Digital Journal. Global microbial and bacterial cellulose market will grow at a CAGR 14.8% and reach USD 570 Million by 2023, from USD 250 Million in 2017. 2018. http://www.digitaljournal.com/pr/3877005.

Douglas, T. Viruses: making friends with old foes. Science 2006, 312, 873–875.

Doyle, L. Lardaceous disease: some early reports by British authors (1722–1879). J. R. Soc. Med. 1988, 81(12), 729–731.

Dry, C. Matrix cracking repair and filling using active and passive modes for smart timed release of chemicals from fibers into cement matrices. Smart Mater. Struct. 1994, 3, 118–123.

Dupraz, C.; Reid, P.; Braissant, O.; Decho, A.; Norman, R.; Visscher, P. Processes of carbonate precipitation in modern microbial mats. Earth-Sci. Rev. 2009, 96, 141–162.

Edvardsen, C. Water permeability and autogenous healing of cracks in concrete. ACI. Mater. J. 1999, 96, 448–454.

Eiben, S. RNA-directed assembly of TMV-like carriers with tunable fractions of differently addressable coat proteins. In Methods in Molecular Biology. Springer: New York, NY, 2018, 35–50.

Eraña, H.; Venegas, V.; Moreno, J.; Castilla, J. Prion-like disorders and transmissible spongiform encephalopathies: an overview of the mechanistic features that are shared by the various disease-related misfolded proteins. Biochem. Biophys. Res. Commun., 2017, 483(4), 1125–1136.

Eş, I.; Khaneghah, A.M.; Barba, F.J.; Saraiva, J.A.; Sant'Ana, A.S.; Hashemi, S.M.B. Recent advancements in lactic acid production—a review. Food Res. Int. 2018, 107, 763–770. doi.org/10.1016/j. foodres.2018.01.001.

Farr, R.; Choi, D.S.; Lee, S.W. Phage-based nanomaterials for biomedical applications. Acta Biomaterialia. 2014, 10, 1741–1750.

Fernandes, S.N.; Almeida, P.L.; Monge, N.; Aguirre, L.E.; Reis, D.; de Oliveira, C.L.P.; Pierranski, P.; Godinho, M.H. Mind the microgap in iridescent cellulose nanocrystal films. Adv. Mater. 2017, 29, 1603560.

Ferrer-Miralles, N.; Rodríguez-Carmona, E.; Corchero, J.; García-Ftuitós, E.; Vázquez, E.; Villaverde, A. Engineering protein self-assembling in protein-based nanomedicines for drug delivery and gene therapy. Crit. Rev. Biotechnol. 2015, 35, 209–221.

Fuenmayor, J.; Gòdia, F.; Cervera, L. Production of virus-like particles for vaccines. New Biotechnol., 2017, 39, 174–180. doi. org/10.1016/j.nbt.2017.07.010.

Galego, N.; Rozsa, C.; Sánchez, R.; Fung, J.; Vázquez, A.; Tomás, J.S. Characterization and application of poly (β-hydroxyalkanoates) family as composite biomaterials. Polym. Test. 2000, 19, 485–492.

Gallardo, R.; Ramakers, M.; De Smet, F.; Claes, F.; Khodaparast, L.; Khodaparast, L.; Rousseau, F. De novo design of a biologically active amyloid. Science, 2016, 354(6313), aah4949.

Gandini, C.; Tarraran, L.; Kalemasi, D.; Pessione, E.; Mazzoli, R. Recombinant Lactococcus lactis for efficient conversion of cellodextrins into L-lactic acid. Biotechnol. Bioeng. 2017, 114, 2807–2817. doi.org/10.1002/bit.26400.

Gao, M.; Li, J.; Bao, Z.; Hu, M.; Nian, R.; Feng, D.; An, D.; Li, X.; Xian, M.; Zhang, H. A natural in situ fabrication method of functional bacterial cellulose using a microorganism. Nat. Commun. 2019, 10, 437.

Gargaud, M. et al. Biopolymer in Encyclopedia of Astrobiology. Springer: Berlin, 2011.

Gazit, E. Self-assembled peptide nanostructures: the design of molecular building blocks and their technological utilization. Chem. Soc. Rev. 2007, 36(8), 1263–1269.

Gerhardt, K.E.; MacNeill, G.J.; Gerwing, P.D.; Greenberg, B.M. Phytoremediation of Salt-Impacted Soils and Use of Plant Growth-Promoting Rhizobacteria (PGPR) to Enhance Phytoremediation. Springer: Berlin, 2017, 19–51.

González, Y.; Meza, J.C.; González, O.; Córdova, J.A. Synthesis and biodegradation of polyhydroxialkanoates: bacterially produced plastics. Rev. Int. Contam. Ambient. 2013, 29, 77–115.

Gonzalez-Montalban, N.; Garcia-Fruitos, E.; Ventura, S. et al. The chaperone DnaK controls the fractioning of functional protein between soluble and insoluble cell fractions in inclusion bodyforming cells. Microb. Cell Fact. 2006, 5, 26.

Goonoo, N.; Bhaw-Luximon, A.; Passanha, P.; Esteves, S.; Jhurry, D. Third Generation Poly(hidroxyacid) composite scaffolds for tissue engineering. Biomed. Mater. Res.

2016, 105B, 1667–1684. doi.org/10.1002/jbm.b.33674.

Greenwald, J.; Riek, R. Biology of amyloid: structure, function, and regulation. Structure 2010, 18(10), 1244–1260.

Guo, X.; Niu, C.; Wu, Y.; Liang, X. Application of an M13 bacteriophage displaying tyrosine on the surface for detection of Fe3 and Fe2 ions. Virol. Sin. 2015, 30, 410–416.

Gustot, A.; Gallea, J.I.; Sarroukh, R.; Celej, M.S.; Ruysschaert, J.M.; Raussens, V. Amyloid fibrils are the molecular trigger of inflammation in Parkinson's disease. Biochem. J. 2015, 471(3), 323–333.

Hammarström, P.; Jiang, X.; Hurshman, A.R.; Powers, E.T.; Kelly, J.W. Sequence-dependent denaturation energetics: a major determinant in amyloid disease diversity. Proceedings of the National Academy of Sciences of the United States of America, 2002, 99(SUPPL. 4), 16427–16432.

Harada, L.K.; Silva, E.C.; Campos, W.F.; Del Fiol, F.S.; Vila, M.; Dąbrowska, K.; Krylov, V.N.; Balcão, V.M. Biotechnological applications of bacteriophages: state of the art. Microbiol. Res. 2018, 212-213, 38–58. doi.org/10.1016/j.micres.2018.04.007.

Hassan, S.E.; Abdel-Rahman, M.A.; Roushdy, M.M.; Salah, M.; Ali, M. Biocatalysis and agricultural biotechnology effective biorefinery approach for lactic acid production based on co- fermentation of mixed organic wastes by Enterococcus durans BP130. Biocatal. Agric. Biotechnol. 2019, 20, 101203. doi.org/10.1016/j.bcab.2019.101203.

Hauser, C.A.E.; Maurer-Stroh, S.; Martins, I.C. Amyloid-based nanosensors and nanodevices. Chem. Soc. Rev. 2014, 43(15), 5326–5345.

Hauser, Charlotte A.E.; Maurer-Stroh, S.; Martins, I.C. Amyloid-based nanosensors and nanodevices. The Royal Society of Chemistry, London, 2014.

Hickey, R.J.; Pelling, A.E. Cellulose biomaterials for tissue engineering. Front. Bioeng. Biotech. 2019, 7, 45.

Hirano, K.; Kurosaki, M.; Nihei, S.; Hasegawa, H.; Shinoda, S. Enzymatic diversity of the Clostridium thermocellum cellulosome is crucial for the degradation of crystalline cellulose and plant biomass. Sci. Rep. 2016, 6, 35709. doi.org/10.1038/srep35709.

Ho, P.T.; Montiel-García, D.J.; Wong, J.J.; Carrillo-Tripp, M.; Brooks, C.L.; Johnson, J.E.; Reddy, V.S. VIPERdb: a tool for virus research. Annu. Rev. Virol. 2018, 5, 477–488.

Hughson, A.G.; Race, B.; Kraus, A.; Sangaré, L.R.; Robins, L.; Groveman, B.R.; Caughey, B. Inactivation of prions and amyloid seeds with hypochlorous acid. PLoS Pathog. 2016, 12 (9), 1–27.

Iconomidou, V.A.; Vriend, G.; Hamodrakas, S.J. Amyloids protect the silkmoth oocyte and embryo. FEBS Lett. 2000, 479 (3), 141–145.

Irwan, J.M.; Anneza, L.H.; Othman, N.; Faisal, A.; Zamer, M.M.; Teddy, T. Calcium lactate addition in bioconcrete: effect on compressive strength and water penetration. MATEC Web of Conferences. 2016, 78, 01027.

Jacek, P.; Dourado, F.; Gama, M.; Bielecki, S. Molecular aspects of bacterial nanocellulose biosynthesis. Microb. Biotech. 2019, 12, 633–649.

Jarosz-Griffiths, H.H.; Noble, E.; Rushworth, J.V.; Hooper, N.M. Amyloid-β receptors: the good, the bad, and the prion protein. J. Biol. Chem. 2016, 291 (7), 3174–3183.

Jeevanandam, J.; Pal, K.; Danquah, M.K. Virus-like nanoparticles as a novel delivery tool in gene therapy. Biochimie 2019, 157, 38–47. doi.org/10.1016/j.biochi.2018.11.001.

Jiménez, C.; Jroundi, F.; Rodríguez, M.; Arias, J.M.; González, M.T. Biomineralization induced by Myxobacteria. In Communicating Current Research and Educational Topics and Trends in Applied Microbiology. A. Méndez-Vilas ed. Formatex: Guadalajara, 2017, 143–154.

Jin, H.E.; Farr, R.; Lee, S.W. Collagen mimetic peptide engineered M13 bacteriophage for collagen targeting and imaging in cancer. Biomaterials 2014, 35, 9236–9245.

Joachim, C.; Mori, H.; Selkoe, D. Amyloid β-protein deposition in tissues other than brain in Alzheimer's disease. Nature 1989, 341, 189–192.

Jolley, C.; Klem, M.; Harrington, R.; Parise, J.; Douglas, T. Structure and photoelectrochemistry of a virus capsid-TiO2nanocomposite. Nanoscale 2011, 3, 1004–1007.

Jonkers, H.M.; Schlangen, E. Crack repair by concrete-immobilized bacteria. In: Proceeding of First International Conference on Self-Healing Materials, Schmets, A.J.M., van der Zwaag, S. eds. Springer: Noordwijk, The Netherlands, 2007.

Juturu, V.; Wu, J.C. Microbial production of lactic acid: the latest development. Crit. Rev. Biotechnol. 2016, 36, 967–977. doi. org/10.3109/07388551.2015.1066305.

Kano, K.; Zhang, Q.; Yoshida, S.; Tanaka, T.; Ogino, C.; Fukuda, H.; Kondo, A. D-lactic acid production from cellooligosaccharides and beta-glucan using L-LDH gene-deficient and endoglucanase secreting *Lactobacillus plantarum*. Appl. Microbiol. Biotechnol. 2010, 85, 643–50. doi. org/10.1007/s00253-009-2111-8.

Kaur, G.; Srivastava, A.K.; Chand, S. Advances in biotechnological production of 1, 3 -propanediol. Biochem. Eng. J. 2012, 64, 106–118.

Kawaguchi, H.; Hasunuma, T.; Ogino, C.; Kondo, A. Bioprocessing of bio-based chemicals produced from lignocellulosic feedstocks. Curr. Opin. Biotechnol. 2016, 42, 30–39. doi.org/10.1016/j.copbio. 2016.02.031.

Kawaguchi, K.; Lijima, M.; Miyakawa, H.; Ohta, M.; Muguruma, T.; Endo, K.; Nakazawa, F.; Mizoguchi, I. Effects of chitosan fiber addition on the properties of polyurethane with termo-responsive shape memory. J. Biomed. Mater. Res. B. Appl.

Biomatter. 2017, 105, 1151–1156. doi 10.1002/jbm.b.33664.

Keating, K.W.; Young, E.M. Synthetic biology for bio-derived structural materials. Curr. Opin. Chem. Eng. 2019, 24, 107–114.

Khalid, A.; Ullah, H.; Ul-Islam, M.; Khan, R.; Khan, S.; Ahmad, F.; Khan, T.; Wahid, F. Bacterial cellulose-TiO2 nanocomposites promote healing and tissue regeneration in burn mice model. RSC Adv. 2017, 7, 47662–47668.

Khanna, S.; Srivastava, A.K. Recent advances in microbial polyhydroxyalkanoates. Process Biochem. 2005, 40, 607–619.

Kim, D.; Rhee, Y. Biodegradation of microbial and synthetic polyesters by fungi. Appl. Microbiol. Biotechnol. 2003, 61, 300–308.

Kim, D.Y.; Kim, H.W.; Chung, M.G.; Rhee, Y.H. Biosynthesis, modification, and biodegradation of bacterial medium-chain-length polyhydroxyalkanoates. J. Microbiol. 2007,45, 87–97.

Kim, I.G.; Jo, B.H.; Kang, D.G.; Kim, C.S.; Choi, Y.S.; Cha, H.J. Biomineralization-based conversion of carbon dioxide to calcium carbonate using recombinant carbonic anhydrase. Chemosphere 2012, 87, 1091–1096.

Knowles, T.P.J.; Mezzenga, R. Amyloid fibrils as building blocks for natural and artificial functional materials. Adv. Mater. (Deerfield Beach, Fla.) 2016, 28 (31), 6546–6561.

Komesu, A.; Rocha de Oliveira, A.J.; da Silva Martins, L.H.; Maciel, M.R.W.; Filho, R.M. Lactic acid production to purification: a review. BioResources 2017, 12, 4364–4383. doi.org/10.15376/ biores.12.2.Komesu.

Konhauser, K. Introduction to Geomicrobiology. Blackwell publishing: Oxford, UK, 2007.

Krupovic, M.; Dolja, V.V.; Koonin, E.V. Origin of viruses: primordial replicators recruiting capsids from hosts. Nat. Rev. Microbiol. 2019, 17, 449–458.

Kuppusamy, S.; Thavamani, P.; Venkateswarlu, K.; Lee, Y.B.; Naidu, R.; Megharaj, M. Remediation approaches for polycyclic aromatic hydrocarbons (PAHs) contaminated soils: technological constraints, emerging trends and future directions. Chemosphere 2017, 168, 944–968.

Lêdo, S., Leite, Â., Souto, T., Dinis, M.A., Sequeiros, J. Mid-and long-term anxiety levels associated with presymptomatic testing of Huntington's disease, Machado-Joseph disease, and familial amyloid polyneuropathy. Revista Brasileira de Psiquiatria, 2016, 38 (2), 113–120.

Lee, C.S.; Aroua, M.K.; Daud, W.M.A.W.; Cognet, P.; Pérès-Lucchese, Y.; Fabre, P.; Reynes, O.; Latapie, L. A review: conversion of bioglycerol into 1,3-propanediol via biological and chemical method. Renew. Sustain. Energy Rev. 2015, 42, 963–972.

Lee, L.A.; Muhammad, S.M.; Nguyen, Q.L.; Sitasuwan, P.; Horvath, G.; Wang, Q. Multivalent ligand displayed on plant virus induces rapid onset of bone differentiation. Mol. Pharm. 2012, 9, 2121–2125.

Lenz, R.W.; Marchessault, R.H. Bacterial polyesters: biosynthesis, biodegradable plastics and biotechnology. Biomacromolecules 2005, 6, 1–8.

Li, V.C.; Herbert, E. Robust self-healing concrete for sustainable infrastructure. J. Adv. Concr. Technol. 2012, 10, 207–218.

Li, Y.; Tian, J.; Tian, H.; Chen, X.; Ping, W.; Tian, C.; Lei, H. Mutation-based selection and analysis of Komagataeibacter hansenii HDM1-3 for improvement in bacterial cellulose production. J. Appl. Microbiol. 2016, 121, 1323–1334.

Lindlbauer, K.; Marx, H.; Sauer, M. Effect of carbon pulsing on the redox household of Lactobacillus diolivorans in order to enhance 1,3-propanediol production. New Biotechnol. 2017, 34, 32–39.

Liovic, M.; Ozir, M.; Bedina, Z.A. et al. Inclusion bodies as potential vehicles for recombinant protein delivery into epithelial cells. Microb. Cell Fact. 2012, 11, 67.

Liu, C.G.; Xiao, Y.; Xia, X.X.; Zhao, X.Q.; Peng, L.; Srinophakun, P.; Bai, F.W. Cellulosic ethanol production: progress, challenges and strategies for solutions. Biotechnol. Adv. 2019, 37, 491–504. doi.org/10.1016/j.biotechadv.2019.03.002.

Liu, G.; Sun, J.; Zhang, J.; Tu, Y.; Bao, J. High titer L-lactic acid production from corn stover with minimum wastewater generation and techno-economic evaluation based on Aspen plus modeling. Bioresour. Technol. 2015, 198, 803–810. doi.org/10.1016/j.biortech.2015.09.098.

Liu, M.; Liu, L.; Jia, S.; Li, S.; Zou, Y.; Zhong, C. Complete genome analysis of Gluconacetobacter xylinus CGMCC 2955 for elucidating bacterial cellulose biosynthesis and metabolic regulation. Sci. Rep. 2018, 8, 6266.

López-Gómez, J.P.; Alexandri, M.; Schneider, R.; Venus, J. A review on the current developments in continuous lactic acid fermentations and case studies utilising inexpensive raw materials. Process Biochem. 2019, 79, 1–10. doi.org/10.1016/j.procbio.2018.12.012.

Lu, K.; Jacob, J.; Thiyagarajan, P.; Conticello, V.P.; Lynn, D.G. Exploiting amyloid fibril lamination for nanotube self-assembly. J. Am. Chem. Soc. 2003, 125 (21), 6391–6393.

Luckanagul, J.; Lee, L.A.; Nguyen, Q.L.; Sitasuwan, P.; Yang, X.; Shazly, T.; Wang, Q. Porous alginate hydrogel functionalized with virus as three-dimensional scaffolds for bone differentiation. Biomacromolecules 2012, 13, 3949–3958.

Luckanagul, J.; Lee, L.A.; You, S.; Yang, X.; Wang, Q. Plant virus incorporated hydrogels as scaffolds for tissue engineering possess low immunogenicity in vivo. J. Biomed. Mater. Res. A. 2014, 103, 887–895.

Maji, S.K.; Schubert, D.; Rivier, C.; Lee, S.; Rivier, J. E.; Riek, R. Amyloid as a depot for the formulation of long-acting drugs. PLoS Biol. 2008, 6 (2), 0240–0252.

Marshall, K.E.; Morris, K. L.; Charlton, D.; O'Reilly, N.; Lewis, L.; Walden, H.; Serpell, L.C. Hydrophobic, aromatic, and electrostatic interactions play a central role in amyloid fibril formation and stability. Biochemistry 2011, 50 (12), 2061–2071.

Maury, C.P.J. The emerging concept of functional amyloid. J. Intern. Med. 2009, 265 (3), 329–334.

Mazzoli, R. Development of microorganisms for cellulose-biofuel consolidated bioprocessings: metabolic engineers' tricks. Comput. Struct. Biotechnol. J. 2012, 3, c201210007. doi.org/10.5936/csbj.201210007.

Mazzoli, R.; Bosco, F.; Mizrahi, I.; Bayer, E.A.; Pessione, E. Towards lactic acid bacteria-based biorefineries. Biotechnol. Adv. 2014, 32, 1216–1236. doi.org/10.1016/j.biotechadv.2014.07.005.

Mazzoli, R.; Lamberti, C.; Pessione, E. Engineering new metabolic capabilities in bacteria: lessons from recombinant cellulolytic strategies. Trends Biotechnol. 2012, 30, 111–119. doi.org/10.1016/j.tibtech.2011.08.003.

McGuinness, M.; Dowling, D. Plant-associated bacterial degradation of toxic organic compounds in soil. Int. J. Environ. Res. Public Health. 2009, 6, 2226–2247.

Mendes, A.C.; Strohmenger, T.; Goycoolea, F.; Chronakis, I.S. Electrostatic self-assembly of polysaccharides into nanofibers. Colloids Surf. A Physicochem. Eng. Asp. 2017, 531, 182–188.

Mendonca, T.T.; Gomez, J.G.; Buffoni, E.; Sánchez, R.J.; Schripsema, J.; Lopes, M.S.; Silva, L.F. Exploring the potential of Burkholderia sacchari to produce polyhydroxyalkanoates. J. Appl. Microbiol. 2014, 116, 815–829.

Metsoviti, M.; Zeng, A.; Koutinas, A.A.; Papanikolaou, S. Enhanced 1,3-propanediol production by a newly isolated Citrobacter freundii strain cultivated on biodiesel-derived waste glycerol through sterile and non-sterile bioprocesses. J. Biotechnol. 2013, 163, 408–418.

Mickoleit, F.; Altintoprak, K.; Wenz, N.L.; Richter, R.; Wege, C.; Schüler, D. Precise assembly of genetically functionalized magnetosomes and tobacco mosaic virus particles generates a magnetic biocomposite. ACS Appl. Mater. Interfaces. 2018, 10, 37898–37910.

Mingardon, F.; Chanal, A.; Tardif, C.; Fierobe, H.P. The issue of secretion in heterologous expression of Clostridium cellulolyticum cellulase-encoding genes in Clostridium acetobutylicum ATCC 824. Appl. Environ. Microbiol. 2011, 77, 2831–2838. doi.org/10.1128/AEM.03012-10.

Montiel-García, D.J.; Mannige, R.V.; Reddy, V.S.; Carrillo-Tripp, M. Structure based sequence analysis of viral and cellular protein assemblies. J. Struct. Biol. 2016, 196, 299–308.

Morais, S.; Shterzer, N.; Grinberg, I.R.; Mathiesen, G.; Eijsink, V.G.; Axelsson, L.; Lamed, R.; Bayer, E.A.; Mizrahi, I. Establishment of a simple Lactobacillus plantarum cell consortium for cellulase-xylanase synergistic interactions. Appl. Environ. Microbiol. 2013, 79, 5242–5249. doi.org/10.1128/AEM.01211-13.

Morell, M.; Bravo, R.; Espargaro, A. et al. Inclusion bodies: specificity in their aggregation process and amyloid-like structure. Biochim. Biophys. Acta, 2008, 1783, 1815–25.

Mozejko-Ciesielska, J.; Kiewisz, R. (2016). Bacterial polyhydroxyalkanoates: Still fabulous? Microbiol. Res. 2016, 192, 271–282.

Muhammadi, S.; Afzal, M.; Hameed, S. Bacterial polyhydroxyalkanoateseco-friendly next generation plastic: production, biocompatibility, biodegradation, physical properties and applications. Green Chem. Lett. Rev. 2015, 8, 56–77.

Müller, U.C.; Deller, T.; Korte, M. Not just amyloid: physiological functions of the amyloid precursor protein family. Nat. Rev. Neurosci. 2017, 18 (5), 281–298.

Muynck, W.; Verbeken, K.; Belie, N.; Verstraete, W. Influence of temperature on the effectiveness of a biogenic carbonate surface treatment for limestone conservation. Appl. Microbiol. Biotechnol. 2013, 97, 1335–1347.

Nagashima, A.; Tsuji, T.; Kondo, T. A uniaxially oriented nanofibrous cellulose scaffold from pellicles produced by Gluconacetobacter xylinus in dissolved oxygen culture. Carbohydr. Polym. 2016, 135, 215–224.

Nakano, S.I.; Miyoshi, D.; Sugimoto, N. Effects of molecular crowding on the structures, interactions, and functions of nucleic acids. Chem. Rev. 2014, 114 (5), 2733–2758.

Nhan, H.S.; Chiang, K.; Koo, E.H. The multifaceted nature of amyloid precursor protein and its proteolytic fragments: friends and foes. Acta Neuropathol. 2015, 129 (1), 1–19.

Nitschke, M.; Costa, S.G.V.A.O.; Contiero, J. Rhamnolipids and PHAs: recent reports on Pseudomonas-derived molecules of increasing industrial interest. Process Biochem. 2011, 46, 621–630.

Nofar, M.; Sacligil, D.; Carreau, P.J.; Kamal, M.R.; Heuzey, M. Poly (lactic acid) blends: processing, properties and applications. Int. J. Biol. Macromol. 2019, 125, 307–360. doi.org/10.1016/j.ijbiomac.2018.12.002.

Okano, K.; Uematsu, G.; Hama, S.; Tanaka, T.; Noda, H. Metabolic engineering of Lactobacillus plantarum for direct L-Lactic acid production from raw corn starch. Biotechnol. J. 2018, 1700 517, 1–6. doi.org/10.1002/biot.201700517.

Othman, N.; Irwan, J.M.; Alshali, A.F.; Anneza, L.H. Effect of ureolytic bacteria on compressive strength and water permeability on bioconcrete. In: Advances in Civil Architectural, Structural and Constructional Engineering: Proceedings of the International Conference on Civil, Architectural, Structural and Constructional Engineering. Kim, D.K., Jung, J., Seo, J. eds. Dong-A University, Busan, South Korea, 2016.

Otzen, D. Functional amyloid: turning swords into plowshares. Prion 2010, 4 (4), 256–264.

Ozkose, E.; Akyol, I.; Kar, B.; Comlekcioglu, U.; Ekinci, M.S. Expression of fungal cellulase gene in Lactococcus lactis to construct novel recombinant silage inoculants. Folia Microbiol. 2009, 54, 335–342. doi.org/10.1007/s12223-009-0043-4.

Pacheco-Torgal, F.; Labrincha, J.A. Biotech cementitious materials: some aspects of an innovative approach for concrete with enhanced durability. Constr. Build. Mater. 2013, 40, 1136–1141.

Pan, C.; Tan, G.A.; Ge, L.; Chen, C.; Wang, J. Two-stage microbial conversion of crude glycerol to 1, 3 -propanediol and polyhydroxyalkanoates after pretreatment. J. Environ. Manage. 2019, 232, 615–624.

Parsons, J.B., Frank, S., Bhella, D. et al. Synthesis of empty bacterial microcompartments, directed organelle protein incorporation, and evidence of filament-associated organelle movement. Mol. Cell. 2010, 38, 305–15.

Patterson, D.P. Encapsulation of active enzymes within bacteriophage p22 virus-like particles. In Methods in Molecular Biology. Springer: New York, NY, 2018, 11–24.

Patwardhan, A. Trends in the Electron Microscopy Data Bank (EMDB). Acta. D Struct. 2017, 73, 503–508.

Pflügl, S.; Marx, H.; Mattanovich, D.; Sauer, M. 1,3-Propanediol production from glycerol with Lactobacillus diolivorans. Bioresour. Technol. 2012, 119, 133–140.

Poli, A., Di Donato, P., Abbamondi, G.R., Nicolaus, B. Synthesis, production, and biotechnological applications of exopolysaccharides and polyhydroxyalkanoates. Archaea 2011, 2011, 693253.

Ponnusamy, V.K.; Nguyen, D.D.; Dharma-raja, J.; Shobana, S.; Banu, J.R.; Saratale, R.G.; Chang, S.W.; Kumar, G. A review on lignin structure, pretreatments, fermenta-tion reactions and biorefinery potential. Bioresour. Technol. 2019, 271, 462–472. doi.org/10.1016/j.biortech.2018.09.070.

Portela, R.; Leal, C.R.; Almeida, P.L.; Sobral, R.G. Bacterial cellulose: a versatile biopolymer for wound dressing applica-tions. Microb. Biotech. 2019, 12, 586–610.

Raha, A.R.; Chang, L.Y.; Sipat, A.; Yusoff, K.; Haryanti, T. Expression of a thermo-stable xylanase gene from Bacillus coag-ulans ST-6 in Lactococcus lactis. Lett. Appl. Microbiol. 2006, 42, 210–214. doi. org/10.1111/j.1472-765X.2006.01856.x.

Reddy, C.S.K.; Ghai, R.; Rashmi y Kalia, V.C. Polyhydroxyalkanoates: an overview. Bioresour. Technol. 2003, 87, 137–146.

Rehm, B. Microbial Production of Biopoly-mers and Polymer Precursors: Applica-tions and Perspectives. Caister Academic Press: Norfolk, 2009.

Rios-González, L.J.; Morales-Martínez, T.K.; Rodríguez-Flores, M.F.; Rodríguez-De la Garza, J.A.; Castillo-Quiroz, D.; Castro-Montoya, A.J.; Martinez, A. Auto-hydrolysis pretreatment assessment in ethanol production from agave bagasse. Bioresour. Technol. 2017, 242, 184–190. doi.org/10.1016/j.biortech.2017.03.039.

Roher, A.E.; Lowenson, J.D.; Clarke, S.; Woods, A.S.; Cotter, R.J.; Gowing, E.; Ball, M.J. β-amyloid-(1-42) is a major component of cerebrovascular amyloid deposits: implications for the pathology of Alzheimer disease. Proceedings of the National Academy of Sciences of the United States of America, 1993, 90 (22), 10836–10840.

Rossi, F.; Rudella, A.; Marzotto, M.; Della-glio, F. Vector-free cloning of a bacterial endo-1,4-β-glucanase in Lactobacillus plantarum and its effect on the acidi-fying activity in silage: use of recombi-nant cellulolytic Lactobacillus plantarum as silage inoculant. Antonie van Leeu-wenhoek. 2001, 80, 139-147. doi. org/10.1023/A:101222322.

Ruan, M.; Luan, H.; Wang, G.; Shen, M. Biopolyols synthesized from bio-based 1,3-propanediol and applications on poly-urethane reactive hot melt adhesives. Ind. Crop. Prod. 2019, 128, 436–444.

Ruth, K.; Grubelnik, A.; Hartmann, R.; Egli, T.; Zinn, M.; Ren, Q. Efficient produc-tion of (R)-3-hydroxycarboxylic acids by biotechnological conversion of polyhy-droxyalkanoates and their purification. Biomacromolecules 2007, 8, 279–286.

Sainsbury, F.; Cañizares, M.C.; Lomonos-soff, G.P. Cowpea mosaic virus: the plant virus–based biotechnology workhorse. Annu. Rev. Phytopathol. 2010, 48(1), 437–455. doi.org/10.1146/annurev-phyto-073009-114242.

Salmond, G.; Fineran, P. A century of the phage: past, present and future. Nat. Rev. Microbiol. 2015, 13, 777–786. doi:10.1038/nrmicro3564.

Schleeger, M.; Vandenakker, C.C.; Deckert-Gaudig, T.; Deckert, V.; Velikov, K.P.; Koenderink, G.; Bonn, M. Amyloids: from molecular structure to mechan-ical properties. Polymer 2013, 54 (10), 2473–2488.

Schlegel, H. General Microbiology. Univer-sity Press Cambridge: Cambridge, 1993.

Scholthof, K. B. G.; Mirkov, T. E.; Scholthof, H. B. Plant virus gene vectors: biotechnology applications in agricul-ture and medicine. In Setlow J.K. (ed) Genetic Engineering. Genetic Engi-neering: Principles and Methods, vol. 24. Springer, USA, 2002, 67–85. doi. org/10.1007/978-1-4615-0721-5_4.

Scott, F.; Keith, C.H.; Rice, M.K.; Warren, C.R. Engineering a living biomaterial via bacterial surface capture of environmental molecules. 2018, ysy017. https://doi. org/10.1093/synbio/ysy017.

Seifan, M.; Samani, A.K.; Berenjian, A. Bioconcrete: next generation of self-healing concrete. Appl. Microbiol. Biotechnol. 2016, 100, 2591–2602.

Seras-Franzoso, J.; Peebo, K.; Corchero, J.L. et al. A nanostructured bacterial bio-scaffold for the sustained bottom-up delivery of protein drugs. Nanomedicine (London) 2013, 8, 1587–1599.

Shahab, R.L.; Luterbacher, J.S.; Brethauer, S.; Studer, M.H. Consolidated bioprocessing of lignocellulosic biomass to lactic acid by a synthetic fungal-bacterial consortium. Biotechnol. Bioeng. 2018, 115, 1207–1215. doi.org/10.1002/bit.26541.

Shammas, S.L.; Knowles, T.P.J.; Baldwin, A.J.; MacPhee, C.E.; Welland, M.E.; Dobson, C.M.; Devlin, G.L. Perturbation of the stability of amyloid fibrils through alteration of electrostatic interactions. Biophys. J. 2011, 100 (11), 2783–2791.

Shukla, S.; Steinmetz, N.F. Virus-based nanomaterials as positron emission tomography and magnetic resonance contrast agents: from technology development to translational medicine. Wiley Interdiscip. Rev. Nanomed. Nanobiotech. 2015, 7, 708–721.

Silva, F.B.; Boon, N.; De Belie, N.; Verstraete, W. Industrial application of biological self-healing concrete: challenges and economical feasibility. J. Commer. Biotechnol. 2015, 21, 31–38.

Singh, R.; Shitiz, K.; Singh, A. Chitin and Chitosan: biopolymers for wound management. Int. Wound. J. 2017, 14, 1276–1289. doi 10.1111/iwj.12797.

Singhvi, M.S.; Zendo, T.; Sonomoto, K. Free lactic acid production under acidic conditions by lactic acid bacteria strains: challenges and future prospects. Appl. Microbiol. Biotechnol. 2018, 102, 5911–5924. doi.org/10.1007/s00253-018-9092-4.

Singhvi, M.S.; Zinjarde, S.S.; Gokhale, D.V. Polylactic acid: synthesis and biomedical applications. J. Appl. Microbiol. 2019, 1–15. doi:10.1111/jam.14290.

Song, L.; Wang, S.; Wang, H.; Zhang, H.; Cong, H.; Jiang, X.; Tien, P. Study on

nanocomposite construction based on the multi-functional biotemplate self-assembled by the recombinant TMGMV coat protein for potential biomedical applications. J. Mater. Sci. Mater. Med. 2015, 26, 97.

Soto, C.M.; Ratna, B.R. Virus hybrids as nanomaterials for biotechnology. Curr. Opin. Biotech. 2010, 21, 426–438.

Speed, M.A.; Wang, D.I.; King, J. Specific aggregation of partially folded polypeptide chains: the molecular basis of inclusion body composition. Nat. Biotechnol. 1996, 14, 1283–1287

Stern, J.; Moraïs, S.; Ben-David, Y.; Salama, R.; Shamshoum, M.; Lamed, R.; Shoham, Y.; Bayer, E.A.; Mizrahi, I. Assembly of synthetic functional cellulosomal structures onto the Lactobacillus plantarum cell surface—a potent member of the gut microbiome. Appl. Environ. Microbiol. 2018, 84, e00282-18. doi.org/10.1128/AEM.00282-18.

Sun, J.; Birnbaum, W.; Anderski, J.; Picker, M.T.; Mulac, D.; Langer, K.; Kuckling, D. Use of light-degradable aliphatic polycarbonate Nanoparticles as drug carrier for photosensitizer. Biomacromolecules. 2018, 19, 4677–4690. doi 10.1021/acs.biomac.8b01446.

Tan, A.C.W.; Polo-Cambronell, B.J.; Provaggi, E.; Ardilla-Suárez, C.; Ramirez-Caballero, G.E.; Baldovino-Medrano, V.G.; Kalaskar, D.M. Design and development of low cost polyurethane biopolymer based on castor oil and glycerol for biomedical applications. Biopolymers 2018, 109, e23078. doi 10.1002/bip.23078.

Tarraran, L.; Mazzoli, R. Alternative strategies for lignocellulose fermentation through lactic acid bacteria: the state of the art and perspectives. FEMS Microbiol. Lett. 2018, 365, fny126. doi.org/10.1093/femsle/fny126.

Thomas, L.; Joseph, A.; Devi, L. Xylanase and cellulase systems of Clostridium sp.: an insight on molecular approaches for strain improvement. Bioresour. Technol. 2014, 158, 343–350.

Tugarova, V.A.; Scheludko, A.V.; Dyatlova, Y.A.; Flipecheva, Y.A.; Kamnev, A.A. FTIR spectroscopic study of biofilms formed by the rhizobacterium Azospirillum brasilense Sp245 and its mutant Azospirillum brasilense Sp245.1610. J. Mol. Struct. 2017, 1140, 142–147.

Tycko, R. Amyloid polymorphism: structural basis and neurobiological relevance. Neuron 2015, 86 (3), 632–645.

Vaghari, H.; Eskandari, M.; Sobhani, V.; Berenjian, A.; Song, Y.; Jafarizadeh-Malmiri, H. Process intensification for production and recovery of biological products. Am. J. Biochem. Biotechnol. 2015, 11, 37–43.

Valbuena, A.; Mateu, M.G. Kinetics of surface-driven self-assembly and fatigue-induced disassembly of a virus-based nanocoating. Biophys. J. 2017, 112, 663–673.

Van Eldijk, M.B.; Schoonen, L.; Cornelissen, J.J.L.M.; Nolte, R.J.M.; van Hest, J.C.M. Metal ion-induced self-assembly of a multi-responsive block copolypeptide into well-defined nanocapsules. Small. 2016, 12, 2476–2483.

Van Tittelboom, K.; De Belie, N. Self-healing in cementitious materials—a review. Materials 2013, 6, 2182–2217.

Vazquez, E.; Corchero, J.L.; Burgueno, J.F. et al. Functional inclusion bodies produced in bacteria as naturally occurring nanopills for advanced cell therapies. Adv Mater, 2012, 24, 1742–1747.

Vieira, P.B.; Kilikian, B.V.; Bastos, R.V.; Perpetuo, E.A.; Nascimento, C.A.O. Process strategies for enhanced production of 1,3 -propanediol by Lactobacillus reuteri using glycerol as a co-substrate. Biochem. Eng. J. 2015, 94, 30–38.

Virchow, R. Congenitale Encephalitis und Myelitis. Archiv Für Pathologische Anatomie Und Physiologie Und Für Klinische Medicin, 1867, 38 (1), 129–138.

Vivek, N.; Aswathi, T.V.; Ri, P.; Pandey, A.; Binod, P. 2017. Self-cycling fermentation for 1,3-propanediol production: comparative evaluation of metabolite flux in cell recycling, simple batch and continuous processes using Lactobacillus brevis N1E9.3.3 strain. J. Biotechnol. 2017, 259, 110–119.

Vivek, N.; Christopher, M.; Kumar, M.K.; Castro, E.; Binod, P.; Pandey, A. Pentose rich acid pretreated liquor as co-substrate for 1,3-propanediol production. Renew. Energy 2018, 129, 794–799.

Vivek, N.; Pandey, A.; Binod, P. Biological valorization of pure and crude glycerol into 1,3-propanediol using a novel isolate Lactobacillus brevis N1E9.3.3. Bioresour. Technol. 2016, 213, 222–230.

Wang, J.; Mignon, A.; Snoeck, D.; Wiktor, V.; Van Vliergerghe, S.; Boon, N.; De Belie, N. Application of modified-alginate encapsulated carbonate producing bacteria in concrete: a promising strategy for crack self-healing. Front. Microbiol. 2015, 6, 1088.

Wang, J.; Soens, H.; De Belie, N.; Verstraete, W. Self-healing concrete by use of micro-encapsulated bacterial spores. Cem. Concr. Res. 2014, 56, 139–152.

Wang, J.Y.; De Belie, N.; Verstraete, W. Diatomaceous earth as a protective vehicle for bacteria applied for self-healing concrete. J. Ind. Microbiol. Biot. 2012, 39, 567–577.

Wege, C.; Eber, F.J. Bottom-up assembly of TMV-based nucleoprotein architectures on solid supports. In Methods in Molecular Biology. Springer: New York, NY, 2018, 169–186.

Wege, C.; Geiger, F. Dual functionalization of rod-shaped viruses on single coat protein subunits. In Methods in Molecular Biology. Springer: New York, NY, 2018, 405–424.

Wen, A.M.; Steinmetz, N.F. Design of virus-based nanomaterials for medicine, biotechnology, and energy. Chem. Soc. Rev. 2016, 45, 4074–4126.

Westwell-Roper, C.; Verchere, C.B. Modulation of innate immunity by amyloidogenic peptides. Trends Immunol. 2019, 1–19.

Wiktor, V.; Jonkers, H.M. Quantification of crack-healing in novel bacteria-based self-healing concrete. Cem. Concr. Compos. 2011, 33, 763–770.

Williams, D.F. Definitions in Biomaterials. Elsevier: Amsterdam, 1987.

Williams, D.F. On the nature of biomaterials. Biomaterials 2009, 30, 5897–5909. doi 10.1016/j.biomaterials.2009.07.027.

Wischral, D.; Fu, H.; Pellegrini Pessoa, F.L.; Pereira, N.; Yang, S. Effective and simple recovery of 1,3-propanediol from a fermented medium by liquid–liquid extraction system with ethanol and K3PO4. Chin. J. Chem. Eng. 2018, 26 :104–108.

Wu, M.; Johannesson, B.; Geiker, M. A review: self-healing in cementitious materials and engineered cementitious composite as a self-healing material. Constr. Build. Mater. 2012, 28, 571–583.

Xafenias, N.; Oluchi, M.; Mapelli, V. Electrochemical startup increases 1,3-propanediol titers in mixed-culture glycerol fermentations. Process Biochem. 2015, 50, 1499–1508.

Xiao, R.; Zheng, Y. Overview of microalgal extracellular polymeric substance (EPS) and their applications. Biotechnol. Adv. 2016, 34, 1225–1244. doi 10.1016/j. biotechadv.2016.08.004.

Xin, F.; Dong, W.; Zhang, W.; Ma, J.; Jiang, M. Biobutanol production from crystalline cellulose through consolidated bioprocessing. Trends Biotechnol. 2018, 37, 1–14. doi.org/10.1016/j.tibtech.2018.08.007.

Xu, X.; Tao, A.; Xu, Y. DNA-templated assembly of viral protein hydrogel. Nanoscale 2014, 6, 14627–14629.

Yamada, R.; Hasunuma, T.; Kondo, A. Endowing non-cellulolytic microorganisms with cellulolytic activity aiming for consolidated bioprocessing. Biotechnol. Adv. 2013, 31, 754–763. doi.org/10.1016/j. biotechadv.2013.02.007.

Yang, B.; Liang, S.; Liu, H.; Liu, J.; Cui, Z.; Wen, J. Metabolic engineering of *Escherichia coli* for 1,3-propanediol biosynthesis from glycerol. Bioresour. Technol. 2018, 267, 599–607.

Yildiz, I.; Shukla, S.; Steinmetz, N.F. Applications of viral nanoparticles in medicine. Curr. Opin. Biotech. 2011, 22, 901–908.

Yoo, L.; Park, J.S.; Kwon, K.C.; Kim, S.E.; Jin, X.; Kim, H.; Lee, J. Fluorescent viral nanoparticles with stable in vitro and in vivo activity. Biomaterials 2012, 33, 6194–6200.

Yoo, N.G.; Dogra, S.; Meinen, B.A.; Tse, E.; Haefliger, J.; Southworth, D.R.; Jakob, U. Polyphosphate stabilizes protein unfolding intermediates as soluble amyloid-like oligomers. J. Mol. Biol. 2018, 430(21), 4195–4208.

Yun, J.; Yang, M.; Magocha, T.A.; Zhang, H.; Xue, Y.; Zhang, G.; Qi, X.; Sun, W. Production of 1,3-propanediol using a novel 1,3-propanediol dehydrogenase from isolated Clostridium butyricum and co-biotransformation of whole cells. Bioresour. Technol. 2018, 247, 838–843.

Zhang, J.; Shishatskaya, E.l.; Volova, T.G.; Da Silva, L.F.; Chen, G.Q. Polyhidroxydealkanoates (PHA) for terapeutical applications. Matter. Sci. Eng. C. 2018, 86, 144–150. doi 10.1016/j.msec.2017.12.035.

Zhang, W.; Zhang, X.E.; Li, F. Virus-based nanoparticles of simian virus 40 in the field of nanobiotechnology. Biotechnol. J. 2018, 13, 1700619.

Zhao, C.; Li, Z.; Li, T.; Zhang, Y.; Bryant, D.A.; Zhao, J. High-yield production of extracellular type-I cellulose by the cyanobacterium Synechococcus sp. PCC 7002. Cell discov. 2015, 1, 15004.

Zhao, Y.; Lukiw, W. Microbiome-generated amyloid and potential impact on amyloidogenesis in Alzheimer's disease. J. Nat. Sci. 2015, 344 (6188), 1173–1178.

Zhou, K.; Eiben, S.; Wang, Q. Coassembly of tobacco mosaic virus coat proteins into nanotubes with uniform length and improved physical stability. ACS Appl. Mater. Interfaces 2016, 8, 13192–13196.

Zhou, L.; Zhao, G.; Feng, Y.; Yin, J.; Jiang, W. Toughening polylactide with polyether-block- amide and thermoplastic starch acetate: influence of starch esterification degree. Carbohydr. Polym. 2015, 127, 79–85. doi 10.1016/j.carbpol.2015.03.022.

Zia, K.M.; Noreen, A.; Zuber, M.; Tabasum, S.; Mujahid, M. Recent developments and future prospects on bio-based polyesters derived from renewable resources: a review. Int. J. Biol. Macromol. 2015, 82, 1028–1040. doi 10.1016/j.ijbiomac.2015.10.040.

CHAPTER 14

Biopolymer Extraction and Its Use in Edible Packaging

THALÍA A. SALINAS-JASSO[1], MARÍA L. FLORES-LÓPEZ[2], J. M. VIEIRA[3],
ANA V. CHARLES-RODRÍGUEZ[4], ARMANDO ROBLEDO-OLIVO[4],
OLGA B. ÁLVAREZ PÉREZ[5], ROMEO ROJAS MOLINA[6], and
MIGUEL A. DE LEÓN-ZAPATA[7*]

[1]Department of Chemical Metrology, Universidad Politécnica de Ramos Arizpe (UPRA), Ramos Arizpe, México

[2]Department of Research and Development, Biocampo S.A. de C.V., Blvd. Dr. Jesús Valdés Sánchez Km. 10, Fracc. Presa de las Casas, Arteaga 25350, México

[3]Department of Food Engineering, Faculty of Food Engineering, University of Campinas (UNICAMP), Campinas, Brazil

[4]Department of Food Science and Technology, Universidad Autónoma Agraría Antonio Narro, Colonia Buenavista, Saltillo, México

[5]Greencorp Biorganiks de México S.A. de C.V., Blvd Luis Donaldo Colosio, Col. San Patricio, Saltillo 25204, México

[6]School of Agronomy, Universidad Autónoma de Nuevo León, Monterrey, México

[7]Research Center for the Conservation of Biodiversity and Ecology of Coahuila (CICBEC), Autonomous University of Coahuila, Saltillo 25280, Mexico

*Corresponding author. E-mail: miguel.leon@uadec.edu.mx

ABSTRACT

This chapter is about methods to obtain biopolymers and their application in edible packaging for foods. Different processes to extract biopolymers and obtain edible coatings and films have been developed to increase the postharvest shelf life of food products and minimally processed. This chapter included a description of processes focused on extracting natural biopolymers, as polysaccharides, proteins, lipids. The properties (e.g., gases and humidity barrier ability) of edible coatings and films to reduce the deterioration of vegetable products are acutely examined. Also, this work reviews the parameters that alter the stability of edible packaging.

14.1 INTRODUCTION

The reduction in the use of nonrenewable synthetic materials has promoted research and development of packaging with lower environmental impact.

Even so, the packaging must provide the physical and mechanical properties necessary for food preservation. To achieve this purpose, the promising materials for the manufacture of packaging from renewable sources are biopolymers (Avérous and Poller, 2012). A fundamental characteristic of biopolymers is that its degradation occurs in short periods, from weeks to a few months (Shimao, 2001).

Edible packaging can encapsulate antioxidant (Cheng et al., 2015), antimicrobial (Arismendi et al., 2013), or nutrients (Vanderroost et al., 2014). Packaging must be effective and economically feasible.

Its characteristics depend on the food product application, which

might be coated. Vegetable products are perishable (Barbosa-Pereira et al., 2014), affecting the products quality, causing losses during storage due to various metabolic reactions that promote the growth of microorganisms (Bosquez-Molina et al., 2003). The biopolymers must present thermal stability, flexibility, a barrier to gases and water, should be nontoxic, and biodegradable (Ali et al., 2010).

In this chapter, the processes for biopolymers extraction and its use in edible packaging to improve the quality and prolong the postharvest shelf life of produce and minimally processed foods are described, as well as its properties, factors influencing the stability, and the elaboration methods of edible coatings and films.

14.2 BIOPOLYMERS-BASED EDIBLE FILMS AND COATINGS

Biopolymers have properties for film formation and can be used as cost-effective constituents in the food industry for encapsulation of natural bioactive compounds. Moreover, they exhibit useful functional properties, including gelation, emulsification, and water-holding properties. These properties enable them to exert specific characteristics, such as appropriate appearance and texture, which promote consumer acceptance.

14.2.1 BIOPOLYMERS

According to European Bioplastics (2018), the term "biopolymer" is synonymous with the term "bioplastic." The oil's resources scarcity and the gradual decline in synthetic plastics have allowed to strive toward environmental sustainability (Bilal and Iqbal, 2019). Renewable biomaterials are safer and green options to reduce waste generation and environmental pollution. The development of innovative products from biopolymers has promoted the growth and development of ecological and sustainable processes allowing a sustainable economy (Bilal and Iqbal, 2019). Biopolymers, like polysaccharides, proteins, and lipids, are an alternative to traditional plastics and to produce edible coating and films (Espitia et al., 2014). Globally, consumers demand high-quality natural products obtained through economic processes that do not affect the environment. Biotechnology companies and researchers have proposed to develop biodegradable technologies with food principles from biopolymers (Mahalik and Nambiar, 2010) as composite films and coatings (Galus et al., 2013; Kurek et al., 2014). Food packagings are biodegradable technologies from hydrophilic and hydrophobic biopolymers with moisture and gas barrier properties.

14.2.2 POLYSACCHARIDES

Polysaccharides are composed of monosaccharides and glycosidic bonds, in addition to insoluble fibers (lignin, galactomannans, cellulose, and xylans) and soluble fibers (arabinoxylans, pectins, and arabinogalactans) (Caprita et al., 2010). Polysaccharides (NSP) without starch are principally without α-glucan polysaccharid.

Table 14.1 describes the parameters of the polysaccharides involved in film-forming.

All these factors (Table 14.1) influence the main functions, such as thickening, gelling, film-forming, foaming, and emulsification, which contributes to obtaining high-quality products with applications in the pharmaceutical and food industry.

Polysaccharides are widely available in nature, are not toxic, and inedible films have selective permeability to gases (Erginkaya et al., 2014).

14.2.3 PROTEINS

Proteins cover a large amount of polymeric type compounds that fulfill specific functions within plants and animals as the contribution of biological activity in addition to providing structure. Proteins are biomolecules made up of carbon, nitrogen, hydrogen, and oxygen,

TABLE 14.1 Factors that Affect the Film-Forming Properties

Factors	Film-Forming Properties
Structural conformation	Polysaccharide ordered conformations depends on the sugar residues and the glycosidic linkages. The linear polysaccharides produce good film because assume conformations ranging from a twofold helix to a sixfold helix that determines how extended the polymer is and how much can be associated to form hydrogen bonds, which are responsible for gelling, film formation, or thickening
Molecular weight	High molecular weight polysaccharides form stronger films
Ionic charge on the molecule	Ionic groups allow polysaccharides to be more polar by providing greater capacity to form hydrogen bonds
Steric groups	OH groups of the polysaccharides are esterified by ether substitution, altering the hydrogen bonds, and influencing the ability of the polysaccharides to form films

Sources: Patra et al. (2012), Sudo (2011).

which can also contain sulfur, phosphorus, iron, magnesium, and copper, among other elements. They are made up of amino acids, alpha-amino and alpha carboxyl groups, and a side chain with different functional groups. Present multiple functional properties due to intermolecular binding capacity to different bond types can be classified based on the ability of proteins to interact with water molecules. Protein structure can be modified through heat, modifying pH, pressure, mechanical treatments, and among others (Chiralt et al., 2017). Biodegradable materials have been sought for inclusion in the food industry, mainly of plant and animal origin, for its biodegradability and film-forming (Calva-Estrada et al., 2019).

Proteins, in general, have various properties, including that they are edible and can carry nutrients. When combined with some other components, they become materials with great advantages due to the networks that manage to develop when undergoing modifications (Gnanasekaran, 2019). Not only can these biopolymers be used in the food area but also in different aspects such as in medicine and engineering to rebuild or reinforce. In the food packaging industry, proteins from milk, gelatin, gluten, egg, zein, soy, among other sources are the most used because they can be consumed directly on food; however, despite the multiple studies that exist in this regard to improve and find the best alternatives that are biodegradable.

14.2.3.1 PLANT PROTEINS

14.2.3.1.1 Soy Protein

Its origin is agricultural and has been known since the 1930s. The main feature it has is its emulsifying and texturing capacity. Its grains contain approximately 40% protein, among which two main fractions have been identified: albumins that are water-soluble and globulins that are soluble in saline solutions, representing 80% of total proteins. Remaining proteins are composed of intracellular enzymes, hemaglutinins, protein inhibitors, and membrane lipoproteins (Kinsella, 1979), and major components are classified according to their sedimentation properties. Due to biodegradability and a nutritional mixture of proteins, soy proteins promote film-forming. Several studies have been reported on its use in the food industry and recently as packaging materials.

14.2.3.1.2 Wheat Proteins

Wheat proteins are classified for their solubility and functionality. These include albumins, globulins, gliadins, and glutenins (De la Vega, 2009). They can also be found classified as proteins belonging to or without gluten. Gluten proteins account for 80%–85% of total wheat proteins. Gluten proteins can be used for the manufacture of transparent and homogeneous packaging with mechanical properties of interest.

14.2.3.1.3 Corn Zein

Zein is a natural protein generated by wet milling as a bioproduct. It is a protein derived from corn endosperm and is a biodegradable component. Zein is formed by polypeptides such as leucine, proline, alanine, serine and glutamine, and Υ-zine that has a high cysteine content. Jornet-Martínez et al. (2016) mentioned the amphipathic structure that is created and combines a helical structure with the polyproline, this results in a hydrophobic material, making it an alternative for food packaging.

14.2.3.2 ANIMAL PROTEINS

14.2.3.2.1 Whey Protein

The FDA reports that milk proteins must have all the proteins found naturally in milk and that they must be in the same relationships, while in the whey protein, isolates and concentrate are obtained by the elimination of nonprotein components and are casein free. These milk proteins have properties that can provide desirable textures and other attributes to the final product. They have multiple applications in traditional food products. Various types of milk protein, such as whey protein concentrates, whey protein isolates, caseins and caseinates, and among others, can be obtained from waste generated by industrial complexes. Whey contains β-Lactoglobulins that are obtained as a residue from the cheese and casein industry (Fox et al., 2015), which provide the properties of solubility, thermal stability, emulsifying, and nutritional value (Walstra et al., 2005). Protein interactions that occur between chains determine the network formation of films and their properties.

14.2.3.2.2 Meat Proteins

They can be classified into three types: sarcoplasmic, stromal, and myofibrillar. Collagen is classified as a fibrous protein and is present in the skin, tendons, bones, and vascular and connective tissue of animals. Due to the presence of covalent bonds, it is said to form intermolecular bridges inside, and disulfide bonds are few due to the low amount of cysteine present. It consists of three parallel chains of the alpha type, which combine to give rise to a triple-stranded superhelical structure (Montalvo et al. 2012). Amino acid sequence is formed by a repeating chain of glycine-proline-hydroxyproline and has been used as packaging for sausages. Gelatin is generated when collagen is subjected to hydrolysis under acidic or alkaline conditions.

Gelatin has large amounts of proline, hydroxyproline, lysine, and hydroxylysine. It is used primarily as a texturing agent. Packaging from this component is usually clear and flexible but very hydrophilic. Moreover, in short, myofibrillar proteins are found in mammals and fish; they have a secondary structure, generating packaging with properties of permeability when subjected to previous denaturation processes (Dangaran et al., 2009).

14.2.4 LIPIDS

Lipids are a set of materials used in the manufacture of food packaging due to their hydrophobic properties. Within the components called hydrophobic substances, we find waxes, oils, and resins. Natural waxes such as candelilla wax, carnauba, bees, and among others are used in the industry. Oil-based waxes such as paraffin, petroleum-derived oils and vegetables, and resins are also found.

14.2.4.1 WAXES

In recent years, waxes have been used to give food, especially those of immediate consumption such as fruits and vegetables, a better look compared to the consumer. Waxes are known to several nonpolar components that can be synthetic or natural. Chemically they are esters of fatty acids and long-chain fatty acids and are characterized by a high hydrocarbon content (approximately 50%) and a relatively low amount of volatile esters. Due to their nonpolar components, they form materials against humidity (Aguirre-Joya et al., 2017).

14.2.4.2 ANIMAL AND INSECT WAXES

They are as the name indicates the waxes that come from the segregation of certain insects or animals. Within this classification, they are distinguished into two subgroups: wax that comes from terrestrial and marine animals. From terrestrial animals, we find the lanolin that comes from wool and marine animals the Spermaceti, which is no longer marketed. However, the wax of greater industrial importance is beeswax. This compound is the final product of bee metabolism and is segregated by worker bee glands (Rhim and Shellhammer, 2005)

14.2.4.3 VEGETABLE WAXES

Vegetable waxes are the result of the climatic conditions in the regions where many plants are found. Generally, plants store waxes in the epidermis of their leaves as protection against water evaporation, especially in drought season. They are classified into waxes of trees and shrubs and in turn from the section where the wax is found as leaves, stems, root, and among others. The most used commercial waxes are the carnaúba wax for the production of preservation agents, cleaning, emulsions, and among others. Another, the widely used waxes for the manufacture of cosmetics, foods, and pharmaceuticals is candelilla wax, which is generated in semidesert plants and is found in stems and leaves (Rhim and Shellhammer, 2005).

14.2.4.4 FATS

Fats are polar and neutral lipids insoluble in water with the ability to form a stable hydrophobic layer on a surface. They are used as emulsifiers and dispersing agents forming interfacial micelles. The classification of this group of lipids is given by two fractions: either by the degree of saturation (saturated and unsaturated) or by the length of the chain (short, medium, and long) (Isabel Castro-González, 2002). Saturated fats have a higher melting point and greater water permeability than unsaturated fats.

14.2.4.5 RESINS

Resin is a viscous substance from vegetable (hydroaromatic structure) or of synthetic origin (higher degree of polymerization) with mechanical, emulsifying, and adhesive properties.

14.3 BIO-BASED POLYMERS EXTRACTION METHODS

Currently, there are several conventional and innovative methods for biopolymers extraction, where the yield depends mainly on geographical conditions of the raw material, physical, and chemical parameters, used solvent, and its polarity. Therefore, the different methods of extracting the main biopolymers as proteins, polysaccharides, and lipids are described below.

14.3.1 POLYSACCHARIDES

Natural polysaccharides such as mucilage, starch, and pectin, present selective permeability to gases (Ergincaya et al., 2014). The extraction of the polysaccharides is based on methods that use solvents, water, salts, and acids (Table 14.2).

14.3.2 PROTEINS

Proteins and peptides are obtained during the technological process, and many extraction methods involving enzymatic hydrolysis, and among others. Protein extraction methods are shown in Table 14.3.

14.3.3 LIPIDS

Lipids such as vegetable oils and waxes are a very heterogeneous family of compounds that minimize moisture loss, provide gloss, and flexibility, in addition to reduce complexity, and cost of films and edible coatings. They present hydrophobicity as a common characteristic and dissolve in organic solvents. In order to carry out the extraction of the most used lipids in films and edible coatings for its application in foods, several methods have been proposed (Tables 14.4–14.9).

14.3.3.1 VEGETABLE OILS

Vegetable oils as sunflower oil, jojoba oil, palm oil, coconut oil, and cocoa butter are mainly composed of triglycerides as palmitic acid, stearic acid, oleic acid, and linolenic acid. They are used in films and edible coatings with the aim of decrease intramolecular forces.

14.3.3.2 SUNFLOWER OIL

Sunflower oil is obtained from the sunflower seed and contains a high concentration of vitamin E and low levels of saturated fat. The pressing method is commonly used to extract sunflower oil. New methods have been developed for the extraction

TABLE 14.2 Main Methods of Polysaccharides Extraction

Polysaccharide	Extraction Method	Parameters (Weight, Volume, Pretreatment, Solvent, pH, Time, Temperature)	Yield (%)	Reference
Starch	Extraction with water	Potato (kg) = 1 Pretreatment process = sliced and crushed The solid residue was washed several times with distilled water to remove all the starch, was filtered and sedimented at the bottom, and finally drying the resulting paste on trays at room temperature	8.6–19.2	Zárate-Polanco et al. (2014)
Starch	Extraction with water	Banana flour (g) = 100 Pretreatment process = shelled, sliced, frozen, lyophilized and milled Solvent = distilled water	56.5	Bello-Lara et al. (2014)
Pectin	Extraction with salts	Banana flour (g) = 100 Pretreatment process = shelled, sliced, frozen, lyophilized and milled Salt = ammonium oxalate (0.05%%) pH = 5.6 in a water bath and constant agitation	9.7	Bello-Lara et al. (2014)
Mucilage	Extraction with solvents	Opuntia ficus-indica (g) = 100 Pretreatment process = washed and milled Centrifuged for the aqueous phase = 5000 rpm for 30 min. Solvent ratio (aqueous phase: solvent (acetone)) = 1: 3	0.8	Abrajān-Villaseñor (2008)
Mucilage	Extraction with solvents	Opuntia ficus-indica (g) = 100 Pretreatment process = washing, grinding and blanching Centrifuged for the aqueous phase = 5000 rpm for 30 min Solvent ratio (aqueous phase: solvent (acetone)) = 1: 3	0.6	Abrajān-Villaseñor (2008)
Mucilage	Extraction with solvents	Opuntia ficus-indica (g) = 100 Pretreatment process = washed, ground and cooked Centrifuged for the aqueous phase = 5000 rpm for 30 min Solvent ratio (aqueous phase: solvent (acetone)) = 1: 3 Washing the precipitate with ethanol and dried under vacuum	0.3	Abrajān-Villaseñor (2008)
Pectin	Extraction with acids and solvents	Pretreatment process = chopped, blanched, filtered, solid washed with ethanol, dried and weighed Dry and ground loquat (g) = 25 Acid = HCl 0.003 N T = 90 °C t = 75 min 98% ethanol was added to the pectic solution to precipitate the pectin, leaving it at rest for 1 h The floating pectin was filtered and washed with ethanol	21–23	Chasquibol-Silva et al. (2008)

TABLE 14.3   Protein Extraction Methods

Extraction Method	Food	Reference
Cell disruption methods		
Homogenization		
Milling and homogenization	Rice	Toldra and Nollet (2013)
	Olive tree seeds	
	Green Alga	Wang et al. (2015)
Ultrasonic homogenization	Peanut	Wang (2017)
	Milk	Ashokkumar et al. (2010)
Pressure homogenization	Milk	(Doona and Feeherry (2008)
	Microalgae	Mulchandani, Kar and Singhal (2015)
Temperature treatments		
	Whey proteins	Schmid and Muller (2018)
Enzymatic treatments	Soybean	Ndlela et al. (2012)
		De Moura et al. (2011a)
	Algae	Wang et al. (2015)
	Lentelis and White beans	Bildstein et al. (2008)
Osmotic and chemical lysis	Spiruline	Hadiyanto and Adetya (2018)
	Ginseng roots	Jiang et al. (2014)
Protein Solubilization/ Precipitation	Bovine serum	McArt et al. (2006)
	Soybean	De Moura et al. (2011b)
Aqueous solutions	Almond	Ge et al. (2016)

of sunflower oil, the most used is the extraction with solvents such as hexane, isopropyl alcohol, and petroleum ether (Table 14.4), where the hexane shows high yield at the 12 h of extraction. However, there are many environmental limitations regarding the use of solvents. It is also highlighting the use of solvents as the hexane and ultrasound irradiation, with the aim of decrease the extraction time in the process (90 min) in comparison with solvents extraction, obtaining high yields similar to those obtained in hexane extraction (Table 14.4).

14.3.3.3   JOJOBA OIL

Jojoba oil is obtained from seeds of the jojoba *Simmondsia chinensis.* Jojoba oil is of light color and contains a mixture of triglycerides

TABLE 14.4 Methods to Extract Sunflower Oil

Extraction Method	Parameters (w:v, Solvent, Temperature, Time)	Yield (%)	Reference
Extraction with solvents	Seed = 5 g Hexane = 200 mL $T = 100\ °C$ $t = 8\ h$	37.9	Fornasari et al. (2017)
Extraction with solvents	Seed = 5 g Petroleum ether = 200 mL $T = 100\ °C$ $t = 8\ h$	36.8	Fornasari et al. (2017)
Extraction with solvents	Seed = 50 g Hexane = 300 mL $T = 50\ °C$ $t = 24\ h$	54.3	Rai et al. (2015, 2016)
Extraction with solvents	Seed = 10 g Hexane = 200 mL T = Normal boiling point $t = 4\ h$	46.2	Ravber et al. (2015)
Extraction with solvents	Seed = 10 g Particle size = 2.0 mm Hexane = 100 mL $t = 12\ h$ $T = 85\ °C$	99.0	Luque-García and Luque de Castro (2004)
Extraction with solvents	Seed = 50 g Hexane = 189 mL $T = 69.5\ °C$	44.4	Gallegos-Infante et al. (2003)
Extraction with solvents	Seed = 50 g Isopropyl alcohol = 159 mL $T = 82\ °C$	43.4	Gallegos-Infante et al. (2003)
Ultrasound-assisted Soxhlet	Seed = 10 g Particle size = 2.0 mm Hexane = 100 mL $T = 85\ °C$ Ultrasound irradiation of the cartridge for 30 cycles (90 min, output amplitude 40% of the nominal amplitude of the converter, applied power 100 W)	99.0	Luque-García and Luque de Castro (2004)

and long-chain esters of unsaturated fatty acids. Commonly, jojoba oil is obtained by pressing the seeds obtaining yields between 30% and 43% (Table 14.5), depending on the number of pressures that are applied during the process. The use of solvents such as hexane, chloroform, and petroleum ether, allows us to obtain yields of 52%–94.2%, 32.5%, and 92.2%, respectively, thus hexane being the most used solvent. The most recent method of extraction is a supercritical fluid, which uses a mixture of solvents as CO_2 + propane, CO_2 + ethanol, and hexane, obtaining yields of 98%, 80%, and 94%, respectively, and showing the highest yields (Table 14.5). CO_2 is also used; however, this method shows lower yields (44%–50.6%) in comparison with the disolvents mixture (Table 14.5). Therefore, the yields obtained by the supercritical fluid method depend on the polarity of the solvent, as well as the temperature and pressure applied in the process.

14.3.3.4   PALM OIL

Palm oil is rich in triglycerides and is extracted from the palm Elaeis guineensis Jacq. Yield depends on the variety of the palm (Table 14.6). Commonly palm oil is extracted by manual pressing with yields of 33%–35% of palmitic acid and 42.2%–43.3% of oleic acid (Table 14.6), which are lower than those obtained by hydraulic and mechanical pressure (Table 14.6). Where the mechanical pressure of the seed pretreated with preheating, cracking, and flaking turned out to be the process with the highest oil yield (Table 14.6) compared to the hydraulic pressure process of the seed pretreated with preheating, cracking, and flaking (Table 14.6).

14.3.3.5   COCONUT OIL

Coconut oil is obtained from *C. nucifera*. Also, this oil does not undergo degradation at high temperatures. Coconut oil is commonly obtained by heating, where the coconut undergoes a process of cracking, flaking, and scratching of the pulp, to subsequently boil the pulp in water, and separate by density, with this process the highest oil yield is obtained (Table 14.7). The solvent extraction method, mainly with hexane, has the lowest oil yield (Table 14.7). Another innovative method for coconut oil extraction is the method of extraction with saline solution (sodium chloride salt) using temperature, which proved to obtain a yield of 60% (Table 14.7).

14.3.3.6   COCOA BUTTER

Cocoa bean (*Theobroma cacao*) consists mainly of cocoa butter,

TABLE 14.5   Main Methods of Jojoba Oil Extraction

Extraction Method	Parameters (Pressure, Solvent, Time, Number of Pressing, Temperature, Particle Size and Pretreatment Process)	Yield (%)	Reference
Supercritical fluid	Solvent = 30%CO_2 + propane P (bar) = 70 T (K) = 313	98	Palla et al. (2014)
Supercritical fluid	Solvent = 30%CO_2 + ethanol P (MPa) = 35 y 45 T (K) = 363	80	Salgin (2007)
Supercritical fluid	Solvent = ScCO$_2$ P (bar) = 450 T (K) = 343	44	Salgin (2007)
Supercritical fluid	Solvent = ScCO$_2$ P (bar) = 600 T (K) = 363	50.6	Salgin et al. (2004)
Extraction with solvents	Particle size (mm) = 0.48 mm Solvent = Hexane	55	Allawzi et al. (2005)
Extraction with solvents	Particle size (mm) ≤ 1 mm Solvent = Chloroform t = 18 h	32.5	Salgin et al. (2004)
Extraction with solvents	Relation Disolvent/solid (L/Kg) = 15 Solvent = Hexane t = 30 min T = 55 °C	94.2	Zaher et al. (2004)
Extraction with solvents	Relation disolvent/solid (L/kg) = 15 Solvent = Petroleum ether t = 30 min T = 55 °C	92.2	Zaher et al. (2004)
Extraction with solvents	Particle size (mm) ≤ 1 mm Solvent = Hexane t = 18 h	52	Abu-Arabi et al. (2000)
Traditional	Type of press = Hydraulic Number of pressing = 1 Pretreatment process = none	35.4	Abu-Arabi et al. (2000)
Traditional	Type of press = Hydraulic Number of pressing = 2 Pretreatment process = Breaking	43.8	Abu-Arabi et al. (2000)
Traditional	Type of press = Rosedowns Number of pressing = 1 Pretreatment process = Preheating	38.2	Rawles, (1978)
Traditional	Type of press = Hander Number of pressing = 1 Pretreatment process = Preheating	35–39	Rawles, (1978)

TABLE 14.5   *(Continued)*

Extraction Method	Parameters (Pressure, Solvent, Time, Number of Pressing, Temperature, Particle Size and Pretreatment Process)	Yield (%)	Reference
Traditional	Type of press = Hander Number of pressing = 2 Pretreatment process = Preheating	40–42	Rawles, (1978)
Traditional	Type of press = Hander Number of pressing = 3 Pretreatment process = Preheating	43	Rawles, (1978)
Traditional	Type of press = Hydraulic Number of pressing = 1 Pretreatment = Cracking and flaking	30.8	Spadaro and Lambou (1972)

TABLE 14.6   Methods to Extract Palm Oil

Extraction Method	Parameters (Variety of Plant Species, Type of Press, Pretreatment)	Yield (%)	Reference
Traditional	Variety of plant species = *D. x Ekona* Type of press = Manual Number of pressing = 1 Pretreatment process = none	Palmitic acid = 33.1 Oleic acid = 43.3	Sandoval-García et al. (2016)
Traditional	Variety of plant species = *D. x Ghana* Type of press = Manual Number of pressing = 1 Pretreatment process = none	Palmitic acid = 35.4 Oleic acid = 42.2	Sandoval-García et al. (2016)
Traditional	Variety of plant species = *D. x Nigeria* Type of press = Manual Number of pressing = 1 Pretreatment process = none	Palmitic acid = 34.2 Oleic acid = 43.3	Sandoval-García et al. (2016)
Pressing	Variety of plant species = *E. guineensis* Jacq. Type of press = Mechanic Number of pressing = 1 Pretreatment process = Preheating, cracking, flaking	97 y 99	Jaimes-M. et al. (2012)
Pressing	Variety of plant species = *Acrocomia aculeata* Type of press = Hydraulic Number of pressing = 1 Pretreatment = Cracking and flaking	Lauric acid = 51.0	Hernández and Mieres-Pitre (2005)

TABLE 14.7 Main Methods of Coconut Oil Extraction

Extraction Method	Extraction Conditions (w: v, Pretreatment Process, Temperature, Time, Solvent)	Yield (%)	Reference
Traditional	Coconut pulp = 1350 g Pretreatment process = cracking, flaking, striped coconut $T = 98\ °C$ $t = 2$ h	222	Da Silva-Rodríguez (2017)
Extraction with solvents	Relation disolvent/solid (mL/g) = 300/80 Solvent = Hexane $t = 1.5$–2 h $T = 62\ °C$	57–64	Rivera-Hernández et al. (2001)
Extraction with saline solution	Relation water/solid (mL/g) = 1000/75 sodium chloride salt (g) = 90 $t = 1$ h $T = 60\ °C$	60	Ramos-Ramírez and Salazar-Montoya (1995)

which is considered as unique, due to its chemical composition. Cocoa beans pretreated by fermentation, drying, and toasted give yields of 50% (Table 14.8). New methods have been implemented for the extraction of cocoa butter, as the supercritical method, that used pressure (35 MPa), temperature (60 °C), and mixture of solvents as CO_2 + 25% ethanol, CO_2 + 25% isopropanol, and CO_2 + 15% acetone obtaining yields of 100%, 96.7%, and 84%, respectively (Table 14.8). Other methods with high yield consist of the extraction with solvents as the *t*-butanol mixed with a solution of ammonium sulfate, where the system adjusts to pH 2, and it is heated at 45 °C for 2 h (Table 14.8). Therefore, there is a demand for green technologies as the ultrasound-assisted supercritical method and supercritical method, using CO_2 as an extraction solvent, obtaining yields of 37.5% and 30.6%, respectively (Table 14.8).

14.3.4 WAXES

Waxes refer to a wide variety of substances of vegetable and animal origin because they comprise a mixture of several long-chain fatty acids and other components. Consumers prefer natural and sustainable

TABLE 14.8   Processes to Extract Cocoa Butter

Extraction Method	Extraction Conditions (w:v, Solvent, Pressure, Temperature, Time, Type of Press, Number of Pressing, Pretreatment Process)	Yield (%)	Reference
Supercritical	Cocoa beans (g) = 100 Solvent = CO_2 P (bar) = 550 T (°C) = 40 t = 360 min	30.6	Rodríguez et al. (2014)
Supercritical	Cocoa beans (g) = 10 Solvent = CO_2 + 25% ethanol P (MPa) = 35 T (°C) = 60 Flow rate = 2 mL/min	100	Asep et al. (2013)
Supercritical	Cocoa beans (g) = 10 Solvent = CO_2 + 25% isopropanol P (Mpa) = 35 T (°C) = 60 Flow rate = 2 mL/min	96.7	Asep et al. (2013)
Supercritical	Cocoa beans (g) = 10 Solvent = CO_2 + 15% acetone P (MPa) = 35 T (°C) = 60 Flow rate = 2 mL/min	84	Asep et al. (2013)
Supercritical	Cocoa beans (g) = 30 Solvent = ethane P (MPa) = 28.3 T (K) = 343.2	100	Saldaña et al. (2002)
Ultrasound-assisted supercritical	Cocoa beans (g) = 100 Solvent = CO_2 P (bar) = 550 T (°C) = 40 t = 360 min Ultrasound irradiation (resonance frequency of 30 kHz and constant energy of 58 W)	37.5	Rodríguez et al. (2014)
Extraction with solvents	kokum kernel powder (g) = 1 Three phase partitioning system = distilled water (16 mL), ammonium sulphate (50% w/v) and t-butanol (0.5:1–3:1) pH = 2 t = 2 h T = 45 °C	95	Vidhate and Singhal (2013)
Traditional	Type of press = Mechanic Number of pressing = 1 Pretreatment process = Fermentation, drying, toasted	50	Codini et al. (2004)

products, which increased the demand for natural waxes global (Attard et al., 2018), mainly the carnauba wax, candelilla wax, and beeswax.

14.3.4.1 CARNAUBA WAX

Carnauba wax is obtained the dried and crushed the palm leaves of *Copernicia prunifera* (Dantas et al., 2013). There are many environmental limitations regarding the use of dangerous solvents. There have been many efforts in order to find cleaner methods, one of these methods is the use of supercritical fluids (Table 14.9), such as CO_2 as a solvent which is inexpensive and completely inert (Palla et al., 2014). Also, obtaining a performance of carnauba wax superior (97%) to that obtained with the use of solvents (3.4%), due to the use of high pressures.

14.3.4.2 CANDELILLA WAX

Candelilla wax is obtained from Euphorbia antisyphilitica Zucc. The common name of the plant "candelilla" comes from the particular form of the stems of the bush, which are long, straight, erect, covered, with wax, and with the appearance of small candles (De León-Zapata, 2008; Rojas-Molina et al., 2013). Traditionally, candelilla wax is

extracted with concentrated sulfuric acid to separate the wax in the form of foam (Table 14.9). Candelilla wax is also obtained by solvent extraction as aliphatic and aromatic hydrocarbons, but it is flammable and not renewable. Hence, an efficient, economical, and eco-friendly method is required as the extraction method with citric acid in combination with temperature and presion obtaining yields of 4–6% of wax (Table 14.9). Another method is the accelerated drying and scraping of candelilla stems obtaining a yield of 4% of wax (Table 14.9). However, it takes a lot of work and time since it is necessary to scrape stem by stem.

14.3.4.3 BEESWAX

The physicochemical properties of the wax depend on the bee species (Reybroeck et al., 2010). Commonly, the wax is extracted by methods that use heat to melt the wax (Table 14.9). In order to produce high-quality wax, it is recommended not to heat at too high temperature and for too long time, because that may damage darken its color. Wax should be heated in containers made of stainless steel. Combs containing fermented honey should not be melted in order to prevent wax off odor, and water with a low mineral content should be used to avoid the formation of water-wax emulsions (Bogdanov, 2004).

TABLE 14.9   Main Methods of Waxes Extraction

Wax	Extraction Method	Extraction Conditions (w:v, Solvent, Pressure, Temperature, Time, Flow Rate)	Yield	Reference
Carnauba wax	Supercritical fluid	Milled date palm leaves (g) = 100 Solvent = CO_2 P (bar) = 400 T (°C) = 100 Flow rate = 40 g/min t = 2 h	97	Al Bulushi et al. (2018)
Carnauba wax	Extraction with solvents	Milled date palm leaves (g) = 10 Solvent = Heptane t = 5 h	3.4	Al Bulushi et al. (2018)
Carnauba wax	Dried and crushed	The leaves are collected, dried and crushed to open the plant tissue and then beaten to separate the wax as a powder	–	Morales-Hernandez (2015)
Candelilla wax	Organic acids	Candelilla stems (kg) = 1 Acid = 0.05% citric acid P (kg/cm²) = 1.05 t = 5 min T = 100 °C	4–6	De León-Zapata et al. (2016)
Candelilla wax	Accelerated drying and scraping	Candelilla stems (kg) = 0.5 Drying = for 48 h at 40 °C	4	Ahumada-Lazo (2012)
Candelilla wax	Traditional	Candelilla stems (kg) = 1 Acid = concentrated sulfuric acid t = 30 min T = 90–100 °C	4	De León-Zapata (2008)
Candelilla wax	Extraction with solvents	Solvent = aliphatic and aromatic hydrocarbons or a mixture of both t = 50 min T = 100 °C	–	Taboada-Reyes (1992)
Candelilla wax	Pressing	Type of press = Mechanic Number of pressing = 1 Pretreatment process = None	–	Treviño-García (1929)
Beeswax	Traditional	Extraction with hot water, steam, heat from electrical or solar power, to melt the wax	–	Bogdanov (2004)

14.4 EDIBLE PACKAGING FOR FOOD APPLICATION

The challenge of the agrifood industry is to maintain the quality and organoleptic properties of the fresh vegetable products. In this context, edible coatings have been incorporated in food processing, because they protect them from water loss during the transpiration in postharvest. Typically, the major constituents are biopolymers, mainly polysaccharides (e.g., chitosan), proteins (e.g., whey protein), and lipids (e.g., beeswax) that can be extracted from products and by-products of the agrifood industry. This section discusses the properties, qualities, and effect on shelf life extension that different materials used in the formation of edible coatings.

14.4.1 POLYSACCHARIDES, PROTEINS, AND LIPIDS FOR VEGETABLE PRODUCTS IN POSTHARVEST

Hydrocolloids (protein and polysaccharides), lipids (fatty acids, waxes, oils, and resins), and composites (interaction between hydrocolloids and lipids) (Dhall, 2013; Valencia-Chamorro et al., 2011) must be selected according to the ripening profile and the surface characteristics of the fruits and vegetables to be coated (Flores-López et al., 2016; Yousuf et al., 2018).

14.4.2 HYDROCOLLOIDS

Whey protein (Schmid et al., 2017), chia protein (Capitani et al., 2016), gelatin (Ahmad et al., 2012), soy protein (Yousuf and Srivastava, 2019), corn-zein (Boyacı et al., 2019), and among others are widely used for food packaging. Protein-based materials form coatings/films (Hassan et al., 2018) pliable, and translucent (Yousuf et al., 2018), and also present mechanical properties due to the possibility of forming different types of linkages; although their hydrophilic nature results in high water vapor permeability (Feng et al., 2018; Flores-López et al., 2016).

Polysaccharides have a selective permeability to gases (O_2 and CO_2); however, their hydrophilicity also influences their water vapor permeability. The most commonly used polysaccharides are alginate (Valero et al., 2013), chitosan (Vieira et al., 2016), pectin and cellulose (Moalemiyan et al., 2012; Pastor et al., 2010), starch (García et al., 2012), carrageenan (Hamzah et al., 2013) and, so on. The research into novel natural sources of polysaccharides for the build of edible coatings/films with improved properties has received world attention. Cerqueira et al. (2009) designed edible coatings based on galactomannans from *Adenanthera pavonina* and *Caesalpinia pulcherrima* seeds, to be applied in tropical fruits. Also,

Dick et al. (2015) investigated the effect of glycerol on the physico-chemical and mechanical properties of chia mucilage-based film. The films were found to have a uniform and transparency appearance as the glycerol concentration increased. The use of residues from the agri-food industry has also allowed us to obtain polysaccharides with prop-erties to form bio-packaging. For instance, Torres-León et al. (2018) developed a new edible film based on mango (var. Ataulfo) by-prod-ucts to extend the shelf life of peach. This improved surface prop-erties and reduced gas transfer rates of the fruit.

14.4.3 LIPIDS

Lipids repel water due to its hydro-phobic property (Hassan et al., 2018), in addition, its low gas permeability and its protective capacity in refrig-eration conditions allow them to be an excellent alternative for noncli-macteric fruits (Flores-López et al., 2016). Within this group, the appli-cation of waxes and oils (synthetic and natural) has been a recurring activity since ancient times (Dhall, 2013). Paraffin wax and beeswax are the most effective lipid materials but alternative naturals waxes have become more acceptable in recent years, as in the case of candelilla wax. Candelilla wax based coating formulations have been shown to

prevent senescence of strawberries (Oregel-Zamudio et al., 2017) and avocados (Saucedo-Pompa et al., 2009). However, lipid-based coat-ings are characterized by improperly adhering to the surface, promote anaerobiosis, and altering the appear-ance and taste of the product to be coated (Hassan et al., 2018; Flores-López et al., 2016; Perez-Gago et al., 2002).

14.4.4 COMPOSITES

Composites are defined as a blend of hydrocolloids (i.e., polysaccharides or proteins) and lipids (Dhall, 2013), in order to improve their character-istics (e.g., mechanical properties or permeability) while minimizing their drawbacks (Tharanathan, 2003; Valencia-Chamorro et al., 2009; Chiumarelli and Hubinger, 2014; Oliveira et al., 2018). Also, the use of composites can reduce the costs of the final coating/film.

14.4.5 USE OF NATURAL ADDITIVES IN EDIBLE PACKAGING

Fresh vegetable products coated with edible films are a reality (Hassan et al., 2018; Yousuf et al., 2018; Zhao, 2019) forming a barrier against microbial attack and growth, and gas exchange control (the main posthar-vest problems) (Ortega-Toro et al.,

2017; Ncama et al., 2018). Promoting the retention of nutritional quality (Figure 14.1). Also, these systems are capable of carrying natural bioactive compounds, for example, antimicrobial, antioxidant, nutrients, and flavorings, from various sources within their matrix (Hassan et al., 2018), limiting the use of synthetic chemicals due to their possible agrotoxicological effects on environment and consumers (Ponce et al., 2008; Vieira et al., 2016). Minimally processed vegetable products are extremely perishable, being more susceptible to the physical, enzymatic, microbiological, and consequently at organoleptic level (Yousuf and Srivastava, 2019; Thakur et al., 2019). The use of edible coatings/films becomes an indispensable alternative in their preservation and this

reality is triggered by the consumers' demand for durable, safe, and stable food without compromising the environment through nonbiodegradable packaging. Table 14.10 shows some recently developed works, with examples of base compositions with and without incorporation of bioactive/antimicrobial agents and their effects when applied to a determined fresh or minimally processed fruit or vegetable.

14.5 PRODUCTION OF EDIBLE COATINGS AND FILMS

Edible packaging can be edible films or coatings, edible coatings are liquid produced by wet methods, and films are solid laminates produced by dry methods.

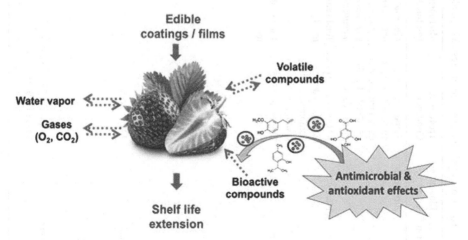

FIGURE 14.1 Edible coatings/films as gas and water vapor barrier and vehicle of natural additives to protect fresh vegetable products.

TABLE 14.10 Coatings/Films Applied in Fresh Vegetable Products

Base Composition	Bioactive Agent	Treatment Conditions	Microorganism Inhibited	Fruit/Vegetable Fresh	Fruit/Vegetable Fresh-cut	Shelf-life Improvement	References
Starch	Aloe vera	Storage at 10 °C/85% RH for 7 days, then stored at 25 °C/85% RH for more 7 days	Fusarium oxysporum; Bipolaris spicifera; Curvularia hawaiiensis	Cherry tomatoes		Weight loss reduction, fungal inhibition, and good appearance promotion	Ortega-Toro et al. (2017)
Isolated soy protein	Honey	4 °C/16 days. Honey was applied in combination with coating	Total yeast and mold		Pineapple	Physicochemical and microbiological quality maintenance, extending shelf life up to two weeks	Yousuf and Srivastava (2019)
Shellac and gelatin	–	Storage at 25 °C for 30 days	Total yeast and mold	Banana		Physicochemical and microbiological quality maintenance, ripening process retarding, weight loss prevention, and good texture promotion	Soradech et al. (2017)
Starch and glucose	–	Storage at 4 °C for 30 days	–	Cucumber		Ripening process retarding, weight loss prevention, partial barrier to O_2 and CO_2, and WVP control. Also, antioxidant activity, proline, and soluble sugar content increased, while catalase and peroxidase activities and protein content decreased	Patel and Panigrahi (2019)
Chitosan and guar gum	–	4 °C/16 days	–	Shiitake mushroom (Lentinus edodes)		Higher tissue firmness, rate of declines in soluble protein and ascorbic acid slowed, increased the total soluble solids and reducing sugars, malondialdehyde and electrolyte leakage. Sensory evaluation confirmed the positive effects of coating	Huang et al. (2019)
Beetroot, corn, and pectin	–	Storage at 25 °C/80–85% RH for 30 days	–	Tomato		Prevents dehydration and senescence. Improves firmness and appearance. Retains the antioxidant content	Sucheta et al. (2019)

TABLE 14.10 *(Continued)*

Base Composition	Bioactive Agent	Treatment Conditions	Microorganism Inhibited	Fruit/Vegetable		Shelf-life Improvement	References
				Fresh	Fresh-cut		
Chitosan	*Aloe vera*	Storage at 5 °C/90% RH for 25 days	*Botrytis cinerea*	Blueberry		Weight loss prevention, additional barrier to reduce fungal contamination and good appearance promotion. The use of coating extended the shelf life for about 5 days	Vieira et al. (2016)
Cactus (*Opuntia dillenii*) polysaccharide	—	Storage at 5 °C for 5 days	Total viable counts		Potato	Browning suppression. Weight loss and microbial growth prevention. Respiration rate and total sugars control	Wu (2019)
Cassava starch and beeswax	—	Storage at 15 °C/90% RH for 15 days	—	Guavas		High WVP reduction. Delay in mass loss and chlorophyll. Physicochemical and organoleptic characterization indicated a quality maintenance and an extended shelf life	Oliveira et al. (2018)
Gellan	Geraniol and pomegranate extract	Storage at 5 °C for 7 days	Mesophilic bacteria, yeast and molds, and psychrophilic bacteria		Strawberry	Geraniol into gellan-based coatings improved the microbiological stability. Gellan-based coatings were found to be good vehicle of natural antimicrobials compatible with organic fresh food	Tomadoni et al. (2018)
Whey protein isolate nanofibrils	Trehalose	Storage at 4 °C for 10 days	—		Apple	Colorless transparent film on the apple surface. It delayed the browning process, the weight loss, and inhibited the total phenolics decreasing. Promotion of greater antioxidant activity and reduction of respiration rate, resulting in an increase of the shelf life.	Feng et al. (2018)

14.5.1 DRY PROCESSING

Biopolymers extracted from biomass, such as proteins, that behaves as thermoplastic materials which they are excellent for processing packaging by dry processing (Table 14.11).

Sustainable use of natural resources for the development of edible coatings and films has taken important global. The most used biopolymers for its elaboration are pectin (Lei et al., 2013; Younis and Zhao, 2019), starch (Galindez et al., 2019; Yildirim-Yalcin et al., 2019), cellulose, zein (Spasojevic et al., 2019), carboxymethyl cellulose (Ruan et al., 2019), methylcellulose (Matta et al., 2019), chitosan (Younis

and Zhao, 2019), agar (Wang et al., 2018), alginate (Salama et al., 2018; Fabra et al., 2018), konjac (Lei et al., 2019), carragenans (Tavassoli-Kafrani et al., 2015), gelatin (Dou et al., 2018), Cassia gum (Cao et al., 2018), mucilage (Gheribi et al., 2018), maltodextrin (Zhang et al., 2019), egg yolk (Fuertes et al., 2017), Tara gum (Ma et al., 2016), and many others. However, the methodology for its preparation is variable from the use of different temperatures, times, stirring, and concentration is according to each polymer used (Table 14.12).

Edible film production consists of solubilizing the base polymer and adding some plasticizer (glycerol)

TABLE 14.11 Main Methods for Obtaining Edible Films by Dry Processing

Methods	Process	Finality	Reference
Thermoforming	The biopolymer is heated and transformed	Elaboration of containers	Hernandez-Izquierdo and Krochta (2008)
Thermopressing	The biopolymer is subjected to high and low temperatures	Elaboration of multilayer materials	Hernandez-Izquierdo and Krochta (2008)
Extrusion	The biopolymer is subjected to shear, high-temperature compression, and cooling	Edible films manufacturing	Ullsten et al. (2006) Barone et al. (2006) Hernandez-Izquierdo et al. (2008) Krishna et al. (2012) Rouilly et al. (2006) Arvanitoyannis and Biliaderis (1999) Fishman et al. (2000) Flores et al. (2010) Li et al. (2011)

TABLE 14.12 Methods for Preparation of Edible Films

Polymer(s)	Concentration	Plasticizer	Time (min)	Temperature	Stirring (rpm)	Drying	Additives	Reference
Ulluco starch	2.0, 2.5, and 3.0% (w/v)	Glycerol (1.0% w/v)	5	95	NR	8 °C	NR	Galindez et al. (2019)
Sodium alginate Carboxymethyl cellulose	1:1 w/w	Glycerol (2.0% v/v)	90	50	NR	50 °C/12 h	Epigallocatechin gallate	Ruan et al. (2019)
Carboxymethyl cellulose	2% (w/v)	NR	NR	NR	NR	50 °C/15 h	Lactobacillus rhamnosus	Singh et al. (2019)
Chitosan	1.5% (w/v) in 1% v/v or acetic acid	Glycerol 18% (w/w)	4–6 h	30	8000	40 °C/36 h	Tween 80	Zheng et al. (2019)
Mucilage	2% (w/v)	Glycerol 25% (w/w)	15	25	500	35 °C/48 h	Tween 80	Ekrami et al. (2019)
Agar	1.5% (w/v)	Glycerol 30% (w/w)	30	95	20,000/2 min	40 °C/72 h	Cellulose	Wang et al. (2018)
Starch and protein	100:0, 90:10, 80:20, 70:30 60:40 and 50:50	Glycerol 20% (w/w)	5	90 °C	NR	35 °C	NR	Chinma et al. (2015)
Chitosan in lactic acid solution	2%; 1% v/v	NR	NR	NR	NR	NR	Essential oil of Thymus zygis	Ballester-Costa et al. (2016)
NaCas	7.5 g/100 g	Glycerol 0.32 (w/w)	2 h	25 °C	NR	23 °C/48 h (50%RH)	NR	Caprioli et al. (2009)
Alyssum homolocarpum seed gum	1.2% (w/v)	Glycerol (25, 30 35 and 45% w/w)	30	50 °C	400 rpm	40 °C/24 h (30% RH)	NR	Mohammadi-Nafchi et al. (2017)

NR: nonreported.

to prevent the film from fracturing once it dehydrates. The agitation and temperature depend on the polymer. The incorporation of additives depends on the use and application of the film and ranges from pure compounds, raw extracts, oils, waxes (to reduce permeability). Finally, they are dehydrated at temperatures not exceeding 50 °C for up to 48 h.

14.5.2   WET PROCESSING

Wet processing consists of dissolving the biopolymers with additives and conditioners; this is applied by immersion in the food promoting its spreading over the surface, and finally, it dried by solvent evaporation (29.13). However, this procedure is not feasible at the industrial scale, mainly by the long drying time (24 h).

An alternative to overcome the various problems of food preservation is the development and application of composite coatings from polysaccharides (Formiga et al., 2019), proteins (Ananey-Obiri et al., 2018), or lipids (Rojas et al., 2015) on their blends. The main production processes of edible coatings are shown in Table 14.14.

In general, the production of edible coatings consists solubilize the polymer by up to 5% (depends on its solubility and its water retention

TABLE 14.13   Wet Processing to Produce Edible Coatings

Methods	Process	Finality	Reference
Tape casting	Film-forming solution is cast as a thin layer on a tape and is dried by heat conduction	Manufacturing of paper, plastic, ceramic, paint and edible, or synthetic films to coat paper, improved the barrier to transmission of oxygen, carbon dioxide, water vapor and UV radiation, in addition to the adherence, transparency, and oil resistance	De Moraes et al. (2013) Guillaume et al. (2010) Farris et al. (2010) Gastaldi et al. (2007) Han and Krochta (2001)
Edible coatings	Consists of dipping the food in the coating solution. The main parameters for the elaboration of the edible coating are the density, viscosity, superficial characteristics of product, and the surface tension	Preserve fresh products of vegetable and animal origin	Cisneros-Zevallos and Krochta (2003) Vargas et al. (2008) Kim et al. (2008)

TABLE 14.14 Methods for Preparation of Edible Coatings

Ingredients	Plasticizer	Conditions	Additives	Reference
Arabic gum, candelilla wax	NR	15,000 rpm/15 min	Tarbush extract (500 ppm)	Rojas et al. (2015)
Laquer wax 2%	Tefose° 2000	Dissolving wax in water at 60 °C/5 min	NR	Hu et al. (2019)
Corn starch 4%	Glycerol 1% v/v	Microwave heating (700 W) 10 min, stirring 5 min and pH 5.6	$CaCl_2$ (2%)	De Oliveira Alves Sena et al. (2019)
Beeswax, hydroxypropyl methylcellulose, and stearic acid 5:1 (BW:SA)	Glycerol 2:1 (HPMC:Gly)	Microwave (90 °C), 1 min at 968 g and 3 min at 3871 g.	NR	Formiga et al. (2019)
Guar gum (0.8%, candelilla wax (0.2%),	Glycerol (0.3%)	Hot water (80 °C), 20,500 rpm/20 min	Gallic acid (0.15%)	Oregel-Zamudio et al. (2017)
Chitosan (1%) and carboxymethyl cellulose (1.5%)	NR	Solubilization at 80 °C (Chitosan) and stirring in acetic acid (0.7% w/v)//2h to CMC. pH 5.6	NR	Yan et al. (2019)
Carnauba wax (10 g), oleic acid (7.5 mL/100g)	NR	90–95 °C at 1200 rpm/3 min, pH 7 and finally dissolved	NR	Singh et al. (2016)
Carnauba and beeswax (20 g), ammonia solution (30%)	NR	Stirring 30 min/90 °C. add 20 mL water (100 °C) and finally stirring 60 min with hot water (150 mL)	NR	Motamedi et al. (2018)
Soy protein 5% (w/v)	Glycerol 3.5% (w/v)	Solubilization in hot water (90 °C)/30 min. pH adjusted at 10	Essential oil from thyme and oregano (1, 2, or 3%)	Yemis and Candogan (2017)
Peach gum (1%)	NR	90 °C/6 h, centrifugued 5000 g/10 min	NR	Li et al. (2017)

NR: nonreported.

capacity). The use of surfactants is more common. In the case of coatings, the particle size of the emulsion is of great importance, as it is homogenizes at up to 25,000 rpm. The use of additives depends on the purpose (antimicrobial and antioxidant).

14.6 STABILITY OF EDIBLE PACKAGING

The short shelf life of fresh and processed foods has been a human concern since food is scarce and population growth. Because of their physical structure (water activity, nutrients as carbohydrates, proteins, and minerals), foods are susceptible to microbial spoilage and degradation. Thus, microbiological and physiological activities play a role in quality degradation during storage. Nowadays there are some serious environmental and health problems due to plastic packaging uses and their disposal conditions (Haward, 2018; Chae and An, 2018; Windsor, 2019; Lebreton and Andrady, 2019). Promoting an environmental mentality and the development of various scientific studies to obtain edible and biodegradable packagings from natural sources like fruits and vegetables peels and kernels (Wu et al., 2019; Nawab et al., 2018) with good sensory attributes; high barrier and mechanical properties; biochemical, physicochemical, and microbial stability (Brody, 2011).

14.6.1 REORGANIZATIONS OF THE FILM MATRIX AND COATING

The contact of the packaging with the food depends on the molecular size of the biopolymer, the chemical nature of the compounds, the temperature and process conditions, and the film structure. When a coating is in contact with food, the film-coating evaporation flux at the interface between the film-coating/environment should be evaluated, as well as the coating absorption flux on the food due to the interface between the food/film-coating (Montero-Garcia, 2016). Edible films-coating needs one or more additives in order to carry functional compounds intended to provide the improved characteristics (sensory, nutritional, microbiological, enzymatic, color, and other chemical reactions). Reorganizations of the matrix depend on the type of polymer and its functional groups, the concentrations of plasticizers and additives added. A high concentration of additives could generate undesirable odors, turbidity, and promote lipid oxidation (Wambura et al., 2011). The dose of the active compounds needs to be relatively low (Silva-Weiss et al., 2013). The selection of natural additives and their application depends on their properties (antioxidant, antimicrobial, antibrowning), cost-effectiveness, and effect on the sensory attributes of the final product

(Perumalla and Hettiarachchy, 2011). The reorganization of the edible matrix is related to the wounding-to response of the biological material (starch crystallization, protein aggregation, and plasticizer migration), chemical degradation (lipid oxidation, nonenzymatic browning, degradation of active compounds), and enzymatic degradation (proteolysis). The starch crystallizes due to the plasticizers content (Perez Sira and Dufour, 2017) and has low stability upon storage due to retrogradation (recrystallization) process of starch. Retrogradation can be minimized by adding cellulose as a matrix filler (Benito-González et al., 2019). Plasticizers provide greater flexibility to the polymeric matrix and reduce the original brittleness of biopolymers. Plasticizers do not chemically interact with the backbone chain but position themselves between the polymer molecules to reduce polymer chain-to-chain interaction (Kadzińska et al., 2019). The most studied and suitable plasticizer is glycerol, due to its marked hydrophilic nature as compared to sorbitol (Jiménez et al., 2018). Nowadays the research has been focused on finding other sources of plasticizers, in order to fulfill the main limitations of glycerol use (high hydrophilicity, low thermal stability, and surface migration over time) (Blanco-Pascual, 2016). Gheribi et al. (2018) developed an edible film with a mixture of mucilage from Opuntia ficus-indica cladodes and sorbitol as plasticizer, reaching water vapor permeability values up to three times lower than the other films evaluated. Other research works have a focus in using different materials, like chia seed mucilage (Dick et al., 2015), chitosan (Sabbah et al., 2019), epigallocatechin gallate and carboxymethyl cellulose (Ruan et al., 2019), and among others. Proteins can also have a similar function as a plasticizer, either both plant and animal sources, the amino acid functional groups in proteins can improve the stability and form an extended structure of the films. Formulations from polysaccharides and proteins show better structural properties than individual proteins (Cerqueira et al., 2011; Jiménez et al., 2018).

14.6.2 CHEMICAL DEGRADATION

Lipids used as hydrophobic material is well-known and is not strange its use as the barrier in foodstuffs; however, the inconvenience lays in the lipid oxidation. Lipid oxidation is a process that results in rancidity and deterioration of fats and progresses via free-radical propagated chain reactions, which yields hydroperoxides that cause a variety of secondary reactions with the evolution of aldehydes, ketones, acids, and other low-molecular-weight volatile substances (Ramis-Ramos, 2003).

14.6.3 ENZYMATIC DEGRADATION

Minimally processed foods as fruits and vegetables (Yoruk and Marshall, 2003), meat and meat products (Hernandez-Hernandez, 2009) rich in phenolic components are subjected to the action of antioxidant enzymes. Postharvest oxidative stress occurs during storage, causing an imbalance between the production and removal of reactive oxygen species, such as H_2O_2, O_2-, and OH- radicals, from the tissues. The protection of fruit or vegetable cells from oxidative injury depends on the enzymes and polyphenols level which scavenge the reactive oxygen species, and prevent harmful effects (Amiot et al., 1992; Zeng et al., 2010). Antioxidants and enzymatic inhibitors are used to prevent browning by the chemical reduction of quinones to colorless ortho-diphenol reduction agents (McEvily and Iyengar, 1992). Millard reactions take place if the temperature is increased and reducing sugars are present. Some reducing agents have been investigated to prevent the antioxidant reaction (Table 14.15). Butylated hydroxyanisole and butylated hydroxytoluene are the most commonly used synthetic antioxidants; therefore, natural antioxidants such as phenolic compounds are alternatives to synthetic antioxidants (Chan et al., 2007; Jongjareonrak et al., 2008; Yen et al., 2008).

14.6.4 MICROBIAL GROWTH

Microbial growth in food products promotes deterioration and reduce the shelf life (Ding et al., 2013). Packaging provides some level of protection to food products from external and internal unfavorable conditions (Mihindukulasuriya and Lim, 2014). Those films have nutrients as substrates for microbial growth. In order to improve the efficiency and stability of edible coating-films, it is essential to find adequate materials (Flores-López, 2015). The incorporation of antimicrobial agents into the used in edible films could enhance its functional properties by retarding microorganism (Soares et al., 2009; Sirelkhatim et al., 2015; Malhotra et al., 2015). Essential oils and polyphenols (tannins, flavonoids, phenolic acids, secondary plant metabolites (Espitia et al., 2014) are listed in Table 14.2, as natural antimicrobials. A concern regarding some functional additives is their flavor and aroma. Sensory evaluation of edible films and foodstuffs packaged is scarce; however, many active compounds are known to be accepted by consumers (Otoni et al., 2017).

TABLE 14.15 Edible Coatings and Films with Natural Antioxidants.

Films Coating	Reduction agent	Application	Reference
Carboximethyl cellulose-based coating	Ascorbic acid	Apple	Saba (2016)
Apple-pectin edible coating		Persimmon	Sanchis (2016)
Aloe vera gel coating		Strawberry	Sogvar (2016)
Carrot puree, carboximethyl cellulose, corn starch, and gelatin edible films	Acetic acid and sodium bicarbonate	Carrot	Wang (2011)
Carrot puree-chitosan-starch	Cinnamaldehyde	Carrot	Wang (2015)
Starch coating	Yerba mate extracts and mango pulp	Mango	Reis (2015)
Xanthan gum nano-coating	Tocopherol	Apple	Zambrano-Zaragoza (2014)
		Apple	Galindo-Perez (2015)
	Cinamic acid	Pear	Sharma (2015)
Nanocapsules-xanthan gum coating	b-carotene	Melon	Zambrano-Zaragoza (2017)
Gum arabic, Aloe vera, chitosan coating	Thyme oil	Avocado	Sivakumar (2014)
Chitosan, carboxymethyl cellulose edible coating	Moringa leaf extract	Avocado	Tesfay (2017)
	Pineapple fruit extract	Apple	Supapvanchi (2012)
Alginate edible coating	Plum extract	*Prunus salicina Lindl.*	Valero (2013)
	Sunflower oil	Kent mangoes	Robles-Sanchez (2013)
	Cinnamon and rosemary essential oils	Apple	Chiabrando (2015)
	Malic acid	Mango	Salinas-Roca (2016)
	Mango peel	Papaya	Valderrain-Rodriguez (2015)
	Lemon grass essential oil	Pineapple	Azaraksh (2014)

TABLE 14.15 *(Continued)*

Films Coating	Reduction agent	Application	Reference
Chitosan coating	chitosan	Strawberry	Petriccione (2015)
	Rosemary extract	Pear	Xiao (2010)
	cinnamon oil	Sweet pepper (Capsicum annum L)	Xing (2011)
	cinnamon oil	Peach	Ayala-Zavala (2013)
	Trans-cinnamaldehyde	Melon	Carvalho (2016)
	Sodium chloride	Pear	Xiao (2011)
	Shrimp waste	Shrimp	Arancibia (2015)
	Pomegranate peel extract	Shrimp	Yuan (2016)
	chitosan	Kiwifruit	Drevinskas (2017)
Chitosan-aloe vera coating	NA	Blueberry (*vaccinium corymbosum*)	Vieira (2016)
Starch-gelatin edible film	NA	Grapes	Fakhouri (2015)
Pectin	Geraniol	Strawberry	Badawy (2016)
Hidroxymethyl cellulose, chitosan	Bergamot essential oil	Grapes	Sanchez-Gonzalez (2011)
Poly (butylene adipate co-terephalate) PBAT	Oregano essential oil	Fish	Cardoso (2017)
Basil seed gum	Oregano essential oil	Apricot	Hashemi (2017)
Soy protein edible coating	Ferulic acid	Apple	Alvez (2017)
	Honey	Melon	Yousuf (2017)
Whitemouth croaker protein isolate	Oregano-clay	Papaya	Cortez-Vega (2014)

TABLE 14.16 Recent Antimicrobial Film-coating Studies

Film or Edible Coating	Antimicrobial Agent	Application	Target Microorganism	Reference
Chitosan-aloe vera coating	Aloe vera	Blueberry	B. cinérea, P. expansum, A. niger	Vieira (2016)
Chitosan, pectin	Trans-cinnamaldehyde	Papaya	Total aerobic, pshycrotrophics, yeast, and molds count	Brasil (2012)
Chitosan edible film	Berberis crataeginas fruit extract and seed oil			
Soy protein film	Thyme and oregano essential oil	Beef	E. coli 0H157:H7, S. aureus, P. Aeruginosa, L. plantarum	Emiroglu (2010)
Gelatin-chitosan edible film	Guarana seeds (Paullinia cupana), leaves of boldolochile (Peumus boldus Molin), cinnamon barks (Cinnamomum sp), leaves of rosemary (Rosmarinus officinalis)	Laboratory level	S. aureus and E. coli	Bonilla (2016)
Zein films	Zataria multiflora Boiss essential oil and monolaurin	Meat	E. coli and L. monocytogenes	Moradi (2016)
Pectin-papaya puree	cinnamaldehyde	Laboratory level	E. coli, S. aureus, L. mono-cytogenes, S. entérica, Choleraesuis	Otoni (2014)
LLDPE	Clove essential oil	Chicken	S. entérica, L. monocytogenes	Mulla (2017)
Chitosan	Green tea extracts	Pork sausages	Total viable count, yeast, mold	Siripatrawa (2012)
	Titanium dioxide	Tomato	fungal	Kaewklin (2018)

KEYWORDS

- **natural polymers**
- **extraction**
- **properties**
- **stability**
- **food packaging**

REFERENCES

Abraján-Villaseñor, M.A. Efecto del método de extracción en las características químicas y físicas del mucílago del nopal (Opuntia ficus indica) y estudio de su aplicación como recubrimiento comestible. Degree Dissertation, Universidad Politécnica de Valencia, Valencia, España. 2008.

Abu-Arabi, M.K.; Allawzi, M.A.; Al-Zoubi, H.S.; Tamimi, A. Extraction of Jojoba oil by pressing and leaching. Chem. Eng. J. 2000, 76, 61–65.

Aguirre-Joya, J.A. et al. Edible coatings and films from lipids, waxes, and resins. In Edible Food Packaging: Materials and Processing Technologies. Taylor & Francis Group: Oxfordshire, 2017, 121–152.

Ahmad, M.; Benjakul, S.; Prodpran, T.; Agustini, T.W. Physico-mechanical and antimicrobial properties of gelatin film from the skin of unicorn leatherjacket incorporated with essential oils. Food Hydrocoll. 2012, 28, 189–199.

Ahumada-Lazo, R., Jr. Secado acelerado y raspado del tallo de candelilla (Euphorbia antisyphilitica) y su relación con el rendimiento en la extracción de cera. Degree Dissertation, Universidad Autónoma de Chihuahua, Chihuahua, Chihuahua, 2012.

Al Bulushi, K.; Attard, T.M.; North, M.; Hunt, A.J. Optimisation and economic evaluation of the supercritical carbon dioxide extraction of waxes from waste date palm (Phoenix dactylifera) leaves. J. Clean. Prod. 2018, 186, 988–996.

Ali, A.; Maqbool, M.; Ramachandran, S.; Alderson, P.G. Gum arabig as a novel edible coating for enhancing shelf life and improving postharvest quality of tomato (*Solanum lycopersicum* L.) fruit. Postharv. Biol. Technol. 2010, 58, 42–47.

Allawzi, M.A.; Abu-Arabi, M.K.; Al-Taher, F.A. Parametric study on the batch leaching process of Jojoba oil. Eur. J. Lipid Sci. Technol. 2005, 107, 469–475.

Alves, M.M.; Gonçalves, M. P.; Rocha, C.M.R. Effect of ferulic acid on the performance of soy protein isolate-based edible coatings applied to fresh-cut apples. LWT—Food Sci. Technol. 2017, 80, 409–415.

Amiot, M.J.; Tacchini, M.; Aubert, S.; Nicolas, J. Phenolic composition and browning susceptibility of various apple cultivars at maturity. J. Food Sci. 1992, 57, 4, 958–962.

Ananey-Obiri, D.; Matthews, L.; Azahrani, M.H.; Ibrahim, S.A.; Galanakis, C.M.; Tahergorabi, R. Application of protein-based edible coatings for fat uptake reduction in deep-fat fried foods with an emphasis on muscle food proteins. Trends Food Sci. Technol. 2018, 80, 167–174.

Arancibia, M.Y.; Lopez-Caballero, M.E.; Gomez-Guillen, M.C.; Montero, P. Chitosan coatings enriched with active shrimp waste for shrimp preservation. Food Control. 2015, 54, 259–266.

Arismendi, C.; Chillo, S.; Conte, A.; Del Nobile, M.A.; Flores, S.; Gerschenson, L.N. Optimization of physical properties of xanthan gum/tapioca starch edible matrices containing potassium sorbate and evaluation of its antimicrobial effectiveness. LWT—Food Sci. Technol. 2013, 53, 290–296.

Arvanitoyannis, I.; Biliaderis, C.G. Physical properties of polyol-plasticized edible blends made of methyl cellulose and soluble starch. Carbohydr. Polym. 1999, 38(1), 47–58.

Asep, E.K.; Jinap, S.; Jahurul, M.H.A.; Zaidul, I.S.M.; Singh, H. Effects of polar cosolvents on cocoa butter extraction using supercritical carbon dioxide. Innov. Food Sci. Emerg. Technol. 2013, 20, 152–160.

Ashokkumar, M. et al. The ultrasonic processing of dairy products—an overview. Dairy Sci. Technol. 2010, 90 (2/3), 147–168.

Attard, T.M.; Bukhanko, N.; Eriksson, D.; Arshadi, M.; Geladi, P.; Bergsten, U.; Budarin, V.L.; Clark, J.H.; Hunt, A.J. 2018. Supercritical extraction of waxes and lipids from biomass: a valuable first step towards an integrated biorefinery. J. Clean. Prod. 2018, 177, 684–698.

Avérous, L.; Poller, E. Biodegradable polymers. In Environmental Silicate Nano-Biocomposites. Springer: New York, NY, 2012; 13–39.

Ayala-Zavala, J.F.; Silva-Espinoza, B.A.; Cruz-Valenzuela, M.R.; Leyva, J.M.; Ortega-Ramírez, L.A.; Carrazco-Lugo, D.K.; Miranda, M.R.A. Pectin–cinnamon leaf oil coatings add antioxidant and antibacterial properties to fresh-cut peach. Flavour. Fragr. J. 2013, 28, 1, 39–45.

Azarakhsh, N.; Osman, A.; Ghazali, H.M.; Tan, C.P.; Adzahan, N.M. Lemongrass essential oil incorporated into alginate-based edible coating for shelf-life extension and quality retention of fresh-cut pineapple. Postharvest Biol. Technol. 2014, 88, 1–7.

Badawy, M.E.; Rabea, E.I.; AM El-Nouby, M.; Ismail, R.I.; Taktak, N.E. Strawberry shelf life, composition, and enzymes activity in response to edible chitosan coatings. Int. J. Fruit Sci. 2017, 17, 2, 117–136.

Ballester-Costa, C.; Sendra, E.; Fernández-López, J.; Viuda-Martos, M. Evaluation of the antibacterial and antioxidant activities of chitosan edible films incorporated with organic essential oils obtained from four Thymus species. J. Food Sci. Technol. 2016, 53, 3374–3379.

Barbosa-Pereira, L.; Angulo, I.; Lagarón, J.M.; Paseiro-Losada, P.; Cruz, J.M. Development of new active packaging films containing bioactive nanocomposites. Innov. Food Sci. Emerg. Technol. 2014, 26, 310–318.

Barone, J.R.; Schmidt, W.F.; Gregoire, N.T. Extrusion of feather keratin. J. Appl. Polym. Sci. 2006, 100(2), 1432–1442.

Bello-Lara, J.E., Balois-Morales, R., Sumaya-Martínez, M.T., Juárez-López, P., Rodríguez-Hernández, A.I., Sánchez-Herrera, L.M., Jiménez-Ruíz, E.I. Extracción y caracterización reológica de almidón y pectina en frutos de plátano 'Pera' (Musa ABB). Rev. Mex. Cienc. Agríc. 2014, 5, 1–6.

Benito-Gonzalez, I.; López-Rubio, A.; Martinez-Sanz, M. High-performance starch biocomposites with celullose from waste biomass: film properties and retrogradation behaviour. Carbohydr. Polym. 2019, 216, 180–188.

Bilal, M.; Iqbal, H.M.N. Naturally-derived biopolymers: potential platforms for enzyme immobilization. Int. J. Biol. Macromol. 2019, 130, 462–482.

Bildstein, M. et al. An enzyme-based extraction process for the purification and enrichment of vegetable proteins to be applied in bakery products. Eur. Food Res. Technol. 2008, 228 (2), 177–186.

Bill, M.; Sivakumar, D.; Korsten, L.; Thompson, A.K. The efficacy of combined application of edible coatings and thyme oil in inducing resistance components in avocado (Persea americana Mill.) against anthracnose during post-harvest storage. Crop Prot. 2014, 64, 159–167.

Blanco-Pascual, N.; Gómez-Estaca, J. Production and processing of edible packaging. Edible Food Packag. Mater. Process Technol. 2016, 36, 153.

Bogdanov, S. Quality standards of pollen and beeswax. Apiacta 2004, 38, 334–341.

Bonilla, J.; Sobral, P.J. Investigation of the physicochemical, antimicrobial and antioxidant properties of gelatin-chitosan edible film mixed with plant ethanolic extracts. Food Biosci. 2016, 16, 17–25.

Bosquez-Molina, E.; Guerrero-Legarreta, I.; Vernon-Carter, E.J. Moisture barrier properties and morphology of mesquite gum: candelilla wax based edible emulsion coatings. Food Res. Int. 2003, 36, 885–893.

Boyacı, D.; Iorio, G.; Sozbilen, G.S.; Alkan, D.; Trabattoni, S.; Pucillo, F.; Farris, S.; Yemenicioğlu, A. Development of flexible antimicrobial zein coatings with essential oils for the inhibition of critical pathogens on the surface of whole fruits: test of coatings on inoculated melons. Food Packag. Shelf Life. 2019, 20, 100316.

Brasil, I.M.; Gomes, C.; Puerta-Gomez, A.; Castell-Perez, M.E.; Moreira, R.G. Polysaccharide-based multilayered antimicrobial edible coating enhances quality of fresh-cut papaya. LWT—Food Sci. Technol. 2012, 47, 1, 39–45.

Brecht, J.K. Physiology of lightly processed fruits and vegetables. HortScience 1995, 30, 1, 18–22.

Brody, A.L. Packaging innovation-past, present, and future. Food Technol. 2011, 65, 80–82.

Calva-Estrada, S.J.; Jiménez-Fernández, M.; Lugo-Cervantes, E. Protein-based films: advances in the development of biomaterials applicable to food packaging. Food Eng. Rev. 2019, 11, 78–92.

Cao, L.; Liu, W.; Wang, L. Developing a green and edible film from Cassia gum: the effects of glycerol and sorbitol. J. Clean. Prod. 2018, 175, 276–282.

Capitani, M.I.; Matus-Basto, A.; Ruiz-Ruiz, J.C.; Santiago-García, J.L.; Betancur-Ancona, D.A.; Nolasco, S.M.; Tomás, M.C.; Segura-Campos, M.R. Characterization of biodegradable films based on Salvia hispanica L. Protein and Mucilage. Food Bioprocess Technol. 2016, 9, 1276–1286.

Caprioli, I.; O'Sullivan, M.; Monahan, F.J. Use of sodium caseinate/glycerol edible films to reduce lipid oxidation in sliced turkey meat. Eur. Food Res. Technol. 2009, 228, 433–440.

Caprita, R.; Caprita, A.; Julean, C. Biochemical aspects of non-starch polysaccharides. J. Anim. Sci. Biotechnol. 2010, 43, 368–375.

Cardoso, L.G.; Santos, J.C.P.; Camilloto, G.P.; Miranda, A.L.; Druzian, J.I.; Guimarães, A.G. Development of active films poly (butylene adipate co-terephthalate)–PBAT incorporated with oregano essential oil and application in fish fillet preservation. Ind. Crops Prod. 2017, 108, 388–397.

Carpiné, D. et al. Development and characterization of soy protein isolate emulsion-based edible films with added coconut oil for olive oil packaging: barrier, mechanical, and thermal properties. Food Bioproc. Tech. 2015, 8(8), 1811–1823.

Carvalho, R.L.; Cabral, M.F.; Germano, T.A.; Carvalho, W.M.; Brasil, I.M.; Gallão, M.I.; Miranda, M.R.A. Chitosan coating with *trans*-cinnamaldehyde improves structural integrity and antioxidant metabolism of fresh-cut melon. Postharvest Biol. Technol. 2016, 113, 29–39.

Cerqueira, M.A.; Bourbon, A.I.; Pinheiro, A.C.; Martins, J.T.; Souza, B.W.S.; Teixeira, J.A.; Vicente, A.A. Galactomannans use in the development of edible films/coatings for food applications. Trends Food Sci. Technol. 2011, 22, 12, 662–671.

Cerqueira, M.A.; Lima, Á.M.; Teixeira, J.A.; Moreira, R.A.; Vicente, A.A. Suitability of novel galactomannans as edible coatings for tropical fruits. J. Food Eng. 2009, 94, 372–378.

Chae, Y.; An, Y.J. Current research trends on plastic pollution and ecological impacts on the soil ecosystem: a review. Environ. Pollut. 2018, 240, 387–395.

Chan, E.W.C.; Lim, Y.Y.; Chew, Y.L. Antioxidant activity of Camellia sinensis leaves and tea from a lowland plantation in Malaysia. Food Chem. 2007, 102, 4, 1214–1222.

Chasquibol Silva, N., Arroyo Benites, E., Morales Gomero, J.C. Extracción y caracterización de pectinas obtenidas a partir de frutos de la biodiversidad peruana. Ing. Ind. 2008 26, 175–199.

Chelikani, P.; Fita, I.; Loewen, P.C. Diversity of structures and properties among catalases. Cell. Mol. Life Sci. 2004, 61, 192–208.

Cheng, S.Y.; Wang, B.J.; Weng, Y.M. Antioxidant and antimicrobial edible zein/chitosan composite films fabricated by incorporation of phenolic compounds and dicarboxylic acids. LWT—Food Sci. Technol. 2015, 63, 115–121.

Chiabrando, V.; Giacalone, G. Effect of essential oils incorporated into an alginate-based edible coating on fresh-cut apple quality during storage. Qual. Assurance Safety Crops Foods 2014, 7, 3, 251–259.

Chinma, C.E.; Ariahu, C.C.; Alakali, J.S. Effect of temperature and relative humidity on the water vapour permeability and mechanical properties of cassava starch and soy protein concentrate based edible films. J. Food Sci. Technol. 2015, 52, 2380–2386.

Chiralt, A. et al. Edible films and coatings from proteins. In Proteins in Food Processing. 2nd ed. Elsevier Ltd.: London, 2017.

Chiumarelli, M.; Hubinger, M.D. Evaluation of edible films and coatings formulated with cassava starch, glycerol, carnauba wax and stearic acid. Food Hydrocoll. 2014, 38, 20–27.

Cisneros-Zevallos, L.; Krochta, J.M. Dependence of coating thickness on viscosity of coating solution applied to fruits and vegetables by dipping method. J. Food Sci. 2003, 68(2), 503–510.

Codini, M.; Díaz-Vélez, F.; Ghirardi, M.; Villavicencio, I. Obtención y utilización de la manteca de cacao. Invenio [Online] 2004, 7. http://www.redalyc.org/articulo.oa?id=87701213 (accessed June 17, 2019).

Cortez-Vega, W.R.; Pizato, S.; Andreghetto De Souza, J.T.; Prentice, C. Using edible coatings from whitemouth croaker (Micropogonias furnieri) protein isolate and

organo-clay nanocomposite for improve the conservation properties of fresh-cut 'Formosa' papaya. Innov. Food Sci. Emerg. Technol. 2014, 22, 197–202.

Cosgrove, J. Emerging edible films: dissolving strips have made minor supplement inroads, but advancing technologies point to progress. [Online 2008]. Available online at: http://www.nutraceu ticalsworld.com/contents/view_online-exclusives/2008-01-01/ emerging-edible-films/ (accessed January 7, 2013).

Cui, Z. et al. Effects of rutin incorporation on the physical and oxidative stability of soy protein-stabilized emulsions, Food Hydrocoll. 2014, 41, 1–9.

Da Silva-Rodríguez, C.P. Extracción de aceite de Coco (Cocos nuciferas) como estrategias de aprovechamiento de los productos locales de Mitú. Vaupes Innova. 2017, 83–89.

Dangaran, K.; Tomasula, P.M.; Qi, P. Edible Films and Coatings for Food Applications. Springer Science & Business Media: Berlin, 2009.

Dantas, A.N.D.S.; Magalhães, T.A.; Matos, W.O.; Gouveia, S.T.; Lopes, G.S. Characterization of carnauba wax inorganic content. J. Am. Oil Chem.' Soc.. 2013, 90, 1475–1483.

De la Vega, G. Proteínas de la harina de trigo. Temas de Ciencia y Tecnología. 2009, 13, 27–32.

De León-Zapata, M.A., Jr. Mejoras tecnológicas al proceso de extracción de cera de candelilla (Euphorbia antisyphilitica Zucc.). Degree Dissertation, Universidad Autónoma de Coahuila, Saltillo, Coahuila, 2008.

De León-Zapata, M.A.; Rojas-Molina, R.; Saucedo-Pompa, S.; Ochoa-Reyes, E.; De La Garza-Toledo, H.; Rodríguez-Herrera, R.; Aguilar-González, C.N. Proceso de extracción de cera de candelilla (Euphorbia antisyphilitica Zucc.). Mexican Patent 347121, 10 February 2016.

De Moraes, J.O.; Scheibe, A.S.; Sereno, A.; Laurindo, J.B. Scale-up of the production of cassava starch based films using tape-casting. J Food Eng. 2013, 119, 4, 800–808.

De Moura, J.M.L.N. et al. Protein extraction and membrane recovery in enzyme-assisted aqueous extraction processing of soybeans. J. Am. Oil Chem. Soc. 2011a, 88, 6, 877–889.

De Moura, J.M.L.N. et al. Protein recovery in aqueous extraction processing of soybeans using isoelectric precipitation and nanofiltration. J. Am. Oil Chem. Soc. 2011b, 88, 9, 1447–1454.

De Oliveira Alves Sena, E.; Oliveira Da Silva, P.S.; De Aragão Batista, M.C.; Alonzo Sargent, S.; Ganassali De Oliveira Junior, L.F.; Almeida Castro Pagani, A.; Gutierrez Carnelossi, M.A. Calcium application via hydrocooling and edible coating for the conservation and quality of cashew apples. Sci. Hortic. (Amsterdam). 2019, 256.

Dhall, R.K. Advances in edible coatings for fresh fruits and vegetables: a review. Crit. Rev. Food Sci. Nutr. 2013, 53, 435–450.

Dick, M.; Costa, T.M.H.; Gomaa, A.; Subirade, M.; De Oliveira Rios, A.; Flores, S.H. Edible film production from chia seed mucilage: effect of glycerol concentration on its physicochemical and mechanical properties. Carbohydr. Polym. 2015, 130, 198-205.

Ding, H.; Fu, T.J.; Smith, M.A. Microbial contamination in sprouts: how effective is seed disinfection treatment. J. Food Sci.. 2013, 78, 4, R495–R501.

Dou, L.; Li, B.; Zhang, K.; Chu, X.; Hou, H. Physical properties and antioxidant activity of gelatin-sodium alginate edible films with tea polyphenols. Int. J. Biol. Macromol. 2018, 118, 1377–1383.

Drevinskas, T.; Naujokaitytė, G.; Maruška, A.; Kaya, M.; Sargin, I.; Daubaras, R.; Česonienė, L. Effect of molecular weight of chitosan on the shelf life and other quality parameters of three different cultivars of Actinidia kolomikta (kiwifruit). Carbohydr. Polym. 2017, 173, 269–275.

Ekrami, M.; Emam-Djomeh, Z.; Ghoreishy, S.A.; Najari, Z.; Shakoury, N. Characterization of a high-performance edible film based on Salep mucilage functionalized with pennyroyal (Mentha pulegium). Int. J. Biol. Macromol. 2019, 133, 529–537.

Emiroğlu, Z.K.; Yemiş, G.P.; Coşkun, B.K.; Candoğan, K. Antimicrobial activity of soy edible films incorporated with thyme and oregano essential oils on fresh ground beef patties. Meat Sci. 2010, 86, 2, 283–288.

Erginkaya, Z.; Kalkan, S.; Unal, E. Use of antimicrobial edible films and coatings as packaging materialsfos food safety. In Food Processing: Strategies for Quality Assessment. Food Engineering Series. Malik, A. et al., eds. Springer: New York, 2014.

Espitia, P.J.P.; Du, W.X.; De Jesús Avena-Bustillos, R.; Soares, N.D.F.F.; McHugh, T.H. Edible films from pectin: physical-mechanical and antimicrobial properties—a review. Food hydrocoll. 2014, 35, 287–296.

European Bioplastics. What are bioplastics. https://www.europeanbioplastics.org/bioplastics/(accessed September 14, 2018).

Fabra, M.J.; Falcó, I.; Randazzo, W.; Sánchez, G.; López-Rubio, A. Antiviral and antioxidant properties of active alginate edible films containing phenolic extracts. Food Hydrocoll. 2018, 81, 96–103.

Fakhouri, F.M.; Martelli, S.M.; Caon, T.; Velasco, J.I.; Mei, L.H.I. Edible films and coatings based on starch/gelatin: film properties and effect of coatings on quality of refrigerated red crimson grapes. Postharvest Biol. Technol. 2015, 109, 57–64.

Farris, S.; Cozzolino, C.A.; Introzzi, L.; Piergiovanni, L. Development and characterization of a gelatin-based coating with unique sealing properties. J. Appl. Polym. Sci. 2010, 118(5), 2969–2975.

Feng, Z.; Wu, G.; Liu, C.; Li, D.; Jiang, B.; Zhang, X. Edible coating based on whey protein isolate nanofibrils for antioxidation and inhibition of product browning. Food Hydrocoll. 2018, 79, 179–188.

Fishman, M.L.; Coffin, D.R.; Konstance, R.P.; Onwulata, C.I. Extrusion of pectin/starch blends plasticized with glycerol. Carbohydr. Polym. 2000, 41(4), 317–325.

Flores, S.K.; Costa, D.; Yamashita, F.; Gerschenson, L.N., Grossmann, M.V. Mixture design for evaluation of potassium sorbate and xanthan gum effect on properties of tapioca starch films obtained by extrusion. Mater. Sci. Eng.: C. 2010, 30(1), 196–202.

Flores-López, M.L.; Cerqueira, M.A.; De Rodríguez, D.J.; Vicente, A.A. Perspectives on utilization of edible coatings and nano-laminate coatings for extension of postharvest storage of fruits and vegetables. Food Eng. Rev. 2016, 8, 3, 292–305.

Formiga, A.S.; Pinsetta, J.S.; Pereira, E.M.; Cordeiro, I.N.F.; Mattiuz, B.-H. Use of edible coatings based on hydroxypropyl methylcellulose and beeswax in the conservation of red guava 'Pedro Sato.' Food Chem. 2019, 290, 144–151.

Fornasari, C.H.; Secco, D.; Ferreira Santos, R.; Benetoli da Silva, T.R.; Galant Lenz, N.B.; Kazue Tokura, L.; Lucian Lenz, M.; Melegari de Souza, S.N.; Zanão Junior, L.A.; Gurgacz, F. Efficiency of the use of solvents in vegetable oil extraction at oleaginous crops. Renew. Sust. Energ. Rev. 2017, 80, 121–124.

Fox, P.F. et al. Dairy chemistry and biochemistry. 2nd ed., Springer International Publishing: Switzerland, 2015, p. 584.

Fuertes, S.; Laca, A.; Oulego, P.; Paredes, B.; Rendueles, M.; Díaz, M. Development and characterization of egg yolk and egg yolk fractions edible films. Food Hydrocoll. 2017, 70, 229–239.

Galindez, A.; Daza, L.D.; Homez-Jara, A.; Eim, V.S.; Váquiro, H.A. Characterization of ulluco starch and its potential for use in edible films prepared at low drying temperature. Carbohydr. Polym. 2019, 215, 143–150.

Galindo-Pérez, M.J.; Quintanar-Guerrero, D.; Mercado-Silva, E.; Real-Sandoval, S.A.; Zambrano-Zaragoza, M.L. The effects of tocopherol nanocapsules/xanthan gum coatings on the preservation of fresh-cut apples: evaluation of phenol metabolism. Food Bioprocess Technol. 2015, 8, 8, 1791–1799.

Gallegos-Infante, J.A.; Rocha-Guzmán, N.; González-Laredo, R.; Zuno-Floriano, F.; Vidaña-Martínez, S.A. Caracterización de dos variedades de girasol con potencial para la producción de aceite extraídos con hexano e isopropanol. Grasas y Aceites. 2003, 54, 245–252.

Galus, S.; Lenart, A.; Voilley, A.; Debeaufort, F. Effect of potato oxidized starch on the physyco-chemical properties of soy protein isolate-based edible films. Food Technol. Biotechnol. 2013, 51, 403–409.

García, L.C.; Pereira, L.M.; De Luca Sarantópoulos, C.I.; Hubinger, M.D. Effect of antimicrobial starch coating on shel-life of fresh strawberries. Packag. Technol. Sci. 2012, 25, 413–425.

Gastaldi, E.; Chalier, P.; Guillemin, A.; Gontard, N. Microstructure of protein-coated paper as affected by physico-chemical properties of coating solutions. Colloids Surf. A Physicochem. Eng. Asp. 2007, 301(1–3), 301–310.

Ge, X.L. et al. Development of an aqueous polyethylene glycol-based extraction and recovery method for almond (*Prunus armeniaca L.*) protein. Food Anal. Methods. 2016, 9 (12), 3319–3326.

Gheribi, R.; Puchot, L.; Verge, P.; Jaoued-Grayaa, N.; Mezni, M.; Habibi, Y.; Khwaldia, K. Development of plasticized edible films from Opuntia ficus-indica mucilage: a comparative study of various polyol plasticizers. Carbohydr. Polym. 2018, 190, 204–211.

Gnanasekaran, D. Green Biopolymers and their Nanocomposites. Materials Horizons: From Nature to Nanomaterials. Springer: New York, 2019.

Guillaume, C.; Pinte, J.; Gontard, N.; Gastaldi, E. Wheat gluten-coated papers for bio-based food packaging: structure, surface and transfer properties. Food Res Int. 2010, 43(5), 1395–1401.

Hadiyanto, H.; Adetya, N.P. Response surface optimization of lipid and protein extractions from Spirulina platensis using ultrasound assisted osmotic shock method. Food Sci. Biotechnol. 2018, 27 (5), 1361–1368.

Hamzah, H.M.; Osman, A.; Tan, C.P.; Ghazali, F.M. Carrageenan as an alternative coating for papaya (Carica papaya L. cv. Eksotika). Postharvest Biol. Technol. 2013, 75, 142–146.

Han, J.H. Edible films and coatings: a review. In Innovations in Food Packaging. Academic Press: Cambridge, MA, 2014; 213–255.

Han, J.H.; Krochta, J.M. Physical properties and oil absorption of Whey-proteincoated paper. J. Food Sci. 2001, 66(2), 294–299.

Han, J.H.; Scanlon, M.C. Mass transfer of gas and solute through packaging materials. In Innovations in Food Packaging. 2 ed. Elsevier: London, 2014; 37–49.

Hashemi, S.M.B.; Khaneghah, A.M.; Ghahfarrokhi, M.G.; Eş, I. Basil-seed gum containing origanum vulgare subsp. viride essential oil as edible coating for fresh cut apricots. Postharvest Biol. Technol. 2017, 125, 26–34.

Hassan, B.; Chatha, S.A.S.; Hussain, A.I.; Zia, K.M.; Akhtar, N. Recent advances on polysaccharides, lipids and protein based edible films and coatings: a review. Int. J. Biol. Macromol. 2018, 109, 1095–1107.

Haward, M. Plastic pollution of the world's seas and oceans as a contemporary challenge in ocean governance. Nat. Commun. 2018, 9, 1, 667.

Hernández, C.; Mieres-Pitre, A. Extracción y purificación del aceite de la almendra del fruto de la palma de corozo (*Acrocomia aculeata*). Rev. Ing. UC. 2005, 12, 68–75.

Hernández-Hernández, E.; Ponce-Alquicira, E.; Jaramillo-Flores, M.E.; Legarreta, I.G. Antioxidant effect rosemary (*Rosmarinus officinalis* L.) and oregano (*Origanum vulgare* L.) extracts on TBARS and colour of model raw pork batters. Meat Sci. 2009, 81, 2, 410–417.

Hernandez-Izquierdo, V.M.; Krochta, J.M. Thermoplastic processing of proteins for film formation—a review. J. Food Sci. 2008, 73, 2, R30–R39.

Hu, H.; Zhou, H.; Li, P. Lacquer wax coating improves the sensory and quality attributes of kiwifruit during ambient storage. Sci. Hortic. (Amsterdam). 2019, 244, 31–41.

Huang, Q.; Qian, X.; Jiang, T.; Zheng, X. Effect of chitosan and guar gum based composite edible coating on quality of mushroom (Lentinus edodes) during postharvest storage. Sci. Hortic. 2019, 253, 382–389.

Isabel Castro-González, M. ÁCIDOS GRASOS OMEGA 3: BENEFICIOS Y FUENTES. Interciencia 2002, 27 (3), 128–136.

Jaimes, W.A.; Rocha, S.; Vesga, J.N.; Kafarov, V. Thermodynamic analysis to a real palm oil extraction process. Prospect 2012, 10, 61–70.

Jiang, R. et al. An Extraction Method Suitable for Two-Dimensional Electrophoresis of Low- abundant Proteins from Ginseng Roots. Lecture Notes in Electrical Engineering, vol. 251, Springer: London, 2014, 3.

Jiménez, A.; Requena, R.; Vargas, M.; Atarés, L.; Chiralt, A. Food hydrocolloids as matrices for edible packaging applications. In Role of Materials Science in Food Bioengineering. Academic Press: London, 2018; 263–299.

Jongjareonrak, A.; Benjakul, S.; Visessaguan, W.; Tanaka, M. Antioxidant activity and properties of fish skin gelatin films incorporated with BHT and α-tocopherol. Food Hydrocoll. 2008, 22, 449–458.

Jornet-Martínez, N.; Campíns-Falcó, P.; Hall, E.A.H. Zein as biodegradable material for effective delivery of alkaline phosphatase and substrates in biokits and biosensors. Biosens. Bioelectron. 2016, 86, 14–19.

Kadzińska, J.; Janowicz, M.; Kalisz, S.; Bryś, J.; Lenart, A. An overview of fruit and vegetable edible packaging materials. Packag. Technol. Sci., 2019, 32, 483–495.

Kaewklin, P.; Siripatrawan, U.; Suwanagul, A.; Lee, Y.S. Active packaging from chitosan-titanium dioxide nanocomposite film for prolonging storage life of tomato fruit. Int. J. Biol. Macromol. 2018, 112, 523–529.

Karbowiak, T.; Debeaufort, F.; Voilley, A.; Trystram, G. From macroscopic to molecular scale investigations of mass transfer of small molecules through edible packaging applied at interfaces of multiphase food products. Innov. Food Sci. Emerg. Technol. 2009, 10, 1, 116–127.

Kim, S.H.; No, H.K.; Prinyawiwatkul, W. Plasticizer types and coating methods affect quality and shelf life of eggs coated with chitosan. J. Food Sci. 2008, 73, 3, S111–S117.

Kinsella, J.E. Functional Properties of Soy Proteins. J. Am. Oil Chem. Soc. 1979, 56, 3, 242–258.

Krishna, M.; Nindo, C.I.; Min, S.C. Development of fish gelatin edible films using extrusion and compression molding. J. Food Eng. 2012, 108, 2, 337–344.

Kumar, R.; Zhang, L. Soy protein films with the hydrophobic surface created through non-covalent interactions. Ind. Crops Prod. 2009, 29, 2/3, 485–494.

Kurek, M.; Galus, S.; Debeaufort, F. Surface, mechanical and barrier properties of bio-based composite films based on chitosan and whey protein. Food. Packag. Shelf Life 2014, 1, 56–67.

Lebreton, L.; Andrady, A. Future scenarios of global plastic waste generation and disposal. Palgrave Commun. 2019, 5, 1, 6.

Lei, Y.; Wu, H.; Jiao, C.; Jiang, Y.; Liu, R.; Xiao, D.; Lu, J.; Zhang, Z.; Shen, G.; Li, S. Investigation of the structural and physical properties, antioxidant and antimicrobial activity of pectin-konjac glucomannan composite edible films incorporated with

tea polyphenol. Food Hydrocoll. 2019, 94, 128–135.

Li, C.; Tao, J.; Zhang, H. Peach gum polysaccharides-based edible coatings extend shelf life of cherry tomatoes. 3 Biotech. 2017, 7, 168.

Li, M.; Liu, P.; Zou, W.; Yu, L.; Xie, F.; Pu, H.; Chen, L. Extrusion processing and characterization of edible starch films with different amylose contents. J. Food Eng. 2011, 106, 1, 95–101.

Luque-García, J.L.; Luque de Castro, M.D. Ultrasound-assisted Soxhlet extraction: an expeditive approach for solid sample treatment. application to the extraction of total fat from oleaginous seeds. J. Chromatogr A. 2004, 1034, 237–242.

Ma, Q.; Hu, D.; Wang, H.; Wang, L. Tara gum edible film incorporated with oleic acid. Food Hydrocoll. 2016, 56, 127–133.

Mahalik, N.O.; Nambiar, A.N. (2010). Trends in food packaging and manufacturing systems and technology. Trends Food Sci. Technol. 2010, 21, 117–128.

Malhotra, B.; Keshwani, A.; Kharkwal, H. Antimicrobial food packaging: potential and pitfalls. Front. Microbiol. 2015, 6, 611–611.

Matta, E.; Tavera-Quiroz, M.J.; Bertola, N. Active edible films of methylcellulose with extracts of green apple (Granny Smith) skin. Int. J. Biol. Macromol. 2019, 124, 1292–1298.

McArt, S.H. et al. A modified method for determining tannin-protein precipitation capacity using accelerated solvent extraction (ASE) and microplate gel filtration. J. Chem. Ecol. 2006, 32, 6, 1367–1377.

McEvily, A.; Iyengar, R.; Otwell, S. Inhibition of enzymic browning in foods and beverages. Crit. Rev. Food. Sci. Nutr. 1992, 32, 3, 253–273.

Mihindukulasuriya, S.D.F.; Lim, L.T. Nanotechnology development in food packaging: a review. Trends Food Sci Technol. 2014, 40, 2, 149–167.

Moalemiyan, M.; Ramaswamy, H.S.; Maftoonazad, N. Pectin-based edible coating for shelf-life extension of Ataulfo mango. J. Food Process Eng. 2012, 35, 572–600.

Mohammadi Nafchi, A.; Olfat, A.; Bagheri, M.; Nouri, L.; Karim, A.A.; Ariffin, F. Preparation and characterization of a novel edible film based on Alyssum homolocarpum seed gum. J. Food Sci. Technol. 2017, 54, 1703–1710.

Montalvo, C.; López Malo, A.; Palou, E. Películas comestibles de proteína: características, propiedades y aplicaciones. Temas Selectos de Ingeniería de Alimentos 2012, 2, 32–46.

Montero-Garcia, M.D.P.; Gómez-Guillén, M.C.; López-Caballero, M.E.; Barbosa-Cánovas, G.V. Edible Films and Coatings: Fundamentals and Applications. CRC Press: Boca Raton, FL, 2016.

Moradi, M.; Tajik, H.; Rohani, S.M.R.; Mahmoudian, A. Antioxidant and antimicrobial effects of zein edible film impregnated with Zataria multiflora Boiss. essential oil and monolaurin. LWT—Food Sci. Technol. 2016, 72, 37–43.

Morales-Hernandez, M., Jr. Evaluación de las propiedades de candelilla y carnauba para su aplicación en emulsiones céreas de uso comercial. Degree Dissertation, Instituto Politécnico Nacional, Mexico City, 2015.

Motamedi, E.; Nasiri, J.; Malidarreh, T.R.; Kalantari, S.; Naghavi, M.R.; Safari, M. Performance of carnauba wax-nanoclay emulsion coatings on postharvest quality of 'Valencia' orange fruit. Sci. Hortic. (Amsterdam). 2018, 240, 170–178.

Mulchandani, K.; Kar, J.R.; Singhal, R.S. Extraction of lipids from chlorella saccharophila using high-pressure homogenization followed by three phase partitioning. Appl. Biochem. Biotechnol. 2015, 176, 6, 1613–1626.

Mulla, M.; Ahmed, J.; Al-Attar, H.; Castro-Aguirre, E.; Arfat, Y.A.; Auras, R. Antimicrobial efficacy of clove essential oil infused into chemically modified LLDPE

film for chicken meat packaging. Food Control. 2017, 73, 663–671.

Nawab, A.; Alam, F.; Haq, M.A.; Haider, M.S.; Lutfi, Z.; Kamaluddin, S.; Hasnain, A. Innovative edible packaging from mango kernel starch for the shelf life extension of red chili powder. Int. J. Biol. Macromol. 2018, 114, 626–631.

Ndlela, S.C. et al. Aqueous extraction of oil and protein from soybeans with subcritical wáter. J. Am. Oil Chem' Soc. 2012, 89, 6, 1145–1153.

Oliveira, V.R.L.; Santos, F.K.G.; Leite, R.H.L.; Aroucha, E.M.M.; Silva, K.N.O. Use of biopolymeric coating hydrophobized with beeswax in post-harvest conservation of guavas. Food Chem. 2018, 259, 55–64.

Oregel-Zamudio, E.; Angoa-Pérez, M.V.; Oyoque-Salcedo, G.; Aguilar-González, C.N.; Mena-Violante, H.G. Effect of candelilla wax edible coatings combined with biocontrol bacteria on strawberry quality during the shelf-life. Sci. Hortic. (Amsterdam). 2017, 214, 273–279.

Otoni, C.G.; Avena-Bustillos, R.J.; Azeredo, H.M.; Lorevice, M.V.; Moura, M.R.; Mattoso, L.H.; McHugh, T.H. Recent advances on edible films based on fruits and vegetables—a review. Compr. Rev. Food Sci. Food Saf. 2017, 16, 5, 1151–1169.

Otoni, C.G.; De Moura, M.R.; Aouada, F.A.; Camilloto, G.P.; Cruz, R.S.; Lorevice, M.V.; Mattoso, L.H. Antimicrobial and physical-mechanical properties of pectin/papaya puree/cinnamaldehyde nanoemulsion edible composite films. Food Hydrocolloids. 2014, 41, 188-194.

Palla. C.; Hegel, P.; Pereda, S.; Bottin, S. Extraction of Jojoba oil with liquid CO2+ propane solvent mixtures. J. Supercrit. Fluids 2014, 91, 37–45.

Pastor, C.; Sánchez-González, L.; Cháfer, M.; Chiralt, A.; González-Martínez, C. Physical and antifungal properties of hydroxypropylmethylcellulose based films containing propolis as affected by moisture content. Carbohydr. Polym. 2010, 82, 1174–1183.

Patra, A.; Verma, P.K.; Mitra, R.K. Slow relaxation dynamics of water in hydroxypropyl cellulose-water mixture traces its phase transition pathway: a spectroscopic investigation. J. Phys. Chem. B. 2012, 116 (5), 1508–1516.

Perez Sira, E.; Dufour, D. Native and modified starches as matrix for edible films and covers. Int. J. Nutr. Food Sci. 2017, 3, 3.

Perez-Gago, M.; Rojas, C.; Del Rio, M. Effect of lipid type and amount of edible composite coatings used to protect postharvest quality of mandarins cv. Fortune. Food Chem. Toxicol. 2002, 67, 2903–2910.

Perumalla, A.V.S.; Hettiarachchy, N.S. Green tea and grape seed extracts—potential applications in food safety and quality. Food Res. Int. 2011, 44, 4, 827–839.

Petriccione, M.; Mastrobuoni, F.; Pasquariello, M.; Zampella, L.; Nobis, E.; Capriolo, G.; Scortichini, M. Effect of chitosan coating on the postharvest quality and antioxidant enzyme system response of strawberry fruit during cold storage. Foods 2015, 4, 4, 501–523.

Ponce, A.G.; Roura, S.I.; Del Valle, C.E.; Moreira, M.R. Antimicrobial and antioxidant activities of edible coatings enriched with natural plant extracts: in vitro and in vivo studies. Postharvest Biol. Technol. 2008, 49, 294–300.

Pushkala, R.; Raghuram, P.K.; Srividya, N. Chitosan based powder coating technique to enhance phytochemicals and shelf life quality of radish shreds. Postharvest Biol. Technol. 2013, 86, 402–408.

Ragaert, P.; Devlieghere, F.; Debevere, J. Role of microbiological and physiological spoilage mechanisms during storage of minimally processed vegetables. Postharvest Biol. Technol. 2007, 44, 3, 185–194.

Rai, A.; Mohanty, B.; Bhargava, R. Modeling and response surface analysis of supercritical extraction of watermelon seed oil using carbon dioxide. Sep. Purif. Technol. 2015, 141, 354–365.

Rai, A.; Mohanty, B.; Bhargava, R. Supercritical extraction of sunflower oil: a central composite design for extraction variables. Food Chem. 2016, 192, 647–659.

Ramis-Ramos, G. Antioxidants. Synthetic Antioxidants. Encyclopedia of Food Sciences and Nutrition. In Finglas, P.; Toldra, F.; Caballero, B. Eds.; 2nd ed.; Academic Press: New Jersey, 2003, pp. 265–275.

Ramos-Ramírez, E.G.; Salazar-Montoya, J.A. Procedimiento para la obtención de aceite de coco a partir de pulpa fresca. Mexican Patent 192953, 1995.

Ravber, M.; Knez, Z.; Škerget, M. Simultaneous extraction of oil- and water-soluble phase from sunflower seeds with subcritical water. Food Chem. 2015, 166, 316–323.

Rawles, L.R. Proceedings of the Third International Conference on Jojoba and Its Uses. University of California, Riverside, CA, 1978, 279–283.

Reis, L.C.B.; De Souza, C.O.; Da Silva, J.B.A.; Martins, A.C.; Nunes, I.L.; Druzian, J.I. Active biocomposites of cassava starch: the effect of yerba mate extract and mango pulp as antioxidant additives on the properties and the stability of a packaged product. Food Bioprod. Process. 2015, 94, 382–391.

Reybroeck, W.; Jacobs, F.J.; De Brabander, H.F.; Daeseleire, E. Transfer of sulfamethazine from contaminated beeswax to honey. J. Agric. Food Chem. 2010, 58, 7258–7265.

Rhim, J.W.; Shellhammer, T.H. Lipid-based edible films and coatings. Innovations in Food Packaging. Elsevier Ltd: London, 2005.

Rivera-Hernández, J.R.; Lomelí-Soto, J.M.; Román-Salinas, L.; Vera-Figueroa, F. Extracción de aceite de coco a partir de la copra por medio de disolventes químicos. Conciencia Tecnológica [Online] 2001, 17. http://www.redalyc.org/articulo.oa?id= 94401703 (accessed June 12, 2019).

434 *Handbook of Research on Bioenergy and Biomaterials*

Robles-Sánchez, R.M.; Rojas-Graü, M.A.; Odriozola-Serrano, I.; González-Aguilar, G.; Martin-Belloso, O. Influence of alginate-based edible coating as carrier of anti-browning agents on bioactive compounds and antioxidant activity in fresh-cut Kent mangoes. LWT—Food Sci. Technol. 2013, 50, 1, 240–246.

Rodríguez, O.; Ortuño, C.; Simala, S.; Benedito, J.; Femenia, A.; Rosselló, C. Acoustically assisted supercritical CO2 extraction of cocoa butter: effects on kinetics and quality. J. Supercrit. Fluids. 2014, 94, 30–37.

Rojas, R.; Contreras-Esquivel, J.C.; Orozco-Esquivel, M.; Muñoz, C.; Aguirre-Joya, J.; Aguilar, C.N.; Castro-López, C.; Ventura-Sobrevilla, J.M.; González-Hernández, M.D.; Rojas, R.; Ascacio-Valdés, J.A.; Aguilar, C.N.; Martínez-Ávila, C.G.; De León-Zapata, M.A.; Sáenz-Galindo, A.; Rojas-Molina, R.; Rodríguez-Herrera, R.; Jasso-Cantú, D.; Aguilar, C.N.; Torres-León, C.; Rojas, R.; Contreras-Esquivel, J.C.; Serna-Cock, L.; Belmares-Cerda, R.E.; Aguilar, C.N.; Cázares, R.; Ruiz, R.B.; Fraga, J.C.; Charles, J.C.; Morales, L.E.; Rojas, R.; Flores, A.; Aranda, J.; Martínez-Avíla, C.; García Carmona, J.; Correa López, M. de J.; Rojas Figueroa, R.A.; Martínez Ibarra, R.; Castro-López, C.; Rojas, R.; Sánchez-Alejo, E.J.; Niño-Medina, G.; Martínez-Ávila, C.G.; Sierra-Hernandez, J.M.; Castillo-Guzman, A.; Selvas-Aguilar, R.; Vargas-Rodriguez, E.; Gallegos-Arellano, E.; Guzman-Chavez, D.; Estudillo-Ayala, J.M.; Jauregui-Vazquez, D.; Rojas-Laguna, R.; Lopez-Contreras, J.J.; Zavala-Garcia, F.; Urias-Orona, V.; Martinez-Avila, C.G.; Rojas, R.; Guillermo, N.-M. Edible candelilla wax coating with fermented extract of tarbush improves the shelf life and quality of apples. Grape Wine Biotechnol. 2015, 57, 1857–1860.

Rojas-Molina, R.; León-Zapata, M.A.; Saucedo-Pompa, S.; Aguilar-González,

M.Á.; Aguilar-González, C.N. Chemical and structural characterization of candelilla (Euphorbia antisyphilitica Zucc.). J. Med. Plants Res. 2013, 7, 702–705.

Rouilly, A.; Meriaux, A.; Geneau, C.; Silvestre, F.; Rigal, L. Film extrusion of sunflower protein isolate. Polym. Eng. Sci. 2006, 46(11), 1635–1640.

Ruan, C.; Zhang, Y.; Wang, J.; Sun, Y.; Gao, X.; Xiong, G.; Liang, J. Preparation and antioxidant activity of sodium alginate and carboxymethyl cellulose edible films with epigallocatechin gallate. Int. J. Biol. Macromol. 2019, 134, 1038–1044.

Ruan, C.; Zhang, Y.; Wang, J.; Sun, Y.; Gao, X.; Xiong, G.; Liang, J. Preparation and antioxidant activity of sodium alginate and carboxymethyl cellulose edible films with epigallocatechin gallate. Int. J. Biol. Macromol. 2019, 134, 1038–1044.

Saba, M.K.; Sogvar, O.B. Combination of carboxymethyl cellulose-based coatings with calcium and ascorbic acid impacts in browning and quality of fresh-cut apples. LWT—Food Sci. Technol. 2016, 66, 165–171.

Sabbah, M.; Di Pierro, P.; Cammarota, M.; Dell'Olmo, E.; Arciello, A.; Porta, R. Development and properties of new chitosan-based films plasticized with spermidine and/or glycerol. Food hydrocoll. 2019, 87, 245–252.

Salama, H.E.; Abdel Aziz, M.S.; Sabaa, M.W. Novel biodegradable and antibacterial edible films based on alginate and chitosan biguanidine hydrochloride. Int. J. Biol. Macromol. 2018, 116, 443–450.

Saldaña, M.D.A.; Mohameda, R.S.; Mazzafera, P. Extraction of cocoa butter from Brazilian cocoa beans using supercritical CO2 and ethane. Fluid Phase Equilibria. 2002, 194–197, 885–894.

Salgin, U. Extraction of Jojoba seed oil using supercritical CO2 + ethanol mixture in green and high-tech separation process. J. Supercrit Fluids 2007, 39, 330–337.

Salgin, U.; Çamili, A.; Zühtü-Uysal, B. Supercritical fluid extraction of Jojoba oil. J. Am. Oil Chem. Soc. 2004, 81, 293–296.

Salinas-Roca, B.; Soliva-Fortuny, R.; Welti-Chanes, J.; Martín-Belloso, O. Combined effect of pulsed light, edible coating and malic acid dipping to improve fresh-cut mango safety and quality. Food Control. 2016, 66, 190–197.

Sánchez-González, L.; Pastor, C.; Vargas, M.; Chiralt, A.; González-Martínez, C.; Cháfer, M. Effect of hydroxypropylmethylcellulose and chitosan coatings with and without bergamot essential oil on quality and safety of cold-stored grapes. Postharvest Biol. Technol. 2011, 60, 1, 57–63.

Sanchís, E.; González, S.; Ghidelli, C.; Sheth, C.C.; Mateos, M.; Palou, L.; Pérez-Gago, M.B. Browning inhibition and microbial control in fresh-cut persimmon (Diospyros kaki Thunb. cv. RojoBrillante) by apple pectin-based edible coatings. Postharvest Biol. Technol. 2016, 112, 186–193.

Sandoval-García, A.M.; Altamirano-Cárdenas, J.R.; Aguilar-Ávila, J.; García-Muñiz, J.G. Chemical characterization of oil obtained by homemade methods from three african palm varieties (Elaeis guineensis Jacq.). Rev. Fitotec. Mex. 2016, 39, 317–322.

Saucedo-Pompa, S.; Rojas-Molina, R.; Aguilera-Carbó, A.F.; Saenz-Galindo, A.; De la Garza, H.; Jasso-Cantú, D.; Aguilar, C.N. Edible film based on candelilla wax to improve the shelf life and quality of avocado. Food Res. Int. 2009, 42, 511–515.

Schmid, M.; Müller, K. Whey Protein-Based Packaging Films and Coatings, Whey Proteins. Elsevier Inc: London, 2018.

Schmid, M.; Pröls, S.; Kainz, D.M.; Hammann, F.; Grupa, U. Effect of thermally induced denaturation on molecular interaction-response relationships of whey protein isolate based films and coatings. Prog. Org. Coat. 2017, 104, 161–172.

Sharma, S.; Rao, T.R. Xanthan gum based edible coating enriched with cinnamic acid

prevents browning and extends the shelf-life of fresh-cut pears. LWT—Food Sci. Technol. 2015, 62, 1, 791–800.

Shimao, M. Biodegradation of plastics. Environ. Biotechnol. 2001, 12, 242–247.

Silva-Weiss, A.; Ihl, M.; Sobral, P.J.A.; Gómez-Guillén, M.C.; Bifani, V. Natural additives in bioactive edible films and coatings: functionality and applications in foods. Food Eng. Rev. 2013, 5, 4, 200–216.

Singh, P.; Magalhães, S.; Alves, L.; Antunes, F.; Miguel, M.; Lindman, B.; Medronho, B. Cellulose-based edible films for probiotic entrapment. Food Hydrocoll. 2019, 88, 68–74.

Singh, S.; Khemariya, P.; Rai, A.; Rai, A.C.; Koley, T.K.; Singh, B. Carnauba wax-based edible coating enhances shelf-life and retain quality of eggplant (Solanum melongena) fruits. LWT—Food Sci. Technol. 2016, 74, 420–426.

Sirelkhatim, A.; Mahmud, S.; Seeni, A.; Kaus, N.H.M.; Ann, L.C.; Bakhori, S.K.M.; Hasan, H.; Mohamad. D. Review on Zinc Oxide Nanopar, 2015.

Siripatrawan, U.; Noipha, S. Active film from chitosan incorporating green tea extract for shelf life extension of pork sausages. Food hydrocoll. 2012, 27, 1, 102–108.

Soares, N.F.F.; Pires, A.C.S.; Camilloto, G.P.; Santiago-Silva, P.; Espitia, P.J.P.; Silva, W.A. Recent patents on active packaging for food application. Recent Pat. Food Nutr. Agric. 2009, 1, 1, 171–178.

Sogvar, O.B.; Saba, M.K.; Emamifar, A. Aloe vera and ascorbic acid coatings maintain postharvest quality and reduce microbial load of strawberry fruit. Postharvest Biol. Technol. 2016, 114, 29–35.

Soliva-Fortuny, R.C.; Martín-Belloso, O. New advances in extending the shelf-life of fresh-cut fruits: a review. Trends Food Sci. Technol. 2003, 14, 9, 341–353.

Soradech, S.; Nunthanid, J.; Limmatvapirat, S.; Luangtana-anan, M. Utilization of shellac and gelatin composite film for coating to extend the shelf life of banana. Food Control 2017, 73, 1310–1317.

Spadaro, J.J.; Lambou, M.G. Proceedings of the first international conference on Jojoba and its uses, Tucson, Arizona. Office of Arid Lands Studies College of Earth Sciences, University of Arizona, Tucson, 1972; 47–60.

Spasojević, L.; Katona, J.; Bučko, S.; Savić, S.M.; Petrović, L.; Milinković Budinčić, J.; Tasić, N.; Aidarova, S.; Sharipova, A. Edible water barrier films prepared from aqueous dispersions of zein nanoparticles. LWT—Food Sci. Technol. 2019, 109, 350–358.

Sucheta; Chaturvedi, K.; Sharma, N.; Yadav, S.K. Composite edible coatings from commercial pectin, corn flour and beet-root powder minimize post-harvest decay, reduces ripening and improves sensory liking of tomatoes. Int. J. Biol. Macromol. 2019, 133, 284–293.

Sudo, S. Dielectric properties of the free water in hydroxypropyl cellulose. J. Phys. Chem. B. 2011, 115 (1), 2–6.

Supapvanich, S.; Prathaan, P.; Tepsorn, R. Browning inhibition in fresh-cut rose apple fruit cv. Taaptimjaan using konjac glucomannan coating incorporated with pineapple fruit extract. Postharvest Biol. Technol. 2012, 73, 46–49.

Taboada-Reyes, F. Procedimiento y equipo para extraccion de cera de candelilla con disolventes selectivos. Mexican Patent MX164638, 1992.

Tavassoli-Kafrani, E.; Shekarchizadeh, H.; Masoudpour-Behabadi, M. Development of edible films and coatings from alginates and carrageenans. Carbohydr. Polym. 2015, 137, 360–374.

Tesfay, S.Z.; Magwaza, L.S. Evaluating the efficacy of moringa leaf extract, chitosan and carboxymethyl cellulose as edible coatings for enhancing quality and extending postharvest life of avocado (Persea americana Mill.) fruit. Food Packag. Shelf Life 2017, 11, 40–48.

Thakur, R.; Pristijono, P.; Scarlett, C.J.; Bowyer, M.; Singh, S.P.; Vuong, Q.V. Starch-based films: major factors affecting their properties. Int. J. Biol. Macromol. 2019, 132, 1079–1089.

Tharanathan, R.N. Biodegradable films and composite coatings: Past, present and future. Trends Food Sci. Technol. 2003, 14, 71–78.

Toivonen, P.M.; DeEll, J.R. Physiology of fresh-cut fruits and vegetables. In Fresh-cut fruits and vegetables. CRC Press: Boca Raton, FL, 2002, 99–131.

Toldrá, F.; Nollet, L.M.L. Proteomics in Foods: Principles and Applications. Springer Science & Business Media: London, 2013, 1–590.

Tomadonia, B.; Moreira, M.R.; Pereda, M.; Ponce, A.G. Gellan-based coatings incorporated with natural antimicrobials in fresh-cut strawberries: microbiological and sensory evaluation through refrigerated storage. LWT—Food Sci. Technol. 2018, 97, 384–389.

Torres-León, C.; Vicente, A.A.; Flores-López, M.L.; Rojas, R.; Serna-Cock, L.; Alvarez-Pérez, O.B.; Aguilar, C.N. Edible films and coatings based on mango (var. Ataulfo) by-products to improve gas transfer rate of peach. LWT—Food Sci. Technol. 2018, 97, 624–631.

Treviño-García, J. Method of extracting candelilla wax. U.S. Patent 1,715,194, 1929.

Ullsten, N.H.; Gallstedt, M.; Johansson, E.; Graslund, A.; Hedenqvist, M.S. Enlarged processing window of plasticized wheat gluten using salicylic acid. Biomacromolecules 2006, 7, 3, 771–776.

Valencia-Chamorro, S.A.; Palou, L.; Del Río, M.A.; Pérez-Gago, M.B. Antimicrobial edible films and coatings for fresh and minimally processed fruits and vegetables: a review. Crit. Rev. Food Sci. Nutr. 2011, 51, 872–900.

Valencia-Chamorro, S.A.; Pérez-Gago, M.B.; Del Río, M.Á.; Palou, L. Effect of antifungal hydroxypropyl methylcellulose (HPMC)-lipid edible composite coatings on postharvest decay development and quality attributes of cold-stored "Valencia" oranges. Postharvest Biol. Technol. 2009, 54, 72–79.

Valero, D.; Díaz-Mula, H.M.; Zapata, P.J.; Guillén, F.; Martínez-Romero, D.; Castillo, S.; Serrano, M. Effects of alginate edible coating on preserving fruit quality in four plum cultivars during postharvest storage. Postharvest Biol. Technol. 2013, 77, 1–6.

Vanderroost, M.; Ragaert, P.; Devlieghere, F.; De Meulenaer, B. Intelligent food packaging: the next generation. Trends Food Sci. Technol. 2014, 39, 47–62.

Vargas, M.; Pastor, C.; Chiralt, A.; McClements, D.J.; González-Martínez, C. Recent advances in edible coatings for fresh and minimally processed fruits. Crit. Rev. Food Sci. Nutr. 2008, 48, 6, 496–511.

Velderrain-Rodríguez, G.R.; Ovando-Martínez, M.; Villegas-Ochoa, M.; Ayala-Zavala, J.F.; Wall-Medrano, A.; Álvarez-Parrilla, E.; González-Aguilar, G.A. Antioxidant capacity and bioaccessibility of synergic mango (cv. Ataulfo) peel phenolic compounds in edible coatings applied to fresh-cut papaya. Food Nutr. Sci. 2015, 6, 3, 365.

Vidhate, G.S.; Singhal, R.S. Extraction of cocoa butter alternative from kokum (Garcinia indica) kernel by three phase partitioning. J. Food Eng. 2013, 117, 464–466.

Vieira, J.M.; Flores-López, M.L.; De Rodríguez, D.J.; Sousa, M.C.; Vicente, A.A.; Martins, J.T. Effect of chitosan–Aloe vera coating on postharvest quality of blueberry (Vaccinium corymbosum) fruit. Postharvest Biol. Technol. 2016, 116, 88–97.

Walstra, P. et al. Dairy Technology-Principles of Milk Properties and Processes. Marcel Dekker Inc.: New York, NY, 2005.

Wambura, P.; Yang, W.; Mwakatage, N.R. Effects of sonication and edible coating containing rosemary and tea extracts on reduction of peanut lipid oxidative rancidity. Food Bioproc. Technol. 2011, 4, 107–115

Wang, D. et al. Combined enzymatic and mechanical cell disruption and lipid extraction of green alga Neochloris oleoabundans. Int. J. Mol. Sci. 2015, 16, 4, 7707–7722.

Wang, Q. Peanut Processing Characteristics and Quality Evaluation. Springer: London, 2017.

Wang, X.; Guo, C.; Hao, W.; Ullah, N.; Chen, L.; Li, Z.; Feng, X. Development and characterization of agar-based edible films reinforced with nano-bacterial cellulose. Int. J. Biol. Macromol. 2018, 118, 722–730.

Wang, X.; Kong, D.; Ma, Z.; Zhao, R. Effect of carrot puree edible films on quality preservation of fresh-cut carrots. Irish J. Agri. Food Res. 2015, 54, 1, 64–71.

Windsor, F.M.; Durance, I.; Horton, A.A.; Thompson, R.C.; Tyler, C.R.; Ormerod, S.J. A catchment-scale perspective of plastic pollution. Global Change Biol. 2019, 25, 4, 1207–1221.

Wu, H.; Lei, Y.; Zhu, R.; Zhao, M.; Lu, J.; Xiao, D.; Li, S. Preparation and characterization of bioactive edible packaging films based on pomelo peel flours incorporating tea polyphenol. Food Hydrocoll. 2019, 90, 41–49.

Wu, S. Extending shelf-life of fresh-cut potato with cactus Opuntia dillenii polysaccharide-based edible coatings. Int. J. Biol. Macromol. 2019, 130, 640–644.

Xiao, C.; Zhu, L.; Luo, W.; Song, X.; Deng, Y. Combined action of pure oxygen pretreatment and chitosan coating incorporated with rosemary extracts on the quality of fresh-cut pears. Food Chem. 2010, 121, 4, 1003–1009.

Xiao, Z.; Luo, Y.; Luo, Y.; Wang, Q. Combined effects of sodium chlorite dip treatment and chitosan coatings on the quality of fresh-cut d'Anjou pears. Postharvest Biol. Technol. 2011, 62, 319–326.

Xing, Y.; Li, X.; Xu, Q.; Yun, J.; Lu, Y.; Tang, Y. Effects of chitosan coating enriched with cinnamon oil on qualitative properties of sweet pepper (Capsicum annuum L.). Food Chem. 2011, 124, 4, 1443–1450.

Yan, J.; Luo, Z.; Ban, Z.; Lu, H.; Li, D.; Yang, D.; Aghdam, M.S.; Li, L. The effect of the layer-by-layer (LBL) edible coating on strawberry quality and metabolites during storage. Postharvest Biol. Technol. 2019, 147, 29–38.

Yemiş, G.P.; Candoğan, K. Antibacterial activity of soy edible coatings incorporated with thyme and oregano essential oils on beef against pathogenic bacteria. Food Sci. Biotechnol. 2017, 26, 4, 1113–1121.

Yen, M.T.; Yang, J.H.; Mau, J.L. Antioxidant properties of chitosan from crab shells. Carbohydr. Polym. 2008, 74, 4, 840–844.

Yıldırım-Yalçın, M.; Şeker, M.; Sadıkoğlu, H. Development and characterization of edible films based on modified corn starch and grape juice. Food Chem. 2019, 292, 6–13.

Yoruk, R.; Marshall, M.R. Physicochemical properties and function of plant polyphenol oxidase: a review. J. Food Biochem. 2003, 27, 5, 361–422.

Younis, H.G.R.; Zhao, G. Physicochemical properties of the edible films from the blends of high methoxyl apple pectin and chitosan. Int. J. Biol. Macromol. 2019, 131, 1057–1066.

Yousuf, B.; Qadri, O.S.; Srivastava, A.K. Recent developments in shelf-life extension of fresh-cut fruits and vegetables by application of different edible coatings: a review. LWT—Food Sci. Technol. 2018, 89, 198–209.

Yousuf, B.; Srivastava, A.K. A novel approach for quality maintenance and shelf life extension of fresh-cut Kajari melon: effect of treatments with honey and soy protein isolate. LWT—Food Sci. Technol. 2017, 79, 568–578.

Yousuf, B.; Srivastava, A.K. Impact of honey treatments and soy protein isolate-based coating on fresh-cut pineapple during storage at 4°C. Food Packag. Shelf Life. 2019, 21, 100361.

Yuan, G.; Lv, H.; Tang, W.; Zhang, X.; Sun, H. Effect of chitosan coating combined with pomegranate peel extract on the quality of Pacific white shrimp during iced storage. Food Control 2016, 59, 818–823.

Zaher, F.A.; El Kinawy, O.S.; El Haron, D.E. Solvent extraction of jojoba oil from pre-pressed jojoba meal. Grasas y Aceites. 2004, 55, 129–134.

Zambrano-Zaragoza, M. L.; Mercado-Silva, E.; Del Real, A.; Gutierrez-Cortez, E.; Cornejo-Villegas, M.A.; Quintanar-Guerrero, D. The effect of nano-coatings with α- tocopherol and xanthan gum on shelf-life and browning index of fresh-cut "Red Delicious" apples. Innov. Food Sci. Emerg. Technol. 2014, 22, 188–196.

Zambrano-Zaragoza, M.L.; Quintanar-Guerrero, D.; Real, A.D.; Pinon-Segundo, E.; Zambrano-Zaragoza, J.F. The release kinetics of β-carotene nanocapsules/xanthan gum coating and quality changes in fresh-cut melon (cantaloupe). Carbohydr. Polym. 2017, 157, 1874–1882.

Zárate-Polanco, L.M.; Ramírez-Suárez, L.M.; Otálora-Santamaría, N.A.; Prieto, L.; Garnica-Holguín, A.M.; Cerón-Lasso, M.S.; Argüelles, J.H. 2014. Extracción y caracterización de almidón nativo de clones promisorios de papa criolla (Solanum tuberosum, Grupo Phureja). Revista Latinoamericana de la Papa 18(1), 2–24.

Zeng, K.; Deng, Y.; Ming, J.; Deng, L. Induction of disease resistance and ROS metabolism in navel oranges by chitosan. Sci. Hortic. 2010, 126, 223–228.

Zhang, R.; Wang, W.; Zhang, H.; Dai, Y.; Dong, H.; Hou, H. Effects of hydrophobic agents on the physicochemical properties of edible agar/maltodextrin films. Food Hydrocoll. 2019, 88, 283–290.

Zhao, Y. Edible coatings for extending shelf-life of fresh produce during postharvest storage. In Encyclopedia of Food Security and Sustainability. Elsevier: Amsterdam, 2019, 506–510.

Zheng, K.; Xiao, S.; Li, W.; Wang, W.; Chen, H.; Yang, F.; Qin, C. Chitosan-acorn starch-eugenol edible film: physico-chemical, barrier, antimicrobial, antioxidant and structural properties. Int. J. Biol. Macromol. 2019, 135, 344–352.

CHAPTER 15

Bioplastics from Plant Oils and Sugars

ALMA BERENICE JASSO-SALCEDO[1], ARACELI MARTÍNEZ PONCE[2],
CHRISTIAN JAVIER CABELLO ALVARADO[3],
MARLENE LARIZA ANDRADE GUEL[4], and
CARLOS ALBERTO ÁVILA ORTA[4,*]

[1]*CONACYT—Centro de Investigación en Química Aplicada, Blvd. Enrique Reyna Hermosillo No. 140 C.P. 25294 Saltillo, Coahuila, México*

[2]*Escuela Nacional de Estudios Superiores, Unidad Morelia, Universidad Nacional Autónoma de México, Antigua Carretera a Pátzcuaro No. 8701, Col. Ex Hacienda de San José de la Huerta, 58190 Morelia, Michoacán, México*

[3]*CONACYT—Consorcio de Investigación Científica, Tecnológica y de Innovación del Estado de Tlaxcala, Calle 1 de mayo No. 22, Col. Centro, C.P. 90000, Tlaxcala de Xicothenccatl Tlaxcala, México*

[4]*Department of Advanced Materials, Centro de Investigación en Química Aplicada, Blvd. Enrique Reyna Hermosillo No. 140 C.P. 25294 Saltillo, Coahuila, México*

Corresponding author. E-mail: carlos.avila@ciqa.edu.mx.

ABSTRACT

Bioplastics can partially participate in the biological carbon cycle which implies less accumulation and less environmental impact. Bioplastics are a type of plastics displaying biodegradability properties because they proceed from a renewable resource like biomass. The renewable biomass sources (vegetable fats and oils, sugars, starch, cellulose, and others) are either chemically synthesized from biological material or entirely biosynthesized by a living organism. In this regard, this chapter shows the different synthetic routes of bioplastics, properties, the process of biodegradation, and its regulations. A particular emphasis on bioplastics synthesized from plant oils and sugars. Indeed, there is an opportunity to move toward renewable sources and more sustainable synthetic and processing methods. A section describes the progress of bioplastics research in Mexico and developing countries. Finally, we display an overview of the potential applications of bioplastics in textiles, health, food product coatings and packing materials, and agriculture.

15.1 INTRODUCTION

Carbon dioxide emissions have reached alarming levels in the atmosphere (Institution of Oceanography, 2019). Carbon capture and utilization technologies could reduce the environmental impact of CO_2 (Cuéllar-Franca and Azapagic, 2015). For instance, the production of plant-based feedstock for bioplastics requires sunlight and some nutrients to capture and transform carbon dioxide into sugar, oil, and biomass. By doing so, the bioplastic industry could reduce its environmental footprint but still face challenges as land usage competition, transportation, upscaling, and plastic industrials perception.

We perceive the lack of international and national standards explicitly designed for bioplastics; therefore, we inquire about the applicability of *commodity* plastic regulations by summarizing them in a structured manner for the reader to select. Additionally, the complexity of bioplastic containing both biological and fossil carbon-based fractions requires an upgrade of production and recycling installations, as well as acceptance in the circular economy of food packaging, pharmaceutical, biomedical, textile, automotive, and agriculture sectors.

In this context, this chapter abridges the latest bioplastics synthetic routes (of the past 15 years) from sugars and oils obtained from plants. The advances in the green processes of bioplastics in non-European countries are highlighted. For instance, researchers from the USA, Brazil, Mexico, Chile, Argentina, and Colombia have shown relevant advances in the enhancement of mechanical properties using nanoparticles, essential oils, and others. Just last year, Mexican and Chilean researchers published a protocol using cheap and less toxic solvents that permitted a molecular weight tunning (García et al., 2019). The relevance of those studies conveys to the use of cost-effective and environmentally friendly precursors in the new bioplastic industry. Recent scientific papers and patents on bioplastics and bio-based plastics of potential use as mulch, pest control, postharvest, and food packaging are revised.

15.2 BIOPLASTICS OVER *COMMODITIES* PLASTICS

The *commodity* plastics have been the most recognized materials worldwide. It estimates that over 250 million tons of these plastics annually are produced (PlasticsEurope, 2015). Some smaller application as materials for electronics or automotive, *commodity* plastics are mainly used in packaging, which is predominantly composed of polyethylene (PE), polypropylene (PP), and polyethylene terephthalate (PET), and as construction materials which is the next largest consuming sector composed mainly of polyvinyl chloride (Geyer et al., 2017). These polymeric materials are long lasting and show good mechanical and thermal properties, which are attractive in many applications (Allen et al., 1994; Edge et al., 1991). However, the increase of packaging and other short-term uses causes a disposal problem for our future generation due to the undefined environmental fate of the materials (Barlow and Morgan, 2013; Eriksen et al., 2013; Ivleva et al., 2017). It is well-known that the huge development of *commodity* plastics made from petroleum-based synthetic polymers unable to degrade in landfills or the natural environment had led to serious environmental issues (Barnes et al., 2009). In response to this increasing awareness, the use of polymers stemming from renewable and sustainable resources to develop bioplastics constitutes a promising alternative to mostly nondegradable *commodity* polymers, to combat the global plastic waste problem (Chanprateep, 2010; Endres, 2017).

Bioplastics cover a wide range of polymers, each of which has attributes in terms of its impact on the environment. According to European Bioplastics e.V., bioplastics are either based on renewable resources, biodegradable, or features both properties (Horvat and Kržan, 2012; Song et al., 2009). IUPAC describes bioplastics as "polymer derived from the biomass or issued from monomers derived from the biomass and which, at some stage in its processing into finished products, can be shaped by flow," these can be renewable or nonrenewable (Vert et al., 2012). Different classifications of various bioplastics have been proposed (Song et al., 2011), and we choose to classify the bioplastics according to their origins and synthesis process (Figure 15.1).

In contrast to the bioplastics that are polymers formed in nature during the growth cycles of living organisms such as natural rubber, starch, proteins, lignin, and chitosan; the ones chemically synthesized from renewable resources are known as bio-based plastics. According to some authors, the term bioplastics includes three types of plastics

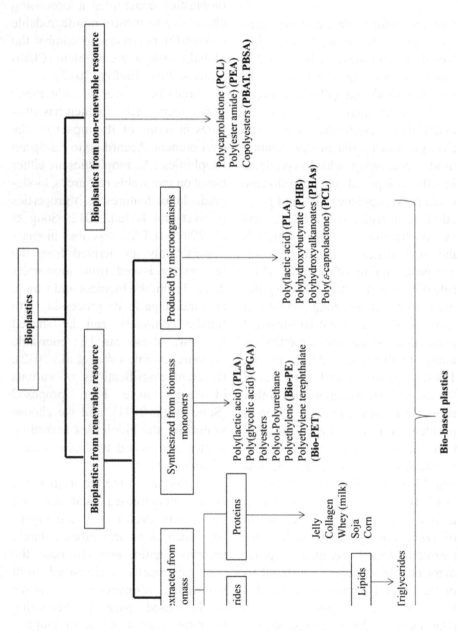

FIGURE 15.1 Classification of bioplastics based on their origins and synthetic process.

of which two types are bio-based (biodegradable and nonbiodegradable) (Spierling et al., 2018); however, the IUPAC only considers bio-based plastics as bioplastic "composed or derived in whole or in part of biological products issued from the biomass (including plant, animal, and marine or forestry materials)." The bio-based plastics are not necessarily environmentally friendly nor biocompatible nor biodegradable, especially if it is similar to oil-based plastics (Vert et al., 2012). The synthesis of the bio-bio-based plastics as bio-PE, bio-polyurethanes, and bio-poly(ethylene terephthalate) (bio-PET) using bioethanol (by the fermentation of glucose), vegetable oils, and bio-based ethylene glycol (30% wt.) as raw material, respectively, there are not biodegradable (Figure 15.2). However, bio-based plastics as poly(lactic acid) (PLA), poly(ε-caprolactone), and polyhydroxyalkanoates (PHAs) are materials containing ester groups in their polymeric chain, which can be degraded by certain enzymes secreted by microorganism and are

thus categorized as bio-based biodegradable plastics, as shown in Figure 15.2 (Iwata, 2015). The sustainability of bio-based plastics and its suitable end-of-life option for particular applications depends on its properties.

15.2.1 BIO-BASED-PLASTICS FROM PLANT OILS

Plants oils are alternative sources for the preparation of bio-based plastics, such as polyesters derivatives, polyurethanes, polycarbonates, polyamides, and other bioplastics. In particular, plant oils are considered renewable feedstocks, cheap, and plentiful application possibilities. It is important to note that the vegetable oils (triglycerides) and the essential oils (terpenes and terpenoids) are part of plant oils. This section will focus on bio-based plastics from vegetable oils.

Vegetable oils are extracted from plants, such as sunflower, corn, soja, *Jatropha curcas* L, *Ricinus communis*, microalgae, and others. Several plants naturally produce

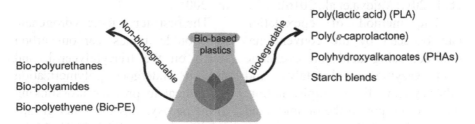

FIGURE 15.2   Nonbiodegradable and biodegradable bio-based plastics.

different oils mainly a mixture of triglycerides, with a different composition of the alkyl chains depending on their origin. Therefore, a great variety of materials can prepare from vegetable oils.

Fatty acid alkyl esters obtained by transesterification of triglycerides from vegetable oils represent the primary feedstock for aliphatic polyesters formation. It is well-known that the ester linkage is hydrolyzable and degradable because of the action of naturally occurring microorganisms such as algae, bacteria, and fungi; thus, polyesters are prone to degradability under specific conditions. Fatty acids alkyl esters are used for chemical conversions, as well as for the direct synthesis of bio-based plastics.

Whether the vegetable oils transesterification synthesis occurs or not, the carbon–carbon double bonds of their fatty acid chains can be reactive and incorporate in a polymerization reaction. Epoxidation and oxy-polymerization are two methods commonly used for the modification of the carbon–carbon double bonds of the triglycerides (Figure 15.3) (Saurabh et al., 2011; Vaidya et al., 2016).

The method of epoxidation can be seen by the conventional method, using acid ion exchange resin, enzyme, metal catalyst, and other systems. For example, authors have been reported the double bond epoxidation of linseed, cottonseed,

soybean, sunflower, corn, and jatropha oils in presence of organic peroxy acid (peroxyacetic acid or hydrogen peroxide), catalyzed by liquid inorganic acids (HCl, H_2SO_4, HNO_3, or H_3PO_4) and carboxylic acid (CH_3COOH or HCOOH) as oxygen carrier (Cai et al., 2008; Dinda et al., 2008; Seniha Güner et al., 2006). These epoxidized triglyceride oils can polymerize via cationic polymerization. When the oils having double bonds are oxidized, they undergo cationic polymerization.

Oxidized oils are widely used in the manufacturing of oil-based binders because they give the final products having high viscosity and excellent film properties (Shahidi and Fereidoon, 2005). Oxy-polymerization is one of the methods commonly used for the modification of triglyceride oils. It has been reported the oxidized soybean oil from permanganate oxidation with sub/supercritical CO_2 (Mercangöz et al., 2004). It found that linseed oil was oxy-polymerized under air at 80 °C, 120 °C, and 200 °C. The oxy-polymerized oils showed plastic liquid type flow behavior (Güler et al., 2004).

The first step is to dehydrogenate the double bonds carbon–carbon found on the triacylglycerol, and then it undergoes polymerization via radical to produce crosslinking of the peroxy, alkyl, or ether groups (Mallégol et al., 2000, 2001; Meier

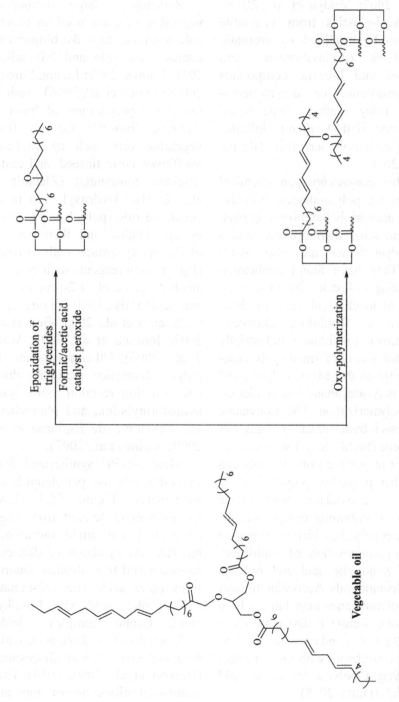

FIGURE 15.3 Modification of carbon–carbon double bonds of the triglycerides by epoxidation and oxy-polymerization.

et al., 2007; Vaidya et al., 2016). Bio-based-plastics from vegetable oils can be modified by aromatic derivatives (divinylbenzene and styrene), and bicyclic compounds (norbornadiene, or dicyclopentadiene) using methyl oleate based trifluoride diethyl etherate initiator via cationic polymerization (Ronda et al., 2011).

Other *commodity* petrochemical plastics are polyurethanes. Polyurethanes are mainly thermosetting plastics, but some thermoplastic grades of polyurethanes are also available. They have found application in coating industries for their wide range of mechanical strength, low-temperature flexibility, chemical, and corrosion resistance. Industrially polyurethanes are formed by the reaction between di-, tri- or polyols, and di- or polyisocyanates via condensation polymerization. The isocyanate synthesis is by the reaction of gaseous phosgene (highly toxic) with amines or their respective salts. Polyols can be either polyether polyol, synthesized by epoxidation with active hydrogen-containing compounds, or polyester polyols, which are prepared by polycondensation of multifunctional carboxylic acid and polyhydroxy compounds. According to their use, polyurethanes may have a high molecular weight (2000 g mol^{-1} to 10,000 g mol^{-1}) and are consider flexible polyurethanes with lower molecular weight polyols are more rigid materials (Dutta, 2018).

Recently, a large number of vegetable oils are used as renewable resources to make biopolyurethanes (Athawale and Nimbalkar, 2011; Endres, 2017; Lu and Larock, 2008; Noreen et al., 2016). Authors report the preparation of biopolyurethane (bio-PU) derived from vegetable oils such as soybean, sunflower, corn, linseed, and castor (Ricinus communis) (Zlatanic et al., 2004). Hydroxyl functional vegetable oils (polyols) can obtain by epoxidation and then reaction of the epoxy groups with different ring-opening reagents such as water, alcohol, glycerol, 1,2-propanediol, and acids (HBr, HCl) (Figure 15.4) (Adhvaryu et al., 2005; Guo et al., 2007; Ionescu et al., 2007; Wang et al., 2009). Other routes for the polyols formation including thiol-ene coupling reaction, ozonolysis, hydroformylation, and photochemical oxidation (de Espinosa et al., 2009; Narine et al., 2007).

Most bio-PU synthesized from vegetable oils use petroleum-based isocyanates. Figure 15.4 shows the isocyanate derived from vegetable oil. For example, the authors reported the synthesis of diisocyanates derived from diacids obtained from oleic acid. The *self*-metathesis of oleic acid with Ru-alkylidene Grubb complex yielded 1,18-octadec-9-enedioic acid, which then converted it into diisocyanate (Hojabri et al., 2009, 2010). Fatty acid-based diisocyanates may also

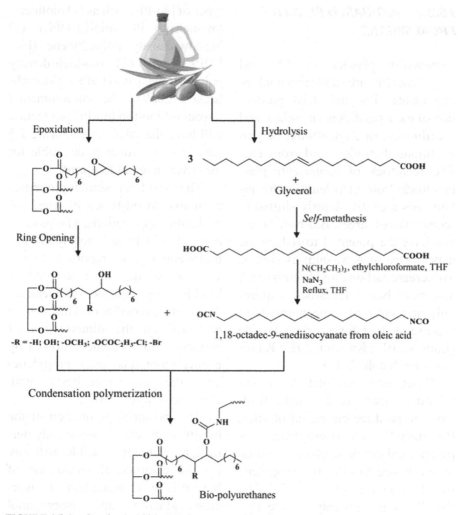

FIGURE 15.4 Synthesis of bio-PU from vegetable oil.

be synthesized from diesters by hydrazinolying dimethyl sebacate in absolute ethanol to generate diacyl hydrazide, which in turn is converted to diacyl azide (More et al., 2013).

It has reported the synthesis of polyols and polyurethanes from linseed seed and passion fruit oils. In situ epoxidation and hydroxylation of vegetable oils in a single step were accomplished using a mixture of hydrogen peroxide and formic acid (de Vasconcelos Vieira Lopes et al., 2013).

15.2.2 BIO-BASED PLASTICS FROM SUGARS

Commodity plastics as PE and propylene (PP) are stable and nonbiodegradable. The industrial production of these plastics is via radical and coordination or Zigler-Natta polymerizations of ethylene and propylene. The feedstock of *commodity* plastics made from petroleum hydrocarbons has recently slightly shifted to renewable resources. Bio-based plastics have the potential to reduce the dependence of petroleum as well as to decrease carbon dioxide emissions. The plant-based feedstock requires only sunlight and some nutrients to capture and transform carbon dioxide (Endres, 2017; Kabasci, 2013; Kamigaito and Satoh, 2015).

Bioethanol obtained from fermenting sugars is the main feedstock to produce bio-based plastics. Bioethanol turns to ethylene, propylene, and ethylene glycol are used to synthesize bio-PE, bio-propylene (bio-PP), and bio-PE terephthalate (bio-PET), respectively. Figure 15.5 shows several bio-based plastics using as feedstocks to the sugarcane.

The Brazilian company Braskem was the first to produce bioethylene by dehydrating ethanol obtained from the fermentation of sugarcane (Bozell and Petersen, 2010; Fuessl et al., 2012; Mathers, 2012; Muellhaupt, 2013). The bioethylene is polymerized and commercialized various types of bio-PE, such as (1) biolinear-low-density PE (bio-LLDPE), (2) bio-low-density polyethylene (bio-LDPE), and (3) bio-high-density polyethylene (bio-HDPE) (Morschbacker, 2009). The environmental impact of one-ton bio-PE production will have the effect of capturing 2.5 tons of CO_2, which is favorable for the environmental footprint.

The bio-PE has similar properties compared to oil-based PE used for packaging applications. The production of bioethanol to synthesize propylene is also reported. Figure 15.5 shows the steps to produce bio-PP using feedstock ethylene by dehydrating bioethanol. This method is based on the dimerization of ethylene into 1-butene followed by isomerization (*cis-*, *trans-*propylene) and the *cross*-metathesis with ethylene (Dwyer, 2007).

The industrial production of the bio-PP is by coordination polymerization. Another possible pathway for the commercial production of bio-PP is by fermentation of sugarcane glucose into isopropanol followed by dehydration or using the glycerol of transesterification of vegetable oils as feedstocks that can be dehydrogenated to produce propylene (Yu et al., 2006). A major market for bio-PP is the automotive industry; it is also used in the packaging and devices like living hinges and textiles.

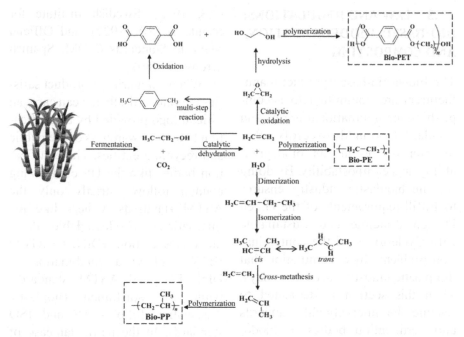

FIGURE 15.5 Bio-based plastics (bio-PE, bio-PP, and bio-PET) using as feedstocks to the sugarcane.

Other bio-based plastics that replaces *commodity* plastics is bio-PET. The bio-PET is widely used as container packaging materials. Bio-PET is produced from biomass using bioethylene glycol that is produced by bioethanol derived from the fermentation sugarcane glucose. The biomass content in bio-PET is approximately 30%. The bio-PET was in 2009, the first generation of PlantBottle™ technology from Coca Cola (Muellhaupt, 2013). The production of bio-PET with 100% biomass was reported as the new generation of PlantBottle™. Bio-PET is produced

from bio-based ethylene glycol and bio-based terephthalic acid (Figure 15.5). The monomer ethylene glycol is produced by bioethanol derived from the fermentation sugarcane glucose and the terephthalic acid is derived from *p*-xylene also obtained by sugarcane (multistep reaction): (1) trimerization of ethylene to hexane, (2) catalytic disproportionation of hexane to 2,4-hexadiene, (3) Diels–Alder reaction between ethylene, and (4) 2,4-hexadiene, and dehydrogenation of 3,6-dimethylcyclohexadiene to *p*-xylene) (Pang et al., 2016).

15.3 LAW AND REGULATIONS: BIO-BASED, BIODEGRADATION, AND COMPOSTING

The bioplastic-based product manufacturers are committed to accomplishing the international norms and standards thoroughly to certify either its biomass origin, its biodegradability, and compostability. By doing so, the bioplastic industry ensures to fulfill requirements of the Green Design Principle and Sustainable Life Cycle to avoid the contamination problems by accumulation that the plastic industry faces nowadays. Then this section is dedicated to resume the international standards and certification bodies of biodegradable and compostable materials and products.

There are at least three international organizations with established guidelines, standards, and definitions for a polymeric material to be considered biodegradable and/or compostable. The ISO (International Organization for Standardization, 1947), EN (European Standards, 1961) and the ASTM (American Society for Testing and Materials, 1898) are those leading standards organizations. Other organizations are the German Institute for Standardization (DIN, German acronym, 1917), Belgian accredited inspection and certification organization (VINÇOTTE, Belgian acronym, 1989), Japanese Standards Association (JSA, 1945), Australian standards

(AS, 1922), Swedish Institute for Standards (SIS, 1922), and Official Mexican Standards (NOM, Spanish acronym, 1980).

When a material or product satisfies the standards, then it can carry an official logo provided by a certification body for awareness of the public and recycling entities. The certification bodies like the US composting council follows strictly only the ASTM standards. Others like the International Biodegradable Polymers Association (DIN CERTCO IBAW) follow a combination of DIN, EN, and ASTM standards, while the Australian Bioplastic Association follows AS and ISO standards. In the particular case of Sweden, the country designed its own set of SPCR 141 standards, an SP Technical Research Institute for accreditation, and a logo. Table 15.1 resumes the international standards used by selected certification rulers in six countries. As can be seen, a certification body can follow more than one international standard. For instance, the ASTM differentiates between biodegradable and degradable *commodity* plastics and can be suitable for bioplastics.

15.3.1 BIO BASED

Maybe it is due to the early stages of bioplastics expansion (<1% of the plastics market) that only a few standards can certify that a material

TABLE 15.1 Certification Institutions, Standards, and Rules Used to Define the Biodegradable and Compostable Status of a Plastic Material or Product

Certification Body	Australian Standards	AIB VINÇOTTE	DIN CERTCO IBAW	Biodegradable Plastics Society	SP Technical Research Institute of Sweden (Sveriges Tekniska Forskningsintitut)	Biodegradable Products Institute and US composting Council
Country	Australia	Belgium	Germany	Japan	Sweden	USA
Test						
Biodegradable	AS 5810	EN 13432 ISO 14851	DIN V 54900 or EN 13432 or ASTM D6400	JIS K 6950 ISO 14851 OECD 301C	SPCR 141 Appendix 4	ASTM D6400
O_2 demand in a respirometer	ISO 14851	ISO 14851	DIN V 54900-2 Method 1	JIS K 6950 ISO 14851		
CO_2 in an aqueous medium	ISO 14852	ISO 14852	DIN V 54900-2 Method 2	JIS K 6951 ISO 14852		
CO_2 in compositing conditions	ISO 14855	ISO 14855	DIN V 54900-2 Method 3	JIS K 6953 ISO 14855		ASTM D5988-03
Compostable	AS 4736		DIN V 54900		SPCR 141 Appendix 1-Industrial scale Appendix 2-Small scale	ASTM D6400-04
Disintegration	ISO 20200	ISO 16929	DIN V 54900-3 pilot scale and full scale			ASTM D6400 with subclause 6.2 or ASTM D6002-96 with subclause 7.2.1
Ecotoxicity	EN 13432 Appendix E	EN 13432 with OECD 208	DIN 54900-4			ASTM D6400 with OECD 208

TABLE 15.1 *(Continued)*

Certification Body	Australian Standards	AIB VINÇOTTE	DIN CERTCO IBAW	Biodegradable Plastics Society	SP Technical Research Institute of Sweden (Sveriges Tekniska Forskningsintitut)	Biodegradable Products Institute and US composting Council
Bioaccumulation	ASTM E1676					OECD 207
Chemical composition	AS 4454	EN 13432	DIN V 54900 or EN 13432 ASTM D6400	GreenPLA certification scheme		ASTM D6400

or product, in part or whole, has a biological origin. A complex carbon 14 isotope (^{14}C) analysis will determine the renewable carbon content derived from biomass (Kunioka et al., 2007). The ASTM D6866, ISO 16620-2, and the European standards EN 16640 and CEN/TS 16137 contain the detailed methods. In the particular case of Mexico, the Mexican Association of Plastic Industry (ANIPAC, Spanish acronym) supported the implementation of an NMX-E law fostered by the Products Certification and Normalization Center (CNCP, Spanish acronym). The NMX-E-267-CNCP-2016 disclaims two methodologies to determine the bio-based content in resin and plastic products in concordance with the mentioned standards. The certification of bodies in Japan differentiates between bio-based and biomass-based polymer. Japan BioPlastics Association certifies bio-based plastics with a content of at least 25%, while Japan Organics Recycling Association certifies composite with a bio-based content of more than 5% (Iwata, 2015). In summary, there is a lot to advance in terms of carbon 14 isotope determination techniques and normativity in many countries.

15.3.2 BIODEGRADABLE

The advantage of a biodegradable bioplastic is that it does not need transport, cleaning, and recycling facilities, nor seek for an end-user of the poor quality product (Song et al., 2009). Biodegradation is the process of decomposition and mineralization in soil carried out by naturally occurring microorganisms in anaerobic or aerobic normalized conditions. Apart from the soil composition, pH, C/N ratio, and water content, there is a time frame to pass, for example, 90% of the organic material is converted into CO_2 within 180 days.

The specific conditions are described in standards such as the ASTM D5988 that determines the degree of aerobic biodegradation by measuring evolved CO_2 as a function of time of plastic is exposed to soil equivalent to the International Standard ISO 17556 for plastic materials in the soil. The European Standard EN 14995 differs on the soil nature since it is on industrial composting and ISO 14855 under controlled composting conditions.

15.3.3 COMPOSTABLE

A compostable bioplastic is decomposed in compost under normalized temperature and humidity producing CO_2, water, and biomass. A bioplastic subjected to a "compostability" test must accomplish several protocols. The first one is the disintegration and the second is a test of the quality of the compost by ecotoxicity that involves plant germination and growth of two different plant species (agronomic test). A third optional test is a bioac-cumulation test based on the survival of earthworms. Finally, the chemical composition of the soil is mandatory in terms of nutrient content, pH, and heavy metal content similar to the applied to the regular compost.

The "compostable" certification is based on the international standards ASTM D6400 (or EN 13432) for compostable packaging, and ISO 17088 (or EN 14995) for compostable plastics. The Australian norm considers the type of soil, for instance, AS 5810 under home compost and AS4736 under municipal and industrial aerobic composting facilities. The Mexican standard NMX-E-273-NYCE-2019 considers aerobic composting based on international standards for compostable packaging distributed on the contrary. A biodegradable bioplastic can meet or not the standards of compostable and it is only a matter of the frame time. For instance, the disintegration test can last 5 weeks (ASTM D6400) or 12 weeks (EN 13432) and remain <10% of the original weight. Therefore, it is said that compostable plastic is always biodegradable but certainly not the opposite.

The report of new tests could change the standards in the near future, especially because of the more accessible tools like plant molecular biology and elemental analysis techniques than 20 years ago. For instance, Serrano-Ruíz et al. proposed in vitro ecotoxicity. Those are designed to test the inhibitory effects on the germination, aerial, and root growth of model plants, that is, tomato and lettuce. The author reported the use of mulch film extracts to measure relative germination inhibition, aerial and root growth and leaves chlorophyll and proline content (Serrano-Ruíz et al., 2018). The mulch extract was obtained by submitting the film to frozen mechanical milling in order to obtain 8 mm pieces that were stored in Murashige and Skoog culture medium for 7 days at room temperature, sterilized using 0.20 μm filter, and finally pH adjusted to 5.7 before use. The pH of culture is important since it contributes to nutrients availability and assimilation by the plant.

15.3.4 RECYCLING

The European industry of plastic has participated enthusiastically in the past two surveys organized by European Plastics Converters Association and has an important position to the recycled Plastics Materials (rPMs) in integrating strategies toward a circular economy. A strong statement says "recycling is key to success in the future of plastics" (Europe Polymer Comply, 2019). They recognize three factors that challenge the growth of the rPM. First, the complexity in quality of the polymers, the requirement for new machinery and technologies for recycling, and finally the need for new applications and target markets of the rPM. The recycling of *commodity* plastics is yet complex so what would we expect of bioplastics recycling?

15.4 CURRENT AND POTENTIAL APPLICATIONS

Bioplastics are gaining ground in industrial applications such as packaging, automotive, health, and agriculture; however, the *commodity* plastics still take up to 40% of the industrial packaging applications and over 21% in the building and construction industry (Schulze et al., 2017). This section compiles applications of the last two decades in different areas which can inspire the reader to come across new ones in their specialty field.

15.4.1 TEXTILES

Three groups are commonly used for textile fibers classification: natural fibers that are either from the animal, vegetal, or mineral origin; synthetic fibers obtained from the chemical industry from synthetic polymers; and artificial fibers that are manufactured by the industrial process, in which material is given

the fiber shape (Ho et al., 2012). Natural fibers have been used as reinforcement to polymeric matrixes of natural origin like PLA, PBA, PHA, and among others, forming biocomposites that can be degraded more easily (Figure 15.6).

Nowadays, environmental degradation is of major concern, for this reason, several solutions have been proposed using biocomposites. An example of this is patent IN2013MU00234A (2013) which describes the usage of conventional plastics with the addition of natural fibers to create biopolymer-based on methacrylated alga oil, this material has diverse applications ranging from safety and defense to the automotive industry. Also, nonwoven fabric based in bioplastics like PLA and a combination of PLA and PHA is available, this kind of fabrics are used as a quilt for the cultivation of vegetables, since they are more resistant than conventional PE quilts and have no tear or degradation against ambient exposure, besides these can be composted and degrade in soil (Dharmalingam et al., 2015).

Ventura et al. (2015) studied palm fibers like reinforcement for a matrix of PLA, where a good balance between lightness and hardness for applications in the automotive industry was detected. Gurukarthik et al. (2018) characterized and identified the physicochemical properties of *Phaseolus vulgaris* plant, best

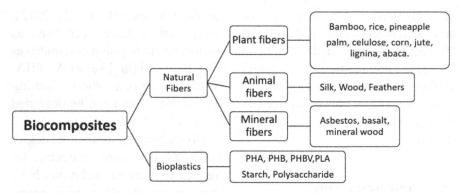

FIGURE 15.6   Constituents of biocomposites.

known in Mexico like bean plant, for its usage like textile fabric in the automotive industry. PLA has also been studied like fabric for manufacturing polyfurfuryl alcohol, PLA tissue is soluble in furfuryl alcohol, in mechanical tests performed on this material showed greater resistance to traction and an increase in bioplastic thermal stability (Sharib et al., 2018).

There are biocomposites reinforced with plant fibers such as flax and jute; these are known as a potential ecological alternative to synthetic fibers. Fibers from nonrenewable resources have drawbacks like high energy consumption during their production and nonbiodegradable (Shaker et al., 2016). In WO 2017122162 A1 (2017), it was patented a functionalized textile with an oligosaccharide that provides repellent properties for its usage in automotive seats. Cellulose fibers have been manufactured with carbon nanotubes like reinforcement for the

detection of polluted water, using the immersion method where carbon nanotubes solution is prepared, and then different immersions in cellulose fibers are preformed, biocomposite showed good reproducibility and detection (Qi et al., 2013). Aliotta et al. (2019) reported the study of cellulosic fibers in the PLA matrix, obtaining biocomposites with a different weight percentage of 5%, 10%, and 20%, then it was defined that these types of biocomposites fulfill with properties for packaging or agriculture. Flax fibers with bioplast films were also studied, biodegradable polymer produced from potato starch, their mechanical and flame retardant properties were assessed and the results showed good resistance to fire (Duquesne et al., 2015).

One of the biomedical applications of flax fabric with PHB is its usage like wounds bandage, Kulma et al. (2015) reported that this fabric inhibits the growth of bacteria and fungi that are common in wounds, besides promoting wound healing, however at molecular level the antimicrobial activity of flax fabric is still not acknowledged. Biofibers like straw, bamboo, and potato shell fibers have been added to the PHB matrix to reduce production costs and improve thermomechanical properties, improved thermal stability of biocomposites after adding fibers was detected (Wei and McDonald, 2016). Such fibers have been compatible with different polymeric matrixes and one of the advantages for using for reinforcement is that reduces plastic concentration, thanks to this feature, the researchers are looking for good adhesion and compatibility between fiber and polymeric matrix. Mathivanan et al. (2016) studied pineapple leaf natural fiber with a tapioca matrix, both fiber and matrix have hydrophilic nature that allows good compatibility; the material was manufactured with 30% of pineapple leaf fiber increasing its mechanical properties.

Borassus (Palmyra Palm) is a fiber belonging to a family of palm trees, which has good mechanical properties, there have been reported PLA/borassus composites prepared by mixture in melt with loads of 5%, 10%, and 15% of the weight of Borassus (Marathe et al., 2019). On the other hand, Dahy (2019) reported the manufacturing of two biocomposites based on PLA/lignin and PLA/rice straw fibers; both are cellulosic fibers recovered from agricultural residues. The biocomposites were prepared with a load of 20% of fibers, the results showed good compatibility between matrix and fibers, the antiflame behavior of biocomposites was also studied, the best antiflame behavior was PLA/lignin biocomposite, due to the relationship between lignin and polymeric matrix. Lignocellulosic materials extracted from plants are considered friendly with environment and with the low energy consumption during their processing, Batista et al. (2010) studied thermal and mechanical properties, as well as biodegradability in the soil of biocomposite of poly (3-hydroxy-butyrate-co-3-hydroxyvalerate and particles of peach palm are degraded faster than neat PHV, since cavities in the matrix are created, which allow the introduction of water and microorganisms for better degradation.

In 1991, Daimler-Mercedes Benz was one of the first companies in manufacturing cars with natural fibers, after companies, like Audi, Fiat, BMW, and among others, were interested in including natural fibers in cars components. In the beginning, these fibers were added to plastics like polyurethanes, nylon, and polyester. However, there were drawbacks due to the incompatibility between the polymeric matrix that only overcame after giving chemical treatments to fibers that had hydrophilic character (Ashori,

2008). Then, biocomposites based on fibers coming from plants that are considered the new green materials for use in automotive textiles were developed. Different research works have addressed to search textile-based biocomposites. An example is cellulose acetate with hemp fibers, the manufacturing process of this biocomposite is based on dust impregnation, extrusion, and injection molding, the biocomposite with best properties was at 30% in weight of fiber (Wibowo et al., 2004). In Figure 15.7, there can be noted that different textile components in the automotive interiors. There have been developed PHB/jute components with the ability of being permeable, allowing its application in cars seats and PLA/kenaf fiber for spare tire covers and car dashboard due to the different properties offered by this combination of materials (Bajpai et al., 2014; Francucci et al., 2010; Nakamura et al., 2009).

The PLA/cotton mixture has characteristics like good resistance to impact making this material proper for different automobile parts, cotton fiber is also considered to give comfort to seats and provides good mechanical stability (Graupner et al., 2009).

15.4.2 HEALTH

15.4.2.1 ANTIMICROBIAL PROPERTIES OF BIOPLASTICS

Bioplastics have applications in different industry sectors like chemistry, biotechnology, food, medicine, and others. Food and medicine industry needs to care environment by producing biodegradable packaging and antimicrobial properties. With the development of PLA films, these can be used as packaging for food and medical products that require being sterile, this type of film has

Celulose/ hemp fiber

PHB/jute

PLA/fiber kenaf

PLA/celulose

FIGURE 15.7   Textiles in automobiles manufactured from biocomposites.

reduced cycle time compared with films produced from fossils origin (Theinsathid et al., 2011). A study has compared PLA with PE terephthalate in the cycle time and antimicrobial properties, the result was that PLA is friendly with the environment and causes less impact on people's health, prevents microbial growth, can also be used in packaging of different types (Lorite et al., 2017).

Chitosan and montmorillonite based bionanocomposites adding rosemary oil have been studied, the obtained results showed antimicrobial activity and good properties of oxygen permeability, such qualities play a key role in defining the shelf-life of package product (Souza et al., 2019). Thanks to different benefits the acceptance of bioplastics at the industrial level have been increased, an alternative is polysaccharides from natural sources like Momordica charantia, known as bitter melon, that after mixing it with plasticizers improves its mechanical properties, besides antioxidant and antimicrobial properties of films produced from this bioplastic were studied, resulting in good activity against bacteria like *E. coli, S. aureus, B. subtilis* (Shafie et al., 2018). Other material conformed by polysaccharide that has antibacterial and antifungal applications is starch with lysine, coordinated with different metals like Mn, Cu, Co, Ni, and Zn, when biologically assessed, polymer with Cu metallic complex obtained greater antimicrobial activity (Nishat and Malik, 2013). Starch mixed with chitosan and lauric acid has shown antimicrobial activity, but when mixing starch with gold or silver nanoparticles then increases its inhibition percentage above 90% against bacteria like *E. coli* and *S. aureus* (Mlalila et al., 2018).

The addition of nanoparticles helps to obtain synergy with bioplastic to increase its antimicrobial properties against different strains. It has been reported that polycaprolactone with nanoclays has antifungal characteristics (Sanchez-Garcia et al., 2010). Nanoparticles used to increase antimicrobial activity were silver, nanoclays, zinc oxide, and titanium dioxide, all of them restrict microbial growth and help to eliminate microorganisms (Jariyasakoolroj et al., 2019). Other additives can be used to increase antimicrobial properties; extracts of plants are an example of this. Spices like cinnamon helped to decrease microbial growth by antibiotic delivery (Perez-Puyana et al., 2017). One of the problems of using these additives is FDA regulations that do not allow some nanoparticles in the packaging of the medical product, due to the potential migration of nanoparticles. ZnO nanoparticles have been approved for use like bioplastics exclusive for cosmetic containers and pharmaceutical products (Lagaron et al., 2012).

15.4.2.2   DRUG DELIVERY

Effective drug delivery has been one of the main problems encountered during disease treatments, which can be achieved thanks to nanopharmaceutical agents that act like vehicles toward action locations, performing diverse mechanisms like adsorption, cell penetration, or endocytosis. With the purpose to solve drug delivery issues, strategies for modifying properties like surface, size, and distribution of materials or vehicles have been designed, with the purpose to adjust delivery kinetics in order to coincide with the action mechanism of active ingredient (Wilczewska et al., 2012). One of the bioplastics that come from biosynthesis is PHA, which is biodegradable and biocompatible, thanks to its properties are possible to consider it as a drug delivery vehicle and actually is also researched as a possible drug against diseases like cancer (Tan et al., 2014). Also, PHA nanoparticles loaded with pyrene have been studied, these were prepared against recombinant *E. coli*, the results of drug delivery in vitro revealed that PHA nanoparticles are safe to transport and these have high load and efficiency capacity (Lam et al., 2017).

PHB (polyhydroxybutyrate) is another bioplastic that offers different drug-eluting advantages among them is that can adapt to diverse shapes like microcapsules, films, and microspheres, the molecule delivery coefficient depends on the microsphere diameter, other advantage is that can be applied for different drugs like anti-inflammatory, hormones, proteins, enzymes, and others (Roohi and Kuddus, 2018). It has been reported that the addition of Sulbactam-cefoperazone medication in PHB for its controlled delivery, the delivery occurs when polymer begins to degrade, this type of medication is used in osteomyelitis chronic disease (Raza et al., 2018).

The bionanocomposites consisting of bioplastic and nanoparticles as polysaccharide and nanoclays, respectively, have drug delivery properties given by their porous structure and can be a candidate for this kind of application (Ruiz-Hitzky et al., 2013). Agarose is a polysaccharide extracted mainly from algae, for synthesizing films glycerol is used as plasticizer, and antibiotics and antiseptic molecules for medicine application have been added to these particles, specifically in wounds and surgical procedures, where fast drug delivery is required (Awadhiya et al., 2017).

On the other hand, bioplastics have the characteristic of adjusting degradation rates allowing to delivery of drugs in the function of degradation time, an example is PLA and PLA copolymers, which are loaded with drugs and targeted to specific organs; delivery is in the function of degradation time (Ge et al., 2000). This delivery can be produced by swelling when water gets in contact with a polymer which has structural bonds, such bonds are broken and deliver drug. PLA has been studied as supply carrier of transepidermal drugs, used PLA nanoparticles are synthesized by the solvent evaporation method (Bano et al., 2018).

15.4.2.3 TISSUE ENGINEERING

Three-dimensional bioprinting is a technique used in tissue engineering, the challenge of this method consists of producing biocompatible materials that help on long-term cell growth as well as on good mechanical properties. Alginate is a polysaccharide that has been used under this method since is compatible with cartilaginous tissue, however has low mechanical properties when compared to other materials (Fahmy et al., 2016).

Other bioplastics used in tissue engineering, in particular for heart valves is PHB, the microspheres had porosity greater to 80% allowing to exist cells growth, besides PHB has been researched for bone tissue regeneration, which has no reject after is grafted and can be used for bone filler (Huang et al., 2010). Biocomposites based in bioplastic matrix and inorganic nanoparticles can help to improve mechanical properties for its application in tissue engineering and nano hydroxylapatite with starch is found as an example, this study showed that this biocomposite has an ideal ratio of 50/50 wt./wt. (Nugraha and Tontowi, 2016). As well as PHB with the addition of hydroxylapatite has been studied for bone tissue regeneration, this kind of composites has been used for hard bone tissue that needed good wear resistance (Porter et al., 2013).

Several research groups have studied scaffold-based tissue engineering with bioplastics, where PHA can be an appropriate option. According to Williams et al. (1999), it is required scaffolds for cellular growth whose degradation leads to nontoxic products. A way to improve mechanical properties, promote cellular growth, and rule biodegradation of bioplastics is the addition of bioceramics like hydroxylapatite, bioglass, tricalcium phosphate, or wollastonite (Misra et al., 2006). On the other hand, in order to define its compatibility with mice fibroblasts for scaffold-based tissue engineering modified cellulose has been studied and it has been defined osteoblasts cellular growth, as well for constructing heart cells based on cellulose acetate and regenerated cellulose (Mishra and Mishra, 2011).

15.4.3 *AGRICULTURE*

15.4.3.1 *MULCHES AND PEST CONTROL*

Agricultural mulches refer to any covering material used to control water and temperature. This physical barrier also can act as weed and pest control. Three bio-based polymers are commonly used for the preparation of mulches: PHAs, PLA, and starch. Several reviews and book chapters had compiled the scientific advances and the benefits of bioplastics for this particular application (Briassoulis, 2004; Kasirajan and Ngouajio, 2012; Malinconico, 2017; Malinconico et al., 2008). Several patents have been published since 2010 on this topic and the Chinese patents are quite abundant. For instance, Si et al. (2015, 2018) reported at least two patents for the preparation of "fully biodegradable" mulching films comprised of PLA and poly-(butylene adipate-co-tere-phthalate) (PBAT). The bioplastic NatureWorks Ingeo™ 4032D is a commercially available PLA of high heat film grade obtained from sugars of annual crops, such as corn, cassava, sugar cane, and sugar beets according to the NatureWorks LLC webpage. On the other side, polyglycolide (PGA) is a bio-based polymer that has been used in a mixture with starch to prepare low-cost thin films with high tensile strength (Wu et al., 2019) and in a mixture with PLA

to prepare low-cost films with high light transmittance (Wu et al., 2018). Both films can potentially replace PE mulch which is the most commonly used polymer in agriculture.

Regarding the weed control, an interesting proposal is the patent reported by Korean researcher is a PLA sheet contain seed embedded, the so-called seeded sheet. The revolutionary sowing seed method contains the seeds in the center, a layer of fertilizer, a layer of nutrients, and finally an herbicide layer (Young, 2016). The PLA sheet acts as mulch after the seeds germination to control weed and advantageously decompose with time. The dosage of a haloxyfop weed killer in biodegradable PHA pellets or films was patented by Rusian authors and granted in 2013 (Volova Tatjana and Shishatskaja Ekaterina, 2013). The postemergent herbicidal has the advantage of avoiding dispersion of the toxic compound in leaves and air and lasts up to 30 days.

Finally, the dosage of pesticides and insecticides in loaded PLA materials is also new and interesting. For instance, Zou et al. (2018) patented the preparation of an insecticidal film containing erucylamide in a biodegradable two-component matrix: PLA and PBAT. The mulch film was designed to exceed the mechanical strength of PE and protect against the corn pest commonly known as corn earthworm (*Helicoverpa armigera*). While Xu et al. (2018)

patented the use of microspheres of PP carbonate and PLA to effectively deliver emamectin benzoate, also a corn earthworm insecticide. All these techniques have the potential of facilitating fieldwork in agriculture.

Preoccupying arguments reached the light last year about the inhibitory effects of released compounds of these biodegradable bioplastics. Serrano-Ruíz et al. (2018) selected a huge range of commercially available bioplastics extracts that were tested in the germination and growth of lettuce and tomato. First, the extract of Mater-Bi (PBAT, Corn TPS, and vegetable oils composite) from the Italian company Novamont had no effect on the germination but showed root inhibition in lettuce. While, Bioplast-SP4 and SP-6 extracts (PBAT, Corn TPS, and PLA composite) from the Spanish company Group Sphere Ibérica Biotech and Mirel extract (PBAT and PHB composite) from Metabolix® in the USA showed negative effects on the germination in tomato. Although the toxic compounds in the extracts were not identified, this study shows the need to conscientiously select the mulch for the crop especially when it is intended to use repeatedly.

15.4.3.2 POSTHARVEST AND FOOD PACKAGING

The bioplastics are having an enormous chance in postharvest and food industry, especially because of the short shelf-life needed for vegetables, substituting single-use high-density PE films. Several reviews are recommended (Arrieta et al., 2017; Calva-Estrada et al., 2019; Cazón et al., 2017) but will focus on a few patent applications. Traditional one component films should be substituted by multiple layer film or sheet. This has the advantage of an adjustable gas permeability due to the multicomponent.

A European patent of Polish authors reported several multilayer materials, for instance, the BIO TRANSPARENT-NK20/F30 composed of 20 μm outer layer cellulose-based film and a 30 μm inner layer of modified starch (Zagrodzki, 2018). The cellulose film provides poor permeability to gases like CO_2 and water vapor, while the modified starch is proven to be compatible with food. The applications of such biodegradable materials are flexible packages like doypack, sachet, and gusset pouch for food storage. A Chinese patent claims a material of three, five, and seven layers with a thickness of 0.018–0.250 mm composed of renewably sourced bioplastics PLA or PHA and nonrenewable PBAT. For instance, the PBAT/PBAT-*co*-PLA/PLA using polyoxyethylene polyoxypropylene block copolymer as chain extender instead of classical isocyanate (Yunxuan et al., 2015) is attractive for food and vegetables sensitive to gas exchange. The challenge still remains in the appropriated chain extender and/or compatibilizing agent.

15.5 BIOPLASTICS RESEARCH IN MEXICO AND LATIN AMERICAN

Bioplastics and bio-based plastics research are rather important in the USA than Brazil, Mexico, Chile, Argentina, and Colombia. These Latin-American counties have shown relevant advances in the study of PHAs, PHBs, and starch and its mechanical properties. This section summarizes the research efforts on this topic and highlights the work of Mexican colleagues.

15.5.1 LATIN AMERICA

Bio-based plastics as PHAs are produced by fermentation of a variety of carbon sources and tend to accumulate in the cytoplasm of the microorganism. Bacteria and microalgae are the two types of microorganisms that are able to metabolize sugars and oils (fatty acids) and through a stress-induced mechanism synthesize PHAs. Two identified groups of bacteria have been extensively studied and reported in the literature (Noar and Bruno-Bárcena, 2018), those are *Pseudomona sp.* (Tortajada et al., 2013) and

Cupriabvidus necator (Lutz Ienczak et al., 2013). Remarkably, Latin-American research groups have been focused in find new PHA accumulating bacteria and the understanding of factors influencing the production of these bio-based plastics. For instance, Batista et al. (2007) from Instituto de Química Biológica in Montevideo, Uruguay reported the root-associated nitrogen-fixing bacteria *Herbaspirillum seropedicae* that produces PHA using disaccharides, such lactose as alone carbon source. We must point out that *H. seropedicae* is unable to metabolize disaccharides. On the other side, Batista et al. (2018) from Federal University of Paraná in Brazil try to study the physiological relevance of PHA production in *H. seropedicae* SmR1 submitted to an oxygen-limiting environment. Interestingly, the author reported that the bacteria was able the use of a huge variety of carbon sources such as organic acids, monosaccharides, and glycerol (Batista et al., 2016) which open the possibility to use multiple feedstocks on a pilot scale.

The second successful and less explored group of PHA-producing microorganisms is the microalgae (Costa et al., 2019). *Nostoc muscorum* (69 wt% dry mass of PHA accumulated), *Synechococcus* sp. (55 wt%), *Spirulina* sp. (30 wt%), and consortium of microalgaes (30–43 wt%) (Bhati and Mallick, 2015; Costa et al., 2019; Nishioka et al., 2001) store

PHA as response to stress conditions of very different nature (dehydration, presence of oxidants, heat, and UV irradiation). Invariably, the stress induced by phosphorus deprivation has shown the best results in *Nostoc* sp. and *Synechococcus* sp. (Bhati and Mallick, 2015; Nishioka et al., 2001). Particularly, Costa et al. (2018) from Federal University of Bahia in Brazil has been studying the influence of nutrient deficiency, for instance nitrogen, on the PHA production by *Synechococcus subsalsus* and *Spirulina sp.* The author also reports the use of native mixotroph microalgaes such as *Chlorella minutissima* finding that it was unable to accumulate PHA which must not discourage to explore others local microorganisms. As the author enumerate in a review that there are still a few challenges before the exploitation of microalgae by the bioplastics industry (Costa et al., 2019; Singh and Mallick, 2017).

15.5.2 MEXICO

Azotobacter vinelandii is the second important PHB producing bacteria in the world and has been extensively studied by Peña and coworkers in UNAM, Mexico (Jiménez et al., 2016; Millán et al., 2016). Early this year, the author reported the upscaling (from 3 to 30 L) of the synthetic protocol to produce ultra-high molecular weight PHB

$(3-11 \times 10^6$ Da). In addition, a two-step protocol for the recovery and purification of PHB (using less toxic ethanol and acetone) was also cheaper than chlorinated solvents (chloroform or sodium hypochlorite) usually employed (García et al., 2019). The relevance of this study relays on the control of PHB molecular weight as well as the use of cost-effective and environmentally friendly solvents.

On the other hand, the biodegradable unsaturated polyesters preparation from lactones such as ω-6-hexadecenlactone and vegetable oils (Martínez et al., 2018; Tlenkopatchev ct al., 2013) in the presence of Ru-alkylidene and Ru-vinylidene complexes have been investigated by Martínez and coworkers in Universidad Nacional Autónoma de México. These polyesters have shown their potential application in the membrane industry for the separation of a gas mixture. Also, the polyesters as the ω-6-polyhexadecenlactone with thermal and mechanical properties similar to that of PE have the advantage of being a biodegradable polymer.

Aguirre-Loredo et al. (2018, 2014) have been studied the effect of sugar and oils (oleic acid and glycerol) content on the water vapor and oxygen permeability, and the mechanical properties of films in Universidad Autónoma del Estado de Hidalgo, Mexico. For instance, high amylose content starch can form more resistant mechanically and less water-permeable film, which is crucial in the shelf life of a product. The biodegradable and edible films have the potential to be used in cut vegetables, meat, and cheese that require to avoid moisture loss.

ACKNOWLEDGMENTS

A.B.J-S. thanks to the Catedras CONACYT (790), A.M.P. the DGAPA-UNAM PAPIIT (IA103620), M.L.A.G. to CONACYT (postdoctoral scholarship 387368), C.J.C. the FOMIX (TLAX-2018-01-01-43129), and FORDECYT (296356).

KEYWORDS

- **bioplastics**
- **sustainable green chemistry**
- **renewable sources**
- **biodegradability**

REFERENCES

Adhvaryu, A., Liu, Z., and Erhan, S. Z. (2005). Synthesis of novel alkoxylated triacylglycerols and their lubricant base oil properties. *Industrial Crops and Products*, *21*(1), 113–119. https://doi.org/10.1016/J.INDCROP.2004.02.001

Aguirre-Loredo, R. Y., Rodríguez-Hernández, A. I., and Chavarría-Hernández, N. (2014). Physical properties of emulsified films based on chitosan and oleic acid. *CyTA-Journal of Food*, *12*(4), 305–312. https://doi.org/10.1080/19476337.2013.853207

Aguirre-Loredo, R. Y., Velazquez, G., Gutierrez, M. C., Castro-Rosas, J., Rangel-Vargas, E., and Gómez-Aldapa, C. A. (2018). Effect of airflow presence during the manufacturing of biodegradable films from polymers with different structural conformation. *Food Packaging and Shelf Life*, *17*, 162–170. https://doi.org/10.1016/J.FPSL.2018.06.007

Aliotta, L., Gigante, V., Coltelli, M. B., Cinelli, P., and Lazzeri, A. (2019). Evaluation of mechanical and interfacial properties of bio-composites based on poly(lactic acid) with natural cellulose fibers. *International Journal of Molecular Sciences*, *20*(4), 960. https://doi.org/10.3390/ijms20040960

Allen, N. S., Edge, M., Mohammadian, M., and Jones, K. (1994). Physicochemical aspects of the environmental degradation of poly(ethylene terephthalate). *Polymer Degradation and Stability*, *43*(2), 229–237. https://doi.org/10.1016/0141-3910(94)90074-4

Arrieta, M., Samper, M., Aldas, M., and López, J. (2017). On the use of PLA-PHB blends for sustainable food packaging applications. *Materials*, *10*(9), 1008. https://doi.org/10.3390/ma10091008

Ashori, A. (2008). Wood-plastic composites as promising green-composites for automotive industries! *Bioresource Technology*, *99*(11), 4661–4667. https://doi.org/10.1016/j.biortech.2007.09.043

Athawale, V. D., and Nimbalkar, R. V. (2011). Polyurethane dispersions based on sardine fish oil, soybean oil, and their interesterification products. *Journal of Dispersion Science and Technology*, *32*(7), 1014–1022. https://doi.org/10.1080/01932691.2010.497459

Awadhiya, A., Tyeb, S., Rathore, K., and Verma, V. (2017). Agarose bioplastic-based drug delivery system for surgical and wound dressings. *Engineering in Life Sciences*, *17*(2), 204–214. https://doi.org/10.1002/elsc.201500116

Bajpai, P. K., Singh, I., and Madaan, J. (2014). Development and characterization of PLA-based green composites: A review. *Journal of Thermoplastic Composite Materials*, *27*(1), 52–81. https://doi.org/10.1177/0892705712439571

Bano, K., Pandey, R., Jamal-e-Fatima, and Roodi. (2018). New Advancements of Bioplastics in Medical Applications. *International Journal of Pharmaceutical Sciences and Research*, *9*(2), 402–416. http://ijpsr.com/bft-article/new-advancements-of-bioplastics-in-medical-applications/?view=fulltext

Barlow, C. Y., and Morgan, D. C. (2013). Polymer film packaging for food: an environmental assessment. *Resources, Conservation & Recycling*, *78*, 74–80. https://doi.org/10.1016/j.resconrec.2013.07.003

Barnes, D. K. A., Galgani, F., Thompson, R. C., and Barlaz, M. (2009). Accumulation and fragmentation of plastic debris in global environments. *Philosophical Transactions of the Royal Society B: Biological Sciences*, *364*(1526), 1985–1998. https://doi.org/10.1098/rstb.2008.0205

Batista, K. C., Silva, D. A. K., Coelho, L. A. F., Pezzin, S. H., and Pezzin, A. P. T. (2010). Soil biodegradation of PHBV/Peach palm particles biocomposites. *Journal of Polymers and the Environment*, *18*(3), 346–354. https://doi.org/10.1007/s10924-010-0238-4

Batista, M. B., Müller-Santos, M., Pedrosa, F. de O., and de Souza, E. M. (2016). Potentiality of Herbaspirillum seropedicae as a Platform for Bioplastic Production. In: *Microbial Models: From Environmental to Industrial Sustainability*. Springer: Singapore, 23–39. https://doi.org/10.1007/978-981-10-2555-6_2

Batista, M. B., Teixeira, C. S., Sfeir, M. Z. T., Alves, L. P. S., Valdameri, G., Pedrosa, F. D. O., Sassaki, G. L., Steffens, M. B. R.,

de Souza, E. M., Dixon, R., and Müller-Santos, M. (2018). PHB biosynthesis counteracts redox stress in *Herbaspirillum seropedicae*. *Frontiers in Microbiology*, *9*(March), 1–12. https://doi.org/10.3389/fmicb.2018.00472

Bhati, R., and Mallick, N. (2015). Poly(3-hydroxybutyrate-co-3-hydroxyvalerate) copolymer production by the diazotrophic cyanobacterium nostoc muscorum agardh: process optimization and polymer characterization. *Algal Research*, *7*, 78–85. https://doi.org/10.1016/j.algal.2014.12.003

Bozell, J. J., and Petersen, G. R. (2010). Technology development for the production of biobased products from biorefinery carbohydrates—the US Department of Energy's "Top 10" revisited. *Green Chemistry*, *12*(4), 539–554. https://doi.org/10.1039/b922014c

Briassoulis, D. (2004). An overview on the mechanical behaviour of biodegradable agricultural films. *Journal of Polymers and the Environment*, *12*(2), 65–81. https://doi.org/10.1023/B:JOOE.0000010052.86786.ef

Cai, C., Dai, H., Chen, R., Su, C., Xu, X., Zhang, S., and Yang, L. (2008). Studies on the kinetics ofin situ epoxidation of vegetable oils. *European Journal of Lipid Science and Technology*, *110*(4), 341–346. https://doi.org/10.1002/ejlt.200700104

Calva-Estrada, S. J., Jiménez-Fernández, M., and Lugo-Cervantes, E. (2019). Protein-based films: advances in the development of biomaterials applicable to food packaging. *Food Engineering Reviews*, *11*, 78–92. https://doi.org/10.1007/s12393-019-09189-w

Catalán, A. I., Ferreira, F., Gill, P. R., and Batista, S. (2007). Production of polyhydroxyalkanoates by Herbaspirillum seropedicae grown with different sole carbon sources and on lactose when engineered to express the lacZlacY genes. *Enzyme and Microbial Technology*, *40*(5), 1352–1357. https://doi.org/10.1016/J.ENZMICTEC.2006.10.008

Cazón, P., Velazquez, G., Ramírez, J. A., and Vázquez, M. (2017). Polysaccharide-based films and coatings for food packaging: a review. *Food Hydrocolloids, 68*, 136–148. https://doi.org/10.1016/J.FOODHYD.2016.09.009

Chanprateep, S. (2010). Current trends in biodegradable polyhydroxyalkanoates. *Journal of Bioscience and Bioengineering, 110*(6), 621–632. https://doi.org/10.1016/J.JBIOSC.2010.07.014

Costa, S. S., Miranda, A. L., Andrade, B. B., Assis, D. de J., Souza, C. O., de Morais, M. G., Costa, J. A. V., and Druzian, J. I. (2018). Influence of nitrogen on growth, biomass composition, production, and properties of polyhydroxyalkanoates (PHAs) by microalgae. *International Journal of Biological Macromolecules, 116*, 552–562. https://doi.org/10.1016/J.IJBIOMAC.2018.05.064

Costa, S. S., Miranda, A. L., de Morais, M. G., Costa, J. A. V., and Druzian, J. I. (2019). Microalgae as source of polyhydroxyalkanoates (PHAs)—a review. *International Journal of Biological Macromolecules, 131*, 536–547. https://doi.org/10.1016/J.IJBIOMAC.2019.03.099

Cuéllar-Franca, R. M., and Azapagic, A. (2015). Carbon capture, storage and utilisation technologies: a critical analysis and comparison of their life cycle environmental impacts. *Journal of CO_2 Utilization, 9*, 82–102. https://doi.org/10.1016/J.JCOU.2014.12.001

Dahy, H. (2019). Efficient fabrication of sustainable building products from annually generated non-wood cellulosic fibres and bioplastics with improved flammability resistance. *Waste and Biomass Valorization, 10*(5), 1167–1175. https://doi.org/10.1007/s12649-017-0135-3

de Espinosa, L. M., Ronda, J. C., Galià, M., and Cádiz, V. (2009). A new route to acrylate oils: crosslinking and properties of acrylate triglycerides from high oleic sunflower oil. *Journal of Polymer Science Part A: Polymer Chemistry, 47*(4), 1159–1167. https://doi.org/10.1002/pola.23225

de Vasconcelos Vieira Lopes, R., Loureiro, N. P. D., Pezzin, A. P. T., Gomes, A. C. M., Resck, I. S., and Sales, M. J. A. (2013). Synthesis of polyols and polyurethanes from vegetable oils–kinetic and characterization. *Journal of Polymer Research, 20*(9), 238. https://doi.org/10.1007/s10965-013-0238-x

Dharmalingam, S., Hayes, D. G., Wadsworth, L. C., Dunlap, R. N., DeBruyn, J. M., Lee, J., and Wszelaki, A. L. (2015). Soil degradation of polylactic acid/polyhydroxyalkanoate-based nonwoven mulches. *Journal of Polymers and the Environment, 23*(3), 302–315. https://doi.org/10.1007/s10924-015-0716-9

Dinda, S., Patwardhan, A. V., Goud, V. V., and Pradhan, N. C. (2008). Epoxidation of cottonseed oil by aqueous hydrogen peroxide catalysed by liquid inorganic acids. *Bioresource Technology, 99*(9), 3737–3744. https://doi.org/10.1016/j.biortech.2007.07.015

Duquesne, S., Samyn, F., Ouagne, P., and Bourbigot, S. (2015). Flame retardancy and mechanical properties of flax reinforced woven for composite applications. *Journal of Industrial Textiles, 44*(5), 665–681. https://doi.org/10.1177/1528083713505633

Dutta, A. S. (2018). Polyurethane foam chemistry. In S. Thomas, A. Vasudeo Rane, K. Kanny, V. K. Abitha, and M. G. Thomas (Eds.), Recycling of Polyurethane Foams. A Volume in Plastics Design Library, 1st edn., pp. 17–27. William Andrew Publishing. https://doi.org/10.1016/B978-0-323-51133-9.00002-4

Dwyer, C. L. (2007). Chapter 6. Metathesis of olefins. In: G. P. Chiusoli and P. M. Maitlis (Eds.), *Metal-Catalysis in Industrial Organic Processes*. Royal Society of Chemistry: London, 201–217. https://doi.org/10.1039/9781847555328-00201

Edge, M., Hayes, M., Mohammadian, M., Allen, N. S., Jewitt, T. S., Brems,

K., and Jones, K. (1991). Aspects of poly(ethylene terephthalate) degradation for archival life and environmental degradation. *Polymer Degradation and Stability*, *32*(2), 131–153. https://doi.org/10.1016/0141-3910(91)90047-U

Endres, H. J. (2017). Bioplastics. In: K. Wagemann and N. Tippkötter (Eds.), *Biorefineries. Advances in Biochemical Enigineering/Biotechnology* (Vol. 166). Springer: Cham, 427–468. https://doi.org/10.1007/10_2016_75

Eriksen, M., Maximenko, N., Thiel, M., Cummins, A., Lattin, G., Wilson, S., Hafner, J., Zellers, A., and Rifman, S. (2013). Plastic pollution in the South Pacific subtropical gyre. *Marine Pollution Bulletin*, *68*(1–2), 71–76. https://doi.org/10.1016/J.MARPOLBUL.2012.12.021

Europe Polymer Comply. (2019). The usage of recycled plastics materials by plastics converters in Europe. A qualitative European industry survey. In: *Polymer Comply Europe*, 2nd ed. British Plastics Federation: London.

Fahmy, M. D., Jazayeri, H. E., Razavi, M., Masri, R., and Tayebi, L. (2016). Three-dimensional bioprinting materials with potential application in preprosthetic surgery. *Journal of Prosthodontics*, *25* (4), 310–318. https://doi.org/10.1111/jopr.12431

Francucci, G., Rodríguez, E. S., and Vázquez, A. (2010). Study of saturated and unsaturated permeability in natural fiber fabrics. *Composites Part A: Applied Science and Manufacturing*, *41*(1), 16–21. https://doi.org/10.1016/j.compositesa.2009.07.012

Fuessl, A., Yamamoto, M., and Schneller, A. (2012). Opportunities in bio-based building blocks for polycondensates and vinyl polymers. In M. Moeller and K. Matyjaszewski (Eds.), *Polymer Science: A Comprehensive Reference* (1st ed., Vol. 5). Elsevier: Amsterdam, 49–70. https://doi.org/10.1016/B978-0-444-53349-4.00132-1

García, A., Pérez, D., Castro, M., Urtuvia, V., Castillo, T., Díaz-Barrera, A., Espín, G., and Peña, C. (2019). Production and recovery of poly-3-hydroxybutyrate [P(3HB)] of ultra-high molecular weight using fed-batch cultures of *Azotobacter vinelandii* OPNA strain. *Journal of Chemical Technology & Biotechnology*, *94*(6), 1853–1860. https://doi.org/10.1002/jctb.5959

Ge, H., Hu, Y., Yang, S., Jiang, X., and Yang, C. (2000). Preparation, characterization, and drug release behaviors of drug-loaded e-caprolactone/L-lactide copolymer nanoparticles. *Journal of Applied Polymer Science*, *75*(7), 874–882. https://doi.org/10.1002/(SICI)1097-4628(20000214)75:7<874::AID-APP3>3.0.CO;2-G

Geyer, R., Jambeck, J. R., and Law, K. L. (2017). Production, use, and fate of all plastics ever made. *Science Advances*, *3*(7), e1700782. https://doi.org/10.1126/sciadv.1700782

Graupner, N., Herrmann, A. S., and Müssig, J. (2009). Natural and man-made cellulose fibre-reinforced poly(lactic acid) (PLA) composites: an overview about mechanical characteristics and application areas. *Composites Part A: Applied Science and Manufacturing*, *40*(6–7), 810–821. https://doi.org/10.1016/j.compositesa.2009.04.003

Güler, Ö. K., Güner, F. S., and Erciyes, A. T. (2004). Some empirical equations for oxypolymerization of linseed oil. *Progress in Organic Coatings*, *51*(4), 365–371. https://doi.org/10.1016/J.PORGCOAT.2004.07.024

Guo, Y., Hardesty, J. H., Mannari, V. M., and Massingill, J. L. (2007). Hydrolysis of epoxidized soybean oil in the presence of phosphoric acid. *Journal of the American Oil Chemists' Society*, *84*(10), 929–935. https://doi.org/10.1007/s11746-007-1126-5

Gurukarthik Babu, B., Princewinston, D., SenthamaraiKannan, P., Saravanakumar, S. S., and Sanjay, M. R. (2019). Study on characterization and physicochemical properties of new natural fiber from Phaseolus vulgaris. *Journal of Natural Fibers*, *16*(7), 1035–1042. https://doi.org/10.1080/15440478.2018.1448318

Ho, M. P., Wang, H., Lee, J. H., Ho, C. K., Lau, K. T., Leng, J., and Hui, D. (2012). Critical factors on manufacturing processes of natural fibre composites. *Composites Part B: Engineering*, *43*(8), 3549–3562. https://doi.org/10.1016/j.compositesb.2011.10.001

Hojabri, L., Kong, X., and Narine, S. S. (2009). Fatty acid-derived diisocyanate and biobased polyurethane produced from vegetable oil: synthesis, polymerization, and characterization. *Biomacromolecules*, *10*(4), 884–891. https://doi.org/10.1021/bm801411w

Hojabri, L., Kong, X., and Narine, S. S. (2010). Novel long chain unsaturated diisocyanate from fatty acid: synthesis, characterization, and application in bio-based polyurethane. *Journal of Polymer Science Part A: Polymer Chemistry*, *48*(15), 3302–3310. https://doi.org/10.1002/pola.24114

Horvat, P., and Kržan, A. (2012). Certification of bioplastics. Plastice: innovative value chain development dor sustainable plastics in Central Europe. European Regional Development Fund.

Huang, W., Shi, X., Ren, L., Du, C., and Wang, Y. (2010). PHBV microspheres—PLGA matrix composite scaffold for bone tissue engineering. *Biomaterials*, *31*(15), 4278–4285. https://doi.org/10.1016/j.biomaterials.2010.01.059

Institution of Oceanography (2019). A daily record of atmospheric carbon dioxide from Scripps Institution of Oceanography at UC San Diego. In J. R. Ehleringer, T. E. Cerling, and M. D. Dearing (Eds.). Scripps Institution of Oceanography. Springer Verlag, Berlin, 83–113. https://scripps.ucsd.edu/programs/keelingcurve/

Ionescu, M., Petrović, Z. S., and Wan, X. (2007). Ethoxylated soybean polyols for polyurethanes. *Journal of Polymers and*

the Environment, *15*(4), 237–243. https://doi.org/10.1007/s10924-007-0065-4

Ivleva, N. P., Wiesheu, A. C., and Niessner, R. (2017). Microplastic in aquatic ecosystems. *Angewandte Chemie International Edition*, *56*(7), 1720–1739. https://doi.org/10.1002/anie.201606957

Iwata, T. (2015). Biodegradable and biobased polymers: future prospects of ecofriendly plastics. *Angewandte Chemie International Edition*, *54*(11), 3210–3215. https://doi.org/10.1002/anie.201410770

Jariyasakoolroj, P., Leelaphiwat, P., and Harnkarnsujarit, N. (2019). Advances in research and development of bioplastic for food packaging. *Journal of the Science of Food and Agriculture*, 100, 5032–5045. https://doi.org/10.1002/jsfa.9497

Jiménez, L., Castillo, T., Flores, C., Segura, D., Galindo, E., and Peña, C. (2016). Analysis of respiratory activity and carbon usage of a mutant of Azotobacter vinelandii impaired in poly-β-hydroxybutyrate synthesis. *Journal of Industrial Microbiology & Biotechnology*, *43*(8), 1167–1174. https://doi.org/10.1007/s10295-016-1774-2

Kabasci, S. (Ed.). (2013). *Bio-Based Plastics. Materials and Applications*. John Wiley & Sons, Ltd: Hoboken, NJ.

Kamigaito, M., and Satoh, K. (2015). Encyclopedia of Polymeric Nanomaterials. In S. Kobayashi and K. Müllen (Eds.), *Encyclopedia of Polymeric Nanomaterials*. Springer-Verlag: Berlin, pp. 109–118. https://doi.org/10.1007/978-3-642-36199-9

Kasirajan, S., and Ngouajio, M. (2012). Polyethylene and biodegradable mulches for agricultural applications: a review. *Agronomy for Sustainable Development*, *32*(2), 501–529. https://doi.org/10.1007/s13593-011-0068-3

Kulma, A., Skórkowska-Telichowska, K., Kostyn, K., Szatkowski, M., Skała, J., Drulis-Kawa, Z., Preisner, M., Zuk, M., Szperlik, J., Wang, Y. F., and Szopa, J. (2015). New flax producing bioplastic fibers for medical purposes. *Industrial Crops and Products*, *68*, 80–89. https://doi.org/10.1016/j.indcrop.2014.09.013

Kulsoom, B., Reetika, P., and Roohi, J-e-F. (2018). New advancements of bioplastics in medical applications. *International Journal of Pharmaceutical Sciences and Research*, *9*(2), 402–416. http://ijpsr.com/bft-article/new-advancements-of-bioplastics-in-medical-applications/?view=fulltext

Kunioka, M., Ninomiya, F., and Funabashi, M. (2007). Biobased contents of organic fillers and polycaprolactone composites with cellulose fillers measured by accelerator mass spectrometry based on ASTM D6866. *Journal of Polymers and the Environment*, *15*(4), 281–287. https://doi.org/10.1007/s10924-007-0071-6

Lagaron, J. M., Ocio, M. J., and Rubio, A. L. (2012). *Antimicrobial Polymers*. Wiley: Hoboken, NJ. https://www.wiley.com/en-us/Antimicrobial+Polymers-p-9780470598221

Lam, W., Wang, Y., Chan, P. L., Chan, S. W., Tsang, Y. F., Chua, H., and Yu, P. H. F. (2017). Production of polyhydroxyalkanoates (PHA) using sludge from different wastewater treatment processes and the potential for medical and pharmaceutical applications. *Environmental Technology*, *38*(13–14), 1779–1791. https://doi.org/10.1080/09593330.2017.1316316

Lorite, G. S., Rocha, J. M., Miilumäki, N., Saavalainen, P., Selkälä, T., Morales-Cid, G., Gonçalves, M. P., Pongrácz, E., Rocha, C. M. R., and Toth, G. (2017). Evaluation of physicochemical/microbial properties and life cycle assessment (LCA) of PLA-based nanocomposite active packaging. *LWT*, *75*(75), 305–315. https://doi.org/10.1016/j.lwt.2016.09.004

Lu, Y., and Larock, R. C. (2008). Soybean-oil-based waterborne polyurethane dispersions: effects of polyol functionality and hard segment content on properties. *Biomacromolecules*, *9*(11), 3332–3340. https://doi.org/10.1021/bm801030g

Luca, L., Marco, R., Solitario, N., Fioravante, B., and Giovanna, C. (2017). *Functionalized Fabric* (Patent No. WO2017122162A1).

Lutz Ienczak, J., Schmidell, W., and Falcão de Aragão, G. M. (2013). High-cell-density culture strategies for polyhydroxyalkanoate production: a review. *Journal of Industrial Microbiology & Biotechnology*, *40*(3–4), 275–286. https://doi.org/10.1007/s10295-013-1236-z

Malinconico, M. (2017). *Soil Degradable Bioplastics for a Sustainable Modern Agriculture* (M. Malinconico (Ed.); Green Chem). Springer: Berlin. https://doi.org/10.1007/978-3-662-54130-2

Malinconico, M., Immirzi, B., Santagata, G., Schettini, E., Vox, G., and Mugnozza, G. S. (2008). An overview on innovative biodegradable. In *Progress in Polymer Degradation and Stability Research* (Issue February 2016). Nova Publishers: Hauppauge, NY.

Mallégol, J., Gonon, L., Lemaire, J., and Gardette, J.-L. (2001). Long-term behaviour of oil-based varnishes and paints 4. Influence of film thickness on the photooxidation. *Polymer Degradation and Stability*, *72*(2), 191–197. https://doi.org/10.1016/S0141-3910(00)00170-1

Mallégol, J., Lemaire, J., and Gardette, J.-L. (2000). Drier influence on the curing of linseed oil. *Progress in Organic Coatings*, *39*(2–4), 107–113. https://doi.org/10.1016/S0300-9440(00)00126-0

Marathe, Y. N., Arun Torris, A. T., Ramesh, C., and Badiger, M. V. (2019). Borassus powder-reinforced poly(lactic acid) composites with improved crystallization and mechanical properties. *Journal of Applied Polymer Science*, *136*(18), 1–11. https://doi.org/10.1002/app.47440

Martínez, A., Tlenkopatchev, M. A., and Gutiérrez, S. (2018). The unsaturated polyester via ring-opening metathesis polymerization (ROMP) of ω-6-Hexadecenlactone. *Current Organic Synthesis*, *15*(4), 566–571. https://doi.org/10.2174/1570179414666171011155831

Mathers, R. T. (2012). How well can renewable resources mimic commodity monomers and polymers? *Journal of Polymer Science Part A: Polymer Chemistry*, *50*(1), 1–15. https://doi.org/10.1002/pola.24939

Mathivanan, D., Norfazilah, H., Siregar, J. P., Rejab, M. R. M., Bachtiar, D., and Cionita, T. (2016). The study of mechanical properties of pineapple leaf fibre reinforced tapioca based bioplastic resin composite. *MATEC Web of Conferences*, *74*, 5–8. https://doi.org/10.1051/matecconf/20167400016

Meier, M. A. R., Metzger, J. O., and Schubert, U. S. (2007). Plant oil renewable resources as green alternatives in polymer science. *Chemical Society Reviews*, *36*, 1788–1802. https://doi.org/10.1039/b703294c

Mercangöz, M., Küsefoğlu, S., Akman, U., and Hortaçsu, Ö. (2004). Polymerization of soybean oil via permanganate oxidation with sub/supercritical CO2. *Chemical Engineering and Processing: Process Intensification*, *43*(8), 1015–1027. https://doi.org/10.1016/J.CEP.2003.10.002

Millán, M., Segura, D., Galindo, E., and Peña, C. (2016). Molecular mass of poly-3-hydroxybutyrate (P3HB) produced by *Azotobacter vinelandii* is determined by the ratio of synthesis and degradation under fixed dissolved oxygen tension. *Process Biochemistry*, *51*(8), 950–958. https://doi.org/10.1016/J.PROCBIO.2016.04.013

Mishra, A. K., and Mishra, S. B. (2011). Cellulose based green bioplastics for biomedical engineering. In *Handbook of Bioplastics and Biocomposites Engineering Applications*. John Wiley & Sons, Inc: Hoboken, NJ, 346–356. https://doi.org/10.1002/9781118203699.ch12

Misra, S. K., Valappil, S. P., Roy, I., and Boccaccini, A. R. (2006). Polyhydroxyalkanoate (PHA)/inorganic phase composites for tissue engineering applications. *Biomacromolecules*, *7*(8), 2249–2258. https://doi.org/10.1021/bm060317c

Mlalila, N., Hilonga, A., Swai, H., Devlieghere, F., and Ragaert, P. (2018). Antimicrobial packaging based on starch, poly(3-hydroxybutyrate) and poly(lactic-co-glycolide) materials and application challenges. *Trends in Food Science and Technology*, *74*(February), 1–11. https://doi.org/10.1016/j.tifs.2018.01.015

More, A. S., Lebarbé, T., Maisonneuve, L., Gadenne, B., Alfos, C., and Cramail, H. (2013). Novel fatty acid based di-isocyanates towards the synthesis of thermoplastic polyurethanes. *European Polymer Journal*, *49*(4), 823–833. https://doi.org/10.1016/J.EURPOLYMJ.2012.12.013

Morschbacker, A. (2009). Bioethanolbased ethylene. *Polymer Reviews*, *49*(2), 79–84. https://doi.org/10.1080/15583720902834791

Muellhaupt, R. (2013). Green polymer chemistry and bio-based plastics: Dreams and reality. *Macromolecular Chemistry and Physics*, *214*(2), 159–174. https://doi.org/10.1002/macp.201200439

Nakamura, R., Goda, K., Noda, J., and Ohgi, J. (2009). High temperature tensile properties and deep drawing of fully green composites. *Express Polymer Letters*, *3*(1), 19–24. https://doi.org/10.3144/expresspolymlett.2009.4

Narine, S. S., Kong, X., Bouzidi, L., and Sporns, P. (2007). Physical Properties of Polyurethanes Produced from Polyols from Seed Oils: I. Elastomers. *Journal of the American Oil Chemists' Society*, *84*(1), 55–63. https://doi.org/10.1007/s11746-006-1006-4

Nishat, N., and Malik, A. (2013). Antimicrobial bioplastics: synthesis and characterization of thermally stable starch and lysine-based polymeric ligand and its transition metals incorporated coordination polymer. *ISRN Inorganic Chemistry*, *2013*, 1–10. https://doi.org/10.1155/2013/538157

Nishioka, M., Nakai, K., Miyake, M., Asada, Y., and Taya, M. (2001). Production of poly-β-hydroxybutyrate by thermophilic cyanobacterium, Synechococcus sp. MA19, under phosphate-limited conditions.

Biotechnology Letters, *23*(14), 1095–1099. https://doi.org/10.1023/A:1010551614648

Noar, J. D., and Bruno-Bárcena, J. M. (2018). Azotobacter vinelandii: the source of 100 years of discoveries and many more to come. *Microbiology*, *164*, 421–436. https://doi.org/10.1099/mic.0.000643

Noreen, A., Zia, K. M., Zuber, M., Tabasum, S., and Zahoor, A. F. (2016). Bio-based polyurethane: an efficient and environment friendly coating systems: a review. *Progress in Organic Coatings*, *91*, 25–32. https://doi.org/10.1016/j.porgcoat.2015.11.018

Nugraha, F. K. A., and Tontowi, A. E. (2016). Shrinkage of [HA/Bioplastic/Sericin] composite part printed by bioprinter. *THE 2016 CONFERENCE ON FUNDAMENTAL AND APPLIED SCIENCE FOR ADVANCED TECHNOLOGY*, 020037. https://doi.org/10.1063/1.4953962

Pang, J., Zheng, M., Sun, R., Wang, A., Wang, X., and Zhang, T. (2016). Synthesis of ethylene glycol and terephthalic acid from biomass for producing PET. *Green Chemistry*, *18*(2), 342–359. https://doi.org/10.1039/c5gc01771h

Perez-Puyana, V., Felix, M., Romero, A., and Guerrero, A. (2017). Development of pea protein-based bioplastics with antimicrobial properties. *Journal of the Science of Food and Agriculture*, *97*(8), 2671–2674. https://doi.org/10.1002/jsfa.8051

PlasticsEurope, E. (2016). Plastics—the facts 2016. An analysis of European plastics production, demand and waste data. In Plastics Europe. https://www.plasticseurope.org/application/files/4315/1310/4805/plastic-the-fact-2016.pdf

Porter, M. M., Lee, S., Tanadchangsaeng, N., Jaremko, M. J., Yu, J., Meyers, M., and McKittrick, J. (2013). Porous Hydroxyapatite-Polyhydroxybutyrate composites fabricated by a novel method via centrifugation. In: *Mechanics of Biological Systems and Materials*. Springer, New York, 63–71. https://doi.org/10.1007/978-1-4614-4427-5_10

Qi, H., Mäder, E., and Liu, J. (2013). Unique water sensors based on carbon nanotube-cellulose composites. *Sensors and Actuators, B: Chemical*, *185*, 225–230. https://doi.org/10.1016/j.snb.2013.04.116

Raza, Z. A., Abid, S., and Banat, I. M. (2018). Polyhydroxyalkanoates: characteristics, production, recent developments and applications. *International Biodeterioration and Biodegradation*, *126*(September 2017), 45–56. https://doi.org/10.1016/j.ibiod.2017.10.001

Ronda, J. C., Lligadas, G., Galià, M., and Cádiz, V. (2011). Vegetable oils as platform chemicals for polymer synthesis. *European Journal of Lipid Science and Technology*, *113*(1), 46–58. https://doi.org/10.1002/ejlt.201000103

Roohi, Z. M. R., and Kuddus, M. (2018). PHB (poly-β-hydroxybutyrate) and its enzymatic degradation. *Polymers for Advanced Technologies*, *29*(1), 30–40. https://doi.org/10.1002/pat.4126

Ruiz-Hitzky, E., Darder, M., Fernandes, F. M., Wicklein, B., Alcântara, A. C. S., and Aranda, P. (2013). Fibrous clays based bionanocomposites. *Progress in Polymer Science*, *38*(10–11), 1392–1414. https://doi.org/10.1016/j.progpolymsci.2013.05.004

Sanchez-Garcia, M. D., Lopez-Rubio, A., and Lagaron, J. M. (2010). Natural micro and nanobiocomposites with enhanced barrier properties and novel functionalities for food biopackaging applications. *Trends in Food Science & Technology*, *21*(11), 528–536. https://doi.org/10.1016/j.tifs.2010.07.008

Saurabh, T., Patnaik, M., Bhagst, S. L., and Renge, V. (2011). Epoxidation of vegetable oils: a review. *International Journal of Advanced Engineering Technology E*, *2*(4), 459–501.

Schulze, C., Juraschek, M., Herrmann, C., and Thiede, S. (2017). Energy analysis of bioplastics processing. *Procedia CIRP*, *61*, 600–605. https://doi.org/10.1016/j.procir.2016.11.181

Seniha Güner, F., Yağcı, Y., and Tuncer Erciyes, A. (2006). Polymers from triglyceride oils. *Progress in Polymer Science*, *31*(7), 633–670. https://doi.org/10.1016/J.PROGPOLYMSCI.2006.07.001

Serrano-Ruíz, H., Martín-Closas, L., and Pelacho, A. M. (2018). Application of an in vitro plant ecotoxicity test to unused biodegradable mulches. *Polymer Degradation and Stability*, *158*, 102–110. https://doi.org/10.1016/J.POLYMDEGRADSTAB.2018.10.016

Shafie, M. H., Samsudin, D., Yusof, R., and Gan, C.-Y. (2018). Characterization of bio-based plastic made from a mixture of momordica charantia bioactive polysaccharide and choline chloride/glycerol based deep eutectic solvent. *International Journal of Biological Macromolecules*, *118*(Pt A), 1183–1192. https://doi.org/10.1016/j.ijbiomac.2018.06.103

Shahidi, F. (2005). *Baileys Industrial Oil And Fat Products* (W. John and Sons Eds.). Wiley: Hoboken, NJ.

Shaker, K., Ashraf, M., Jabbar, M., Shahid, S., Nawab, Y., Zia, J., and Rehman, A. (2016). Bioactive woven flax-based composites: development and characterisation. *Journal of Industrial Textiles*, *46*(2), 549–561. https://doi.org/10.1177/1528083715591579

Sharib, M., Kumar, R., and Kumar, K. D. (2018). Polylactic acid incorporated polyfurfuryl alcohol bioplastics: thermal, mechanical and curing studies. *Journal of Thermal Analysis and Calorimetry*, *132*(3), 1593–1600. https://doi.org/10.1007/s10973-018-7087-0

Shrinivas, Y. O., Sadashiv, D. B., Vitthal, G. V., and Dharamsing, R. S. (2013). *A Process for the Preparation of Bio-Polymer Composite from Algae Oil* (Patent No. IN 2013MU00234 A).

Si, P., Zhang, Y., Hao, N., Liu, Y., Qi, Y., and Zou, J. (2015). *Controllable bio-based full-degradable mulching film* (Patent No. CN104559087A). https://worldwide.espacenet.com/publicationDetails/biblio?CC=CN&NR=104559087A&KC=A&FT=D&ND=3&date=20150429&DB=EPODOC&locale=en_EP

Singh, A. K., and Mallick, N. (2017). Advances in cyanobacterial polyhydroxyalkanoates production. *FEMS Microbiology Letters*, *364*(20), 1–13. https://doi.org/10.1093/femsle/fnx189

Song, J., Kay, M., and Coles, R. (2011). Bioplastics. In R. Coles and M. Kirwan (Eds.), *Food and Beverage Packaging Technology*. Wiley-Blackwell: Hoboken, NJ, 295–319. https://doi.org/10.1002/9781444392180.ch11

Song, J., Murphy, R. J., Narayan, R., and Davies, G. B. H. (2009). Biodegradable and compostable alternatives to conventional plastics. *Philosophical Transactions of the Royal Society B: Biological Sciences*, *364*(1526), 2127–2139. https://doi.org/10.1098/rstb.2008.0289

Souza, V. G. L., Pires, J. R. A., Vieira, É. T., Coelhoso, I. M., Duarte, M. P., and Fernando, A. L. (2019). Activity of chitosan-montmorillonite bionanocomposites incorporated with rosemary essential oil: from in vitro assays to application in fresh poultry meat. *Food Hydrocolloids*, *89*(October 2018), 241–252. https://doi.org/10.1016/j.foodhyd.2018.10.049

Spierling, S., Röttger, C., Venkatachalam, V., Mudersbach, M., Herrmann, C., and Endres, H. J. (2018). Bio-based plastics—a building block for the circular economy? *Procedia CIRP*, *69*, 573–578. https://doi.org/10.1016/j.procir.2017.11.017

Tan, G. Y. A., Chen, C. L., Li, L., Ge, L., Wang, L., Razaad, I. M. N., Li, Y., Zhao, L., Mo, Y., and Wang, J. Y. (2014). Start a research on biopolymer polyhydroxyalkanoate (PHA): a review. *Polymers*, *6*(3), 706–754. https://doi.org/10.3390/polym6030706

Theinsathid, P., Chandrachai, A., Suwannathep, S., and Keeratipibul, S. (2011). Lead users and early adoptors of bioplastics: a market-led approach to innovative food packaging films. *Journal of Biobased Materials and Bioenergy*, *5*(1), 17–29. https://doi.org/10.1166/jbmb.2011.1128

The International Organization for Standardization, 1947, European Standards, 1961, American Society for Testing and Materials, 1898, JSA, 1945, AS, 1922, and SIS, 1922 are acronym and year of creation of each standard organization and the text was modified accordingly as follow: "The ISO (International Organization for Standardization) established in 1947, EN (European Standards) in 1961 and the ASTM (American Society for Testing and Materials) in 1898 are those leading standards organizations. Other organizations are the German Institute for Standardization (DIN, German acronym, 1917), Belgian accredited inspection and certification organization (VINÇOTTE, Belgian acronym), Japanese Standards Association (JSA), Australian standards (AS), Swedish Institute for Standards (SIS) and Official Mexican Standards (NOM, Spanish acronym)."

Tlenkopatchev, M. A., Gutiérrez, S., and Martínez, A. (2013). *Química sostenible: metátesis de hules y aceites naturales: metátesis sostenible*. Editorial Académica Española.

Tortajada, M., da Silva, L. F., and Prieto, M. A. (2013). Second-generation functionalized mediumchain-length polyhydroxyalkanoates: the gateway to high-value bioplastic applications. *International Microbiology*, *16*(1), 1–15. https://doi.org/10.2436/20.1501.01.175

Vaidya, R., Chaudhari, G. N., and Raut, N. (2016). Synthesis pathways for biocomposites from vegetable oils. *International Journal of Advance Research in Science and Engineering*, *5*(2), 533–543.

Ventura, H., Morón, M., and Ardanuy, M. (2015). Characterization and treatments of oil palm frond fibers and its suitability for technical applications. *Journal of Natural Fibers*, *12*(1), 84–95. https://doi.org/10.1080/15440478.2014.897670

Vert, M., Doi, Y., Hellwich, K., Hess, M., Hodge, P., Kubisa, P., Rinaudo, M., and Schué, F. (2012). Terminology for biorelated polymers and applications (IUPAC Recommendations 2012). *Chemistry International—Newsmagazine for IUPAC*, *84*(2), 377–410. https://doi.org/10.1515/ci.2011.33.2.25c

Volova Tatjana, G., and Shishatskaja Ekaterina, I. (2013). *Herbicidal Long-acting composition for use in soil* (Patent No. RU2494621C1). https://patents.google.com/patent/RU2494621C1/en?oq=RU2494621C1

Wang, C.-S., Yang, L.-T., Ni, B.-L., and Shi, G. (2009). Polyurethane networks from different soy-based polyols by the ring opening of epoxidized soybean oil with methanol, glycol, and 1,2-propanediol. *Journal of Applied Polymer Science*, *114*(1), 125–131. https://doi.org/10.1002/app.30493

Wei, L., and McDonald, A. G. (2016). A Review on Grafting of Biofibers for biocomposites. *Materials*, *9*(4), 303. https://doi.org/10.3390/ma9040303

Wibowo, A. C., Mohanty, A. K., Misra, M., and Drzal, L. T. (2004). Chopped industrial hemp fiber reinforced cellulosic plastic biocomposites: thermomechanical and morphological properties. *Industrial and Engineering Chemistry Research*, *43*(16), 4883–4888. https://doi.org/10.1021/ie030873c

Wilczewska, A. Z., Niemirowicz, K., Markiewicz, K. H., and Car, H. (2012). Nanoparticles as drug delivery systems. *Pharmacological Reports*, *64*(5), 1020–1037. http://www.ncbi.nlm.nih.gov/pubmed/23238461

Williams, S. F., Martin, D. P., Horowitz, D. M., and Peoples, O. P. (1999). PHA applications: addressing the price performance issue: I. Tissue engineering. *International Journal of Biological Macromoleculeslogical Macromolecules*, *25*(1–3), 111–121. http://www.ncbi.nlm.nih.gov/pubmed/10416657

Wu, X., Si, P., Wu, W., Weng, W., and Liu, Z. (2018). *Low-cost bio-based fully-degradable high-transmittance film and preparation method thereof* (Patent No. CN107652641A). JIANGSU JINJU ALLOY MAT CO LTD. https://worldwide.espacenet.com/publicationDetails/biblio?DB=EPODOC&II=1&ND=3&adjacent=true&locale=en_EP&FT=D&date=20180202&CC=CN&NR=107652641A&KC=A

Wu, X., Si, P., Wu, W., Weng, W., and Liu, Z. (2019). *Low-cost bio-based fully-degradable thin film and preparation method therefor* (Patent No. WO2019052150A1). JIANGSU GOLDEN POLY ALLOY MAT CO LTD. https://worldwide.espacenet.com/publicationDetails/biblio?DB=EPODOC&II=0&ND=3&adjacent=true&locale=en_EP&FT=D&date=20190321&CC=WO&NR=2019052150A1&KC=A1

Xu, S., Wu, G., Wu, Z., and Wei, T. (2018). *Biodegradable material-loaded pesticide microsphere suspension agent and preparation method thereof* (Patent No. CN108684690A). http://210.34.85.210:8219/fjnldx/item/itemDetail/67236.shtml

Young, K. J. (2016). *The multiple use degradable seed sheet for agriculture* (Patent No. KR20160123756A). https://worldwide.espacenet.com/publicationDetails/biblio?II=1&ND=3&adjacent=true&locale=en_EP&FT=D&date=20161026&CC=KR&NR=20160123756A&KC=A

Yu, C., Ge, Q., Xu, H., and Li, W. (2006). Propane dehydrogenation to propylene over Pt-based catalysts. *Catalysis Letters,* *112*(3–4), 197–201.https://doi.org/10.1007/s10562-006-0203-y

Yunxuan, W., Xiaoqian, D., Zhigang, H., Min, Z., Yujuan, J., and Jing, H. (2015). *Biodegradable multilayer material with adjustable gas transmission rate and preparation method and applications thereof* (Patent No. CN104691067(A)). Univ Beijing Tech & Business. https://worldwide.espacenet.com/publicationDetails/biblio?II=0&ND=3&adjacent=true&locale=en_EP&FT=D&date=20150610&CC=CN&NR=104691067A&KC=A

Zagrodzki, A. (2018). *Multi-layer biodegradable laminated structures intended for the production especially of food packaging and food packaging obtained therefrom* (Patent No. EP3266608(A1)). https://worldwide.espacenet.com/publicationDetails/biblio?II=0&ND=3&adjacent=true&locale=en_EP&FT=D&date=20180110&CC=EP&NR=3266608A1&KC=A1

Zlatanic, A., Lava, C., Zhang, W., and Petrovic, Z. S. (2004). Effect of structure on properties of polyols and polyurethanes based on different vegetable oils. *Journal of Polymer Science Part B: Polymer Physics,* *42*(5), 809–819. https://doi.org/10.1002/polb.10737

Zou, J., Si, P., and Gao, C. (2018). *Cycle-controllable fully-biodegradable insecticidal mulching film and preparation method thereof* (Patent No. CN107955338A). https://patents.google.com/patent/CN107955338A/en

CHAPTER 16

Biofibers for Polymer Reinforcement: Macro- and Micromechanical Points of View

F. J. ALONSO-MONTEMAYOR[1], R. I. NARRO-CÉSPEDES[2,*],
M. G. NEIRA-VELÁZQUEZ[3], D. NAVARRO-RODRÍGUEZ[3], and
C. N. AGUILAR[4]

[1]PhD Program in Materials Science and Technology, School of Chemistry (FCQ), Autonomous University of Coahuila (UAdeC), 25280 Saltillo, México

[2]Departament of Polymers, FCQ-UAdeC, 25280 Saltillo, México

[3]Department of Polymers Synthesis, Research Center for Applied Chemistry (CIQA) 25294, Saltillo, Coahuila, México

[4]Bioprocesses and Bioproducts Research Center, FCQ-UAdeC, 25280 Saltillo, México

*Corresponding author. Email: rinarro@uadec.edu.mx.

ABSTRACT

Despite the use of bio-based fibers to reinforce materials is not recent, the performance of resulting materials is still relatively poor. However, the current deterioration of the environmental conditions is making us turn our attention to them. The bio-based fibers come from renewable sources; therefore, their environmental impact has minor negative effects as compared with those produced by the mineral ones, which are still widely used. Moreover, the bio-based fibers outstrip the minerals ones in that they are not harmful, produce little damage to machines, allow the production of lightweight, and eco-friendly composites, among others. However, not all bio-based fibers are able to reinforce polymer materials, and those able to do it, sometimes require specific treatment to improve the fiber–polymer interfacial adhesion (compatibility).

Some other important factors determining the level of reinforcement are fiber loading, dispersion, orientation, and aspect ratio, among others. In this chapter, the reinforcement of polymers with bio-based fibers is described from the macro- and micro-mechanical points of view.

16.1 INTRODUCTION

Social trends toward green consumption practices and the implementation of rigorous environmental policies are motivating engineers and scientists to develop sustainable materials and processes. For many years, a great variety of products have been developed without thinking about the environmental impact like, for instance, reinforced polymers with synthetic and mineral fibers (Espinach et al., 2016). Some examples of such fibers are aramid, carbon and glass fibers, which are commonly used in industry for polymer reinforcement. Especially, glass fibers are intensely used in polymer reinforcement as they provide remarkable mechanical properties to plastics, around 85% of fiber-reinforced plastics incorporate glass fiber (E-glass) (Shah et al., 2013). However, glass fibers are harmful to humans (lung and skin diseases), cause significant damage to processing machines, reduce recyclability, require high-energy consumption, among other disadvantages (Delgado-Aguilar et al., 2018a,

2018b; Fernandes et al., 2015). It is to point out that reinforced plastics and green products are manufactured with <2% of bio-based fibers (Shah et al., 2013). Some other attractive aspects of bio-based fibers are biodegradability, abundant availability, and low cost. Flax, hemp, jute, coir, pineapple, abaca, cotton, kenaf, and sisal fiber are now increasingly investigated for their potential use in polymer reinforcement. All these fibers are less dense ($1.25–1.5$ g/cm^3) and softer than E-glass (2.54 g/cm^3), carbon ($1.8–2.1$ g/cm^3) fibers, and many other natural fibers (Alhuthali et al., 2013). Although lignocellulosic fibers show suitable intrinsic properties, some processing is needed to obtain industrially attractive fibers (Del Rey et al., 2017). This chapter explains the mechanical behavior of bio-based reinforced polymers from the macro- and micro-mechanical points of view.

16.2 REINFORCEMENT OF POLYMERS WITH FIBERS

It is a typical industry practice to reinforce polymers with fibers to render them mechanically stronger and stiffer, and thus improve their structural performance to be used, for instance, in tribological applications (Archaya et al., 2008), building products, transportation (automotive, railroad, and trucking), and furniture, among others. The polymer matrix works as an adhesive to maintain fibers in place (Mohanty et al., 2005); stress is transferred from the polymer matrix to fibers through the interface. Meanwhile, fibers hinder the polymer chain mobility, decreasing the matrix deformation capacity (Ogunsona et al., 2018).

In reinforced polymers, the interfacial adhesion is a crucial factor determining the mechanical strength, but it has a little impact on the tensile and flexural modulus (López et al., 2012a; Oliver-Ortega et al., 2018a, 2018b). The interfacial adhesion depends on mechanical interlocking and chemical interactions (de Farias et al., 2017). Other factors to consider are dimensional thermal stability, orientation, loading, length, and dispersion of fibers (Kim et al., 2001; Mulinari et al., 2016; Oliver-Ortega et al., 2018a, 2018b; Savetlana et al., 2017). Short polymer chains fill the empty spaces at the interface, improving compatibility (Lu et al., 2006).

Lignocellulosic fiber composition and properties depend on the source, extraction process, and treatment (de Melo et al., 2017). Its main chemical components are cellulose, hemicellulose, and lignin, although the most abundant is cellulose in a semicrystalline phase. Cellulose is a linear chain of polysaccharides comprising D-anhydrous glucopyranose units joined by β-1,4 glycosidic bonds. It imparts mechanical strength and stiffness to fibers. Meanwhile, lignin and hemicellulose are amorphous phases that work as adhesive for the cellulose microfibrils network. They exhibit a branched structure consisting of a mixture of polysaccharides with lower molecular weight as compared with those of cellulose. All three phases preserve fiber structural integrity. A schematic representation of the structure of lignocellulose fibers is depicted in Figure 16.1 (Lau et al., 2018; Takhur and Singha, 2015).

16.3 BIO-BASED FIBERS

All bio-based fibers from vegetable biomass are lignocellulosic. Most of them come from sources like wood, straw (rice, wheat, and corn), stalks (kenaf, flax, and pineapple), seeds (cotton and coconut), and herbs (bamboo and grass) (Mohanty et al., 2005).

16.4 TREATMENT OF BIO-BASED FIBERS: TYPICAL METHODS

Bio-based fibers usually show poor interfacial adhesion with nonpolar polymers (Gurunathan et al., 2015). In this sense, much research work has been performed to improve the fiber–matrix interfacial adhesion

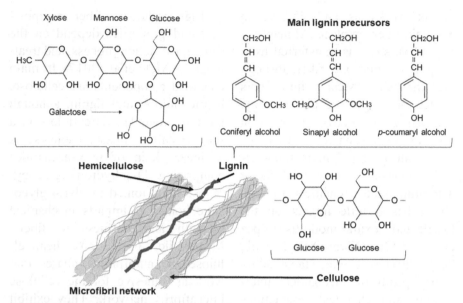

FIGURE 16.1   Structure of lignocellulose fibers.

by modification of both fibers and polymers through physical and chemical treatments (Venkatachalam et al., 2016), like heating, alkalization, silanization, acetylation, etherification, amidation, among others (Gurunathan et al., 2015).

Chemical treatments are widely used for mercerizing the bio-based fibers, making them more reactive and thermally stable. Alkali treatment is perhaps the most usual chemical method. It allows the extraction of lignin, hemicellulose, oils, waxes, and other low molar mass components that cover the surface of fibers. This treatment results in an improved fiber-matrix interfacial adhesion. The mechanism behind such improvement is related to the cleavage of hydrogen bonds and an

increase in roughness (de Melo et al., 2017). Microfibrils individualization is another consequence of alkalization (Santos et al., 2018).

16.5   MACROMECHANICS OF BIO-BASED FIBER-REINFORCED POLYMERS

In polymers reinforced with bio-based fibers, the reinforcement mechanism depends on electrostatic interactions between fibers and the polymer matrix (Takhur and Singha, 2015). The chemical affinity promotes a strong enough interfacial adhesion between both phases (López et al., 2012b). The stress transference can occur through weak or strong interfaces, although weak interactions produce weak interfaces that are inefficient to transfer stress (Takhur and Singha, 2015). Figure 16.2 schematically represents the stress transference in bio-based fiber-reinforced polymers.

A linear behavior of the macro-mechanical properties concerning the fiber content suggests good fiber dispersion and interfacial interaction (Oliver-Ortega et al., 2018b). Table 16.1 shows the symbols of the macromechanical properties described in this chapter. The linear behavior also allows the application of micromechanical models (Del Rey et al., 2017).

Figure 16.3 shows typical bio-based fiber-reinforced composite stress–strain curves. The elastic modulus is defined as the slope of the elastic deformation strain–stress curves in the elastic deformation zone and refers to the opposition of materials to be deformed. However, the mechanical strength is the ability of materials to withstand an applied load without failure.

16.6 MICROMECHANICS OF BIO-BASED FIBER-REINFORCED POLYMERS

Different types of micromechanical tests such as single fiber fragmentation, single fiber pull out, single fiber compression, and microdebond are commonly used to examine the interfacial bonding properties of fibers and other micromechanical

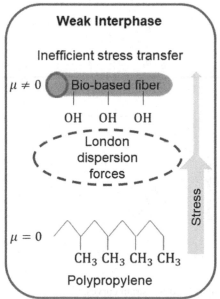
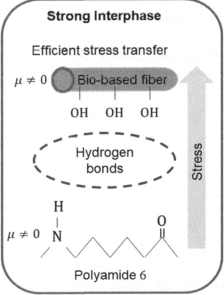

FIGURE 16.2   Stress transference in bio-based fiber reinforced polymers.

TABLE 16.1   Symbols of the Macromechanical Properties

Macromechanical property	Symbol
Composite tensile strength	σ_t^C
Composite tensile modulus	E_t^C
Composite tensile strain	ε_t^C
Composite flexural strength	σ_f^C
Composite flexural modulus	E_f^C
Composite flexural strain	ε_f^C
Matrix tensile strength	σ_t^M
Matrix tensile modulus	E_t^M
Matrix tensile strain	ε_t^M
Matrix flexural strength	σ_f^M
Matrix flexural modulus	E_f^M
Matrix flexural strain	ε_f^M
Matrix contribution to σ_t^C at ε_t^C	σ_t^{M*}

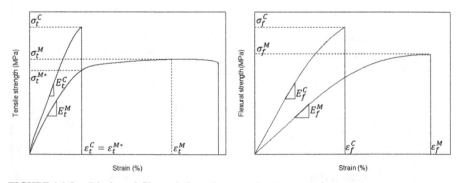

FIGURE 16.3   Bio-based fiber-reinforced composite stress–strain curves.

properties (Beckermann and Pickering, 2009; Lau et al., 2018). For instance, nanoindentation can be used to find the thickness range of the interface (Uribe et al., 2017). However, the extrapolation of results from these experimental tests remains highly controversial (Tripathi and Jones, 1998; Gu et al., 1995). These tests are not applied in bio-based fibers because of the high variability in their mechanical properties. The use of some mathematical models to calculate the micromechanical properties is recommended (Espinach et al., 2013).

There is a variety of micromechanical models for fiber-reinforced polymers; Table 16.2 summarizes some of them. It is to point out that X^r and Y^r refer to the contributions of subcritical and supercritical lengths expressed in terms of interfacial shear strength; meanwhile X^{LC} and Y^{LC} refer to the same contributions but in terms of critical length. The expressions for σ_t^C and the fiber tensile strength factor (FTSF) have the same algebraic structure with respect to those for E_t^C and the fiber tensile modulus factor (FTMR), which are not mentioned in this table (Aruan Efendy and Pickering, 2019; Kalaprasad et al., 1997). Although, this chapter does not include equations describing the flexural micromechanics of nontimber fibers because they are soft and, therefore, their flexural properties cannot be measured (Cao et al., 2013).

Micromechanical models describe the matrix and reinforcing phase contributions to the composite mechanical strength and modulus. The following assumptions validate these models. The assumptions of Voigt and Reuss state that the matrix, fiber, and composite elastic strains (on a small scale) are equal to each other $(\varepsilon_t^M = \varepsilon_t^F = \varepsilon_t^C)$ (Cao et al., 2013; Facca et al., 2006; Tucker III and Liang, 1999). Some other relevant assumptions, especially for the Bowyer–Bader solution, are the absence of fiber debonding, the composite porosity is negligible, τ does not depend on the load direction (Beckermann and Pickering, 2009), the stress transfer across the interface increases linearly from the fiber tips inwards, fiber diameter is constant, and χ_1 only depends on the processing framework (Thomason, 2001).

16.6.1 MEASUREMENT OF INPUTS IN MICROMECHANICAL MODELS

Some physical properties of single fibers and the fiber–matrix interface are inputs in micromechanical models. They can be determined by specific equations of the micromechanical models, like those of Cox–Krenchel and Halpin–Tsai–Pagano equations. Their calculation involves some easy-to-measure parameters like X^F, fibers length distribution (L_i and L_j), D, and L^W.

TABLE 16.2

Model	Strength	Eq.
Rule of mixtures "longitudinal load"	$\sigma_t^C = \sigma_t^F X^F + \sigma_t^M \left(1 - X^F\right)$	16.1
Rule of mixtures "transversal load"	$\sigma_t^C = \dfrac{\sigma_t^F \sigma_t^M}{\sigma_t^M X^F + \sigma_t^F \left(1 - X^F\right)}$	16.2
Modified rule of mixtures	$\sigma_t^C = \chi_1 \chi_2 \sigma_t^F X^F + \sigma_t^M \left(1 - X^F\right)$	16.3
	$FTSF = f_C \sigma_t^F = \dfrac{\sigma_t^C - \sigma_t^M \left(1 - X^F\right)}{X^F}$	16.4
	$f_C = \chi_1 \chi_2$	16.5
Cox-Krenchel equations	$\chi_1 = \cos^4 \alpha$ for all kind of arrangements in modulus calculus	16.6
	$\chi_1 = \dfrac{\sin \alpha}{\alpha} \sqrt{\dfrac{(3-v)\sin \alpha}{4\alpha} + \dfrac{(1+v)\sin 3\alpha}{12\alpha}}$ only for a square arrangement	16.7
	$\chi_2 = 1 - \dfrac{2 \tanh \dfrac{\beta L^W}{2}}{\beta L^W}$	16.8
	$\beta = \sqrt{\dfrac{2\pi G_m}{E_t^F A^F \ln \dfrac{2R}{D}}} = \dfrac{2}{D} \sqrt{\dfrac{E_t^M}{E_t^F (1-v) \ln \sqrt{\dfrac{\pi}{4X^F}}}}$	16.9
	$A^F = \dfrac{\pi D^2}{4}$	16.10
	$R = \sqrt{\dfrac{\pi D^2}{2 X^F \sqrt{3}}}$ for a hexagonal arrangement	16.11
	$R = \dfrac{D}{2} \sqrt{\dfrac{\pi}{4X^F}}$ for a square arrangement	16.12
	$\chi_2 = 1 - \dfrac{2 \tanh ns}{ns}$ is a Cox simplified expression	16.13
	$n = \sqrt{-\dfrac{2E_t^M}{E_t^F \left(2 + X^F\right) \ln X^F}}$	16.14
	$s = \dfrac{L^W}{D}$	16.15
Hirsch	$\sigma_t^C = \beta \left(\sigma_t^M \left(1 - X^F\right) + \sigma_t^F X^F\right) + \dfrac{(1-\beta)\sigma_t^F \sigma_t^M}{\sigma_t^M X^F + \sigma_t^F \left(1 - X^F\right)}$	16.16

TABLE 16.2 *(Continued)*

Model	Strength	Eq.
	$\sigma_t^F = E_t^F \varepsilon_t^C$ only for elastic deformation of the composite	16.17
Halpin–Tsai–Pagano equations	$\sigma_t^C = \dfrac{3\sigma^1 + 5\sigma^2}{8}$	16.18
	$\sigma^1 = \dfrac{\sigma_t^M \left(1 + \zeta \eta_1 X^F\right)}{1 - \psi \eta_1 X^F}$	16.19
	$\sigma^2 = \dfrac{\sigma_t^M \left(1 + 2\eta_2 X^F\right)}{1 - \eta_2 X^F}$	16.20
	$\zeta = \dfrac{2L^W}{D}$ for fibers with circular section	16.21
	$\psi = 1 + \dfrac{X^F(1-\phi)}{\phi^2}$	16.22
	$\eta_1 = \dfrac{\dfrac{\sigma_t^F}{\sigma_t^M} - 1}{\dfrac{\sigma_t^F}{\sigma_t^M} - \zeta}$ have the same algebraic structure than module expressions	16.23
	$\eta_2 = \dfrac{\dfrac{\sigma_t^F}{\sigma_t^M} - 1}{\dfrac{\sigma_t^F}{\sigma_t^M} - 2}$ have the same algebraic structure than module expressions	16.24
Kelly–Tyson model with Bowyer–Bader solution	Level $\varepsilon_{t\,min}^C$: $\sigma_{t\,min}^C - Z_{min} = \chi_1\left(X_{min} + Y_{min}\right)$	16.25
	Level $\varepsilon_{t\,max}^C$: $\sigma_{t\,max}^C - Z_{max} = \chi_1\left(X_{max} + Y_{max}\right)$	16.26
	$X^{L_C} = \dfrac{\sigma_t^F}{2L_C}\sum_{i=0}^{L_C} X_i^F L_i$ the contribution of the L_{sub} fibers	16.27
	$Y^{L_C} = \sigma_t^F \sum_{L_C}^{j=\infty} X_j^F - \dfrac{\sigma_t^F L_C}{2}\sum_{L_C}^{j=\infty} \dfrac{X_j^F}{L_j}$ the contribution of the L_{sup} fibers	16.28
	$Z = \sigma_t^{M*}\left(1 - X^F\right)$	16.29
	$L_C = \dfrac{\sigma_t^F D}{2\tau}$	16.30
	$X^\tau = \dfrac{\tau}{D}\sum_{i=0}^{L_C} X_i^F L_i$ the contribution of the L_{sub} fibers	16.31

TABLE 16.2 *(Continued)*

Model	Strength	Eq.
	$Y^{\tau} = \sigma_t^F \sum_{L_C}^{j=\infty} X_j^F - \dfrac{\sigma_t^F D}{4\tau} \sum_{L_C}^{j=\infty} \dfrac{X_j^F}{L_j}$ the contribution of the L_{sup} fibers	16.32
	$R_{exp} = \dfrac{\sigma_{t\,min}^C - Z_{min}}{\sigma_{t\,max}^C - Z_{max}}$	20.33
	$R_{the} = \dfrac{X_{min} + Y_{min}}{X_{max} + Y_{max}}$	20.34

σ_t^C : Composite tensile strength

σ_t^M : Matrix tensile strength

σ_t^{M*} : Matrix contribution to the σ_t^C at rupture

σ_t^F : Fiber tensile strength

E_t^C : Composite tensile module

E_t^M : Matrix tensile module

E_t^F : Fiber tensile module

ε_t^C : Composite tensile strain

ε_t^M : Matrix tensile strain

ε_t^F : Fiber tensile strain

τ: Interfacial shear strength

φ: Fibers fraction with most relevant arrangement

G_m: Interfacial shear modulus

X^F: Fibers volume fraction

α: Average orientation fibers angle

χ_1: Fibers orientation factor

χ_2: Fibers interphase-length factor

f_c: Reinforcement efficiency factor

$FTSF$: Fiber tensile strength factor

$FTMF$: Fiber tensile modulus factor

β: Load concentration factor at the fiber tips

ζ: Geometric adjustment parameter

ψ: Fibers arrangement-concentration factor

σ^1: Composite longitudinal tensile strength

σ^2: Composite transversal tensile strength

η_1: Longitudinal efficiency factor

η_2: Transversal efficiency factor

L^W: Fiber weighted average length

D: Fiber average diameter

A^F: Fiber transversal section

R: Average distance between reinforcement fibers

v: Poisson's coefficient

n: Cox simplified factor

s: Length-diameter ratio

L_C: Critical length

L_{sub}: Subcritical length fibers

L_{sup}: Supercritical length fibers

min: Minimum level

max: Maximum level

X: Contribution of the L_{sub} fibers to the σ_t^C

Y: Contribution of the L_{sup} fibers to the σ_t^C

Z: Contribution of the matrix to the σ_t^C

L: Fiber length measure

R_{exp}: Experimental ratio

R_{the} : Theoretical ratio

ISO 1183-1 allows calculating X^F from the composite, matrix, and fibers density represented by ρ^C, ρ^M, and ρ^F, respectively. The process described in this standard testing consists of determining the weight and volume of the composite (using distilled water as reference liquid). The identity employed to measure ρ^C is defined by Eq. (16.35), and it can be used to measure ρ^M in the same way, where w^C and w^{H_2O} are the composite, and water weight, respectively, V is the pycnometer volume, and ρ^{H_2O} is the density of distilled water at room temperature. Eq. (16.36) allows determining ρ^F from ρ^C and ρ^M, where w^F and w^M are the weight of fibers and matrix, respectively. Finally, Eq. (16.37) is used to obtain X^F from ρ^M and ρ^F, where x^F is the fiber's mass fraction.

$$\rho^C = \frac{w^C}{V - w^{H_2O} \rho^{H_2O}} \quad (16.35)$$

$$\rho^F = \frac{w^F \rho^C \rho^M}{w^C \rho^M - w^M \rho^C} \quad (16.36)$$

$$X^F = \frac{\rho^M x^F}{1 + \rho^F (1 - x^F)} \quad (16.37)$$

On the other hand, the dimensions and length distribution of fibers in the composite needs to be determined because the length of fibers decreases during processing. Therefore, it necessary to extract the fibers from the composite to determine their dimensions, and it is normally done by the Soxhlet

extraction described elsewhere (Del Rey et al., 2017).

ISO/FDIS 160652 allows determining the dimensions and length distribution of the extracted fibers by means of a Morfi analyzer. This analysis requires the use of an aqueous suspension of 25 mg/L of the reinforcing fibers. The mean length (L), D, and L^W are calculated according to Eqs. (16.38)–(16.40), respectively (Delgado-Aguilar et al., 2018b; Lafranche et al., 2013, 2015).

$$L = \frac{\sum_y n_y L_y}{n_{\text{Total}}} \quad (16.38)$$

$$D = \frac{\sum_y n_y D_y}{n_{\text{Total}}} \quad (16.39)$$

$$L^W = \frac{\sum_y n_y L_y^2}{\sum_y n_y L_y} \quad (16.40)$$

In these equations, L_y and D_y are specific y length and diameter, respectively, n_y is the number of fibers with specific y length and diameter, and n_{Total} is the total number of measurements.

16.6.2 ORIENTATION AND INTERFACE FACTORS

The micromechanical models also involve some micromechanical properties that cannot be experimentally measured (theoretical), such as χ_1 [Eqs. (16.6) and (16.7)] and χ_2 [Eqs. (16.8) and (16.13)]. According to the Cox–Krenchel model, χ_1 takes values of 1, 0.375, and 0.2 for axial, aleatory, and tridimensional orientation, respectively (Aruan Efendy and Pickering, 2019), but these values are only approximations. To precisely know both χ_1 and χ_2 factors, it is necessary to determine some other theoretical properties like β, v, σ_t^F, and E_t^F. Cox–Krenchel equations allow calculating β by using Eq. (16.9), although β values near 0.4 or 0.4335 allow reproducing the experimental results faithfully (Cao et al., 2013; Del Rey et al., 2017; Espigulé et al., 2013; Espinach et al., 2013, 2016; Jiménez et al., 2017; Kalaprasad et al., 1997; López et al., 2012b; Oliver et al., 2016; Vilaseca et al., 2010, 2018). Moreover, β allows calculating G_m through Eq. (16.8), provided that E_t^F or σ_t^F, A^F, R, and v are known.

The Hirsch model lets us calculate E_i^F and σ_t^F by an iterative method [Eq. (16.16)]. For doing so, an initial value of E_t^F (or σ_t^F) is iterated up to fit the experimental E_t^C (or σ_t^C). The determined E_t^F can then be used to calculate σ_t^F through Eq. (16.17). Equation (16.10) allows calculating A^F from D while for Eqs. (16.11) and (16.12) let us estimate R from D and X^F. However, according to Eqs. (16.11) and (16.12), there will be two possible values for R, and in consequence two possible values for G_m too, unless the ordering of fibers corresponds to a hexagonal or square-like arrangement, respectively. G_m needs to be constant because it is related to the elastic deformation of the material. The determination of the arrangement of fibers is relevant for the Halpin–Tsai–Pagano equations [Eqs. (16.18)–(16.24)] because they depend on it, where φ takes values of 0.785, 0.907, and 0.82 for square, hexagonal, and random arrangements, respectively (Aruan Efendy and Pickering, 2019).

On the other hand, experimental tests can provide v values or we can find them in literature for many polymer matrices. For instance, v values for polyamide 6 ranges between 0.35 and 0.48. It should be pointed out that v values vary very little between polymer matrices and composites (Arhant et al., 2016, 2019; Chen et al., 2019; Fornes and Paul, 2003; Meddad and Fisa, 1994; Piao et al., 2019; Zhu et al., 2014).

Finally, once χ_1 is known, it is possible to estimate α by Eq. (16.6) for all kinds of fiber arrangements, but only applicable to moduli data, or by Eq. (16.7), which provides valid values only to square fiber arrangements (Vilaseca et al., 2018). However, the calculated α values may considerably differ from those experimentally obtained (Espinach et al., 2013). The same occurs with the simplified Cox expressions [Eqs. (16.13)–(16.15)].

16.6.3 DIRECTIONAL STRENGTHS AND MODULI OF COMPOSITES

Equations (16.1) and (16.19) help calculating the longitudinal strength and modulus; meanwhile, Eqs. (16.2) and (16.20) allow estimating the transversal strength and modulus. The predicted directional values are limit values for the experimental strength and modulus (Facca et al., 2006).

While the directional strength and modulus are straightforward calculated by the rule of mixtures, they are rather difficult to calculate by the Halpin–Tsai–Pagano. First, it is necessary to determine ζ [Eq. (16.21)], which depends only on experimental values, and ψ [Eq. (16.22)], which only depends on φ. Once these two parameters are known, η_1 and η_2 can be calculated through Eqs. (16.23) and (16.24), respectively. Halpin–Tsai–Pagano equations, which considers the contribution of the orientation and length of the fibers, offer better-adjusted values to the experimental data, then the rule of mixtures, which involves more ideal equations.

16.6.4 REINFORCEMENT EFFICIENCY FACTORS

Equations involved in the modified rule of mixtures allow calculating nondirectional reinforcement efficiency factors. FTSF and FTMF are factors that refer to the fiber contribution to the composite tensile strength and modulus, respectively, and can be determined by Eq. (16.4) (Delgado-Aguilar et al., 2018a, 2018b; 2018c). Another commonly reported efficiency factor is f_c [Eqs. (16.3) and (16.5)]. This factor ranges from 0.17 (or 0.18) to 0.20 (Espinach et al., 2016; Del Rey et al., 2017; Jiménez et al., 2017; Vilaseca et al., 2010) for the strength-based calculations, and from 0.4 (or 0.4334) to 0.6 (Espinach et al., 2013; Jiménez et al., 2017; Kalaprasad et al., 1997; López et al., 2012b; Vilaseca et al., 2010) for the modulus-based calculations.

16.6.5 INTERFACIAL SHEAR STRENGTH AND CRITICAL FIBER LENGTH

Other important micromechanical properties [both defined by Eq. (16.30)] are the interfacial shear strength (τ) and the fiber's critical length (L_C). There are two criterions to determine the range of τ values. These are the Von Mises criterion [Eq. (16.41)], which is the upper bound, and the Tresca criterion, which is the lower bound [Eq. (16.42)] (Adams and Coppendale, 1979; Jiménez et al., 2016; Vallejos et al., 2012).

$$\tau = \frac{\sigma_t^C}{\sqrt{3}} \qquad (16.41)$$

$$\tau = \frac{\sigma_t^C}{2} \qquad (16.42)$$

To determine τ is crucial to first obtain a precise value of L_C, and this is critical because both τ and L_C are inputs for the Kelly–Tyson model. In this sense, Bowyer–Bader offered an iterative solution to the Kelly–Tyson model, which allows obtaining precise values of τ and L_C. Such a solution implies the obtention of two Kelly–Tyson expressions [Eqs. (16.25) and (16.26)] each one related to an arbitrary ε_t^C level. The left side of these two equations considers experimentally determinable terms like σ_t^c and the contribution of the matrix to the σ_t^c, which is denoted by Z [Eq. (16.29)]. On the other hand, the right side considers theoretically determinable terms, which are the contribution of the subcritical and supercritical length fibers, which are denoted by X [Eqs. (16.27) and (16.31)] and Y [Eqs. (16.28) and (16.32)], respectively.

One of the two Kelly–Tyson strain level expressions is divided by the other to discard χ_1 and obtain two constant ratios; one is based on experimental data while the other based on theoretical ones, being represented by R_{exp} [Eq. (16.33)] and R_{the} [Eq. (16.34)], respectively. R_{the} depends only on τ and L_C and needs to be adjusted to R_{exp} by iteration. Once τ and L_C are known, Eqs. (16.25) or (16.26) can be used to obtain a precise value of χ_1, while only for Eq. (16.5) let us estimate χ_2.

16.6.6 CONTRIBUTION OF SUBCRITICAL AND SUPERCRITICAL FIBERS

The Kelly–Tyson model also allows determining the specific contribution of the subcritical fiber length (L_{sub}) and supercritical fiber length (L_{sup}) to the strength and modulus of the composite. Figure 16.4 offers a schematic representation of L_{sub} and L_{sup}. Subcritical fibers tend to debond from the matrix without breaking because their interfaces are not strong enough to resist transfer loads. On the other hand, the supercritical fibers must break inside the matrix because their interface is strong enough to transfer loads higher than σ_t^F (Beckermann and Pickering, 2009; Vallejos et al., 2012).

The identities $\chi_1 X$, $\chi_1 Y$, and $(1 - X^F)Z$ allow calculating the contribution of the subcritical fibers, supercritical fibers, and matrix to the strength and modulus of the composite, respectively. On the other hand, the volumetric concentration of each reinforcing phase can be calculated by the cumulative X_i^F, from $i = 0$ to $i = L_C$, for the subcritical fibers, and the cumulative X_j^F, from $j = L_C$ to $j = \infty$, for the supercritical fibers. In comparing stress and volumetric contributions, fiber's supercritical contribution to σ_t^c is higher

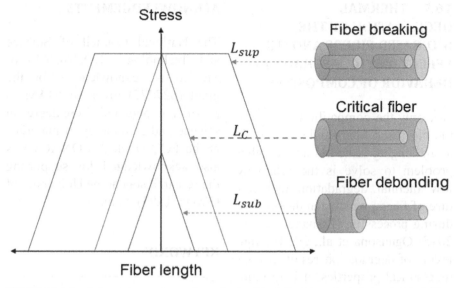

FIGURE 16.4 Schematic representation of subcritical fiber length and supercritical fiber length.

than that of the subcritical one, despite the concentration of supercritical fibers is lower. This difference between both contributions is because the supercritical fibers interface is more capable of transferring load from the matrix to fibers.

16.6.7 *MICROMECHANICAL MODELS VALIDATION*

Finally, after obtaining micromechanical predictions, it is necessary to validate them by comparing the theoretical composite stress in the function of X^F, with the experimental one. Some models like Halpin–Tsai–Pagano [Eq. (16.18)] and Cox–Krenchel fit better to the

experimental data than the ideal Cox model and rule of mixtures. The discrepancy between the calculated and experimentally values would come from the lack of adequate adjustment factors that consider orientation, length–interface relation, compatibility, and dimensional irregularity of the fibers (Lafranche et al., 2013).

On the other hand, the micromechanical parameters obtained by the modified rule of mixtures, Hirsch models, and Kelly–Tyson model cannot be validated in this way, although validation seems to be implicit in calculations. This is because the strength and modulus of the composite and matrix are references and inputs in these models.

16.7 THERMAL DEGRADATION OF THE BIO-BASED FIBERS AND ITS EFFECT ON THE MECHANICAL BEHAVIOR OF COMPOSITES

Not only the compatibility between the bio-based fibers and polymers is a challenging issue but also another problem to solve is the relatively low thermal degradation temperature of fibers, which can decompose during processing (Fernandes et al., 2015; Ogunsona et al., 2018). High levels of degradation result in poor mechanical properties. It is to point out that lignin, hemicellulose, and cellulose exhibit thermal degradation above 200 °C, 250–290 °C, and 350–375 °C, respectively (Baltazary-Jimenez et al., 2008; El-Sabbagh et al., 2014; Kiziltas et al; 2016). Therefore, the more lignin and hemicellulose is present in bio-based fibers, the less is the thermal resistance of fibers.

16.8 FINAL REMARK

In the fiber-reinforced polymer field, the current ecological concerns are motivating engineers and scientists to use bio-based fibers, specifically lignocellulosic ones. This chapter explores the mechanical behavior of polymers reinforced with bio-based fibers from both macro- and micromechanical perspectives, including different micromechanical models.

ACKNOWLEDGEMENTS

The National Council of Science and Technology (CONACYT) of Mexico is acknowledged for the grant (627977) given to FJAM to pursuit and obtain his MSc degree in science and technology of materials at the FCQ-UAdeC. CONACYT is also acknowledged for supporting the research stay at the University of Girona (UdG), Spain.

KEYWORDS

- **natural fibers**
- **reinforced composite**
- **micromechanical model**

REFERENCES

Adams, R.D.; Coppendale, J. The stress-strain behaviour of axially-loaded butt joints. J. Adhes. 1979, 10, 49–62.

Alhuthali, A.; Low, M. Mechanical properties of cellulose fibre reinforced vinyl-ester composites in wet conditions. J. Mater. Sci. 2013, 48, 6331–6340.

Archaya, S.K.; Mishra, P.; Mehar, S.K.; Dikshit, V. Weathering behavior of bagasse fiber reinforced polymer composite. J. Reinf. Plast. Compos. 2008, 27, 1839–1846.

Arhant, M.; Briançon, C.; Burtin, C.; Davies, P. Carbon/polyamide 6 thermoplastic composite cylinders for deep sea applications. Compos. Struct. 2019, 212, 535–546.

Arhant, M.; Le Gac, P.-Y.; Le Gall, M.; Burtin, C.; Briançon, C.; Davies, P. Effect of sea water and humidity on the tensile and compressive properties of carbon-polyamide 6 laminates. Compos. Part A Appl. Sci. Manuf. 2016, 91, 250–261.

Aruan Efendy, M.G.; Pickering, K.L. Comparison of strength and Young moduli of aligned discontinuous fibre PLA composites obtained experimentally and from theoretical prediction models. Compos. Struct. 2019, 208, 566–573.

Baltazar-y-Jimenez, A.; Juntaro, J.; Bismarck, A. Effect of atmospheric air pressure plasma treatment on the thermal behaviour of natural fibers and dynamical mechanical properties of randomly-oriented short fibre composites. J. Biobased Mater. Bioenergy. 2008, 2, 264–272.

Beckermann, G.W.; Pickering, K.L. Engineering and evaluation of hemp fibre reinforced polypropylene composites: micro-mechanics and strength prediction modeling. Compos. Part A Appl. Sci. Manuf. 2009, 40, 210–217.

Cao, Y.; Wang, W.; Wang, Q.; Wang, H. Application of mechanical models to flax fiber/wood fiber/plastic composites. BioResources. 2013, 8, 3276–3288.

Chen, J.; Han, J.; Xu, D. Thermal expansion properties of the polycaprolactam nanocomposites reinforced with single walled carbon nanotubes. Results Phys. 2019, 13, 1645–1652.

de Farias, J.G.G.; Cavalcante, R.C.; Canabarro, B.R.; Viana, H.M.; Scholz, S.; Simão, R.A. Surface lignin removal on coir fibers by plasma treatment for improved adhesion in thermoplastic starch composites. Carbohydr. Polym. 2017, 165, 429–436.

de Melo, R.P.; Marques, M.F.V.; Navard, P.; Duque, N.P. Degradation studies and mechanical properties of treated curauá fibers and microcrystalline cellulose in composites with polyamide 6. J. Compos. Mat. 2017, 51, 1–9.

Del Rey, R.; Serrat, R.; Alba, J.; Perez, H.; Mutje, P.; Espinach, F.X. Effect of sodium hydroxide treatments on the tensile strength and the interphase quality of hemp core fiber-reinforced polypropylene composites. Polymers 2017, 9, 377–391.

Delgado-Aguilar, M.; Ortega, H.O.; Méndez, J.A.; Camps, J.; Espinach, F.X.; Mutjé, P. The role of lignin on the mechanical performance of polylactic acid and jute composites. Int. J. Biol. Macromol. 2018a, 116, 299–304.

Delgado-Aguilar, M.; Reixach, R.; Tarrés, Q.; Espinach, F.X.; Mutjé P.; Méndez, J.A. Bleached kraft eucalyptus fibers as reinforcement of poly(lactic acid) for the development of high performance biocomposites. Polymers 2018b, 10, 1–15.

Delgado-Aguilar, M.; Vilaseca, F.; Tarrés, Q.; Julian, F.; Mutjé, P.; Espinach, F.X. Extending the value chain of corn agriculture by evaluating technical feasibility and quality of the interphase of chemo-thermomechanical fiber from corn stover reinforced polypropylene biocomposites. Compos. Part B Eng. 2018c, 137, 16–22.

El-Sabbagh, A.; Steuernagel, L.; Ziegmann, G.; Meiners, D.; Toepfer, O. Processing parameters and characterisation of flax fibre reinforced engineering plastic composites with flame retardant fillers. Compos. Part B Eng. 2014, 62, 12–18.

Espigulé, E.; Vilaseca, F.; Puigvert, X.; El Mansouri, N.-E.; Espinach, F.X.; Verdaguer, N.; Mutjé, P. Biocomposites from starch-based biopolymer and rape fibers. Part I: interfacial analysis and intrinsic properties of rape fibers. Curr. Org. Chem. 2013, 17, 1633–1640.

Espinach, F.X.; Granda, L.A.; Tarres, Q.; Duran, J.; Fullana-i-Palmer, P; Mutjé, P. Mechanical and micromechanical tensile strength of eucalyptus bleached fibers reinforced polyoxymethylene composites. Compos. Part B Eng. 2016, 116, 333–339.

Espinach, F.X.; Julian, F.; Verdaguer, N.; Torres, Ll.; Pelach, M.A.; Vilaseca, F.; Mutje, P. Analysis of tensile and flexural moduli in hemp strands/polypropylene composites. Compos. Part B Eng. 2013, 47, 339–342.

Facca, A.G.; Kortschot, M.T.; Yan, N. Predicting the elastic modulus of natural fibre reinforced thermoplastics. Compos. Part A Appl. Sci. Manuf. 2006, 37, 1660–1671.

Fernandes, F.C.; Gadioli, R.; Yassitepe, E.; De Paoli, M.-A. Polyamide-6 composites reinforced with cellulose fibers and fabricated by extrusion: effects of fiber bleaching on mechanical properties and stability. Polym. Compos. 2015, 38, 299–308.

Fornes, T.D.; Paul, D.R. Modeling properties of nylon 6/clay nanocomposites using composite theories. Polymer 2003, 44, 4993–5013.

Gu, X.H.; Young, R.J.; Day, R.J. Deformation micromechanics in model carbon fibre-reinforced composites. J. Mater. Sci. 1995, 30, 1409–1419.

Gurunathan, T.; Mohanty, S.; Nayak, S.K. A review of the recent developments in biocomposites based on natural fibres and their application perspectives. Compos. Part A Appl. Sci. Manuf. 2015, 77, 1–25.

Jiménez, A.M.; Delgado-Aguilar, M.; Tarrés, Q.; Quintana, G.; Fullana-i-Palmer, P.; Mutjé, P.; Espinach, F.X. Sugarcane bagasse reinforced composites: studies on the Young's modulus and macro and micro-mechanics. BioResources 2017, 12, 3618–3629.

Jiménez, A.M.; Espinach, F.X.; Granda, L.A.; Delgado-Aguilar, M.; Quintana, G.; Fullana-i.Palmer, P.; Mutjé, P. Tensile strength assessment of injection-molded high yield sugarcane bagasse-reinforced polypropylene. BioResources 2016, 11, 6346–6361.

Kalaprasad, G.; Joseph, K.; Thomas, S.; Pavithran, C. Theoretical modelling of tensile properties of short sisal fibre-reinforced low-density polyethylene composites. J. Mater. Sci. 1997, 32, 4261–4267.

Kim, E.G.; Park, J.K.; Jo, S.H. A study on fiber orientation during the injection molding of fiber-reinforced polymeric composites (comparison between image processing results and numerical simulation). J. Mater. Process. Technol. 2001, 111, 225–232.

Kiziltas, E.E.; Yang, H.-S.; Kiziltas, A.; Boran, S.; Ozen, E.; Gardner, D.J. Thermal analysis of polyamide 6 composites filled by natural fiber blend. BioResources 2016, 11, 4758–4769.

Lafranche, E.; Martins, C.I.; Oliveira, V.M.; Krawczak, P. Prediction of tensile properties of injection moulding flax fibre reinforced polypropylene from morphology analysis. Key Eng. Mater. 2013, 554–557, 1573–1582.

Lafranche, E.; Oliveira, V.M.; Martins, C.I.; Krawczak, P. Prediction of injection-moulded flax fibre reinforced polypropylene tensile properties through a micro-morphology analysis. J. Compos. Mater. 2015, 49, 113–128.

Lau, K.t.; Hung, P.-y.; Zhu, M.-H.; Hui, D. Properties of natural fibre composites for structural engineering applications. Compos. Part B Eng. 2018, 136, 222–233.

López, J.P.; Mutje, P.; Pelach, M.A.; El Mansouri, N.E.; Boufi, S.; Vilaseca, F. Analysis of the tensile moduli of PP composites reinforced with stone ground-wood fibers from softwood. BioResources. 2012a, 7, 1310–1323.

López, J.P.; Vilaseca, F.; Barbera, L.; Bayer, R.J.; Pèlach, M.A.; Mutjé P. Processing and properties of biodegradable composites based on Mater-Bi® and hemp core fibres. Resour. Conserv. Recycl. 2012b, 59, 38–42.

Lu, J.Z.; Negulescu, I.I.; Chen, Y. The influences of fiber feature and polymer melt index on mechanical properties of sugarcane fiber/polymer composites. J. Appl. Polym. Sci. 2006, 102, 5607–5619.

Meddad, A.; Fisa, B. Stress-strain behavior and tensile dilatometry of glass bead-filled polypropylene and polyamide 6. J. Appl. Polym. Sci. 1996, 64, 653–665.

Mohanty, A.K.; Misra, M.; Drzal, L.T. Natural Fibers, Biopolymers, and

Biocomposites. Taylor & Francis: New York, NY, 2005; 1–232.

Mulinari, D.R.; Voorwald, H.J.C.; Cioffi, M.O.H.; da Silva, M.L.C.P. Cellulose fiber-reinforced high-density polyethylene composites—mechanical and thermal properties. J. Compos. Mater. 2016, 51, 1–9.

Ogunsona, E.O.; Codou, A.; Misra, M.; Mohanty, A.K. Thermally stable pyrolytic biocarbon as an effective and sustainable reinforcing filler for polyamide biocomposites fabrication. J. Polym. Environ. 2018, 26, 3574–3589.

Oliver-Ortega, H.; Granda, L.A.; Espinach, F.X.; Delgado-Aguilar, M.; Duran, J.; Mutjé P. Stiffness of bio-based polyamide 11 reinforced with soft wood stone ground-wood fibres as an alternative to polypropylene glass fibre composites. Eur. Polym. J. 2016, 84, 481–489.

Oliver-Ortega, H.; Llop, M.F.; Espinach, F.X.; Tarrés, Q.; Ardauny, M.; Mutjé P. Study of the flexural moduli of lignocellulosic fibers reinforced bio-based polyamide11 green composites. Compos. Part B Eng. 2018a, 152, 126–132.

Oliver-Ortega, H.; Méndez, J.A.; Reixach, R.; Espinach, F.X.; Ardanuy, M.; Mutjé, P. Towards more sustainable material formulations: a comparative assessment of PA11-SGW flexural performance versus oil-based composites. Polymers 2018b, 10, 1–16.

Piao, H.; Kiryu, Y.; Chen, L.; Yamashita, S.; Ohsawa, I.; Takahashi, J. Influence of water absorption on the mechanical properties of discontinuous carbon fiber reinforced polyamide 6. J. Polym. Res. 2019, 26, 63–70.

Santos, E.B.C.; Moreno, C.B.; Barros, J.J.P.; Moura, D.A.; Carvalho, F.; Ries, A.; Wellen, R.M.R.; Silva, L.B. Effect of alkaline and hot water treatments on the structure and morphology of piassava fibers. Mater. Res. 2018, 21, 1–11.

Savetlana, S.; Mulvaney-Johnson, L.; Gough, T.; Kelly, A. Properties of nylon-6-based composite reinforced with coconut shell particles and empty fruit bunch fibres. Plast. Rubber Compos. 2017, 47, 77–86.

Shah, D.U.; Schubel, P.J.; Clifford, M.J. Can flax replace E-glass in structural composites? A small wind turbine blade case study. Compos. Part B Eng. 2013, 52, 172–181.

Takhur, V.K.; Singha, A.S. Surface Modification of Biopolymers. John Wiley & Sons: Hoboken, NJ, 2015; 1–44.

Thomason, J.L. Micromechanical parameters from macromechanical measurements on glass reinforced polyamide 6,6. Compos. Sci. Technol. 2001, 61, 2007–2016.

Tripathi, D.; Jones, F.R. Single fibre fragmentation test for assessing adhesion in fibre reinforced composites. J. Mater. Sci. 1998, 33, 1–16.

Tucker III, C.L.; Liang, E. Stiffness prediction for unidirectional short-fiber composites: review and evaluation. Compos. Sci. Technol. 1999, 59, 655–671.

Uribe, B.E.B.; Chiromito, E.M.S.; Carvalho, A.J.F.; Tarpani, J.R. Low-cost, environmentally friendly route for producing CFRP laminates with microfibrillated cellulose interphase. eXPRESS Polym. Lett. 2017, 11, 47–59.

Vallejos, M.E.; Espinach, F.X.; Julián, F.; Torres, Ll.; Vilaseca, F.; Mutjé, P. Micromechanics of hemp strands in polypropylene composites. Compos. Sci. Techol. 2012, 72, 1209–1213.

Venkatachalam, N.; Navaneethakrishnan, P.; Rajsekar, R.; Shankar, S. Effect of pretreatment methods on properties of natural fiber composites: a review. Polym. Polym. Compos. 2016, 24, 555–566.

Vilaseca, F.; Del Rey, R.; Serrat, R.; Alba, J.; Mutje, P.; Espinach, F.X. Macro and micro-mechanics behavior of stiffness in alkaline treated hemp core fibers polypropylene-based composites. Compos. Part B Eng. 2018, 144, 118–125.

Vilaseca, F.; Valadez-González, A.; Herrera-Franco, P.J.; Pèlach, M.À.; López, J.P.; Mutjé, P. Biocomposites from abaca strands and polypropylene. Part I: evaluation of tensile properties. Bioresour. Technol. 2010, 101, 387–395.

Zhu, Q.; Burtin, C.; Binetruy, C. Acousto-elastic effect in polyamide 6: linear and nonlinear behavior. Polym. Test. 2014, 40, 178–186.

CHAPTER 17

Trends in the Modification and Obtaining of Biomaterials Used in Physical Rehabilitation and Tissue Engineering

L. F. MORA-CORTES[1], R. I. NARRO-CÉSPEDES[2*], A. SÁENZ-GALINDO[2], J. C. CONTRERAS-ESQUIVEL[3], M. G. NEIRA-VELÁZQUEZ[4], and F. ÁVALOS-BELMONTES[2]

[1]PhD Program in Materials Science and Technology, School of Chemistry (FCQ), Autonomous University of Coahuila (UAdeC), 25280 Saltillo, Mexico

[2]Departament of Polymers, FCQ-UAdeC, 25280 Saltillo, Mexico

[3]Glycobiology Research Center, FCQ-UAdeC, 25280 Saltillo, Mexico

[4]Polymer Synthesis Department, Research Center of Applied Chemistry (CIQA), 25294 Saltillo, México

*Corresponding author. E-mail: rinarro@uadec.edu.mx.

ABSTRACT

This chapter describes the characteristics, development, current status, and future trends of biomaterials used in two fields of medicine: physical rehabilitation and tissue engineering, as well as the importance of the properties of each biomaterial to determine its application in the medical area. The work presented here is mainly dedicated to biomaterials made from metals, ceramics, and synthetic polymers, in addition to mentioning the importance of the combination of biomaterials, such as compounds and hybrids and the importance and tendency to use various types of modification, in the improvement of the properties of the different existing biomaterials.

17.1 INTRODUCTION

In the 1980s and 1990s, numerous surgical interventions related to joints, skin injuries, the nervous system, tendons, implants, prostheses, and organ replacement were considered of high risk, mainly due to the scarce medical technology available, the lack of tissues and/or organs of a donor and also, due to the effects of repulsion of the immune system to the material introduced into the body (O'Brien, 2011; Zavaglia and Prado, 2016; Saini et al., 2015; Limongi et al., 2017); all this entailed sometimes to a second operation. However, today medicine has evolved and is helping to counteract this problem more effectively by replacing or regenerating structurally and functionally damaged tissues or organs (Saini et al., 2015; Limongi et al., 2017; Huebsch and Mooney, 2009); and this without the need for a donor or surgery of post-implantation. Such a medical–scientific contribution that medicine is having is mainly due to its connection with other fields of science, such as material engineering and tissue engineering (IT), which consists primarily of the combination of biomaterials with advanced cellular technology (O'Brien, 2011; Zavaglia and Prado, 2016; Saini et al., 2015; Limongi et al., 2017).

This interdisciplinarity is known as regenerative medicine. Biomaterials are the main component of IT (Limongi et al., 2017); however, they are also widely used in the specific medical field of physical rehabilitation (Saini et al., 2015; Huebsch and Mooney, 2009). Today, the design of

biomaterials aims to create biocompatible options for replacement, rehabilitation, and/or therapeutic regeneration of a tissue or function ready to be used at the time required. For this, the biomaterials are elaborated with the capacity to give structural and functional support to the damaged tissue or organ according to the time tissue needs; they can also be designed to provide a surface that allows the adhesion, proliferation, and differentiation of specific cell populations in regeneration processes (O'Brien, 2011; Zavaglia and Prado, 2016; Limongi et al., 2017); and finally, if required, biomaterials can be prepared to be reabsorbed as a source of energy for cells while regenerating damaged tissue or organ (Zavaglia and Prado, 2016; Limongi et al., 2017). In addition, the byproducts of this bioabsorption proved to be nontoxic, being capable of leaving the body without interference with other organs (O'Brien, 2011). A little over 30 years ago, this was probably seen as science fiction but nowadays it is already a fact that only needs to be perfected (Limongi et al., 2017).

However, the use of materials in order to improve or repair the function of a tissue or organ of the human body dates back to ancient times (Saini et al., 2015; Huebsch and Mooney, 2009). For example, when natural materials such as wood were used in an attempt to structurally replace, the limbs lost due to diseases or traumas (Huebsch and Mooney, 2009). The discovery of the alloys was the next step in improving properties regarding pure metals. Among the alloys that stand out are those of stainless steel (SS) and titanium (Ti), for the medical and material engineering applications they have today (Saini et al., 2015; Ciardelli and Chiono, 2006; Pérez et al., 2008). Likewise, ceramic materials have been used by men for more than 5000 years, and today they have a great boom in biomedicine since they have mechanical and thermal properties that are used to improve the bone system (Saini et al., 2015). From the mid-19th century to the beginning of the 20th century, significant progress was made in new materials with the peek of plastic synthesis and resins, to obtain various polymers (Pina et al., 2019; Bhat and Kumar, 2013). The properties and easy handling of these have placed them as replacement materials for metals, alloys, and ceramics in many applications of daily life, including those related to interaction within the human body (Saini et al., 2015; Huebsch and Mooney, 2009).

Today, thanks to the currently available technology, it can be seen that metals, ceramics, and polymers are studied, synthesized, modified, and improved with precision to introduce them into the human body, to optimize the quality of life significantly (O'Brien, 2011; Zavaglia and Prado, 2016; Saini et al.,

2015; Limongi et al., 2017; Huebsch and Mooney, 2009; Ciardelli and Chiono, 2006). Consequently, what is known at this time as biomaterials can be considered as the branch of material research with the greatest boom today. Biomaterials outline to be one of the most important research and development areas of the following years along with the search and use of renewable energy sources (Huebsch and Mooney, 2009; Bhat and Kumar, 2013).

Day by day research is being carried out on new biomaterials, which is why this work has been given the task of discussing recent trends and advances, as well as the challenges and future directions in the evolution and application of biomaterials for rehabilitation surgery and regenerative medicine. In particular, this review focuses on the importance and exploration of the structural, mechanical, biochemical, and biological information of biomaterials to provide what their current applications are and where the design of future biomaterials is directed, as well as the trend of modification of biomaterials, for the improvement of properties, and creation of custom materials.

17.2 BIOMATERIALS IN THE MEDICAL AREA

Medicine is a field that becomes more interdisciplinary day by day.

Currently, some of its primary objectives are the replacement or regeneration of nerve tissue, joints, and organs damaged structurally and functionally. To achieve this, different technological approaches are used, based on IT, 3D printing, and biomaterials that are known as regenerative medicine, as Pina et al. (2019) mention in their review of strategies for IT and regenerative medicine (Pina et al., 2019). A biomaterial from the perspective of medical care can be defined as any material, whether metallic, ceramic, polymeric, or compound that has properties which make it appropriate to interact with living tissue without causing adverse reactions of immune rejection (Bhat and Kumar, 2013) or according to the National Institute of Health of the USA and the Food and Drug Administration, is defined as "any substance or combination of substances, other than medicines, of synthetic or natural origin, which may be used for any period of time, which partially or totally reinforces or replaces any tissue, organ or function, in order to maintain the quality of life of the individual" (Zavaglia and Prado, 2016; Bhat and Kumar, 2013). Such definitions converge in what we call biocompatibility of a material. Studies to consider that a biomaterial is medically applicable, involve at least considering its source of obtention, chemical, mechanical, thermal, cytotoxic, and response properties in vivo and

in vitro in animals and/or humans (O'Brien, 2011; Zavaglia and Prado, 2016; Bhat and Kumar, 2013).

The first biomaterial that met the minimum necessary biocompatible characteristics was developed in the 1920s. Reiner Erdle and Charles Prange joined their knowledge of dentistry and metallurgy, respectively, to develop the first metallic biomaterial, the Vitallium Cr alloy (65%)—Co (30%)—Mo (5%) (Venable and Stuck, 1943), used in dental implants to date. Today, more than 2 million tons per year of all types of biomaterials are commercialized for different sustainable, technological, and biomedical purposes, it is not surprising to know that, therefore, biomaterial technology is the branch of materials with the greatest boom today. Because it is considered a priority issue for health, therefore it is expected to be one of the largest and most important research and development areas worldwide in the following years (Huebsch and Mooney, 2009; Bhat and Kumar, 2013). According to the aforementioned definitions, an essential characteristic that must be met by all biomaterials is biocompatibility (Huebsch and Mooney, 2009; Bhat and Kumar, 2013; Kappel, 2014). Other characteristics that must be taken into account in the vast majority of biomaterials are their biodegradability or if materials are discussed in biological terms bioabsorption, as well as its bioactivity (Turnbull et al., 2018). Biodegradability can be present in ceramics, plastics, glasses, and many other materials. Although almost everything can be biodegradable, the time at which its decomposition occurs is not the same. In this sense, the biodegradability referred to here is the decomposition of a material or substance in the chemical elements that comprise it by cellular action, enzymes, or microorganisms in a short-time; this time must be £4 years for materials used in the cellular environment (Shah, 2008; Song et al., 2009; Laycock et al., 2017; Du et al., 2018). The greater the biodegradability of the biomaterial, the easier its decomposition (Sivan, 2011; Satti, 2017).

Bioactivity, on the other hand, is the ability of a material to chemically interact with the tissues of the organism to stimulate a positive effect and, normally, leads to regeneration in a shorter period of time than the organism alone can have (Turnbull et al., 2018; Du et al., 2018). A clear example can be found in the review by Zhong et al. (2010) about scaffolds for skin healing, where thanks to the biodegradability of certain material; in this case, a biomaterial based on polylactic acid) (PLA) and polyglycolic acid) (PGA) (Zhong, 2010) can be used as a temporary substitute for damaged tissue while it regenerates, due to the bioactivity of this compound specifically (Figure 17.1a). The PLA/PGA graft (known as polylactic-*co*-glycolic

acid (PLGA) and named polyglactin in this application) was cultured with human neonatal fibroblasts and collagen leading to the development of a commercial cryopreserved Dermagraft® product (Figure 17.1b). It is important to recognize that the modification by means of the combination between biomaterial cells and collagen results in a biodegradable and bioactive material. In the analyzed evaluations of this material, the PLGA mass was gradually decreasing due to cell action, metabolism, as well as physicochemical mechanisms such as hydrolysis (Figure 17.1c). Grafts of this type are normally completely reabsorbed by the body in a time not exceeding 28 days and are currently common to heal deep wounds on the skin of patients with diabetes, or with the second- and third-degree burns (Zhong, 2010).

17.3 CLASSIFICATION OF BIOMATERIALS, TYPES OF MATERIALS, AND THEIR APPLICATIONS

Biomaterials, therefore, may have only the biocompatibility property, or two properties together, the biocompatibility and biodegradability properties, or three properties: biocompatibility, biodegradability, and bioactivity. However, according to the medical definitions of biomaterial mentioned above,

 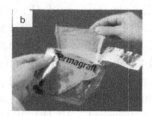 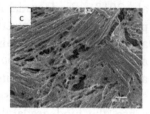

FIGURE 17.1 (a) Polyglactin mesh, (b) a person receiving a Dermagraft® package, (c) dermal fibroblasts and collagen cultured in the polyglactin mesh, the collagen promotes adhesion and growth of fibroblasts after 5 days of culture. The cells take as a source of energy the mass of the biomaterial. (From Marston, 2004; Zhong and Nanomed, 2010; adapted with permission.).

there is an ambiguity that deserves to be clarified in depth. Mainly because of the evocation that refers to the fact that it is only enough that the organism does not present a rejection for a material so that it is considered as biologically compatible. In that sense, there is one more branch of biomaterials that, although it does not generate immunological rejection, has very limited interaction within the organism, so they are considered as inert materials.

Consequently, there are four types of biomaterials according to their interaction with the organism: biocompatibles, biodegradables. Bioactives, and bioinerts. Biocompatibles meet the minimum acceptance for not presenting immunological rejection (Saini et al., 2015; Huebsch and Mooney, 2009; Bhat and Kumar, 2013); biocompatibles and biodegradable, which will be named in this chapter only as biodegradable, are fully recognized by the organism, and its degradation

products are used as a source of energy by the cells (O'Brien, 2011; Laycock et al., 2017; Satti, 2017; Pajarinen et al., 2014). Bioactives, in addition to being biodegradable, have biological interaction; these materials help to regenerate tissues more quickly and effectively (O'Brien, 2011; Zavaglia and Prado, 2016; Saini et al., 2015; Turnbull et al., 2018; Du et al., 2018). Finally, bioinerts not only present rejection but also do not present significant biological activity in the organism (Saini et al., 2015; Huebsch and Mooney, 2009; Bhat and Kumar, 2013; Du et al., 2018). In addition, there are three individual groups of biomaterials from the perspective of their origin, which can be divided as follows:

(1) synthetics (metals, ceramics, polymers, and compounds),
(2) derivatives of nature (natural polymers obtained from living beings), and

(3) hybrid materials (combinations of synthetic and natural materials).

Synthetic biomaterials are characterized by the intervention of man, from the obtaining of the raw material to the final elaboration of the material. In addition, they are mostly obtained from nonrenewable resources. On the contrary, those are derived from nature renewable raw materials obtained from fruits, plants, and animals; however, the finished biomaterial still depends on the man. Synthetic and natural biomaterials have strengths and weaknesses, so the use of hybrid combinations becomes increasingly common in order to obtain customized materials according to the needs of the patient. All this opens up a very wide range of possibilities in terms of the composition and final properties of the material (Zavaglia, 2011; Saini et al., 2015; Limongi et al., 2017; Huebsch and Mooney, 2009; Turnbull et al., 2018; Du et al., 2018; Zhong, 2010). Therefore, there is a general classification for medical devices that are introduced to the human body, which is presented in Table 17.1. This classification is based on the official information of the governmental regulatory authorities of different countries.

TABLE 17.1 Classification of Medical Devices

Regulatory authority	Class			
TGA[a]	I	IIa, IIb	III	
FDA[b]	I	II	III	
ECHCP[c]	I	IIa	IIIb	III
MDB[d]	I	II	III	IV
General description	Noninvasive and / or transitory use (≤ 6 months)	Minimally invasive, short-term (6 months to 4 years)	Short and medium-term contact (de 6 months a 10 years)	Medium and long-term contact (≥ 10 years)
Restrictions	Minimum before commercialization	Specific before commercialization	Rigorous control before approval and commercialization	Rigorous control before approval and commercialization
Potential health risk	Low	Low/Moderate	Moderate/High	High

[a] Therapeutic Products Administration of Australia

[b] Food and Drug Administration

[c] European Commission. Health and Consumer Protection

[d] Medical Devices Bureau

Finally, the biomaterials in this chapter have been classified according to the type of material used and the biological response they cause when they are introduced into the human body. As a synthesis, Table 17.2 is a summary classification proposed by the authors regarding the source of obtaining the material and the type of biological response that they have in the organism. This is a classification that generally involves some of the materials most commonly used in regenerative medicine and physical rehabilitation, both natural and synthetic and combinations since the possibilities to obtain biomaterials are wide enough. Therefore, some examples will be presented gradually throughout this investigation and there will be a general mention of the new trends with composite and hybrid biomaterials.

Normally, biocompatible biomaterials and bioinerts are intended to be formulated to permanently replace an organ or function within the human body (Figure 17.2). They are not intended to be replaced (O'Brien, 2011; Saini et al., 2015; Huebsch and Mooney, 2009; Pajarinen et al., 2014) unless incompatibility problems occur, or the material wears prematurely. A large part of these biomaterials is used in implants or prostheses as a replacement material, mainly in physical rehabilitation surgeries, such as hip, knee, and spine, while others are formulated to replace arteries, trachea, liver, heart, between other organs and functions of the human body. On the other hand, it is not intended that biodegradable and bioactive biomaterials remain as permanent substitutes within the human body (Figure 17.2) (O'Brien, 2011; Zavaglia and Prado, 2016; Limongi et al., 2017; Moshiri et al., 2016). Their use is mainly aimed at tissue regeneration and, therefore, it is only required that they serve as a means for the tissue to recover more efficiently. Likewise, due to their degradation qualities, such biomaterials can be used as a vehicle for the controlled drug release.

Although a large number of biodegradable biomaterials can also help in physical rehabilitation therapies, mainly in muscle tear and cartilage regeneration treatments, most of these (except epidermis healing treatments) are designed to regenerate deeper tissues of the body and to fulfill the role of temporary support maintaining a structural stability in the place where the new tissue is generated. In these biomaterials, the degradation products are used by the body's cells as a source of energy during the regeneration of the damaged tissue or function (Limongi et al., 2017; Moshiri et al., 2016). In this way, the damaged tissue cells have an adequate environment to grow until the tissue is completely restored while the biomaterial is reabsorbed.

TABLE 17.2 Classification of Biomaterials Based on Their Composition and Biological Activity in the Organism

Biological Activity of the Materials	Metals	Ceramics	Synthetic		Naturals	Hybrids
			Polymers	**Compounds**		
Biodegradable (Classes I and II)			PLA[Turnbull (2018)] PVA[Zavaglia (2016), Turnbull (2018)] PCL[Zavaglia (2016), Limongi (2017), Ciardelli (2006), Du (2018)] PGA[Zavaglia (2016), Turnbull (2018)] PEG[Bhat (2013), Turnbull (2018)] PGMA[Bhat (2013), Turnbull (2018)] PLLA[O'Brien (2011), Turnbull (2018)] PDLA[Turnbull (2018)] PLGA[O'Brien (2011), Turnbull (2018)] PHEMA[Bhat (2013)]			
Bioactive (Classes I, II, and III)	Fe[Chen (2015)] Mg[Chen (2015)] Ca[Chen (2015)] Na[Chen (2015)] Cu[Chen (2015)] Zn[Chen (2015)]	sHA[O'Brien (2011), Saini (2015), Turnbull (2018), Du (2018)] CP[Turnbull (2018), Du (2018)]		Carbon/silicon[Saini (2015)] Fe-Mg[Turnbull (2018)] BG/PCL[Turnbull (2018)] BG/ PLLA[Turnbull (2018)] PLA/sHA[Turnbull (2018)] PLLA/sHA[Turnbull (2018)]	*Polysaccharides* Chitosan[Sensharma (2017), Turnbull (2018)] GAGs[Sensharma (2017)] Alginate[Sensharma (2017), Turnbull (2018)]	PCL/Q[Ciardelli (2006), Turnbull (2018)] PCL/A[Turnbull (2018)] PET/Col[Bhat (2013)] BG/G[Turnbull (2018)] PCL/G[Turnbull (2018)] PLLA/G[Turnbull (2018)]

TABLE 17.2 *(Continued)*

Biological Activity of the Materials	Metals	Ceramics	Synthetic Polymers	Compounds	Naturals	Hybrids
		BCP[Turnbull (2018), Du (2018)] TCP[O'Brien (2011), Saini (2015), Turnbull (2018), Du (2018)] CS[Saini (2015), Du (2018)] BG[Saini (2015), Turnbull (2018)]		BG/sHA[Turnbull (2018)] BG/TCP[Turnbull (2018)] BG/PU[Bhat (2013)] PET/PU[Bhat (2013)] BG/PS[Bhat (2013)] BG/Mg[Turnbull (2018)] CFs/PS[Bhat (2013)] PET/PHEMA[Bhat (2013)] KF/PMA[Bhat (2013)] KF/PE[Bhat (2013)] GF/PU[Bhat (2013)] PCL/TCP[Turnbull (2018)] PTFE/PU[Bhat (2013)] CF/PTFE[Bhat (2013)] PLA/TCP[Turnbull (2018)] Silicon/sHA[Turnbull (2018)] Ni-Ti/sHA[Turnbull (2018)] CS/sHA[Turnbull (2018)] PCL/sHA[Turnbull (2018)] PLGA/sHA[Turnbull (2018)] PCL/sHA/CNTs[Turnbull (2018)] PCL/PLA/SiO$_2$[Turnbull (2018)] PCL/PLGA/sHA[Turnbull (2018)] RGD-PLA[Turnbull (2018)] PEG/CM/sHA[Turnbull (2018)]	Agarose[Turnbull (2018)] Proteins Colllagen[O'Brien (2011), Sensharma (2017), Turnbull (2018)] Elastin[Sensharma (2017), Turnbull (2018)] Silk[Turnbull (2018)] Gelatin[Sensharma (2017), Turnbull (2018)] Fibrinogen[Sensharma (2017), Turnbull (2018)] Col/SC/HLA[Turnbull (2018)] C/Col/OMPs[Turnbull (2018)] CNF/A[Turnbull (2018)] CNF/HLA[Turnbull (2018)] Glicol-Q/nHA/HLA[Turnbull (2018)] A/HLA[Turnbull (2018)] Col/HLA/FB[Turnbull (2018)]	PCL/silk[Turnbull (2018)] PGA-paclitaxel[Bhat (2013)] PEG-irinotecan[Bhat (2013)] PEG-camptothecin[Bhat (2013)] PHMHM-camptothecin[Bhat (2013)] Cells/PTFE[Bhat (2013)] Cells/PET[Bhat (2013)] PET/Col[Bhat (2013)] TCP/Col[Turnbull (2018)] PGMA/Q[Turnbull (2018)] TCP/A[Bhat (2013)] C/PLGA[Bhat (2013), Turnbull (2018)] Sr/nHA/C[Bhat (2013), Turnbull (2018)] Col/sHA[Bhat (2013), Turnbull (2018)] Silk/sHA[Turnbull (2018)] Col/Mg-sHA[Turnbull (2018)] PLLA/C/sHA/AA[Turnbull (2018)] C/nHA/CFs[Turnbull (2018)] HLA/sHA/TCP[Turnbull (2018)] nHA/PDA/PCL[Turnbull (2018)] PCL/cell-laden/pluronic F127[Turnbull (2018)] PECE/Col/nHA[Turnbull (2018)]

TABLE 17.2 *(Continued)*

Biological Activity of the Materials	Metals	Ceramics	Synthetic Polymers	Compounds	Naturals	Hybrids
Bioinerts (Classes III and IV)	Ti puro[Saini (2015), Turnbull (2018), Du (2018)] Zr puro[Turnbull (2018), Du (2018), Chen (2015)] Ta puro[Saini (2015), Gee (2016)]	Alumina[Saini (2015), Du (2018), Chen (2015)] Zirconium dioxide[Turnbull (2018), Du (2018), Chen (2015)] TiO_2[Turnbull (2018)]		Ti-SiO_2[Turnbull (2018)] Ti-Co-Cr[Turnbull (2018)] (Ti-6Al-4V)[Saini (2015)] *Vitallium*[Venable (1943) (Navarro (2008)] Co-Cr-Mo[Prakasam (2017)] Ni-Ti[Andani (2014), Chen (2015)] Alloys Co-Cr/nHA[Prakasam (2017)]		
Biocompatible (Classes III y IV)	Au[Saini (2015), Chen (2015)] SS[Saini (2015), Turnbull (2018)] Nb[Saini (2015), Chen (2015)]		PS[O'Brien (2011), Bhat (2013)] PMMA[Saini (2015), Bhat (2013)] PE[Saini (2015), Bhat (2013)] PADs[Saini (2015)] PET[Bhat (2013)] PTFE[Saini (2015), Bhat (2013)] PU[Carcielli (2006), Saini (2015), Bhat (2013)] PDMS[Bhat (2013)] Silicon[Saini (2015), Bhat (2013)] PVC[Bhat (2013)]			

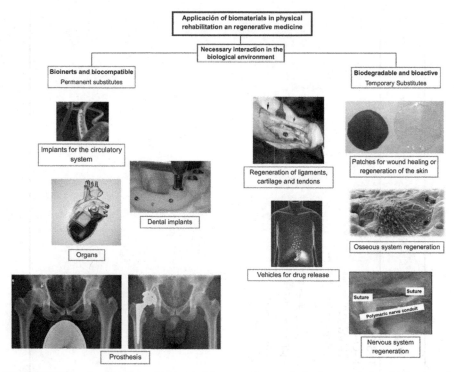

FIGURE 17.2 Application of biomaterials according to their characteristics: bioinert and biocompatibles; biodegradable and bioactive. (From Pajarinen, 2014; Moshiri, 2016; Yang, 2011; Hong, 2017; Thamarai-Selvi et al., 2018; adapted with permission.).

This type of materials generally focuses on the repair of nerves, tendons, cartilage, and bone, where the material is commonly made from biodegradable synthetic polymers or natural polymers (Bettinger and Borenstein, 2010; Sang, 2015; Camarero, 2016; Fukushima, 2016; Grémare, 2018; Rao, 2018; Tabasum, 2018; Shi, 2018). Today, research on biodegradable, bioactive, or synthetic polymers is becoming more frequent. These materials lose their weight percentage in a period generally <3 years, although the vast majority can do so in <1 year and a half (Shah, 2008; Song et al., 2009; Laycock et al., 2017; Du et al., 2018).

Biodegradable synthetic polymers are obtained from natural molecules such as lactic acid, through reactions (organic synthesis) that are not spontaneous in nature. These polymers only in some cases have slight incompatibility reactions (Limongi et al., 2017; Ciardelli and Chiono, 2006; Bettinger and Borenstein, 2010; Sang, 2015; Camarero, 2016; Fukushima, 2016; Grémare,

2018; Rao, 2018; Tabasum, 2018; Shi, 2018). Such materials are designed to present flexibility, good porosity for cellular nutrition (in case it is a cell regeneration device, which includes growth factors), and mechanical resistance. The most commonly used biopolymers for this purpose are aliphatic polyesters, such as PLAs (Hsu, 2011; Serra, 2013), PGA (Nakamura, 2004; Chaudhury, 2012; Manavitehrani et al., 2016), PEG (Leach and Schmidt, 2005; Mokarizadeh et al., 2015), PCL (Patrício, 2013; Malikmammadov et al., 2018), copolyesters derived from these such as PLGA, CPLA, and other copolyesters (Chaudhury, 2012; Manavitehrani et al., 2016; Bini et al., 2004). Some examples of different applications of biodegradable synthetic polymers in the body depending on their morphology are presented in Figure 17.3.

17.4 TENDENCIES AND ALTERNATIVES OF MODIFICATION AND IMPROVEMENT OF BIOMATERIALS

Among biomaterials with modification alternatives for improvement, bioactive biomaterials are the most complex to obtain because they not only provide a favorable environment to regenerate damaged tissue and then reabsorb but also influence the regeneration process so that it is performed in a better way (O'Brien, 2011; Limongi et al., 2017; Yang, 2011). This is being achieved thanks to biodegradable biomaterials that can be synthetic polymers that are modified by adding any of the following biological stimulations, such as specific cells, stem cells, growth factors/specific cells, growth factors/stem cells, or extracellular matrix (ECM) (O'Brien, 2011;

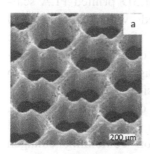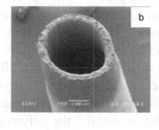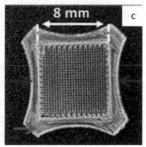

FIGURE 17.3 Grafts and implants made with biodegradable synthetic biopolymers: (a) honeycomb-shaped glycerol-co-sebacato (PGS) that provides structural and mechanical signals to guide cardiac regeneration, (b) cryofractured sections of PCL ducts for the regeneration of peripheral nerve tissues, and (c) PLA polylactic acid mesh for bone regeneration. (From Ciardelli, 2006; Bettinger, 2010; Grémare, 2018; adapted with permission.)

Zavaglia and Prado, 2016; Turnbull et al., 2018; Du et al., 2018). These biodegradable synthetic polymers offer the possibility of being modified by joining the cells or bioactive molecules on their surface, gradually releasing them during biodegradation. It is here that these polymers can be converted into bioactive biomaterials, and this combination of properties results in what is known as scaffolding (O'Brien, 2011; Serra, 2013).

Having this in mind, the growth factors are a set of proteins presented in plasma and in the platelets, which play an essential role in tissue repair and regeneration processes. ECM is a set of macromolecules such as collagen and elastin that are essential for the adhesion of cells in tissues and also regulate the growth factors. Therefore, if bioactive biomaterials contain ECM, they can influence regeneration processes because they have physical and chemical properties that allow cell adhesion, proliferation, and differentiation (O'Brien, 2011; Zavaglia and Prado, 2016; Turnbull et al., 2018; Du et al., 2018).

Adhesion is an important property in biomaterials since it corresponds to the ability of adhesion and bioabsorption of the material under standard cell culture conditions (immune system cells), as well as proliferation, cell phenotype, and differentiation, to specifically regenerate the required damaged tissue or organ. These principles, both without (biodegradable) and

with culture (bioactive), have been applied to nervous, bone, connective tissue, etc., since biodegradable and bioactive biomaterials must mimic soft bone tissues or neural nerve tissues, to mineralized hard tissues (Zhong, 2010; Moshiri et al., 2016; Yang, 2011). Some examples of bioactive biomaterials are Ciardelli et al. (2006) who performed in vitro cell adhesion and proliferation tests on PCL and PEU gelatin-coated ducts using S5Y5 neuroblast cell lines. The researchers chose this cell line since it shows homologous characteristics regarding human neurons. The results showed that the gelatin coating increases the adhesion and proliferation of cells in the hybrid ducts, which represents a considerable advance in a line of research that had a little approach to the PEU as a nerve conduit (Ciardelli and Chiono, 2006).

Likewise, Serra et al. (2013) worked with 3D printed PLA scaffolds coated with collagen and stem cells (Serra, 2013). These researchers obtained surface coatings of collagen in the PLA in two ways: physically and covalently. In the case of physical absorption, the polymer was directly immersed in a solution containing collagen (100 μg/mL), while, to achieve a covalent bond, the surface of the PLA was pretreated with NaOH to activate the necessary functional groups and subsequently the support was immersed in the collagen solution (100 µg/mL). To quantify the amount of protein (collagen) on the surface of the PLA scaffolds, the bicinconinic acid test (BCA test), also known as the Smith test, was performed. After surface modifications of the PLA with collagen, the researchers' cultured mesenchymal stem cells (rMSCs), and performed the cell viability analysis on the scaffold surface by the lactate dehydrogenase assay (LDH assay). Cellular studies showed a positive cellular response in collagen-functionalized PLA scaffolds but in greater proportion for scaffolds that were modified with collagen covalently bound to PLA. In fact, after 72 h of culture, mesenchymal stem cells in the scaffolds functionalized with collagen covalently bound to the PLA spread very well and completely covered the surface of the scaffold (Figure 17.4b and d). Serra et al. added that 3D-printed PLA scaffolds are not only for an application in specific regenerative medicine but can be used to regenerate any tissue, as necessary (Serra, 2013). The scaffolds used in this research and their results are shown in Figure 17.4.

The combination of properties between biomaterial (biocompatible and biodegradable) and biological-cellular stimuli would be the broadest definition of what is known as IT. Figure 17.5 shows the areas that involve IT (O'Brien, 2011; Limongi et al., 2017; Du et al., 2018). As has been seen so far in this research, its application is direct in regeneration medicine and, therefore, usually the term IT and regenerative medicine is usually interchangeable or equivalent (O'Brien, 2011; Limongi et al., 2017).

In the evolution of regenerative medicine, bioactive biomaterials, or scaffolding are considered the most advanced to date (Turnbull et al., 2018; Du et al., 2018; Thamarai-Selvi et al., 2018). Typically there are three individual groups of scaffolds for IT, which are ceramics, synthetic polymers (biodegradable and nonbiodegradable), and natural polymers (O'Brien, 2011; Limongi et al., 2017; Turnbull et al., 2018; Du et al., 2018). Each of these groups has specific advantages and disadvantages so that the use of scaffolding composed of different phases are becoming increasingly common, as mentioned by Du et al. (2010) and Thamarai-Selvi et al. (2018), in their work on ceramic-based 3D scaffolds and wound healing patches, respectively (Du et al., 2018; Thamarai-Selvi et al., 2018). In their review, Du et al. (2010) clearly mention that to be able to formulate 3D ceramic scaffolding, ceramic (powder, raw material) and a polymeric matrix where the ceramic material is dispersed are required, thus providing flexibility to the scaffolding and facilitating 3D printing. The polymer can be natural or synthetic, while the ceramic can be bioinert or bioactive. On the other hand, Thamarai-Selvi et

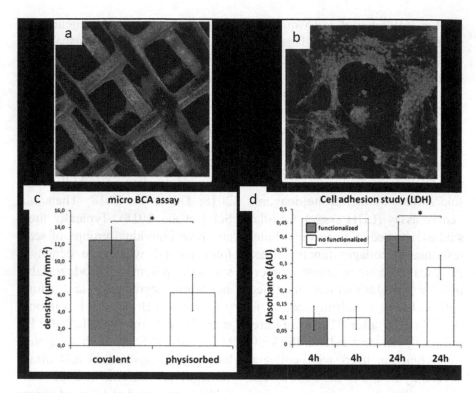

FIGURE 17.4 Surface functionalization of 3D printed PLA scaffolds: (a) collagen-covered scaffold completely and homogeneously; (b) rMSCs grown on functionalized scaffolds after 72 h; (c) quantification of the amount of collagen on the scaffold surface. Covalently functionalized scaffolds showed a significantly higher protein density than those treated with physical absorption; (d) amount of rMSCs adhered after 4 and 24 h on the scaffolds of PLA-collagen-covalent and nonfunctionalized PLA; functionalized scaffolds showed a greater number of viable cells after 24 h. (From Serra, 2013; with permission.).

al. (2018) obtained a patch based on sodium alginate, TiO$_2$ nanoparticles, curcumin, and PVA, that is, a combination of natural polymers with a biodegradable synthetic polymer and a metal oxide. For each of the materials that make up the patch was formulated, it can be noted that the patch is effectively designed to provide antioxidant (curcumin) and antibacterial (TiO$_2$) properties, stabilizing properties (sodium alginate), and finally easy molding properties and biodegradable (PVA) (Thamarai-Selvi et al., 2018).

Scaffolding can be designed in different morphologies depending on its application, such as cube, membrane, mesh, adhesive gel or graft of nerve, tendon or cartilage, among other morphologies. All these morphologies have in common porosity,

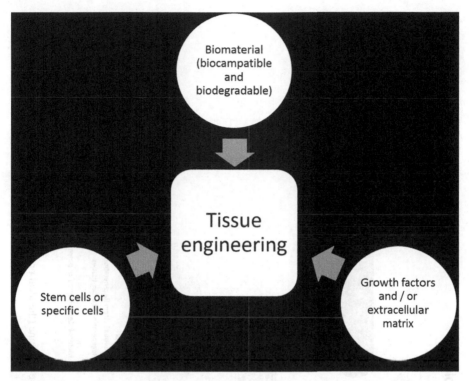

FIGURE 17.5 Tissue engineering (IT): cells, biosignals (chemically provided by growth factors and/or one or more macromolecules of the ECM), and the biomaterial that acts as the support for tissue formation by allowing cells migrate, adhere, and produce tissue.

which is essential because it provides the greatest amount of surface area for cell adhesion (Limongi et al., 2017; Turnbull et al., 2018; Du et al., 2018; Thamarai-Selvi et al., 2018). Membranes and meshes are generally designed for wound healing and tissue regeneration. Therefore, they are normally formulated based on biodegradable or resorbable polymers and/or ceramics, as well as the hybrid combination of collagen/hydroxyapatite or PLA/hydroxyapatite (Shah, 2008). On the other hand, cylindrical ducts are ideally designed to reconstruct nerve tissue, cartilage, or tendons, so its formulation should range from biodegradation or reabsorption through natural or synthetic polymers, to bioactive, with the surface modification of its properties either physically, chemically, or biologically. The latter can be done, for example, with the culture of stem cells. In Figure 17.6, the morphologies and ideal properties that a scaffold must have been described.

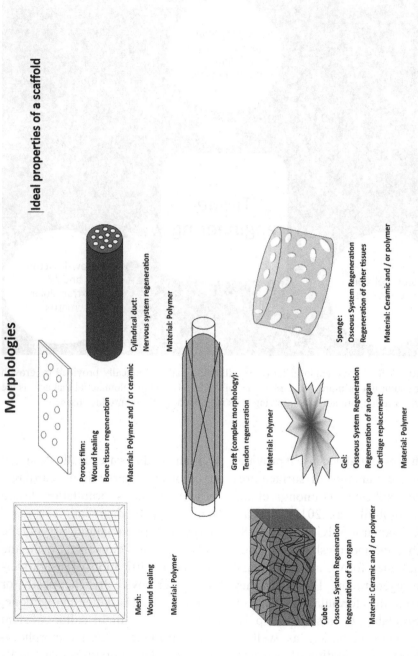

FIGURE 17.6 Design and properties of a scaffold for regenerative medicine (the ideal properties will always depend on both the biomaterial and the culture of cells and biosignals added to the scaffold).

Suture materials (surgical point) can also be considered as scaffolding, as well as controlled drug release (O'Brien, 2011; Venable and Stuck, 1943; Shah, 2008; Sivan, 2011; Bettinger and Borenstein, 2010). The latter refers to formulations, technologies, and systems for transporting an active compound in the body to safely achieve the desired therapeutic effect in the affected area (Bhat and Kumar, 2013; Turnbull et al., 2018; Thamarai-Selvi et al., 2018). The effectiveness of regenerative therapies with active compounds depends on both the ingredient and the vehicle. In this sense, the biomaterial is the vehicle that is responsible for carrying the active ingredient, but it may also have the ability to be part of the treatment (O'Brien, 2011; Thamarai-Selvi et al., 2018; Bettinger and Borenstein, 2010). Ye et al. (2019), conducted a review about scaffolds for drug release in wounds, obtained using electro-spinning. The authors mentioned that the most important active compounds in wound healing are vitamins A, C, and E, zinc, iodine, silver nanoparticles, and copper minerals. Which must be supplied by controlled release. In other studies, also, Ye et al. found that there was an increasing tendency in other investigations to incorporate growth factors or antibacterial molecules in electrospun materials to improve the quality of wound healing (Ye,

2019). Therefore, adequate interaction with the biological environment is likewise the most important characteristic that the vehicle must fulfill within the controlled release of drugs. Additionally, it is clarified that in regenerative applications of suture and controlled drug release, the materials are necessarily required to be fully biodegradable (Bhat and Kumar, 2013; Turnbull et al., 2018; Thamarai-Selvi et al., 2018).

In addition to the multiple biochemical and bioelectric signals that can be added to a biomaterial, IT has also used the technology currently available to make additional physical and chemical modifications to the biomaterials. These modifications are made based on the needs that the material requires and can improve or change the chemical, physical, and mechanical properties of the biomaterials. For example, inert biomaterials are sometimes filled with a bioactive material (Ciardelli and Chiono, 2006), while many natural polymers must be crosslinked to improve their mechanical properties, due to their rapid dissolution in environments, such as those of the internal tissue of human beings (Yang, 2011). Also, a plus of the biomaterials designed for regenerative medicine using these modifications is that the problems of incompatibility, corrosion, and degradability between the interface of the biomaterial and living tissue are corrected.

Another route to prevent all these difficulties is also to obtain biomaterials from renewable sources, mainly talking about polymers (Miculescu, 2017). An example of the modification of biomaterials is that of Li and Liu (2015), in their book chapter, where they speak about the chemical modification of metal biomaterials. In this chapter, they mention that there are generally five types of chemical modification of metal biomaterials or alloys: treatment with hydrogen peroxide (H_2O_2), acid treatment, alkaline treatment, thermal oxidation, and enzymatic modification. Such modifications influence the surface properties of the metal at the micrometric or nanometric level, providing them with porosity, changing their crystallization properties, or adding a coating (layer, spheres, cubes). Each of these modifications is aimed at improving its characteristics within the human body, as well as its thermal, mechanical stability, and mainly, at reducing toxic waste from the metal to the surrounding tissue (Li, 2015).

Another example is that of Minati et al. (2017), who address the issue of surface modification by plasma of biomaterials, such as polyether ketone (PEEK). This advanced engineering material is one of the most outstanding synthetic polymers because it has excellent mechanical and thermal properties (Tensile strength: 90–100 MPa; $T_f = 343$ °C)

and also resistance to aggressive sterilization procedures. PEEK is widely used in medical applications, for the treatment of spinal disorders, to repair orthopedic injuries, for joint arthroplasty, dentistry, and prostheses to replace metal parts (Minati, 2017). However, like most synthetic polymers, PEEK is hydrophobic, so it must be superficially modified for applications such as scaffolding, where it is desired to induce cells to adhere to PEEK to promote tissue regeneration. Plasma enhanced chemical vapor deposition (PECVD) is the process normally used for these purposes in PEEK, which refers to the deposition of thin films that have selected functionalities. Particular treatment is developed using PECVD polymerization processes based on CH_4 ($CH_4 + O_2$) and ($CH_4 + N_2$) or simply O_2, which improves the adhesion of cells to PEEK (Minati, 2017).

A further example is to perform functional grafts on the surface of the biopolymers by means of ATRP or RAFT polymerizations, as mentioned by Liang et al. (2017). A great contribution in this regard is the polymerization of unnatural amino acids (polypeptides or polyamino acids), which help both to mimic the functions of proteins and to interact with cellular systems. The strategies have even offered an economical and convenient approach to the production of high molecular weight polypeptides, with precise structures and custom-made functionalities, not only as a graft on another biopolymer but also as an independent biomaterial (Liang, 2017). The modifications that are used according to the needs and want to be reinforced in terms of biocompatibility, biodegradability, or improvement of the physical and mechanical properties of the biomaterials are presented in Figure 17.7.

17.5 ADVANTAGES AND DISADVANTAGES OF SYNTHETIC BIOMATERIALS

Many types of biomaterials are used in the replacement of artificial organs, prostheses, and cell regeneration, among other biochemical and biomedical applications (Eliaz, 2012; Alvarez, 2009; Xi, 2010;

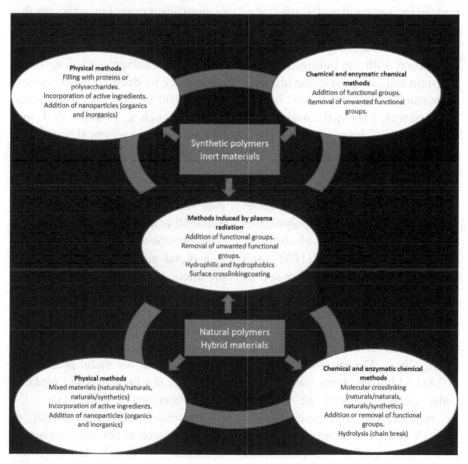

FIGURE 17.7 Scheme of the modifications that are used in biomaterials to improve their physicochemical properties. (From Saini et al., 2015; Bhat and Kumar, 2013; Li, 2015; Minati, 2017; Liang, 2017; Gübitz, 2003.).

Montero, 2012; Sabree, 2015; Monteiro, 2015; Salmoria, 2016; Oliveira, 2016; Meyer, 2016; Tamaddon, 2017; Chen, 2017). Initially, most of these devices were manufactured from synthetic nondegradable biomaterials. But little by little, it was evidenced that the good interaction with certain specific cells within the organism was difficult. For this reason, they had to be filled, modified, or impregnated with some material derived from nature, which was compatible with promoting a better interaction with the biological environment and attenuating the immune response of the organism in the affected area. Among the main materials that have been used for these applications are synthetic polymers (polymers that are mostly derived from nonrenewable sources such as petroleum), such as acrylics (Uzman and Villegas, 1983), polyolefins (Cordeiro et al., 1989), elastomeric hydrogels (Keeley et al., 1991), and silicones (Chen, 2000). Other sets of materials used for these purposes are metals, different types of SS (Minati, 2017), alloys of Ti, Co, among others (Saini et al., 2015; Pajarinen et al., 2014; Thamarai-Selvi et al., 2018). Although, with these materials, there has been an advance in regenerative treatments and a growth in the manufacture of artificial organs, grafts, and prostheses, it has been found that many nondegradable materials tend to cause an

inflammatory response, which leads to chronic compression around the treated area. In addition, many could present a series of additional inconveniences, such as poor porosity, which hinders cellular respiration, and finally, poor or no flexibility (Ciardelli and Chiono, 2006). As a result, its use was gradually limited to body parts where the immune system does not present an alarming or significant rejection, although this also depends on the type of polymer used. However, many of the plastics have an advantage regarding resistance since its decomposition is very slow (>100 years). Therefore, some polymers such as PMMA, PET, and silicone, among others (Uzman and Villegas, 1983; Cordeiro et al., 1989; Keeley et al., 1991; Chen, 2000) can be used as artificial organs, thanks to their durability and modifications. Some other applications may include bone prostheses where the main part is a metal and the adhesive and/or prosthetic joints are a polymer (Saini et al., 2015; Bhat and Kumar, 2013; Pajarinen et al., 2014). For example, UHMWPE one of the most stable and low friction polymers that is used as a joint for metal prostheses, mainly in the knee and hip replacement (Pajarinen et al., 2014).

However, the manipulation of properties such as porosity and biodegradability of metals and polymers has been achieved, thanks to new research on composite biomaterials, biodegradable synthetic biomaterials (as discussed above), and hybrid biomaterials (O'Brien, 2011; Turnbull et al., 2018; Du et al., 2018; Thamarai-Selvi et al., 2018), as well as to surface modifications such as those mentioned by Li and Liu (2015) in their chapter on metal biomaterials (Li, 2015), Minati et al. (2017) by means of plasma technology for the modification of synthetic polymers (Minati, 2017), which can also be applicable for metals or ceramics, and Gübitz et al. (2003) also modified synthetic polymers but enzymatically (Gübitz, 2003), such as the modification of polyamides as Nylon with different types of proteases. One of the great advantages of these last two biomaterial modification (plasma and enzymatic) techniques is that they are considered sustainable chemical processes.

An example of obtaining a hybrid material is the research that is carried out by Vashisth and Bellare (2018), who designed a 3D biomimetic scaffolding considering the natural architecture of the bone. The scaffolding consists of a favorable interconnected porous structure, with a characteristic nanomicroscale level of great mechanical resistance. The main components of the hybrid scaffold were collagen nanofibers (natural polymer) and a hydrogel made from gellan gum (natural polymer) and hydroxyapatite (bioactive synthetic ceramic), properly arranged to create a bone imitation. Morphological analysis using SEM and 4D X-ray microscopy revealed that the hybrid scaffold consists of rolled nanofiber rings in a highly porous hydrogel matrix that shows structural similarity to osteons (basic structural elements of the bone matrix through which blood flows). The reinforcement of electrospun nanofibers in hydrogel influenced the mechanical properties of the scaffold. The potential application of the biomimetic hybrid scaffold, and the role of its specific architecture, was investigated using a human osteosarcoma fibroblast cell line. These hybrid biomaterials presented great potential for bone tissue regeneration, and although it is a development at the level of tissue engineering, it also has direct application in bone regeneration so that it can be considered as part of the new physical rehabilitation therapies (Vashisth, 2018).

As for metals, very few pure metallic materials are accepted biologically by the tissues that are in contact with them, so that most of these biomaterials are alloys (Saini et al., 2015; Pajarinen et al., 2014; Li, 2015). In regenerative medicine, gradually it was discovered the types of materials based on metals and ceramics that can interact better with the cellular environment. As explained in the work done by Pajarinen et al. (2014) that is based on obtaining biomaterials for total hip replacement and that of Li and Liu (2015) about metallic materials for biomedical applications (Pajarinen et al., 2014; Li, 2015). Pajarinen et al. (2014), for example, mention that many of the first attempts for total hip replacement between the 1960s and 1990s were made with polymeric materials but these presented great problems of incompatibility and mechanical fracture. However, although it was clear that the mechanical characteristics of joint replacement are very important, Pajarinen et al. (2014) mentioned that many of these problems have already been resolved with the use of superalloys and exhaustive preclinical mechanical tests.

The advantage of metal and ceramic alloys is that they are rigid, resistant and have no rejection response from the body, so most of them are used as prostheses to replace parts of the bone system (Saini et al., 2015; Turnbull et al., 2018; Du et al., 2018; Pajarinen et al., 2014; Li, 2015). Although most metals are already known to formulate an alloy with a safe application within the human body, biocompatibility studies are essential for new alloys that are intended to be used as replacement biomaterials. While implants and ceramic prostheses are a little more noble in terms of their direct application in the human body, and their physicochemical similarity with hard tissues can make them

more proper to have not only biocompatibility but also bioabsorption and bioactivity, so that in the long-term, they can become part of living tissue (Turnbull et al., 2018; Du et al., 2018; Pajarinen et al., 2014). In this sense, both metallic and ceramic biomaterials can be inert or bioabsorbable (but few metal alloys are bioabsorbable) (Eliaz, 2012; Montero, 2012). As specified above, the inert metals or ceramics would be those that do not chemically or biologically bond with the tissue; therefore, the organism cannot absorb them but they are not rejected by the tissue and, therefore, do not produce an allergy or secondary reaction (Saini et al., 2015; Du et al., 2018; Liu, 2002; McGinnis et al., 1998; Mattioli et al., 2003; Bellón et al., 2005; Goharian and Abdullah, 2017; Navarro, 2008; Stevens, 2008; Mano, 2004). Bioabsorbable and bioactive biomaterials would lose their mass progressively while the tissue regenerates, by action of the cells, metabolism, as well as by physical–chemical processes, like the hydrolysis, until disappearing completely in a certain time (O'Brien, 2011; Limongi et al., 2017; Turnbull et al., 2018; Du et al., 2018; Montero, 2012).

17.6 FINAL REMARK

This chapter clearly described the relationship between the type of biomaterial and its biological activity within the human body. This allowed us to know that, according to the chemical, physical, and mechanical characteristics of metals, ceramics, and polymers, the application of a biomaterial within the organism can be deducted. In this sense, in this chapter, it was found that the majority of metal and ceramic biomaterials are used for bone regeneration or rehabilitation, prostheses, and implants. Polymers are being used mainly in tendon reconstruction, nerve tissue, organ replacement, and wound healing, while both polymers and ceramics are becoming increasingly important as vehicles for controlled drug release. It should be clarified that the artificial organs in regenerative medicine are manufactured mainly from nonbiodegradable synthetic polymers because they have biocompatibility and flexibility similar to that of the organ but mainly because they have a long-term chemical and dimensional resistance (>100 years). However, trends are evolving toward the use of hybrid materials (combinations of synthetic and natural biomaterials), mainly due to the fact that these materials can be chemically, physically, or biologically modified to adapt and combine all their properties. Thanks to the modifications that can be made in biomaterials, the possibilities of using ceramics and metals in combination with polymers for applications in implants, prostheses, and scaffolds are amplified, both for applications in physical rehabilitation, as in regenerative medicine. Therefore, the trend indicates that, as IT improves the knowledge and adaptation of these biomaterials and their combinations in the human body, the results will continue to be narrowed in the investigations carried out with the actual application. In this way, it is very likely that in the following years, more opportunities will be found to correct all types of tissue, or function damaged within the organism reliably and effectively.

ACKNOWLEDGMENTS

The National Council of Science and Technology (CONACyT) is thanked for the grant number 447932 given to MSc Luis Fernando Mora Cortes in the PhD Program in Materials Science and Technology of the FCQ.

KEYWORDS

- **biodegradable**
- **bioactive**
- **bioinert**
- **biocompatible**
- **scaffold**
- **prostheses**
- **metals**
- **ceramics**
- **polymers**

REFERENCES

Alvarez, K.; Nakajima, H. Metallic scaffolds for bone regeneration. Materials, **2009**, 2, 790–832.

Andani, M.; Moghaddam, N.; Haberland, C.; et al. Metals for bone implants. Part 1: powder metallurgy and implant rendering. Acta Biomater., **2014**, 10, 4058–4070.

Bellón, J.; García, N.; Serrano, N.; et al. Composite prostheses for the repair of abdominal wall defects: effect of the structure of the adhesion barrier component. Hernia, **2005**, 9, 338–343.

Bettinger, C.; Borenstein, J. Biomaterials-based microfluidics for engineered tissue constructs. Soft Matter., **2010**, 6, 4999–5015.

Bhat, S.; Kumar, A. Biomaterials and bioengineering tomorrow's healthcare. Biomaterials, **2013**, 3, 1–12.

Bini, T.; Gao, S.; Xu, X.; et al. Peripheral nerve regeneration by microbraided poly(L-lactide-co-glycolide) biodegradable polymer fibers. J. Biomed. Mater. Res. A, **2004**, 68, 286–295.

Camarero, S.; Rothen, B.; Fostera, E.; Weder, C. Articular cartilage: from formation to tissue engineering. Biomater. Sci., **2016**, 4, 734–767.

Chaudhury, S. Mesenchymal stem cell applications to tendon healing. Musc. Lig. Tend. J., **2012**, 2, 222–229.

Chen, Q.; Thouas, G. Metallic implant biomaterials. Mat. Sci. Eng. R, **2015**, 87, 1–57.

Chen, Y.; Hsieh, C.; Tsai, C.; et al. Peripheral nerve regeneration using silicone rubber chambers filled with collagen, laminin and fibronectin. Biomaterials, **2000**, 21, 1541–1547.

Chen, Z.; Song, Y.; Zhang, J.; et al. Laminated electrospun nHA/PHB-composite scaffolds mimicking bone extracellular matrix for bone tissue engineering. Mater. Sci. Eng. C, **2017**, 72, 341–351.

Ciardelli, G.; Chiono, V. Materials for peripheral nerve regeneration. Macromol. Biosci., **2006**, 6, 13–26.

Cordeiro, P.; Seckel, B.; Lipton, S.; et al. Acidic fibroblast growth factor enhances peripheral nerve regeneration in vivo. Plast. Reconstr. Surg., **1989**, 83, 1013–1019.

Du, X.; Fua, S.; Zhu, Y. 3D printing of ceramic-based scaffolds for bone tissue engineering: an overview. J. Mater. Chem. B, **2018**, 6, 4397–4412.

Eliaz, N.; Hakshur, K. Chapter-10, Fundamentals of tribology and the use of ferrography and bio-ferrography for monitoring the degradation of natural and artificial joints. In Ddegradation of Implant Materials. Springer: Berlin, **2012**, 286–293.

Fukushima, K. Poly(trimethylene carbonate)-based polymers engineered for biodegradable functional biomaterials. Biomater. Sci., **2016**, 4, 9–24.

Gee, E.; Jordan, R.; Hunt, J.; Saithna, A. Current evidence and future directions for research into the use of tantalum in soft tissue re-attachment surgery. J. Mater. Chem. B, **2016**, 4, 1020–1034.

Goharian, A.; Abdullah, M. Chapter 7, Bioinert metals (stainless steel, titanium, cobalt chromium). In Trauma plating systems. Elsevier: Amsterdam, **2017**, 115–142.

Grémare, A.; Guduric, V.; Bareille, R.; Heroguez, V.; et al. Characterization of printed PLA scaffolds for bone tissue engineering. Biomed. Mater. Res. Part A, **2018**, 106, 887–894.

Gübitz, G.; Paulo, A. New substrates for reliable enzymes: enzymatic modification of polymers. Curr. Opin. Biotech., **2003**, 14, 577–582.

Hong, D.; Oh, J. Recent advances in dental implants. Maxillofac. Plast. Reconstr. Surg., **2017**, 39, 1–10.

Hsu, S.; Chan, S.; Chiang, C.; Chen, C.; Jiang, C. Peripheral nerve regeneration using a microporous polylactic acid asymmetric conduit in a rabbit long-gap sciatic nerve transection model. Biomaterials, **2011**, 32, 3764–3775.

Huebsch, N.; Mooney, D. Inspiration and application in the evolution of biomaterials. Nature, **2009**, 462, 426–432.

Kappel, R.; Cohen, J.; Pruijn, G. Autoimmune/inflammatory syndrome induced by adjuvants (ASIA) due to silicone implant incompatibility syndrome in three sisters. Clin. Exp. Rheu., 2014, 32, 256–258.

Keeley, R.; Nguyen, K.; Stephanides, M.; Padilla, J.; Rosen, J. The artificial nerve graft: a comparison of blended elastomer-hydrogel with polyglycolic acid conduits. J. Reconstr. Microsurg., 1991, 7, 93–100.

Laycock, B.; Nikolic, M.; Colwell, J.; Gauthier, E.; et al. Lifetime prediction of biodegradable polymers. Prog. Polym. Sci., 2017, 71, 144–189.

Leach, J.; Schmidt, C. Characterization of protein release from photocrosslinkable hyaluronic acid-polyethylene glycol hydrogel tissue engineering scaffolds. Biomaterials, 2005, 26, 125–135.

Li, J.; Liu, X. Chapter-5, Chemical surface modification of metallic biomaterials. In Surface coating and modification of metallic biomaterials. Elsevier: Amsterdam, 2015, 159–183.

Liang, Y.; Li, L.; Scott, R.; Kiick, K. Polymeric biomaterials: diverse functions enabled by advances in macromolecular chemistry. Macromolecules, 2017, 50, 483–502.

Limongi, T.; Lizzul, L.; Giugni, A.; Tirinato, L.; et al. Laboratory injection molder for the fabrication of polymeric porous poly-epsilon-caprolactone scaffolds for preliminary mesenchymal stem cells tissue engineering applications. Microelec. Eng., 2017, 175, 12–16.

Liu, D.; Yang, Q.; Troczynski, T. Sol–gel hydroxyapatite coatings on stainless steel substrates. Biomaterials, 2002; 23, 691–698.

Malikmammadov, E.; Tanir, T.E.; Kiziltay, A.; Hasirci, V.; Hasirci, N. PCL and PCL-based materials in biomedical applications. J. Biomater. Sci. Polym. Ed., 2018, (7-9), 863–893.

Manavitehrani, I.; Fathi, A.; Badr, H.; Dehghani, F.; et al. Biomedical applications of biodegradable polyesters. Polymers, 2016, 8, 20–52.

Mano, J.; Sousa, R.; Boesel, L.; Neves, N.; Reis, R. Bioinert, biodegradable and injectable polymeric matrix composites for hard tissue replacement: state of the art and recent developments. Compos. Sci. Technol., 2004, 64, 789–817.

Marston, W. Dermagraft (R), a bioengineered human dermal equivalent for the treatment of chronic nonhealing diabetic foot ulcer. Exp. Rev. Med. Devices, 2004, 1, 21–31.

Mattioli, M.; Giavaresi, G.; Biagini, G.; et al. Tailoring biomaterial compatibility: in vivo tissue response versus in vitro cell behavior. Int. J. Artif. Organs, 2003, 26, 1077–1085.

McGinnis, M.; Larsen, P.; Miloro, M.; Beck, F. Comparison of resorbable and nonresorbable guided bone regeneration materials: a preliminary study. Int. J. Oral Maxillofac. Implants, 1998, 13, 30–35.

Meyer, C.; Stenberg, L.; Gonzalez, F.; et al. Chitosan-film enhanced chitosan nerve guides for long-distance regeneration of peripheral nerves. Biomaterials, 2016; 76, 33–51.

Miculescu, F.; Maidaniuc, A.; Ioan, S.; Kumar, V; et al. Progress in hydroxyapatite–starch based sustainable biomaterials for biomedical bone substitution applications. ACS Sustain. Chem. Eng., 2017, 5, 8491–8512.

Minati, L.; Migliaresi, C.; Lunelli, L.; Viero, G.; Dalla, M.; Speranza, G. Plasma assisted surface treatments of biomaterials. Biophys. Chem., 2017, 229, 151–164.

Mokarizadeh, A.; Mehrshad, A.; Mohammadi, R. Local polyethylene glycol in combination with chitosan based hybrid nanofiber conduit accelerates transected peripheral nerve regeneration. J. inves. surg., 2015, 29, 167–174.

Monteiro, N.; Martins, A.; Reis, R.; Neves, N. Nanoparticle-based bioactive agent release systems for bone and cartilage tissue engineering. Regen. Ther., 2015, 1, 109–118.

Montero, J.; Lorda, C.; Hurlé, J. Regenerative medicine and connective tissues:

cartilage versus tendon. J. Tissue Eng. Regen. Med., **2012**, 6, 337–347.

Moshiri, A.; Oryan, A.; Meimandi, A.; et al. Effectiveness of hybridized nano- and microstructure biodegradable, biocompatible, collagen-based, three-dimensional bioimplants in repair of a large tendon-defect model in rabbits. J. Tissue Eng. Regen. Med., **2016**, 10, 451–465.

Nakamura, T.; Inada, Y.; Fukuda, S.; et al. Experimental study on the regeneration of peripheral nerve gaps through a polyglycolic acid–collagen (PGA–collagen) tube. Brain Res., **2004**, 1027, 18–29.

Navarro, M.; Michiardi, A.; Castaño, O.; Planell, J. Biomaterials in orthopaedics. J. R. Soc. Interface, **2008**, 5, 1137–1158.

O'Brien, F. Biomaterials & scaffolds for tissue engineering. Materialstoday, **2011**, 14, 88–95.

Oliveira, C.; Pereira, F.; Oliveira, Z.; et al. Evaluation of biodegradable polymer conduits—poly(L-lactic acid)—for guiding sciatic nerve regeneration in mice. Methods, **2016**, 99, 28–36.

Pajarinen, J.; Lin, T.; Sato, T.; Yao, Z.; Goodman, S. Interaction of materials and biology in total joint replacement-successes, challenges and future directions. J. Mater. Chem. B, **2014**, 2, 7094–7108.

Patrício, T.; Domingos, M.; Gloria, A.; Bártolo, P. Characterisation of PCL and PCL/PLA scaffolds for tissue engineering. Procedia CIRP, **2013**, 5, 110–114.

Pérez, J.; Terán, J.; Herrera, M.; Martínez, M.; Genescá. J. Assessment of stainless steel reinforcement for concrete structures rehabilitation. J. Const. Steel Res., **2008**, 64, 1317–1324.

Pina, S.; Ribeiro, V.; Marques, C.; Silva, T.; et al. Scaffolding strategies for tissue engineering and regenerative medicine applications. Materials, **2019**, 12, 1824–1866.

Prakasam, M.; Locs, J.; Salma, K.; et al. Biodegradable materials and metallic implants—a Review. J. Funct. Biomater., **2017**, 8, 44, https://doi:10.3390/jfb8040044.

Rao, S.; Harini, B.; Kumar, R.; Balagangadharan, K.; Selvamurugan, N. Natural and synthetic polymers/bioceramics/bioactive compounds-mediated cell signalling in bone tissue engineering. Int. J. Biol. Macromol., **2018**, 110, 88–96.

Sabree, I.; Gough, J.; Derby B. Mechanical properties of porous ceramic scaffolds: influence of internal dimensions. Ceram. Int., **2015**, 41, 8425–8432.

Saini, M.; Singh, Y.; Arora, P.; Arora, V.; Jain, K. Implant biomaterials: a comprehensive review. World J. Clin. Cases., **2015**, 16, 52–57.

Salmoria, G.; Paggi, R.; Castro, F.; et al. Development of PCL/Ibuprofen tubes for peripheral nerve regeneration. Procedia CIRP, **2016**, 49, 193–198.

Sang, L.; González, C.; Kwang, K.; Kyoung, Kim.; Kuroda, K. Catechol-functionalized synthetic polymer as a dental adhesive to contaminated dentin surface for a composite restoration Biomacromolecules, **2015**, 16, 2265–2275.

Satti, S.; Shah, A.; Auras, R.; Marsh, T. Isolation and characterization of bacteria capable of degrading poly(lactic acid) at ambient temperature. Polym. Degrad. Stab., **2017**, 144, 392–400.

Thamarai-Selvi, R.; Prasanna, A.; Niranjan, R.; Kaushik, M.; et al. Metal oxide curcumin incorporated polymer patches for wound healing. Appl. Surf. Sci., **2018**, 449, 603–609.

Sensharma, P.; Madhumathi, G.; Jayant, R.; Jaiswal, A. Biomaterials and cells for neural tissue engineering: current choices. Mater. Sci. Eng. C, **2017**, 77, 1302–1315.

Serra, T.; Mateos-Timoneda, M.; Planell, J.; Navarro, M. 3D printed PLA-based scaffolds. Organogenesis, **2013**, 9, 239–244.

Shah, A.; Hasan, F.; Hameed, A.; Ahmed, S. Biological degradation of plastics: a comprehensive review. Biotechnol. Adv., **2008**, 26, 246–265.

Shi, Y.; Liu, J.; Yu, L.; Zhong, L.; Jiang, H. β-TCP scaffold coated with PCL as biodegradable materials for dental applications. Ceram. Int., **2018**, 44, 15086–15091.

Sivan, A. New perspectives in plastic biodegradation. Curr. Opi. Biotech., **2011**, 22, 422–426.

Song, J.; Murphy, R.; Narayan, R.; Davies, G. Biodegradable and compostable alternatives to conventional plastics. Phil. Trans. R. Soc. B, **2009**, 364, 2127–2139.

Stevens, M. Biomaterials for bone tissue engineering. Materialstoday, **2008**; 11, 18–25.

Tabasum, S.; Noreen, A.; Farzam M.; Umar, H.; et al. A review on versatile applications of blends and composites of pullulan with natural and synthetic polymers. Int. J. Biol. Macromol. Part A, **2018**, 120, 603–632.

Tamaddon, M.; Samizadeh, S.; Wang, L.; Blunn, G.; Liu, C. Intrinsic osteoinductivity of porous titanium scaffold for bone tissue engineering. Int. J. Biomater., **2017**, 5093063, https://doi.org/10.1155/2017/5093063.

Turnbull, G.; Clarke, J.; Picard, F.; Riches, P.; et al. 3D bioactive composite scaffolds for bone tissue engineering. Bioact. Mater., **2018**, 3, 278–314.

Uzman, B.; Villegas, G. Mouse sciatic nerve regeneration through semipermeable tubes: a quantitative model. J. Neurosci. Res., **1983**, 9, 325–338.

Vashisth, P.; Bellare, J. Development of hybrid scaffold with biomimetic 3D architecture for bone regeneration. Nanomedicine, **2018**, 14, 1325–1336.

Venable, C.; Stuck, W. Clinical uses of vitallium. Ann. Surg., **1943**, 117, 772–782.

Xi, Y.; Fei, L.; Yuan, X.; Chang, W. In vitro study in the endothelial cell compatibility and endothelialization of genipin-crosslinked biological tissues for tissue-engineered vascular scaffolds. J. Mater. Sci. Mater. Med., **2010**, 21, 777–785.

Yang, Y.; Shen, C.; Cheng, H.; Liu, B. Sciatic nerve repair by reinforced nerve conduits made of gelatin-tricalcium phosphate composites. J. Biomed. Mater. Res. Part A, **2011**, 96, 288–300.

Yang, Y.; Zhao, W.; He, J.; Zhao, Y.; et al. Nerve conduits based on immobilization of nerve growth factor onto modified chitosan by using genipin as a crosslinking agent. Eur. J. Pharm. Biopharm., **2011**, 79, 519–525.

Ye, K.; Kuang, H.; You, Z.; Morsi, Y.; Mo, X. Electrospun nanofibers for tissue engineering with drug loading and release. Pharmaceutics, **2019**, 11, 182–199.

Zavaglia, C.; Prado, M. Feature article: biomaterials. Mater. Sci. Mater. Eng., **2016**, 1–5. https://doi: 10.1016/B978-0-12-803581-8.04109-6

Zhong, S.; Zhang, Y.; Lim, C. Tissue scaffolds for skin wound healing and dermal reconstruction. Nanomed. Nanobiotechnol., **2010**, 2, 510–525.

CHAPTER 18

Hydrogel Systems Based on Collagen and/or Fibroin for Biomedical Applications

JESÚS A. CLAUDIO-RIZO[1], CLAUDIA M. LÓPEZ-BADILLO[2], and BRENDA R. CRUZ-ORTIZ[2*]

[1]Advanced Materials Department, School of Chemical Sciences, University Autonomous of Coahuila, Saltillo, Mexico

[2]Ceramic Materials Department, School of Chemical Sciences, University Autonomous of Coahuila, Saltillo, Mexico

*Corresponding author. E-mail: b.cruz@uadec.edu.mx.

ABSTRACT

Fibrillar proteins such as collagen and fibroin can produce highly hydrated networks generating hydrogel systems with potential applications in the area of biomedicine. Hydrogels based on collagen and/or fibroin have high biocompatibility, but have poor mechanical properties and rapid degradation rates, limiting their applicability in the area of tissue engineering. Due to these limitations, various strategies for the design of collagen and/or fibroin based on hydrogels coupled with natural polymers and/or inorganic phases have been developed, improving their performance in biomedical applications. The study of biomaterials in the hydrogel state requires to know detailing relevant aspects of the processing conditions, the characterization techniques employed to study their properties, and the main advantages offered by these biomaterials in the area of biomedicine and tissue engineering. Thus, this chapter aims to review the main contributions of collagen and/or fibroin-based hydrogels systems detailing the innovative results in the biomedical field.

18.1 INTRODUCTION

Hydrogels are highly crosslinked polymeric networks that can absorb water and show promising applications in the field of biomedicine. Besides, their swelling and degradation properties allow the controlled diffusion of exudates and/or essential biomolecules in the tissue healing processes, providing moisture to the wound bed; and on the other hand, diverse strategies allow to design of hydrogels with adapted properties to improve their performance in these fields of research. Hydrogels based on natural polymers, such as proteins, specifically collagen, and fibroin, have been widely studied due to their high biocompatibility. These proteins can polymerize at physiological conditions (pH 7.4 and 37 °C) to form fibers that have the capacity to absorb water and form materials in the hydrogel state (Figure 18.1).

Hydrogels derived from collagen and fibroin have excellent biocompatibility but show mechanical weakness, and they biodegrade quickly, limiting their applicability in tissue engineering strategies, where hydrogels with tailored properties are required. Among the main approaches to improve these properties, there are (1) the formation of interpenetrated networks (IPNs) with other types of polymers and (2) the chemical crosslinking of the proteins with exogenous molecules with reactive groups.

The formation of IPN-based hydrogels occurs when different polymeric chains are physically mixed by continuous cooling-drying processes, producing hybrid matrices with improved mechanical properties, without a significant change in the rate of degradation. The Chemical crosslinking process generates covalent bonds among the reactive functional groups of the polymeric chains with the functional groups related to exogenous molecules. The main protein crosslinkers used for the design of hydrogels are glutaraldehyde, carbodiimide,

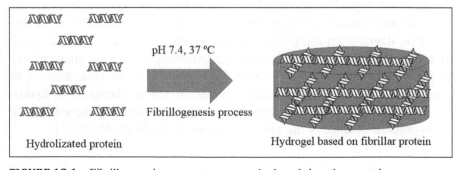

pH 7.4, 37 °C

Fibrillogenesis process

Hydrolizated protein

Hydrogel based on fibrillar protein

FIGURE 18.1 Fibrillogenesis process to generate hydrogels based on proteins.

genipin, adipic acid, citric acid, oligourethanes, and among others. Hydrogels produced by chemical crosslinking can control their physicochemical properties according to the degree of crosslinking, allowing them to control the mechanical properties and degradation of hydrogels. In general, low crosslinking indices produce an improvement in the properties of the materials and do not show a significant decrease in the biocompatibility of the modified proteins; by increasing the crosslinking index, biocompatibility decreases significantly.

Other strategies involve the generation of hybrid hydrogels that include the coupling of collagen and/or fibroin with other chemical agents, allowing them to tailor their properties for successful applications in tissue engineering and regenerative medicine. Mainly, the coupling with natural polymers and inorganic phases has allowed generating systems of hydrogels with improved performance in the biomedical field. The natural polymers (polysaccharides and proteins) permit to adapt the properties of the collagen hydrogels in direct relation with the concentration, taking advantage of the high biocompatibility present in these biopolymers. Besides, different inorganic phases, such as nanoparticles of Si, Ag, Au, Fe, and hydroxyapatite, show improvement in the properties of the

materials and control of the biological response required in tissue engineering applications. Therefore, the objective of this chapter is to reveal the main contributions of collagen and fibroin-based hydrogel systems coupled with natural polymers and inorganic phases that are in the literature applied in biomedical aspects. The present work focuses on the processing conditions of the systems, the characterization techniques used to study their physicochemical properties, and the in vitro and in vivo evaluation of innovative results presented by these biomaterials.

18.2 COLLAGEN HYDROGELS FOR TISSUE ENGINEERING

Collagen-based hydrogels have shown to provide properties that make them excellent candidates for application in tissue engineering and biomedicine; however, the performance of collagen hydrogels is low due to their ease of degradation and poor mechanical stability. The generation of synthesis strategies of new hydrogels that allow modulating their properties and their structure is an area of interest in current research. Collagen has been chosen as a model protein to generate biomaterials with high biocompatibility since this protein constitutes the connective and skeletal tissues of animals (Tang et al., 2009). Cellular tissues have a support matrix that allows them to perform their fundamental functions, this extracellular matrix (ECM) contains collagen as a scaffolding protein, and is formed by biochemical processes performed by fibroblasts and epithelial cells (Theocharis et al., 2016). Depending on the molecular aggregates that form the quaternary structures of collagen, about 20 specific forms of collagen exists; each form or type of collagen shows significant variation in their primary and quaternary structures (Eyre, 1980). The primary structure of collagen depends on the sequence of amino acids that constitute it, while the quaternary structure depends on the interactions that form with other collagen molecules. The most abundant amino acids in collagen are glycine (Gly), proline (Pro), and hydroxyproline (Hyp). The presence of five-member aromatic rings in the Pro and Hyp amino acids gives collagen its characteristic secondary structure in the form of alpha-helix (Gould, 1968).

Collagen molecules can be self-assembled to generate high-order hierarchical structures in the form of fibrils; the fibrils homogeneously generate fibers, and these fibers constitute the scaffold base of the extracellular matrices that make up the tissues (Kastelic et al., 1978). The process of collagen fiber formation depends on ionic strength, pH, and temperature. Collagen oligopeptides that have amino and carboxylic acid functional groups at their ends, initiate self-assembly reactions to generate amide bonds, whereby the collagenic structure grows. These molecules capable of polymerizing are called tropocollagens and are made up of about 240 amino groups and 230 carboxylic acids, with an average length of 300 nm (Kastelic et al., 1978). The stability of collagen fibrils is due to hydrogen bridge interactions, and molecular interactions that involve short-range bonds such as van der Waals and London forces, as well as interactions between dipoles (Harris et al., 2013). These types of interactions are also associated with the cross-linking of collagen fibrils to produce fibers with a high hierarchical structure. Collagen fibers hydrolyze in the presence of dilute acids to produce soluble oligopeptides, which reassemble to generate fibers after modifying the pH and temperature (Harris et al., 2013). Collagen regions that do not have a helical structure are known as telopeptides, which are associated with producing adverse reactions when in contact with exogenous agents or hosts (Lowther, 1963; Whitesides and Boncheva, 2002). During the process of assembling the collagen fibers, the physical crosslinking processes determine the maturity of the fiber and its mechanical strength. The elimination of telopeptidic regions of collagen fibers significantly improves their mechanical strength because the final structure has a homogeneous helical character (Kadler et al., 1996).

Collagen fibers can absorb water to produce hydrogels with high structural stability. Thus, various techniques involving hydrolysis assisted by proteolytic enzymes, such as pepsin, and papain, have been developed to extract helical collagen oligopeptides that can polymerize to generate materials in the hydrogel state. Collagen hydrogels have been extensively studied in biomedicine and tissue engineering applications. The structural stability and the similarity with natural scaffolds that collagen fibers present are two of the main factors for their use. These factors permit the encapsulation of cells, allowing them to grow, migrate, differentiate, and proliferate, as well as regulate the metabolism (Kadler et al., 1996; Overstreet et al., 2012). The water content and the swelling capacity of collagen hydrogels are fundamental properties that allow the modulation of the cellular phenotype for applications in tissue engineering. It is also possible to encapsulate exogenous molecules that allow improving the application of the hydrogel for the generation of new tissue and avoiding bacterial infections (Tan and Marra, 2010).

Recently, the use of various tissues for the extraction of collagen has been investigated in order to observe the relationship that the ECM molecules have in the collagen hydrogels that are generated. Matrix proteins such as laminin, fibrin, elastin, fibroin, among others, and ECM polysaccharides such as peptidoglycans and glycosaminoglycans have been shown to have a potential advantage when they are present in the designed collagen hydrogels. These biomolecules have typical biocompatibility and regulate cellular metabolic processes that result in successful biomedical applications (Badylak, 2007). Because of this, it is essential that the protocols for the extraction and purification of collagen are not harmful to these components of the ECM to generate hydrogels with improved biocompatibility. Different types of tissues have been used to generate hydrolyzed collagen oligopeptides, among which are pig, cattle, fish, mice, and jellyfish (Baldwin et al., 2014; Banerjee and Shanthi, 2012; Claudio-Rizo et al., 2017; Li et al., 2017; Meng et al., 2015; Wang et al., 2016; Wolf et al., 2012). The tissues formed mostly by tendons are rich in type I collagen; skin-derived tissues show high collagen type I, type III, fibronectin, laminin, and elastin. It is essential to know the biochemical composition of the starting tissues for the generation of collagen hydrogels with improved properties. The importance of tissue and the properties of the collagen hydrogels generated, as well as their innovation in the biomedical field, has been recently reported (Rangel-Argote et al., 2018).

The collagen hydrogels also have a desired moist environment, have an interconnected pore structure allowing the correct diffusion of nutrients and/or critical exogenous molecules in cell dynamics, their swelling capacity is related to their potential to absorb wound exudates, and its controlled degradation allows new tissue to be generated avoiding scar formation. Collagen hydrogels have been used for medical repairs of skin, bone, connective, and ophthalmological tissue, as well as to study cell–matrix extracellular interaction (Hinderer et al., 2016). Despite the excellent properties that hydrogels have in terms of biocompatibility and biomedical success, their application is limited for processes that involve high mechanical load and prolonged use times. The reduced mechanical stability and high biodegradation of collagen hydrogels are properties that must be improved to increase their applicability in tissue engineering. Increasing the mechanical stability of collagen hydrogels allows encapsulating exogenous molecules and/ or supporting exogenous polymer networks that give specific properties to hydrogels while controlling their degradation rate allows modulating the rate of formation of new tissue (Claudio-Rizo et al., 2016). Better mechanical stability can be achieved by controlling the structure of collagen hydrogels, generating strategies that involve chemical

functionalization with biopolymers and inorganic phases.

With this in mind, the purpose of the following section is to provide strategies to design collagen-based hydrogels with several coupling phases, highlighting the conditions of processing, the characterization techniques, the novel characteristics of the systems, and the main results in vitro/in vivo of application.

18.2.1 COLLAGEN HYDROGELS COUPLED WITH POLYSACCHARIDES

18.2.1.1 ALGINATE

Alginate is a polymer comprised of D-mannuronic and L-guluronic monosaccharides; the alginate is extracted generally from brown algae belonging to the phylogenic class *Phaeophyceae*. Collagen derived from soft coral *Sarcophyton* was utilized in combination con alginate to generate hydrogels with tailored properties, using 1-ethyl-3-(3-dimethyl aminopropyl) carbodiimide (EDC)-*N*-hydroxysuccinimide (NHS) as a crosslinker system. The synthesis of collagen-alginate hydrogels was at room temperature (RT) during a reaction time of 48 h, and they were characterized by DIC to verify the fibrillar patron in materials, scanning electron microscope (SEM) to reveal the microstructure, and mechanical analysis to study the improvement in the storage modulus of hydrogels. Alginate coupled hydrogels show improved mechanical properties, exhibiting a similar elastic behavior present in human epithelial tissues. This improvement in the storage module of this type of material was associated with the concentration of collagen, and how the fibers are formed and oriented by the presence of alginate, forming a double network hydrogel (IPNs) (Sharabi et al., 2014). In another work, alginate-collagen hydrogels were reinforced with $CaSO_4$ and Na_2HPO_4, generating novel biocomposites. The preparation of the composite was at 4 °C for a crosslinking time of 20 min, maintaining the pH at 7, then maturation of composites was at 37 °C for 1.5 h. Dynamic mechanical analysis (DMA) allowed to evaluate the mechanical properties depending on temperature, Fourier transform infrared spectroscopy (FTIR) to study the functional groups and the modifications during the formation of the hydrogel, thermogravimetric analysis (TGA) to determine the thermal stability, X-ray diffraction (XRD) to evaluate the semicrystalline properties, and SEM to observe the microstructure. Composite hydrogels show satisfactory improvement in their mechanical properties and benefiting the growth of bone tissue cells. The innovation offered is that it is an injectable formulation with high biocompatibility that allows controlling the gelation

capacity, from 5 to 10 min, depending on the concentration of alginate used, which is very applicable for bone fillings avoiding complicated surgeries (Bendtsen and Wei, 2015).

18.2.1.2 CELLULOSE

Cellulose is a biopolymer that has β-glucose molecules. Cellulose is the most abundant organic biomolecule since it forms the majority of terrestrial biomass. Plants and microorganisms of the protist kingdom are the producers of cellulose. Cellulose derivatives can be used as coupling agents to synthesize collagen hydrogels. In a first approach, collagen extracted of the bovine Achilles tendon is mix with dialdehyde carboxymethyl cellulose (DCMC), a derivative with collagen crosslinking capacity. The hydrogels were obtained centrifuging the component mix at 2500 rpm, using a temperature of 4 °C in a time of 20 min, then they were frozen at -20 °C for five days and thawed at RT. The materials were characterized using LLS spectroscopy to generate reliefs that reveal the cell-hydrogel interaction, FTIR to study the chemical structure of materials, SEM to observe the hybrid fibrillar structure, differential scanning calorimetry (DSC) to assay the thermal stability of hydrogels, and swelling analysis. Hybrid hydrogels show improvement in their swelling capacity in lower hydrodynamic equilibrium time, improved thermal properties, and excellent hemocompatibility. The swelling capacity is directly related to the concentration of DCMC and the variation in the pH. At high concentrations of DCMC, a decrement in the hemolysis capacity of hydrogels is observed; besides, an increment in antithrombotic activity is reported too (Huan Tan et al., 2015).

In another approach, collagen obtained from porcine dermis was functionalized using microcrystalline cellulose and 1-butyl-3-methylimidazolium chloride. The hydrogels were synthesized using a temperature of 100 °C under N_2 atmosphere for the reaction time of 2 h, and the novel materials were characterized by SEM to evaluate their microstructure, FTIR to determinate the changes in the functional groups that comprise the hydrogels, mechanical analysis to get the storage modules of matrices, and absorption/desorption assays for Cu(II) quantification. The hydrogels have a 3D microstructure with interconnected pores; the chemical structure of the materials high in reactive amino groups benefits coordination with Cu(II) ions, generating hydrogels with an improved adsorbent capacity. Hydrogels exhibit high degradation at acidic conditions and can be used up to four times, indicating that these materials represent an innovative strategy in

areas of bioremediation of contaminated water. The adsorption capacity of Cu(II) ions is dependent on the composition of the hydrogel, specifically on the collagen/cellulose mass ratio. The maximum adsorption capacities were recorded at pH 6, and the materials had a 95% desorption of Cu(II) (Wang et al., 2013). In the biomedical area, these hydrogels could be used as injectable systems in patients suffering from contamination by Cu(II), representing an easy strategy to remove copper from the body.

Another system using cellulose derivatives includes the combination of collagen derived from rat tail tendon with bacterial derived cellulose in the presence of the system EDC/NHS as a crosslinking agent. The hydrogels were prepared at 25 °C in a reaction time of 24 h; then, the hydrogels were dried to 37 °C. The characterization techniques used were FTIR to study the chemical structure of materials, ^{31}P NMR to evaluate the cell metabolism by the identification of phosphorus species, SEM to study the fibrillar structure in the hybrid hydrogels, TGA for the evaluation of thermal resistance, mechanical analysis to assay the hydrogel behavior and cytotoxicity, mutagenic and genotoxicity assays which allowed to verify the improved biocompatibility of the materials. This new material shows favorable performance for bone tissue growth, indicating that there are no cytotoxic, mutagenic, and geno-toxic effects on bone cells. Hydro-gels encapsulated with osteogenic growth peptide provide to stimulate early development and modulation of the osteoblastic phenotype, significantly improving cell growth and migration. (Figure 18.2) (Saska et al., 2017).

18.2.1.3 CHITOSAN

Chitosan is a macromolecule comprised of units of β-(1→4) D-glucosamine and N-acetyl-D-glucosamine. Commercial collagen (*BD Biosciences*) was combined with chitosan (4% v/v acetic acid) to produce novel hydrogels with controlled properties. The synthesis

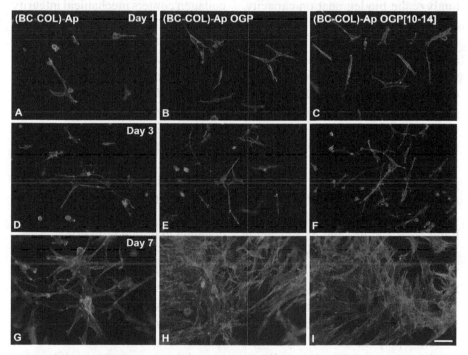

FIGURE 18.2 Epifluorescence of osteoblastic cells cultured on (BC-COL)-Ap (A, D, G), (BC COL)-Ap OGP (B, E, H), and (BC-COL)-Ap OGP(10-14) (C, F, I) at 1 (A-C), 3 (D-F) and 7 (G-I) days. Green fluorescence (Alexa Fluor 488-conjugated phalloidin) reveals actin cytoskeleton. At day 1, cells were adhered and exhibited fusiform shape on all membranes (A-C). At day 3, cell spreading was clearly noticed for all membranes (D–F). At day 7, a reduced number of cells was observed for cultures grown on (BC-COL)-Ap, while cells on other membranes showed areas of cell multilayering (compare G to H and I). Bar: A–F = 100 m. (For interpretation of the references to colour in this figure legend, the reader is referred to the web version of this article).

was at 37 °C for a reaction time of 30 min in the presence of an atmosphere containing 5% CO_2. The hydrogels were characterized by SEM to study the microstructure, rheological analysis to reveal the mechanical behavior, live/dead assay which allowed to verify the improved biocompatibility of the materials and the evaluation of the cytotoxicity of materials, enzymatic degradation to analyze the biodegradation capacity, and LDH cytotoxicity assay and histological and immunofluorescent staining evidencing the variation in the cell phenotype. The results indicate that chitosan-coupled hydrogels have significantly improved their mechanical properties, due to the formation of a double network hydrogel; this network allows the growth, proliferation, and maturation of cardiomyocytes for up to 7 days in vivo conditions. On the other hand, these hydrogels encapsulated with angiopoietin-1 peptide allow the cardiomyocytes to live inside the hydrogel, and the hydrogels remained at the injection site after seven days (Reis et al., 2012).

The preparation of hydrogels has been reported based on collagen derived from skin fish using chitosan, gelatin, and glutaraldehyde as a crosslinker agent. The hydrogels were synthesized employing a temperature of −20 °C for a crosslinking time of 24 h, then they were lyophilized until dried, and they were characterized using SEM to reveal the microstructure, oscillatory rheology to study the viscous modulus, mechanical analysis to determinate Young's modulus and biodegradation assay to study their proteolytic stability. The composition of the hydrogels is decisive for their physicochemical properties, such that higher swelling capacities were observed in materials that include higher chitosan content; collagen confers mechanical integrity and flexibility to the hybrid network while gelatin decreases flexibility and generates brittle materials. On the other hand, it was observed that hydrogels derived from salmon skin collagen have a higher solubility and less turbidity, while those derived from African catfish skin collagen show higher rigidity, less flexibility, and better antioxidant capacity (Tylingo et al., 2016), exhibiting the role of the type of ECM on the collagen hydrogel properties.

Another research involves the coupling of collagen (Mian Scientific suppliers, Lahore) with chitosan and 2-deoxy-D-ribose to generate hybrid hydrogels. Triethyl orthoformate was used as a crosslinker. The materials were synthesized, keeping the pH at 10, with a temperature of −20 °C for a reaction time of 24 h, then the materials were washed in absolute ethanol and dried at 25 °C. FTIR was used to evaluate the chemical structure of hydrogels, SEM to evaluate the fibrillar structure, microbiological and biochemistry analysis, and chick chorionic allantoic membrane assay was the characterization techniques used to understand the variation in the biological response. The hydrogels presented a controlled and sustainable release of 2-deoxy-D-ribose, improving angiogenesis, and wound healing. Also, superficial wounds in rats were completely closed, and the skin was regenerated entirely after 17 days of treatment (Yar et al., 2017).

Another research involves collagen derived from rat tail tendon combined with chitosan and silica particles of 210 nm and 438 nm, employing genipin as a crosslinking agent. The hydrogels were prepared at pH 7.4 using a phosphate buffer, then all components were vortexed and then were incubated at 37 °C for a maturation time of 24 h. The materials were characterized by SEM to elucidate the microstructure, cell viability assay to study the cytotoxicity of hydrogels, biochemical, and molecular biology analysis to reveal the improvement in the cell behavior on the novel materials. The results show that these innovative biocomposites in the hydrogel state have improved pro-osteogenic capacity without the need to encapsulate osteogenic inducers; also, these materials promote osteogenic differentiation of stem cells derived from bone marrow in vitro conditions. The process of differentiation of these cells is significantly influenced by the size of the silica particles (Figure 18.3) (Filipowska et al., 2018).

Other strategy referents to biofunctional hydrogels comprise of collagen derived from dermis porcine, chitosan, alginate, and curcumin. The EDC–NHS system was used as a crosslinker agent. The hybrid materials were prepared to maintain constant the pH at 7, using a temperature of 4 °C for a reaction time of 2 h. The characterization of hydrogels was taken out by SEM to reveal the structure, XRD to determinate crystallinity, TGA to study the thermal stability, mechanical analysis to know the storage modules, and cell viability/morphology assays to evaluate the cytotoxic effect of materials. The results show that this type of biocomposite allows the controlled release of curcumin, significantly improving the healing process of diabetic wounds. The wounds in rats with induced diabetes treated with this innovative biomaterial healed entirely and much faster than the control, evidencing inhibition of inflammatory response in the wounds (Karri et al., 2016).

18.2.1.4 DEXTRAN AND CYCLODEXTRINS

Dextran is a branched and complex polysaccharide comprised of glucose molecules. Dextran can have average molecular weights from 10 to 150 kDa. It is formed by α-(1→6)

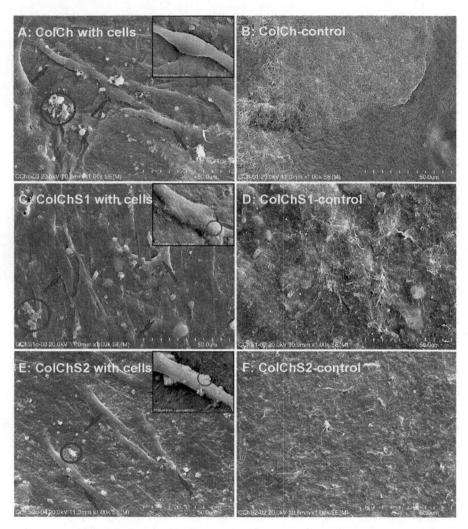

FIGURE 18.3 SEM microphotographs of materials after 20 days of cells culturing (A, C, E) and control materials (B, D, F) (without cells but treated with the same manner, kept in incubator with medium changed at the same time point).

glycosidic linkages between glucose molecules, while the ramifications are presented at α-(1→4) linkages (in some cases also at α-(1→2) and α-(1→3) linkages). Cyclodextrins are a family of compounds formed by monosaccharide molecules forming a cyclic structure (cyclic oligosaccharides). In total six to eight glucopyranoside units constitute the cyclodextrins. Cyclodextrins are produced from starch by enzymatic conversion.

Novel hydrogels with improved properties have been designed by the coupling of bovine Achilles tendon collagen and β-cyclodextrins (β-CD) and polyrotaxane as a crosslinking agent. The hydrogels were prepared under a temperature of 4 °C for a reaction time of 48 h; then, the mixture was stirred at a temperature of 37 °C for 15 min. Finally, the resultant mixture was frozen at a temperature of −20 °C for a time of 48 h. ^{1}H NMR revealed the crosslinking of hydrogels, XRD provided crystal regions of materials, FTIR elucidated the modification in the chemical structure of materials, SEM exhibited the microstructure, rheology determined the mechanical behavior and biochemical analysis, enzymatic degradation, and cytotoxicity assay were employed to characterize the biological response of the hydrogels.

The materials shown improved mechanical properties, the hydrogels exhibited good swelling, with both improved cell adhesion and proliferation. The biodegradation rate has also been improved and shows lower cytotoxicity, similar to hydrogels with carbodiimide. Interestingly, the crosslinking produced by modified cyclodextrin has a direct relationship with the tailoring of the properties of collagen, allowing to L929 fibroblasts cell culturing in hydrogel showing excellent adhesion, growing and proliferation after five days (Liu et al., 2015).

Dextran has been used to generate hydrogels with high application in biomedicine and tissue engineering. Hybrid hydrogels comprised of bovine Achilles tendon collagen and aldehyde-functionalized dextran were synthetized. The processing conditions include a temperature of 4 °C, keeping the pH at 7.0, and then the gelation process was taken out with a temperature of 37 °C with an aging time of 24 h. The materials were characterized by NMR to elucidate the crosslinking effects, SEM to evaluate the microstructure, rheological, and mechanical analysis to determine the storage modulus of materials, oxidized degree, and cytotoxicity assay to know the effect of the composition of hydrogels on the biological response. The results indicate that the novel hybrid hydrogel has an improved mechanical property, the collagen shown resistance to denaturalization, and the fibrillar structure is homogenous. Besides, the materials exhibited enhanced biocompatibility with fibroblast and Hela cells (Zhang et al., 2014).

18.2.1.5 HYALURONIC ACID

Hyaluronic acid (HA) is a polysaccharide of the type of glycosaminoglycans composed of repetitive polymeric disaccharides of D-glucuronic acid and *N*-acetyl-D-glucosamine linked by β-(1 → 3) bonds. It is widely distributed in both prokaryotic and eukaryotic cells. In the human, it is abundant in the skin and is present in the vitreous humor, umbilical cord, synovial fluid, skeletal tissue, heart valves, lung, aorta, prostate, tunica albuginea, cavernous bodies, and fluffy penis. HA is produced mainly by mesenchymal cells in addition to other types. Collagen-based hydrogels coupled with HA have been prepared using as the source of collagen bovine Achilles tendon, and pullulan (PL) as collagen crosslinkers; the component mixing takes out to pH 7 during 60 min with constant stirring at RT, then is changed the pH at 8 to generate the hybrid hydrogels at 37 °C for 30 min. The materials were characterized using SEM to study their fibrillar structure, ^{1}H NMR to determine the chemical modification, fluorescence microscopy to evaluate the cytotoxicity, crosslinking analysis for photocolorimetry, rheological analysis to know the mechanical behavior and stability. These hydrogels show excellent viability of eukaryotic cells, and this type of cell proliferates when they are encapsulated in these types of materials. By using fibroblasts derived from adipose tissue, these hydrogels also show excellent biocompatibility to this cell line, allowing satisfactory encapsulation. Hydrogels crosslinked with PL exhibit high cell viability; however, hydrogels indicate a decrease in metabolic activity (Luca et al., 2017).

In other work, it has been suggested the coupling of collagen derived from bovine tendon with thiolated HA; the 3D hybrid matrices were prepared at pH 7.4, a temperature of 37 °C, and with an atmosphere containing 5% of CO_2 during a reaction time of 1 h. Confocal reflectance microscopy allowed the study of the cell–matrix interaction, mechanical analysis for the determination of the module of storage of materials, SEM to evaluate the microstructure, migrations experiments on hydrogels with cells grown in vitro, and cell spreading assay were taken out to biological characterizing the materials. The mechanical properties of the hydrogels were adapted directly by controlling the concentration of HA, and, besides that, HA directly influences the phenotype of glioblastoma cells. The microstructure of this novel material generates a 3D environment that facilitates the study of tumor cell interactions with the ECM. Also, the concentration of HA induces changes in the morphology, dispersion, and migration of cancer cells within the hydrogel. This work represents a possibility to study the behavior of cancer cells and the modification of their phenotype by changing the composition of the hydrogel (Rao et al., 2013).

18.2.1.6 PULLULAN

PL is a polysaccharide comprised of maltotriose units, also known as α-(1→4);α-(1→6)-glucan. PL is produced from starch by the fungus *Aureobasidium* PLs. Human-like collagen expressed in *Escherichia coli* was coupled with carboxyl PL, using 1,4-butanediol diglycidyl ether as a crosslinker for the design of novel hydrogels with tailored properties. The hydrogels were prepared at a constant pH of 8, 25 °C for a reaction time of 30 min; then the gelation of materials was induced promoting the crosslinking at the temperature of 50 °C for a reaction time of 3 h; the materials were washed in a hot bath containing water and stirred for 72 h, refreshing water every 2 h. FTIR was used to evaluate the chemical structure of materials, SEM to observe the fibrillar microstructure, mechanical analysis to determine the hydrogel behavior, and swelling behavior to know the crosslinking effects. The hydrogels crosslinked with PL had a significant improvement in their storage module. The crosslinking index does not have a direct relationship with the swelling capacity of the materials; also, it improved both cellular adhesion and cell viability, evidencing a decrement in the inflammatory response. In vitro results indicated that epithelial cells could grow, migrate, and proliferate within hydrogels without cytotoxic

effects. In vivo results revealed anti-inflammatory effects, biocompatibility, and anti-biodegradability in mice for 24 weeks. The adapted properties of these hydrogels make them excellent candidates for potential applications in tissue engineering and biomedicine (Li et al., 2015).

In another approach, collagen derived from skin fish coupled with PL under the action of sodium trimetaphosphate is used as a crosslinking agent to synthesize hybrid hydrogels. The reaction conditions involved a constant pH of 9 at 25 °C for a reaction time of 90 min. Then, the pH was modified at 7, and the temperature at 50 °C for promoting the crosslinking during a time of 30 min. The materials were analyzed by SEM to study the microstructure, FTIR to evaluate the chemical modification, TGA to determine the thermal stability, DSC to evaluate the resistance to the collagen denaturalization, compatibility assay by cytotoxic effects, swelling, and biochemical analysis to evidence the variation on the biological response. The results indicate that these novel hydrogels were transparent, showing a super absorbent capacity, and having improved storage modules and enhanced biocompatibility for wound healing. The swelling capacity of the hydrogels is improved by up to 320%. Hybrid materials show excellent biocompatibility, improved cell adhesion, and controlled proliferation of fibroblasts depending on the degree of cross-linking of the materials, evidenced that these cells can regulate the blood vessel formation process when using wounded chickens; on the other hand, rats with wounds exhibited total cure after 12 days in contact with this novel material (Iswariya et al., 2016).

18.2.2 COLLAGEN HYDROGELS COUPLED WITH AMINO ACIDS AND PROTEINS

18.2.2.1 FIBRIN

Fibrin is a fibrillar protein with the ability to form 3D networks of blood vessels and tendons. This protein acts as a kind of glue or thread between the platelets that are exposed in a wound. The fibrin keeps the scab glued to the wound until a new layer of skin appears. Novel hydrogels based on rat tail tendon collagen and fibrin were prepared using PEG (poly(ethylene glycol))-ether tetra-succinimidyl glutarate. The hydrogels were synthesized, maintaining the pH at 7 with a regulate temperature of 37 °C and in the presence of an atmosphere of 5% of CO_2. FTIR was used to determine the changes in the chemical structure, SEM to visualize the fibrillar structure, cell viability/motility assays to evaluate the cytotoxicity of materials, and molecular biology techniques were employed to characterize the genotypic alteration of cells encapsulated. Hydrogels based on this biopolymer system show high hydrolytic stability while maintaining a swelling capacity dependent on the degree of crosslinking dependent on the modified PEG. These also show excellent viability to astrocytes. Astrocytes can easily migrate to the fibrillar network that constitutes hydrogels. On the other hand, no alterations in the integrity of the hydrogels were observed for up to 9 days of cellular encapsulation, thus demonstrating resistance to cellular contraction. The importance of fibrin is associated with the control of astrocyte migration within the hybrid matrix (Seyed-hassantehrani et al., 2016). Another research involves the combination of collagen derived from porcine articular joint and fibrin, to generate hydrogels with controlled properties. Fibrin amino groups function as reactive sites of chemical crosslinking with collagen. These hydrogels were prepared at pH 7.4 at a temperature of 37 °C and with a reaction time of 30 min. The materials were characterized by SEM to assess the fibrillar structure, histological, biochemical, and immunohistochemical analysis to probe the enhancement of the biological response. The formulation of this hybrid hydrogel allows the encapsulation of hydrolysates derived from the ECM and growth factors. This system with encapsulated bio components exhibits improved chondrogenesis and short-term time stimulation of chondroinductivity; these improvements could be exploited in strategies for bone and cartilage tissue regeneration. The results in vitro and in vivo indicate that hydrogels show a sustainable release of growth factors and ECM components that improve and accelerate the metabolic processes related to chondrogenesis, generating new cartilage in 15 days (Almeida et al., 2016).

18.2.2.2 POLY(γ-GLUTAMIC ACID)

Polyglutamic acid (γ-PGA) is a polymer comprised of amino glutamic acid (GA). It is formed by bacterial fermentation. Bioinspired hydrogels have been designed using commercial-type I collagen (Biolad) combined with γ-PGA. The cross-linking of the chemical moieties was taken out under the presence of fibronectin. The hydrogels were prepared at a temperature of 4 °C under constant stirring. Circular dichroism (CD) and UV–vis (ultraviolet–visible) spectroscopy were used to study the fibrillar patron and fibrillogenesis process, respectively; optical and fluorescence microscopy to evaluate the cytotoxicity of materials, SEM to study the microstructure, and rheological analysis were used to characterize the mechanical behavior. The results indicate that this system has the properties of an injectable scaffold since the formulation in a liquid state can gel in contact with the skin in a time of 2 min. Being innovative and offering excellent potential for application in regenerative medicine. This type of hydrogel was used as an encapsulation material for skin mesenchymal cells and drugs with antioxidant capacity, revealing controlled release when they were applied to wounds in rat kidneys. Interestingly, the composition of this hydrogel does not affect the growth and proliferation of mesenchymal cells and allows increased viability for up to 4 days of encapsulation. In the presence of renal defects produced in rats, the cells can leave the hydrogel and occupy the site of the lesions promoting constructive effects and healing (Cho et al., 2017).

18.2.2.3 SILK

Silk is a natural fiber formed by proteins. It is produced by several groups of arthropod animals, such as spiders and various types of insects; currently, only silk produced by the larvae of the *Bombyx mori* butterfly has been used in biomedical research. Novel bioinspired hydrogels have been synthetized combining porcine dermis collagen and silk, under the action of glutaraldehyde as a crosslinker. The components were stirred to gel formation at 25 °C for a reaction time of 24 h, maintaining a constant pH of 7.0. SEM revealed the fibrillar microstructure, XRD identified the crystal regions in the materials, FTIR determined the chemical modification of the components that comprise the hydrogel, mechanical analysis detailed the storage module of materials, cell viability assays exhibited the cytotoxicity of the hybrid materials. The double polymer network system allowed the correct dispersion and support of hydroxyapatite particles having diameters from 30 to 100 nm. The improvement in mechanical properties was related to the silk content. However, the presence of the inorganic phase and collagen allowed improved biocompatibility of osteoblasts encapsulated in the composite hydrogels (Chen et al., 2014).

Another exciting approach involves the use of genetically engineered silk–collagen-like copolymer produced by *Pichia pastoris*, to form bioinspired hydrogels. To the generation of the hydrogels, the copolymer was mixed for a reaction time of 4 h, and the gelation process was induced for 12 h at 25 °C, maintaining a constant pH of 7.4. The hydrogels were characterized by FTIR to evaluate the chemical structure, rheological analysis to determine the enhancement mechanical, oxidized degree to estimate the degradation capacity, cell viability/proliferation assay, and molecular biology to demonstrate the influence of the composition of novel materials on the biological response. The results indicate that these bioinspired materials show stability at long periods of contact with acidic and proteolytic media. The mechanical properties are dependent on the degree of crosslinking produced by the interaction of silk with collagen. Mesenchymal cells derived from mouse bone were encapsulated in these matrices, presenting controlled viability for up to 21 days. However, the proliferation and mineralization processes of the cells were affected due to the high degree of crosslinking they present (Włodarczyk-Biegun et al., 2014).

18.2.3 COLLAGEN HYDROGELS COUPLED WITH INORGANIC PHASES

18.2.3.1 HYDROXYLAPATITE

Hydroxylapatite (HAp), also called hydroxyapatite or apatite, is a mineral and a biological component formed by crystalline calcium phosphate $Ca_5(PO_4)_3(OH)$. HAp particles have been dispersed in collagen hydrogels derived from rat tail tendon. The composites were prepared at a constant pH of 7.0, the temperature of 25 °C for a reaction time of 1 h, then the aging of materials was taken out at 40 °C for a time of 22.5 h. Transmission electron microscopy (TEM) was used to observe the dispersion of HAp particles inside materials, SEM to observe the modifications in fibrillar structure, XRD to assess the alteration in the crystallinity, TGA to determinate the improvement in thermal resistance, DMA and oscillatory rheology was used to study the mechanical material properties. The dispersion of HAp shows favorable mechanical improvement due to the increase in the biocomposites storage module, also without observation in the modification of the structural integrity of the collagen, maintaining a fibrillar morphology of interconnected pores. The content of HAp maintains a direct relationship with the alignment of the pores generated in the composites, and both improved the mechanical and thermal properties. Osteoblasts, seeded on these hydrogels, show improved viability and excellent adhesion and migration in the hydrogel (Xia et al., 2014). Another strategy involves the use of fish dermis collagen, coupled with HAp and alendronate (sodium bisphosphonate). These novel hydrogels were crosslinked by genipin. The materials were prepared at a constant pH of 7.4, and at a temperature of 37 °C for a polymerization time of 2 h; and they were characterized by ^{1}H NMR and FTIR to elucidate the chemical crosslinking, XRD to evaluate the crystal regions in the biocomposite, and TGA to determinate the thermal resistance. The content of HAp shows a direct

dependence on the polymerization rate of the composites, recording gelation times from 5 to 36 min. The materials exhibited notable improvement in their swelling capacity and controlled degradation behavior using collagenase as a proteolytic enzyme. Mouse osteoblasts were cultured on these hydrogels, and the results indicate that the cells improved the viability and proliferation when alendronate is present. This hydrogel system represents a material with potential application in bone tissue repair (Ma et al., 2016).

18.2.3.2 MAGNETIC NANOPARTICLES

Magnetic nanoparticles are a class of inorganic phases that have the property to be manipulated through magnetic fields. These particles are mainly synthetized from magnetic elements such as iron, nickel, and cobalt and their chemical compounds. The magnetic nanoparticles based on Fe have been thoroughly researched for the generation of hydrogels with potential applications in biomedicine. Biocomposite hydrogels with modulated properties have been synthetized from bovine Achilles tendon collagen and magnetic nanoparticles based on ferrite. The hydrogels were prepared at a constant pH of 7.0, using a temperature of 25 °C and under the application of variant magnetic fields from 162 to 2110 G. The materials were studied by oscillatory rheology to determinate their mechanical properties, TEM to observe the dispersion and size of magnetic nanoparticles and SEM to evaluate the changes in the fibrillar collagen structure. This methodology allows controlling the morphology and orientation of the collagen fibers according to the intensity of the magnetic field used. The results of experiments in vivo using neurons indicate that these novel biocomposites have the ability to regulate the electrical activity of these nerve cells, significantly improving their viability (Antman-Passig and Shefi, 2016).

In this same approach, hydrogels based on rat tail tendon collagen coupled with superparamagnetic iron oxide particles have also been reported. These hybrid materials were synthetized under a pH of 7.0, at a temperature of 37 °C during a polymerization time of 20 min. Magnetic resonance imaging was used to get 3D contours of materials, histological analysis, and molecular biology were used to evaluate the material properties on the biological response of human mesenchymal cells. This type of biocomposite allows modulating the cellular behavior by varying the intensity of the magnetic field, observing both enhanced viability, and migration of the cells after a transplant for the repair of articular cartilage. The proliferation of mesenchymal

cells is dependent on the orientation of the collagen fibers, which can be controlled by varying the magnetic field. The presence of iron nanoparticles prevents mesenchymal cells from changing their cellular phenotype; that is, they inhibit osteogenic, adipogenic, or chondrogenic differentiation (Heymer et al., 2008).

Other research has been focused on the design of magnetic hydrogels comprised of collagen derived from chicken sternum cartilage and iron oxide nanoparticles. The materials were generated using the electrodeposition process under a constant pH of 5.0 and a temperature of 37 °C. The characterization was assessed by SEM to visualize the fibrillar morphology, a vibrating sample magnetometer to measure the magnetic properties, inductively coupled plasma mass spectrometer (ICP-MS) to evaluate the chemical composition, confocal laser scanning microscopy to observe the interaction matrix-cell, and mechanical analysis using oscillatory rheology. The concentration of nanoparticles shows a direct relationship with the change in the swelling capacity and flexibility of the composites in the hydrogel state. In vitro results using mesenchymal cells indicate that these novel biocomposites can regulate the osteoblastic differentiation of this cell line being dependent on the improvement in the mechanical properties of the fibrillar network.

Interestingly, the intensity of the applied magnetic field can regulate the mechanical-transducer signaling pathways of the encapsulated cells on these materials, observing the increment in the biosynthesis of genes related to these metabolic pathways (Zhuang et al., 2018).

18.2.3.3 SILICA NANOPARTICLES

Hydrogels based on rat tail tendon collagen with silica nanoparticles dispersed inside the 3D matrix have been designed. These materials were prepared, adjusting the pH at 7.0 and 25 °C. Oscillatory rheology was employed to determine the storage modules of biocomposites, inductively coupled plasma atomic emission *spectroscopy* to evaluate the chemical composition, SEM to observe the fibrillar structure, DSC was used to evaluate the thermal material properties. Hydrogels show a direct relationship in both mechanical and thermal improvement with the concentration of silica nanoparticles. Fibroblasts encapsulated on these biocomposites exhibit significant improvement in their viability and proliferation, and hydrogel contraction is not observed. When the materials were transplanted into mice with skin wounds, it was observed that encapsulated fibroblasts could be released from the material to colonize the injured tissue, significantly improving the angiogenesis process without inducing significant inflammatory response (Desimone et al., 2011). Mesoporous silica nanoparticles have been trapped in a 3D collagen matrix derived from rat tail tendon. These novel biocomposites were synthetized with a reaction temperature of 37 °C for a time of 30 min and under the presence of an atmosphere of CO_2. The material properties were evaluated by TEM to know the dispersion of nanoparticles inside the 3D fibrillar matrix, N_2 adsorption-desorption and Brunauer–Emmett–Teller method to evaluate the surficial area and pore size of biocomposite, FTIR to study the chemical structure, laser Doppler electrophoresis, and confocal microscopy to study the cytotoxicity of materials, cell viability assay, and molecular biology to assess the modification in the biological response. The results indicate that this type of biocomposite in the hydrogel state has ideal properties for the loading and controlled release of nerve growth factors (NGFs). Hydrogels with NGF show a controlled and sustainable release of NGF for up to one week, which is regulated by the content of silica nanoparticles. On the other hand, human neurites encapsulated in this bioinspired material show improved bioactivity, producing accelerated neuritogenesis. These results indicate that these novel materials could be used as candidates in nerve tissue repair (Lee et al., 2013).

In other work, studied the coupling of silica nanoparticles with either mineral phase HAp, calcium carbonate ($CaCO_3$), strontium phosphate ($Sr_3(PO_4)_2$), or strontium carbonate ($SrCO_3$) inside matrix hydrogel based on bovine Achilles tendon collagen. These biocomposites were synthetized using orbital stirring at a temperature of 37 °C, in the presence of an atmosphere containing 95% relative humidity during a polymerization time of 3 days. SEM allowed knowing the fibrillar modifications, confocal microscopy to study the cell-material interactions, fluorescence spectroscopy to evaluate the cytotoxicity, histological, and biochemical analyses were applied to study the biocompatibility of the hydrogels. These biocomposites exhibit a rough fibrillar morphology dependent on the nature of the inorganic phase. On the other hand, the coupling with the inorganic phase significantly improves the production of architectures with interconnected pores useful for applications in tissue engineering. Mouse osteoblasts seeded on hydrogels show excellent adhesiveness and controlled viability. Also, in vitro studies done with monocytes indicate that these can be differentiated to osteoclasts on the surface when the biocomposites are enriched with osteoclastic differentiation factors, where the concentration and presence of the Ca(II) ion has a significant impact on this process. However, materials that include Sr(II) show inhibition in the osteoclastic differentiation process. This work evidences the importance of the inorganic composition on the biological response modulated in biocompatible hydrogels (Rößler et al., 2018).

The evaluation of the influence of nanoparticle size on the hydrogel properties was taken out using 12 or 80 nm silica nanoparticles inside hydrogels based on collagen derived from rat tail tendon. The conditions to generate these materials were maintaining a constant pH of 7, at 25 °C. The pregel solution can encapsulate cell suspension and medium with phosphate-buffered saline for then to polymerize at 37 °C for 3 h. The biocomposite was characterized by SEM to evaluate the microstructure, TEM, to observe the dispersion of silica nanoparticles inside the fibrillar matrix, DSC, to evaluate the denaturalization temperature of collagen, oscillatory rheology to determinate the storage modules, immunodetection techniques, biochemical and histological analysis to study the biocompatibility. According to the results, hydrogels show excellent biocompatibility for the encapsulation of mouse fibroblasts with controlled viability. Fibroblasts can carry out their vital functions within these biocomposites, allowing them to proliferate and divide. The 12 nm silica particle size exhibited a higher cytotoxic character toward fibroblasts because they can be adsorbed on nuclear membrane proteins limiting their cellular metabolism. The 80 nm particles show improved biocompatibility in fibroblasts (Figure 18.4) (Mebert et al., 2018). In this approach, 300 nm silica nanoparticles have been trapped in hydrogels comprised of rat tail tendon collagen also have been reported. These novel materials were prepared at a constant pH of 3, 25 °C exposed to an atmosphere of ammonium vapors for 12 h, then pH was changed at 7.4 for the aging of materials for 24 h. SEM was used to observe the fibrillar microstructure of hydrogels, FTIR to elucidate the chemical modifications and microbiological analysis was used to study the biocompatibility of composites. The

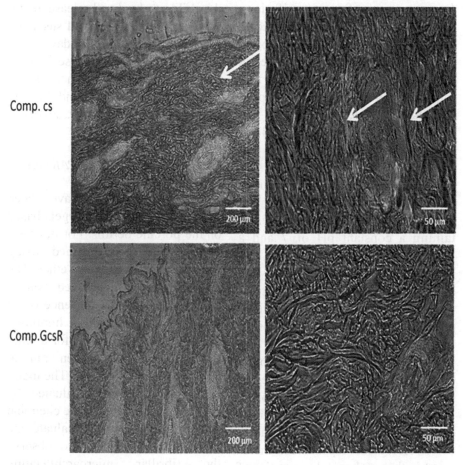

FIGURE 18.4 Immunodetection inflammatory macrophages (phenotype M1) CD-68 (white arrows, red staining) in animal treated with unloaded (comp.cs) and double-loaded (comp. GcsR) composites. Cells nuclei are stained in blue by DAPI.

presence of this type of nanoparticles allows the adsorption and encapsulation of biomolecules with antibacterial capacity, showing controlled and sustained release. Hydrogels loaded with rifampicin and streptomycin were placed in rat skin wounds, not showing inflammatory responses or irritation. Besides, the composition of the materials exhibited a prolonged antibacterial effect toward *Staphylococcus aureus* for up to 10 days, reducing infection in wounds infected with said bacteria. These hydrogels can be explored by coupling different molecules with biological interest to improve their performance in applications of tissue engineering and biomedicine (Alvarez et al., 2014).

Other hydrogels based on collagen derived from rat tail tendons were coupled with silica nanoparticles of several sizes. The materials were obtained, maintaining a constant pH of 7 and a reaction temperature of 37 °C. The hydrogels were studied by SEM to evaluate the fibrillar microstructure, TEM the morphology of silica particles in hybrid materials, oscillatory rheology to determine the mechanical properties, enzymatic degradation to assess the biodegradation, histological, microbiological, biochemical, and cytotoxicity analysis to evaluate the modification of the biological response. An increment in the mechanical properties of the hydrogels was observed, which is dependent on the content of silica nanoparticles. On the other hand, improvement in the rate of proteolytic degradation under the action of collagenase and improved viability for mouse fibroblasts are obtained in hydrogels composed of 500 nm nanoparticles. Increasing the size of the silica particles reports a decrement in cytotoxicity toward fibroblasts. Besides, hydrogels composed of 500 nm nanoparticles also show the ability to load and release gentamycin in a controlled and sustained manner for up to 10 days. The adapted properties of these innovative biomaterials make them excellent candidates for wound healing strategies (Desimone et al., 2010).

18.2.3.4   SILVER NANOPARTICLES

Silver nanoparticles have been trapped in collagen hydrogels based on the porcine dermis. For it, crosslinking was accomplished using 1,4-butanediol diglycidyl ether. The hydrogels were synthetized using a pH of 11, under the presence of an atmosphere with 100% humidity, the temperature of 25 °C for a reaction time of 24 h, and then aging at 37 °C for a time of 1 day. The material properties were evaluated by ICP-MS to determinate the chemical composition, DSC to evaluate the thermal stability, SEM to observe the fibrillar microarchitecture, mechanical analysis by oscillatory rheology, enzymatic degradation to assess the biodegradation, biocompatibility, cytotoxicity, antimicrobial, and biochemical analysis to verify the modification on the biological response. Collagen hydrogels, coupled with silver nanoparticles, have an improvement in their mechanical properties, registering a storage module similar to that of epidermal tissues, also presenting the antibacterial property. In vivo experiments in chronic wounds in mice indicate that this type of hydrogels with encapsulated keratinocytes show adapted viability and proliferation. Keratinocytes can migrate to injured tissue by decreasing inflammation processes; is, excellent antibacterial capacity is observed for both Gram (+) and Gram (–) bacteria. The novel hydrogels are biocompatible and safe for use in the regeneration of chronically infected/inflamed tissues (Chatzistavrou et al., 2016).

In another approach, hydrogels based on bovine Achilles tendon were coupled with silver doped bioactive glass. The collagen crosslinking was taken out with bovine fibrinogen. For the preparation of the hydrogels, all solution components were mixed and centrifuged at a temperature of 4 °C, using 600 rpm for 5 min and the gelation process was taken out to 37 °C for 24 h. Reflectance confocal microscopy allowed to study the cell-matrix interactions, and antimicrobial analysis were evaluated on the materials to evaluate their antimicrobial capacity. The composition of these innovative materials is optimal to allow the encapsulation of stem cells derived from dental pulp, showing high resistance to bacterial growth. The encapsulated cells showed excellent migration, growth, viability, and proliferation within the composite matrix; in addition, the composition of the hydrogel allowed the differentiation of the cells toward fibroblastic phenotype for in vitro and in vivo results. The presence of silver in the matrix is linked with improvement in the mechanics of the fibrillar network by modifying the cellular phenotype, also significantly increasing microbial inhibition. These novel materials could be used to generate wound-dressings for successful applications in infected wounds (Alarcon et al., 2015).

18.3 SILK FIBROIN HYDROGELS FOR TISSUE ENGINEERING

Silk has been used as biomaterial due to its excellent mechanical resistance, elasticity, low or null toxicity, biocompatibility, and adjustable biodegradability. The leading silk producers are silkworms, spiders, bees, and scorpions. Being the most important, the silkworm *Bombyx mori*, which can produce up to 800–1500 m of silk per worm.

Silk is composed of two main proteins, fibroin and sericin (Tamada and Kojima, 2014), according to different studies the sericin can produce an immunogenic response in the host when it is used as a biomaterial and is preferable its elimination through alkaline or enzymatic treatments, this process is named degummed. After the degumming process, the fibroin is solubilized in chaotropic salts, such as LiBr, $CaCl_2$/ethanol/water, LiSCN, etc. Fibroin is soluble for days at neutral pH without shearing forces. Table 18.1 shows the most employed methods to obtain silk fibroin (SF) free of sericin (Gao et al., 2008; Hu et al., 2016; Kim et al., 2014; Kumar et al., 2018; Mehrabani et al., 2018; Ming et al., 2015; Yan et al., 2018). For instance, fibroin is the silk protein more employed in tissue engineering and drug delivery systems (Gaviria Arias and Caballero Mendez, 2015).

Fibroin has hydrophobic and hydrophilic fibers, also known as crystalline and amorphous states, respectively. The crystalline state is formed by the amino acids Gly, Ala, and Ser, and the amorphous state by Thr, Gly, Ser, Phe, Pro, Tyr, Val, Ala, Asp, and Glu. The solubility of the fibroin, in aqueous solutions, is higher when it is in an amorphous state, formed mainly by random coils. Fibroin can suffer a transition from the amorphous state to the crystalline state by physical or chemical treatments. In the crystalline state, the fibroin is composed of α-helix and β-sheets structures, also known as silk I and silk II, respectively

(Galateanu et al., 2019). Fibroin has terminal carboxylic groups that can covalently bind peptides, hormones, and growth factors. The crystalline state can cover up to 60%–70% of the total fibroin, giving the characteristic strength to the silk fibers, and the amorphous state contributes to the elasticity. Another classification that receives since 1976 is based on that the fibroin is composed of three components a heavy chain, a light chain, and a glycoprotein P25 in a ratio of 6:6:1, respectively, and showing a disulfide bridge between the heavy and the light fibroin chains. The glycoprotein helps to keep the silk fibers stable. The heavy fibroin is constituted by a sequence of amino acids that provides hydrophobic character and β-sheets, presenting excellent mechanical properties and crystallinity. Finally, the light fibroin is formed by amino acids with hydrophilic character, showing good elasticity and null crystallinity. According to this classification, neither the heavy nor the light fibroin are toxic, and both present good cell adhesion (Zafar et al., 2015). In Figure 18.5, the amorphous and crystalline states with their amino acids composition are shown (Koperska et al., 2014).

Fibroin hydrogels for biomedical applications is an area of great interest due to the functional properties that the fibroin presents. A hydrogel absorbs a high quantity of water; this can be measured as the

TABLE 18.1 Methodologies Used to Obtain Silk Fibroin

			Stages of the Process				
Degumming	**Washed**	**Drying**	**Solubilization**	**Dialysis**	**Powder Recovery**	**Yield**	**Reference**
Na_2CO_3 0.5 wt.% at 100 °C for 1 h	With distilled water	–	$CaCl_2/H_2O/C_2H_5OH$ molar ratio 1:8:2 at 80 °C for 30 min	Against distilled water for 3 days	By filtration	30–40 mg/mL	Gao et al. (2008)
Na_2CO_3 0.2 wt.% and Marseille soup 0.3 wt.% at 100 °C for 3 h	Distilled water	–	$CaCl_2/C_2H_5OH/H_2O$ molar ratio 1:2:8 at 80 °C for 5 min	Against distilled water for 3 days	By lyophilization	–	Kim et al. (2014)
Na_2CO_3 0.02 M at 100 °C for 30 min	With distilled water	At 50 °C overnight	$CaCl_2/C_2H_5OH/H_2O$ molar ratio 1:2:8 at 70 °C for 4–5 h	Against distilled water for 3 days	By centrifugation	2.5 % wt./ vol.	Mehrabani et al. (2018)
Na_2CO_3 0.05 wt.% at 100 °C for 30 min	With deionized water	–	Formic acid/LiBr at room temperature	Against distilled water for 72 h	By drying	1 wt. %	Ming et al. (2015)
Na_2CO_3 0.5 wt.% at 100 °C for 20 min by triplicate	With deionized water	Air dried	LiBr 9 M	Against distilled water	By drying	1 wt. %	Hu et al. (2016)
Na_2CO_3 0.005 % wt./ vol. at 100 °C for 30 min	–	–	$CaCl_2/C_2H_5OH/H_2O$ molar ratio 1:2:8 at 70 °C for 2 h	Against distilled water	By filtration	–	Yan et al. (2018)
Na_2CO_3 0.02 M at 4 °C for 20 min	–	Dried overnight	LiBr 9.3 M at 60 °C for 4 h	Against distilled water for 3 days	By centrifugation	10 % wt./ vol.	Kumar et al. (2018)

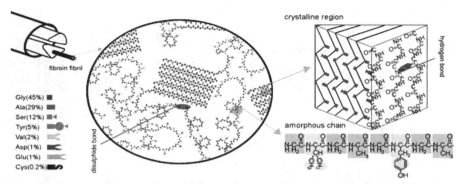

FIGURE 18.5   Amino acids composition in amorphous and crystalline fibroin.

swelling capacity and can keep its structural integrity. Other properties that are measured in hydrogels are porosity, water content, size and shape, rheologic and mechanical properties, biocompatibility, and biodegradability. The hydrophobic amino acids such as glycine, serine, and alanine, permit the gelation without a gelling agent. The gel formation is due to the conversion from coils to β-sheets (Kapoor and Kundu, 2016). The hydrophilic zones produce the gel swelling and the hydrophobic zones maintain the mechanical properties and manage the swelling rate of the hydrogel. The intrinsic elastic properties of the hydrogel control the water absorption. The hydrogels can be obtained with different sizes from micrometers to nanometers; for hydrogels with a size of 5 μm, the administration should be via oral, from 10 to 100 nm can be through injection. The excretion when the hydrogel is below 10 nm is through kidneys, and between 0.5 and 10 nm is by phagocytosis. The hydrogels can be used in tissue engineering, controlled release of drugs and other active substances in regenerative medicine. In Figure 18.6, a scheme of typical hydrogel composition is shown.

Exist advanced hydrogels with self-healing properties and shape memory can be seen in the work of Mahinroosta et al. (2018). Also, the hydrogels are employed in food, water treatment, and pharmaceutical products (Hasan and Abdel-Raouf, 2019).

18.3.1 SILK HYDROGELS COUPLED WITH POLYSACCHARIDES

18.3.1.1 AGAROSE

Different authors had published works using agarose and SF gels that later were freeze-dried to obtain scaffolds for bone repair. Fibroin/agarose

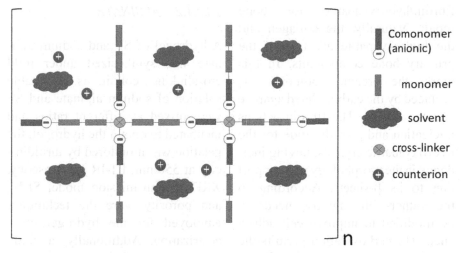

FIGURE 18.6   Scheme of typical hydrogel composition.

scaffolds with different fibroin gelation degrees were obtained using sodium dodecyl sulfate. The agarose helped to maintain the water stability in the hydrogel after the freeze drying. Additionally, it is believed that the drying process modulates the properties of the scaffold depending on its gelation degree. The gels had an opaque appearance, and the rate of gelation decreased when the sodium dodecyl sulfate concentration increased, with an optimal of 5.5 mM within the range from 4.5 to 6.5 mM. The agarose changed the solubility of the fibroin gel, increasing the β-sheet content, increasing its insolubility in aqueous solutions, and maintaining its shape at high temperature (around 100 °C). At low temperatures (−20 °C), the scaffolds form a hydrogel, and without agarose, it becomes liquid.

The rest of the sodium dodecyl sulfate showed cytotoxicity, so it is necessary to remove it from the scaffolds. Physicochemical characterization of the scaffolds was carried out using the following techniques FT-IR spectroscopy, SEM, determination of porosity using cyclohexane as displacement liquid, and evaluation of mechanical properties through the compressive modulus. The results showed that the fibroin/agarose scaffolds are biocompatible. At a high degree of gelation, the porosity and elasticity increased, as well as cell proliferation. However, the compression modulus decreases in the scaffolds with high porosity, which affects the mineralization and the osteogenic differentiation (Lu et al., 2017).

Other work published the fabrication of a composite of agarose/

fibroin/hydroxyapatite for bone repair. Naturally, the collagen and the hydroxyapatite are two of the primary bone components. In this work, the organic component is replaced by the carboxylated agarose and the fibroin. The fibroin acts as nucleation and growth guide for the hydroxyapatite crystals, having incidence in the morphology and support due to its β-sheets. According to the authors, the agarose needs to be modified to improve cell adherence. The carboxylation permits the crosslinking between the fibroin and the composite. The agarose was carboxylated to increase its properties, showing better thermal stability than the unmodified agarose. The characterization was performed by FT-IR spectroscopy, FE-SEM, TEM, EDS, TGA, and DSC. According to the FT-IR results, it was possible to corroborate the transition from random coils to β-sheets. The EDS confirmed an excellent distribution of hydroxyapatite in the composite through a Ca and P mapping. Also, amorphous and crystalline areas or peaks that correspond to the organic polymers and hydroxyapatite were observed from the XRD results. The composite showed good mechanical properties for body support. The results showed that the acetylated agarose improves the cellular adhesion, the viability, and slows the degradation, thus helping bone regeneration (Hu et al., 2016).

18.3.1.2 ALGINATE

A hydrogel of SF and sodium alginate was synthesized under mild crosslinking conditions. A stable solution of sodium alginate and SF were mixed at different ratios and incubated to obtain the hydrogel; the gelation was monitored by turbidimetry at 550 nm. FT-IR spectroscopy, XRD in transmission mode, SEM, and porosity were the techniques employed for the hydrogel characterization. Additionally, a study of its mechanical properties was performed. The hydrogel obtained showed characteristics similar to those of the ECM in which the cells are immersed. The results showed a decrease in the compressive stress as the alginate content increased, a porosity of around 96% in the hydrogels with different ratios of sodium alginate and SF tested, and the SEM micrographs displayed that the hydrogel morphology is of nanofibers, which did not have a dependence with the temperature of synthesis. XRD results showed that the addition of sodium alginate decreases the generation of β-sheets and promotes the presence of random coils. The FT-IR bands at 1660 cm^{-1} indicate the presence of random coils, and the bands at 1521 and 1631 cm^{-1} the presence of β-sheets, both types of silks (I and II) are present in the composite. The degradation essays were positive using proteases, with 53% more degradation compared to the control. The proliferation assays did not show cytotoxicity, and the alginate content did not influence cell proliferation. The possible application of this hydrogel for tissue engineering and drug delivery is demonstrated by Ming et al. (2015).

For the treatment of burn wounds, a hydrogel of fibroin/carboxymethyl cellulose/calcium alginate, abbreviated as SF/CC/CA, was fabricated. First, a solution of carboxymethyl cellulose and calcium alginate at different ratios was prepared in deionized water; then, the SF was added to obtain the hydrogel. Adhesion test, moisture content, cytotoxicity, and histological observation of the wound were determined during the study. The results showed that the material with better adhesion was the Purilon Gel®, a hydrogel of CC and sodium alginate used as a reference, then the hydrogel SF/CC/CA and finally a conventional medical gauze, this test was employed due to it can be considered proportional to the intrinsic viscosity of the hydrogel, which is essential to know the capacity that has the hydrogel to absorb the fluids exudated by the wound. The moisture content percentage was similar for the SF/CC/CA and the Purilon Gel® and ≈10% less for the medical gauze. The SF/CC/CA and the Purilon Gel® had a similar healing effect, and both

materials were better than a medical gauze, neither were cytotoxic. The primary purpose of the SF/CC/CA hydrogel is to keep hydrated and absorb the residues produced by the wound, helping to a rapid wound healing (Ju et al., 2014).

18.3.1.3 CHITIN OR CHITOSAN

A scaffold of chitin, fibroin, and TiO_2 was synthesized. First, the chitin was dissolved in a solution of NaOH and urea, then SF was added, and glycerol was used as a plasticizer. The following step was the addition of TiO_2 at different proportions after 24 h of stirring the crosslinking was initialized with the incorporation of glutaraldehyde. Finally, the bandages were obtained after a freezing and freeze-drying process. They determined by FT-IR spectroscopy of the bands associated with OH⁻ and C=O of chitin at 3436 and 1633 cm^{-1}, and the vibrations related to silk I components, random coils, and α-helix, at 1637, 1533, and 1264 cm^{-1}, respectively. Using SEM and the Archimedes' principle was possible to calculate the porosity; the composites showed up to 90% of porosity; however, the increment in porosity produces a decrease in the mechanical properties. Also, they observed that a high content of TiO_2 decreased the swelling percentage and the porosity of the material. The degradation of the material increased with time until reach 60%–70% at 3 weeks, also they noticed that the addition of TiO$_2$ above 0.5% had a negative effect on the interaction between polymeric chains showing a complete dissolution of the material in 3 weeks. The mechanical properties such as elongation and tensile strength were improved due to the H-bonding between amide groups of chitin and fibroin.

Additionally, the chitin increased the crystallinity of fibroin.

Interestingly, high amounts of TiO$_2$ decreased the mechanical properties of the material. The positive charge of chitin and fibroin benefit the coagulation, considering the negative charge of blood. The TiO$_2$ did not affect coagulation. The antibacterial properties of chitin/fibroin increased with the addition of TiO$_2$, also the cytotoxicity increased with the increasing addition of TiO$_2$, being the optimal concentration between 0.5% and 1.5% (Mehrabani et al., 2018).

For skin and cartilage repair, hydrogels based on fibroin, chitosan, agarose, and silver nanoparticles with antibacterial properties were synthesized. The addition of chitosan increased notably the porosity and the compression modulus compared to the hydrogel made only of fibroin and agarose. The increase of agarose content from 0.25% to 0.5% w/v also improved the porosity, the swelling percentage, and the compression

modulus; the last parameter also increases, until a value of 1371 Pa, with the presence and increment of silver nanoparticles in the hydrogel. They tested the antibacterial activity of the hydrogels against *E. coli* and *S. aureus*, finding that the hydrogels with Ag have good antibacterial activity, and the hydrogels without silver presented low antibacterial activity. The hydrogels showed good cytocompatibility in the tests using fibroblasts, but the hydrogels with fibroin and agarose were the ones that presented higher cytocompatibility. According to the rheology results, the hydrogels are steady in solid-state; this is associated with chemical bonds or physical–chemical interactions between the chitosan and the fibroin. Additionally, the XRD patterns showed that the addition of chitosan induces the transition from β-sheets to random coils in the fibroin, leading to an amorphous hydrogel due to the hydrogen bonds formed between the fibroin and the chitosan (Chen et al., 2019).

18.3.1.4 DEXTRAN

Biocomposites, biomaterials composed of at least two different components of dextran with SF, had been reported to improve the moist environment necessary for bandages to heal skin wounds. As it is well-known, the fibroin suffers a transition from random coils, which are soluble to a β-sheet conformation, which is crystalline and insoluble in water, producing that the hydrogel films for wound bandages cannot maintain an adequate environment for skin regeneration. The alternative is bonding covalently the dextran, previously oxidized, to the fibroin and increased the water retention of the fibroin/dextran film. The results showed a swelling increment of up to 80% in the film with dextran. However, at the early stage of the biomaterial evaluation, low cell adhesion and proliferation was observed; this was attributed to the wettability produced by the dextran. At a later stage, no significant difference in cell proliferation was detected between the mats with or without dextran, showing good collagen production (Kim et al., 2014).

18.3.1.5 FUCOIDAN

Current information related to hydrogels of fibroin and fucoidan for biomedical applications is scarce. A work from 2008 studied the films of SF and fucoidan obtained after mixing and reducing the volume at 50 °C; later, the concentrated was poured in Petri dishes and let dry at 60 °C to obtain films. The materials were characterized by ATR FT-IR, XPS, SEM, contact angle, clotting time, cell attachment, and proliferation. According to the FT-IR results, the addition of fucoidan did not interfere with the β-sheet formation in fibroin, which, as we mentioned before, gives excellent mechanical properties to the hydrogel. XPS showed an increase in S and O ratio in SF/fucoidan hydrogel compared to the SF hydrogel, attributing to the presence of sulfate and hydroxyl groups of fucoidan that increases the hydrophilicity, these results are in agreement with the contact angle result, which is lower for the hydrogel with fucoidan. SEM micrographs showed a rough surface in the fibroin and fucoidan hydrogel that helps the cell adhesion and proliferation. Also, the hydrogels with fucoidan presented good anticoagulant response, cell attachment, and proliferation, using endothelial cells (Gao et al., 2008).

18.3.1.6 HYALURONIC ACID

In order to create biocompatible hydrogels that have good mechanical properties with short gelation times, and that could be used as biomaterials for tissue regeneration, have been developed SF/HA hydrogels crosslinked with EDC (1-ethyl-3-(3-dimethyl aminopropyl) carbodiimide hydrochloride), NHS (*N*-hydroxysuccinimide), and MES (2-morpholinoethanesulfonic acid) as assisted agent. XRD, SEM, FTIR, histological evaluation, control release, and degradation analysis were applied to study the hydrogel (Yan et al., 2018).

The formation times of the composite gel were 80, 27, and 20 min at RT, 37, and 55 °C, respectively. The increase in the ratio of HA in the composite gel led to a decrease in the average size, shrinkage, elastic modulus, resilience deformation ability while favoring an increase in porosity, reaching maximum values of compressive failure strength of 207 kPa using a 6:4 ratio of SF: HA. As for the drug release (Rhodamine B), during the first seven days, about 60% of the drug was released; subsequently, there was a cumulative release of 80% at 38 days. These types of SF-HA hydrogels also showed excellent histocompatibility and biodegradation in vivo.

In other words, to improve the mechanical properties of silk hydrogels, silk-HA hydrogel composite has been created, also, to have a biologically friendly process and improve the polymer's limitations, the composite hydrogel has been enzymatically crosslinked with horseradish peroxidase (HRP) and H_2O_2. The hydrogels were prepared for 3–4 h at 37 °C (Raia et al., 2017). They were characterized by FTIR, rheological analysis, mechanical properties, swelling measurements, enzymatic degradation, and cell viability.

After 8 days of immersion of the hydrogels in an enzyme solution, silk-HA (10%) hydrogels showed the degradation of 50% by weight, silk hydrogels degraded by 30%, and HA hydrogel entirely degraded after 6 days. For in vitro cell response using human mesenchymal stem cells (hMSCs) from bone marrow aspirates, SF-HA hydrogels showed cellular proliferation and cellular growth, revealing cytocompatibility, otherwise with HA hydrogels which inhibited hMSCs growth (Raia et al., 2017).

These hydrogels presented a wide range of compression modulus since they presented values ranging from 10 kPa to almost 1MPa, which makes it very versatile in tissue engineering applications.

18.3.1.7 PULLULAN

Based on the fact that the PL presents low mechanical properties and the SF hydrogel lack cell-specific "epitopes," present in sericin linked with immune responses, investigations have been carried out where the SF/PL is combined (Li et al., 2018) into a composite hydrogel to obtain an injectable hydrogel. Carboxymethyl pullulan (CMPL) conjugated with tyramine was used for the preparation of the SF/CMPL hydrogel, polymerizing the hydrogel by enzymes, which was developed under physiological conditions with HRP as a catalyst and hydrogen peroxide as an oxidant. The gelation times of the SF/CMPL hydrogels were ~12–60 min. Hydrogels were characterized by rheological measurements, SEM, NMR, mechanical tests, mesenchymal stem cells test, cell encapsulation, and cell viability. CMPL-TA hydrogels gelled in 6 min; with the increase in CMPL content, the gelation time decreased in such a way that by adjusting the content of CMPL biomedical applications are expanded. In hydrogels composition SF/CMPL (20%), 90% of cells lived after nine days of cell culture, and the MSCs were incorporated homogeneously. The hydrogels obtained presented a good cytocompatibility (encapsulation of rabbit MSCs) and a compressive modulus with potential uses in musculoskeletal tissue engineering.

18.3.1.8 STARCH

Currently, there are only some studies related to silk fiber-starch composites. Nano biocomposites scaffolds have been developed as starch-based hydrogels and silk nanofibers for potential bone tissue regeneration. The silk nanofibers were elaborated by wet electrospinning and the hydrogels were prepared with constant stirring, maintaining a pH of 2, at 80 °C (Hadisi et al., 2015). The composites were analyzed by FTIR-ATR, SEM, XRD, swelling ratio test, and cell culture.

In order to favor the bioactivity, the scaffolds were subjected to immersion in saturated calcium and phosphate solutions. When the amount of silk nanofibers in the starch hydrogel is increased, the biocomposite has a lower capacity for the formation of calcium phosphate and a decrease in the average pore size and percentage of porosity of the composite. In cell culture experiments of SF/starch (Hadisi et al., 2015), human osteoblast-like cells (MG63), it has been observed that when silk nanofibers are added within the hydrogel, cell viability is improved as well as its attachment and proliferation.

Because rice starch (RS) films are easily deformable, the SF/RS hydrogels have been created and modified with trisodium trimetaphosphate (STMP) (Racksanti et al., 2014); since STMP reacts quickly with polysaccharides and is a crosslinking agent that is cheap and has no toxicity. The hydrogel was prepared by adjusting the pH at 12 of the SF/Starch (1% STMP) solution, constantly stirring at RT for 30 min, then the pH was adjusted to 7. The hydrogels were characterized by FTIR, NMR, and crosslinking density tests were performed. The resulting hydrogels presented better flexibility and were more translucent at a 5:95 weight ratio (SF/RS), which are probably viable as absorbable hydrogels, which can be biocompatible and biodegradable.

18.3.2 SILK HYDROGELS COUPLED WITH AMINO ACIDS AND PROTEINS

18.3.2.1 POLY (γ-GLUTAMIC ACID)

Poly(γ-glutamic acid) g-PGA has been used in the biomedical area since it does not present toxicity; it is biodegradable and biocompatible. However, it has low mechanical properties, which limits the field of application. Although silk/g-PGA hydrogels have not been well researched, have only been reported works focused on wound dressing applications using silk sericin (SS), to improve the mechanical and biological properties (Shi et al., 2015).

Previously, allergic reactions had been attributed to SS, which was probably due to the combination of SS and SF (Gilotra et al., 2018). However, there is currently evidence that it can be used in the biomedical area and tissue regeneration due to the presenting bioactivity (Gilotra et al., 2018). g-PGA/SS hydrogel has been evaluated by FTIR, SEM, mechanical analysis, fluorescent microscopy, animal experiments, in vitro degradation, cell culture, histological examination.

The PGA and SS solution was crosslinked using ethylene glycol diglycidyl ether, with a pH of 5.0, and after injection, it was maintained at 60 °C for 5 h. The incorporation of SS in the PGA/SS hydrogel favored the increase in the swelling ratio (49.2), increasing the degradation rate of the hydrogel where soluble nutrients favored cell proliferation as well as the formation of new skins. These increases in the SS content allowed that a more significant amount of naproxen was released in the solution, releasing up to a maximum of 80% after 3 days in buffer solutions.

As for in vivo tests, using rats, the hydrogel reduced the risk of dehydration in the wounds and was easily removed without damaging the wound, the wound is almost closed on day 15, compared to the control, see Figure 18.7.

In the cytocompatibility studies conducted in PGA/SS hydrogels with L929 cells, cell adhesion and cell proliferation were possible, and according to the MTT tests, the hydrogel did not show cytotoxicity.

18.3.3 SILK HYDROGELS COUPLED WITH INORGANIC PHASES

18.3.3.1 GOLD AND SILVER NANOPARTICLES

Hydrogels that exhibit antibacterial activity have been developed for the SF-HAp system, using silver (AgNp) and gold (AuNp) nanoparticles (Ribeiro et al., 2017). In general, for the formation of the HAp/SF hydrogel with AgNps, $AgNO_3$ was

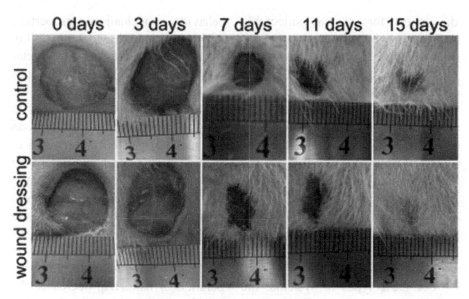

FIGURE 18.7 Photographs of the wound taken postoperatively.

mixed with the SF solution; subsequently, the nanoHAP powder was mixed with ethanol and then added to the aqueous solution of SF kept at 37 °C, and finally carried out the formation of AgNps within the hydrogel to RT. As for the hydrogels containing AuNps, $HAuCl_4 \cdot 3H_2O$ was used, which was mixed with nanoHAp in the presence of ethanol, subsequently were slowly mixed with SF solution at 37 °C where the formation of AuNPs within the hydrogel was performed at 60 °C. The properties of the materials were evaluated by UV–visible, TEM, TGA, mechanical analysis, cytotoxicity, and antimicrobial analysis. The results show that the storage moduli were favored with the increase in the concentration of NPs in hydrogels,

in cytocompatibility studies using osteoblastic cells, cell viability was favored in hydrogels with concentrations up to 0.5% of AgNps and for all concentrations of AuNPs. Using 1% AgNps, a reduction in cell viability was observed. On the other hand, these hydrogels had antibacterial activity against gram-positive and gram-negative bacteria; this effect was stronger in hydrogels containing 0.5% of AgNPs. However, it is worth mentioning that for hydrogels with AuNPs, there was no antimicrobial activity toward *S. epidermidis* (Ribeiro et al., 2017).

18.3.3.2 HYDROXYLAPATITE

Silk nanofibers hydrogels with silk-HAp nanoparticles have been developed for use as injectable systems to improve osteogenesis. Where the HAp nanoparticles were coated with a concentration of 6% (w/v) of SF solution, incubated for 24 h at 60 °C, the SF-HAp composite hydrogels contained up to 60% (w/w) of HAp (Ding et al., 2017).

They were characterized by the techniques of SEM, XRD, FTIR, carrying out rheological studies, as well as determinations of viscosity, injectability, in vitro biocompatibility, cell differentiation, and in vivo animal studies.

For these types of systems, the maximum content of HAp particles to avoid their aggregation and not present structural deterioration was 50%, the module reached of the composite was 10 times higher compared to silk hydrogel. According to cell culture tests, the addition of HAp particles had no adverse effect on cytocompatibility. These hydrogels showed good osteogenesis both in vitro and in vivo tests (Ding et al., 2017).

In recent years, a new method was developed to accelerate the formation process of the SF/nanoHAp hydrogel, which uses ethanol as a gelling agent (Ribeiro et al., 2015). Since these hydrogels conventionally take 48–65 h, at 50 °C and 35 °C, respectively, in their gelation process with this method, their gelation process was 6.3–7.6 min. These hydrogels were characterized by SEM, XRD, DTA, their swelling

behavior, mechanical properties, ALP activity, and cytocompatibility were also determined. The metabolic activity of these hydrogels was evaluated using osteoblast-like cell MG63. It was observed that in the non-frozen hydrogels, there was less metabolic activity compared to the frozen ones, probably due to the larger pore size of the frozen material. However, both frozen/nonfrozen hydrogel favored the attachment and spreading of cells, the hydrogels showed the ability to swell rapidly in the first 10 min, although the incorporation of nanoHAp decreased the swelling percentage. On the other hand, in both types of hydrogels, ALP activity increased with culture time, favoring the functional activity of the bone-derived cells with nanoHAp aggregates (Ribeiro et al., 2015).

18.3.3.3 MAGNETIC PARTICLES

Currently, there are some studies where hydrogels have the main component of silk with magnetic nanoparticles.

One of these works is the hydrogel based on regenerated silk fibroin (RSF) and magnetic nanoparticles; the RSF-based hydrogel was made with 90% RSF and 10% HPMC (hydroxypropyl methylcellulose). Because the formation of Fe_2O_3 was via in situ, this allowed the process to be carried out in a minimum number of steps (Luo and Shao, 2017).

The RSF: HPMC solution had a 9:1 ratio, they were mixed at RT, subsequently heated at 70 °C for 2 h. To obtain the composite, the fibroin hydrogels were mixed in a solution of $FeCl_2$ and $FeCl_3$, were left for 72 h at 4 °C, subsequently washed and placed in ammonium hydroxide for 2 h. The characterization was carried out by XRD, SEM, TGA, magnetic measurements, and UV-vis.

This hydrogel with Fe_2O_3 nanoparticles showed catalytic activity, maintaining the peroxidase-like activity since the Fe_2O_3 nanoparticles were present in the hydrogel; the prevention of the oxidation of such nanoparticles was achieved.

18.4 CONCLUSION

Hydrogels based on fibrillar proteins such as collagen and silk have shown to be systems with modulated properties for improved applications in tissue engineering and biomedicine. Their high biocompatibility and easiness to generate 3D hydrated matrices in the hydrogel state are advantages that have been used to design innovative material systems that can modulate the associated biological response in wound healing processes, cancer therapies, tissue growth, and controlled release of biofunctional molecules. However, hydrogels based on these proteins require strategies to improve their physicochemical characteristics, such as degradation rate and mechanical properties. In this sense, hydrogel systems have been designed by coupling biomolecules such as carbohydrates and proteins, as well as including inorganic phases. The hydrogels of IPNs and/or biocomposites studied exhibit adaptation in their properties allowing us to control biocompatibility improving their performance in biomedicine strategies. Finally, it is important to mention that the development of this type of hydrogel system represents a promising research opportunity where materials with potential applications in biotechnological areas can be designed, where the chemical composition of hydrogels will be decisive in modulating their properties.

KEYWORDS

- **hydrogels**
- **collagen**
- **fibroin**
- **biomaterials**
- **composites**

REFERENCES

Alarcon, E.I.; Udekwu, K.I.; Noel, C.W.; Gagnon, L.B.P.; Taylor, P.K.; Vulesevic, B.; Simpson, M.J.; Gkotzis, S.; Islam, M.M.; Lee, C.-J.; Richter-Dahlfors, A.; Mah, T.-F.; Suuronen, E.J.; Scaiano, J.C.; Griffith, M. Safety and efficacy of composite collagen–silver nanoparticle hydrogels as tissue engineering scaffolds. *Nanoscale* 2015, *7*, 18789–18798.

Almeida, H.V.; Eswaramoorthy, R.; Cunniffe, G.M.; Buckley, C.T.; O'Brien, F.J.; Kelly, D.J. Fibrin hydrogels functionalized with cartilage extracellular matrix and incorporating freshly isolated stromal cells as an injectable for cartilage regeneration. *Acta Biomater.* 2016, *36*, 55–62.

Alvarez, G.S.; Hélary, C.; Mebert, A.M.; Wang, X.; Coradin, T.; Desimone, M.F. Antibiotic-loaded silica nanoparticle–collagen composite hydrogels with prolonged antimicrobial activity for wound infection prevention. *J. Mater. Chem. B* 2014, *2*, 4660.

Antman-Passig, M.; Shefi, O. Remote magnetic orientation of 3D collagen hydrogels for directed neuronal regeneration. *Nano Lett.* 2016, *16*, 2567–2573.

Badylak, S.F. The extracellular matrix as a biologic scaffold material. *Biomaterials* 2007, *28*, 3587–3593.

Baldwin, S.J.; Quigley, A.S.; Clegg, C.; Kreplak, L. Nanomechanical mapping of hydrated rat tail tendon collagen I fibrils. *Biophys. J.* 2014, *107*, 1794–1801.

Banerjee, P.; Shanthi, C. Isolation of novel bioactive regions from bovine Achilles tendon collagen having angiotensin I-converting enzyme-inhibitory properties. *Process Biochem.* 2012, *47*, 2335–2346.

Bendtsen, S.T.; Wei, M. Synthesis and characterization of a novel injectable alginate–collagen–hydroxyapatite hydrogel for bone tissue regeneration. *J. Mater. Chem. B* 2015, *3*, 3081–3090.

Claudio-Rizo, J.A.; Mendoza-Novelo, B.; Delgado, J.; Castellano, L.E.; Mata-Mata, J.L. A new method for the preparation of biomedical hydrogels comprised of extracellular matrix and oligourethanes. *Biomed. Mater.* 2016, *11*, 035016.

Claudio-Rizo, J.A.; Rangel-Argote, M.; Castellano, L.E.; Delgado, J.; Mata-Mata, J.L.; Mendoza-Novelo, B. Influence of residual composition on the structure and properties of extracellular matrix derived hydrogels. *Mater. Sci. Eng. C* 2017, *79*, 793–801.

Chatzistavrou, X.; Rao, R.R.; Caldwell, D.J.; Peterson, A.W.; McAlpin, B.; Wang, Y.-Y.; Zheng, L.; Fenno, J.C.; Stegemann, J.P.; Papagerakis, P. Collagen/fibrin microbeads as a delivery system for Ag-doped bioactive glass and DPSCs for potential applications in dentistry. *J. Non-Cryst. Solids* 2016, *432*, 143–149.

Chen, L.; Hu, J.; Ran, J.; Shen, X.; Tong, H. Preparation and evaluation of collagen-silk fibroin/hydroxyapatite nanocomposites for bone tissue engineering. *Int. J. Biol. Macromol.* 2014, *65*, 1–7.

Chen, S.-H.; Li, Z.; Liu, Z.-L.; Cheng, L.; Tong, X.-L.; Dai, F.-Y. Antimicrobial hydrogels with controllable mechanical properties for biomedical application. *J. Mater. Res.* 2019, *34*, 1–11.

Cho, S.-H.; Noh, J.-R.; Cho, M.Y.; Go, M.-J.; Kim, Y.-H.; Kang, E.S.; Kim, Y.H.; Lee, C.H.; Lim, Y.T. An injectable collagen/poly(γ-glutamic acid) hydrogel as a scaffold of stem cells and α-lipoic acid for enhanced protection against renal dysfunction. *Biomater. Sci.* 2017, *5*, 285–294.

Desimone, M.F.; Hélary, C.; Quignard, S.; Rietveld, I.B.; Bataille, I.; Copello, G.J.; Mosser G.; Giraud-Guille, M.-M.; Livage, J.; Meddahi-Pellé, A.; Coradin, T. In vitro studies and preliminary in vivo evaluation of silicified concentrated collagen hydrogels. *ACS Appl. Mater. Interfaces* 2011, *3*, 3831–3838.

Desimone, M.F.; Hélary, C.; Rietveld, I.B.; Bataille, I.; Mosser, G.; Giraud-Guille, M.-M.; Livage, J.; Coradin, T. Silica–collagen bionanocomposites as three-dimensional scaffolds for fibroblast immobilization. *Acta Biomater.* 2010, *6*, 3998–4004.

Ding, Z.; Han, H.; Fan, Z.; Lu, H.; Sang, Y.; Yao, Y.; Cheng, Q.; Lu, Q.; Kaplan, D.L. Nanoscale silk–hydroxyapatite hydrogels for injectable bone biomaterials. *ACS Appl. Mater. Interfaces* 2017, *9*, 16913–16921.

Eyre, D.R. Collagen: molecular diversity in the body's protein scaffold. *Science* 1980, *207*, 1315–1322.

Filipowska, J.; Lewandowska-Łańcucka, J.; Gilarska, A.; Niedźwiedzki, Ł.; Nowakowska, M. In vitro osteogenic potential of collagen/chitosan-based hydrogels-silica particles hybrids in human bone marrow-derived mesenchymal stromal cell cultures. *Int. J. Biol. Macromol.* 2018, *113*, 692–700.

Galateanu, B.; Hudita, A.; Zaharia, C.; Bunea, M.C.; Vasile, E.; Buga, M.R.; Costache, M. Silk-based hydrogels for biomedical applications. In: Cellulose-Based Superabsorbent Hydrogels. Polymers and Polymeric Composites: A Reference Series; Ed: Springer: Cham, 2019; vol. 1, p. 1794.

Gao, Z.; Wang, S.; Zhu, H.S.; Zhao, D.X.; Xu, J.C. Improvements of anticoagulant activities of silk fibroin films with fucoidan. *Front. Mater. Sci. China* 2008, *2*, 221–227.

Gaviria Arias, D.; Caballero Mendez, L.C. Uso de biomateriales a partir de la fibroína de la seda de gusano de seda (Bombyx mori L.) Para procesos de medicina regenerativa basada en ingeniería de tejidos TT—Fibroin from silkworm (Bombyx mori L) as biomaterial used in regenerative medicine proc. *Revista Médica de Risaralda* 2015, *21*, 38–47.

Gilotra, S.; Chouhan, D.; Bhardwaj, N.; Nandi, S.K.; Mandal, B.B. Potential of silk sericin based nanofibrous mats for wound dressing applications. *Mater. Sci. Eng. C* 2018, *90*, 420–432.

Gould, B.S. Collagen biosynthesis at the ribosomal level. *Int. Rev. Connect. Tissue Res.* 1968, *4*, 33–65.

Hadisi, Z.; Nourmohammadi, J.; Mohammadi, J. Composite of porous starch-silk fibroin nanofiber-calcium phosphate for bone regeneration. *Ceram. Int.* 2015, *41*, 10745–10754.

Harris, J.R.; Soliakov, A.; Lewis, R.J. In vitro fibrillogenesis of collagen type I in varying ionic and pH conditions. *Micron (Oxford, England: 1993)* 2013, *49*, 60–68.

Hasan, A.M.A.; Abdel-Raouf, M.E.-S. *Cellulose-Based Superabsorbent Hydrogels*. Springer: New York, NY, 2019.

Heymer, A.; Haddad, D.; Weber, M.; Gbureck, U.; Jakob, P.M.; Eulert, J.; Nöth, U. Iron oxide labelling of human mesenchymal stem cells in collagen hydrogels for articular cartilage repair. *Biomaterials* 2008, *29*, 1473–1483.

Hinderer, S.; Layland, S.L.; Schenke-Layland, K. ECM and ECM-like materials—biomaterials for applications in regenerative medicine and cancer therapy. *Adv. Drug Del. Rev.* 2016, *97*, 260–269.

Hu, J.X.; Ran, J.B.; Chen, S.; Jiang, P.; Shen, X.Y.; Tong, H. Carboxylated Agarose (CA)-silk fibroin (SF) dual confluent matrices containing oriented hydroxyapatite (HA) crystals: biomimetic organic/inorganic composites for tibia repair. *Biomacromolecules* 2016, *17*, 2437–2447.

Iswariya, S.; Bhanukeerthi, A.V.; Poornima, V.; Uma, T.S.; Paramasivan Thirumalai, P. Design and development of a piscine collagen blended pullulan hydrogel for skin tissue engineering. *RSC Adv.* 2016, *6*, 57863–57871.

Ju, H.W.; Lee, O.J.; Moon, B.M.; Sheikh, F.A.; Lee, J.M.; Kim, J.H.; Park, H.J.; Kim, D.W.; Lee, M.C.; Kim, S.H.; Park, H.R.; Lee, H.R. Silk fibroin based hydrogel for regeneration of burn induced wounds. *Tissue Eng. Regener. Med.* 2014, *11*, 203–210.

Kadler, K.E.; Holmes, D.F.; Trotter, J.A.; Chapman, J.A. Collagen fibril formation. *Biochem. J.* 1996, *316(Pt 1)*, 1–11.

Kapoor, S.; Kundu, S.C. Silk protein-based hydrogels: promising advanced materials for biomedical applications. *Acta Biomater.* 2016, *31*, 17–32.

Karri, V.V.S.R.; Kuppusamy, G.; Talluri, S.V.; Mannemala, S.S.; Kollipara, R.;

Wadhwani, A.D.; Malayandi, R. Curcumin loaded chitosan nanoparticles impregnated into collagen-alginate scaffolds for diabetic wound healing. *Int. J. Biol. Macromol.* 2016, *93*, 1519–1529.

Kastelic, J.; Galeski, A.; Baer, E. The multicomposite structure of tendon. *Connect. Tissue Res.* 1978, *6*, 11–23.

Kim, M.K.; Kwak, H.W.; Kim, H.H.; Kwon, T.R.; Kim, S.Y.; Kim, B.J.; Park, Y.H.; Lee, K.H. Surface modification of silk fibroin nanofibrous mat with dextran for wound dressing. *Fibers Polym.* 2014, *15*, 1137–1145.

Koperska, M.A.; Pawcenis, D.; Bagniuk, J.; Zaitz, M.M.; Missori, M.; Łojewski, T.; Łojewska, J. Degradation markers of fibroin in silk through infrared spectroscopy. *Polym. Degradation Stab.* 2014, *105*, 185–196.

Kumar, M.; Gupta, P.; Bhattacharjee, S.; Nandi, S.K.; Mandal, B.B. Immunomodulatory injectable silk hydrogels maintaining functional islets and promoting anti-inflammatory M2 macrophage polarization. *Biomaterials* 2018, *187*, 1–17.

Lee, J.H.; Park, J.-H.; Eltohamy, M.; Perez, R.; Lee, E.-J.; Kim, H.-W. Collagen gel combined with mesoporous nanoparticles loading nerve growth factor as a feasible therapeutic three-dimensional depot for neural tissue engineering. *RSC Adv.* 2013, *3*, 24202.

Li, Q.; Mu, L.; Zhang, F.; Sun, Y.; Chen, Q.; Xie, C.; Wang, H. A novel fish collagen scaffold as dural substitute. *Mat. Sci. Eng. C-Mater. Biol. Appl.* 2017, *80*, 346–351.

Li, T.; Song, X.; Weng, C.; Wang, X.; Wu, J.; Sun, L.; Gong, X.; Zeng, W.-N.; Yang, L.; Chen, C. Enzymatically crosslinked and mechanically tunable silk fibroin/pullulan hydrogels for mesenchymal stem cells delivery. *Int. J. Biol. Macromol.* 2018, *115*, 300–307.

Li, X.; Xue, W.; Zhu, C.; Fan, D.; Liu, Y.; Xiaoxuan, Ma. Novel hydrogels based on carboxyl pullulan and collagen crosslinking with 1, 4-butanediol diglycidylether for use as a dermal filler: initial in vitro and in vivo investigations. *Mater. Sci. Eng. C* 2015, *57*, 189–196.

Liu, S.; Xie, R.; Cai, J.; Wang, L.; Shi, X.; Ren, L.; Wang, Y. Crosslinking of collagen using a controlled molecular weight bio-crosslinker: β-cyclodextrin polyrotaxane multi-aldehydes. *RSC Adv.* 2015, *5*, 46088–46094.

Lowther, D.A. Chemical aspects of collagen fibrillogenesis. *Int. Rev. Connect. Tissue Res.* 1963, *1*, 63–125.

Lu, Y.; Zhang, S.; Liu, X.; Ye, S.; Zhou, X.; Huang, Q.; Ren, L. Silk/agarose scaffolds with tunable properties: Via SDS assisted rapid gelation. *RSC Adv.* 2017, *7*, 21740–21748.

Luca, A.; Maier, V.; Maier, S.S.; Butnaru, M.; Danu, M.; Ibanescu, C.; Pinteala, M.; Popa, M. Biomacromolecular-based ionic-covalent hydrogels for cell encapsulation: the atelocollagen—oxidized polysaccharides couples. *Carbohydr. Polym.* 2017, *169*, 366–375.

Luo, K.; Shao, Z. A novel regenerated silk fibroin-based hydrogels with magnetic and catalytic activities. *Chin. J. Polym. Sci.* 2017, *35*, 515–523.

Ma, X.; He, Z.; Han, F.; Zhong, Z.; Chen, L.; Li, B. Preparation of collagen/hydroxyapatite/alendronate hybrid hydrogels as potential scaffolds for bone regeneration. *Colloids Surf. B. Biointerfaces* 2016, *143*, 81–87.

Mahinroosta, M.; Jomeh Farsangi, Z.; Allahverdi, A.; Shakoori, Z. Hydrogels as intelligent materials: a brief review of synthesis, properties and applications. *Mater. Today Chem.* 2018, *8*, 42–55.

Mebert, A.M.; Alvarez, G.S.; Peroni, R.; Illoul, C.; Hélary, C.; Coradin, T.; Desimone, M.F. Collagen-silica nanocomposites as dermal dressings preventing infection in vivo. *Mater. Sci. Eng. C* 2018, *93*, 170–177.

Mehrabani, M.G.; Karimian, R.; Rakhshaei, R.; Pakdel, F.; Eslami, H.; Fakhrzadeh, V.;

Rahimi, M.; Salehi, R.; Kafil, H.S. Chitin/ silk fibroin/TiO$_2$ bio-nanocomposite as a biocompatible wound dressing bandage with strong antimicrobial activity. *Int. J. Biol. Macromol.* 2018, *116*, 966–976.

Meng, F.W.; Slivka, P.F.; Dearth, C.L.; Badylak, S.F. Solubilized extracellular matrix from brain and urinary bladder elicits distinct functional and phenotypic responses in macrophages. *Biomaterials* 2015, *46*, 131–140.

Ming, J.; Pan, F.; Zuo, B. Structure and properties of protein-based fibrous hydrogels derived from silk fibroin and sodium alginate. *J. Sol-Gel Sci. Technol.* 2015, *74*, 774–782.

Overstreet, D.J.; Dutta, D.; Stabenfeldt, S.E.; Vernon, B.L. Injectable hydrogels. *J. Polym. Sci., Part B: Polym. Phys.* 2012, *50*, 881–903.

Racksanti, A.; Janhom, S.; Punyanitya, S.; Watanesk, R.; Watanesk, S. Crosslinking density of silk fibroin—rice starch hydrogels modified with trisodium trimetaphosphate. *Appl. Mech. Mater.* 2014, *446–447*, 366–372.

Raia, N.R.; Partlow, B.P.; McGill, M.; Kimmerling, E.P.; Ghezzi, C.E.; Kaplan, D.L. Enzymatically crosslinked silk-hyaluronic acid hydrogels. *Biomaterials* 2017, *131*, 58–67.

Rangel-Argote, M.; Claudio-Rizo, J.A.; Mata-Mata, J.L.; Mendoza-Novelo, B. Characteristics of collagen-rich extracellular matrix hydrogels and their functionalization with Poly(ethylene glycol) derivatives for enhanced biomedical applications: a review. *ACS Appl. Bio Mater.* 2018, *1*, 1215–1228.

Rao, S.S.; DeJesus, J.; Short, A.R.; Otero, J.J.; Sarkar, A.; Winter, J.O. Glioblastoma behaviors in three-dimensional collagen-hyaluronan composite hydrogels. *ACS Appl. Mater. Interfaces* 2013, *5*, 9276–9284.

Reis, L.A.; Chiu, L.L.Y.; Liang, Y.; Hyunh, K.; Momen, A.; Radisic, M. A peptide-modified chitosan–collagen hydrogel for cardiac cell culture and delivery. *Acta Biomater.* 2012, *8*, 1022–1036.

Ribeiro, M.; de Moraes, M.A.; Beppu, M.M.; Garcia, M.P.; Fernandes, M.H.; Monteiro, F.J.; Ferraz, M.P. Development of silk fibroin/nanohydroxyapatite composite hydrogels for bone tissue engineering. *Eur. Polym. J.* 2015, *67*, 66–77.

Ribeiro, M.; Ferraz, M.P.; Monteiro, F.J.; Fernandes, M.H.; Beppu, M.M.; Mantione, D.; Sardon, H. Antibacterial silk fibroin/ nanohydroxyapatite hydrogels with silver and gold nanoparticles for bone regeneration. *Nanomed. Nanotechnol. Biol. Med.* 2017, *13*, 231–239.

Rößler, S.; Heinemann, C.; Kruppke, B.; Wagner, A.S.; Wenisch, S.; Wiesmann, H.P.; Hanke, T. Manipulation of osteoclastogenesis: bioactive multiphasic silica/collagen composites and their effects of surface and degradation products. *Mater. Sci. Eng. C* 2018, *93*, 265–276.

Sapru, S.; Das, S.; Mandal, M.; Ghosh, A.K.; Kundu, S.C. Nonmulberry silk protein sericin blend hydrogels for skin tissue regeneration—in vitro and in vivo. *Int. J. Biol. Macromol.* 2019, *137*, 545–553.

Saska, S.; Teixeira, L.N.; de Castro Raucci, L.M.S.; Scarel-Caminaga, R.M.; Franchi, L.P.; dos Santos, R.A.; Santagneli, S.H.; Capela, M.V.; de Oliveira, P.T.; Takahashi, C.S.; Gaspar, A.M.M.; Messaddeq, Y.; JL Ribeiro, S.; Marchetto, R. Nanocellulose-collagen-apatite composite associated with osteogenic growth peptide for bone regeneration. *Int. J. Biol. Macromol.* 2017, *103*, 467–476.

Seyedhassantehrani, N.; Li, Y.; Yao, L. Dynamic behaviors of astrocytes in chemically modified fibrin and collagen hydrogels. *Integr. Biol.* 2016, *8*, 624–634.

Sharabi, M.; Mandelberg, Y.; Benayahu, D.; Benayahu, Y.; Azem, A.; Haj-Ali, R. A new class of bio-composite materials of unique collagen fibers. *J. Mech. Behav. Biomed. Mater.* 2014, *36*, 71–81.

Shi, L.; Yang, N.; Zhang, H.; Chen, L.; Tao, L.; Wei, Y.; Liu, H.; Luo, Y. A novel poly(γ-glutamic acid)/silk-sericin hydrogel for wound dressing: synthesis, characterization and biological evaluation. *Mater. Sci. Eng. C* 2015, *48*, 533–540.

Tamada, Y.; Kojima, K. (2014). Silk fibers as smart materials toward medical textiles. In X. Tao (Ed.), *Handbook of Smart Textiles* (pp. 1–13). Singapore: Singapore.

Tan, H.; Marra, K.G. Injectable, biodegradable hydrogels for tissue engineering applications. *Materials* 2010, *3*, 1746–1767.

Tan, H.; Wu, B.; Li, C.; Mu, C.; Li, H.; Lin, W. Collagen cryogel cross-linked by naturally derived dialdehyde carboxymethyl cellulose. *Carbohydr. Polym.* 2015, *129*, 17–24.

Tang, H.; Buehler, M.J.; Moran, B. A constitutive model of soft tissue: from nanoscale collagen to tissue continuum. *Ann. Biomed. Eng.* 2009, *37*, 1117–1130.

Theocharis, A.D.; Skandalis, S.S.; Gialeli, C.; Karamanos, N.K. Extracellular matrix structure. *Adv. Drug Del. Rev.* 2016, *97*, 4–27.

Tylingo, R.; Gorczyca, G.; Mania, S.; Szweda, P.; Milewski, S. Preparation and characterization of porous scaffolds from chitosan-collagen-gelatin composite. *React. Funct. Polym.* 2016, *103*, 131–140.

Wang, J.; Wei, L.; Ma, Y.; Li, K.; Li, M.; Yu, Y.; Wang, L.; Qiu, H. Collagen/cellulose hydrogel beads reconstituted from ionic liquid solution for Cu(II) adsorption. *Carbohydr. Polym.* 2013, *98*, 736–743.

Wang, W.; Zhang, X.; Chao, N.-N.; Qin, T.-W.; Ding, W.; Zhang, Y.; Sang, J.-W.; Luo, J.-C. Preparation and characterization of pro-angiogenic gel derived from small intestinal submucosa. *Acta Biomater.* 2016, *29*, 135–148.

Whitesides, G.M.; Boncheva, M. Beyond molecules: self-assembly of mesoscopic and macroscopic components. *Proc. Natl. Acad. Sci. USA* 2002, *99*, 4769–4774.

Włodarczyk-Biegun, M.K.; Werten, M.W.T.; de Wolf, F.A.; van den Beucken, J.J.J.P.; Leeuwenburgh, S.C.G.; Kamperman, M.; Cohen Stuart, M.A. Genetically engineered silk–collagen-like copolymer for biomedical applications: production, characterization and evaluation of cellular response. *Acta Biomater.* 2014, *10*, 3620–3629.

Wolf, M.T.; Daly, K.A.; Brennan-Pierce, E.P.; Johnson, S.A.; Carruthers, C.A.; D'Amore, A.; Nagarkar, S.P.; Velankar, S.S.; Badylak, S.F. A hydrogel derived from decellularized dermal extracellular matrix. *Biomaterials* 2012, *33*, 7028–7038.

Xia, Z.; Villa, M.M.; Wei, M. A biomimetic collagen–apatite scaffold with a multi-level lamellar structure for bone tissue engineering. *J. Mater. Chem. B* 2014, *2*, 1998.

Yan, S.; Wang, Q.; Tariq, Z.; You, R.; Li, X.; Li, M.; Zhang, Q. Facile preparation of bioactive silk fibroin/hyaluronic acid hydrogels. *Int. J. Biol. Macromol.* 2018, *118*, 775–782.

Yar, M.; Shahzadi, L.; Mehmood, A.; Raheem, M.I.; Román, S.; Chaudhry, A.A.; ur Rehman, I.; Douglas, C.W.I.; MacNeil, S. Deoxy-sugar releasing biodegradable hydrogels promote angiogenesis and stimulate wound healing. *Mater. Today Commun.* 2017, *13*, 295–305.

Zafar, M.S.; Belton, D.J.; Handy, B.; Kaplan, D.L.; Perry, C.C. Functional material features of bombyx mori silk light vs. heavy chain proteins. *Biomacromolecules* 2015, *16*, 606–614.

Zhang, X.; Yang, Y.; Yao, J.; Shao, Z.; Chen, X. Strong collagen hydrogels by oxidized dextran modification. *ACS Sustain. Chem. Eng.* 2014, *2*, 1318–1324.

Zhuang, J.; Lin, S.; Dong, L.; Cheng, K.; Weng, W. Magnetically actuated mechanical stimuli on Fe_3O_4/mineralized collagen coatings to enhance osteogenic differentiation of the MC_3T_3-E1 cells. *Acta Biomater.* 2018, *71*, 49–60.

CHAPTER 19

Understanding Consolidated Processes in the Design of Sustainable Biomaterials as Thermal Insulators

LUIS FERNANDO SÁNCHEZ TERÁN, SALVADOR CARLOS HERNÁNDEZ, ARTURO I. MARTÍNEZ ENRÍQUEZ, and LOURDES DÍAZ JIMÉNEZ*

Centro de Investigación y Estudios Avanzados del Instituto Politécnico Nacional, CINVESTAV-Saltillo, 25900 Ramos Arizpe, México

Corresponding author. E-mail: lourdes.diaz@cinvestav.edu.mx.

ABSTRACT

Climatization systems use around 40% of the total energy consumption in buildings and household activities, contributing to 15%–20% of greenhouse gas emissions. The environmental impact is due to the pumping of cold or hot air to compensate for high or low externals temperatures and to regulate it for internal comfort. Besides the energy rationalization and passive solar energy, the use of materials that allow efficient heat transfer between the external and internal environment in buildings could drastically reduce energy consumption. The study of such materials is currently a widespread research topic. The objective of this chapter is the analysis the application of the consolidation theory in the development of insulation materials for buildings. First, a literature analysis to determine the status of biomaterials as thermal insulators in buildings is introduced. After, a brief description of the evolution of the consolidation approach is presented. Under the perspective of consolidated theory, a methodology to design and assess thermal insulator materials based on organic wastes is proposed. Finally, concluding remarks, challenges, and future research directives are presented.

19.1 INTRODUCTION

The problems facing humanity with the development of society (environmental pollution, global warming, and depletion of natural resources) have forced actions to help mitigate those issues. Among the actions considered, the maximum use of resources, specifically their waste, stands out as a promising alternative.

Solid wastes are those derived from human activities that have lost their use for what they were created (Tadesse, 2004). The revalorization of solid waste is a cornerstone within the framework of contemporary sustainability because it raises the development of novel materials from wastes of human activities or industrial processes. One way to revalue a waste is through the manufacture of thermal insulation for buildings.

In addition to traditional thermal insulators such as wood (Cetiner and Shea, 2018), straw, and cork, those of synthetic origins such as polystyrene, expanded polystyrene (EPS), or polyurethane have stood out. The latter, despite their low costs and easy application in the construction industry, have the disadvantage of a high environmental impact during their manufacture. In addition to that, in case of fire, they emit toxic substances (Jelle, 2011). In that sense, there is an increasingly large number of thermal insulators made from recycled materials. Materials derived from natural fibers are a viable alternative to replace synthetic fiber derivatives, especially in lightweight applications. They have also

marked advantages over synthetic materials because, in addition to being environmentally friendly, they offer competitive results in their resistance to stress, impact, and interlaminar cutting, besides its thermal properties and water absorption (Sanjay et al., 2018).

On the other hand, consolidation theory, originally proposed to describe the phenomenon of soil compaction, is a tool that can be applied in the design of thermal insulators. This is because during the performance of an insulator, various elements, as air, rain, sun, and pollutants, can affect the insulator efficiency throughout its life cycle. Then, one of the applications to be explored is the design and manufacturing of thermal insulators. The interest of this chapter is to analyze the generated knowledge of consolidated materials and to evaluate the application of this theory on the synthesis of thermal insulators.

19.2 THERMAL INSULATION IN BUILDINGS

19.2.1 THERMAL COMFORT

One of the first definitions of thermal comfort was proposed by the ASHRAE in the early 1960s, in which it was defined as "that condition of mind which expresses satisfaction with the thermal environment and is assessed by subjective evaluation" (Parsons, 2010). Although this meaning has been accepted worldwide, there are significant variations in people, both physiologically and psychologically aspect (ASHRAE, 2015). In this way, different scales of thermal comfort have been developed in which the ideal state of thermal comfort is indicated at the center of a seven-level scale (Parsons, 2010). Those defined a thermal comfort as a bipolar phenomenon ranging from "too cold" to "too hot" in which the neutral sensation (or thermal comfort) falls in the middle of the scales (Epstein and Moran, 2006).

Another way to define thermal comfort could be that zone delimited by a set of temperature thresholds in which most people manifest feeling good (Fernández García, 1994; Nicol and Humphreys, 2002). Thus, thermal comfort scales have been related to the environment temperature and physiological response of the human body. A comparison between numeral scale designations and temperature thresholds range for thermal sensations is expressed in Table 19.1. It seems plausible that the behavioral activity stimulated by discomfort provides the human being with his principal means of long-term natural thermoregulation (Gagge et al., 1967) like the autonomic and hormonal systems (Fleisher et al., 1996).

Since in industrialized countries, people spend more than 90% of their time indoors (Höppe, 2002), urban building planning should consider the different possible microclimates.

TABLE 19.1 Effective Temperature Thresholds and Comfort Scales for Thermal Sensations

Effective Temperature (°C)	Sensation		Numerical Equivalent	Physiological Response
	Thermal	Comfort		
40—Upper temperatures	Very hot	Very uncomfortable	3	Regulatory problems
35–39	Hot		2	Increased tension by sweating and increased blood flow
30–34	Slightly warm		1	Normal regulation due to sweating and vascular change
25–29	Neutral	Comfortable	0	Vascular regulation
20–24	Slightly cool		−1	Increasing in losses of dry heat
15–19	Chill	Slightly uncomfortable		
10–14	Cool		−2	Vasocontraction in hands and feet
Lower temperatures	Too cold	Uncomfortable	−3	

Sources: Fernández García (1994) and Nicol and Humphreys (2002).

For that reason, one of the fundamentals of urban planning is to understand that it is not sensible to evaluate the bioclimate of a city as a whole, but it is necessary to distinguish between the microclimates of different urban building structures (Mayer and Höppe, 1987).

As a response attempt, comfort standards are intended to optimize the thermal acceptability of indoor environments with scales based on static models of thermal comfort that view occupants as passive recipients of the thermal stimulus (Table 19.1) (de Dear and Brager, 1998). This analysis generates the necessity of developing computational adaptive systems from databases developed from the survey answers of many people from all around the world to implement the best possible solutions (de Dear and Brager, 1998).

The adaptive principle recognizes that a person is not a passive receiver of sense impressions. However, it is an active participant in dynamic equilibrium with the thermal environment (Humphreys and Nicol, 1998). Thus, the empirical findings of field surveys would be used as a guide for informing the design of buildings to provide comfortable conditions (Nicol, 2004).

19.2.2 ENVIRONMENTAL EFFECTS

According to Kaswan et al. (2019), environmental sustainability is the

human ability "to satisfy its current needs without anyway compromising the quality of environment/ ecosystem." More precisely, "environmental sustainability could be defined as a condition of balance, resilience, and interconnectedness that allows human society to satisfy its needs while neither exceeding the capacity of its supporting ecosystems to continue to regenerate the services necessary to meet those needs nor by our actions diminishing biological diversity" (Morelli, 2011).

Either way, environmental sustainability is an issue that has been recognized as strategic to reach socioeconomic development in any country. In the situation of environmental deterioration, global warming is the element that represents the most severe environmental impact (Zhou et al., 2015). The concept of sustainability is embraced by a wide variety of industries and businesses, such as mining and minerals, and building. It has been over 20 years since environmental issues became critical in those areas (Inyim et al., 2014).

In any city, buildings are the largest consumers of energy, mainly due to the operation of their thermal conditioning systems. As climate modifiers, buildings have been designed to shelter occupants and achieve thermal comfort in the occupied space, backed up mechanical heating and air-conditioning systems as necessary (Al-Homoud, 2005).

The initiatives to improve the performance of energy in buildings resulted initially from the necessity to reduce the energy consumed by these mechanical systems (Papadopoulos and Giama, 2007). Today, world energy consumption contributes to pollution, environmental deterioration, and global greenhouse emissions (Ozel, 2012). Future predictions by the International Energy Agency indicate that the energy-related emissions of carbon dioxide (CO_2) will double by 2050 without decisive action (Cuce et al., 2014).

As part of the actions to face increasingly significant environmental problems, buildings have to become more energy efficient (Dimoudi and Tompa, 2008). In that sense, different proposals have emerged focused on increasing the energy efficiency of buildings. For example, the Zero Energy Building (ZEB) is a strategy that tries to reach an efficient use of energy sources in buildings. A ZEB is either a residential or a commercial building that uses both conventional and renewable energy with zero net energy consumption. That means the total amount of energy used by the building is equivalent to the amount of renewable energy that is produced on the site. Typically, a ZEB uses all cost-effective measures to reduce energy usage through energy efficiency systems and includes the application of renewable energy systems to produce enough energy

to meet their remaining energy demand. This concept is no longer perceived as a concept of a distant future, but as a realistic solution to reach a reduction of CO_2 emissions and the reduction of energy use in buildings (Marszal et al., 2011).

Historical definitions of zero energy are based mainly on annual energy use for the building's operation like heating, cooling, ventilation, lighting, and among others (Hernandez and Kenny, 2010). Currently, several countries have adopted or considering the ZEBs as their future building energy targets, such as the Building Technology Program of the US Department of Energy and the EU Directive on Energy Performance of Buildings (Li et al., 2013).

To achieve more efficient energy performance of buildings is essential to understand that buildings are dynamic and nonlinear systems, which include a series of components, where each component has a crucial influence on the total energy consumption of the building (Han et al., 2014).

That is why it would be beneficial to analyze which factors influence the environmental impact of a building the most (Haapio and Viitaniemi, 2008), without forgetting that each element affects the others like the cost-effective, and incomplete environmental performance criterion, the mismatch between the models of optimization and design practice, and the scope definition (Wang et al., 2005).

In synthesis, the main goal is to create a low-energy design of the urban environment and buildings so the population should and could be accommodated with a minimum worsening of the environmental quality (Omer, 2008).

19.2.3   THERMAL INSULATORS

Thermal insulation is considered the best strategy for saving energy used in heating and cooling buildings (Dylewski and Adamczyk, 2011). A proper amount of thermal insulation in the building envelope helps to reduce climatization energy demands and works in the reduction of CO_2 and SO_2 and other greenhouse gas emissions (Kaynakli, 2012). Since when the concentration of greenhouse gases in the atmosphere remains constant, the temperature of the earth keeps unchanged (Çomakl and Yüksel, 2004); it is essential to achieve efficient use of energy, avoiding, or at least minimizing greenhouse gas emissions.

The efficiency of thermal insulation is determined by its thermal conductivity (λ), which can be understood as the capability of any material to transfer heat. According to the American Society for Testing and Materials (ASTM, 2013), thermal conductivity is the "the time rate of heat flow, under steady conditions, through the unit area, per unit temperature gradient in the direction perpendicular to the area." The thermal conductivity is one of the transport properties. Other properties are the viscosity associated with the transport of momentum, diffusion coefficient associated with the transport of mass.

Then, thermal conductivity acts as an indicator of the rate at which heat energy is transferred through a medium by a conduction phenomenon. According to Fourier's Law, thermal conductivity can be defined as shown in the following equation:

$$\lambda = -q_\lambda/(A(\mathrm{d}T/\mathrm{d}x)) \qquad (19.1)$$

where λ is the thermal conductivity, q_λ is the exchanged heat flow, A is a cross-sectional area, and $\mathrm{d}T/\mathrm{d}x$ is the temperature gradient (Kreith et al., 2012). In the equation, the physical meaning of the negative sign is related to the heat flow, which is in the direction of negative gradient temperature, and that serves to make heat flow positive.

A thermal insulator is considered efficient when its thermal conductivity is <0.10 W/m K (Jelle, 2011). Currently, some of the most common materials used as thermal insulators in buildings possess thermal conductivities ranged between 0.033 and 0.04 W/m K. Among these materials that stand out are mineral and glass wool, EPS, extruded polystyrene (XPS), and cellulose (Cetiner and Shea, 2018; Jelle, 2011; Jelle et al.,

2010). Table 19.2 remarks on the main advantages and disadvantages of this kind of insulators. On the other side, novel materials as vacuum insulation panels have typical thermal conductivities between 0.004 and 0.007 W/m K (Wakili et al., 2004), while products like phase change materials take advantage of latent heat of the material through its storage on release by application of temperature; besides, this kind of materials possess the ability to change their state with a specific temperature range (Mondal, 2008).

Among the thermal conductivity of buildings, the quality of an insulating material depends on its adaptability to the local edifications (Papadopoulos, 2005), considering the life cycle assessment (LCA) of materials (extraction, processing, transportation, and implementation). For example, the energy used to manufacture and transport buildings material represents nearly 8% of all primary energy used in the UK, whereas 50% of all energy consumed is attributable to the occupation of the dwellings (Morel et al., 2001). The LCA of materials should be promoted since this analysis allows identifying the stages representing more challenges related to energy consumption, greenhouses gas emissions, and other aspects.

TABLE 19.2 Common and Novel Insulators Used in Buildings

Insulator	TC (W/m K)	Disadvantages	Advantages
Mineral and glass wool	0.03–0.04	TC increases with moisture. High environmental impact. Adverse health effects	Easy handling in buildings
EPS	0.029–0.04	TC increases with moisture. High environmental impact	Easy to adjust in-site. Lightweight
Extruded polystyrene	0.03–0.04	TC increases with moisture. High environmental impact	Easy to adjust in-site. Lightweight
Cellulose	0.04–0.05	TC increases with moisture	Easy to adjust in-site
Polyurethane	0.02–0.03	In the case of fire, it releases harmful substances	Easy to adjust in-site
Perlite	0.04	Nonrenewable mineral resource	Not expensive. Lightweight. Improve fire resistance of the building's envelope. Resistant to degradation
Aerogels	0.004–0.013	Very expensive	
Vacuum Insulating Panels	0.001–0.009	Very expensive	

Sources: Jelle (2011) and Jelle et al. (2010)

In the LCA context and in realizing the goal of adequate buildings for "all people living" the utilization of local materials could make a crucial contribution to the environment and human habitability (Mathur, 2006). Local materials reduce costs and minimize environmental effects since the transport is minimal; also, it is an alternative to promote the local economy and then social benefits. Besides, it is supposed that the characteristics of materials are better adapted to the local conditions. Studies demonstrated how thermally insulating buildings are a much cost-effective measure of intervention when compared to others.

In fact, the integration of passive and active renewable technologies could optimize building behavior to eliminate discomfort and to improve energy efficiency (Bekkouche et al., 2014). Among the materials which can be considered for thermal insulators are wastes produced from local economic activities. The use of these materials is a route to promote sustainability and the implementation of ZEBs. The next section focuses on the description of organic wastes in the context of raw materials for thermal insulators.

19.2.4 ORGANIC WASTES AS MATERIALS

Organic wastes are typically by-products derivate from farming, industrial, or municipal activities, and they are usually called "wastes" because they are not the target product (Westerman and Bicudo, 2005) and do not have a systematic application. In a world limited by resources with an increasing population, recovery of materials is fundamental to achieve sustainable development (Narayana, 2009). In the context of efficient use of energy and resources, the creation of novel materials with low thermal conductivity values produced from solid waste is a motivating alternative (Lertsutthiwong et al., 2008).

There are several propositions for the use of different products derived from organic waste to processing in many different presentations (Pinto et al., 2012). For example, Alabdulkarem et al. (2018) have developed hybrid specimens derived from wood waste (WW) fibers of different densities; the materials presented thermal conductivity values between 0.042 and 0.052 W/m K, which are in the range of adequate insulators. The WW can be applied to timber frame wall construction by manual or mechanical filling between the studs of the frame without any binder (Cetiner and Shea, 2018).

Other accessible organic materials are the ones made from the fiber of jute, flax, or hemp, which generally present low thermal conductivity values. These materials are water vapor permeable and accumulate moisture adsorption from the air, mechanisms that could favorably influence the indoor humidity (Korjenic et al., 2011). Usually, the fibers are mixed with fire retardants, and their range of thermal conductivity is between 0.038 and 0.060 W/m K (Schiavoni et al., 2016).

Natural fibers can also be excellent reinforcing materials for preparing polymer matrix-based composites. By adding Agave fiber on nonorganic polymer material, the resulting composite exhibits higher mechanical strength incrementing the percentage of fiber (Singha and Rana, 2012). The tensile strength of the hybridization of Agave americana with copper, aluminum, and fiber has been analyzed, showing results in a tensile strength range between 16.66 and 99.73 MPa, depending on fiber orientations in composite design (Geethika and Rao, 2017).

An issue of irresponsible management of organic waste is that the decomposition of biodegradable textile waste generates methane, a potent greenhouse gas contributing to global warming (Dissanayake et al., 2018). Besides, the dumping of postconsumer textile waste is a substantial urban waste problem, and there is a common practice of diverse organizations collecting discarded textiles (Luiken and Bouwhuis, 2015). However, there exist different alternatives to revalorize textile wastes; the combination of these materials with other ones can be used to synthesize thermal insulators as described in the next lines.

Cellulose aerogels are a practical solution to approach textile waste because the resultant materials regularly have high porosity and low density and, as such, find use in a vast range of applications (Zeng et al., 2019). Cellulose fibrils are the principal reinforcing constituent of the plant cell walls so the microfibrillated cellulose, with an average size of 5–50 nm, has a Young Modulus between 1800 and 5400 MPa (Svagan et al., 2008). These are adequate properties to promote the desired heat transfer in buildings.

As can be seen in presented applications, replacing (or hybridizing) synthetic fibers with natural fibers can be an alternative to replace common construction materials to achieve cost savings and to provide some desirable properties (Sanjay et al., 2018).

19.3 CONSOLIDATED THEORY

19.3.1 A BRIEF HISTORY

The consolidation theory was first proposed to explain the interaction between different physical properties of soil that allows the compaction process. The consolidation process of saturated soil is understood as an interaction of the solid displacement field with a flow field of pore fluid (Aubram, 2019). Between 1921 and

1925, Karl Terzaghi published six classic papers related to this topic (Clayton et al., 1995), but the one of 1923 proposed a theory for 1D consolidation, and this theory has since played a mainstream role in soil mechanics (Cryer, 1963).

The theory proposed by Terzaghi presents a numerical approach for consolidation analysis utilizing the finite element method and a procedure that solves all governing equations simultaneously at each step (Hwang et al., 1971).

Conformity between the 1D Terzaghi theory and experimental results is not ideal (Taylor and Merchant, 1940); the theoretical solution is an accurate approximation during the initial phase of consolidation, but there is a second phase called secondary consolidation, during which consolidation continues at a higher rate than theoretical predictions (Morgenstern and Nixon, 1971).

Terzaghi acknowledged the limitations of his proposal when the secondary consolidation exceeds 20% of the total settlement (Ahmed and Siddiqua, 2014); for that reason, some other studies were developed to deal with that limitation. For the large-strain consolidation, Monte and Krizekt (1976) developed a mathematical model where the water content in soil determines the fluid limit, and it is directly related to a "stress-free" condition of the soil and is taken as a reference state from which strains are measured.

As far back as 1941, Biot presented an analysis of the transient flow problem in saturated soils. He suggested two constitutive relations for soil and changes in pore-water pressure with time (Fredlund and Hasan, 1979).

In 1942, Taylor was one of the first authors to extend and elaborate on consolidation theory with the square foot of time method for estimating the primary consolidation in an odometer test. This technique is now a part of the standard laboratory procedure (Christian and Baecher, 2014). As well, Carrillo (1942) studied the 2D and 3D consolidation for an infinite rectangular prism of saturated soil, which is loaded suddenly and is permitted to drain at all its faces.

By 1963, Biot (Barden, 1963) considered that the type of anisotropy existing in various undisturbed soils contributes to the overestimation of settlements in overconsolidated clays, and is partially responsible for the observed stress distribution in the sand. In 1977, Fredlund and Morgensten determined that for unsaturated soil are three sets of possible stress tensors, of which only two are independent (Fredlund, 2006).

Zienkiewicz et al. (1980) formulated a method to determine some limits of validity of consolidation assumptions; to reach this, a set of dynamic equations was presented for a nonlinear two-phase medium, following the work of Biot.

In recent times, Shi et al. (2017) present a consolidation model for lumpy composite soils within a homogenization framework. This model considers the nonuniform stress (strain) distribution and evolution of interlump porosity during the consolidation process. Currently, the theory of consolidation is surely now part of basic knowledge in the fields of elementary soil mechanics and is of significant interest in the analysis of deformations of porous media (Verruijt, 2005).

Besides soil mechanics, the consolidation theory has found application in other areas. In recent years, this theory has evolved into the processing of materials; the aim is to reduce the number of steps in the obtaining of some specific product. Among the applications of consolidated theory, stand out: the synthesis of materials for manufacturing applications; the fabrication of spare parts for electric and electronic equipment; the formation of concrete for civil engineering applications; the manufacturing of construction blocks from different materials; the restoration of historic buildings; and the study of compaction of sludges from wastewater treatment plants. Since the management of materials is involved in several applications, the consolidation principles can be used to improve the synthesis of thermal insulators for buildings. These insulators require a combination of two or more materials with

different properties to allow efficient heat transfer to achieve a comfortable temperature. In this chapter, this topic is explored to determine the feasibility of synthesizing consolidated thermal insulators.

19.3.2 CONSOLIDATED THEORY IN THERMAL INSULATOR PHENOMENON WITH BIOMATERIALS

This section describes the principals involved in consolidation theory and its relationship with the biomaterials as insulators.

19.3.2.1 DARCY'S LAW

Darcy's law describes the dynamic of water (or another fluid) through a porous medium, introducing the concept of hydraulic conductivity. When a fluid passes through a porous media, the corresponding flow depends on the geometrics of the medium and its "affinity" to the fluid. This "affinity" is associated with the ability of the fluid to pass through the medium or the resistance of the medium to the fluid. From these relationships, it is possible to study the interaction between solids and liquids and to determine the behavior of materials in face of some environmental conditions. The flow of water on porous materials was first studied by Darcy in 1856 when he was searching for water filtering systems. He found that the water flow through a bed of sand is directly proportional to the cross-sectional area of the porous material and to the hydraulic gradient, which is defined as the relation between the difference of water head loss and the difference of flow length. This phenomenon, or hydraulic conductivity, is analogous to the flow of electrons through an electrical conductor, which is expressed by Ohm's law: the electrical current is proportional to the voltage and inversely proportional to the electrical resistance of the conductor material.

Darcy formulated an empirical equation to represent the water flow through porous media. After that, the mathematical expression has been verified experimentally on numerous occasions (Neuman, 1977). The equation, which is known today as Darcy's Law, is presented in the following equation:

$$Q = -KA\, \Delta h/\Delta L \qquad (19.2)$$

where Q = flow rate of water fluid, K = coefficient of permeability, A = cross-sectional area to the water flow, Δh = difference between the final and initial head of water, ΔL = length of the flow path (Shekhar, 2017).

Since the water flow produces a readjustment of particles in porous media, Darcy's law is used to study filtration processes where a solid/liquid separation is desired. This separation leads to the compaction of solids, which is an important element to study the consolidation phenomena.

Also, considering the readjustment of particles from the pass of fluid through porous materials, Darcy's law can be helpful to study the integration of several biomaterials into a single one. If each material presents specific permeability to face a determined fluid, it is possible to determine the required conditions to promote the biomaterial consolidation. These conditions should be related to the size of the particle, the flow of an agglomerating agent, the geometry of biomaterial, and other characteristics. That means Darcy's law could be used as an indicator of the level of integration of biomaterials in the early phase of consolidation.

19.3.2.2 THE TERZAGHI–TAYLOR CONSOLIDATION MODEL

In the Terzaghi theory for 1D consolidation during primary consolidation, the settlement is controlled by the dissipation of excess pore pressures and Darcy's law. During secondary consolidation, the rate of settlement is controlled by soil viscosity (Leroueil, 2006).

Secondary consolidation takes place once the excess pore water pressure is dissipated from the consolidating material; this situation can occur due to the plastic deformation of materials in their integration and also due to the readjustment of particles.

Different works were performed in order to study the disagreement between the Terzaghi consolidation theory and the behavior of the laboratory consolidation tests (Šuklje, 1957; Crawford, 1997); from the obtained results a complementary theory was developed. This theory assumes that the speed of occurrence of secondary compression ($\partial e/(\partial t)$) depends directly on the undeveloped secondary compression (ρ_s) and the viscosity constant of the fluid (μ), as shown in the following equation:

$$\partial e/\partial t = -\mu\rho_s. \qquad (19.3)$$

Several tests were performed by Taylor, focusing principally on the differences between the rates of compression observed in laboratory samples and the rates given by theory (Crawford, 1997).

From the consolidated biomaterials synthesis viewpoint, the phenomenon of primary and secondary consolidation is important since it improves the integration of materials, and the interactions between them during all the consolidation process can be studied. The primary consolidation can be understood as a first stage in the integration of different organic materials;

it can be used to study the elimination of pores and water through a compression process in order to promote adequate characteristics for heat transfer and/or thermal isolation. The secondary consolidation is useful to improve or "consolidate" the characteristics of biomaterials achieved in the first stage. Both steps of consolidation are useful tools to determine the operating conditions required to promote the integration of materials, for example, the pressure, the fluid which allows improving materials integration, the influence of the material geometry, and among others. Then, the theory developed by Terzaghi and complemented by several authors could be extended to the application of consolidated biomaterials.

19.3.2.3 THE BIOT EQUATIONS

In Terzaghi's model, the porous media flow is uncoupled from the deformation process of the complete material, and then, the mathematical expression describing the compaction is reduced to a diffusion equation.

In order to include a more integrated representation of consolidation, the Biot model considers a coupling condition between the soil and fluid stresses and strains; this coupling implies a mathematical representation more complex, but more complete to study the consolidation process (Cheng and Liggett, 1984). Specifically, instead of particles, the Biot equations represent a fraction of material, usually a cube; then, it is possible to include the interconnection of pores and the physical phenomena derived from the flow of fluids and the application of pressures. Then, from this representation, the soil or material studied can be considered as an elastic system respecting conservation properties, such as components compressibility and shearing rigidity. The displacement or readjustment of particles is then considered as an average of the components affected in the whole material, and different scenarios can be studied: traverse isotropy, anisotropy, incompressible and compressible materials, and multiple strata.

Biot's work furnishes general solutions for the case of consolidation for a porous material. The stress field of porous material is denoted by the next tensor:

$$\{\blacksquare(\sigma_{xx} + \sigma \& \sigma_{xy} \& \sigma_{xz} @ \sigma_{yz} \& \sigma_{yy} \& \sigma_{yz} @ \sigma_{zx} \& \sigma_{zy} \& \sigma_{zz} + \sigma)\}$$

where entries σ_{ij} represent the forces acting on the soil part of the faces of a unit cube of bulk material, and σ is the force applied to the fluid part.

Introducing the porosity f, the pressure p in the fluid is related to σ, as shown in Eq. (19.4) (Biot, 1956):

$$\sigma = -fp \qquad (19.4)$$

For heterogeneous soils with multiple strata and nonlinear behavior, the Biot model allows being discretized. This discretization aims to evaluate validity intervals and then, further integration is performed.

Concerning biomaterials, the Biot model allows studying the consolidation as for soils with multiple strata. This approximation is feasible since it is expected that biomaterials are composed of several organic substrates with different properties: compressibility, permeability, and elasticity. This implies heterogeneous initial conditions in the biomaterial consolidation; after that, the magnitude of pressure and even the necessity of binders as a fluid to remove porosity and promote consolidation can be determined from the solution of the Biot model.

19.3.3 PERSPECTIVES

The existing literature about soil consolidation is heavily addressed toward laboratory tests and theoretical analyses without corresponding full-scale performance studies (Crawford and Morrison, 1996). One probable reason is that laboratory consolidation tests could take only 2 or 3 h to be done (Aboshi et al., 1970). Instead, the case histories are time-consuming, taking several days or even weeks to be completed (Crawford and Morrison, 1996).

Another reason is that the laboratory oedometer test results and the consolidation behavior of case histories were not exactly convergent because even with the most rigorous reproductions on laboratory conditions there was a significant difference against the field tests (Chai et al., 2005).

In comparison, the laboratory tests regularly underestimate the in-situ results (Kabbaj et al., 1988), especially in the effective stress after the application of preconsolidation pressure. However, the laboratory methods are still a useful tool for measuring the soil stress s-strain-strength behavior to modeling a more accurate selection of design parameters to understand the mechanical conditions of soils and clays (DeGroot, 2004).

Otherwise, the implementation of the digital computational systems has revolutionized the way that soil mechanics is now applied in engineering practice (Fredlund et al., 2012). The computer simulations of processes with the involvement of finite elements allow a better interpretation of physical phenomena (Wong et al., 1998).

Presenting verified formulation and implementation, in conjunction with a validated material model of the system, provides the absolute confidence to realistically predict the mechanical behavior of soils during dynamic loading events (Jeremić et al., 2008). The simulations are capable of reproducing almost all main characteristics of saturated–unsaturated soil mechanics problems (Ye et al., 2007) that it would not be possible without the use of digital computational systems.

It is now a fact that engineering has moved into a new paradigm, in which computer software is no longer luxury but a necessity for an efficient practice to provide new data about systems that are hard or impossible to obtain (Barberousse et al., 2009). In this context, with the use of simulations platforms, it is possible to extend the consolidation applications with less economic and technical risks. For this reason, the formulation of robust mathematical models is a great necessity.

19.4 CONSOLIDATED THEORY IN THERMAL INSULATION

19.4.1 RAW MATERIALS

Several products are used as raw material for thermal insulation. The selection of raw materials is based on many factors like recyclable and renewable materials, low resource production techniques, and the potential of applicability (Kymäläinen and Sjöberg, 2008; Panyakaew and Fotios, 2011).

Roux and Espuna (2012) present a list of parameters that must be considered when selecting an organic material for building applications. They proposed it for the design and fabrication of compressed earth blocks (CET) added with coconut fibers. For didactic reasons, here is a generalized interpretation of the method described by the authors, adapted to define the main characteristics of the selected raw material:

- Material background
- Definition of the selected material
- Typical applications of the material
- Why use the selected material
- Extraction process
- Description of the extraction process
- Type of waste generated
- Cleaning of materials
- Definition of material contaminants
- Establishment of the cleaning process
- Extra treatments
- Which kind of treatments the material needs for its correct functionality
- Raw material management
- The management that the material must take to ensure its most prolonged preservation

It is prudent to clarify that the current social concerns about the environmental impacts of construction practices and materials have been expressed through an increasing demand, production, and use of sustainable building products (Corscadden et al., 2014). This practice allows having a methodologic control of the life cycle of the material since the beginning of its implementation is an advantage to create an ecologic architecture (Huberman and Pearlmutter, 2008).

19.4.2 DESIGN OF THERMAL INSULATORS

As expressed in Section 19.2.3, the efficiency of a thermal insulator is provided from the estimation of its thermal conductivity. This physical property measures the insulating capacity of a material; the lower the thermal conductivity, the better properties as thermal insulator material. This factor does not have a fixed value but depends on many factors, such as temperature, density, moisture, and deterioration of the material (Velázquez Rodríguez, 2015). So, the correct design of thermal insulations must consider these parameters besides the calculation of thermal conductivity:

Thermal conductivity depending on temperature. The thermal conductivity of materials presents variations with the environmental temperature. These variations over a few rates of temperature are negligible for some materials, but it is highly significative for other ones. The thermal conductivities of determinate solids show a surprising increase at closer temperatures to absolute zero when these solids become superconductors. The dependency of temperature from thermal conductivity causes considerable complexity in conduction analysis. Thus, it is a common practice to evaluate the thermal conductivity at the average temperature and treat it as constant in the calculations (Çengel, 2007).

19.4.2.1 DENSITY

For solid materials, density is defined as the relationship between mass and volume. The density concept could present the following two definitions.

Bulk density: defined as the quotient of mass over the volume of a sample without considering pores in the material. In the case of granular material, the terms particle density and particle volume are used.

Apparent density: defined as the relationship between the mass and volume of the material, including pores and water. The terms bulk density and bulk volume are used for granular materials (Rodríguez Ramíres et al., 2012).

19.4.2.2 MOISTURE ABSORPTION

When a material (it could be homogeneous or composite) is exposed to a moist environment, depending on the environmental and material conditions, the material absorbs or loses moisture as manifests by weight loss. This capacity is defined as moisture absorption, expressed in percent moisture content (percent weight gain) of the material as a function of time (Shen and Springer, 1976).

19.4.2.3 FACTORS OF DEGRADATION

The degradation of materials is the result of different physical, chemical, and biological effects. Deterioration also depends on the type of materials and products used and is influenced by specific conditions of processing and use (Moncmanová, 2007).

19.4.3 HEAT TRANSFER STUDY

The heat transfer studies seek to predict the transport of energy that can occur between material bodies, and it is represented as a temperature gradient (Holman, 1999).

ASTM proposed a methodology technique to evaluate heat transfer: the ASTM standard E1215-13 (ASTM, 2013). This method works with a test specimen inserted under load between two specimens of the same geometry of known thermal properties. It must be established a temperature gradient along with the system with a heating power source. At equilibrium conditions (steady-state), the temperatures are recollected with the use of temperature sensors assuming a linear heat distribution (Figure 19.1).

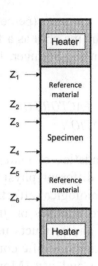

FIGURE 19.1 Thermal conductivity determination apparatus. Z1, Z2, Z3, etc., are the position of temperature sensor along with the system.

The approximated specimen thermal conductivity can be calculated using the following equations:

$$[q']T = \lambda \, M(T_2 - T_1)/(Z_2 - Z_1) \quad (19.5)$$

for the upper reference material and

$$[q']B = \lambda \, M(T_6 - T_5)/(Z_6 - Z_5) \quad (19.6)$$

for the lower reference material, where $q'T$, and $q'B$ are heat exchange in reference specimens; λ is the known thermal conductivity of reference materials; $T1$, $T2$, $T5$, and $T6$ are the temperatures registered by temperature sensors, and $Z1$, $Z2$, $Z5$, and $Z6$ are the position of temperatures sensors measured from the beginning of the upper reference material. From this formulation, it is possible to calculate the thermal conductivity with the following equation:

$$[\lambda']s = (([q']T + [q']_b)$$
$$(Z_4 + Z_3))/2(T_4 + T_3) \quad (19.7)$$

where λ's is the thermal conductivity of specimen, Z_3, and Z_4 are the position of temperatures sensors measured from the beginning of the upper reference material, and T_3, and T_4 are temperatures registered by temperature sensors.

19.4.4   A SIMULATION ANALYSIS

Using molecular dynamics (MD) simulations, it could be studied the influence of harmonic contributions to the atomic interaction potential on thermal conductivity and heat transport in 2D structures.

One fundamental problem in determining experimental thermal conductivity is the heat loss surroundings. In computational simulations, this problem can be solved by the imposition of periodic boundary conditions.

The next equation shows the development of a steady nano-equilibrium state, based on the use of Fourier's law of heat flow:

$$J = \langle \Delta E \rangle / A\Delta t = -\lambda \, 2\Delta T/((L/2)) \quad (19.8)$$

where, j = heat flux necessary to maintain the temperature difference, $2\Delta T$ = temperature difference, Δt = step used in the MD simulation, A = interface area of sample perpendicular to heat flux, $\langle \Delta E \rangle$ = average energy per time step Δt which is added and subtracted, respectively, in the layers representing heat contacts, and λ = thermal conductivity.

Based on the principle of Eq. (19.8) is possible to simulate several systems sizes L and to extrapolate thermal conductivity $L \rightarrow \infty$. Resuming, the goal of this kind of simulations is to find an agreement between theory and experimental thermal conductivity multiplying the analysis possibilities (Oligschleger and Schön, 1999).

19.4.5 THE CONSOLIDATION APPROACH

The thermal insulation from the consolidation theory viewpoint can be addressed in two topics: the study of the phenomena involved in thermal insulation, and the design and synthesis of thermal insulator materials.

The buildings are exposed to different internal and external conditions due to the differences between indoor and outdoor environments, as illustrated in Figure 19.2. Solar radiation, airflow, dust, moisture, and noise are the elements influencing the global outdoor environment; meanwhile, heat and carbon dioxide from human activities, oxygen from plants, dust, and particles from matter degradation, noise, and gaseous elements from kitchens are some of the elements determining the indoor environment. The internal and internal conditions should contribute to providing comfort sensation inside the buildings. Since these conditions

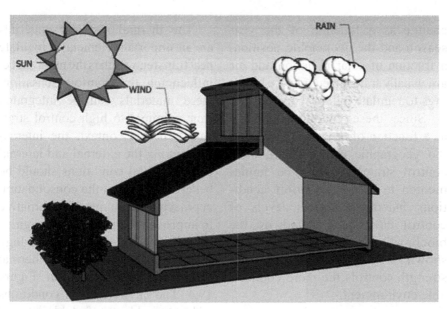

FIGURE 19.2 Internal and external conditions in the building.

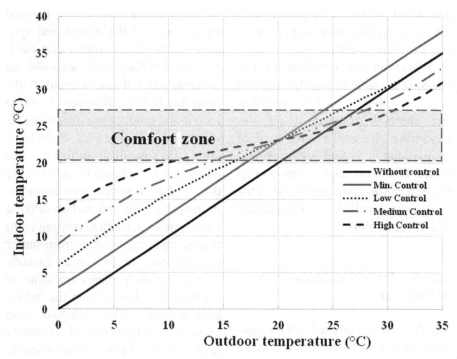

FIGURE 19.3   Indoor comfort zone as a function of a climate control strategy.

change as a function of the year season and the geographic position, a fraction of solar radiation and air are usually transferred into the buildings to regulate indoor temperature.

Since these conditions change as a function of the year season and the geographic position, a climate control strategy should be implemented to achieve comfort conditions indoors. Different levels of control have been stated by the bioclimatic design perspective, as shown in Figure 19.3, the more strength controls the more comfortable environment.

The thermal insulator materials are an important element to regulate heat transfer, which is the main aspect influencing the comfort sensation; these materials can be integrated from minimal to high control strategies. In this context, the interaction among the external and internal environmental conditions should be better studied, and the consolidation approach is an interesting alternative to improve the bioclimatic design.

A consolidated methodology to evaluate heat transfer in thermal insulators is presented in Figure 19.4. The environmental conditions and a desirable comfortable interval

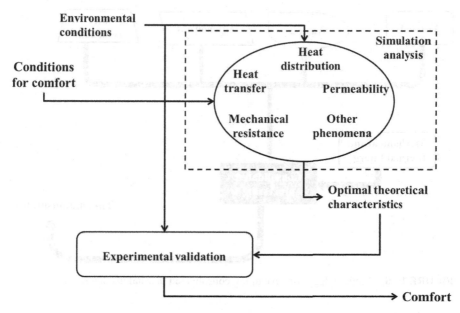

FIGURE 19.4 A strategy based on the consolidation theory to evaluate heat transfer in thermal insulators.

are used as parameters to perform a theoretical study to determine the best characteristics of a thermal insulator. This study is based on the consolidated theory fundamentals; such flow dynamic through insulator materials by Darcy's law, the Terzaghi–Taylor consolidation model and the Biot equations; the thermal conductivity by the Fourier law considering micro- and nano-conditions; the heat transfer by the ASTM standard methodology. From computer simulations, a series of parameters should be found, these are considered as the optimal theoretical characteristics which the insulator material must achieve in order to contribute to the climate control strategy. Among these characteristics, it can be mentioned the required transfer coefficient, the thermal conductivity, the density, viscosity, mechanical resistance, porosity, the consolidation constant, and others. The advantage of this theoretical study is that several scenarios can be evaluated without the need for experiments minimizing time, resources, and costs. Once the theoretical optimal characteristics are obtained, the next step is the experimental validation considering similar environmental conditions as for the theoretical studies; in this step, maybe some parameters need to be adjusted to achieve the comfort condition.

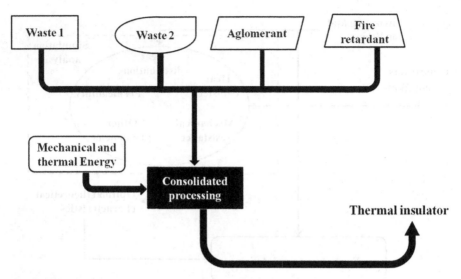

FIGURE 19.5   Methodology for developing consolidated thermal insulators.

Concerning the synthesis of thermal insulators, consolidation theory can also be applied. A general synthesis strategy is presented in Figure 19.5. The strategy is based on the combination of materials and components, promoting an optimal integration to obtain a new material that achieves efficient thermal insulation. In this case, it is proposed to use waste materials as key elements of the thermal insulator, for example, textile wastes, lignocellulosic materials, wasted polymers, and among others. Besides, at least the other two components are required, a fire retardant and an agglomerate promoter; the former is required to modify the combustion point of the whole material and then to increase the resistance against fire; the last component allows to promote the

integration, by an agglomerating effect, of all the materials involved in the thermal insulator synthesis. The integration is influenced by the controlled application of thermal and mechanical energy by following the principles of processes such as densification, composite production, sintering, and among others.

19.5 CONCLUSIONS AND FUTURES RESEARCH DIRECTIVES

Understanding of consolidation process in solid materials is essential to understanding the intrinsic behavior of solid materials. Making a regular practicing the consolidation tests in thermal insulation for buildings would scientifically extend its life cycle and allows its implementation as external covering systems that may be prepared to affront the constant variations in outdoor climatological conditions.

The use of organic waste as thermal insulation should be seen as an alternative solution. This kind of solution must always be analyzed and considered as the first option that must be implemented. To know the technical characteristics of raw materials is essential to ensure its correct disposal and its efficient approach.

The disagreements between theoretical and experimental results, in both cases for consolidation and heat transfer experiments, must always be taken into account, so the ideal situation is to have the possibility of realizing this double checkup to know with less uncertainty the characteristics of working materials.

Together with theoretical and experimental developments, the software simulations are a tremendous third-party mechanism to ensure a correct determination of physical properties.

KEYWORDS

- **thermal conductivity**
- **sustainable insulator**
- **heat transfer**
- **thermal comfort**

REFERENCES

Aboshi, H., Yoshikuni, H., Maruyama, S. Constant loading rate consolidation test. Soils Found. 1970, X, 43–56.

Ahmed, S.I., Siddiqua, S. A review on consolidation behavior of tailings. Int. J. Geotech. Eng. 2014, 8, 102–111. https://doi.org/10.1179/1939787913y.0000000012

Al-Homoud, M.S. Performance characteristics and practical applications of common building thermal insulation materials. Build. Environ. 2005, 40, 353–366. https://doi.org/10.1016/j.buildenv.2004.05.013

Alabdulkarem, A., Ali, M., Iannace, G., Sadek, S., Almuzaiqer, R. Thermal analysis, microstructure and acoustic characteristics of some hybrid natural insulating materials. Constr. Build. Mater. 2018, 187, 185–196. https://doi.org/10.1016/j.conbuildmat.2018.07.213

ASHRAE. Thermal environmental conditiosn for human occupancy, ASHRAE Standard, 2015. https://doi.org/1041-2336

ASTM. Standard test method for thermal conductivity of solids using the guarded-comparative-longitudinal heat flow technique. 2013. https://doi.org/10.1520/E1225-13.2.

Aubram, D. Explicitly coupled consolidation analysis using piecewise constant pressure. Acta Geotech. 2019, 15, 1–7. https://doi.org/10.1007/s11440-019-00792-z

Barberousse, A., Franceschelli, S., Imbert, C. Computer simulations as experiments. Synthese 2009, 169, 557–574. https://doi.org/10.1007/s11229-008-9430-7

Barden, L. Discussion: stresses and displacements in a cross-anisotropic soil. Géotechnique 1963, 13, 198–210. https://doi.org/10.1680/geot.1963.13.3.198

Bekkouche, S.M.A., Benouaz, T., Cherier, M.K., Hamdani, M., Benamrane, N., Yaiche, M.R. Thermal resistances of local building materials and their effect upon the interior temperatures case of a building located in Ghardaïa region. Constr. Build.

Mater. 2014, 52, 59–70. https://doi.org/10.1016/j.conbuildmat.2013.10.052

Biot, M.A. General solutions of the equations of elasticity and consolidation for a porous material. J. Appl. Mech. 1956, 23, 91–96.

Carrillo, N. Simple two and three dimensional case in the theory of consolidation of soils. Stud. Appl. Math. 1942, 21, 1–5. https://doi.org/10.1002/sapm19422111

Çengel, Y.A. Transferencia de calor y masa. McGrall-Hill/Interamericana Editores, S. A. de C. V., Madrid, Mexico, 2007.

Cetiner, I., Shea, A.D. Wood waste as an alternative thermal insulation for buildings. Energy Build. 2018, 168, 374–384. https://doi.org/10.1016/j.enbuild.2018.03.019

Chai, J.C., Carter, J.P., Hayashi, S. Ground deformation induced by vacuum consolidation. J. Geotech. Geoenvironmental Eng. 2005, 131, 1552–1561. https://doi.org/10.1061/(asce)1090-0241(2005)131:12(1552)

Cheng, A.H., Liggett, J.A. Boundary integral equation method for linear porous-elasticity with applications to soil consolidation. Int. J. Numer. Methods Eng. 1984, 20, 255–278. https://doi.org/10.1002/nme.1620200207

Christian, J.T., Baecher, G.B. D. W. Taylor and the foundations of modern soil mechanics. J. Geotech. Geoenviron. Eng. 2014, 141, 1–22. https://doi.org/10.1061/(asce)gt.1943-5606.0001249

Clayton, C.R.I., Müller Steinhagen, H., Powrie, W. A method of calculating the coefficient of permeability of clay from the variation of hydrodynamic stress with time. Procedia Inst. Civ. Eng. Geotech. Eng. 1995, 113, 191–205.

Çomakl, K., Yüksel, B. Environmental impact of thermal insulation thickness in buildings. Appl. Therm. Eng. 2004, 24, 933–940. https://doi.org/10.1016/j.applthermaleng.2003.10.020

Corscadden, K.W., Biggs, J.N., Stiles, D.K. Sheep's wool insulation: a sustainable alternative use for a renewable resource? Resour. Conserv. Recycl. 2014, 86, 9–15. https://doi.org/10.1016/j.resconrec.2014.01.004

Crawford, C.B. Soil consolidation—from theory to practice. Department of Civil Engineering, University of British Columbia, Vancouver, 1997.

Crawford, C.B., Morrison, K.I. Case histories illustrate the importance of secondary-type consolidation settlements in the Fraser River delta. Can. Geotech. J. 1996, 33, 866–878.

Cryer, C.W. A comparison of the three-dimensional consolidation theories of Biot and Terzaghi. Q. J. Mech. Appl. Math. 1963, XVI, 401–412. https://doi.org/10.1093/qjmam/16.4.401

Cuce, E., Cuce, P.M., Wood, C.J., Riffat, S.B. Optimizing insulation thickness and analysing environmental impacts of aerogel-based thermal superinsulation in buildings. Energy Build. 2014, 77, 28–39. https://doi.org/10.1016/j.enbuild.2014.03.034

de Dear, R., Brager, G.S. Developing an adaptive model of thermal comfort and preference. ASHRAE Trans. 1998, 104, 1–18.

DeGroot, D.J. Laboratory measurement and interpretation of soft clay mechanical behavior. In: Soil Behavior and Soft Ground Construction. ASCE: Reston, VA, 2004, 167–200.

Dimoudi, A., Tompa, C. Energy and environmental indicators related to construction of office buildings. Resour. Conserv. Recycl. 2008, 53, 86–95. https://doi.org/10.1016/j.resconrec.2008.09.008

Dissanayake, D.G.K., Weerasinghe, D.U., Wijesinghe, K.A.P., Kalpage, K.M.D.M.P. Developing a compression moulded thermal insulation panel using postindustrial textile waste. Waste Manag. 2018, 79, 356–361. https://doi.org/10.1016/j.wasman.2018.08.001

Dylewski, R., Adamczyk, J. Economic and environmental bene fits of thermal insulation of building external walls. Build. Environ. 2011, 46, 2615–2623. https://doi.org/10.1016/j.buildenv.2011.06.023

Epstein, Y., Moran, D.S. Thermal comfort and the heat stress indices. Ind. Health 2006, 44, 388–398.

Fernández García, F. Clima y confortabilidad humana. Aspectos metodológicos. Ser. Geográfica 1994, 4, 109–125.

Fleisher, L.A., Frank, S.M., Sessler, D.I., Cheng, C., Matsukawa, T., Vannier, C.A. Thermoregulation and heart rate variability. Clin. Sci. 1996, 90, 97–103. https://doi.org/10.1042/cs0900097

Fredlund, D.G. Unsaturated soil mechanics in engineering practice. J. Geotech. Geoenviron. Eng. 2006, 132, 286–321.

Fredlund, D.G., Hasan, J.U. One-dimensional consolidation theory: unsaturated soils. Can. Geotech. J. 1979, 16, 521–531.

Fredlund, D.G., Rahardjo, H., Fredlund, M.D. Unsaturated soil mechanics in engineering practice. John Wiley & Sons: Hoboken, NJ, 2012.

Gagge, A.P., Stolwijk, J.A.J., Hardy, J.D. Comfort and thermal sensations and associated physiological responses at various ambient temperatures. Environ. Res. 1967, 1, 1–20. https://doi.org/10.1016/ 0013-9351 (67)90002-3

Geethika, V.N., Rao, V.D.P. Study of tensile strength of Agave americana fibre reinforced hybrid composites. Mater. Today Proc. 2017, 4, 7760–7769. https://doi.org/ 10.1016/j.matpr.2017.07.111

Haapio, A., Viitaniemi, P. Environmental effect of structural solutions and building materials to a building. Environ. Impact Assess. Rev. 2008, 28, 587–600. https://doi.org/10.1016/j.eiar.2008.02.002

Han, G., Srebric, J., Enache-Pommer, E. Variability of optimal solutions for building components based on comprehensive life cycle cost analysis. Energy Build. 2014, 79, 223–231. https://doi.org/10.1016/j.enbuild.2013.10.036

Hernandez, P., Kenny, P. From net energy to zero energy buildings: defining life cycle zero energy buildings (LC-ZEB). Energy Build. 2010, 42, 815–821. https://doi.org/10.1016/j.enbuild.2009.12.001

Holman, J.P. Transferencia de calor, 10th ed. Continental, S. A. de C. V., México, D. F. 1999.

Huberman, N., Pearlmutter, D. A life-cycle energy analysis of building materials in the Negev desert. Energy Build. 2008, 40, 837–848. https://doi.org/10.1016/j.enbuild.2007.06.002

Humphreys, M.A., Nicol, J.F. Understanding the adaptive approach to thermal comfort. ASHRAE Trans. 1998, 104, 1–14.

Hwang, C.T., Morgenstern, N.R., Murray, D.W. On solutions of plane strain consolidation problems by finite element methods. Can. Geotech. J. 1971, 8, 109–118. https://doi.org/10.1139/t71-009

Inyim, P., Rivera, J., Zhu, Y. Integration of building information modeling and economic and environmental impact analysis to support sustainable building design. J. Manag. Eng. 2014, 31, A4014002-1. https://doi.org/10.1061/(asce)me.1943-5479.0000308

Jelle, B.P. Traditional, state-of-the-art and future thermal building insulation materials and solutions—properties, requirements and possibilities. Energy Build. 2011, 43, 2549–2563. https://doi.org/10.1016/ j.enbuild.2011.05.015

Jelle, B.P., Gustavsen, A., Baetens, R. The path to the high performance thermal building insulation materials and solutions of tomorrow. J. Build. Phys. 2010, 34, 99–123. https://doi.org/10.1177/1744259110372782

Jeremić, B., Cheng, Z., Taiebat, M., Dafalias, Y. Numerical simulation of fully saturated porous materials. Int. J. Numer. Anal. Methods Geomech. 2008, 32, 1635–1660. https://doi.org/10.1002/nag.687

Kabbaj, M., Tavenas, F., Leroueil, S. In situ and laboratory stress–strain relationships. Géotechnique 1988, 38, 83–100.

Kaswan, V., Choudhary, M., Kumar, P., Kaswan, S., Bajya, P. Green production strategies. In: Encyclopedia of Food Security and Sustainability. Elsevier: Amsterdam. 2019. https://doi.org/10.1016/ b978-0-08-100596-5.22292-0

Kaynakli, O. A review of the economical and optimum thermal insulation thickness for building applications. Renew. Sustain. Energy Rev. 2012, 16, 415–425. https://doi.org/10.1016/j.rser.2011.08.006

Korjenic, A., Petránek, V., Zach, J., Hroudová, J. Development and performance evaluation of natural thermal-insulation materials composed of renewable resources. Energy Build. 2011, 43, 2518–2523. https://doi.org/10.1016/j.enbuild.2011.06.012

Kreith, F., Manglik, R.M., Bohn, M.S. Principios de transferencia de calor, 7th ed. Cengage Learning, Inc.: Mexico City, 2012.

Kymäläinen, H., Sjöberg, A. Flax and hemp fibres as raw materials for thermal insulations. Build. Environ. 2008, 43, 1261–1269. https://doi.org/10.1016/j.buildenv.2007.03.006

Leroueil, S. The isotache approach. Where are we 50 years after its development by Professor Šuklje? in: Proceedings of the 13th Danube-European Conference on Geotechnical Engineering. Ljubljana, 2006, pp. 13–46.

Lertsutthiwong, P., Khunthon, S., Siralertmukul, K., Noomun, K., Chandrkrachang, S. New insulating particleboards prepared from mixture of solid wastes from tissue paper manufacturing and corn peel. Bioresour. Technol. 2008, 99, 4841–4845. https://doi.org/10.1016/j.biortech.2007. 09.051

Li, D.H.W., Yang, L., Lam, J.C. Zero energy buildings and sustainable development implications—a review. Energy 2013, 54, 1–10. https://doi.org/10.1016/j.energy.2013.01.070

Luiken, A., Bouwhuis, G. Recovery and recycling of denim waste. In: Denim: Manufacture, Finishing and Applications. Elsevier: Amsterdam, 2015. https://doi.org/10.1016/B978-0-85709-843-6.00018-4

Marszal, A.J., Heiselberg, P., Bourrelle, J.S., Musall, E., Voss, K., Sartori, I., Napolitano, A. Zero Energy Building—a review of definitions and calculation methodologies. Energy Build. 2011, 43, 971–979. https://doi.org/10.1016/j.enbuild.2010.12.022

Mathur, V.K. Composite materials from local resources. Constr. Build. Mater. 2006, 20, 470–477. https://doi.org/10.1016/j.conbuildmat.2005.01.031

Mayer, H., Höppe, P. Thermal comfort of man in different urban environments. Theor. Appl. Climatol. 1987, 38, 43–49. https://doi.org/10.1007/BF00866252

Moncmanová, A. Environmental deterioration of materials, 1st ed. WIT Press, Southhampton. 2007.

Mondal, S. Phase change materials for smart textiles—an overview. Appl. Therm. Eng. 2008, 28, 1536–1550. https://doi.org/10.1016/j.applthermaleng.2007.08.009

Monte, J.L., Krizekt, R.J. One-dimensional mathematical model for large-strain consolidation. Géotechnique 1976, 26, 495–510.

Morel, J.C., Mesbah, A., Oggero, M., Walker, P. Building houses with local materials: means to drastically reduce the environmental impact of construction. Build. Environ. 2001, 36, 1119–1126.

Morelli, J. Environmental Sustainability: a definition for environmental professionals. J. Environ. Sustain. 2011, 1, 1–10. https://doi.org/10.14448/jes.01.0002

Morgenstern, N.R., Nixon, J.F. One-dimensional consolidation of thawing soils. Can. Geotech. J. 1971, 8, 558–565. https://doi.org/10.1139/t71-057

Narayana, T. Municipal solid waste management in India: from waste disposal to recovery of resources? Waste Manag. 2009, 29, 1163–1166. https://doi.org/10.1016/j.wasman.2008.06.038

Neuman, S.P. Theoretical derivation of Darcy's law. Acta Mech. 1977, 25, 153–170. https://doi.org/10.1007/BF01376989

Nicol, F. Adaptive thermal comfort standards in the hot-humid tropics. Energy Build. 2004, 36, 628–637. https://doi.org/10.1016/j.enbuild.2004.01.016

Nicol, J.F., Humphreys, M.A. Adaptive thermal comfort and sustainable thermal standards for buildings. Energy Build. 2002, 34, 563–572. https://doi.org/10.1016/S0378-7788(02)00006-3

Oligschleger, C., Schön, J.C. Simulation of thermal conductivity and heat transport in solids. Phys. Rev. B 1999, 59, 4125–4133.

Omer, A.M. Renewable building energy systems and passive human comfort solutions. Renew. Sustain. Energy Rev. 2008, 12, 1562–1587. https://doi.org/10.1016/j.rser.2006.07.010

Ozel, M. Cost analysis for optimum thicknesses and environmental impacts of different insulation materials. Energy Build. 2012, 49, 552–559. https://doi.org/10.1016/j.enbuild.2012.03.002

Panyakaew, S., Fotios, S. New thermal insulation boards made from coconut husk and bagasse. Energy Build. 2011, 43, 1732–1739. https://doi.org/10.1016/j.enbuild.2011.03.015

Papadopoulos, A.M. State of the art in thermal insulation materials and aims for future developments. Energy Build. 2005, 37, 77–86. https://doi.org/10.1016/j.enbuild.2004.05.006

Papadopoulos, A.M., Giama, E. Environmental performance evaluation of thermal insulation materials and its impact on the building. Build. Environ. 2007, 42, 2178–2187. https://doi.org/10.1016/j.buildenv.2006.04.012

Parsons, K. Thermal comfort in buildings. In: Materials for Energy Efficiency and Thermal Comfort in Buildings, Hall, M.R. (Ed.). CRC, Cornwall, 2010, 127–147.

Pinto, J., Cruz, D., Paiva, A., Pereira, S., Tavares, P., Fernandes, L., Varum, H. Characterization of corn cob as a possible raw building material. Constr. Build. Mater. 2012, 34, 28–33. https://doi.org/10.1016/j.conbuildmat.2012.02.014

Rodríguez Ramíres, J., Méndez Lagunas, L., López Ortiz, A., Sandoval Torres, S. True density and apparent density during the drying process for vegetables and fruits: a review. J. Food Sci. 2012, 77, 145–154. https://doi.org/10.1111/j.1750-3841.2012.02990.x

Roux Gutierres, R.S., Espuna Mújica, J.A. Bloques de tierra comprimida adicionados con fibras naturales. Universidad Autónoma de Tamaulipas, Ciudad Victoria, 2012.

Sanjay, M.R., Madhu, P., Jawaid, M., Senthamaraikannan, P., Senthil, S., Pradeep, S. Characterization and properties of natural fiber polymer composites: a comprehensive review. J. Clean. Prod. 2018, 172, 566–581. https://doi.org/10.1016/j.jclepro.2017.10.101

Schiavoni, S., D'Alessandro, F., Bianchi, F., Asdrubali, F. Insulation materials for the building sector: a review and comparative analysis. Renew. Sustain. Energy Rev. 2016, 62, 988–1011. https://doi.org/10.1016/j.rser.2016.05.045

Shekhar, S. Darcy's law. Hydrogeology & Engineering Geology, 2017.

Shen, C., Springer, G.S. Moisture absorption and desorption of composite materials. J. Compos. Mater. 1976, 10, 1–20. https://doi.org/10.1177/002199837601000101

Shi, X.S., Herle, I., Wood, D.M. A consolidation model for lumpy composite soils in open-pit mining. Géotechnique 2017, 68, 1–16. https://doi.org/10.1680/jgeot.16.p.054

Singha, A.S., Rana, R.K. Natural fiber reinforced polystyrene composites: effect of fiber loading, fiber dimensions and surface modification on mechanical properties. Mater. Des. 2012, 41, 289–297. https://doi.org/10.1016/j.matdes.2012.05.001

Svagan, A.J., Samir, A.S.A., Berglund, L.A. Biomimetic foams of high mechanical performance based on nanostructured cell walls reinforced by native cellulose nanofibrils. Adv. Mater. 2008, 20, 1263–1269. https://doi.org/10.1002/adma.200701215

Tadesse, T. Solid waste management—for environmental and occupational health students. The Ethiopian Ministry of Health, and the Ethiopian Ministry of Education, Addis Ababa, 2004.

Taylor, D.W., Merchant, W. A theory of clay consolidation accounting for secondary compression. J. Math. Phys. 1940, 19, 167–185.

Velázquez Rodríguez, M. Materiales aislantes sostenibles. Universidad de Extrremadura, Extremadura. 2015.

Verruijt, A. Consolidation of soils. In: Encyclopedia of Hydrological Sciences. Wiley: Hoboken, NJ, 2005. https://doi.org/10.5860/choice.47-6894

Wakili, K.G., Bundi, R., Binder, B. Effective thermal conductivity of vacuum insulation panels. Build. Res. Inf. 2004, 32, 293–299. https://doi.org/10.1080/0961321042000189644

Wang, W., Zmeureanu, R., Rivard, H. Applying multi-objective genetic algorithms in green building design optimization. Build. Environ. 2005, 40, 1512–1525.https://doi.org/10.1016/j.buildenv.2004.11.017

Westerman, P.W., Bicudo, J.R. Management considerations for organic waste use in agriculture. Bioresour. Technol. 2005, 96, 215–221. https://doi.org/10.1016/j.biortech.2004.05.011

Wong, T.T., Fredlund, D.G., Krahn, J. A numerical study of coupled consolidation in unsaturated soils. Can. Geotech. J. 1998, 35, 926–937. https://doi.org/10.1139/t98-065

Ye, B., Ye, G., Zhang, F., Yashima, A. Experiment and numerical simulation of repeated liquefaction-consolidation of sand. Soils Found. 2007, 47, 547–558. https://doi.org/10.3208/sandf.47.547

Zeng, B., Wang, X., Byrne, N. Development of cellulose based aerogel utilizing waste denim—a Morphology study. Carbohydr. Polym. 2019, 205, 1–7. https://doi.org/10.1016/j.carbpol.2018.09.070

Zhou, Q., Zhou, H., Zhu, Y., Li, T. Data-driven solutions for building environmental impact assessment, in: Proceedings of the 2015 IEEE 9th International Conference on Semantic Computing, 2015, pp. 316–319. https://doi.org/10.1109/ICOSC.2015.7050826

Zienkiewicz, O.C., Chang, C.T. and, Bettess, P. Drained, undrained, consolidating and dynamic behaviour assumptions in soils. Géotechnique 1980, 30, 385–395. https://doi.org/10.1680/geot.1980.30.4.385

Catabolic Regulation of CCM in Bacteria for the Accumulation of Products of Commercial Interest

M. C. TAMAYO-ORDOÑEZ[1], B. A. AYIL-GÚTIERREZ[2],
F. A. TAMAYO-ORDOÑEZ[3], E. A. DE LA CRUZ-ARGUIJO[4],
J. C. CONTRERAS-ESQUIVEL[5], E. CÁZARES-SÁNCHEZ[6], and
Y. J. TAMAYO-ORDOÑEZ[7*]

[1]Laboratorio de Ingeniería Genética, Departamento de Biotecnología, Facultad de Ciencias Químicas, Universidad Autónoma de Coahuila, Ing J. Cardenas Valdez S/N, República, 25280 Saltillo, Mexico

[2]CONACYT—Centro de Biotecnología Genómica, Instituto Politécnico Nacional, Blvd. del Maestro, s/n, Esq. Elías Piña, 88710 Reynosa, Mexico

[3]Facultad de Química, Dependencia Académica de Ciencias Química y Petrolera, Universidad Autónoma del Carmen, Ciudad del Carmen, Carmen 47480, México

[4]Centro de Biotecnología Genómica, Instituto Politécnico Nacional, Blvd. del Maestro, s/n, Esq. Elías Piña, Reynosa 88710, México

[5]Laboratorio de Glicobiotecnologia Aplicada, Departamento de Investigación en Alimentos, Facultad de Ciencias Químicas, Universidad Autónoma de Coahuila, Ing J. Cardenas Valdez S/N, República, 25280 Saltillo, Mexico

[6]Instituto Tecnológico de la Zona Maya, carretera Chetumal Escárcega Km. 21.5. C.P. 77960 Ejido Juan Sarabia, México

[7]Estancia Posdoctoral Nacional-CONACyT, Posgrado en Ciencia y Tecnología de Alimentos, Facultad de Ciencias Químicas, Universidad Autónoma de Coahuila, Ing J. Cardenas Valdez S/N, República, 25280 Saltillo, Mexico

*Corresponding author. E-mail: yahatamayo@hotmail.es.

ABSTRACT

For centuries, people have used enzymes and other compounds produced by several model organisms, like microorganisms, plants, and animals, to obtain biotechnological products. The production of enzymes and different products of commercial interest can be obtained through the genetic manipulation of the metabolic pathways. The products obtained by the catabolic regulation of the metabolism of these microorganisms are widely applied in the industry for the production of food, beverages, drugs, animal feed, personal hygiene, products, pulp and paper, diagnosis, and therapy. This chapter describes the advances made in the genetic manipulation for catabolic regulation for the accumulation of commercial interest products in bacteria.

20.1 INTRODUCTION

In microorganisms, for an efficient metabolism to be carried out in a correct way, it is necessary to have a balance between catabolism and anabolism (Commichau et al., 2006; Papagianni, 2012). The bacteria to affect an efficient metabolism and synthesis of their cellular components need carbon, nitrogen, and other nutrients (Park et al., 2012; Marzluf, 1981).

Coordinated regulation of gene expression has long been studied for genes involved in the metabolism of compounds that are directly related to different branches of metabolism. Genetic engineering is a biotechnological tool that has helped to clarify basic processes of genetics and regulation of different metabolic pathways present in microorganisms (e.g., bacteria, yeast, fungi) (Papagianni, 2012; Wang et al., 2019). From the regulation of different metabolic pathways such as those of the central carbon metabolism (CCM) and nitrogen metabolism, it has been possible to obtain different products of commercial importance, such as proteins and amino acids, organic acids, metabolites for the food industry, and chemicals with use in therapy (Papagianni, 2012; Nielsen and Nielsen, 2017).

The understanding of CCM has been widely studied in most bacteria, and *E. coli* is a model of reference (Park et al., 2012). The CCM involves the synergy among several processes that involve the interaction of metabolic pathways, such as the phosphotransferase (PTS) system, pentose phosphate (PP) pathway, glycolysis, and the tricarboxylic acid cycle (TCA). Due to the complexity of processes that are unconnected and that are important for the regulation of CCM, the goals of researchers has been to select metabolic steps to reroute carbon fluxes toward precursors. This has been a difficult task difficult and many times leads to the results not expected because there is still ignorance of all the factors involved in the regulation of CCM.

On the other hand, the metabolism of nitrogen and carbon in microorganisms is interconnected by two major players that are the PII proteins and the PTS system. Many DNA-binding transcription has been described to regulate both carbon and nitrogen metabolism, suggesting that certain transcription factors could be the next goal to regulate the metabolism of nitrogen and carbon.

This chapter of the book addresses the genetic regulation advances that have been made in understanding the regulatory pathways of carbon in microorganisms important and biological models such as bacteria. Focusing on the regulation of carbon catabolite repression (CCR) in biological models, such as *Escherichia Coli*, *Bacillus subtilis*, *Pseudomonas*, and *Shinorhizobium meliloti*; the importance of the evolution of the genes of the CCR and its impact on the regulation of this process in bacteria and genetic engineering in the CCR to obtain different chemicals and fuels.

20.2 ADVANCES IN THE REGULATORY MECHANISMS RESPONSIBLE FOR CCR (CATABOLIC REPRESSION) IN BACTERIA

20.2.1 CCR: REPRESSION OF CARBON CATABOLITES IN BACTERIA

The repression of catabolites is a universal phenomenon, which is found in practically all living organisms, ranging from unicellular organisms (bacteria) and multicellular organisms (fungi, plants, and animals), although in bacteria there is a greater scientific advance in the phenomenon. The repression of the carbon catabolite (CCR) controls carbon metabolism and its interaction with other related pathways; for example, in a cell, the CRR is regulated by carbon sources used for the production of energy, and the conversion to biomass. Because of the complexity of this system, and the importance of regulatory mechanisms that affect metabolism, gene transcription, and signaling, interest has been aroused in knowing the basic molecular aspects of how CCR works (Gorke and Stulke, 2008; Kremling et al., 2015).

The repression of the carbon catabolite (CCR) is an important global regulator system in several bacteria that allows them to preferably use the most energy-efficient carbon source in a mixture (Gorke and Stulke, 2008). Catabolite repression carbon (CCR) is a phenomenon that involves preferential utilization of glucose as a carbon source. For years in *E. coli*, a mechanism was established that involves cyclic AMP and its receptor protein (CRP) known as cAMP–CRP complex; from this model, it has been assumed that this mechanism serves as a prototype for the repression of catabolic in all complex organisms (Karimova et al., 2004). However, recent studies have shown that this mechanism is restricted to enteric bacteria and their nearby organisms. The mechanisms of the repression of independent catabolites AMP cyclic are produced in other bacteria, yeasts, plants, and even in *E. coli*. Unicellular organisms such as *E. coli*, *B. subtilis*, and *Saccharomyces cerevisiae* exhibit multiple mechanisms of catabolite repression, and most of them are independent cyclic AMP (Stülke and Hillen, 1999; Inada et al., 1996).

Recently, the second hierarchy of CCR has been identified that leads to the preferential consumption of arabinose on xylose, mediated by regulator protein AraC is sensitive to arabinose. This CCR mechanism causes the preferential use of sugars such as arabinose, galactose, glucose, mannose, and xylose on a short-chain fatty acid (propionate)

(Park et al., 2012; Kang et al., 1998). Because other operons that metabolize sugar are positively regulated by the cAMP–CRP complex, their catabolic genes are expressed only after glucose depletion (Karimova et al., 2004). Although the repression of catabolite by glucose has been studied extensively, CCR, among other sugars other than glucose is still incomplete.

For example, the order of preference between pentose sugars has been established (arabinose followed by xylose followed by ribose). The complex transcriptional regulation of proteins such as AraC and XylR ensures the preferential use of arabinose and xylose on ribose (Kang et al., 1998). Recently, the regulatory mechanisms that lead to the preferential use of arabinose on xylose have been identified. Arabinose interacts with AraC and represses the xylose promoter, which causes cells to use arabinose first and then xylose. Xylose, in turn, inhibits the expression of genes that regulate the metabolism of D-ribose through the transcriptional repressor, XylR (Kang et al., 1998). These results demonstrate that the expression of genes that regulate the metabolism of a carbon source is regulated by other carbon sources in the culture medium. The hierarchy for the use of the carbon source is not only restricted to the glucose domain but also includes other sugars.

20.2.2 REGULATION OF CCR IN E. COLI, B. SUBTILIS, PSEUDOMONAS, AND SHINORHIZOBIUM MELILOTI MODELS

To date, it is believed that the CCR in *E. coli* is mediated by a complex of cyclic adenosine monophosphate (cAMP) and the receptor protein (CRP) (Karimova et al., 2004). The preferential use of glucose over other sugars is a central phenomenon of CCR in this microorganism. When glucose is the predominant sugar in the medium, the synthesis of cAMP is inhibited. When the substrates that induce catabolic repression are the sugars (mainly glucose) and the control over the catabolic genes for the metabolism of other compounds (e.g., lactose) is avoided the transcriptional activation of those genes. The key parts of this system are the IIA component (EIIAGlc or Crr) of the PEP–PTS system for the specific transport of sugars, the adenylate cyclase (AC) enzyme, the cAMP metabolite, and the transcriptional activator CRP or catabolite gene activator protein (CAP) (from "cAMP receptor protein" or "catabolite gene—activator protein") (Gorke and Stulke, 2008).

The PEP–PTS system (phosphoenolpyruvate (PEP) phosphotransferase (PTS)) is a multiprotein system that couples the phosphorylation of sugars with their transport through the membrane. This system is composed of three types of enzymes that carry out the phosphorylation chain; enzyme I (EI), histidine protein (HPr), and enzyme II (EII), the latter formed by domains A, B, and C. In the presence of glucose and some other sugars in the medium, the cytosolic EI component is autophosphorylated using PEP as a donor, and then transfer that phosphate group to a histidine residue of the (also soluble) HPr protein, which in turn donates it to the "IIA" domain (substrate-specific) that is part of the EII transporters. The phosphate group is transferred from the "A" domain of the EII transporter to the "B" domain and, from the latter, directly to the sugar, which is thus phosphorylated, and transported through the "C" domain of the membrane (Harwood et al., 1976; Feucht et al., 1980), Figure 20.1A.

In the phenomenon of the catabolic repression of *E. coli*, the state of phosphorylation of the EIIA component is fundamental. For the expression of catabolic genes from nonpreferential carbon sources (e.g., lactose), it is essential that phosphorylated-EIIA activates the enzyme AC, which generates cAMP. This metabolite binds to the CRP protein, which only then promotes the binding of the RNA polymerase to the promoters or facilitates the formation of the open complex, acting as well as an activator of the catabolic genes and operons (e.g., the *lac* operon for the use of lactose) for the use of carbon sources other than the "PTS sugars" (Harwood et al., 1976; Tagami and Aiba, 1998).

In the presence of glucose or other PTS sugars, the EIA component is mostly dephosphorylated because of the transfer of the phosphate group to the transported sugar. However, it has been shown that the phosphorylation status of EIIA depends not only on transport through PTS but on the pyruvate ratio: PEP present in the cell. If the amount of pyruvate is high concerning that of PEP, the EIIA component will be mostly dephosphorylated and vice versa. That explains that the EIIA component is mostly dephosphorylated not only when *E. coli* is grown in the presence of PTS sugars (glucose or mannitol) but also when the sources of carbon available to them (e.g., glucose-6-phosphate, lactose, or melibiose), despite access to the cell through alternative transporters, they are easily metabolizable to pyruvate and increase the pyruvate: PEP ratio (Hogema et al., 1997; Hogema et al., 1998), see Figure 20.1B.

Traditionally, it was thought that the use in *E. coli* of sugars PTS (as the glucose) would generate a majority of dephosphorylated EIIA and, therefore, low levels of cAMP and expression of alternative catabolic operons (e.g., *lac* operon).

Alternatively, the use of nonpreferential sugars (e.g., lactose) would cause a most phosphorylated EIIA (due to the absence of PTS transport), high levels of cAMP, and increased activation mediated by CRP. However, it has been shown that percentage of dephosphorylated EIIA is very similar in the presence of sugars that transport (mannitol, fructose) or not (lactose, melibiose) through PTS (Hogema et al., 1998) and, furthermore, that cAMP levels do not always have a direct relationship with the phosphorylation levels of EIIA or with the type of transport of sugars. Thus, the cells have lower concentrations of cAMP in the presence of glucose-6-phosphate, glucose, and gluconate cAMP levels are almost identical to the characteristics of the growth in glucose, but EIIA remains mostly phosphorylated (Hogema et al., 1997).

The repression of catabolic operons (e.g., the *lac* operon) is not, therefore, mainly driven by cAMP levels since levels of this metabolite similarly low to those recorded in the presence of glucose still allow for activation mediated by CRP (Inada et al., 1996). Instead, the mechanism of "inductor exclusion" is mediated by dephosphorylated EIIA. In that state, EIIA interacts, among others, with the LacY transporter and inactive, preventing the entry of lactose.

Since lactose functions as an inducer of its degradation path, the interruption of its transport implies repression of the *lac* genes (Inada et al., 1998). This mechanism of "exclusion of the inducer" is also applicable to the metabolism of the maltose, melibiose, arabinose, and galactose (Titgemeyer et al., 1994; Misko et al., 1987). The importance of the cAMP–CRP complex is not, however, nonexistent, since it is required for the expression of the catabolic genes, and it contributes to the catabolic repression through the activation of the expression of the genes for the EIIB and EIIC components of the glucose transporter (enhancing its entry into the cell if possible) (Kimata et al., 1997).

Finally, the cAMP–CRP complex not only regulates the expression of proteins, also regulates genes that code for small regulatory RNAs, as in the case of CyaR or Spot42 (De Lay and Gottesman 2009; Polayes et al., 1988). Spot42 is involved in the regulation of genes necessary for primary and secondary metabolism, the redox balance, and the consumption of various nonpreferential carbon sources (Beisel and Storz, 2011). For its part, CyaR is a small RNA that it binds to the Hfq protein, and it regulates the post-transcriptional expression of many genes.

For this part, *B. subtilis*, glucose constitutes the preferential carbon source, generating the response of catabolic repression in cells through the PEP–PTS system. In this case, the control over the catabolic genes for other compounds is exerted by a mechanism of transcriptional repression of these genes (Figure 20.1C).

In *B. subtilis*, the key elements of the whole process are the glycolytic intermediaries' fructose-1,6-bisphosphate and glucose-6-phosphate, the HPr protein of the PTS system, and the transcriptional repressor CcpA (from "catabolite e control protein A"). In this bacterium, the Hpr protein can be phosphorylated of two different residues in histidine-15 by the action of EI (as in *E. coli*) or in serine-46 thanks to the interaction with the protein bifunctional HPr kinase/phosphorylase (HPrK/P). Phosphorylation at histidine is part of the phosphate transfer chain up to the EII component and finally the transported sugar. Phosphorylation in serine has a purely regulatory function since it triggers the catabolic repression response. In a situation of carbon limitation, the cells accumulate inorganic phosphate, which enhances the phosphorylase activity of the HprK protein on HPr-Ser46-P (Mijakovic et al., 2002). The decrease in phosphorylated HPr in serine relaxes the catabolic repression process and allows greater phosphorylation of the same protein in histidine-15 for sugar transport mediated by the PTS system.

On the contrary, after the entry of glucose into the cells and their metabolization, fructose-1,6-bisphosphate is produced, which enhances the kinase activity of HPrK, resulting

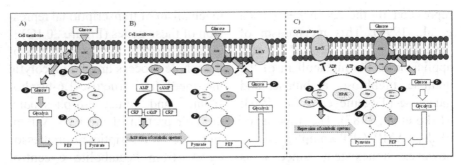

FIGURE 20.1 Global mechanisms of CCR in *E. coli* and *B. subtilis*. In *E. coli* and *B. subtilis*, the molecular mechanisms of global CCR regulation are different. In *E. coli*, CCR is mediated by the prevention of transcriptional activation of catabolic genes in the presence of glucose. By contrast, in *B. subtilis*, CCR is mediated by negative regulation through a repressor protein in the presence of glucose. Although the mechanisms of CCR differ in these two organisms, the PEP–carbohydrate PTS is important in both organisms in the signal-transduction pathways that lead to CCR. (A) The PTS is a multiprotein phosphorylation system that couples the transport of carbohydrates across the cytoplasmic membrane with their simultaneous phosphorylation. This type of active transport is characteristic of bacteria. The PTS is composed of at least three distinct proteins called: EI, HPr, and EII. EI initiates the phosphorylation chain by autophosphorylating with PEP, and the phosphoryl group is subsequently transferred to the His15 residue in HPr. HPr then donates the phosphoryl group to a histidine residue in the (A) domains of the various substrate-specific transporters or EIIs. Finally, the phosphoryl group is transferred to a residue in the EIIB domain and from there to the carbohydrate during its uptake through the membrane domain (EIIC). All phosphoryl transfer reactions between PTS proteins are reversible and the response to nutritional conditions and the metabolic state of the cell provides the basis for PTS-mediated signaling and regulation. In addition, HPr (His-P) controls the activities of metabolic enzymes and transcriptional regulators by modulating their phosphorylation. EI and some EIIs have regulatory functions: nonphosphorylated EI mediates chemotaxis toward PTS substrates and some EIIs regulate the activities of their cognate transcriptional regulatory proteins by phosphorylation. (B) General mechanism of the PEP carbohydrate PTS system in the CRR of *E. coli*. In the presence of high glucose levels, the transport of other metabolites is inhibited (e.g., LacY lactose transporter). The EIIA domain of the glucose transporter is the starting point of the CCR in *E. coli*. When phosphorylated, EIIA binds, and activates AC, which leads to the synthesis of cyclic AMP (cAMP). For the expression of genes involved in the metabolism of nonpreferential carbon sources, the presence of the route inducer (lactose) and the activation by CRP in the presence of cAMP produced by the enzyme AC are essential. High concentrations of cAMP trigger the formation of cAMP–CRP complexes, which bind and activate catabolic gene promoters. In its nonphosphorylated form, EIIAGlc cannot activate AC. (C) General mechanism of the PEP carbohydrate PTS system in the CRR of *B. subtilis*. In the presence of high glucose levels, the HPrK protein phosphorylates HPr in its serine residue (HPr Ser-P), a state that triggers the process of catabolic repression. HPr Ser-P binds to the CcpA protein, the formation of this complex binds to the *cre* site in the DNA, repressing the transcription of catabolic genes. The transcriptional regulator CcpA plays a fundamental role in the repression of genes for the metabolism of secondary carbon sources (Modified from Görke and Stülke, 2008; Stülke and Hillen, 1999).

in the phosphorylation of HPr in the serine residue (Reizer et al., 1998; Jault et al., 2000). In this state, the protein HPr-Ser46-P interacts with the regulator CcpA, acting as a cofactor and promoting its binding to the promoters of the regulated catabolic genes, whose expression is repressed. The interaction between HPr-Ser46-P and CcpA is enhanced by the binding of the glycolytic metabolites fructose-1,6-bisphosphate and glucose-6-phosphate (Schumacher et al., 2007). The CcpA regulator joins palindromic operators in the gene promoters called *cre* sites ("catabolite responsive elements"). Generally, these sites are located in the transcription initiation regions or superimposed on the consensus sequences of the promoters (Miwa et al., 2000). Also, sites upstream of the promoter can be encoded, with the CcpA protein acting as a transcriptional activator rather than as a repressor. For example, the genes *ackA* and *pta* whose protein products are involved in the excretion of acetate when the cells grow more than carbon (Grundy et al., 1993).

In addition to the control exercised by CcpA through direct binding to the *cre* sites, there are some examples of indirect regulation mediated by CcpA (Blencke et al., 2003). For example, some genes are induced in the presence of glucose in the glycolytic gap operon (involved in the production of PEP from glyceraldehyde-3-phosphate). It is known, in this case, that the deletion of the gene for CcpA generates an overactivity of HPrK in its kinase modality, producing an accumulation of HPr phosphorylated in serine. The absence of phosphorylated Hpr in the histidine residue prevents the transport of sugars through the PTS system, which ultimately ends up affecting the expression of numerous operons (such as the operon gap) that depend on glucose induction (Ludwig et al., 2002).

Finally, there are also some described cases of catabolic repression in *B. subtilis* where simultaneously, but independently of the action of CcpA, the process of "exclusion of the inducer" plays a fundamental role. For example, in the regulation of the expression of the glpFK operon (for the transport and phosphorylation of glycerol), it is known that the HPr-His15-P protein promotes the phosphorylation (and therefore the activity) of GlpK, the kinase necessary for glycerol-3P production (the operon inducer). In the presence of glucose, phosphorylation of GlpK mediated by HPr-His15-P does not occur, preventing the expression of the glpFK operon and, therefore, the transport of glycerol to the cell interior (Darbon et al., 2002).

With respect to the CCR regulation in the genus *Pseudomonas*, the molecular basis of catabolic repression in this genus differs substantially from that known for enterobacteria or *B. subtilis*. The first difference is established hierarchy in the use of various carbon sources: bacteria of this genus preferentially employ certain organic acids (e.g., succinate, pyruvate, and acetate) or amino acids instead of glucose. The CSV86 strain of *P. putida*, for example, metabolizes naphthalene in the first place even in the presence of glucose (Basu et al., 2006) and the only sugar transported through of PTS sugar transport system in *Pseudomonas* is fructose (Velazquez et al., 2007). In *P. putida*, a global role in the assimilation of certain compounds and the regulation of carbon and nitrogen metabolism has been demonstrated (Daniels et al., 2010; Milanesio et al., 2011) but it has not been possible to confirm any function in the phenomenon of catabolic repression. Among the elements involved in catabolic repression in *Pseudomonas* that are known are the Crc regulator (whose activity is in turn controlled by the CbrA and CbrB proteins indirectly, by the small crcZ RNA and, in some strains, also by the small RNA crcY), the terminal oxidase Cyo and the PTSNtr system (homologous to the classical PTS system).

The Crc regulator is a protein with a fundamental role in the control of the transport and metabolism of amino acids and sugars (Moreno et al., 2009) being essential in the global organization of metabolism.

In some strains, their involvement in mobility and biofilm formation has also been demonstrated (O'Toole et al., 2000). The mechanism of action of Crc (Moreno et al., 2012) consists of the post-transcriptional regulation of genes through the interaction with messenger RNAs. Recently it has been demonstrated that this interaction is not direct, but requires the formation of a ribonucleoprotein complex in which the presence of the RNA-binding protein Hfq is fundamental (Moreno et al., 2014). In *Pseudomonas*, the regulation of Crc depends on a small RNA, crcZ (and in the case of *P. putida* of also small crcY RNA) (Moreno et al., 2012; Filiatrault et al., 2013), and this interaction depends of Hfq (Moreno et al., 2014).

In addition, to the control by Crc in *P. putida*, it has been possible to demonstrate the participation of a terminal ubiquinol oxidase of the electron transport chain, called Cyo, in the process of catabolic repression associated with the degradation of phenol and alkanes (Petruschka et al., 2001). The inactivation of this oxidase, but not that of any of the other four-terminal oxidases posed by *P. putida*, attenuates the level of catabolic repression in the presence of succinate or amino acids.

Finally, there is in many bacteria (including *E. coli*), a system homologous to the classical PTS system has been described and was named

PTSNtr for its relation to the assimilation of nitrogen in *Klebsiella pneumoniae* (Reizer et al., 1992; Merrick and Coppard, 1989). This system, composed of the proteins PtsP, PtsO (or NPr), and PtsN, homologous to EI, HPr, and EII. In *P. putida*, this PTSNtr system is involucrate in the catabolic repression associated with the degradation of toluene and xylene in the presence of glucose or succinate. The inactivation of ptsN produces relaxation in the catabolic repression phenomenon without affecting the glucose consumption (Cases et al., 1999), while that of ptsO generates the opposite effect, repressing the catabolism of toluene even in the absence of glucose. The dephosphorylated form of PtsN (in His-68) has shown are involved in catabolic repression and the PtsO regulation.

Finally, in *Shinorhizobium meliloti* the carbon sources preferentially used by this bacterium are the dicarboxylic acids of four carbon atoms, principally the succinate, to the detriment of the sugars such as glucose, fructose, raffinose, or lactose. In the simultaneous presence of these two types of carbon sources, the bacterial cultures of *S. meliloti* manifest a diauxic growth generated by the use of the dicarboxylic acid. In the growth of *S. meliloti* are observed phase lag is set to the absence of net growth caused by metabolic and physiological changes required for adaptation to the use of the second carbon source. It is during the lag phase that the catabolic repression of certain routes relaxes to allow the use of secondary nutrients.

S. meliloti has a PTS system of type Ntr composed of the proteins EINtr (gene smc02437), Hpr (gene smc02754), and EIINtr (gene smc 01141) and the implication of these PTS components in the process of catabolic repression of the raffinose (agp) and lactose (lac) catabolic genes has been described (Pinedo et al., 2008; Pinedo and Gage, 2009). Also, the components EINtr and HPr are controlling the expression of the metabolism routes of α- and β-galactosides.

The Hpr is regulated by phosphorylation (in histidine-22) mediate by HprK protein. It is assumed that the function of HPrK is to decrease the phosphorylation in His-22 by EINtr since it is described that the serine residue is important for the interaction between HPr and EI (Deutscher et al., 2006).

In systems of mutants in HPrK show poor growth in several different carbon sources (succinate, fructose, lactose, raffinose, glucose, or glycerol), although not in rich medium, and induction of the *lac* or *gap* genes well below the normal even in the presence of lactose as only carbon sources (Pinedo et al., 2008; Pinedo and Gage, 2009; Muhammed et al., 2017). The same phenotype manifests itself in mutants before HPrK in which Ser-53 is replaced by an alanine suggesting that it is the accumulation of HPrK phosphorylated in His-22 responsible for these phenotypes. The authors propose that the HPrK protein of *S. meliloti* must be involved in the regulation of carbon metabolism through the control of the genes or enzymes necessary for transport and catabolism (Pinedo et al., 2008; Pinedo and Gage, 2009; Muhammed et al., 2017).

Also, deletion mutants in HPrK or those in which His-22 is substituted by an alanine show a decrease in intensity (not total elimination) of catabolic repression and a deficiency in growth based only on some nutrients (raffinose or maltose) (Pinedo et al., 2008; Pinedo and Gage, 2009; Muhammed et al., 2017).

Through the analysis of RNA-seq and promoter fusion data, it has been demonstrated that the PckR protei are a key regulator of CCM in *Sinorhizobium meliloti*; during growth with gluconeogenic substrates, PckR represses expression of the complete Entner–Doudoroff glycolytic pathway and induces expression of the *pckA* and *fbaB* gluconeogenic genes (Muhammed et al., 2017). Also, *in silico* studies predicted that PckR may regulate as many as 81 genes in *S. meliloti* (Galardini et al. 2015), suggesting PckR may be a global regulator of carbon metabolism.

20.3 PARTICIPATION OF REGULATORY PROTEINS IN THE CARBOHYDRATE TRANSPORT SYSTEM AND CAMP IN THE CCR

20.3.1 MECHANISMS FOR THE SPECIFIC INDUCTION OF CARBOHYDRATES IN BACTERIA

The mode of absorption of carbohydrates in bacteria can be diverse depending on the proteins involved in transport. The phosphorylation of carbohydrates is important for CCR. In the process of carbohydrate transport and concomitant phosphorylation participate PEP-dependent carbohydrate PTS system (PEP) (Marzluf, 1981). These systems included sugar-specific PTS permeases, also known as EII, and two general PTS proteins, EI and histidine-containing proteins (HPr), which participate in the phosphorylation of all PTS-carbohydrate transport.

Through processes such as the exclusion of the inductor, the expulsion of the inductor (Saier et al., 1996) or the control of the activity of the regulators by phosphorylation (Stülke et al., 1998), the presence of preferred substrates results in lack of expression of alternative routes. The exclusion of the inducer is a regulatory phenomenon by which a carbohydrate inhibits the uptake of another carbon source. In *E. coli*, the exclusion of the inducer is mediated by the enzyme IIA (EIIAglc) of the glycosic specification of the PTS (Table 20.1). When a PTS substrate is present (e.g., glucose), the phosphate group of the PTS proteins is drained to the incoming sugar. Consequently, EIIAglc exists predominantly in its nonphosphorylated form. This form of EIIAglc binds to sugar permeases without PTS, which are specific for lactose, maltose, melibiose, and glucose. As a result, the transport of these sugars is inhibited and is avoided by the formation of inductors (e.g., allolactose for the induction of lac).

Due to the lack of specific intracellular inducers, the respective operons are still poorly expressed. Interestingly, the interactions of EIIAglc permease are improved in the presence of related substrates, which guarantees economical use of the regulatory protein (Postma et al., 1993). Also, EIIAglc can bind to glycerol kinase, preventing the production of glycerol 3-phosphate, the inducer of the glycerol catabolic genes. Since glycerol enters *E. coli* via facilitated digestion and glycerol-3- phosphate is not a substrate for the glycerol facilitator, phosphorylation is required to trap glycerol within the cell. Consequently, the repression of the activity of glycerol kinase by EIIAglc also reduces glycerol uptake.

In Gram-positive bacteria, another component of the PTS was identified, the HPr (phospho-prosthetic transport protein) participates in the exclusion of the inducer. It was demonstrated in vivo that a form of HPr (PSer-HPr), phosphorylated at serine at position 46 by the HPr kinase, is essential for the repression of transport of sugar without PTS in *Lactobacillus casei* (Dossonnet et al., 2000) and *Lactococcus lactis* (Monedero et al., 2001). Another form of inducer control involving glycerol kinase identified in *B. subtilis* (Darbon et al., 2002). The activity of glycerol kinase is stimulated by PEP-dependent phosphorylation by HPr phosphohistidine (P-His-HPr) and the lack of this phosphorylation, for example, in the presence of PTS substrates, results in the prevention of uptake of glycerol and the repression of the use of glycerol.

In addition to the exclusion of the inducer, it is believed that the expulsion of the inducer contributes to the CCR in some Gram-positive bacteria (Saier et al., 1996). The phenomenon was detected during studies of sugar consumption using nonmetabolizable carbohydrates. This has been described in *L. lactis* and *L. casei* were showed that P-Ser-HPr does not mediate the expulsion of the inducer in these organisms (Monedero et al., 2001; Darbon et al., 2002).

Another way of controlling the specific induction of carbohydrates depends on the regulatory domain of

TABLE 20.1 Proteins Involved in Carbon Catabolite Repression (CCR).

Organism	Protein	Function	Target/s	Reference
Enteric bacteria	CAP (catabolite activator protein)	Activation	CAP sites	Brückner and Titgemeyer, 2002
Enteric bacteria	Cya (adenylate cyclase)	Production of cAMP	—	Brückner and Titgemeyer, 2002
Enteric bacteria	Crp (cAMP receptor protein)	Transcription activator	cAMP and Cya	Kremling et al., 2015
Enteric bacteria	EI^{glc} (glucose specific enzyme I)	Glucose phosphorylation, interaction with regulator(s)	Transport and phosphorylation of glucose	Kremling et al., 2015
Enteric bacteria	EIIA^{glc} (glucose specific enzyme IIA)	Stimulation	Sugar permeases, glycerol kinase	Brückner and Titgemeyer, 2002
Enteric bacteria	EIIBC^{glc} (glucose specific enzyme IIB, IIC)	Glucose phosphorylation, interaction with regulator(s)	Transport and phosphorylation of glucose	Kremling et al., 2015
Enteric bacteria	P-EIIA^{glc} (glucose specific enzyme IIA phosphorylation)	Transcripcional corepresor	Cya	Brückner and Titgemeyer, 2002
Enteric bacteria	LacI (Lactose operon repressor protein)	Transcription repressor	Operon lactose, response to the availability of the carbon source	Görke and Stülke, 2008
Enteric bacteria	AraC (Arabinose operon repressor protein)	Transcription repressor	Operon arabinose, response to the availability of the carbon source	Görke and Stülke, 2008
Enteric bacteria	Mlc (repressor protein)	Transcription repressor	Inhibit the synthesis of EIIBC	Kremling et al., 2009
Gram positive bacteria	HPrk/P (histidine containing phosphocarrier protein)	Phosphorylation of HPr at serine-46 allosteric inhibition	HPr	Brückner and Titgemeyer, 2002
Gram positive bacteria	CcpA (carbon catabolite protein)	Repression/activation	CcpA sites (cre: glucose kinase)	Brückner and Titgemeyer, 2002

TABLE 20.1 *(Continued)*

Organism	Protein	Function	Target/s	Reference
Gram positive bacteria	P-Ser-Hpr[a] (histidine containing phosphocarrier protein, phosphorylated serine residue)	Control of protein activities	CcpA, sugar permeases	Brückner and Titgemeyer, 2002
Enteric bacteria, Gram positive bacteria	P-His-HPr[a] (histidine containing phosphocarrier protein, phosphorylated histidine residue)	Glucose phosphorylation, interaction with regulator(s)	PRD regulators, glycerol kinase	Brückner and Titgemeyer, 2002
Gram negative bacteria	Crc (Catabolite repression control)	Post-transcriptional regulation	crcZ, Hfq and control of the transport and metabolism of amino acids and sugars	Moreno et al., 2012; Filiatrault et al., 2013
Gram negative bacteria	PtsP (intermediate of PTS sugar transport system)	EI homologous	Degradation of toluene and xylene in the presence of glucose or succinate	Cases et al., 1999
Gram negative bacteria	PtsO (intermediate of PTS sugar transport system)	Phosphocarrier protein NPr homologous	Toluene catabolism repressor protein	Cases et al., 1999
Gram negative bacteria	PtsN (intermediate of PTS sugar transport system)	EII homologous	Catabolic repression and in the PtsO regulation	Cases et al., 1999
Gram negative bacteria	EINtr (PTS component in the process of catabolic repression of the raffinose and lactose	Protein control the expression of the metabolism of galactosides α and β.	agp, lac catabolic genes	Pinedo et al., 2009
Gram negative bacteria	EIINtr (PTS component in the process of catabolic repression of the raffinose and lactose)	PtsN homologous	agp, lac catabolic genes	Pinedo et al., 2009

Enteric bacteria (*E. coli*), Gram positive bacteria (*Bacillus subtilis, Lactobacillus casei, Lactococcus lactis*), Gram negative bacteria *Pseudomonas putida, Klebsiella pneumoniae, Shinorhizobium meliloti*). PRD: PTS regulation domain.

[a]In *B. subtilis* and, most likely, in other bacilli, the HPr homologue Crh participates as P-Ser-Crh in CcpA corepression.

PTS (PRD), which is found in anti-terminators and activators (Stülke, 1998). PRD proteins mediate the specific induction of sugar, mainly from the catabolic carbohydrate operon PTS. The RNA or DNA binding activity of their N-terminal vector domains is controlled through two PRDs by multiple phosphorylations mediated by PTS in a complex manner. The related PTS permeases, which act negatively on the regulators, are necessary for the specific induction of sugar. Most, but not all, regulators containing PRD require phosphorylation mediated by P-His-HPr for their activity.

This positive regulation provides the opportunity to control the induction process by CCR. If phosphorylation of PRD is avoided, for example, by transporting another PTS substrate, induction does not occur. Therefore, the final result of the regulation mediated by PRD is the same as for the exclusion of the inductor, the prevention of the specific induction of sugar.

S. thermophilus is a specialized regulation mediated by PTS through a protein domain similar to PTS EIIA that is fused with a lactose transporter, not PTS. Phosphorylation of this PTS domain by P-His-HPr is not involved in the establishment of a sugar utilization hierarchy but modulates the transport of lactose in a self-regulating control circuit (Deutscher, 2008).

20.3.2 REGULATION OF CATABOLITES OF THE OPERON IN BACTERIA

In addition to the global mechanisms of CCR, there are specific mechanisms for the induction of the operon. These mechanisms involve either the formation or uptake of the specific inducer of the operon (exclusion of the inducer) or the activity of the specific transcription factors of the operon (prevention of ion induction). Both mechanisms have been described in *E. coli* and *B. subtilis*. For example, the mechanism of exclusion of the inducer in the lactose operon of *E. coli* and the prevention of induction in the bglPH operon of *B. subtilis* (Deutscher, 2008).

20.3.2.1 CONTROL OF CATABOLITES OF SUGAR TRANSPORTERS

The *lac* operon of *E. coli* is expressed only if one isomer lactose consisting of b-galactosidase binds and inactivates the repressor lac. The formation of this isomer requires the absorption of lactose that only is transported into the cell in the presence of glucose, that inactive the lactose permease (LacY) (Smirnova et al., 2007). EIIA Glc has a key role in the control of LacY activity: in the absence of glucose, EIIA Glc is phosphorylated and does not interact with LacY. In the presence of glucose, it is not phosphorylated. EIIA Glc can inactivate LacY (Nelson et al., 1983; Hogema et al., 1999). Interestingly, this interaction only occurs if lactose is present (Smirnova et al., 2007). Also, this mechanism involves the transport of the other secondary carbohydrate sources (maltose, melibiose, raffinose, and galactose) (Titgemeyer et al., 1994; Misko et al., 1987).

Control of the inhibition interaction between EIIAGlc and specific permeases has been described for Gram-positive bacteria. In *Lactobacillus brevis*, HPr (Ser-P) is formed in the presence of glucose binds and inactivates galactose permease (Djordjevic et al., 2001). For this part, *S. thermophilus* lactose permease is controlled by HPr (His-P)-dependent phosphorylation. This non-PTS permease contains to the domain that is similar to the EIIA domain of the PTS. In the absence of glucose, HPr (His-P) can phosphorylate this PTS-like domain, which activating the permease for lactose transport (Poolman et al., 1995). If glucose is present, HPr becomes available in Ser46 and can no longer activate the lactose permease (Gunnewijk and Poolman, 2000).

20.3.2.2 *CATABOLITE CONTROL OF TRANSCRIPTION FACTORS*

In both, *E. coli* and *B. subtilis,* several operons for the catabolism of PTS substrates are controlled by transcription regulators that contain duplicated PTS regulatory domains (PRDs). These regulators can act as transcription activators or as RNA-binding antitermination proteins. Their activity is modulated by the availability of their specific substrate and by the superimposed mechanism of CCR. The PRD-containing regulators perceive the information on substrate and glycose availability by PTS-dependent phosphorylation of their PRDs (Stülke et al., 1998). This type of regulation allows the hierarchical use of PTS sugars. Each of these regulatory systems consists of a PRD-containing transcription regulator and the sugar-specific cognate EII of the PTS. The PDR regulatory system for the *B. subtilis* has been described LicT antiterminator, which controls the expression of the bglPH operon for the use of aryl-β-glucosides. In the absence of β-glucosides, LicT is phosphorylated by the β-glucoside-specific EII on its first PRD (PRD1). This phosphorylation results in the inactivation of LicT. In the presence of β-glucosides, the incoming sugars become phosphorylated, and LicT is dephosphorylated and, therefore, activated (Tortosa et al., 2001). However, the activity of LicT is not only controlled by the specific substrate but also by the availability of glucose or other PTS sugars. In the presence of glucose, LicT is inactive (Krüger et al., 1996). To become active, LicT not only needs to be dephosphorylated on its PRD1, but it must also be phosphorylated on its PRD2. The phosphorylation of PRD2 is catalyzed by HPr (His-P), which is only available in the absence of glucose (Ludwig et al., 2002; Lindner et al., 2002). Thus, the phosphorylation state of HPr links the availability of PTS substrates to the activity of LicT.

For this part, glucose-regulated phosphorylation of PRDs has also been observed for other PRD-containing regulators, including the *E. coli* BglG antiterminator and the *B. subtilis* GlcT antiterminator that controls glucose transport does not require HPr-dependent phosphorylation for activity and is therefore active in the presence of glucose (Schmalisch et al., 2003).

20.3.3 TRANSCRIPTIONAL CONTROL OF THE SUGAR TRANSPORT SYSTEM AND THE CAMP IN THE CATABOLIC REPRESSION BY GLOBAL REGULATORS

One of the best-studied consequences of the availability of carbohydrates is the activation of global transcriptional control systems that are mechanistically different in enteric bacteria and Gram-positive bacteria.

In *E. coli*, CAP activates transcription at more than 100 promoters and is in some cases also involved in repression CAP needs the allosteric effector cAMP in order to bind efficiently to DNA. Global regulation by CAP responds to the intracellular amount of cAMP and using auto-regulation to CAP levels (Busby and Ebright, 1999; Ishizuka et al., 1994). The intracellular cAMP level, in turn, is adjusted by AC, whose activity depends on the phosphorylated form of EIIAglc (P-EIIAglc) (Nasser et al., 2001; Postma et al., 1993). Besides controlling carbohydrate catabolic genes, CAP is directly involved in the modulation of a large number of other cellular processes. It also exerts indirect control by influencing the expression of global regulators such as FIS (Saier et al., 1996) or by contributing to c factor selectivity (Colland et al., 2000).

In Gram-positive bacteria, CcpA is of central importance for global transcription control in CCR. Genome-wide analysis of CCR in *B. subtilis* estimated about 300 genes to be regulated by CcpA (Moreno et al., 2001). CcpA functions mainly as a repressor of transcription but activation is also documented. The regulator requires the corepressor P-Ser-HPr to bind efficiently to its operator sequence *cre* (catabolite responsive element). Consequently, CcpA activity responds to the level of P-Ser-HPr. In *B. subtilis* the proteins, ccpB and ccpC, participate in glucose repression (Jourlin-Castelli et al., 2000).

20.3.4 EFFICIENT REGULATION OF THE REPRESSION OF CARBON CATABOLITE (CCR) IN BACTERIA

Repressing carbon sources implicate that other carbohydrates who used genes are repressed in the presence of the favored substrate would not be able to trigger CCR. However, any carbohydrate entering glycolysis in Gram-positive bacteria will inevitably activate HPrK/P and turn on CcpA-dependent regulation and inducer exclusion. In *E. coli,* carbohydrates provoke an equivalent response by modulating cAMP and P-EIIAglc levels. Therefore, CCR results in autoregulatory restriction of sugar utilization (Brückner and Titgemeyer, 2002).

When *E. coli* cells grow on lactose, one could assume that lac operon expression would be at its maximum. Two autoregulatory ways, however, restrict lac expression under these conditions.

First, in contrast to a widespread belief, the level of cAMP in lactose-consuming *E. coli* remains even slightly lower than in glucose-grown cells, leading to an only moderate activation of the lac promoter by CAP (Inada et al., 1996). Secondly, since about half of EIIAglc exists in nonphosphorylated form in lactose-consuming cells, inducer exclusion is operative, and reduced lactose permease activity is the consequence (Hogema et al., 1999).

Gram-positive bacteria since CcpA-dependent regulation is triggered by glycolytic intermediates, all catabolic genes for glycolytic sugars harboring *cre* for CcpA regulation are subject to CcpA-mediated autoregulation. Also, inducer control may be exerted, provided the cognate sugar permease is sensitive to inhibition by P-Ser-HPr.

A third autoregulatory consequence of HPrK/P-mediated HPr phosphorylation has recently been described (Monedero et al., 2001). Mutant of *L. casei,* were hprK/P gene encoding an HPrK/P enzyme with strongly reduced and was expressed in *B. subtilis,* showed not growth on PTS sugars. The elevated level of P-Ser-HPr in the mutant strain apparently blocked PTS phosphate transfer. These results obtained with mutant HPrK/P strongly suggest that phosphorylation of HPr by HPrK/P reduces PTS phosphotransfer also in the wild-type. Also, regulatory phosphorylations by P-His-HPr enhancing the activity of glycerol kinase and PRD-containing regulators are reduced. Autoregulatory CCR seems to be important to protect cells from adverse effects caused by the uptake of excess carbohydrates.

20.4 THE EVOLUTIONARY IMPORTANCE OF THE CCR IN BACTERIA

The repression of carbon catabolite (CCR) in bacteria is generally considered as a regulatory mechanism to ensure the sequential use of carbohydrates. The selection of carbon sources is mainly carried out at the level of the specific induction of carbohydrates. Since, all carbohydrates catabolic genes or operons are regulated by specific control proteins and require inducers for high-level expression, direct control of the activity of the regulators or control of the formation of inductors is an effective measure repressed. In bacteria can establish a hierarchy of sugar utilization. In addition to the control of the processes of induction by CCR, the bacteria have developed global circuits of transcriptional regulation, in which the pleiotropic regulators are activated. These global control proteins, the CAP, also known as cAMP receptor protein in *E. coli* or the catabolite control protein (CcpA) in Gram-positive bacteria with low GC content, act on a large number of catabolic genes/operons. Since virtually any carbon source is capable of triggering a global transcriptional control, the expression of sugar utilization genes is restricted even in the exclusive presence of its related substrates. Consequently, the repression of catabolites dependent on CAP or CcpA serves as a self-regulating device to maintain sugar utilization at a certain level instead of establishing preferential utilization of certain carbon sources. Along with other mechanisms of self-regulation that do not act at the level of gene expression, CCR helps bacteria to adjust the use of sugar to their metabolic capacities (Brückner and Titgemeyer, 2002).

The CCR occurs in microorganisms in which the presence of a carbon source in the medium can suppress the expression of certain genes and operons, whose genetic products are often related to the use of alternative carbon sources (Stülke and Hillen, 1999). The classic example of *E. coli*, which uses glucose/lactose diauxie as the preferred carbon source, has been widely discussed. In contrast, the lactic acid bacterium *Streptococcus thermophilus* prefers lactose over glucose (Van Den Bogaard et al., 2000), indicating that adaptation to special ecological niches can result in the ability to utilize the variety of substrates. The final result of CCR is a uniform and reduced expression of certain genes and operons, the mechanisms that lead to repression can be very diverse. The presence of a repressor carbon source may result in lower concentrations of specific inductors for alternative routes of catabolism, altered activities of specific regulators, or the activation of global control proteins, among others.

The regulation of catabolic repression by carbon (CCR) can be achieved by different mechanisms, including activation/repression transcriptional and translational control by protein RNA bindings in different bacteria. In addition, the CCR regulates the expression of virulence factors in many pathogenic bacteria (Gorke and Stulke, 2008; Liu et al., 2005). The bacteria can cometabolize different carbon sources or can preferably use the most sugar accessible allowing faster growth. CCR is one of the most important regulatory phenomena in many bacteria (Blencke et al., 2003; Liu et al., 2005). The CCR is important for competition in natural environments since the selection of the preferred carbon source is a determining factor or in the rate of growth and, therefore, a competitive success with other microorganisms. On the other hand, CCR has a critical role in the expression of virulence genes, which often allow bacteria to access new sources of nutrients. The ability to select the carbon source that allows faster growth is the driving force for the evolution of CCR both in free-living bacteria and in pathogenic bacteria (Moreno et al., 2001; Yoshida et al., 2001).

CCR is observed in the majority of free-living heterotrophic bacteria, including autotrophic bacteria that repress genes for the fixation of carbon dioxide in the presence of organic carbon sources (Bowien and Kusian, 2002). However, some pathogenic bacteria, such as *Chlamydia trachomatis* and *Mycoplasma pneumoniae*, which are very well adapted to nutrient-rich environments, appear to lack CCR (Bowien and Kusian, 2002). These organisms have a small genome and are adapted to a few habitats and, therefore, lack the most regulatory phenomena. Another feature is the cofermentation of glucose, and other carbon sources produced in *Corynebacterium*, although the cofermentation is highly regulated (Wendisch et al., 2001; Frunzke et al., 2008). Finally, for some bacteria, such as *S. thermophilus, Bifidobacterium longum,* and *Pseudomonas aeruginosa,* reverse RCC in which glucose is used as a secondary carbon source, and genes for the utilization of glucose are repressed whenever preferred carbon sources are available (Collier, 1996; Van Den Bogaard et al., 2000).

20.5 ENGINEERING GENETICS IN THE CCR TO OBTAIN DIFFERENT CHEMICALS AND FUELS

Microbial production of chemicals and fuels from renewable resources such as land- and marine-derived biomass has attracted considerable attention in recent decades (Stülke and Arnaud, 1998; Tagami and Aiba, 1998). One of the challenges in realizing the potential of this emerging technology is designing microbial biocatalysts that can utilize multiple carbohydrate substrates (e.g., xylose, arabinose, galactose, mannose, sucrose) from these various renewable resources to facilitate the production of chemicals and fuels (Bothast et al., 1999; Chan et al., 2012).

Increased breakthrough in genetic engineering strategies has been the CCR regulation in bacteria, with objectives that have a higher affinity to use a type of sugar concerning another. For example, the rebuilt the Leloir pathway of galactose metabolism in *E. coli*, mediated the introduction of synthetic components, such as promoters and 50-UTR, to control the expression of each gene for eliminated the regulation arising from CCR at transcriptional and translational levels, has resulted in that the mutant strains resulting to enhance galactose uptake and assimilation rates and facilitate simultaneous utilization of glucose and galactose (Lim et al., 2013). The application of these strains would facilitate the production of various chemicals and fuels through the cofermentation of glucose and galactose.

Also, during the metabolism of natural microorganisms, glucose is converted to pyruvate through glycolytic pathways, such as the Embden-Meyerhof-Parnas pathway, Entner–Doudoroff pathway, and their variations, and is further converted to two molecules of acetyl-CoA (AcCoA) by oxidative decarboxylation. AcCoA is the building block of butanol, fatty acids, isoprenoids, amino acids, and a variety of complex compounds (Cordova and Alper, 2016; Matsumoto et al., 2017), so it obtains more AcCoA molecules that have been a primary objective for different products of commercial interest. Bogorad et al. (2013), designed the nonoxidative glycolytic pathway, which can transform 1 glucose into 3 AcCoA without loss of CO_2 during the metabolism, thus maximizing the carbon yield, and Wang et al. (2019), built a new pathway named EP-bifido pathway in *E. coli* by combining Embden-Meyerhof-Parnas Pathway, PP Pathway and "bifid shunt," to generate high yield acetyl-CoA from glucose. The engineered strains showed greatly improved PHB (poly-β-hydroxybutyrate) yield (from 26.0% to 63.7 mol%) and fatty acid yield (from 9.17 to 14.36%), suggesting that this strain can be used efficiently for the production of chemicals that use acetyl-CoA as a precursor.

20.6 CONCLUSION

The preferential use of the best available nutrition can be an adaptation that allows bacteria to survive in a competitive environment. However, the elimination of the repression of catabolites in industrial hosts is an important mechanism for increasing yields, particularly when lignocellulosic biomass is used as a substrate. To date, there is knowledge of the proteins that could be regulating carbon metabolism, as discussed in this chapter CAP, CcpA, Cya, HPrK/P, EIIAglc, HPr, GlkA, and CRE, among others. Although there is basic knowledge about the role that they play in carbon regulation, there are still bottlenecks to be overcome to generate mutant strains where the metabolic pathways are correctly regulated to obtain a specific product. Although most of the metabolic pathway engineering practices have focused on the modulation of the metabolic pathways of interest, the importance of directing studies to understand the overall main metabolic regulation in response to the specific pathway mutation for efficient metabolic engineering stands out. It may be also considered to modulate transcription factors for significant strain improvement (Millard et al., 2017; Matsuoka and Shimizu et al., 2012).

Futurity, the regulation of enzymes involved in the CCR could help improve the use of sugar varieties that can increase the yield of products obtained by different metabolic pathways, for example, degradation of different biomass, yields of different biotechnological products, such as biomass degradation, generation of high yield strains of different products of biotechnological interest.

KEYWORDS

- bacteria
- catabolic regulation
- enzymes
- carbohydrates

REFERENCES

Basu, A.; Apte, S.K.; Phale, P.S. Preferential utilization of aromatic compounds over glucose by *Pseudomonas putida* CSV86. *Appl. Environ. Microbiol.* **2006**, 72(3), 2226–2230.

Beisel, C.L.; Storz, G. The base-pairing RNA spot 42 participates in a multioutput feedforward loop to help enact catabolite repression in *Escherichia coli. Mol. Cell.* **2011**, 41(3), 286–297.

Blencke, H.M.; Homuth, G.; Ludwig, H.; Mäder, U.; Hecker, M.; Stülke, J. Transcriptional profiling of gene expression in response to glucose in *Bacillus subtilis*: regulation of the central metabolic pathways. *Metab. Eng.* **2003**, 5(2), 133–149.

Bogorad, I.W.; Lin, T.S.; Liao, J.C. Synthetic non-oxidative glycolysis enables complete carbon conservation. *Nature* **2013**, 502, 693–697.

Bothast, R.J.; Nichols, N.N.; Dien, B.S. Fermentations with new recombinant organisms. *Biotechnol. Prog.* **1999**, 15(5), 867–875.

Bowien, B.; Kusian, B. Genetics and control of CO2 assimilation in the chemoautotroph *Ralstonia eutropha. Arch. Microbiol.* **2002**, 178, 85–93.

Brückner, R.; Titgemeyer, F. Carbon catabolite repression in bacteria: choice of the carbon source and autoregulatory limitation of sugar utilization. *FEMS Microbiol. Lett.* **2002**, 209(2), 141–148.

Busby, S.; Ebright R.H. Transcription activation by catabolite activator protein (CAP). *J. Mol. Biol.* **1999**, 293, 199–213.

Cases, I.; Perez-Martin, J.; De Lorenzo, V. The IIANtr (PtsN) protein of *Pseudomonas putida* mediates the C source inhibition of the sigma54-dependent Pu promoter of the TOL plasmid. *J. Biol. Chem.* **1999**, 274(22), 15562–15568.

Commichau, F.M.; Forchhammer, K.; Stülke, J. Regulatory links between carbon and nitrogen metabolism. *Curr. Opin. Microbiol.* **2006**, 9(2), 167–172.

Colland, F.; Barth, M.; Hengge-Aronis, R.; Kolb A. C factor selectivity of *Escherichia coli* RNA polymerase: role for CRP, IHF and Lrp transcription factors. *EMBO J.* **2000**, 19, 3028–3037.

Collier, D.N.; Hager, P.W.; Phibbs, P.V. Jr. Catabolite repression control in the *Pseudomona. Res. Microbiol.* **1996**, 147, 551–561.

Cordova, L.T.; Alper, H.S. Central metabolic nodes for diverse biochemical production. *Curr. Opin. Chem. Biol.* **2016**, 35, 37–42.

Chan, S.; Kanchanatawee, S.; Jantama, K. Production of succinic acid from sucrose and sugarcane molasses by metabolically engineered *Escherichia coli. Biores. Technol.* **2012**, 103(1), 329–336.

diCenzo, G.; Checcucci, A.; Bazzicalupo, M. et al. Metabolic modelling reveals the specialization of secondary replicons for niche adaptation in *Sinorhizobium meliloti. Nat. Commun.* **2016**, 7, 12219.

De Lay, N.; Gottesman, S. The Crp-activated small noncoding regulatory RNA CyaR (RyeE) links nutritional status to group

behavior. *J. Bacteriol.* **2009**, 191(2), 461–476.

Darbon, E.; Servant, S.P.; Jamet, E.; Deutscher J. Antitermination by GlpP, catabolite repression via CcpA, and inducer exclusion elicited by P-GlpK dephosphorylation control *Bacillus subtilis* glpFK expression. *Mol. Microbiol.* **2002**, 43, 1039–1052.

Daniels, C.; Godoy, P.; Duque, E. et al. Global regulation of food supply by *Pseudomonas putida* DOT-T1E. *J. Bacteriol.* **2010**, 192(8), 2169–2181.

Deutscher, J.; Francke, C.; Postma, P.W. How phosphotransferase system-related protein phosphorylation regulates carbohydrate metabolism in bacteria. *Microbiol. Mol. Biol. Rev.* **2006**, 70(4), 939–1031.

Deutscher J. The mechanisms of carbon catabolite repression in bacteria. *Curr. Opin. Microbiol.* **2008**, 11, 87–93.

Djordjevic, G.M.; Tchieu, J.H.; Saier, M.H. Jr. Genes involved in control of galactose uptake in *Lactobacillus brevis* and reconstitution of the regulatory system in *Bacillus subtilis. J. Bacteriol.* **2001**, 183, 3224–3236.

Dossonnet, V.; Monedero, V.; Zagorec, M.; Galinier, A.; Pérez- Martínez, G.; Deutscher, J. Phosphorylation of HPr by the bifunctional HPr Kinase/P-Ser-HPr phosphatase from *Lactobacillus casei* controls catabolite repression and inducer exclusion but not inducer expulsion. *J. Bacteriol.* **2000**, 182, 2582–2590.

Feucht, B.U.; Saier, M.H. Jr. Fine control of adenylate cyclase by the phosphoenolpyruvate: sugar phosphotransferase systems in *Escherichia coli* and *Salmonella typhimurium. J. Bacteriol.* **1980**, 141, 603–610.

Filiatrault, M.J.; Stodghill, P.V.; Wilson, J. et al. CrcZ and CrcX regulate carbon source utilization in *Pseudomonas syringae* pathovar tomato strain DC3000. *RNA Biol.* **2013**, 10(2), 245–255.

Frunzke, J.; Engels, V.; Hasenbein, S.; Gätgens, C.; Bott, M. Co-ordinated regulation of gluconate catabolism and glucose uptake in *Corynebacterium glutamicum* by two functionally equivalent transcriptional regulators, GntR1 and GntR2. *Mol. Microbiol.* **2008**, 67, 305–322.

Grundy, F.J.; Waters, D.A.; Takova, T.Y. et al. Identification of genes involved in utilization of acetate and acetoin in *Bacillus subtilis*. *Mol. Microbiol.* **1993**, 10(2), 259–271.

Gorke, B; Stulke, J. Carbon catabolite repression in bacteria: many ways to make the most out of nutrients. *Nat. Rev. Microbiol.* **2008**, 6(8), 613–624.

Gunnewijk, M.G.; Poolman, B. Phosphorylation state of HPr determines the level of expression and the extent of phosphorylation of the lactose transport protein of *Streptococcus thermophilus*. *J. Biol. Chem.* **2000**, 275, 34073–34079.

Gunnewijk, M.G.; van den Bogaard, P.T.; Veenhoff, L.M.: Heuberger, E.H.; de Vos W.M.; Kleerebezem, M.; Kuipers, O.P.; Poolman, B. Hierarchical control versus autoregulation of carbohydrate utilization in bacteria. *J. Mol. Microbiol. Biotechnol.* **2001**, 3, 401–413.

Harwood, J.P. et al. Involvement of the glucose enzymes II of the sugar phosphotransferase system in the regulation of adenylate cyclase by glucose in *Escherichia coli*. *J. Biol. Chem.* **1976**, 251, 2462–2468.

Hogema, B.M.; Arents, J.C.; Inada, T. et al. Catabolite repression by glucose 6- phosphate, gluconate and lactose in *Escherichia coli*. *Mol. Microbiol.* **1997**, 24(4), 857–867.

Hogema, B.M.; Arents, J.C.; Bader, R. et al. Inducer exclusion in *Escherichia coli* by non-PTS substrates: the role of the PEP to pyruvate ratio in determining the phosphorylation state of enzyme IIAGlc. *Mol. Microbiol.* **1998**, 30(3), 487–498.

Hogema, B.M.; Arents, J.C.; Bader, R.; Postma, P.W. Autoregulation of lactose uptake through the LacY permease by enzyme IIAGlc of the PTS in *Escherichia coli* K-12. *Mol. Microbiol.* **1999**, 31, 1825–1833.

Inada, T.; Kimata, K.; Aiba, H. Mechanism responsible for glucose–lactose diauxie in *Escherichia coli*: challenge to the cAMP model. *Genes Cells.* **1996**, 1, 293–301.

Ishizuka, H.; Hanamura, A.; Inada, T.; Aiba H. Mechanism of the down-regulation of cAMP receptor protein by glucose in *Escherichia coli*: role of autoregulation of the crp gene. *EMBO J.* **1994**, 13, 3077–3082.

Jault, J.M.; Fieulaine, S.; Nessler, S. et al. The HPr kinase from *Bacillus subtilis* is a homo oligomeric enzyme which exhibits strong positive cooperativity for nucleotide and fructose 1,6-bisphosphate binding. *J. Biol. Chem.* **2000**, 275(3), 1773–1780.

John, R.P., Anisha, G.S., Nampoothiri, K.M.; Pandey, A. Micro and macroalgal biomass: a renewable source for bioethanol. *Biores. Technol.* **2011**, 102(1), 186–193.

Jourlin-Castelli, C.; Mani, N.; Nakano, M.M.; Sonenshein, A.L. CcpC, a novel regulator of the LysR family required for glucose repression of the citB gene in *Bacillus subtilis*. *J. Mol. Biol.* **2000**, 295, 865–878.

Kang, H.Y.; Song, S.; Park, C. Priority of pentose utilization at the level of transcription: arabinose, xylose and ribose operons. *Mol. Cells.* **1998**, 8, 318–323.

Karimova, G.; Ladant, D.; Ullmann, A. Relief of catabolite repression in a cAMP independent catabolite gene activator mutant of *Escherichia coli*. *Res. Microbiol.* **2004**, 155, 76–79.

Kimata, K.; Takahashi, H.; Inada, T. et al. cAMP receptor protein-cAMP plays a crucial role in glucose-lactose diauxie by activating the major glucose transporter gene in *Escherichia coli*. *Proc. Natl. Acad. Sci. USA* **1997**, 94(24), 12914–12919.

Kremling, A.; Geiselmann, J.; Ropers, D.; de Jong, H. Understanding carbon catabolite repression in *Escherichia coli* using quantitative models. *Trends Microbiol.* **2015**, 23(2), 99–109.

Kremling, A.; Kremling, S.; Bettenbrock, K. Catabolite repression in *Escherichia coli*—a comparison of modelling approaches. *FEBS J.* **2009**, 276(2), 594–602.

Krüger, S.; Gertz, S.; Hecker, M. Transcriptional analysis of bglPH expression in *Bacillus subtilis*: evidence for two distinct pathways mediating carbon catabolite repression. *J. Bacteriol.* **1996**; 178, 2637–2644.

Lim, H.G.; Seo, S.W.; Jung, G.Y. Engineered *Escherichia coli* for simultaneous utilization of galactose and glucose. *Biores. Technol.* **2013**, 135, 564–567.

Lindner, C.; Hecker, M.; Le Coq, D; Deutscher, J. *Bacillus subtilis* mutant LicT antiterminators exhibiting enzyme I- and HPrindependent antitermination affect catabolite repression of the bglPH operon. *J. Bacteriol.* **2002**, 184, 4819–4828.

Liu, M. et al. Global transcriptional programs reveal a carbon source foraging strategy by *Escherichia coli*. *J. Biol. Chem.* **2005**, 280, 15921–15927.

Ludwig, H.; Rebhan, N.; Blencke, H.M.; Merzbacher, M.; Stülke J. Control of the glycolytic gapA operon by the catabolite control protein A in *Bacillus subtilis*: a novel mechanism of CcpA-mediated regulation. *Mol. Microbiol.* **2002**, 45, 543–553.

Marzluf, G.A. Regulation of nitrogen metabolism and gene expression in fungi. *Microbiol. Rev.* **1981**, 45(3), 437.

Matsumoto, T.; Tanaka, T.; Kondo, A. Engineering metabolic pathways in *Escherichia coli* for constructing a "microbial chassis" for biochemical production. *Biores. Technol.* **2017**, 245, 1362–1368.

Matsuoka, Y.; Shimizu, K. Importance of understanding the main metabolic regulation in response to the specific pathway mutation for metabolic engineering of *Escherichia coli*. *Comput. Struct. Biotechnol. J.* **2012**, 3(4), e201210018.

Merrick, M.J.; Coppard, J.R. Mutations in genes downstream of the rpoN gene (encoding sigma 54) of *Klebsiella pneumoniae* affect expression from sigma 54-dependent promoters. *Mol. Microbiol.* **1989**, 3(12), 1765–1775.

Mijakovic, I.; Poncet, S.; Galinier, A. et al. Pyrophosphate-producing protein dephosphorylation by HPr kinase/phosphorylase: a relic of early life? *Proc. Natl. Acad. Sci. USA* **2002**, 99(21), 13442–13447.

Milanesio, P.; Arce-Rodriguez, A.; Munoz, A. et al. (**2011**). Regulatory exaptation of the catabolite repression protein (Crp)-cAMP system in *Pseudomonas putida*. *Environ. Microbiol.* 13(2), 324–339.

Millard, P.; Smallbone, K.; Mendes, P. Metabolic regulation is sufficient for global and robust coordination of glucose uptake, catabolism, energy production and growth in Escherichia coli. *PLoS Comput. Biol.* **2017**, 13(2), e1005396.

Misko, T.P.; Mitchell, W.J.; Meadow, N.D.; Roseman, S. Sugar transport by the bacterial phosphotransferase system. Reconstitution of inducer exclusion in *Salmonella typhimurium* membrane vesicles. *J. Biol. Chem.* **1987**, 262(33), 16261–16266.

Miwa, Y.; Nakata, A.; Ogiwara, A. et al. Evaluation and characterization of catabolite responsive elements (*cre*) of *Bacillus subtilis*. *Nuc. Acids Res.* **2000**, 28(5), 1206–1210.

Monedero, V.; Kuipers, O.P.; Jamet, E.; Deutscher, J. Regulatory functions of serine-46-phosphorylated HPr in *Lactococcus lactis*. *J. Bacteriol.* **2001**, 183, 3391–3398.Moreno, M.S.; Schneider, B.L.; Maile, R.R.; Weyler, W.; Saier, M.H. Jr. Catabolite repression mediated by the CcpA protein in *Bacillus subtilis*: novel modes of regulation revealed by whole-genome analyses. *Mol. Microbiol.* **2001**, 39, 1366–1381.

Moreno, R.; Fonseca, P.; Rojo, F. Two small RNAs, CrcY and CrcZ, act in concert to sequester the Crc global regulator in *Pseudomonas putida*, modulating catabolite repression. *Mol. Microbiol.* **2012**, 83(1), 24–40.

Moreno, R.; Hernandez-Arranz, S.; La Rosa, R. et al. The Crc and Hfq proteins of *Pseudomonas putida* co-operate in catabolite repression and formation of RNA complexes with specific target motifs. *Environ. Microbiol.* **2014**, 17(1), 105–118.

Moreno, R.; Martinez-Gomariz, M.; Yuste, L. et al. The *Pseudomonas putida* Crc global regulator controls the hierarchical assimilation of amino acids in a complete medium: evidence from proteomic and genomic analyses. *Proteomics* **2009**, 9(11), 2910–2928.

Muhammed, Z.; Østerås, M.; O'Brien, S. A.; Finan, T.M. A key regulator of the glycolytic and gluconeogenic central metabolic pathways in *Sinorhizobium meliloti*. *Genetics* **2017**, 207(3), 961–974.

Nasser, W.; Schneider, R.; Travers, A.; Muskhelishvili, G. CRP modulates fis transcription by alternate formation of activating and repressing nucleoprotein complexes. *J. Biol. Chem.* **2001**, 276, 17878–17886.

Nelson, S.O.; Wright, J.K.; Postma P.W. The mechanism of inducer exclusion. Direct interaction between purified IIIGlc of the phosphoenolpyruvate: sugar phosphotransferase system and the lactose carrier of *Escherichia coli*. *EMBO J.* **1983**, 2, 715–720.

Nielsen, J.C.; Nielsen, J. Development of fungal cell factories for the production of secondary metabolites: linking genomics and metabolism. *Synth. Syst. Biotechnol.* **2017**, 2(1), 5–12.

O'Toole, G.A.; Gibbs, K.A.; Hager, P.W. et al. The global carbon metabolism regulator Crc is a component of a signal transduction pathway required for biofilm development by *Pseudomonas aeruginosa*. *J. Bacteriol.* **2000**, 182(2), 425–431.

Papagianni, M. Recent advances in engineering the central carbon metabolism of industrially important bacteria. *Microb. Cell Fact.* **2012**, 11(1), 50.

Park, J.M.; Vinuselvi, P.; Lee, S.K. The mechanism of sugar-mediated catabolite repression of the propionate catabolic genes in *Escherichia coli*. *Gene* **2012**, 504(1), 116–121.

Petruschka, L.; Burchhardt, G.; Muller, C. et al. The cyo operon of *Pseudomonas putida* is involved in carbon catabolite repression of phenol degradation. *Mol. Genetic. Genom.* **2001**, 266(2), 199–206.

Pinedo, C.A.; Bringhurst, R.M.; Gage, D.J. *Sinorhizobium meliloti* mutants lacking phosphotransferase system enzyme HPr or EIIA are altered in diverse processes, including carbon metabolism, cobalt requirements, and succinoglycan production. *J. Bacteriol.* **2008**, 190(8), 2947–2956.

Pinedo, C.A.; Gage, D.J. HPrK regulates succinate-mediated catabolite repression in the gram-negative symbiont *Sinorhizobium meliloti*. *J. Bacteriol.* **2009**, 191(1), 298–309.

Polayes, D.A.; Rice, P.W.; Garner, M.M. et al. Cyclic AMP-cyclic AMP receptor protein as a repressor of transcription of the spf gene of *Escherichia coli*. *J. Bacteriol.* **1988**, *170*(7), 3110–3114.

Poolman, B.; Knol, J.; Mollet, B.; Nieuwenhuis, B.; Sulter G. Regulation of bacterial sugar-H+ symport by phosphoenolpyruvate-dependent enzyme I/HPr-mediated phosphorylation. *Proc. Natl. Acad. Sci. USA* **1995**, 92, 778–782.

Postma, P.W.; Lengeler, J.W.; Jacobson G.R. Phosphoenolpyruvate: carbohydrate phosphotransferase systems of bacteria. *Microbiol. Rev.* **1993**, 6, 543–594.

Reizer, J.; Hoischen, C.; Titgemeyer, F.; Rivolta, C.; Rabus, R.; Stülke, J.; Karamata, D.; Saier, M.H.J.; Hillen W. A novel protein kinase that controls carbon catabolite repression in bacteria. *Mol. Microbiol.* **1998**, 27(6), 1157–1169.

Reizer, J.; Reizer, A.; Saier, M.H. Jr.; et al. A proposed link between nitrogen and carbon metabolism involving protein phosphorylation in bacteria. *Protein Sci.* **1992**, 1(6), 722–726.

Saier, Jr. M.H.; Chauvaux, S.; Cook, G.M.; Deutscher, J.; Paulsen, I.T.; Reizer, J.; Ye, J.J. Catabolite repression and inducer control in Gram-positive bacteria. *Microbiol.* **1996**, *142*, 217–230.

Schmalisch, M.; Bachem, S.; Stülke, J. Control of the *Bacillus subtilis* antiterminator protein GlcT by phosphorylation: elucidation of the phosphorylation chain leading to inactivation of GlcT. *J. Biol. Chem.* **2003**, 278, 51108–51115.

Schumacher, M.A.; Seidel, G.; Hillen, W. et al. Structural mechanism for the fine-tuning of CcpA function by the small molecule effectors glucose 6-phosphate and fructose 1,6-bisphosphate. *J. Mol. Biol.*, **2007**, 368(4), 1042–1050.

Smirnova, I. et al. Sugar binding induces an outward facing conformation of LacY. *Proc. Natl. Acad. Sci. USA* **2007**, 104, 16504–16509.

Stülke, J.; Hillen, W. Carbon catabolite repression in bacteria. *Curr. Opin. Microbiol.* **1999**, 2, 195–201.

Stülke, J.; Arnaud, M.; Rapoport, G.; Martin-Verstrate, I. PRD—a protein domain involved in PTS-dependent induction and carbon catabolite repression of catabolic operons in bacteria. *Mol. Microbiol.* **1998**, 28, 865–874.

Tagami, H.; Aiba, H. A common role of CRP in transcription activation: CRP acts transiently to stimulate events leading to open complex formation at a diverse set of promoters. *EMBO J.* **1998**, 17(6), 1759–1767.

Titgemeyer, F.; Mason, R.E.; Saier, M.H. Jr. Regulation of the raffinose permease of *Escherichia coli* by the glucose-specific enzyme IIA of the phosphoenolpyruvate: sugar phosphotransferase system. *J. Bacteriol.* **1994**, 176(2), 543–546.

Tortosa, P. et al. Sites of positive and negative regulation in the *Bacillus subtilis* antiterminators LicT and SacY. *Mol. Microbiol.* **2001**, 41, 1381–1393.

Van Den Bogaard, P.T.; Kleerebezem, M.; Kuipers, O.P.; de Vos, W.M. Control of lactose transport, L-galactosidase activity, and glycolysis by CcpA in *Streptococcus thermophilus*: evidence for carbon catabolite repression by a non-phosphoenolpyruvate-dependent phosphotransferase system sugar. *J. Bacteriol.* **2000**, 182, 5982–5989.

Velazquez, F.; Pfluger, K.; Cases, I. et al. The phosphotransferase system formed by PtsP, PtsO, and PtsN proteins controls production of polyhydroxyalkanoates in *Pseudomonas putida. J. Bacteriol.* **2007**, 189(12), 4529–4533.

Wang, Q.; Xu, J.; Sun, Z.; Luan, Y.; Li, Y.; Wang, J. et al. Engineering *in vivo* EP-bifido pathway in *Escherichia coli* for high-yield acetyl-CoA generation with low CO2 emission. *Metab. Eng.* **2019**, 51, 79–87.

Wendisch, V.F.; de Graaf, A.A.; Sahm, H.; Eikmanns, B.J. Quantitative determination of metabolic fluxes during coutilization of two carbon sources: comparative analyses with *Corynebacterium glutamicum* during growth on acetate and/or glucose. *J. Bacteriol.* **2000**, 182, 3088–3096.

Yoshida, K.I. et al. Combined transcriptome and proteome analysis as a powerful approach to study genes under glucose repression in Bacillus subtilis. *Nuc. Acids Res.* **2001**, 29, 6683–6692.

CHAPTER 21

Limonene Biotransformation: An Efficient Strategy for the Production of Industrial Value Compounds

H. A. LUNA-GARCÍA[1,2], P. VILLARREAL QUINTERO[1,2], A. ILINÁ[1],
J. L. MARTÍNEZ-HERNÁNDEZ[1], E. P. SEGURA-CENICEROS[1],
C. N. AGUILAR[2], and M. L. CHÁVEZ-GONZÁLEZ[1,2*]

[1]*Nanobioscience Group, School of Chemistry, Universidad Autónoma de Coahuila, Blvd. V. Carranza esquina con José Cárdenas Valdés s/n Col, República Oriente, 25280 Saltillo, México*

[2]*Bioprocesses & Bioproducts Group, Food Research Department, School of Chemistry, Universidad Autónoma de Coahuila, Blvd. V. Carranza esquina con José Cárdenas Valdés s/n Col, República Oriente, 25280 Saltillo, México*

Corresponding author. E-mail: monicachavez@uadec.edu.mx.

ABSTRACT

The use of complete cells to carry out chemical biotransformations is a viable and efficient alternative for obtaining fine chemical compounds with applications in various industrial areas. Limonene is a monoterpene that has a very particular chemical structure and is considered as a substrate molecule for the synthesis of different compounds with extensive biological activities. This chapter deals with the use of microbial biotransformations for the production of limonene derivatives as a way of making compound production processes more efficient.

21.1 INTRODUCTION

Biotransformation is a process in which an organism changes a chemical substance by transforming it into a different one through a reaction or set of reactions in a complete biological system. These chemical modifications serve to microorganism as an adaptation mechanism to change that occurs in environmental conditions (Singh, 2017). If the chemical conversion of the substance is carried out with the help of an isolated enzyme, it is known as biocatalysis.

Many chemicals with applications in different industrial areas are originated by chemical synthesis. The chemical processes of obtaining traditional compounds usually involve multiple stages of transformation in which each of these stages uses a big amount of solvents that are corrosive and/or toxic to human health; also the chemical synthesis generates a large number of effluents that are discarded and constitute a serious problem of environmental pollution. In terms of efficiency, chemical synthesis processes have low recovery yields of the products of interest, due to the multiple stages of synthesis and how difficult it is to control each one of them (Lima et al., 2016).

Therefore, biotransformation processes offer advantages over traditional synthesis processes; biotransformation uses a complete cell and due to the enzymatic machinery that possesses the chemical transformation processes to occur with greater enanto- and regio-specificity. The high catalytic specificity of enzymes is reflected in the high yields of the desired products. In addition, the growth conditions of the microorganisms used are of low severity, being in most cases nontoxic. Biotransformation is a good way to the production of different compounds derived from a wide variety of substrates.

21.2 TERPENES

Terpenes are an interesting group of organic compounds that are derived from isoprene, that basically constituted by a basic unit of five carbon atoms. Terpenes are classified based on the number of isoprene units, terpenes with one unite of isoprene are called hemiterpenes, monoterpenes (2) sesquiterpenes (3), diterpenes (4), sesterpenes (5), triterpenes (6), tetraterpenes (8), and polyterpenes (more than 8 isoprene units) (Hegazy et al., 2020). Although these compounds present a common biosynthesis and structure, they do not perform the same functions (Lima et al. 2016).

When the chemical structure of terpenoid is modified, it is called a terpenoid (Tetali, 2019). Some of the modifications that can be involved are oxidation in carbon skeletons, ciclations, elimination, or displacements of protons and methyls. These modifications can give rise to variations in their structures, giving rise to modified terpenes, with other functional groups, such as alcohols, aldehydes, or ketones (Lima et al. 2016).

Terpenes and terpenoids are the main constituents in the essential oils of plants, which are produced as a mechanism of defense against predators (Ludwiczuk and Asakawa, 2019). These types of molecules can be extracted from plants through different methodologies. These methodologies are classified in classic or conventional methods, in where heat is used to evaporate the volatile fraction that contains the essential oils. Essential oils are also obtained through the use of alternative technologies of recovery, where many of these techniques have resulted in increasing the recovery yields of the oils, thanks to the high selectivity they present as well as the reduction of energy consumption or the emission of gases making these techniques a better economic and environmental option (De Matos et al., 2019). Terpenes can be located in different parts of plants such as leaves, fruits, flowers, roots, and trunk (Singh and Sharma, 2015) (Table 21.1).

21.3 TERPENES CLASSIFICATION

Terpenes have been classified according to the isoprene unit number, below are the terpene subclasses (Figure 21.1).

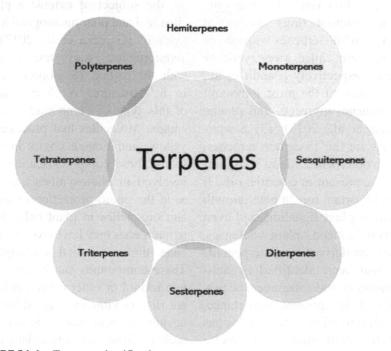

FIGURE 21.1 Terpenes classification.

(1) Hemiterpenes are the smallest terpenes, with 1 isoprene unit (Lima et al., 2016). There are about 50 compounds that are classified as hemiterpenes, of which the most prominent is the isoprene obtained from plant leaves. Some other known hemiterpenes are the acids: angelic, isovaleric, and senecioic (Ben Salha et al., 2019); another subclass of monoterpenes is (2) monoterpenes that are characterized by having 10 carbons. Monoterpenes and their derivatives are better known as components of volatile flower essences and as part of the essential oils of herbs and spices (Zebec et al., 2016). Monoterpenes can be classified according to the presence of carbon rings, for example, terpenes without rings are called "acyclic," while terpenes with one or two rings are called monocyclic or bicyclic, respectively (Pandit et al., 2015). Some of the most important are limonene, myrcene, and pinene (Materić et al., 2017). (3) Sesquiterpenes are the 15-carbon terpenes. Like monoterpenes, many sesquiterpenes are present in essential oils. It has an important role in plant growth regulators, plant signaling, and even herbivore-induced plant defenses. Some plant-derived sesquiterpenoids have also been identified as anti-inflammatory and anticancer species. Despite their apparent importance, the limited number of trade rules has hampered their study and precise

quantification (Duhamel et al., 2016). (4) Diterpenes contain 20 carbons in its structure. Diterpenes and their derivatives are known for their highly diverse structures and strong bioactivities, such as antitumor, anti-inflammatory, and neurotoxic activity (Zhou et al., 2014). (5) Sesterpenes: they are 25-carbon terpenes. Sesterpenes derivatives exhibit a variety of biological activities, such as cytotoxic, anti-inflammatory, and platelet aggregation inhibition effects (Meng et al. 2009). (6) Triterpenes: these are 30-carbon terpenes. Most triterpenes are colorless crystalline compounds characterized by low reactivity and high melting point. They are widely distributed in the plant kingdom and are the subject of extensive phytochemical and pharmacological investigations (Gajęcka et al., 2017). (7) Tetraterpenes: they are terpenes with 40 carbon (8 isoprene units) in its structure. A clear example of this type of compound is carotenoids. Molecules that produce red, yellow, and orange colors in some foods; these compounds are widely involved in photosynthesis, as well as in the photo-protection of important organelles in plant cells. Some tetraterpenes may have oxygen molecules that are called xanthophylls. These compounds can be converted into retinol or other compounds that are rich in vitamin A in addition to having different biological activities such as being an antioxidant agent

(Habtemariam, 2019); and, finally, (8) polyterpenes contain more than 8 units of isoprene, the most important are rubber and latex.

21.4 MONOTERPENES

Monoterpenes are a diverse group of volatile compounds of low molecular weight, great economic value, and they had been tested by its clinical and nutrition interest. Due to its chemical properties, monoterpenes are frequently used as aromatic additives in foods and perfumes (Rico Molins, 2015). From the classification of terpenes, monoterpenes stand out for being the most abundant in the essential oils of plants, flowers, seeds, etc.

The monoterpenes have strong antimicrobial activity, for example, the monoterpenes menthol, thymol, and linalyl acetate have shown strong activity against gram-positive bacteria (Bhatti et al., 2014).

Monoterpenes are the main constituents of essential oils found in many plants, fruits, vegetables, and flowers including stems, leaves, and roots. Essential oils are complex mixtures, usually liquid and volatile, and are extracted by water vapor steam from plants (Vásquez et al., 2001). In addition to natural sources, monoterpenes can be synthesized industrially through cell cultures, fermentation, and chemical synthesis (Rubulotta and Quadrelli, 2019).

21.5 LIMONENE AND ITS APPLICATIONS

Limonene is the most abundant monoterpene in nature. It is a natural substance that is extracted from the oil of citrus peels and confers the characteristic smell to citrus and is characterized by a colorless liquid (Chávez, 2015). Limonene is cyclohexene, with a methyl group on carbon 1, and an isopropenyl group on carbon 4 (Figure 21.2). The name of the structure corresponds to 4-isopropenyl-1-methyl-cyclohexene or 1-methyl-4-(1-methylethenyl)-cyclohexene. It has an asymmetric chiral carbon, giving rise to two optical isomers: R-limonene and S-limonene. Thus, it finds many applications both in the food industry and in cosmetics. However, due to its lipophilic character, it has low absorption, high oxidation, and hydrophobic character (Evageliou and Saliari, 2017).

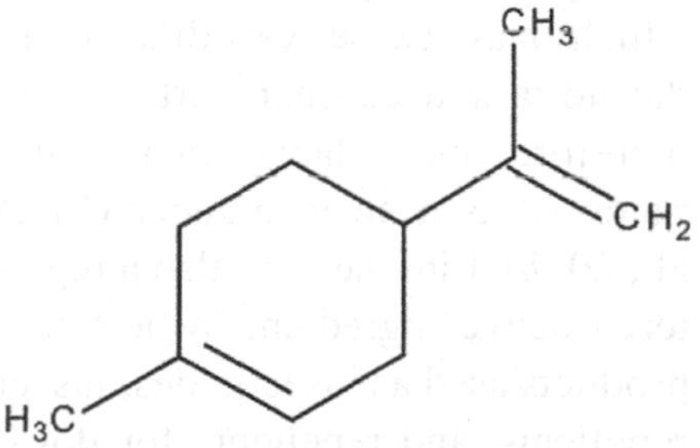

FIGURE 21.2 Chemical structure of limonene.

D-limonene has been considered as a green alternative potential to replace nonpolar organic compounds such as n-hexane for the extraction of bioactive compounds (Aissou et al., 2017). In addition to their therapeutic applications, limonene is of great industrial interest, being used in the pharmaceutical industry, as aromatizers and/or flavorings in the food and beverage industry, in perfumery, cosmetics, personal hygiene products, among others (Chávez, 2015) and is widely used in aromatherapy (Li et al., 2018). It is also used as a cleaning and degreasing agent in industrial and domestic applications (Aissou et al., 2017). Limonene has been used to prevent gastric diseases, for gallstones and exerts antiproliferative effects on various types of cancer cells. It has antioxidant, antigenotoxic (Bacanlı et al., 2015), antimicrobial (Hạc-Wydro et al., 2017), antiallergic and anti-inflammatory properties, is also an agonist ligand for alpha 2 receptors of adenosine, which may cause vasodilation in the aorta and coronary artery, and therefore may have therapeutic potential as a sleep inducer (Li et al., 2018). Limonene is also a registered active ingredient in pesticide products used as insecticides, insect repellents, and repellents for dogs, cats, and ornamental plants, as it is not toxic to humans and domestic animals, nor harmful to gardening, indoor plants, or the environment (Hebeish et al. 2008).

Limonene can be founded as the main component of oils obtained from citrus fruits such as orange, lemon, and grapefruit (Bacanli et al., 2015) and can also be found in the flowers of some plants, such as rosemary, eucalyptus, lavender, pine, cardamom, and among others (Li et al., 2018; Zahi et al., 2015). Limonene can be obtained through different methodologies. The classic methodologies are based on the use of solvents such as hexane among others, which can be potentially dangerous for the environment. Therefore, new ways are currently being sought to obtain limonene from other types of solvents that are not hazardous to the environment, such as isopropyl alcohol, cyclopentyl methyl ether, and among others (Ozturk et al., 2019).

Due to its chemical structure, limonene has been widely used as substrate molecules for the generation of other compounds with wide industrial applications. In general, these processes of obtaining derivative compounds take place through chemical synthesis (Wüst and Croteau et al., 2002). Among the most important compounds derived from limonene can be found: carveol, α-terpineol perylaldehyde, carvone, perylic acid, menthol, and perillyl alcohol. The chemical synthesis of these important compounds derived from limonene presents a series of drawbacks such as low yields in the synthesis processes due to the low specificity of the reactions (Tsao et al., 1995; Giguere et al., 1982).

21.6 LIMONENE BIOTRANSFORMATION

The use of biological systems, specifically the use of fungi, to carry out the biotransformation of monoterpenes such as limonene promises to be an alternative to organic chemical synthesis. A large number of microorganisms are capable of growing on a wide range of organic substrates (Sharma and Tripathi, 2008). It has been reported that genera such as *Aspergillus* and *Penicillium* can transform monoterpenes into interesting products (Chávez, 2015). Another advantage of biotransformation is that microorganisms have the capacity to grow at environmental temperatures, so the reactions carried out are in average conditions, so costs are reduced by lowering energy costs.

Biotransformation of limonene has disadvantages related to different factors such as (1) chemical instability, (2) low levels of solubility, (3) high volatilization capacity, (4) toxicity between substrate and product. Besides a possible low yield and the costs generated by the production, which can affect the production of compounds derived from terpenes, so optimization processes are necessary (Sales et al., 2019).

The products obtained from the biotransformation of terpenes are used in different industries such as cosmetics, food, pharmaceuticals, and among others. This makes this biotechnology flourish due to the advances that have been achieved recently. Microorganisms can chemically modify a wide variety of organic compounds in a specific way. An organic compound is modified in one or more reactions that are part of biological processes. For example, some vitamins, antibiotics, amino acids, and steroid hormones are obtained by biotransformation processes (Sales et al., 2019).

The biotransformation of different compounds such as terpenes is possible through biocatalysis carried out by enzymes, extract from different cells, or even complete cells, cyanobacteria, fungi, and yeasts. The use of these biological systems for the biotransformation of these compounds has its advantages and disadvantages, for example, the use of enzymes can be expensive but has the advantage of being the most sensitive method of all so that you can obtain the desired compound with greater security, on the other hand, the use of whole cells is more economical but has a risk of contamination (de Carvalho and da Fonseca, 2006).

The biotransformation of limonene results in oxidation, epoxidation, and hydroxylation modifications, obtaining derivatives, many of the commercial, and diverse utilities. Bier et al. (2017) carried out biotransformation work on limonene in synthetic and natural fermentation media using orange peel. The endophyte fungus *Phomopsis sp.* was used for fermentation. The main biotransformation products obtained in this work were a-terpineol, carvone and limonene-1,2-diol, and terpinen-4-ol (Table 21.1).

Molina et al. (2015) succeeded in biotransforming the *R*-(+)- and *S*-(−)-limonene isomers into products such as a-terpineol and limonene-1,2-epoxide, respectively, using the fungus *Fusarium oxysporum*. The use of microorganisms has been the object of study since the conditions for their optimal growth can be manipulated. There are reports of limonene biotransformation using bacteria, yeasts, and fungi; it has been seen that the fungal systems are much more effective, a multiplicity of biotransformation products (Table 21.2). Also, the culture media used and process control are much simpler.

During the limonene biotransformation process, different derivatives are produced and its application is varied, some of them are given below.

Perillyl alcohol is an oxygenated derivative that is appreciated for its application in the pharmaceutical industry and has a potent anticancer activity (Belanger, 1998; Gupta and Myrdal, 2004). It has

TABLE 21.1 Classification of Terpenes and Natural Sources

Terpenes	Sources	References
Hemiterpenes	Plants of family *Convolvulaceae* as *Convolvulus glomeratus* Choisy, *Evolvulus alsinoides (L.) L* y *Bonamia* sp.	Lopes et al. (2015)
Monoterpenes	Trees & plants of gender *Pinus*, fruits of gender *Citrus*, flowers *Hemerocallis minor* Mill, grapes (*Vitis vinifera* L.)	Chávez (2015), Zhao et al. (2017), Zhang et al. (2017)
Sesquiterpenes	*Inula montana* L., *Grangea maderaspatana*	Garayev et al. (2017), Chang et al. (2016)
Diterpenes	*Grangea maderaspatana, Persea indica*	Chang et al. (2016), Fraga et al. (2017)
Sesterterpenes	*Salvia dominica*	Dal Piaz et al. (2010)
Triterpenes	Fruits and vegetables as, apples, *Grangea maderaspatana*	Gajęcka et al. (2017), Chang et al. (2016)
Tetraterpenes	*Abies fabri, Botryococcus braunii*	Li et al. (2012), Delahais and Metzger (1997)

TABLE 21.2 Studies of Limonene Biotransformation Through Fungal Systems Reported in the Literature

Fungi	Biotransformation Products	References
Pleurotus sapidus	Carveol	Onken and Berger (1999)
	Carvone	
Phomopsis sp.	Limonene-1,2-diol	Bier et al. (2017)
	α-terpineol	
	terpinen-4-ol	
	(*R*)-(−)-Carvone	
Penicillium sp.	(+)α—terpineol	Demyttenaere et al. (2001)
	γ-terpinene	
Penicillium digitatum	(+)a-terpineol	Demyttenaere et al. (2001)
	óxido de *cis*-limonene	
	1,3,5-*p* mentatrien	
	exo-2-Hidroxicineol	
Corynespora cassiicola	(1*S*, 2*S*, 4*R*)-Limoneno-1,2-diol	Demyttenaere et al. (2001)
	R-(+)-a-Terpineol	
	cis-p-Ment-2-en-1-ol	
	trans-p-Ment-2-en-1-ol	
	neo-dihidrocarveol	
	óxido *cis*-limoneno	
	piperitona	
Aspergillus niger	Carveol	Chávez (2015)
Fusarium oxysporum	α-Terpineol	Molina et al. (2015)
	limoneno-1,2-epóxido	
Fusarium oxysporum	α-Terpineol	Maróstica and Pastore (2007)

been characterized by therapeutic properties against a wide variety of cancers (Chávez, 2015; Xu et al., 2004). Another important monoterpene is Perillaldehyde that can be found it in abundance in *Perilla frutescens* and citrus peel (Tian et al., 2015); this compound is used as a flavoring agent to add spice and citrus flavor to foods due to its characteristic smell and is being investigated for potential hypolipidemic, anti-inflammatory, neuroprotective, antidepressant, and antifungal effects (Hobbs et al., 2016).

Perillic acid is a monoterpenoid acid that it is present as a glycoside in *Perilla frutescens* and as a minor component in the essential oils of lemongrass, Citrus, and Perilla, exerts a strong growth-inhibiting effect on bacteria and moulds, making it an attractive candidate for natural use as a preservative in the cosmetics and food industry. Also, it has been found that perillic acid has acted as a therapeutic agent in the treatment of different cancer (Mirata et al., 2009; Yeruva et al. 2007). Carveol is another compound reported to be a fragrance ingredient used in the production of cosmetics, fine fragrances, shampoos, antiperspirants, soaps, body lotions, bath products, etc. It has been reported that carveol does not promote mutagenicity (Bhatia et al. 2008). Carvone is a flavoring substance widely used in pharmaceuticals, chewing gum, toothpaste, mouthwashes, foods, and beverages. Additionally, carvone is known as a fungicide, insecticide, and antibacterial agent (Barrera et al., 2008). It has been recognized as a chemopreventive agent against different types of cancer (Thamizharasi et al. 2019). α-terpineol is the most important commercially monocyclic terpenic alcohol, it is one of the 30 flavoring compounds with the highest world demand. α-terpineol can be found in two enantiomeric forms, each with a different odor, R-(+)-α-terpineol has a floral lilac aroma while S-(−)-terpineol has a coniferous wood aroma. Due to its characteristics, it is not only used as a perfume, insect repellent, antifungal, disinfectant, metal flotation agent but also has been reported effective as an analgesic (Gouveia et al., 2018; Dagne et al., 2000). Two limonene epoxides can be obtained from limonene, epoxy-1,2-limonene, and epoxy-8,9-limonene, which have been found naturally in some foods, such as mandarin and ginger (van der Werf et al., 2000). The terminal epoxide-8,9-limonene can be transformed into aldehyde, p-menten-9-al, which is used in the flavoring industry (Pinto et al., 2008; Santa et al., 2008;). *p*-cimene is another monoterpenoid widely used as an intermediate in the synthesis of fine chemicals for aromas, fragrances, perfumes, fungicides, pesticides, etc. Also, it is used as a starting material for the synthesis of p-cresol and nonnitrified musks (Kamitsou et al., 2014). Linalool is found in essential oils of plants and herbs that produce multiple effects in the mevalonate pathway and antiproliferative activity in cancer cells. Linalool acts as a sedative, anxiolytic, analgesic, anesthetic, anti-inflammatory, antioxidant, neuroprotective, antimicrobial, and antitumor (Rodenak-Kladniew et al., 2017). And finally, menthol has stood out the varied biological activities such as antimicrobial, analgesic effect, immunomodulatory, and chemopreventive properties (Luo et al., 2019; Suchodolsk et al., 2017). Menthol is well-known for its cooling effect or sensation when inhaled, chewed, consumed, or applied to the skin (Wright et al., 2019; Kamatou et al., 2013).

KEYWORDS

- terpenes
- terpenoids
- biotransformation
- microbial catalysis

REFERENCES

Aissou, M., Chemat-Djenni, Z., Yara-Varón, E., Fabiano-Tixier, A.-S., & Chemat, F. (2017). Limonene as an agro-chemical building block for the synthesis and extraction of bioactive compounds. *Comptes Rendus Chimie*, *20*(4), 346–358. https://doi.org/10.1016/j.crci.2016.05.018

Bacanli, M., Başaran, A. A., & Başaran, N. (2015). The antioxidant and antigenotoxic properties of citrus phenolics limonene and naringin. *Food and Chemical Toxicology*, *81*, 160–170. https://doi.org/10.1016/j.fct.2015.04.015

Barrera, R. de J., Alarcón, E. A., González, L. M., Villa, A. L., & Montes de Correa, C. (2008). Síntesis de carveol, carvona, verbenol y verbenona. *Ingeniería y Competitividad*, *10*(1), 43–63.

Belanger, J. T. (1998). Perillyl alcohol: applications in oncology. *Alternative Medicine Review*, *3*(6), 448–457.

Ben Salha, G., Abderrabba, M., & Labidi, J. (2019). A status review of terpenes and their separation methods. *Reviews in Chemical Engineering*, https://doi.org/10.1515/revce-2018-0066

Bhatti, H. N., Khan, S. S., Khan, A., Rani, M., Ahmad, V. U., & Choudhary, M. I. (2014). Biotransformation of monoterpenoids and their antimicrobial activities. *Phytomedicine*, *21* (12), 1597–1626. https://doi.org/10.1016/j.phymed.2014.05.011

Bhatia, S. P., McGinty, D., Letizia, C. S., & Api, A. M. (2008). Fragrance material review on carveol. *Food and Chemical Toxicology*, *46*(11), S85–S87. doi:10.1016/j.fct.2008.06.032

Bier, M. C. J., Medeiros, A. B. P., & Soccol, C. R. (2017). Biotransformation of limonene by an endophytic fungus using synthetic and orange residue-based media. *Fungal Biology*, *121*(2), 137–144. https://doi.org/10.1016/j.funbio.2016.11.003

Chang, F.-R., Huang, S.-T., Liaw, C.-C., Yen, M.-H., Hwang, T.-L., Chen, C.-Y., Hou, M.-F., Yuan, S.-S., Cheng, Y.-B., & Wu, Y.-C. (2016). Diterpenes from Grangea maderaspatana. *Phytochemistry*, *131*, 124–129. https://doi.org/10.1016/j.phytochem.2016.08.009

Chávez, M. L. (2015). *Aprovechamiento de residuos de la industria citrícola para la obtención de limoneno y su biotransformación*. Universidad Autónoma de Coahuila, Saltillo.

Dagne, E., Bisrat, D., Alemayehu, M., & Worku, T. (2000). Essential oils of twelve Eucalyptus species from Ethiopia. *Journal of Essential Oil Research*, *12*(4), 467–470. doi:10.1080/10412905.2000.9699567

Dal Piaz, F., Imparato, S., Lepore, L., Bader, A., & De Tommasi, N. (2010). A fast and efficient LC–MS/MS method for detection, identification and quantitative analysis of bioactive sesterterpenes in Salvia dominica crude extracts. *Journal of Pharmaceutical and Biomedical Analysis*, *51*(1), 70–77. https://doi.org/10.1016/j.jpba.2009.08.006

de Carvalho, C. C. C. R., & da Fonseca, M. M. R. (2006). Biotransformation of terpenes. *Biotechnology Advances*, *24*(2), 134–142. https://doi.org/10.1016/j.biotechadv.2005.08.004

De Matos, S. P., Teixeira, H. F., De Lima, Á. A. N., Veiga-Junior, V. F., & Koester, L. S. (2019). Essential oils and isolated terpenes in nanosystems designed for topical administration: a review. *Biomolecules*, *9*(4), 1–19. https://doi.org/10.3390/biom9040138

Delahais, V., & Metzger, P. (1997). Four polymethylsqualene epoxides and one acyclic tetraterpene epoxide from Botryococcus braunii. *Phytochemistry*, *44*(4), 671–678. https://doi.org/10.1016/S0031-9422 (96)00563-8

Demyttenaere, J. C., Van Belleghem, K., & De Kimpe, N. (2001). Biotransformation of (R)-(+)- and (S)-(−)-limonene by fungi and the use of solid phase microextraction for screening. *Phytochemistry*, *57*(2), 199–208. https://doi.org/10.1016/S0031-9422(00)00499-4

Duhamel, N., Martin, D., Larcher, R., Fedrizzi, B., & Barker, D. (2016). Convenient synthesis of deuterium labelled sesquiterpenes. *Tetrahedron Letters*, *57*(40), 4496–4499. https://doi.org/10.1016/j.tetlet.2016.08.079

Evageliou, V., & Saliari, D. (2017). Limonene encapsulation in freeze dried gellan systems. *Food Chemistry*, *223*, 72–75. https://doi.org/10.1016/j.foodchem.2016.12.030

Fraga, B. M., Terrero, D., Bolaños, P., & Díaz, C. E. (2017). Diterpenes with new isoryanodane derived skeletons from Persea indica. *Tetrahedron Letters*, *58*(23), 2261–2263. https://doi.org/10.1016/j.tetlet.2017.04.081

Gajęcka, M., Przybylska-Gornowicz, B., Zakłos-Szyda, M., Dąbrowski, M., Michalczuk, L., Koziołkiewicz, M., Babuchowski, A., Zielonka, Ł., Lewczuk, B., & Gajęcki, M. T. (2017). The influence of a natural triterpene preparation on the gastrointestinal tract of gilts with streptozocin-induced diabetes and on cell metabolic activity. *Journal of Functional Foods*, *33*, 11–20. https://doi.org/10.1016/j.jff.2017.03.019

Garayev, E., Herbette, G., Di Giorgio, C., Chiffolleau, P., Roux, D., Sallanon, H., Ollivier, E., Elias, R., & Baghdikian, B. (2017). New sesquiterpene acid and inositol derivatives from Inula montana L. *Fitoterapia*, *120*, 79–84. https://doi.org/10.1016/j.fitote.2017.05.011

Giguere, R. J., Von Ilsemann, G., & Hoffmann, H. M. R. (1982). Terpenes and terpenoid compounds. 9. Homologs of monocyclic monoterpenes. Tetramethylated derivatives of carvone, carveol.beta.-terpineol, sobrerol, and related compounds. *The Journal of Organic Chemistry, 47*(25), 4948–4954. doi:10.1021/jo00146a024

Gouveia, D. N., Costa, J. S., Oliveira, M. A., Rabelo, T. K., Silva, A. M. de O. e., Carvalho, A. A., Miguel-Dos-Santos, R., Lauton-Santos, S., Scotti, L., Scotti, M. T., Viana Dos Santos, M. R., Quintans-Júnior, L. J., De Albuquerque Junior, R. L. C., & Guimarães, A. G. (2018). α-Terpineol reduces cancer pain via modulation of oxidative stress and inhibition of iNOS. *Biomedicine & Pharmacotherapy*, *105*, 652–661. doi:10.1016/j.biopha.2018.06.027

Gupta, A., & Myrdal, P. B. (2004). Development of a perillyl alcohol topical cream formulation. *International Journal of Pharmaceutics*, *269*(2), 373–383. https://doi.org/10.1016/j.ijpharm.2003.09.026

Habtemariam, S. (2019). Introduction to plant secondary metabolites—from biosynthesis to chemistry and antidiabetic action. In: *Medicinal Foods as Potential Therapies for Type-2 Diabetes and Associated Diseases* (pp. 109–132). Elsevier: Amsterdam. https://doi.org/10.1016/B978-0-08-102922-0.00006-7

Hąc-Wydro, K., Flasiński, M., & Romańczuk, K. (2017). Essential oils as food ecopreservatives: model system studies on the effect of temperature on limonene antibacterial activity. *Food Chemistry*, *235*, 127–135. https://doi.org/10.1016/j.foodchem.2017.05.051

Hebeish, A., Fouda, M. M. G., Hamdy, I. A., EL-Sawy, S. M., & Abdel-Mohdy, F. A. (2008). Preparation of durable insect repellent cotton fabric: limonene as insecticide. *Carbohydrate Polymers*, *74*(2), 268–273. https://doi.org/10.1016/j.carbpol.2008.02.013

Hegazy, M.-E. F., Elshamy, A. I., Mohamed, T. A., Hussien, T. A., Helaly, S. E., Abdel-Azim, N. S., Shams, K. A., Shahat, A. A., Tawfik, W. A., Shahen, A. M., Debbab, A., El Saedi, H. R., Mohamed, A. El-H. H., Hammouda, F. M., Sakr, M., Paré, P. W., & Efferth, T. (2020). Terpenoid biotransformations and applications via cell/organ cultures: a systematic review. *Critical Reviews in Biotechnology*, *40*(1), 64–82. https://doi.org/10.1080/07388551.2019.1681932

Hobbs, C. A., Taylor, S. V., Beevers, C., Lloyd, M., Bowen, R., Lillford, L., Maronpot, R., & Hayashi, S. (2016). Genotoxicity assessment of the flavouring agent, perillaldehyde. *Food and Chemical Toxicology*, *97*, 232–242. https://doi.org/10.1016/j.fct.2016.08.029

Kamatou, G. P. P., Vermaak, I., Viljoen, A. M., & Lawrence, B. M. (2013). Menthol: a simple monoterpene with remarkable biological properties. *Phytochemistry*, *96*, 15–25. https://doi.org/10.1016/j.phytochem.2013.08.005

Kamitsou, M., Panagiotou, G. D., Triantafyllidis, K. S., Bourikas, K., Lycourghiotis, A., & Kordulis, C. (2014). Transformation of α-limonene into p-cymene over oxide catalysts: A green chemistry approach. *Applied Catalysis A: General*, *474*, 224–229. https://doi.org/10.1016/j.apcata.2013.06.001

Li, L. J., Hong, P., Jiang, Z. D., Yang, Y. F., Du, X. P., Sun, H., Wu, L. M., Ni, H., & Chen, F. (2018). Water accelerated transformation of d-limonene induced by ultraviolet irradiation and air exposure. *Food Chemistry*, *239*, 434–441. https://doi.org/10.1016/j.foodchem.2017.06.075

Lima, P. S. S., Lucchese, A. M., Araújo-Filho, H. G., Menezes, P. P., Araújo, A. A. S., Quintans-Júnior, L. J., & Quintans, J. S. S. (2016). Inclusion of terpenes in cyclodextrins: Preparation, characterization and pharmacological approaches. *Carbohydrate Polymers*, *151*, 965–987. https://doi.org/10.1016/j.carbpol.2016.06.040

Ludwiczuk, A., & Asakawa, Y. (2019). Bryophytes as a source of bioactive volatile terpenoids—a review. *Food and Chemical Toxicology*, *132*, 110649. *110649*. doi:10.1016/j.fct.2019.110649

Luo, Y., Sun, W., Feng, X., Ba, X., Liu, T., Guo, J., Xiao, L., Jiang, J., Hao, Y., Xiong, D., & Jiang, C. (2019). (−)-menthol increases excitatory transmission by activating both TRPM8 and TRPA1 channels in mouse spinal lamina II layer. *Biochemical and Biophysical Research Communications*, *516*, 825–830. doi:10.1016/j.bbrc.2019.06.135

Lopes, E. V., Dias, H. B., Torres, Z. E. dos S., Chaves, F. C. M., Siani, A. C., & Pohlit, A. M. (2015). Coumarins, triterpenes and a hemiterpene from Bonamia ferruginea (Choisy) Hallier f. *Biochemical Systematics and Ecology*, *61*, 67–69. https://doi.org/10.1016/j.bse.2015.04.034

Maróstica, M. R., & Pastore, G. M. (2007). Biotransformação de limoneno: Uma revisão das principais rotas metabólicas. *Quimica Nova*, *30*(2), 382–387. https://doi.org/10.1590/S0100-40422007000200027

Materić, D., Lanza, M., Sulzer, P., Herbig, J., Bruhn, D., Gauci, V., Mason, N., & Turner, C. (2017). Selective reagent ion-time of flight-mass spectrometry study of six common monoterpenes. *International Journal of Mass Spectrometry*, *421*, 40–50. https://doi.org/10.1016/j.ijms.2017.06.003

Meng, X.-J., Liu, Y., Fan, W.-Y., Hu, B., Du, W., & Deng, W.-P. (2009). The first synthesis of marine sesterterpene (+)-scalarolide. *Tetrahedron Letters*, *50*(35), 4983–4985. https://doi.org/10.1016/j.tetlet.2009.06.064

Mirata, M. A., Heerd, D., & Schrader, J. (2009). Integrated bioprocess for the oxidation of limonene to perillic acid with Pseudomonas putida DSM 12264. *Process Biochemistry*, *44*(7), 764–771. https://doi.org/10.1016/j.procbio.2009.03.013

Molina, G., Bution, M. L., Bicas, J. L., Dolder, M. A. H., & Pastore, G. M. (2015). Comparative study of the bioconversion process using R-(+)- and S-(−)-limonene as substrates for Fusarium oxysporum 152B. *Food Chemistry*, *174*, 606–613. https://doi.org/10.1016/j.foodchem.2014.11.059

Onken, J., & Berger, R. (1999). Effects of R-(+)-limonene on submerged cultures of the terpene transforming basidiomycete Pleurotus sapidus. *Journal of Biotechnology*, *69*(2/3), 163–168. https://doi.org/10.1016/S0168-1656(99)00040-1

Ozturk, B., Winterburn, J., & Gonzalez-Miquel, M. (2019). Orange peel waste valorisation through limonene extraction using bio-based solvents. *Biochemical Engineering Journal*, *151*, 107298. https://doi.org/10.1016/j.bej.2019.107298

Pandit, J., Aqil, M., & Sultana, Y. (2015). Terpenes and Essential Oils as Skin Penetration Enhancers. In *Percutaneous Penetration Enhancers Chemical Methods in Penetration Enhancement* (pp. 173–193). Springer: Berlin. https://doi.org/https://doi.org/10.1007/978-3-662-47039-8_11

Pinto, L. D., Dupont, J., de Souza, R. F., & Bernardo-Gusmão, K. (2008). Catalytic asymmetric epoxidation of limonene using manganese Schiff-base complexes immobilized in ionic liquids. *Catalysis Communications*, *9*(1), 135–139. doi:10.1016/j.catcom.2007.05.025

Rico Molins, J. (2015). Producción heteróloga de monoterpenos en Saccharomyces cerevisiae: selección y mejora de cepas mediante técnicas de ingeniería metabólica. https://doi.org/10.4995/Thesis/10251/49359

Rodenak-Kladniew, B., Islan, G. A., de Bravo, M. G., Durán, N., & Castro, G. R. (2017). Design, characterization and in vitro evaluation of linalool-loaded solid lipid nanoparticles as potent tool in cancer therapy. *Colloids and Surfaces B: Biointerfaces*, *154*, 123–132. https://doi.org/10.1016/j.colsurfb.2017.03.021

Rubulotta, G., & Quadrelli, E. A. (2019). Terpenes: a valuable family of compounds for the production of fine chemicals. *Studies in Surface Science and Catalysis*, *178*, 215–229. https://doi.org/10.1016/B978-0-444-64127-4.00011-2

Santa A., A. M., Vergara G., J. C., Palacio S., L. A., & Echavarría I., A. (2008). Limonene epoxidation by molecular sieves zinc-ophosphates and zincochromates. *Catalysis Today*, *133-135*, 80–86. doi:10.1016/j.cattod.2007.12.025

Sales, A., Moreira, R. C., Pastore, G. M., & Bicas, J. L. (2019). Establishment of culture conditions for bio-transformation of R-(+)-limonene to limonene-1,2-diol by Colletotrichum nymphaeae CBMAI 0864. *Process Biochemistry*, *78*, 8–14. https://doi.org/10.1016/j.procbio.2019.01.022

Sales, A., Pastore, G. M., & Bicas, J. L. (2019). Optimization of limonene biotransformation to limonene-1,2-diol by Colletotrichum nymphaeae CBMAI 0864. *Process Biochemistry*, *86*, 25–31. https://doi.org/10.1016/j.procbio.2019.07.022

Sharma, N., & Tripathi, A. (2008). Effects of Citrus sinensis (L.) Osbeck epicarp essential oil on growth and morphogenesis of Aspergillus niger (L.) Van Tieghem. *Microbiological Research*, *163*(3), 337–344. https://doi.org/10.1016/j.micres.2006.06.009

Singh, R. (2017). Microbial biotransformation : a process for chemical alterations. *Journal of Bacteriology & Mycology: Open Access*, *4*, 47–51. https://doi.org/10.15406/jbmoa.2017.04.00085.

Singh, B., & Sharma, R. A. (2015). Plant terpenes: defense responses, phylogenetic analysis, regulation and clinical applications. *3 Biotech*, *5*(2), 129–151. https://doi.org/10.1007/s13205-014-0220-2

Suchodolski J, Feder-Kubis J, Krasowska A. (2017). Antifungal activity of ionic liquids based on (−)-menthol: a mechanism study. *Microbiol Res. 197,* 56–64. doi: 10.1016/j.micres.2016.12.008.

Tetali, S. D. (2019). Terpenes and isoprenoids: a wealth of compounds for global use. *Planta.* 249(1), 1–8. doi: 10.1007/s00425-018-3056-x.

Thamizharasi, G., Sindhu, G., Veeravarmal, V., & Nalini, N. (2019). Preventive effect of D-carvone during DMBA induced mouse skin tumorigenesis by modulating xenobiotic metabolism and induction of apoptotic events. *Biomedicine & Pharmacotherapy, 111*, 178–187.

Tian, J., Wang, Y., Zeng, H., Li, Z., Zhang, P., Tessema, A., & Peng, X. (2015) Efficacy and possible mechanisms of perillaldehyde in control of Aspergillus niger causing grape decay. *International Journal of Food Microbiology, 202*, 27–34. doi:10.1016/j.ijfoodmicro.2015.02.022

Tsao, R., Lee, S., Rice, P. J., Jensen, C., & Coats, J. R. (1995). Monoterpenoids and their synthetic derivatives as leads for new insect-control agents. *Synthesis and Chemistry of Agrochemicals IV, 584*, 312–324. doi:10.1021/bk-1995-0584.ch028

van der Werf, M. J., Keijzer, P. M., & van der Schaft, P. H. (2000). Xanthobacter sp. C20 contains a novel bioconversion pathway for limonene. *Journal of Biotechnology, 84*(2), 133–143. https://doi.org/10.1016/S0168-1656(00)00348-5

Vásquez R. O., Alenguer, A., & Marreros V, J. (2001). Extracción y Caracterización del Aceite Esencial de Jengibre (Zingiber officinale). *Revista Amazónica de Investigación Alimentaria, 1*(1), 38–42. Retrieved from http://www.unapiquitos.edu.pe/pregrado/facultades/alimentarias/descargas/vol1/6.pdf

Velasco B., R., Montenegro M., D. L., Vélez S., J. F., García P., C. M., & Durango R., D. L. (2009). Biotransformación de compuestos aromáticos sustituidos mediante hongos filamentosos fitopatógenos de los géneros Botryodiplodia y Collecotrichum. *Revista de la Sociedad Química del Perú, 75*(1), 94–111.

Wüst, M., & Croteau, R. B. (2002). Hydroxylation of Specifically Deuterated Limonene Enantiomers by Cytochrome P450 Limonene-6-Hydroxylase Reveals the Mechanism of Multiple Product Formation†. *Biochemistry, 41(6)*, 1820–1827. doi: 10.1021/bi011717h

Wright, A., Benson, H. A. E., & Moss, P. (2019). Development of a topical menthol stimulus to evaluate cold hyperalgesia. *Musculoskeletal Science and Practice, 41*, 55–63. doi:10.1016/j.msksp.2019.03.010

Xu, M., Floyd, H. S., Greth, S. M., Chang, W.-C. L., Lohman, K., Stoyanova, R., Kucera, G. L., Kute, T. E., Willingham, M. C., Miller, M. S. (2004). Perillyl alcohol-mediated inhibition of lung cancer cell line proliferation: potential mechanisms for its chemotherapeutic effects. *Toxicology and Applied Pharmacology, 195*(2), 232–246. https://doi.org/10.1016/j.taap.2003.11.013

Yeruva, L., Pierre, K. J., Elegbede, A., Wang, R. C., & Carper, S. W. (2007). Perillyl alcohol and perillic acid induced cell cycle arrest and apoptosis in non small cell lung cancer cells. *Cancer Letters, 257*(2), 216–226. doi:10.1016/j.canlet.2007.07.020

Zahi, M. R., Liang, H., & Yuan, Q. (2015). Improving the antimicrobial activity of d-limonene using a novel organogel-based nanoemulsion. *Food Control, 50*, 554–559. https://doi.org/10.1016/j.foodcont.2014.10.001

Zebec, Z., Wilkes, J., Jervis, A. J., Scrutton, N. S., Takano, E., & Breitling, R. (2016). Towards synthesis of monoterpenes and derivatives using synthetic biology. *Current Opinion in Chemical Biology, 34*, 37–43. https://doi.org/10.1016/j.cbpa.2016.06.002

Zhang, E., Chai, F., Zhang, H., Li, S., Liang, Z., & Fan, P. (2017). Effects of sunlight exclusion on the profiles of monoterpene biosynthesis and accumulation in grape exocarp and mesocarp. *Food Chemistry, 237*, 379–389. https://doi.org/10.1016/j.foodchem.2017.05.127

Zhao, X., Guo, Y., Zhang, Y., Xie, Y., Yan, S., Jin, H., & Zhang, W. (2017). Monoterpene derivatives from the flowers of the Hemerocallis minor Mill. *Phytochemistry Letters*, *21*, 134–138. https://doi.org/10.1016/j.phytol.2017.06.006

Zhou, S.-Z., Yao, S., Tang, C., Ke, C., Li, L., Lin, G., & Ye, Y. (2014). Diterpenoids from the flowers of *Rhododendron molle*. *Journal of Natural Products*, *77*(5), 1185–1192. https://doi.org/10.1021/np500074q

CHAPTER 22

Mangiferin: Biological Diversity, Properties, and Biotechnological Applications

J. C. TAFOLLA-ARELLANO[1], R. ROJAS MOLINA[2],
J. M. TIRADO-GALLEGOS[3], E. OCHOA-REYES[4], R. BAEZA-JIMÉNEZ[4] and
J. J. BUENROSTRO-FIGUEROA[4*]

[1]Departamento de Ciencias Básicas, Laboratorio de Biotecnología y Biología Molecular. Núcleo Básico del Doctorado en Ciencias en Recursos Fitogenéticos para Zonas Áridas, Universidad Autónoma Agraria Antonio Narro, 25315. Buenavista, Saltillo, Coahuila, México.

[2]Universidad Autónoma de Nuevo León, School of Agronomy, Research Center and Development for Food Industries, General Escobedo, 66050, Nuevo León, México

[3]School of Animal Science and Ecology, Universidad Autónoma de Chihuahua. 31453, Chihuahua, Chihuahua, México

[4]Research Center in Food and Development, A.C. Av. 4ta Sur 3820, Fracc. Vencedores del Desierto, 33089, Cd. Delicias, Chihuahua, México

*Corresponding author. E-mail: jose.buenrostro@ciad.mx

ABSTRACT

Mangiferin (1,3,6,7-tetrahydroxyxanthone-C2-β-D-glucoside) is a xanthone present in mango (*Mangifera indica* L.). It was first isolated more than 100 years ago as a coloring matter from the mango tree. Also, it has been found in different parts of mango fruit, such as peel, stalks, leaves, barks, kernel, and seed. Nevertheless, it is also present in many angiosperm plants. Mangiferin exhibits a full spectrum of biological properties such as anticancer, antiviral, anti-inflammatory, antidiabetic, cardiovascular protective role, and among others, which are attributed to its antioxidant property. The latter turns out in an increased global demand for mangiferin-based food supplements, having considered mangiferin as a "super antioxidant" or a "natural miracle bioactive compound." The technical literature related to mangiferin biosynthesis at molecular and biochemical levels, as well as biotechnological processes for its recovery and applications, is limited, and in consequence, not well understood. Then, in this chapter, a detailed review of biosynthesis, obtention sources, and biological properties of mangiferin, and also extraction methods, and biotechnological processes for its applications are given for this promising phytochemical compound.

22.1 INTRODUCTION

Mangiferin (1,3,6,7-tetrahydroxyxanthone-C2-β-D-glucoside) is a natural polyphenol, considered as a "super antioxidant" or a "natural miracle bioactive compound" that belongs to xanthones, one of the several classes of natural compounds known as polyphenols (Imran et al., 2017). Xanthone is a C-glycoside occurring in many plant species, such as angiosperms and ferns. Mangiferin was first isolated more than 100 years ago as a coloring matter from the stem bark of mango tree (*Mangifera indica* L.). It has also been found in different parts of mango fruit, such as the peel, stalks, leaves, barks, kernel, and seed (Wu et al., 2010). The first glances

of mangiferin structure and sugar component showed a C-glucosyl structure, the first C-glucosyl xanthone (Iseda, 1957; Bhatia et al., 1967). Through nuclear magnetic resonance studies and chemical degradations, Haynes and Taylor (1966) reported that mangiferin is a C-glycosyl derivative 2-*p*-D-glucopyranosyl-1,3,6,7-tetrahydroxyxanthone. Its structure consisted of two aromatic rings, nonaromatic secondary hydroxyl groups, one lactonic carbonyl group, and one primary glycosidic hydroxyl group (Jyotshna et al., 2016). Recently, after extensive research, mangiferin data are fully identified and known, is a C-glycosyl compound consisting of 1,3,6,7-tetrahydroxyxanthen-9-one having a β-D-glucosyl residue at 6-position (Table 22.1) (Sekar, 2015; Kim et al., 2018).

This chapter is divided into four sections: Section 22.1 deals with biosynthesis and illustrates the putative biosynthetic pathway. Section 22.2 is related to extraction methods which determine the quality and quantity of mangiferin, including its occurrence in several sources. Section 22.3 details its biological properties, namely anticancer, antimicrobial, antiviral, anti-inflammatory, antidiabetic, cardiovascular protective role, and among others. Finally, Section 22.4 refers to the applications of mangiferin in recent years.

TABLE 22.1 Mangiferin Data Identification

Name	Mangiferin
IUPAC Name	1,3,6,7-tetrahydroxy-2-[(2S,3R,4R,5S,6R)-3,4,5-trihydroxy-6-(hydroxymethyl) oxan-2-yl]xanthen-9-one
Synonyms:	• Alpizarin
	• Chinomin
	• Hedysarid
	• 1,3,6,7-Tetrahydroxyxanthone C_2-β-D-glucoside
	• 2-β-D-glucopyranosyl-1,3,6,7-tetrahydroxy-9H-xanthen-9-one
	• 9H-xanthen-9-one, 2-beta-D-glucopyranosyl-1,3,6,7-tetrahydroxy-
PubChem CID:	5281647
KEEG compound:	C10077
CAS number:	4773-96-0
Structure:	
Molecular formula:	$C_{19}H_{18}O_{11}$
Molecular weight:	422.3 g/mol
Melting point	269-270 °C

22.2 MANGIFERIN BIOSYNTHESIS

Xanthones are yellow pigments of phenolic origin; the majority of xanthones are rarely glucosylated, as mangiferin (Caspi et al., 2011). Little is known about the molecular basis for mangiferin pathway regulation and the genes involved in its biosynthesis, which is an opportunity for plant biology. There are some biochemical advances in the knowledge of biosynthetic pathways; however, they are not entirely described due to gaps and redundancy of the routes proposed (Kumar et al., 2015). This is a bottleneck for the development of strategies for its production and biological applications. However, some approaches in chemical synthesis have been carried out. Wu et al. (2010) synthesized mangiferin from a C-glycosylation of a xanthene derivative with perbenzyl- glucopyranosyl N-phenyltrifluoroacetimidate (for detailed information refer to Jyotshna et al., 2016 and Wei et al., 2016).

Mangiferin biosynthesis is a complex process and may differ among plants. Three different routes have been proposed (Kumar et al., 2015) (Figure 22.1). Xanthones and flavonoids share structural similarities, the coexistence of mangiferin with other polyhydroxylated metabolites and experimental evidence suggest that its biosynthesis includes a mixed shikimate-acetate pathway (Ehianeta et al., 2016). In the first route (1), the mangiferin is biosynthesized from phenylalanine, cinnamic acid, *p*-coumaric acid, and malonic acid (acetate pathway), with iriflophenone and maclurin. This was deduced by the incorporation of C_6-C_3 unit from shikimate (Fujita and Inoue, 1977; Sekar, 2015). On the other hand, the second route was proposed by Fujita and Inoue (1981) with labeling ^{14}C isotope incorporation studies, consisting of caffeic acid degradation into phloroglucinol with malonic acid, after C-glycosylation and cyclization to produce mangiferin. Finally, the third route includes phenylalanine through *m*-hydroxybenzoic acid to produce benzophenone maclurin arises as a critical intermediate where C-glycosylation occurs, and then mangiferin is formed (Kumar et al., 2015; Ehianeta et al., 2016).

In *Swertia chirayita*, Kumar et al. (2015) reported through the detection of intermediate metabolites

iriflophenone, maclurin, deoxyloganic acid, loganic acid, and 1,3,6,7-tetrahydroxy-9H-xanthen-9-one, the biosynthetic pathways for mangiferin and other compounds such as amarogentin, amaroswerin, and amaronitidin. In this sense, the presence of iriflophenone, maclurin, and 1,3,6,7-tetrahydroxy-9H-xanthen-9-one suggests that mangiferin biosynthesis is mainly carried out through iriflophenone, and glucosylation occurs after the formation of xanthone moiety (as in the first route, Figure 22.1). Rai et al. (2016) also proposed a mangiferin pathway based on a transcriptomic analysis in *Swertia japonica* showing the unigenes expression levels in leaf and root. Results exhibited 16 enzyme coding genes to be involved in mangiferin biosyntheses, such as tyrosine ammonia-lyase, p-coumarate-3-hydroxylase (C3H), trans-cinnamate- 4-hydroxylase (C4H), and phenylalanine ammonia-lyase. Mangiferin is biosynthesized through coumaric acid derived from phenylalanine, which then is converted into caffeic acid or to benzophenone intermediates, such as iriflophenone, and combined with malonyl-CoA results in the final product. Although these efforts are essential contributions to the elucidation of mangiferin biosynthesis, further analysis is required for the complete elucidation and understanding of the pathways.

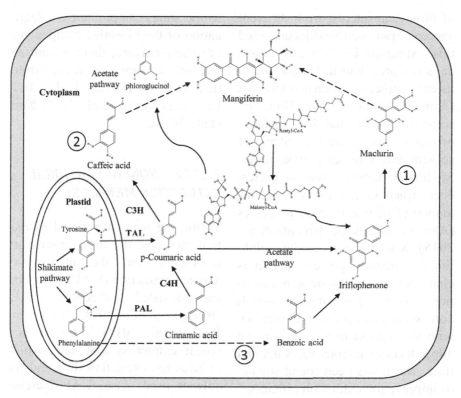

FIGURE 22.1 Proposed mangiferin biosynthesis pathway. (Adapted from: Chouhan et al., 2017; Kumar et al., 2015, and Rai et al., 2016. The compound structures were obtained from MolView V2.4 (molview.org). C4H: trans-cinnamate 4-hydroxylase; C3H: *p*-coumarate 3-hydroxylase.

22.3 EXTRACTION METHODS

22.3.1 PHYSICOCHEMICAL PROCESSES

22.3.1.1 CONVENTIONAL EXTRACTION METHODS

Extraction methods determine the quality and quantity of the mangiferin recovered. There are different extraction methods applied to obtain mangiferin and other bioactive compounds from plant matrices. The most common conventional extraction techniques include Soxhlet, maceration, infusion (sample is allowed to stand in boiling water), decoction (boiling for some time at 100 °C using direct flame), and among other (Brusotti et al., 2014; Medina Ramírez et al., 2016). Soxhlet extraction method is the most used process to extract molecules of biological interest from different sources. Franz von Soxhlet proposed this extraction in 1879 and nowadays is a prevalent technique due to its easy implementation and low cost (Armenta et al., 2015). However, some of the principal disadvantages of this technique are the volume of solvent and long extraction times. Soxhlet extraction serves as a reference when comparing the performance of new extraction processes (Azmir et al., 2013; Armenta et al., 2015). Maceration is another solid–liquid extraction process, and it is characterized by its smooth execution and low implementation costs. It consists of mixing the solid material with the solvent in a container and allow them to interact for a defined time, which has been found stirring to improve the extraction efficiency. Once the process is finished, the solvent containing the molecules of interest is separated from the mixture by filtration. Among its main disadvantages are the long extraction times and large liquid/solid ratios (Azmir et al., 2013). In general, the conventional organic solvent-based extraction methods are techniques with low extraction yields, the use of organic solvents, with a consequent environmental impact.

Moreover, extraction times can be so long, and the extracts obtained may contain toxic residues (Dorta et al., 2012; Kim et al., 2010). On the same hand, the use of relatively high temperatures can promote the degradation of the bioactive components. For these reasons, there is constant research for new and novel extraction methods (Dorta et al., 2012; Fernández-Ponce et al., 2012; Kaur et al., 2019).

22.3.1.2 NONCONVENTIONAL EXTRACTION METHODS

Within the nonconventional extraction methods, the most promissory techniques are the microwave-assisted extraction (MAE) and ultrasound-assisted extraction (UAE) (Brusotti et al., 2014; Fernández-Ponce et al., 2012). The UAE is a recent extraction technique, which reduces the extraction time and the solvent used. Also, UAE increase extraction yield with low energy consumption and higher purity in the final product (Forero and Pulido, 2016; Chemat et al., 2017). The ultrasonic waves break the cell walls by the high shear forces generated in the media by the cavitation phenomena. The implosion of cavitation bubbles leads to microjetting process and turbulences in the solid/liquid media, which turns out in particle breakdown of solid material, increasing the contact surface area between the solid and liquid phase, facilitating the release of the molecules of interest to the solvent (Brusotti et al., 2014; Chemat et al.,

2017). On another hand, MAE is a highly effective green extraction method with low energy and less solvent consumptions, conducting to automation, and shorter extraction times (Ekezie et al., 2017). The power generated by the microwaves (electromagnetic waves, usually operated at a frequency of 2.45 GHz) is converted into heat turning out in ionic conduction and dipole rotation mechanisms (Wijngaard et al., 2012; Azmir et al., 2013). The most critical variables that affect the efficiency of this method are irradiation time, microwave power, and the liquid-to-solid ratio (L/S) (Ekezie et al., 2017). The combination of this phenomenon causes cell breakdown due to internal overheating allowing the consequent penetration of the solvent in the matrix and transport of molecules into the liquid (Wijngaard et al., 2012; Azmir et al., 2013; Ekezie et al., 2017).

The nonconventional methods also include the use of supercritical fluid extraction (SFE), which is an interesting option with great potential for obtaining biological molecules from plant matrix (Brusotti et al., 2014; da Silva et al., 2016). The low viscosity and high diffusivity of supercritical fluids provide them with better transport properties than liquid solvents, and therefore, they can be diffused easily through the solid matrix, improving the efficiency of the extraction process. Supercritical

fluids acquire these properties by changing their density because of the application of different pressures and temperatures above their critical point (da Silva et al., 2016). One of the compounds widely used in SFE is the supercritical carbon dioxide (CO_2), which is eco-friendly and not a residual toxic remnant in the recovered extracts (Kim et al., 2010; Brusotti et al., 2014). However, the use of CO_2 as a unique solvent for SFE did not work for the extraction of mangiferin, but when it was combined with ethanol or methanol (MeOH), the extraction process efficiency was significantly improved, due to changes in polarity (Zou et al., 2013). On the other hand, in the case of a subcritical fluid, the pressure is maintained above the critical value while the temperature is below a critical value. These types of fluids have been used to extract mangiferin of mango leaves with good efficiency performance (Fernández-Ponce et al., 2012; Kim et al., 2010). The subcritical water extraction (SWE) uses subcritical water as a solvent, which allows obtaining phytoconstituents (polar or slightly polar) efficiently without generating toxic waste, with shorter extraction time and it is eco-friendly (Kim et al., 2010). Another nonconventional extraction method is the three phases partitioning (TPP), which is a simple one-step process for the separation, purification, and concentration of biomolecules from complex mixtures. In TPP, ammonium sulfate is added to the water containing the extract, and then butanol is added further allows the formation of three phases, with the upper phase (butanol) containing nonpolar compounds, the lower aqueous phase contains the polar compounds, both phases are separated through a third phase formed by a protein precipitate (Kulkarni and Rathod, 2014, 2015). Mangiferin extraction from mango leaves by TPP and TPP coupled with ultrasound (UTPP) was first proposed by Kulkarni and Rathod (2014). Under optimal conditions (25 min, pH 6, ammonium sulfate saturation 40% (w/v), slurry to t-butanol ratio of 1:1, solute to solvent ratio of 1:40, frequency 25 kHz, power 180 W, duty cycle of 50%, soaking time of 5 min and temperature of 30 ± 2 °C), a yield of 41 mg/g was attained with UTPP, which it was higher than the one obtained for TPP (28 mg/g after 2 h). Soxhlet extraction method allowed to achieve the highest yield of mangiferin (57 mg/g) with water, after 5 h (Kulkarni and Rathod, 2014). In another study, the same authors increased the efficiency of the extraction from leaves using microwave-assisted three-phase partitioning extraction (MTPP) (Kulkarni and Rathod, 2015). Under optimal conditions (microwave irradiation time of 5 min, 40% w/v $(NH_4)_2SO_4$, 272 W, solute to solvent ratio of 1:20, slurry to t-butanol ratio of 1:1,

soaking time of 5 min and duty cycle of 50%), the mangiferin yield with MTPP was 54 mg/g, which is similar to that obtained by Soxhlet extraction in water for 5 h (57 mg/g) (Kulkarni and Rathod, 2015). Following the above mentioned, TPP assisted with UAE, and MAE reduced the extraction times significantly and offered good yields. In Table 22.2, some studies about the extraction of mangiferin using conventional and nonconventional techniques during the last years are listed.

According to the detailed information in Table 22.2, most of the studies conducted to optimize mangiferin extraction have mainly focused on mango leaves. The best yields using this material were attained when nonconventional methods such as UAE. Regarding other sources, Forero and Pulido (2016) and Kshirsagar et al. (2017) isolated mangiferin from different mango barks and Swertia sp., respectively. These authors observed that mangiferin yield was strongly influenced by the variety (Table 22.2). The marked differences in the yield between biological materials, even among on, can be determined by genetics, ontogenetic, and environmental factors, since the production of bioactive compounds occurs at the secondary level of metabolism and this, in turn, is a function of gene expression (Forero and Pulido, 2016). It is to note that another parameter of special importance in the extraction process is the grinding stage (particle

TABLE 22.2 Extraction of Mangiferin From Different Plant Materials Using Conventional and Nonconventional Extraction Techniques

Source	Extraction Method	Main Results	Reference
Mango barks	Soxhlet extraction with EtOH/H$_2$O (70:30 v/v, 250 mL, 8 h)	The mangiferin yield was affected by the bark variety. The highest mangiferin yield was reported in the varieties Pig 693, Rosa and Julie with 5.257, 4.906,2 y 4.879,1 mg/100 g.	Forero and Pulido (2016)
P. macrocarpa fruits	SWE (4–6 h and 90–110 °C, L/S of 40:1 mL/g)	The optimal extraction conditions found by RSM were 105 °C and 6 h extraction time 38.7 with a mangiferin yield of 38.7 mg/g.	Alara et al. (2017)
Mango leaves	MAE (45% EtOH, L/S of 30:1 mL/g, extraction time of 123 s at 474 W), Maceration in EtOH (40% v/v) during 30–120 min	The mangiferin yield was 36.10 mg/g under optimal conditions in MAE. The yield with Soxhlet method was 15.24 and 25.17 mg/g at 30 and 120 min, respectively	Zou et al. (2013)
Mango leaves	SC-CO$_2$ plus cosolvents (20% EtOH or 20% MeOH, 100 and 400 bar, 35 and 55 °C), SWE (40 bar and 100 °C)	The highest mangiferin yield was observed with SC-CO$_2$ + 20% MeOH at 100 bar and 55 °C (0.939 mg/g) and SC-CO$_2$ + 20% EtOH at 100 bar and 35 °C (1.918 mg/g). However, these values were lower than those obtained with SWE (13.659 mg/g)	Fernández-Ponce et al. (2012)
Mango leaves	Infusion in water (letting stand for 5 min in boiling water) UAE (water, ultrasound power of 35 W, frequency of 40 kHz)	The mangiferin yield was higher at the highest concentration of leaves (5%, mv) in the solvent. Mangiferin yield obtained by UAE in young and mature leaves was 0.77 and 0.114 mg/mL. The higher yield was observed with infusion process in young (0.285 mg/mL) and mature leaves (0.57 mg/mL)	Medina Ramírez et al. (2016)
Mango peels	Maceration (200 rpm, 25 °C, 24 h), UAE (42 kHz, 25 °C, 30 min) and HHP extraction (150 MPa, 25 °C, 20 min). All extractions with EtOH/H$_2$O (8:2 v/v) and L/S: 10:1 mL/g.	UAE method obtained the best yield of mangiferin ($\approx$13.5 mg/g) than those obtained with HHP ($\approx$11 mg/g) and maceration ($\approx$5.8 mg/g)	Ruiz-Montañez et al. (2014)

TABLE 22.2 *(Continued)*

Source	Extraction Method	Main Results	Reference
Mango leaves	UAE (ultrasound power of 200 W and frequency of 40 kHz)	EtOH concentration, L/S, and extraction time had significant effects on the yield of mangiferin extraction. The best response found by RSM (44% EtOH, L/S of 38:1, time of 19.2 min at 60 °C) reach a mangiferin yield of 58.46 mg/g	Zou et al. (2014)
Swertia sp.	Static maceration (L/S: 100/1 mL/g, 28 °C at 6, 12, and 24 h), CSM (L/S: 100/1 mL/g, constant stirring of 100 rpm, 28 °C at 6, 12, and 24 h). Both processes were performed in MeOH	The mangiferin yield with static maceration was highest in *S. chirayita* (155.76 mg/g) at 24, which was lower than those observed at 12 h in *S. nervosa* (80 mg/g). The mangiferin yields in CSM were lower than those registered static maceration, ranging from 0.53 to 64.93 mg/g during extraction time of 12 and 24 h in *Swertia angustifolia* var. *pulchella* and *S. chirayita* (PRK-20), respectively	Kshirsagar et al. (2017)

EtOH: Etanol; RSM: response surface methodology; SC-CO_2: subcritical fluid extraction with CO_2; HHP: high hydrostatic pressure extraction; CSM: continuous shaking maceration.

size), since the contact surface, the recirculation, and solvent penetration depends on the particle size (Forero and Pulido, 2016). In general, the literature suggests that the solvent, biological material, method, and extraction time have a strong influence on the amount of mangiferin recovered. Moreover, the technical literature refers that recovery is also affected by the extraction method.

22.3.1.3 BIOTECHNOLOGICAL PROCESSES

The enzymatic extraction of molecules of biological interest is a novel technique of world interest. In this method, the enzymes added to degrade the cell wall of the plant material, facilitate the extraction of the biomolecules of interest, and improve the efficiency (Gullón et al., 2017; Nadar et al., 2018). Moreover, this technique is considered as an eco-friendly process. The main enzymes involved in these extraction methods are pectinases, cellulases, and hemicellulases (Gullón et al., 2017). Enzyme-assisted biomolecule extraction has been reported for obtaining phenolic compounds from unripe apples (Zheng and Chung, 2014), grape residues (Gómez-García et al., 2012), and cauliflower (Huynh et al., 2014). In the particular case of mango, there are only reports that describe the extraction of oil from mango seed

kernel using extraction assisted by enzymes (Gaur et al., 2007; Womeni et al., 2008). Therefore, mangiferin extraction using this method may offer great opportunities.

22.4 BIOLOGICAL PROPERTIES OF MANGIFERIN

The natural and heat-stable C-glucosyl xanthone mangiferin has been identified in several parts of mango: leaves, fruits, stem bark, heartwood, and roots. Then, it is easy to conclude that mangiferin is a phenolic compound. Due to its chemical nature, mangiferin exhibits antioxidant, as well as antiviral, antilipid peroxidation, antibacterial, antidiabetic, immunomodulatory, antitumor, cardiotonic, hypotensive, wound healing, antidegenerative, and anti-inflammatory activities. In Cuba and Sri Lanka, it is sold under the brand names Vimang® and Salaretin®, respectively. In this section, some of its bioactivities are reviewed.

22.4.1 ANTIOXIDANT ACTIVITY

The four hydroxyl groups in mangiferin demonstrate remarkable antioxidant properties by scavenging the noxious oxygen-free radicals. The free-radical scavenging activity of mangiferin is attributed to the C-glucosyl linkage along with the presence of polyhydroxy groups

(Saha et al., 2016). For centuries mangiferin, it had been used in traditional medicine in countries such as India and China for its health benefits. For instance, it is considered to be one of the main active constituents in more than 40 polyherbal formulations in traditional Chinese medicine (Franklin et al., 2009; Jyotshna et al., 2016; Matkowski et al., 2013).

Sánchez et al. (2000) carried out an in vivo study with OF1 mice in order to compare Vimang® (50, 110, 250 mg/kg), mangiferin (50 mg/kg), vitamin C (100 mg/kg), vitamin E (100 mg/kg), vitamin E plus vitamin C (100 mg/kg each), and β-carotene. The inductor of oxidative damage, TPA (12-O-tetradecanoylphorbol-13-acetate), was administered (0.1 µg, i.p.). According to their findings, these authors indicated that Vimang® is comparable or superior to nutritional antioxidants in protecting mice from oxidative stress, concluding that the effect was dose-dependent. Mangiferin showed the same pattern of impact as Vimang® except for GPx (no restoration of glutathione peroxidase levels was observed).

Dar et al. (2005) assayed the potent antioxidant activity of mangiferin (EC50 5.8 ± 0.96 µg/mL). They observed that it lowers hydrogen peroxide-induced lipid peroxidation in human peripheral blood lymphocytes (this occurs in a dose-dependent manner). In another study, Nishigaki et al. (2007) evaluated in vitro the protective effect of mangiferin on human umbilical vein endothelial cells against glycated protein-iron chelate induced toxicity. A protective effect of mangiferin was measured through enhanced levels of antioxidant enzymes concerning lipid peroxidases. These observations allow concluding that mangiferin prevents the development of vascular complications induced by glycated protein-iron chelate.

A biochemical study was conducted by Rajendran et al. (2008b), where they mentioned decreased activities of electron transport chain complexes and some Krebs cycle enzymes: isocitrate dehydrogenase, succinate dehydrogenase, malate dehydrogenase, and α-ketoglutarate dehydrogenase, in lung cancer-bearing animals. This study confirms the chemopreventive and chemotherapeutic effect of mangiferin against cancer via oxidative stress. Karuppanan et al. (2014) referred to the hepatoprotective and antioxidant effects of mango leaf extracts against mercury chloride-induced liver toxicity in mice. Two alcohol-based extracts were obtained, and the one that was capable of increasing hepatic catalase, glutathione *S*-transferase, and glutathione peroxidase levels was methanolic extract. Meneses et al. (2015) obtained an extract from mango by-products, rich in mangiferin, quercetin, and kaempferol. The authors recovered a dry powder with antioxidant activity of 852 μmol TE/g and an IC_{50} of DPPH radical of 90 μg/mL.

22.4.2 IMMUNOMODULATORY EFFECT

Some of the biochemical roles of mangiferin are related to (1) increasing the intracellular glutathione (GSH) levels, (2) suppressing NF-κB activation induced by inflammatory agents, including tumor nuclear factor (TNF), and (3) mediating the downregulation of NF-κB.

In that sense, the immunomodulatory activity of alcoholic extract containing mangiferin was evaluated for its effect on cell-mediated and humoral components of the immune system in mice (Makare et al., 2001). The authors mentioned that mangiferin produced an increase in humoral antibody titer and delayed-type hypersensitivity (DTH) in mice, which is essential in terms of immunostimulant properties. Mangiferin exerted inhibitory effects on rat macrophage functions, including phagocytic activity and the respiratory burst (García et al., 2002). Their findings indicated the depressor effects of mangiferin on the phagocytic and ROS production activities of rat macrophages. This can be highlighted for the treatment of diseases of immunopathological origin characterized by the hyperactivation of phagocytic cells.

On the other hand, García et al. (2003) determined the effects of orally administered mangiferin on mouse antibody responses induced by inoculation with spores of microsporidian parasites. They observed that mangiferin did not affect either IgM or IgG2a, but improved the production of IgG1 and IgG2b. These results suggest a potential application in modulating the humoral response in different immunopathological disorders.

The use of mangiferin on an ischemia-reperfusion (IR) induced model of myocardial injury in rats was evaluated (Suchal et al., 2016). A significant effect against cardioprotective activity against IR injury was observed at 40 mg/kg/day. The author attributed this effect by modulating mitogen-activated protein-kinases (MAPK) mediated inflammation and apoptosis. In another study, Leiro et al. (2004) performed in vivo the immunomodulatory activity of mangiferin on thioglycollate-elicited mouse macrophages, which were stimulated with lipopolysaccharide (LPS) and gamma interferon (IFN-γ). A total of 10 μM of mangiferin was enough to confirm that mangiferin modulates the expression of a large number of genes critical for the regulation of apoptosis, viral replication, tumorigenesis, inflammation, and various autoimmune diseases. This is remarkable for the treatment and prevention of inflammatory diseases and/or cancer. It is worth noting the work of Garrido et al. (2005), where the authors described some biological activities, at a molecular level, of an aqueous stem bark extract of *M. indica.* They found that mangiferin inhibited early and late events in T cell activation, including CD25 cell surface expression, progression to the S-phase of the cell cycle, and proliferation in response to T cell receptor stimulation.

22.4.3 ANTI-INFLAMMATORY ACTIVITY

Inflammatory processes involve a broad spectrum of chemical mediators, in this sense, Pardo-Andreu et al. (2006) affirmed for both in vitro and in vivo models, the anti-inflammatory potential of mangiferin is a dose-dependently inhibition of the inflammatory cytokines, TNFα, NO (nitric oxide), and NF-Kβ. Gong et al. (2013) reported that mangiferin could attenuate cecal ligation and puncture-induced mortality as well as acute lung injury. On the other hand, mangiferin inhibited sepsis-activated MAPK and NF-Kβ signaling. In another study, mangiferin improved kidney function, increased CD73 expression in kidneys (from IR injury), improved adenosine production, and inhibited proinflammatory responses and tubular apoptosis (Wang et al., 2015).

For the protective role of mangiferin, Agustini et al. (2016) assayed it against doxorubicin (DOX)-induced cardiotoxicity and inflammatory responses in rats, via downregulation of proapoptotic and proinflammatory gene expressions, and the upregulation of SERCA2a gene expression, as well as normalization of cytosolic calcium levels. Further, a molecular mechanism of mangiferin against LPS and D-galactosamine-induced acute liver injury and inflammation was carried out by Pan et al. (2016). They found that mangiferin activates the Nrf2 pathway and regulates NLRP3 inflammasome activation.

On the other hand, Szandruk et al. (2018) analyzed the protective role of mangiferin in colon tissues against 2,4,6-trinitrobenzensulfonic acid (TNBS)-induced colitis in rodents. According to their findings, mangiferin decreased macroscopic and microscopic damage, as well as reduced MDA levels in the colon at 30 and 100 mg/kg body weight doses. The TNF-α and IL-17 levels and SOD activity in the colon tissues were low in the presence of mangiferin (100 mg/kg body weight). The researchers concluded that mangiferin significantly protects against TNBS-induced inflammatory manifestations and colitis in rats.

To establish a comparison, Garrido et al. (2001) demonstrated the analgesic and anti-inflammatory effects of Vimang®. Analgesia was produced by using acetic acid-induced abdominal constriction and formalin-induced licking. *M. indica* extract exerted a potent and dose-dependent antinociceptive influence against acetic acid test in mice. This study was the first report dealing with the antinociceptive and anti-inflammatory actions of *M. indica* extract. Conversely, Bhatia et al. (2008) evaluated the ability of mangiferin to reduce both prostaglandin $F_(2)$ ($PGF_(2)$)) and 8-iso-prostaglandin F(2alpha) (8-iso-PGF(2alpha)) production by LPS-activated primary rat microglia. Mangiferin was able to reduce both productions and decreased LPS-induced COX-2 protein synthesis in a dose-dependent manner. From their findings, the authors suggested that mangiferin may play a vital role in the mechanism of cerebral protection.

22.4.4 ANTIDIABETIC ACTIVITY

Diabetes mellitus comprises a series of metabolic conditions associated with hyperglycemia, derived from defects in insulin secretion, and/or insulin action. In this sense, Rolo and Palmeira (2006) mentioned oxidative damage in the heart and kidney due to the formation of advanced glycosylated end-products (which turns out in the generation of ROS) in long-standing hyperglycemia with diabetes mellitus.

Muruganandan et al. (2002, 2005) demonstrated the effects of mangiferin on type I diabetes in terms of hyperglycemia, atherogenicity, and oxidative damage to cardiac and renal tissues in streptozotocin-induced diabetic rats. The authors found reduced catalase and SOD activities in the kidney, increased in heart, and unaltered in erythrocytes.

Concerning type II diabetes, KK-Ay mice were fed mangiferin (90 mg/kg), and after 7 h after oral administration, the baseline glucose level was reduced 56% (Miura et al., 2001a). In the same model, mangiferin (30 mg/kg, p.o., once daily followed 30 min later by exercise (120 min motorized treadmill) for two weeks) reduced both blood cholesterol (~40%) and triglyceride levels (~70%) (Miura et al., 2001b).

Apontes et al. (2014) reported that mangiferin protected against high-fat diet-induced weight gain increased aerobic mitochondrial capacity and thermogenesis, and improved glucose and insulin profiles. They found that mangiferin significantly increased glucose oxidation in the muscle of high-fat diet-fed mice without changing fatty acid oxidation. These results indicate that mangiferin redirects fuel utilization toward carbohydrates. Mangiferin also inhibited anaerobic metabolism of pyruvate to lactate but enhanced pyruvate oxidation, being pyruvate dehydrogenase a key target. These results imply a potential carbohydrate utilization.

Hu et al. (2007) demonstrated the significant antidiabetic activity of mangiferin by establishing its protein tyrosine phosphatase1B (PTP1B) inhibitory activity, which plays a vital role in the treatment of diabetes mellitus. On the other hand, Prashanth et al. (2001) conducted an in-vitro study for the inhibition of α-glucosidase using an ethanolic extract of *M. indica* (mangiferin), and they concluded it has a potential against obesity and diabetes. Yoshikawa et al. (2001) reported that mangiferin inhibits sucrase, isomaltase, and maltase in rats. In consequence, mangiferin reduces blood glucose levels and possesses both pancreatic and extrapancreatic mechanisms. They also found that mangiferin inhibited body weight gain in experimental rats.

22.4.5 LIPOLYTIC

This significant biological effect has been studied for triglycerides and cholesterol. Yoshikawa et al. (2001) studied this important biological effect on rat epididymal fat-derived cultured adipocytes, and they found that mangiferin (100 mg/L) reduced 35% triglycerides in those adipocytes. On the other hand, Muruganandan et al. (2005) observed that mangiferin significantly reduced total plasma, cholesterol, triglycerides,

and LDL-cholesterol associated with a concomitant increase in HDL-cholesterol levels and a decrease in the atherogenic index in diabetic rats. This suggests a potential antihyperlipidemic and antiatherogenic activity.

22.4.6 ANTIMICROBIAL AND ANTIVIRAL

Stoilova and Ho (2004) analyzed the mangiferin role against bacterial (*Bacillus pumilus, B. cereus, Staphylococcus aureus, Staphylococcus, E. coli, Salmonella agona, K. pneumoniae*, yeast (*S. cerevisiae*)) and fungi species (*Thermoascus aurantiacus, Trichoderma reesei, Aspergillus flavus,* and *A. fumigatus*).

Perrucci et al. (2006) compared mangiferin to paromomycin (an active drug), in a neonatal mouse model, at a dose of 100 mg/kg. They found that mangiferin exhibited a similar inhibitory activity on *Cryptosporidium parvum*. On the other side, mangiferin has also been assayed against Herpes simplex virus type 2. Mangiferin does not directly inactivate HSV-2 but inhibits the late event in HSV-2 replication (Zhu et al., 1993). In vitro, mangiferin was also able to inhibit HSV-1 virus replication within cells (Zheng and Lu, 1989) and to antagonize the cytopathic effects of HIV (Guha et al., 1996).

The antiviral potential has been explored by Rechenchoski et al.

(2018), who evaluated the activity of mangiferin against poliovirus type-1 (PV-1). They reached 50% cytotoxic concentration (CC_{50}) > 2000 μg/mL in HEp-2 cell cultures, by the dimethylthiazolyl-diphenyl-tetrazolium bromide method. The 50% inhibitory concentration (IC_{50}) was 53.5 μg/mL. The authors also referred to the inhibition of viral protein synthesis by immunofluorescence assay, concluding that mangiferin is a potential candidate for the control of PV infection.

22.4.7 ANTITUMORAL

Guha et al. (1996) reported that mangiferin had in vivo growth-inhibitory activity against ascitic fibrosarcoma in Swiss mice. The authors mentioned that mangiferin induced antitumor effect irrespective of the size of tumor inoculum, as well as a cytotoxic effect on tumor cells. Mangiferin was also found to antagonize in vitro the cytopathic effect of HIV in MT-2 cells and prevented cell death.

On the other hand, Yoshimi et al. (2001) investigated the effects of mangiferin in rat colon carcinogenesis induced by chemical carcinogen and azoxymethane. The presence of mangiferin observed a reduction in cell proliferation in colonic mucosa. The authors point out mangiferin as a chemopreventive agent and a promising cancer controller.

22.5 POTENTIAL APPLICATIONS OF MANGIFERIN

Knowing the potential of mangiferin due to its bioactivities, it is essential to its incorporation in foods (Luo et al., 2011). Nanotechnology has taken an important role in incorporating bioactive compounds into food matrices, such as nutraceuticals and pharmacological applications (Jahanshahi and Babaei, 2008). Nanometric-scale systems have taken on great importance, such as high-water solubility, resistance to chemical degradation, high absorption in the human body, prolonged circulation, controlled release, high cell uptake, and specific target delivery (Baán et al., 2019; Zuidam and Shimoni, 2010). Some of the essential applications of mangiferin in recent years are shown in Table 22.3.

It is important to mention that mangiferin needs a protection system to be able to incorporate it into a system (either food and/or pharmaceutical) to ensure its bioavailability inside the body. All these systems are of great importance since they can not only be incorporated into food matrices but can also be used in the pharmaceutical industry for the human and animal areas (Chowdhury et al., 2019; Samadarsi and Dutta, 2019). It should be noted that today the applications are found at the laboratory level both in vitro and in vivo. The recommendation of uses

TABLE 22.3 Recent Studies About the Potential Use of Mangiferin

Compound	Source	Method	Main Results	Application	Reference
Mangiferin/β-lactoglobulin	Curcuma amada	Desolvation method	Spherical and uniform size (70 nm), resistance to pepsin digestion, release in the colon of 80%	Oral delivery	Baán et al. (2019)
Chitosan/mangiferin	M. indica L.	Spray-drying	Adsorption capacity of Cr(VI)	Preventive material in cases of human or animal contamination with Cr(VI)	Sampaio et al. (2015)
Chitosan/pectin/mangiferin	M. indica L.	Microspheres	Spheres (650–680 μm), highest release (7.8 mg mangiferin/g bead)	As nutraceutical to be added in functional foods, and also to improve gastroprotection [6]	de Souza et al. (2009)
Mangiferin/β-cyclodextrin	M. indica	Co-evaporation	Increased antioxidant activity	Encapsulation to use a release in situ of enhancement of therapeutic effects	Ferreira et al. (2013)
Mangiferin	Sigma Chemical company St. Louis, MO, USA	Direct with corn oil as a vehicle	Increase in the levels of glycoproteins, membrane ATPases and membrane lipid peroxidation were observed in animals with lung carcinoma	Incorporation into foods to anticancer effect (chemoprevention)	Rajendran et al. (2008a)
Mangiferin	Belamcanda chinensis	Direct consumption	Reduce the intensity of damage (colitis) in rats	Can be used into foods to prevent colitis	Szandruk et al. (2018)

TABLE 22.3 *(Continued)*

Compound	Source	Method	Main Results	Application	Reference
Mangiferin	Carbosynth Limited (UK)	Nanoemulsion (10,000 rpm/5 min)	Droplets size 295 nm, −30 mV, pseudoplastic, the release depends on the molecular weight of the polymer	Topic (not consumption)	Pleguezuelos-Villa et al. (2019)
Mangiferin	Nanjing Nutri Herb BioTech Co., Ltd, (Nanjing, China)	Direct application	The combination of mangiferin and oral hypoglycemic agents metformin and gliclazide improved the overall diabetic conditions and can be translated in the clinical practice also.	Combined oral consumption is suggested	Sekar et al. (2019)
Mangiferin/ polyamine-β- cyclodextrin	Kunming Pharmaceutical Co. (Yunnan, PR China.	Suspension	Improved the water solubility and cytotoxicity	As herbal medicine or healthcare products	Liang et al. (2019)
Mango peel extract	*M. indica* L. var. Ataulfo	Water-in-oil-in-water	0.96 ± 0.02 mg/g mangiferin. This technique is an alternative to encapsulated and maintain stable the emulsion	New functional food products containing bioactive ingredients from agroindustrial by-products	Velderrain-Rodriguez et al. (2019)

is made either in food or as a complement of medicines in the so-called phytomedicine (Sajid et al., 2019; Sekar et al., 2019). However, foods distributed in which mangiferin is integrated have not been reported.

22.6 CHALLENGES AND PERSPECTIVES

People around the world are increasingly worried about their health and nutrition. Mangiferin is a bioactive compound with an important role in human health. Several biological properties of mangiferin have been reported at the laboratory level both in vitro and in vivo, with a potential application in the pharmaceutical industry for the human and animal areas. For that, future research should be focused on the study of the reaction system, including the type of system, solvent, and operational conditions. On the other hand, evaluate biotechnological processes as the environmental process to release the mangiferin and improve their efficiency of extraction, using a fermentation process to produce the needed enzyme to release the compound from a matrix complex. Omics tools are necessary to build the complete metabolic analysis of the compounds and their relation about them into the mangiferin biosynthesis and biotransformation. Once obtained the metabolite

(mangiferin), stability and storage methods are needed, as well as evaluate its production on a large scale with their technological and economic assays. Finally, evaluate the incorporation of mangiferin on functional food development.

KEYWORDS

- **mangiferin**
- **biosynthesis**
- **sources**
- **biological properties**
- **biotechnological processes**

REFERENCES

Agustini, F. D.; Arozal, W.; Louisa, M.; Siswanto, S.; Soetikno, V.; Nafrialdi, N.; Suyatna, F., Cardioprotection mechanism of mangiferin on doxorubicin-induced rats: focus on intracellular calcium regulation. *Pharm. Biol.* **2016**, *54* (7), 1289–1297.

Alara, O. R.; Abdul Mudalip, S. K.; Olalere, O. A., Optimization of mangiferin extracted from *Phaleria macrocarpa* fruits using response surface methodology. *J. Appl. Res. Med. Aromat. Plants.* **2017**, *5*, 82–87.

Apontes, P.; Liu, Z.; Su, K.; Benard, O.; Youn, D. Y.; Li, X.; Li, W.; Mirza, R. H.; Bastie, C. C.; Jelicks, L. A.; Pessin, J. E.; Muzumdar, R. H.; Sauve, A. A.; Chi, Y., Mangiferin stimulates carbohydrate oxidation and protects against metabolic disorders induced by high-fat diets. *Diabetes.* **2014**, *63* (11), 3626–3636.

Armenta, S.; Garrigues, S.; de la Guardia, M., The role of green extraction techniques in green analytical chemistry. *Trends Anal. Chem.* **2015**, *71*, 2–8.

Azmir, J.; Zaidul, I. S. M.; Rahman, M. M.; Sharif, K. M.; Mohamed, A.; Sahena, F.; Jahurul, M. H. A.; Ghafoor, K.; Norulaini, N. A. N.; Omar, A. K. M., Techniques for extraction of bioactive compounds from plant materials: a review. *J. Food Eng.* **2013**, *117* (4), 426–436.

Baán, A.; Adriaensens, P.; Lammens, J.; Delgado Hernandez, R.; Vanden Berghe, W.; Pieters, L.; Vervaet, C.; Kiekens, F., Dry amorphisation of mangiferin, a poorly water-soluble compound, using meso-porous silica. *Eur. J. Pharm. Biopharm.* **2019**, *141*, 172–179.

Bhatia, H. S.; Candelario-Jalil, E.; de Oliveira, A. C. P.; Olajide, O. A.; Martínez-Sánchez, G.; Fiebich, B. L., Mangiferin inhibits cyclooxygenase-2 expression and prostaglandin E2 production in activated rat microglial cells. *Arch. Biochem. Biophys.* **2008**, *477* (2), 253–258.

Bhatia, V. K.; Ramanathan, J. D.; Seshadri, T. R., Constitution of mangiferin. *Tetrahedron* **1967**, *23* (3), 1363–1368.

Brusotti, G.; Cesari, I.; Dentamaro, A.; Caccialanza, G.; Massolini, G., Isolation and characterization of bioactive compounds from plant resources: the role of analysis in the ethnopharmacological approach. *J. Pharm. Biomed. Anal.* **2014**, *87*, 218–228.

Caspi, R.; Altman, T.; Dreher, K.; Fulcher, C. A.; Subhraveti, P.; Keseler, I. M.; Kothari, A.; Krummenacker, M.; Latendresse, M.; Mueller, L. A.; Ong, Q.; Paley, S.; Pujar, A.; Shearer, A. G.; Travers, M.; Weerasinghe, D.; Zhang, P.; Karp, P. D., The MetaCyc database of metabolic pathways and enzymes and the BioCyc collection of pathway/genome databases. *Nucleic Acids Res.* **2011**, *40* (D1), D742–D753.

Chemat, F.; Rombaut, N.; Sicaire, A.-G.; Meullemiestre, A.; Fabiano-Tixier, A.-S.; Abert-Vian, M., Ultrasound assisted extraction of food and natural products. Mechanisms, techniques, combinations, protocols and applications. A review. *Ultrason. Sonochem.* **2017**, *34*, 540–560.

Chouhan, S.; Sharma, K.; Zha, J.; Guleria, S.; Koffas, M. A. G., Recent advances in the recombinant biosynthesis of polyphenols. *Front. Microbiol.* **2017**, *8*, 2259-2259.

Chowdhury, A.; Lu, J.; Zhang, R.; Nabila, J.; Gao, H.; Wan, Z.; Adelusi Temitope, I.; Yin, X.; Sun, Y., Mangiferin ameliorates acetaminophen-induced hepatotoxicity through APAP-Cys and JNK modulation. *Biomed. Pharmacother.* **2019**, *117*, 109097.

da Silva, R. P. F. F.; Rocha-Santos, T. A. P.; Duarte, A. C., Supercritical fluid extraction of bioactive compounds. *Trends Anal. Chem.* **2016**, *76*, 40–51.

Dar, A.; Faizi, S.; Naqvi, S.; Roome, T.; Zikr-ur-Rehman, S.; Ali, M.; Firdous, S.; Moin, S. T., Analgesic and antioxidant activity of mangiferin and its derivatives: the structure activity relationship. *Biol. Pharm. Bull.* **2005**, *28* (4), 596–600.

de Souza, J. R. R.; de Carvalho, J. I. X.; Trevisan, M. T. S.; de Paula, R. C. M.; Ricardo, N. M. P. S.; Feitosa, J. P. A., Chitosan-coated pectin beads: characterization and *in vitro* release of mangiferin. *Food Hydrocolloid.* **2009**, *23* (8), 2278–2286.

Dorta, E.; Lobo, M. G.; Gonzalez, M., Reutilization of mango byproducts: study of the effect of extraction solvent and temperature on their antioxidant properties. *J. Food. Eng.* **2012**, *77* (1), C80–C88.

Ehianeta, T. S.; Laval, S.; Yu, B., Bio- and chemical syntheses of mangiferin and congeners. *BioFactors* **2016**, *42* (5), 445–458.

Ekezie, F.-G. C.; Sun, D.-W.; Cheng, J.-H., Acceleration of microwave-assisted extraction processes of food components by integrating technologies and applying emerging solvents: a review of latest developments. *Trends Food Sci. Technol.* **2017**, *67*, 160–172.

Fernández-Ponce, M. T.; Casas, L.; Mantell, C.; Rodríguez, M.; Martínez de la Ossa,

E., Extraction ofantioxidant compounds from different varieties of *Mangifera indica* leaves using green technologies. *J. Supercrit. Fluids.* **2012**, *72*, 168–175.

Ferreira, F. d. R.; Valentim, I. B.; Ramones, E. L. C.; Trevisan, M. T. S.; Olea-Azar, C.; Perez-Cruz, F.; de Abreu, F. C.; Goulart, M. O. F., Antioxidant activity of the mangiferin inclusion complex with β-cyclodextrin. *LWT—Food Sci. Technol.* **2013**, *51* (1), 129–134.

Forero, L. F.; Pulido, D. A. P., Extracción, purificación y cuantificación de mangiferina en la corteza de algunos cultivares de mango (*Mangifera indica* L.). *Rev. Colomb. Cienc. Hortic.* **2016**, *10* (2), 292–300.

Franklin, G.; Conceição, L. F. R.; Kombrink, E.; Dias, A. C. P., Xanthone biosynthesis in hypericum perforatum cells provides antioxidant and antimicrobial protection upon biotic stress. *Phytochemistry* **2009**, *70* (1), 60–68.

Fujita, M.; Inoue, T., Biosynthesis of mangiferin in *Anemarrhena asphodeloides*: intact incorporation of C6-C3 precursor into xanthone. *Tetrahedron Lett.* **1977**, *18* (51), 4503–4506.

Fujita, M.; Inoue, T., Further studies on the biosynthesis of mangiferin in *Anemarrhena asphodeloides*: hydroxylation of the shikimate-derived ring. *Phytochemistry* **1981**, *20* (9), 2183–2185.

García, D.; Delgado, R.; Ubeira, F. M.; Leiro, J., Modulation of rat macrophage function by the *Mangifera indica* L. extracts Vimang and mangiferin. *Int. Immunopharmacol.* **2002**, *2* (6), 797–806.

García, D.; Leiro, J.; Delgado, R.; Sanmartín, M. L.; Ubeira, F. M., *Mangifera indica* L. extract (Vimang) and mangiferin modulate mouse humoral immune responses. *Phytother. Res.* **2003**, *17* (10), 1182–1187.

Garrido, G.; Blanco-Molina, M.; Sancho, R.; Macho, A.; Delgado, R.; Muñoz, E., An aqueous stem bark extract of *Mangifera indica* (Vimang®) inhibits T cell proliferation and TNF-induced activation of nuclear transcription factor NF-κB. *Phytother. Res.* **2005**, *19* (3), 211–215.

Garrido, G.; González, D.; Delporte, C.; Backhouse, N.; Quintero, G.; Núñez-Sellés, A. J.; Morales, M. A., Analgesic and anti-inflammatory effects of *Mangifera indica* L. extract (Vimang). *Phytother. Res.* **2001**, *15* (1), 18–21.

Gaur, R.; Sharma, A.; Khare, S. K.; Gupta, M. N., A novel process for extraction of edible oils: enzyme assisted three phase partitioning (EATPP). *Bioresour. Technol.* **2007**, *98* (3), 696–699.

Gómez-García, R.; Martínez-Ávila, G. C. G.; Aguilar, C. N., Enzyme-assisted extraction of antioxidative phenolics from grape (*Vitis vinifera* L.) residues. *3 Biotech.* **2012**, *2* (4), 297–300.

Gong, X.; Zhang, L.; Jiang, R.; Ye, M.; Yin, X.; Wan, J., Anti-inflammatory effects of mangiferin on sepsis-induced lung injury in mice via up-regulation of heme oxygenase-1. *J. Nutr. Biochem.* **2013**, *24* (6), 1173–1181.

Guha, S.; Ghosal, S.; Chattopadhyay, U., Antitumor, immunomodulatory and anti-HIV effect of mangiferin, a naturally occurring glucosylxanthone. *Chemotherapy* **1996**, *42* (6), 443–451.

Gullón, B.; Lú-Chau, T. A.; Moreira, M. T.; Lema, J. M.; Eibes, G., Rutin: a review on extraction, identification and purification methods, biological activities and approaches to enhance its bioavailability. *Trends Food Sci. Technol.* **2017**, *67*, 220–235.

Haynes, L. J.; Taylor, D. R., C-glycosyl compounds. Part V. Mangiferin; the nuclear magnetic resonance spectra of xanthones. *J. Chem. Soc. C: Org.* **1966**, 1685–1687.

Hu, H. G.; Wang, M. J.; Zhao, Q. J.; Yu, S. C.; Liu, C. M.; Wu, Q. Y., Synthesis of mangiferin derivates and study their potent PTP1B inhibitory activity. *Chin. Chem. Lett.* **2007**, *18* (11), 1323–1326.

Huynh, N. T.; Smagghe, G.; Gonzales, G. B.; Van Camp, J.; Raes, K., Enzyme-assisted extraction enhancing the phenolic release from cauliflower (*Brassica oleracea* L. var. *botrytis*) outer leaves. *J. Agri. Food Chem.* **2014**, *62* (30), 7468–7476.

Imran, M.; Arshad, M. S.; Butt, M. S.; Kwon, J.-H.; Arshad, M. U.; Sultan, M. T., Mangiferin: a natural miracle bioactive compound against lifestyle related disorders. *Lipids Health Dis.* **2017**, *16* (1), 84–84.

Iseda, S., On Mangiferin, the coloring matter of mango (*Mangifera indica* Linn.). V. Identification of sugar component and the structure of mangiferin. *Bull. Chem. Soc. Jpn.* **1957**, *30* (6), 629–633.

Jahanshahi, M.; Babaei, Z., Protein nanoparticle: a unique system as drug delivery vehicles. *Afr. J. Biotechnol.* **2008**, *7*, 4926–4934.

Jyotshna; Khare, P.; Shanker, K., Mangiferin: a review of sources and interventions for biological activities. *BioFactors.* **2016**, *42* (5), 504–514.

Karuppanan, M.; Krishnan, M.; Padarthi, P.; Namasivayam, E., Hepatoprotective and antioxidant effect of *Mangifera indica* leaf extracts against mercuric chloride-induced liver toxicity in mice. *Euroasian J. Hepatogastroenterol.* **2014**, *4* (1), 18–24.

Kaur, P.; Kumar Pandey, D.; Gupta, R. C.; Dey, A., Simultaneous microwave assisted extraction and HPTLC quantification of mangiferin, amarogentin, and swertiamarin in *Swertia* species from Western Himalayas. *Ind. Crops Prod.* **2019**, *132*, 449–459.

Kim, S.; Chen, J.; Cheng, T.; Gindulyte, A.; He, J.; He, S.; Li, Q.; Shoemaker, B. A.; Thiessen, P. A.; Yu, B.; Zaslavsky, L.; Zhang, J.; Bolton, E. E., PubChem 2019 update: improved access to chemical data. *Nucleic Acids Res.* **2018**, *47* (D1), D1102–D1109.

Kim, W.-J.; Veriansyah, B.; Lee, Y.-W.; Kim, J.; Kim, J.-D., Extraction of mangiferin from *Mahkota Dewa* (*Phaleria macrocarpa*) using subcritical water. *J. Ind. Eng. Chem.* **2010**, *16* (3), 425–430.

Kshirsagar, P. R.; Gaikwad, N. B.; Pai, S. R.; Bapat, V. A., Optimization of extraction techniques and quantification of swertiamarin and mangiferin by using RP-UFLC method from eleven *Swertia* species. *S. Afr. J. Bot.* **2017**, *108*, 81–89.

Kulkarni, V. M.; Rathod, V. K., Extraction of mangiferin from *Mangifera indica* leaves using three phase partitioning coupled with ultrasound. *Ind. Crops Prod.* **2014**, *52*, 292–297.

Kulkarni, V. M.; Rathod, V. K., A novel method to augment extraction of mangiferin by application of microwave on three phase partitioning. *Bioresour. Technol.* **2015**, *6*, 8–12.

Kumar, V.; Sood, H.; Chauhan, R. S., Detection of intermediates through high-resolution mass spectrometry for constructing biosynthetic pathways for major chemical constituents in a medicinally important herb, Swertia chirayita. *Nat. Prod. Res.* **2015**, *29* (15), 1449–1455.

Leiro, J.; Arranz, J. A.; Yáñez, M.; Ubeira, F. M.; Sanmartín, M. L.; Orallo, F., Expression profiles of genes involved in the mouse nuclear factor-kappa B signal transduction pathway are modulated by mangiferin. *Int. Immunopharmacol.* **2004**, *4* (6), 763–778.

Liang, J.; Li, F.; Lin, J.; Song, S.; Liao, X.; Gao, C.; Yang, B., Host-guest inclusion systems of mangiferin and polyamine-β-cyclodextrins: preparation, characterization and anti-cancer activity. *J. Mol. Struct.* **2019**, *1193*, 207–214.

Luo, Y.; Zhang, B.; Whent, M.; Yu, L.; Wang, Q., Preparation and characterization of zein/chitosan complex for encapsulation of α-tocopherol, and its in vitro controlled release study. *Colloids Surf. B Biointerfaces.* **2011**, *85* (2), 145–152.

Makare, N.; Bodhankar, S.; Rangari, V., Immunomodulatory activity of alcoholic extract of *Mangifera indica* L. in mice. *J. Ethnopharmacol..* **2001**, *78* (2), 133–137.

Matkowski, A.; Kus, P.; Wozniak, E. G.; Dorota, Mangiferin—a Bioactive Xanthonoid, not only from Mango and not just Antioxidant. *Mini.-Rev. Med. Chem.* **2013**, *13* (3), 439–455.

Medina Ramírez, N.; Monteiro Farias, L.; Apolonio Santana, F.; Viana Leite, J. P.; De Souza Dantas, M. I.; Lopes Toledo, R. C.; De Queiroz, J. H.; Stampini Duarte Martino, H.; Machado Rocha Ribeiro, S., Extraction of mangiferin and chemical characterization and sensorial analysis of teas from *Mangifera indica* L. Leaves of the Ubá variety. *Beverages* **2016**, *2* (4), 33.

Meneses, M.A.; Caputo, G.; Scognamiglio, M.; Reverchon, E.; Adami, R., Antioxidant phenolic compounds recovery from *Mangifera indica* L. by-products by supercritical antisolvent extraction. *J. Food Eng.* **2015**, *163*, 45–53.

Miura, T.; Ichiki, H.; Iwamoto, N.; Kato, M.; Kubo, M.; Sasaki, H.; Okada, M.; Ishida, T.; Seino, Y.; Tanigawa, K., Antidiabetic activity of the Rhizoma of *Anemarrhena asphodeloides* and active components, mangiferin and its glucoside. *Biol. Pharm. Bull.* **2001a**, *24* (9), 1009–1011.

Miura, T.; Iwamoto, N.; Kato, M.; Ichiki, H.; Kubo, M.; Komatsu, Y.; Ishida, T.; Okada, M.; Tanigawa, K., The suppressive effect of mangiferin with exercise on blood lipids in Type 2 diabetes. *Biol. Pharm. Bull.* **2001b**, *24* (9), 1091–1092.

Muruganandan, S.; Gupta, S.; Kataria, M.; Lal, J.; Gupta, P. K., Mangiferin protects the streptozotocin-induced oxidative damage to cardiac and renal tissues in rats. *Toxicology* **2002**, *176* (3), 165–173.

Muruganandan, S.; Srinivasan, K.; Gupta, S.; Gupta, P. K.; Lal, J., Effect of mangiferin on hyperglycemia and atherogenicity in streptozotocin diabetic rats. *J. Ethnopharmacol.* **2005**, *97* (3), 497–501.

Nadar, S. S.; Rao, P.; Rathod, V. K., Enzyme assisted extraction of biomolecules as an approach to novel extraction technology: a review. *Food Res. Int.* **2018**, *108*, 309–330.

Nishigaki, I.; Venugopal, R.; Sakthisekaran, D.; Rajkapoor, B., *In vitro* protective effect of mangiferin against glycated protein-iron chelate induced toxicity in human umbilical vein endothelial cells. *J. Biol. Sci.* **2007**, *7* (7), 1227–1232.

Pan, C.-w.; Pan, Z.-z.; Hu, J.-j.; Chen, W.-l.; Zhou, G.-y.; Lin, W.; Jin, L.-x.; Xu, C.-l., Mangiferin alleviates lipopolysaccharide and D-galactosamine-induced acute liver injury by activating the Nrf2 pathway and inhibiting NLRP3 inflammasome activation. *Eur. J. Pharmacol.* **2016**, *770*, 85–91.

Pardo-Andreu, G. L.; Delgado, R.; Núñez-Sellés, A. J.; Vercesi, A. E., Dual mechanism of mangiferin protection against iron-induced damage to 2-deoxyribose and ascorbate oxidation. *Pharmacol. Res.* **2006**, *53* (3), 253–260.

Perrucci, S.; Fichi, G.; Buggiani, C.; Rossi, G.; Flamini, G., Efficacy of mangiferin against *Cryptosporidium parvum* in a neonatal mouse model. *Parasitol. Res.* **2006**, *99* (2), 184.

Pleguezuelos-Villa, M.; Nácher, A.; Hernández, M. J.; Ofelia Vila Buso, M. A.; Ruiz Sauri, A.; Díez-Sales, O., Mangiferin nanoemulsions in treatment of inflammatory disorders and skin regeneration. *Int. J. Pharm.* **2019**, *564*, 299–307.

Prashanth, D.; Amit, A.; Samiulla, D. S.; Asha, M. K.; Padmaja, R., α-Glucosidase inhibitory activity of *Mangifera indica* bark. *Fitoterapia* **2001**, *72* (6), 686–688.

Rai, A.; Nakamura, M.; Takahashi, H.; Suzuki, H.; Saito, K.; Yamazaki, M., High-throughput sequencing and de novo transcriptome assembly of *Swertia japonica* to identify genes involved in the biosynthesis of therapeutic metabolites. *Plant Cell Rep.* **2016**, *35* (10), 2091–2111.

Rajendran, P.; Ekambaram, G.; Magesh, V.; Sakthisekaran, D., Chemopreventive efficacy of mangiferin against benzo(a)pyrene induced lung carcinogenesis in experimental animals. *Environ. Toxicol. Pharmacol.* **2008a**, *26* (3), 278–282.

Rajendran, P.; Ekambaram, G.; Sakthisekaran, D., Effect of mangiferin on benzo(a) pyrene induced lung carcinogenesis in experimental Swiss albino mice. *Nat. Prod. Res.* **2008b**, *22*, 672–680.

Rechenchoski, D.; Galhardi, L.; Cunha, A.; Ricardo, N.; Nozawa, C.; Linhares, R., Antiviral potential of mangiferin against poliovirus. *Int. J. Pharmacol. Res.* **2018**, *8*, 34.

Rolo, A. P.; Palmeira, C. M., Diabetes and mitochondrial function: role of hyperglycemia and oxidative stress. *Toxicol. Appl. Pharmacol.* **2006**, *212* (2), 167–178.

Ruiz-Montañez, G.; Ragazzo-Sánchez, J. A.; Calderón-Santoyo, M.; Velázquez-de la Cruz, G.; Ramírez de León, J. A.; Navarro-Ocaña, A., Evaluation of extraction methods for preparative scale obtention of mangiferin and lupeol from mango peels (*Mangifera indica* L.). *Food Chem.* **2014**, *159*, 267–272.

Saha, S.; Sadhhukhan, P.; Sil, P.C., Mangiferin: a xanthonoid with multipotent anti-inflammatory potential. *BioFactors* **2016**, *42* (5), 459–474.

Sajid, M.; Cameotra, S. S.; Ahmad Khan, M. S.; Ahmad, I., Chapter 23, Nanoparticle-based delivery of phytomedicines: challenges and opportunities. In: *New Look to Phytomedicine*. Ahmad Khan, M. S.; Ahmad, I. and Chattopadhyay, D., Eds. Academic Press: Cambridge, **2019**, 597–623.

Samadarsi, R.; Dutta, D., Design and characterization of mangiferin nanoparticles for oral delivery. *J. Food. Eng.* **2019**, *247*, 80–94.

Sampaio, C. d. G.; Frota, L. S.; Magalhães, H. S.; Dutra, L. M. U.; Queiroz, D. C.; Araújo, R. S.; Becker, H.; de Souza, J. R. R.; Ricardo, N. M. P. S.; Trevisan, M. T. S., Chitosan/mangiferin particles for Cr(VI) reduction and removal. *Int. J. Biol. Macromol.* **2015**, *78*, 273–279.

Sánchez, G. M.; Re, L.; Giuliani, A.; Núñez-Sellés, A. J.; Davison, G. P.; León-Fernández, O. S., Protective effects of *Mangifera indica* L. extract, mangiferin and selected antioxidants against

TPA-induced biomolecules oxidation and peritoneal macrophage activation in mice. *Pharmacol. Res.* **2000**, *42* (6), 565–573.

Sekar, M., Molecules of interest-mangiferin-a review. *Ann. Res. Rev. Biol.* 2015, *5* (4), 307–307.

Sekar, V.; Mani, S.; Malarvizhi, R.; Nithya, P.; Vasanthi, H. R., Antidiabetic effect of mangiferin in combination with oral hypoglycemic agents metformin and gliclazide. *Phytomedicine* **2019**, *59*, 152901.

Stoilova, I.; Ho, L., Antimicrobial and anti-oxidant activity of the polyphenol mangiferin. *Herba Polonica.* **2004**, *51*, 37–43.

Suchal, K.; Malik, S.; Gamad, N.; Kumar, M.R.; Goyal, S.N.; Ojha, S.; Kumari, S.; Bhatia, J.; Singh, D., Singh Arya, S. D., Mangiferin protect myocardial insults through modulation of MAPK/TGF-β pathways. *Eur. J. Pharmacol.* **2016**, *776*, 34–43.

Szandruk, M.; Merwid-Ląd, A.; Szeląg, A., The impact of mangiferin from *Belamcanda chinensis* on experimental colitis in rats. *Inflammopharmacology* **2018**, *26* (2), 571–581.

Velderrain-Rodríguez, G. R.; Acevedo-Fani, A.; González-Aguilar, G. A.; Martín-Belloso, O., Encapsulation and stability of a phenolic-rich extract from mango peel within water-in-oil-in-water emulsions. *J. Funct. Foods.* **2019**, *56*, 65–73.

Wang, B.; Wan, J.; Gong, X.; Kuang, G.; Cheng, X.; Min, S., Mangiferin attenuates renal ischemia-reperfusion injury by inhibiting inflammation and inducing adenosine production. *Int. Immunopharmacol.* **2015**, *25* (1), 148–154.

Wei, X.; Liang, D.; Wang, Q.; Meng, X.; Li, Z., Total synthesis of mangiferin, homomangiferin and neomangiferin. *Org. Biomol. Chem.*, **2016**, *14* (37), 8821–8831.

Wijngaard, H.; Hossain, M. B.; Rai, D. K.; Brunton, N., Techniques to extract bioactive compounds from food by-products of plant origin. *Food Res. Int.* **2012**, *46* (2), 505–513.

Womeni, H. M.; Ndjouenkeu, R.; Kapseu, C.; Mbiapo, F. T.; Parmentier, M.; Fanni, J., Aqueous enzymatic oil extraction from Irvingia gabonensis seed kernels. *Eur. J. Lipid Sci. Technol.* **2008,** *110* (3), 232–238.

Wu, Z.; Wei, G.; Lian, G.; Yu, B., Synthesis of mangiferin, isomangiferin, and homomangiferin. *J. Org. Chem.* **2010,** *75* (16), 5725–5728.

Yoshikawa, M.; Nishida, N.; Shimoda, H.; Takada, M.; Kawahara, Y.; Matsuda, H., Polyphenol constituents from Salacia species: quantitative analysis of mangiferin with alpha-glucosidase and aldose reductase inhibitory activities. *YAKUGAKU ZASSHI.* **2001,** *121* (5), 371–378.

Yoshimi, N.; Matsunaga, K.; Katayama, M.; Yamada, Y.; Kuno, T.; Qiao, Z.; Hara, A.; Yamahara, J.; Mori, H., The inhibitory effects of mangiferin, a naturally occurring glucosylxanthone, in bowel carcinogenesis of male F344 rats. *Cancer Lett.* **2001,** *163* (2), 163–170.

Zheng, H. Z.; Chung, S. K., Case study: optimization of enzyme-aided extraction of polyphenols from unripe apples by response surface methodology. In: *Mathematical and Statistical Methods in Food Science and Technology.* John Wiley & Sons: Hoboken, **2014,** 31–42.

Zheng, M.; Lu, Z., Antiviral effect of mangiferin and isomangiferin on herpes simplex virus. *Acta Pharmacol. Sin.* **1989,** *10* (1), 85–90.

Zhu, X.; Song, J.; Huang, Z.; Wu, Y.; Yu, M., Antiviral activity of mangiferin against herpes simplex virus type 2 in vitro. *Acta Pharmacol. Sin.* **1993,** *14* (5), 452–454.

Zou, T.-B.; Xia, E.-Q.; He, T.-P.; Huang, M.-Y.; Jia, Q.; Li, H.-W., Ultrasoundassisted extraction of mangiferin from mango (*Mangifera indica* L.) leaves using response surface methodology. *Molecules* **2014,** *19* (2), 1411–1421.

Zou, T.; Wu, H.; Li, H.; Jia, Q.; Song, G., Comparison of microwave-assisted and conventional extraction of mangiferin from mango *(Mangifera indica* L.) leaves. *J. Sep. Sci.* **2013,** *36* (20), 3457–3462.

Zuidam, N. J.; Shimoni, E., Overview of microencapsulates for use in food products or processes and methods to make them. In *Encapsulation Technologies for Active Food Ingredients and Food Processing.* Zuidam, N. J. and Nedovic, V., Eds. Springer: New York, NY, **2010**; 3–29.

CHAPTER 23

Release and Production of Phenolic Biomolecules by Fungal Enzymes From Biomass

ERIKA ACOSTA-CRUZ[1], LUIS VÍCTOR RODRÍGUEZ-DURÁN[2],
LEONARDO SEPÚLVEDA[3], FERNANDO JASSO-JUAREZ[1],
MARISOL CRUZ-REQUENA[4], JOSÉ ANTONIO RODRÍGUEZ-DE LA GARZA[1],
LEOPOLDO J. RÍOS-GONZÁLEZ[1], and MIGUEL A. MEDINA-MORALES[1*]

[1]Departamento de Biotecnología, Facultad de Ciencias Químicas, Universidad Autónoma de Coahuila, Blvd. Venustiano Carranza, 25280 Saltillo, México

[2]Departamento de Ingeniería Bioquímica, Unidad Académica Multidisciplinaria, Universidad Autónoma de Tamaulipas, Cd. Mante, 87000 Cd Victoria, México

[3]Grupo de Bioprocesos y Bioquímica Microbiana, Facultad de Ciencias Químicas, Universidad Autónoma de Coahuila, 25280 Saltillo, México

[4]Departamento de Investigación en Alimentos, Facultad de Ciencias Químicas, Universidad Autónoma de Coahuila, 25280 Saltillo, México

*Corresponding author. E-mail: miguel.medina@uadec.edu.mx.

ABSTRACT

The biotechnological importance at industrial levels is increasing everyday. By combining the availability of residual materials or responsible exploitation of biomass with biotechnological means, the overall process becomes safer and environmentally friendly. There is an extensive array of compounds of natural origin, which have been attributed with several bioactivities, many of which are of phenolic nature. These compounds are found in nature and most of them are glycosylated in plant tissues. The most common way to obtain them is by physicochemical extraction, which can be costly and may produce pollutants. By the use of microorganism that is able to degrade these types of compounds, it is possible to produce and recover the enzymes responsible to release bioactive aglycones and accumulate said compounds in the same bioprocess or in a separate system. An approach to bioprocess resources for enzyme and bioactive production is fermentations with microorganisms that produce the necessary enzymes for the purpose of interest. Also, examples will be addressed for biomolecules isolation, recovery, and β-glucosidase structure simulations from microbial sources.

23.1 INTRODUCTION

In plants, polyphenolic compounds are produced by the shikimic acid metabolic pathway and there is a wide array of compounds derived from polyphenols such as lignans, stilbenoids, flavonoids, and phenolic acids (Brillouet et al., 2013). Currently, there is a renewed interest in natural alternatives for the production or release of bioactive compounds with several industrial applications (Duan et al., 2019). For the effects of this chapter, special attention will be paid to flavonoids and phenolic acids. Among polyphenols, flavonoids constitute a large part of these molecules. Its structure corresponds to a carbon backbone of C_6-C_3-C_6 (Kawabata et al., 2015). There are

several types of flavonoids, which are anthocyanins, flavanones, flavonols, flavones, isoflavones, and flavanols. Depending on the substitution in its rings, flavonoids can be classified as flavones or isoflavones (Pandey et al., 2016). Phenolic acids are derivates from cinnamic and hydroxybenzoic acids with a C_6-C_1 and C_6-C_3 basic conformation (Zhang et al., 2019). Molecules of these types can be found in nature polymerized and/or glycosylated. Examples of phenolic acids are caffeic, chlorogenic, ellagic, and gallic acids. From these last two, polymerized forms exist such as ellagitannins and gallotannins (Aguilera-Carbó et al., 2008; Chávez-González et al., 2018). As previously mentioned, glycosylated phenolic molecules also are present in plants. Naturally, phenolic molecules may be water-insoluble due to their structure and glycosylation favors solubility and transport in plant tissues such as anthocyanins and glycosides of caffeic acid, gallotannins, and ellagitannins (Barnaba et al., 2018; Núñez-López et al., 2019). This type of compound exhibits bioactivities such as anticarcinogenic, antimicrobial, antiviral, and antioxidant and they promote benefits to human health. Most of the bioactivities are linked to the ability of polyphenols to bond to several molecules or free radicals and it is by this action that these molecules can act as antimicrobial or antioxidant agents (Yang et al., 2018). As these compounds can be found in nature, several methods for the extraction or assist extraction exist for polyphenols recovery, such as mechanical and/or physicochemical techniques such as pressing, milling, solvent extraction, ultrasound, microwave, among others. For some of these extraction methods, solvents and equipment are needed which may increase production costs (Wong-Paz et al., 2015; Zhang et al., 2013; Pinheiro do Prado et al., 2014).

Biotechnological means of extraction or release is also an option, many researchers are considering as a viable for it has very little or none repercussions in the environment and is more cost-effective. There are several studies where, by the effect of fermentation or enzymatic degradation, biomolecules have been released and recovered (Medina-Morales et al., 2017; Gligor et al., 2019). If this is the case, for effective exploitation of available resources by biotechnological means, a bioprocess configuration must be applied to determine the most suitable microorganism and conditions for biodegradation to take place and biomolecules release detection and recovery (Figure 23.1). According to the microorganism and the substrate, the type of culture must be considered. Firstly, the microorganism must be suited to produce the enzymes required to degrade the plant structures that are bound to the

molecules of interest. Microorganisms are capable of degrading plant tissue bound to bioactive molecules, but special emphasis is given to fungi due to its higher adaptability to recalcitrant growth conditions and higher resistance to polyphenolic content in plants or biomass (Mutabaruka et al., 2007; Martins et al., 2011; Cao et al., 2015). The microorganisms, adapted to growth conditions, are likely able to produce glycosidases responsible for biomolecules release, as well as cellulolytic enzymes to favor degradation and access to nutrients and energy. There are studies where agro-wastes and biomass have been used to release bioactives through a fermentative process with promising results, and it has given a pattern in the use of these resources for bioactive release and recovery (López-Trujillo et al., 2017). Secondly, the substrate available is of the utmost importance. Many wastes or by-products, from agro-industries and biomass, that are available for biotechnological revalorization and the bioprocessing of these materials is, as previously mentioned, by fermentation (de Oliveira Rodrigues et al., 2017). Being considered a problem the amount of residue or by-product generated by agro-industries, the idea is to utilize high amounts of substrate in the bioprocess. For this reason, solid-state fermentation is an adequate alternative to use high amounts of solids, low water content, and higher adaptation by microorganism (fungi

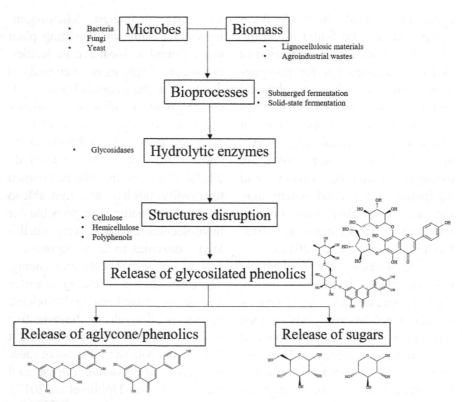

FIGURE 23.1 Overall scheme of biomass biodegradation to release phenolic biomolecules.

in particular) (Soccol et al., 2017). By using agro-food wastes, ellagic acid has been produced from pomegranate residues by solid-state fungal fermentation (Sepúlveda-Torre et al., 2018), soybean isoflavones by yeast and fungal fermentation (Queiroz-Santos et al. 2018), gallic and digallic acid from mango seeds (Torres León et al., 2019), among other research works. Concerning the use of biomass, the use of *Flourensia cernua* biomass by fungal solid-state bioprocess, glycosides of luteolin and apigenin were obtained (Medina-Morales et al., 2017), ellagic and gallic acid have been produced from *Larrea tridentata* (Aguilera-Carbó et al., 2008), just to name a few examples. Now, for the release of bioactive phenolics, enzymes must be produced by the microorganisms which must be coded into their DNA. In-plant tissues, the compounds of interest often are bound to the polysaccharides present in their structure (Quirós-Sauceda et al., 2014). The enzymes required to degrade polysaccharides, such as cellulose are endoglucanases, cellobiohydrolases, and β-glucosidases; for hemicellulose, xylanases, β-xylosidases, arabino-furanosidases, among many others for other substrates (Juturu and Wu, 2012, Cao et al., 2015, Marín et al., 2019). Considering these enzymes degrade polysaccharides and polyphenols are bound to these fibers, fibrolytic enzymes must play a role in the release of the bioactive compounds (Lopez-Trujillo et al., 2017). Enzymes can often interact with more than one type of substrate; it may be possible that the bonds between phenolic compounds and polysaccharides can be cleaved by enzymes such as β-glucosidases and it may possible that glycosides can be degraded by these enzymes as well (Yeoman et al., 2010). This chapter will deal with fermentation processes that release bioactive phenolics and the enzymes responsible for that action.

23.2 EXTRACTION AND BIOTECHNOLOGICAL APPROACH

The most common way to obtain phenolic extracts readily for use from any given source is aqueous extraction. These compounds have been used for a long time for several purposes, much of it unknowingly, as infusions for several ailments' relief, or in leather tanning. As these molecules possess bioactivities that help hinder or inhibit the causing agents of diseases, the conformation, concentration, and availability of the bioactive molecules could be improved by biotechnological means. The sources of these molecules are plant materials from which a certain amount is considered residues, others are food and vegetation. Several methods have been employed such as Soxhlet, microwave, organosolv, ultrasound, among others (Gómez-Mejía et al., 2019; Pavlić et al., 2019; Ravindran et al., 2018). Although these methods are effective, there are still molecules bound to lignocellulose fiber which can be further processed for their extraction. Concerning fiber-phenolics association, biodegradation is a suitable alternative because the enzymes produced by microorganisms, depolymerize fibers, thus leaving phenolic molecules bioavailable.

Polyphenols have been known to act as antagonists to microbial growth. Several derivates of these compounds have been reported as antimicrobials, such as flavonoids, procyanidins, lignans, phenolic acids, and hydrolysable tannins (Daglia, 2012). However, there are other microorganisms that are able to grow in the presence of these types of molecules and also, produce enzymes that are resistant to the ability of the polyphenols to form covalent bonds with proteins. Bhat et al. (1998) reported several microorganisms that are able to grow in the presence of polyphenols, such as tannins, that are known to have antimicrobial activity. Several bacteria and fungi are able to grow and, therefore, degrade those compounds, from which fungi such as *Aspergillus* and *Penicillium* are mentioned. Many research papers mention *Aspergillus* as an effective biodegradator and/or producer of enzymes that catalyzes hydrolytic reactions on polyphenols (Medina-Morales et al., 2017; López-Trujillo et al., 2017; Chávez-González et al., 2014). Given the fact that lignocellulosic fibers and polyphenols are closely related, enzymes such as cellulases, hemicellulases, and glycosidases are needed to release polyphenols, being flavonoids, phenolic acids, and/or hydrolyzable tannins.

23.3 MICROBIAL RELEASE OF FLAVONOIDS AND ITS GLYCOSIDES

These molecules hold great importance in several aspects. As they promote natural product addition in food and cosmetics, its extraction has required research in biotechnological options. The main interest of these molecules is in human health application. These molecules have beneficial effects, such as antioxidant, antiviral, antifungal, antibacterial, anticancer, antiallergenic, and anti-inflammatory (Sordon et al., 2016). Flavonoids may be present as glycosylated or as an aglycone and its absorption has not been understood in its totality. It is important to study its release, in this case by microbiological action, to later process polyphenolic sources to accumulate these types of molecules and apply or scale it (Figure 23.2).

There are several methodologies for the extraction of flavonoids where solvents such as acetone, ethanol, and methanol have been used (Yahia et al., 2017). Other options are microwave extraction, which rapidly heats solvents and plant material in an electromagnetic field (Švarc et al., 2013) and ultrasound-assisted extraction has been used for the same purpose (Irakli et al., 2018). As effective as these techniques, some aspects are enhanced by biotechnological approaches. If a biological degradation process takes place in a substrate, be it biomass or agro-industrial wastes, bioactive flavonoids can be released, as well as the enzymes responsible for its degradation. This is most interesting due to the ability of microorganisms to produce several types of enzymes depending on the substrate (Lu et al., 2010).

Fungi, bacteria, and yeasts are able to degrade material and increase phenolics content in fermented material and its subsequent recovery if so desired. Yeasts such as *Debaryomyces hansenii* is a common find in high-protein content fermented plant grains, such as soybean. This grain has isoflavones which its

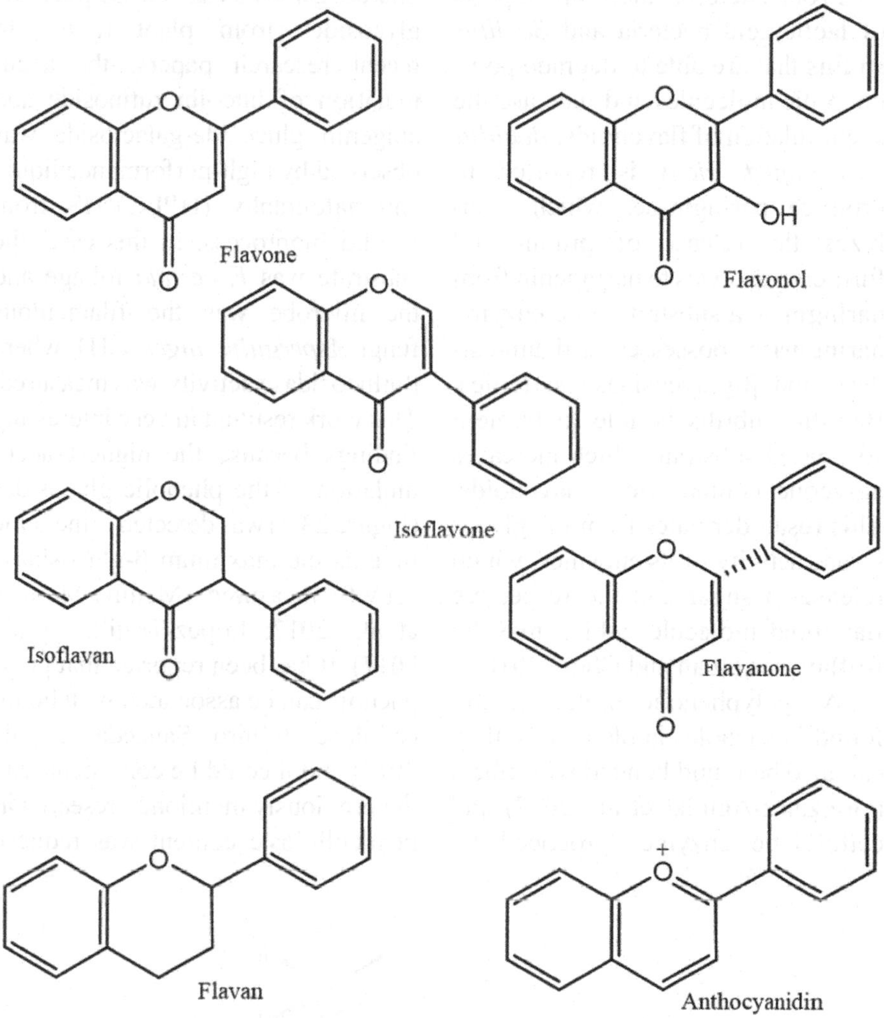

FIGURE 23.2 Basic carbon-skeletons of flavonoids found in nature.

natural form as a glycoside. It has been found that this yeast produces a β-glucosidase enzyme that releases sugars from the glycosylated form of the isoflavones from soy molasses. Since there are several types of isoflavones in soybean, one of these is β-glycosyl isoflavones such as b-glucosides of genistin, glycitin, and daidzin; this enzyme was able to release aglycones in the free and immobilized form (Maitan-Alfenas et al., 2014; Falcão et al., 2018).

From bacteria, there are reports of lactic acid bacteria and *Bacillus* strains that are able to degrade polyphenolic molecules and increase the accumulation of flavonoids. *Bacillus amyloliquefasciens* is reported to produce naringinase, which catalyzes the release of prunin and further hydrolysis to naringenin from naringin as a substrate. The enzyme naringinase possesses a-rhamnosidase and β-gluycosidase activities. Bacillus subtilis is able to ferment soy-based substrate which increases aglycone content from flavonoids. This result derivates from the glycosidase activity of its enzymes, which releases a sugar and the respective flavonoid molecule, which may be isoflavones (Juan and Chou, 2010).

As polyphenolic molecules are found in vacuoles in plant cells, they can also be found bonded with fibers (González-Aguilar et al., 2017) and cellulolytic enzymes produced by

microorganisms can release phenolic glycosides from plant tissue. In recent research papers, the accumulation of luteolin rutinoside and apigenin glucoside-galactoside was observed by high-performance liquid chromatography (HPLC)/MS from a solid bioprocess. In this case, the substrate was *F. cernua* foliage and the microbe was the filamentous fungi *Aspergillus niger* GH1 where β-glucosidase activity was measured. This work resulted in very interesting findings because the highest accumulation of the phenolic glycosides (Figure 23.3) was detected at the same time as the maximum β-glucosidase activity measured (Medina-Morales et al., 2017; Lopez-Trujillo et al., 2017). It has been reported that polyphenols can be associated with hemicellulose (Quiros-Sauceda et al., 2014) and it could be consistent with the previously mentioned research as hemicellulosic content was reduced

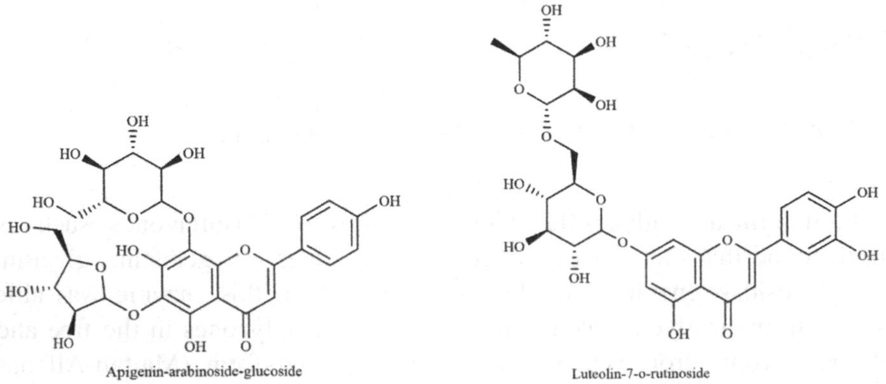

Apigenin-arabinoside-glucoside

Luteolin-7-o-rutinoside

FIGURE 23.3   Glycosilated flavones found in *Flourensia cernua* were released by fungal solid-state fermentation.

and may be correlated to polyphenol-fiber interaction (López-Trujillo et al., 2017).

To our knowledge, these are more reports involving fungi in flavonoid degradation than other microbes. Several reports also mention that the growth cultures for fungi to degrade these molecules are solid-state fermentation. Fungi, being mostly aerobic organisms, are commonly found in decaying wood or in unstored bread. Solid-state fermentation in laboratory conditions allows the emulation of the habitat of fungi which in turn

also allows a better observation and study of its degradation mechanisms in that growth system. In the case of fungi, microbes, such as *Aspergillus niger*, are able to produce several hydrolytic enzymes for plant tissue depolymerization. As it has been mentioned, glycosyl hydrolases are capable of cleaving the bonds between phenolics and sugars (Figure 23.4); considering that, it is possible that the enzymes for cellulose and hemicellulose degradation can also hydrolyze glycosylated phenolics to release aglycones (Lu et al., 2010).

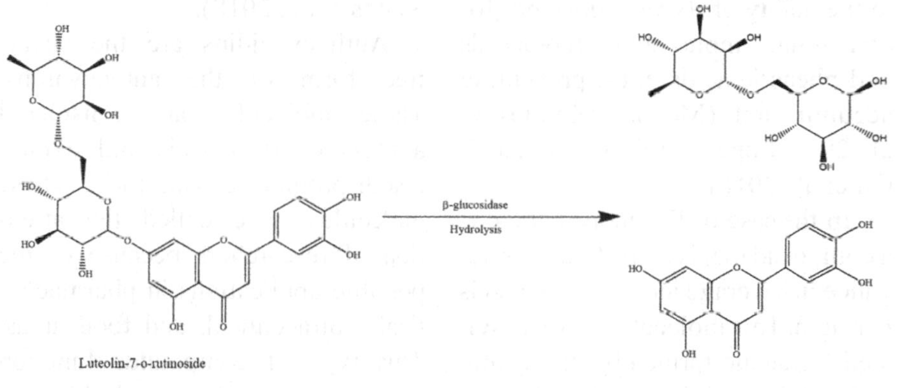

β-glucosidase
Hydrolysis

Luteolin-7-o-rutinoside

FIGURE 23.4 Proposed diagram for enzymatic degradation of Luteolin-7-o-rutinoside by fungal β-glucosidase.

23.3.1 IMPORTANCE OF FLAVONOID BIOMOLECULES AND ITS GLYCOSIDES

The importance of these molecules is that they exhibit several bioactivities. Flavonols have been studied for their bioactivities from which they

are several reported such as a beneficial role in cardiovascular diseases, neuroprotective activities as well as anti-Alzheimer, anti-inflammatory, and anti-arthritic activities; also, anticancer, antiaging, antioxidant, among many others (Semwal et al., 2019). Flavone glycosides also have been evaluated for bioactivities similar to the already mentioned (Duan et al., 2019).

Glycosylated flavones have been detected in fermented *F. cernua* foliage in solid-state. The discussion in that research is that glycoside hydrolases, such as β-glucosidases, are involved in the release of these types of compounds. In the same system, it is also possible that the same enzymes promote the aglycones liberation from sugar molecules due to the detection, although at very low levels, the presence of free apigenin and luteolin. The absence of accumulation may be attributed to the ability of *Aspergillus niger* for consuming monomeric flavonoids and phenolic acids, thus, preventing accumulation (Medina Morales et al., 2017; Lopez-Trujillo et al., 2017; Oh et al., 2018).

In the case of flavanones, there is a compound that is considered of relevance in beverage industries, which is naringin. This molecule is a glycosylated flavanone formed by naringenin and neohesperidiose. The enzyme responsible for its degradation is known as naringinase. This enzyme has been produced by bacteria and fungi such as *Bacillus amyloliquefaciens*, *Aspergillus sojae*, and *Penicillium decumbens*. Also, another example is hesperidin, which is the glycoside of hesperetin and rutinose. It has been reported that naringinase is able to release aglycones from

hesperidin by a glycosidase from *A. sojae*, which is also a naringinase (Cui et al., 2016; Lee et al., 2012; Zhu 2017).

Isoflavones are another type of flavonoids that can possess antioxidant activity and immunomodulatory and estrogenic activity as well. These compounds are highly notorious because of these biological activities and their extraction and recovery of a subject of research. There is research where it's showed that *Saccharomyces cerevisiae* β-glucosidases while fermenting soybean, released aglycones from b-glucosides isoflavones (Queiroz-Santos et al., 2018).

Anthocyanidins are the sugar-free form of the anthocyanins. These molecules are considered a class of flavonoids and exhibit health-promoting activities. These molecules have called the attention of researchers because of the possible applications in pharmaceutical, nutraceutical, and food areas. This type of compounds function as pigments in plants and also as a defense mechanism against potential plant pathogens which also poses as another alternative as a natural dye in products, such as drugs and food (Veitch and Grayer, 2011). In this case, there are research reports where bacteria are able to degrade anthocyanins to anthocyanidins. Lactic acid bacteria are able to produce enzymes such as β-glucosidase, β-galactosidase, and α-galactosidase, which may be related to anthocyanin degradation. In the same cases, the degradation of such molecules goes forward to produce chalcones and protocatechuic acid (Braga 2017; Shin 2018).

23.4 ENZYMATIC MECHANISMS FOR THE RELEASE OF HYDROLYSABLE POLYPHENOLIC COMPOUNDS

Tannins are a wide range of polyphenolic compounds that are distributed in nature in leaves, stems, flowers, and fruits. At the moment, tannins are classified into three great classes as proanthocyanidins or condensed tannins, gallotannins and ellagitannins or hydrolysable tannins, and finally in phlorotannins that are found in the brown algae (Quideau, et al. 2011). Currently, information about fungal enzymes involved in hydrolysis for the release of these polyphenolic compounds is very scarce. This release of molecules can be using bioprocesses. Next, some research papers on the conditions of solid and/or submerged fermentation processes for the release of polyphenols and the possible enzymatic mechanism involved are described.

The process of discoloration in mill waters of the olive residue and the biodegradation of phenolic compounds by *Geotrichum candidum* was investigated. The results of this study suggest that the culture conditions produce a high activity of lignin peroxidase and manganese peroxidase and as a result lead to high levels of decolorization of the residue by *G. candidum* due to the depolymerization of phenolic compounds with high molecular weight, which are the most recalcitrant compounds (Asses et al., 2009). On the other hand, they studied the biotransformation of quercetin by *Gliocladium deliquescens* NRRL 1086. The results showed that the formation of these polyphenols can take two metabolic pathways: regio-selectivity glycosylation and 2,3,dioxygenation of quercetin coexisting in the culture. In addition, they describe that the formation of these compounds is influenced by the amount of glucose and some trace elements in the system. The authors concluded that *G. deliquescens* was able to glycosylate in a highly selective manner to phenolic hydroxyl in carbon 3 of the flavonols and observed the presence of enzymes involved in the degradation process (Xu et al., 2017). In another study, they developed a bioprocess using theaflavins in an immobilized polyphenol oxidase system for the conversion of green tea leaf catechins. The authors concluded that this system was able to adequately activate hydroxyl groups for the development of the biotransformation of tea catechins in theaflavins. These molecules are important antioxidants that can be used in the food, pharmacy, and cosmetic industry (Sharma et al. 2009).

They studied the effect of a sequential addition of substrate on the tannase activity for the increase of epigallocatechin and gallic acid. They observed that the increase in the percentage of the substrate can inhibit the formation of the molecules. The release of epigallocatechin and gallic acid is mainly due to the action of tannase. This enzyme can hydrolyze the ester bonds of substrates, such as tannic acid, epicatechin gallate, chlorogenic acid, etc. The authors concluded that the biotransformation of these molecules from green tea extracts can be used particularly in the food and beverage industry as additives (Noh et al., 2014). In another study, they evaluated the chemopreventive potential of a green tea extract and epigallocatechin gallate after tannase-mediated hydrolysis. The results showed that biotransformed compounds retained the beneficial properties of the original compounds. The authors highlighted the benefit of biotechnological modifications of natural food molecules (Macedo et al., 2012).

Grape pomace is a residue, the process of making wine that is rich in polyphenolic compounds. They studied the effect of tannase, pectinase, cellulase, and the combination of these for the release of polyphenols from different residues of the wine industry. The increase in quercetin content is attributed to the enzyme tannase, where the content increases by 3.8 times. In addition, the increase in transresveratrol content is due to the hydrolysis of pectinase, cellulase, and β-glucosidase. The authors concluded that due to the high content of resveratrol and gallic acid from the hydrolysis of grape residues, they have potential application as foods functional, nutraceutical, and/or cosmetic (Martins et al., 2016). The juice industry generates a lot of waste. However, these residues are rich in phenolic glycosides. In another study, enzyme-assisted extraction and biotransformation of phenolic compounds from citrus juice by-products were evaluated. They evaluated the individual and combined effect of pectinase, cellulase, tannase, and β-glucosidase for the biotransformation of phenolic compounds. The best results showed that β-glucosidase promotes phenolic conversion mainly to hesperetin. In addition, the mixture between β-glucosidase and tannase generates bioactive compounds, such as hesperetin, naringenin, and diosmetin. These compounds can be used as functional ingredients and natural antioxidants (Roggia-Ruviaro et al., 2019).

On the other hand, they evaluated the solid-state fermentation process with a combination of different microorganisms from guava leaves to promote the release of polyphenolic compounds. The best results showed that cofermentation with *Monascus anka* and *S. cerevisiae* led to a significant increase in the content of quercetin and total polyphenols, including gallic acid, chlorogenic acid, quercetin, kaempferol. The authors concluded that this method is a feasible application for microbial conversion to improve the nutritional or medicinal values of natural herbal products (Wang et al., 2016). Green tea and yerba mate are plants rich in polyphenolic compounds. In another study, they evaluated the enzymatic biotransformation reaction catalyzed by the *Paecilomyces variotii* tannase. The results showed that the tannase could probably hydrolyse the substrates contained in the tea, and the hydrolysis products apparently improve the antioxidant activity (Macedo et al., 2011). The production of cellulases was evaluated by *Trichoderma reesei* for the release of polyphenols from rice straw as a carbon source. The main phenolic compounds released were phenolic acids and tannins. However, the high concentration of coumaric acid and ferulic acid can inhibit the production of cellulases and decrease their activity by up to 23%. The authors concluded that it was the first report on the influence of the release of polyphenolic compounds from rice straw for the production of cellulases (Zheng et al., 2017).

Finally, the biodegradation of tannic acid by *Aspergillus niger* GH1 in submerged fermentation and solid-state fermentation was evaluated by liquid chromatography coupled to mass spectrometry. They marked the differences in the chemical composition of tannic acid, in addition to achieving substrate consumption and identifying biodegradation intermediaries. The authors demonstrated that the hydrolysis profile is the same in both systems evaluated but with differences in substrate absorption time and product release (Chávez-González et al., 2014). Figure 23.5 shows the possible route of biodegradation of tannic acid and some intermediates. On the other hand, they developed a process of biotransformation of phenolic compounds from citrus residues by solid-state fermentation with the microorganism *P. variotii*. The best fermentation conditions were 10 g of waste as a source of carbon and energy, 20 mL of volume at 32 °C for 48 h of incubation. The authors mention that one of the possible enzymes involved in the hesperidin and naringin release may be the tannase. The authors concluded that the use of agro-industrial citrus wastes is important due to their availability, low cost, and characteristics that allow to obtain bioactive phenolic compounds (Madeira Jr et al., 2014).

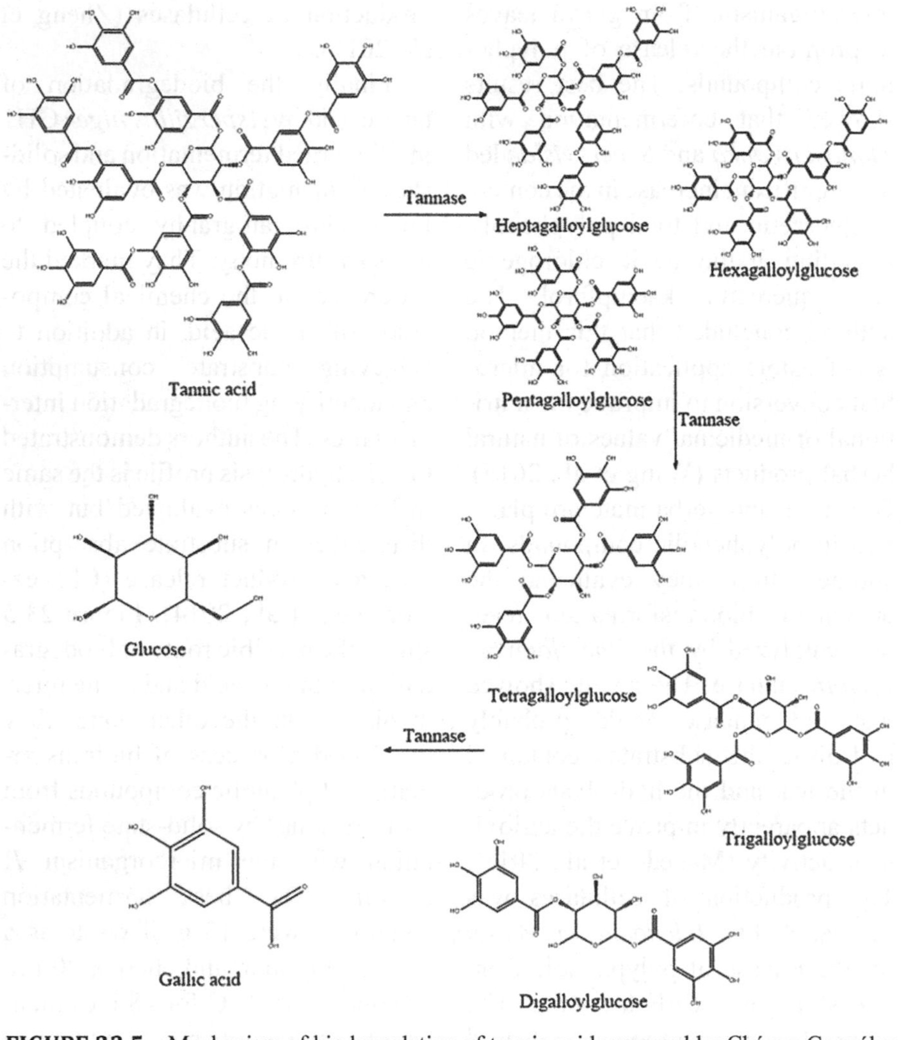

FIGURE 23.5 Mechanism of biodegradation of tannic acid proposed by Chávez-González et al. (2014) with some modifications. (Author image).

23.5 ISOLATION AND PURIFICATION OF PHENOLIC BIOMOLECULES

The polyphenol-rich extracts obtained from natural sources are complex mixtures whose composition depends on the source and the extraction methods used. In general, polyphenol-rich extracts are dilute solutions that contain the phenolic compound, carbohydrates, lipids, and other molecules. In addition, phenolic compounds are present in nature in groups of molecules with related chemical structures.

Many techniques have been developed for the analytical separation of bioactive phenolic compounds. Most of them are based on chromatographic techniques such as HPLC and ultra-performance liquid chromatography coupled to ultraviolet, photodiode array, or mass spectrometry detectors (Cavaliere et al., 2018). However, relatively few protocols have been developed for the isolation and purification of bioactive compounds on a preparative scale. Table 23.1 summarizes some procedures for the isolation of phenolic compounds from different sources.

Several researchers have developed protocols for the separation of phenolic-rich extracts in order to obtain specific phenolic compounds with the degree of purity required for certain applications (Jampani and Raghavarao, 2015; Lu et al., 2004; Wang et al., 2014). In these cases, the presence of the target compound is monitored in the chromatography fractions by analytical techniques, such as thin layer chromatography and HPLC. On the other hand, other researchers have fractionated polyphenol-rich extracts in order to isolate and identify the compound responsible for certain biological activity, in the so-called bioassay-guided isolation method (Erenler et al., 2017; Oldoni et al., 2016; Sudha and Srinivasan, 2014; Wang et al., 2018). In the bioassay-guided isolation method,

the researcher measures a biological activity of interest, such as antioxidant, antiproliferative, or anticancer activities. The fractions that exhibit the desired biological activity are collected, analyzed, and, if necessary, fractionated by another preparative technique. In either case, the separation protocol depends on the target molecule or molecules, the composition of the polyphenol-rich extract, and the degree of purity desired.

Some classic methods consist of a series of liquid–liquid extractions, precipitations, and crystallizations. For example, Hulme (1953) isolated chlorogenic acid from immature apples by Soxhlet extraction, washing with ethyl ether, precipitation with lead acetate, extraction with ethyl acetate, precipitation with light petroleum, and crystallization from water. Jurd (1956) isolated ellagic acid from walnut pellicles by Soxhlet extraction with methanol, evaporation, solubilization in ether, washing with water, and liquid–liquid extraction with ethyl acetate and crystallization. These procedures, although effective, are slow, tedious, and require the use of large amounts of organic solvents. Therefore, faster and more efficient protocols have been developed for the isolation and purification of bioactive polyphenols.

A common strategy to isolate bioactive phenolics is to use liquid–liquid extraction to eliminate the most hydrophilic and the most hydrophobic

TABLE 23.1 Recovery Examples of Biomolecules Citing Extraction Agents and Isolation Methods

Compounds	Source	Extraction	Isolation	Purity	Reference
Flavonoids	Apple peels	Aqueous acetone extraction	Column chromatography (Diaion HP-20, silica gel) and preparative HPLC (C18 column)	Pure quercetin-3-O-β-D-glucopyranoside and quercetin-3-O-β-D-galactopyranoside	He and Liu (2008)
Caffeic acid, chlorogenic acid and luteolin	L. japonica stem	Ethanolic extraction	HSCCC	95.55% caffeic acid, 97.24% chlorogenic acid and 98.11% luteolin	Wang et al. (2008)
Chlorogenic acid	Honeysuckle (L. japonica)	Aqueous microwave-assisted extraction	Column chromatography (HPD-850 resin)	50%	Zhang et al. (2008)
Anthocyanins	Blood oranges juice	Squeezed into a centrifugal extractor	Column chromatography (NKA-9 and TSK HW-40S)	Not reported	Cao et al. (2010)
Anthocyanins	Sweet cherries	Microwave assisted extraction	Semipreparative reversed-phase liquid chromatography (C18 column)	98%, cyanidin-3-O-glucoside 97%, cyanidin-3-O-rutinoside	Grigoras et al. (2012)
Anthocyanins	Coffee pulp	Methanolic extraction	Adsorption (C18 cartridge) and column chromatography (Sephadex LH-20)	89% cyanidin 3-rutinoside	Murthy et al. (2012)
Anthocyanins	Jamun (Syzygium cumini L.) fruits	Aqueous solid–liquid extraction	Adsorption on Amberlite XAD-7HP	Sugar-free extract	Jampani et al. (2014)
Anthocyanins	Wild blueberries	Maceration	Liquid–liquid extraction, Column chromatography (Amberlite XAD-7HP and Sephadex LH20) and semipreparative HPLC (C18 column)	97.7% malvidin-3-O-glucoside 99.3% petunidin-3-O-glucoside 95.4% delphinidin-3-O-glucoside	Wang et al. (2014)

TABLE 23.1 *(Continued)*

Compounds	Source	Extraction	Isolation	Purity	Reference
Anthocyanins	Red cabbage	Aqueous extraction	Aqueous two phase extraction and forward osmosis	0.3 % total anthocyanins*	Jampani and Raghavarao (2015)
Proanthocyanidins	Peanut skin	Aqueous acetone extraction	Column chromatography (Amberlite XAD-2, Sephadex LH20) and semipreparative HPLC (C18 column))	High purity proanthocyanidin A1 and proanthocyanidin A2	Oldoni et al. (2016)
Punicalagin	Pomegranate husk	Aqueous extraction	Column chromatography (Amberlite XAD 16) and semi-preparative medium pressure liquid chromatography (C18 column)	97%	Aguilar-Zárate et al. (2017)
Phenolic compounds	*O. rotundifolium*	Aqueous extraction	Liquid-liquid extraction and column chromatography on silica	Highly pure Apigenin, ferulic acid, vitexin, caprolactam, rosmarinic acid, and globoidnan	Erenler et al. (2017)
Proanthocyanidins	Red wine	Evaporation	Column chromatography (Diaion HP-2MG and Sephadex LH-20) Preparative and semipreparative HPLC (C18 column)	Highly pure: Malvidin 3-*O*-glucoside, petunidin 3-*O*-glucoside, Malvidin 3-*O*-glucoside, Trimer A, B, and C, procyanidin C1 and epicatechin 3-*O*-gallate	Fujimaki et al. (2018)
Flavonoid alkaloids	*S. moniliorrhiza*	Ethanolic extraction	Extracción líquido-líquido con CHCl₃, columna chromatography (silica gel), preparative HPTLC, column chromatography (Sephadex LH-20) and chiral chromatography (chiralpak AS-H)	Pure scumoniline A, B, C and D	Han et al. (2018)

TABLE 23.1 *(Continued)*

Compounds	Source	Extraction	Isolation	Purity	Reference
Anticancer and antioxidant compounds	*G. pentaphyllum*	Solvent extraction (ethyl acetate and buthanol)	Column chromatography (silica gel, polyamide column, Sephadex LH20 and ODS columns)	Pure 3,4-dihydroxy phenyl-O-β-D-glucoside, gypenoside XLVI, gypenoside L and ginsenoside Rd.	Wang et al. (2018)
Flavone glucosides	*A. juncea*	Ethanolic extraction	Column chromatography (silica gel and sephadex LH20)	Pure Eupatilin (5,7-dihydroxy-3′,4′,6-trimethoxyflavone) and 4′,5-dihydroxy-3′,5′,6-trimethoxyflavone-7-O-β-D-glucoside	Okhundedaev et al. (2019)
Flavonoids	Mulberry leaves	Ultrasonic-assisted extraction	HSCCC	93.8%	Zhang et al. (2019)

compounds from polyphenol-rich extracts. For example, De Colmenares et al. (1998) isolated two proanthocy-anidin-containing preparations from aqueous acetone extract of coffee pulp. After evaporation of acetone, aqueous extracts were extracted with petroleum ether to remove lipids and fat-soluble pigments, with ethyl acetate to solubilize simple phenols, and with dichloromethane to eliminate the caffeine. The aqueous extract was fractionated into a Sephadex LH-20 column. A faster alternative is the use of solid-phase extraction cartridges for the concentration of phenolic compounds and the elimination of sugars and other polar compounds. For example, Murthy et al. (2012) purified anthocyanins from an aqueous methanol extract of coffee pulp. After solvent evaporation, polyphenols were concentrated in C18 Sep-Pak cartridges, washed with acidified water, and eluted with acidified methanol. Methanol was vacuum evaporated, and pigments were separated into a Sephadex LH-20 column. However, solid-phase extraction cartridges are expensive and with limited capacity. Therefore, it is more common to use column chromatography with macroporous resins, such as the Amberlite XAD resins. In this technique, a column packed with a macroporous resin is equilibrated with water, loaded with an aqueous polyphenol-rich extract, and washed with water to remove the most polar compounds. Then, the

phenolic compounds are eluted with some polar organic solvent, often aqueous ethanol or aqueous acetone. The polyphenol-enriched fraction, commonly called sugar-free polyphenolic extract, can then be purified to isolate specific compounds. Amberlite has been used for the isolation of anthocyanins from Jamun fruits (Jampani et al., 2014) and wild blueberries (Wang et al., 2014), proanthocyanidins from peanut skin (Oldoni et al., 2016), and ellagitannins from pomegranate husks (Aguilar-Zárate et al., 2017).

The purification of phenolic compounds is usually carried out by gravity or flash column chromatography. The most commonly used stationary phases are silica gel and Sephadex LH-20. Silica gel is the preferred stationary phase for normal phase chromatographs. Sephadex LH-20 is a hydroxypropylated derivative of Sephadex G25 that has both hydrophobic and hydrophilic natures. Thus, separation on Sephadex LH-20 combines adsorption with gel filtration chromatography (Zhang et al., 2018). Low-pressure column chromatography has been used for purification of anthocyanins from the coffee pulp (Murthy et al., 2012) and wild blueberries (Wang et al., 2014), proanthocyanidins from peanut skin (Oldoni et al., 2016) and red wine (Fujimaki et al., 2018), phenolic compounds from *Origanum rotundifolium* (Erenler et al., 2017), flavonoid alkaloids from

Scutellaria moniliorrhiza (Han et al., 2018), flavone glucosides from *Artemisia juncea* (Okhundedaev et al., 2019), and anticancer and antioxidant compounds from *Gynostemma pentaphyllum* (Wang et al., 2018), among others.

Low-pressure column chromatography is an effective technique that has been used to isolate many phenolic compounds; however, these procedures can laborious and time-consuming. An alternative for the purification of phenolic compounds is the use of semipreparative or preparative HPLC systems. Semipreparative HPLC chromatography is a high-resolution technique useful for obtaining pure compounds in the order of several milligrams per run. Separation in semipreparative or preparative HPLC columns can be carried out in the normal or reverse phase. However, the reverse phase is by far the most used technique. The preferred stationary phase for separations in preparative or semipreparative in HPLC is the octadecylsilane (ODS) bonded silica, commonly known as C18. Semipreparative HPLC in C18 column has extensively used for the purification of many phenolic compounds, such as quercetin-3-O-β-D-glucopyranoside and quercetin-3-O-β -D-galactopyranoside 98%, cyanidin-3-O-glucoside (He and Liu, 2008) cyanidin-3-*O*-rutinoside (Grigoras et al., 2012), malvidin-3-O-glucoside,

petunidin-3-O-glucoside, delphinidin-3-*O*-glucoside (Fujimaki et al., 2018; Wang et al., 2014), proanthocyanidin A1, proanthocyanidin A2 (Oldoni et al., 2016), procyanidin C1 (Fujimaki et al., 2018), punicalagin (Aguilar-Zárate et al., 2017), and among others.

High-speed counter-current chromatography (HSCCC) is a liquid–liquid partition chromatography process in which both the mobile and the stationary phase are liquids. The "column" is simply a long length of tubing wound on a drum (Valls et al., 2009). HSCCC has several advantages compared to conventional column chromatography with a solid stationary phase, including high loading capacity, high sample recovery, and low solvent consumption. The main limitation of HSCCC is that it only separates the compounds in a relatively narrow polarity window (Zhang et al., 2018). HSCCC has been used to simultaneously purify caffeic acid, chlorogenic acid, and luteolin from a crude extract of *Lonicera japonica* (Wang et al., 2008). HSCCC has also been used to purify flavonoids from mulberry leaves (Zhang et al., 2019).

23.6 STRUCTURE SIMULATION

Flavonoids, being the most widespread group of phenolic compounds, are produced in plants as secondary metabolites. The extraction of this type of compounds by chemical, physicochemical, and physical techniques generally leads to low yield, since their union to the cell wall material through a –OH group (O-glycosides) or carbon–carbon bonds (C-glycosides) (Acosta-Estrada et al., 2014). Therefore, a hydrolysis treatment before the extraction is needed, and fermentation or enzymatic treatments are the best options, since they avoid deleterious transformations of the extracted compounds and they are environment-friendly processes (Puri et al., 2012; Martins et al., 2011).

As previously discussed, the release of flavonoids in fermentation relies on two crucial steps: cell wall degradation and the hydrolysis of glycosydic bond (Figure 23.1). Among the several enzymes involved in these processes, β-glucosidase has been widely reported as responsible for catalyzing the hydrolysis of glycosidic linkages in alkyl and aryl-β-D-glucosides to release phenolic aglycone moieties. Huynh et al. (2014) reviews reports of this kind of enzymes in several fungi like *Aspergillus awamori nakazawa, Aspergillus niger, Aspergillus oryzae, Lentinus edodes, Pleurotus ostreatus, Rhizopus oligosporus, Phanerochaete chrysosporium, Rhizopus oryzae,* the yeast *Rhodotorula glutimis, Sacharomyces cerevisiae, Wickerhamomyces anomalus,* and bacteria like *Lactobacillus plantarum* and *Lactobacillus rhamnosus.*

The enzyme is conserved in the genus *Aspergillus* but the primary structure of these proteins does not show a high similarity among the species compared here. BLASTp showed over 90% coverage but only around 30% of identity between *L. plantarum* and *A. niger,* and 78% coverage and 48% identity among *R. oligosporus* and *A. niger* (Atschul et al., 1990).

To our knowledge, there are no crystallizations of a β-glucosidase from any of these species available in the Protein Data Bank (rscb.org, Berman et al., 2000), some amino acid sequences can be found on the National Center for Biotechnology Information database for proteins. For this chapter, we made a prediction of the 3D structures of these proteins using I-TASSER server (Zhang et al., 2017; Yang et al., 2015) from *Aspergillus niger, Lactobacillus plantarum,* and *Rhizopus oligosporus* (Figure 23.6a–c, respectively) (Accession numbers RDH20571.1, WP_114648652 and XP_023469485.1, respectively).

These structures show a TM-align score of 0.75–0.90 with over 80% of the chain being aligned in each computation (Zhang et al., 2005), which is very interesting, due to the difference in the primary structure mentioned above and the fact that prokaryotes and eukaryotes were compared. Moreover, there are

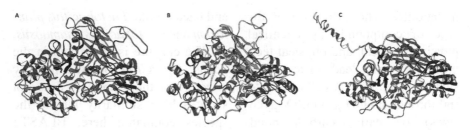

FIGURE 23.6 β-glucosidase from (A) *Aspergillus niger*, (B) *Lactobacillus plantarum*, and (C) *Rhizopus oligosporus*. Molecular graphics and analyses performed with UCSF ChimeraX, developed by the resource for biocomputing, visualization, and informatics at the University of California, San Francisco, with support from NIH R01-GM129325 and P41-GM103311 (Goddard et al., 2018).

conserved residues in the predicted ligand binding, particularly alkaline and acidic residues which, chemically, can interact with glucose through hydrogen bonds. The similarity in these structures suggests a strong evolutionary conservation across different domains and the relevance of this enzyme activity.

23.7 CONCLUDING REMARKS

These molecules, as are widely found in nature, and possess an also ample array of bioactivities. It is important to establish that there are many sources for compound extraction and biotechnological aspects represent an environmentally friendly option for its processing. One of the advantages of the bioprocessing of these sources is that, besides the high added-value of these compounds' extraction, also the enzymes responsible for its release can be produced and therefore, its recovery. Even though polyphenols are increasingly being noticed due to their importance, new developments are still needed to design better recovery strategies and improve the bioprocesses to increase yields in molecule release, isolation, and their bioactive features. In the case of the particular case, the enzyme addressed in this chapter, there is still a wide range of research options for enzyme production, biomolecules release, and the interaction among these metabolites, specially β-glucosidase. The hydrolysis of glycosylated polyphenols, its microbial production, and the possibility of producing an enzyme that can both participate in cellulose hydrolysis and the cleavage of glycosidic bonds among flavonoids and phenolic acids. For its recovery, several works have been reported which serves as alternatives for further evaluation of these types of enzymes along with the entailed molecular aspects available.

KEYWORDS

- bioactive compounds
- biodegradation
- bioprocess
- glycosides
- aglycones

REFERENCES

Acosta-Estrada, B.A.; Gutiérrez-Uribe, J.A.; Serna-Saldívar, S.O. Bound phenolics in foods: a review. *Food Chem.* **2014**, 152, 46–55.

Aguilar-Zárate, P.; Wong-Paz, J.E.; Michel, M.; Buenrostro-Figueroa, J.; Díaz, H.R.; Ascacio, J.A.; Contreras-Esquivel, J.C.; Gutiérrez-Sánchez, G.; Aguilar, C.N. Characterisation of pomegranate-husk polyphenols and semi-preparative fractionation of punicalagin. *Phytochem. Anal.* **2017**, 28, 433–438.

Altschul, S.F.; Gish, W.; Miller, W.; Myers, E.W.; Lipman, D.J. Basic local alignment search tool. *J. Mol. Biol.* **1990**, 215, 403–410.

Aguilera-Carbo, A.; Hernández, J.S.; Augur, C.; Prado-Barragan, L.A.; Favela-Torres, E.; Aguilar, C.N. Ellagic acid production from biodegradation of creosote bush ellagitannins by *Aspergillus niger* in solid state culture. *Food Bioprocess Technol.* **2008**, 2, 208–212. https://doi.org/10.1007/s11947-008-0063-0.

Asses, N.; Ayed, L.; Bouallagui, H.; Sayadi, S.; Hamdi, M. Biodegradation of different molecular-mass polyphenols derived from olive mill wastewaters by *Geotrichum candidum*. *Int. Biodeter. Biodegr.* **2009**, 63, 407–413.

Barnaba, C.; Larcher, R.; Nardin, T.; Dellacassa, E.; Nicolini, G. Glycosylated simple phenolic profiling of food tannins using high resolution mass spectrometry (Q-Orbitrap). *Food Chem.* **2018**, 267, 196–203. https://doi.org/10.1016/j.foodchem.2017.11.048.

Berman, H.M.; Westbrook, J.; Feng, Z.; Gilliland, G.; Bhat, T.N.; Weissig, H.; Shindyalov, I.N.; Bourne, P.E. The protein data bank. *Nuc. Acids Res.* 2000, 28, 235–242.

Bhat, T.; Singh, B.; Sharma, O. Microbial degradation of tannins—a current perspective. *Biodegradation* **1998**, 9, 343–357.

Braga, A.R.C.; Mesquita, L.M. de S.; Martins, P.L.G.; Habu, S.; de Rosso, V.V. *Lactobacillus* fermentation of jussara pulp leads to the enzymatic conversion of anthocyanins increasing antioxidant activity. *J. Food Compos. Anal.* **2018**, 69, 162–170. https://doi.org/10.1016/j.jfca.2017.12.030.

Brillouet, J.M.; Romieu, C.; Schoefs, B.; Solymosi, K.; Cheynier, V.; Fulcrand, H.; Verdeil, J.L.; Conéjéro, G. The tannosome is an organelle forming condensed tannins in the chlorophyllous organs of *Tracheophyta. Ann. Bot.* **2013**, 112, 1003–1014. https://doi.org/10.1093/aob/mct168.

Cao, H.; Chen, X.; Reza, A.; Xiao, J. Microbial biotransformation of bioactive flavonoids. *Biotechnol. Adv.* **2015**, 33, 214–223. https://doi.org/10.1016/j.biotechadv.2014.10.012.

Cao, S.-q.; Pan, S.-y.; Yao, X.-l.; Fu, H.-f. Isolation and purification of anthocyanins from blood oranges by column chromatography. *Agric. Sci. Chin.* 2010, 9, 207–215.

Cavaliere, C.; Capriotti, L.A.; La Barbera, G.; Montone, M.C.; Piovesana, S.; Laganà, A. Liquid chromatographic strategies for separation of bioactive compounds in food matrices. *Molecules* **2018**, 23, 3091. https://doi.org/10.3390/molecules23123091

Chávez-González, M.L.; Guyot, S.; Rodríguez-Herrera, R.; Prado-Barragán, A.; Aguilar, C.N. Production profiles of phenolics from fungal tannic acid biodegradation in submerged and solid-state fermentation. *Process Biochem.* **2014**, 49, 541–546.

Chávez-González, M.L.; Guyot, S.; Rodrí-guez-Herrera, R.; Prado-Barragán, A.; Aguilar, C.N. Exploring the degradation of gallotannins catalyzed by tannase produced by *Aspergillus niger* GH1 for ellagic acid production in submerged and solid-state fermentation. *Appl. Biochem. Biotechnol.* **2018**, 185, 476–483.

Cui, P.; Li, T.D.S.; Ping, L.Z.; Sun, W.Y. Highly selective and efficient biotransformation of linarin to produce tilianin by naringinase. *Biotechnol. Lett.* **2016**, 38, 1367–1373. https://doi.org/10.1007/s10529-016-2116-1.

De Colmenares, N.G.; Ramírez-Martínez, J.R.; Aldana, J.O.; Ramos-Niño, M.E.; Clifford, M.N.; Pékerar, S.; Méndez, B. Isolation, characterisation and determination of biological activity of coffee proanthocyanidins. *J. Sci. Food Agric.* **1998**, 77, 368–372.

de Oliveira Rodrigues, P.; dos Santos, B.V.; Costa, L.; Henrique, M.A.; Pasquini, D.; Baffi, M.A. Xylanase and β-glucosidase production by *Aspergillus fumigatus* using commercial and lignocellulosic substrates submitted to chemical pre-treatments. *Ind. Crops Prod.* **2017**, 95, 453–459. https://doi.org/10.1016/j.indcrop.2016.10.055.

Duan, L.; Zhang, W.H.; Zhang, Z.H.; Liu, E.H.; Guo, L. Evaluation of natural deep eutectic solvents for the extraction of bioactive flavone C-glycosides from *Flos Trollii*. *Microchem. J.* **2019**, 145, 180–186. https://doi.org/10.1016/j.microc.2018.10.031.

Erenler, R.; Meral, B.; Sen, O.; Elmastas, M.; Aydin, A.; Eminagaoglu, O.; Topcu, G. Bioassay-guided isolation, identification of compounds from *Origanum rotundifolium* and investigation of their antiproliferative and antioxidant activities. *Pharm. Biol.* **2017**, 55, 1646–1653.

Falcão, H.G.; Handa, C.L.; Silva, M.B.R.; de Camargo, A.C.; Shahidi, F.; Kurozawa, L.E.; Ida, E.I. Soybean ultrasound pre-treatment prior to soaking affects β-glucosidase activity, isoflavone profile and soaking time. *Food Chem.* **2018**,

269, 404–412. https://doi.org/10.1016/j.foodchem.2018.07.028.

Fujimaki, T.; Mori, S.; Horikawa, M.; Fukui, Y. Isolation of proanthocyanidins from red wine, and their inhibitory effects on melanin synthesis in vitro. *Food Chem.* **2018**, 248, 61–69.

Gligor, O.; Mocan, A.; Moldovan, C.; Locatelli, M.; Crişan, G.; Ferreira, I.C.F.R. Enzyme-assisted extractions of polyphenols—a comprehensive review. *Trends Food Sci. Technol.* **2019**, 88, 302–315. https://doi.org/10.1016/j.tifs.2019.03.029.

Goddard, T.D.; Huang, C.C.; Meng, E.C.; Pettersen, E.F.; Couch, G.S.; Morris, J.H.; Ferrin, T.E. UCSF ChimeraX: meeting modern challenges in visualization and analysis. *Protein Sci.* **2018**, 27(1), 14–25.

Gómez-Mejía, E.; Rosales-Conrado, N.; León-González, M.E.; Madrid, Y. Citrus peles waste as a source of value-added compounds: extraction and quantification of bioactive polyphenols. *Food Chem.* **2019**, 295, 289–299.

González-Aguilar, G.A.; Blancas-Benítez, F.J.; Sáyago-Ayerdi, S.G. Polyphenols associated with dietary fibers in plant foods: molecular interactions and bioaccessibility. *Curr. Opin. Food Sci.* **2017**, 13, 84–88. https://doi.org/10.1016/j.cofs.2017.03.004.

Grigoras, C.G.; Destandau, E.; Zubrzycki, S.; Elfakir, C. Sweet cherries anthocyanins: an environmental friendly extraction and purification method. *Sep. Purif. Technol.* **2012**, 100, 51–58.

Han, Q.-T.; Ren, Y.; Li, G.-S.; Xiang, K.-L.; Dai, S.J. Flavonoid alkaloids from *Scutellaria moniliorrhiza* with anti-inflammatory activities and inhibitory activities against aldose reductase. *Phytochemistry* **2018**, 152, 91–96.

He, X.; Liu, R.H. Phytochemicals of apple peels: isolation, structure elucidation, and their antiproliferative and antioxidant activities. *J. Agric. Food Chem.* **2008**, 56, 9905–9910.

Hulme, A.C. The isolation of chlorogenic acid from the apple fruit. *Biochem. J.* **1953**, 53, 337–340.

Huynh, N.T.; Van Camp, J.; Smagghe, G.; Raes, K. Improved release and metabolism of flavonoids by steered fermentation processes: a review. *Int. J. Mol. Sci.* **2014**, 15, 19369–19388.

Irakli, M.; Chatzopoulou, P.; Ekateriniadou, L. Optimization of ultrasound-assisted extraction of phenolic compounds: oleuropein, phenolic acids, phenolic alcohols and flavonoids from olive leaves and evaluation of its antioxidant activities. *Ind. Crops Prod.* **2018**, 124, 382–388. https://doi.org/10.1016/j.indcrop.2018.07.070.

Jampani, C.; Naik, A.; Raghavarao, K.S.M.S. Purification of anthocyanins from jamun (*Syzygium cumini* L.) employing adsorption. *Sep. Purif. Technol.* **2014**, 125, 170–178.

Jampani, C.; Raghavarao, K.S.M.S. Process integration for purification and concentration of red cabbage (Brassica oleracea L.) anthocyanins. *Sep. Purif. Technol.* **2015**, 141, 10–16.

Juan, M.Y.; Chou, C.C. Enhancement of antioxidant activity, total phenolic and flavonoid content of black soybeans by solid state fermentation with *Bacillus subtilis* BCRC 14715. *Food Microbiol.* **2010**, 27, 586–591. https://doi.org/10.1016/j.fm.2009.11.002.

Jurd, L. Plant polyphenols. I. The polyphenolic constituents of the pellicle of the walnut (*Juglans regia*). *J. Am. Chem. Soc.* **1956**, 78, 3445–3448.

Juturu, V.; Wu, J.C. Microbial xylanases: engineering, production and industrial applications. *Biotechnol. Adv.* **2012**, 30, 1219–1227. https://doi.org/10.1016/j.biotechadv.2011.11.006.

Kawabata, K.; Mukai, R.; Ishisaka, A. Quercetin and related polyphenols: new insights and implications for their bioactivity and bioavailability. *Food Funct.* **2015**, 6, 1399–1417. https://doi.org/10.1039/c4fo01178c.

Lee, Y.S.; Huh, J.Y.; Nam, S.H.; Moon, S.K.; Lee, S.B. Enzymatic bioconversion of citrus hesperidin by *Aspergillus sojae* naringinase: enhanced solubility of hesperetin-7-O-glucoside with in vitro inhibition of human intestinal maltase, HMG-CoA reductase, and growth of *Helicobacter pylori*. *Food Chem.* **2012**, 135, 2253–2259. https://doi.org/10.1016/j.foodchem.2012.07.007.

Lopez-Trujillo, J.; Medina-Morales, M.A.; Sanchez-Flores, A.; Arevalo, C.; Ascacio-Valdes, J.A.; Mellado, M.; Aguilar, C.N.; Aguilera-Carbo, A.F. Solid bioprocess of tarbush (*Flourensia cernua*) leaves for β-glucosidase production by *Aspergillus niger*: initial approach to fiber–glycoside interaction for enzyme induction. *3 Biotech.* **2017**, 7, 271. https://doi.org/10.1007/s13205-017-0883-6.

Lu, H.T.; Jiang, Y.; Chen, F. Application of preparative high-speed counter-current chromatography for separation of chlorogenic acid from *Flos Lonicerae*. *J. Chromatogr. A.* **2004**, 1026, 185–190.

Lu, X.; Sun, J.; Nimtz, M.; Wissing, J.; Zeng, A.; Rinas, U. The intra- and extracellular proteome of *Aspergillus niger* growing on defined medium with xylose or maltose as carbon substrate. *Microb. Cell Fact.* **2010**, 9, 1–13.

Macedo, J.A.; Battestin, V.; Ribeiro, M.L.; Macedo, G.A. Increasing the antioxidant power of tea extracts by biotransformation of polyphenols. *Food Chem.* **2011**, 126, 491–497.

Macedo, J.A.; Ferreira, L.R.; Camara, L.E.; Santos, J.C.; Gambero, A.; Macedo, G.A.; Ribeiro, M.L. Chemopreventive potential of the tannase-mediated biotransformation of green tea. *Food Chem.* **2012**, 126, 133, 358–365.

Madeira Jr, J.V.; Mayumi-Nakajima, V.; Alves-Macedo, V.; Alves-Macedo, G. Rich bioactive phenolic extract production by microbial biotransformation of Brazilian Citrus residues. *Chem. Eng. Res. Des.* **2014**, 92, 1802–1810.

Maitan-Alfenas, G.P.; Lage, L.G.D.A.; Almeida, M.N.; Visser, E.M.; Rezende,

S.T. De; Guimarães, V.M. Hydrolysis of soybean isoflavones by *Debaryomyces hansenii* UFV-1 immobilised cells and free β-glucosidase. *Food Chem.* **2014**, 146, 429–436. https://doi.org/10.1016/j.foodchem.2013.09.099.

Marín, M.; Sánchez, A.; Artola, A. Production and recovery of cellulases through solid-state fermentation of selected lignocellulosic wastes. *J. Clean. Prod.* **2019**, 209, 937–946. https://doi.org/10.1016/j.jclepro.2018.10.264.

Martins, S.; Mussatto, S.I.; Martínez-Avila, G.; Montañez-Saenz, J.; Aguilar, C.N.; Teixeira, J.A. Bioactive phenolic compounds: production and extraction by solid-state fermentation. A review. *Biotechnol. Adv.* **2011**, 29, 365–373. https://doi.org/10.1016/j.biotechadv.2011.01.008.

Medina-Morales, M.A.; López-Trujillo, J.; Gómez-Narváez, L.; Mellado, M.; García-Martínez, E.; Ascacio-Valdés, J. A.; Aguilar, C. N.; Aguilera-Carbó, A. Effect of growth conditions on β-glucosidase production using *Flourensia cernua* leaves in a solid-state fungal bioprocess. *3 Biotech.* **2017**, 7, PMC5622878. https://doi.org/10.1007/s13205-017-0990-4.

Murthy, P.S.; Manjunatha, M.R.; Sulochannama, G.; Naidu, M.M. Extraction, characterization and bioactivity of coffee anthocyanins. *Eur. J. Biol. Sci.* **2012**, 4, 13–19.

Mutabaruka, R.; Hairiah, K.; Cadisch, G. Microbial degradation of hydrolysable and condensed tannin polyphenol-protein complexes in soils from different land-use histories. *Soil Biol. Biochem.* **2007**, 39, 1479–1492. https://doi.org/10.1016/j.soilbio.2006.12.036.

Noh, D.; Choi, H.; Suh, H.; Catechine biotransformation by tannase with sequential addition of substrate. *Process Biochem.*, **2014**, 49, 271–276.

Núñez-López, G.; Herrera-González, A.; Hernández, L.; Amaya-Delgado, L.; Sandoval, G.; Gschaedler, A.; Arrizon, J.; Remaud-Simeon, M.; Morel, S.

Fructosylation of phenolic compounds by levansucrase from *Gluconacetobacter diazotrophicus. Enzyme Microb. Technol.* **2019**, 122, 19–25. https://doi.org/10.1016/j.enzmictec.2018.12.004.

Pavlic, B.; Kaplan, M.; Bera, O.; Olgun, E.; Canli, O.; Milosavljevic, N.; Antic, B.; Zekovic, Z.; Microwave-assisted extraction of peppermint polyphenols—artificial neural networks approach. *Food Bioprod. Process.* **2019**, 118, 258–269.

Oh, J.M.; Lee, J.P.; Baek, S.C.; Kim, S.G.; Jo, Y.; Kim, J.; Kim, H. Characterization of two extracellular β-glucosidases produced from the cellulolytic fungus *Aspergillus* sp. YDJ216 and their potential applications for the hydrolysis of flavone glycosides. *Int. J. Biol. Macromol.* **2018**, 111, 595–603. https://doi.org/10.1016/j.ijbiomac.2018.01.020.

Okhundedaev, B.S.; Bacher, M.; Mukhamatkhanova, R.F.; Shamyanov, I.J.; Zengin, G.; Böhmdorfer, S.; Mamadalieva, N.Z.; Rosenau, T. Flavone glucosides from *Artemisia juncea. Nat. Prod. Res.* **2019**, 33, 2169–2175.

Oldoni, T.L.C.; Melo, P.S.; Massarioli, A.P.; Moreno, I.A.M.; Bezerra, R.M.N.; Rosalen, P.L.; da Silva, G.V.J.; Nascimento, A.M.; Alencar, S.M. Bioassay-guided isolation of proanthocyanidins with antioxidant activity from peanut (*Arachis hypogaea*) skin by combination of chromatography techniques. *Food Chem.* **2016**, 192, 306–312.

Pandey, R.P.; Parajuli, P.; Koffas, M.A.G.; Sohng, J.K. Microbial production of natural and non-natural flavonoids: pathway engineering, directed evolution and systems/synthetic biology. *Biotechnol. Adv.* **2016**, 34, 634–662. https://doi.org/10.1016/j.biotechadv.2016.02.012.

Pinheiro do Prado, A.C.; da Silva, H.S.; da Silveira, S.M.; Barreto, P.L.M.; Vieira, C.R.W.; Maraschin, M.; Ferreira, S.R.S.; Block, J.M. Effect of the extraction process on the phenolic compounds profile and the antioxidant and antimicrobial activity of extracts of pecan nut [*Carya illinoinensis* (Wangenh) C. Koch] Shell. *Ind. Crops Prod.* **2014**, 52, 552–561. https://doi. org/10.1016/j.indcrop.2013.11.031.

Protein [Internet]. National Center for Biotechnology Information (US). **2010**, Bethesda, MD. https://www.ncbi.nlm.nih. gov/protein/.

Puri, M.; Sharma, D.; Barrow, C.J. Enzyme-assisted extraction of bioactives from plants. *Trends Biotechnol.* **2012**, 30, 37–44.

Queiroz Santos, V.A.; Nascimento, C.G.; Schimidt, C.A.P.; Mantovani, D.; Dekker, R.F.H.; da Cunha, M.A.A. Solid-state fermentation of soybean okara: isoflavones biotransformation, antioxidant activity and enhancement of nutritional quality. *LWT* **2018**, 92, 509–515. https:// doi.org/10.1016/j.lwt.2018.02.067.

Quideau, S.; Deffieux, D.; Douat-Casassus, C.; Pouységu, L. Plant polyphenols: chemical properties, biological activities, and synthesis. *Angew Chem Int Ed Engl.*, **2011**, 50, 586–621.

Quirós-Sauceda, A.E.; Palafox-Carlos, H.; Sáyago-Ayerdi, S.G.; Ayala-Zavala, J.F.; Bello-Perez, L.A.; Álvarez-Parrilla, E.; De La Rosa, L.A.; González-Córdova, A.F.; González-Aguilar, G.A. Dietary fiber and phenolic compounds as functional ingredients: interaction and possible effect after ingestion. *Food Funct.* **2014**, 5, 1063–1072. https://doi.org/10.1039/c4fo00073k.

Ravindran, R.; Desmond, C.; Jaiswal, S.; Jaiswal, A. Optimisation of organosolv pretreatment for the extraction of polyphenols from spent coffee waste and subsequent recovery of fermentable sugars. *Bioresour. Technol. Rep.* **2018**, 3, 7–14.

Roggia-Ruviaro, A.; Menezes-Barbosa, P.; Alves-Macedo, G. Enzyme-assisted biotransformation increases hesperetin content in citrus juice by-products. *Food Res. Int.*, **2019**, 124, 213–221.

Semwal, R.B.; Semwal, D.K.; Combrinck, S.; Trill, J.; Gibbons, S.; Viljoen, A. Acacetin—a simple flavone exhibiting diverse pharmacological activities. *Phytochem. Lett.* **2019**, 32, 56–65. https://doi. org/10.1016/j.phytol.2019.04.021.

Sepúlveda, L.; Wong-Paz, J.E.; Buenrostro-Figueroa, J.; Ascacio-Valdés, J.A.; Aguilera-Carbó, A.; Aguilar, C.N. Solid state fermentation of pomegranate husk: recovery of ellagic acid by SEC and identification of ellagitannins by HPLC/ESI/MS. *Food Biosci.* **2018**, 22, 99–104. https://doi. org/10.1016/j.fbio.2018.01.006.

Sharma, K.; Bari, S.; Singh, H. Biotransformation of tea catechins into theaflavins with immobilized polyphenol oxidase. *J. Mol. Catal., B Enzym.* **2009**, 56, 253–258.

Shin, H.Y.; Kim, S.M.; Lee, J.H.; Lim, S.T. Solid-state fermentation of black rice bran with *Aspergillus awamori* and *aspergillus oryzae*: effects on phenolic acid composition and antioxidant activity of bran extracts. *Food Chem.* **2019**, 272, 235–241. https://doi.org/10.1016/j. foodchem.2018.07.174.

Soccol, C.R.; Scopel, E.; Alberto, L.; Letti, J.; Karp, S.G.; Woiciechowski, A.L.; de Souza, L. Recent developments and innovations in solid state fermentation. *Biotechnol. Res. Innov.* **2017**, 1, 52–71. https://doi. org/10.1016/j.biori.2017.01.002.

Sordon, S.; Popłoński, J.; Huszcza, E. Microbial glycosylation of flavonoids. *Polish J. Microbiol.* **2016**, 65, 137–151. https://doi. org/10.5604/17331331.1204473.

Sudha, A.; Srinivasan, P. Bioassay-guided isolation and antioxidant evaluation of flavonoid compound from aerial parts of *Lippia nodiflora* l. *BioMed Res. Int.* **2014**, 2014, 10.

Švarc, J.; Stojanović, Z.; Segura Carretero, A.; Arráez Román, D.; Borrás, I.; Vasiljević, I. Development of a microwave-assisted extraction for the analysis of phenolic compounds from *Rosmarinus officinalis*. *J. Food Eng.* **2013**, 119, 525–532. https:// doi.org/10.1016/j.jfoodeng.2013.06.030.

Torres-León, C.; Ramírez-Guzmán, N.; Ascacio-Valdés, J.; Serna-Cock, L.; dos

Santos Correia, M.T.; Contreras-Esquivel, J.C.; Aguilar, C.N. Solid-state fermentation with *Aspergillus niger* to enhance the phenolic contents and antioxidative activity of Mexican mango seed: a promising source of natural antioxidants. *LWT* **2019**, 112, 108236. https://doi.org/10.1016/j.lwt.2019.06.003.

Valls, J.; Millán, S.; Martí, M.P.; Borràs, E.; Arola, L. Advanced separation methods of food anthocyanins, isoflavones and flavanols. *J. Chromatogr.* **2009**, 1216, 7143–7172.

Veitch, N.C.; Grayer, R.J. Flavonoids and their glycosides, including anthocyanins. *Nat. Prod. Rep.* **2011**, 25, 555–611. https://doi.org/10.1039/c1np00044f.

Wang, E.; Yin, Y.; Xu, C.; Liu, J. Isolation of high-purity anthocyanin mixtures and monomers from blueberries using combined chromatographic techniques. *J. Chromatogr.* **2014**, 1327, 39–48.

Wang, L.; Wei, W.; Tian, X.; Shi, K.; Wu, Z. Improving bioactivities of polyphenol extracts from *Psidium guajava* L. leaves through co-fermentation of *Monascus anka* GIM 3.592 and *Saccharomyces cerevisiae* GIM 2.139. *Ind. Crops Prod.* **2016**, 94, 206–215.

Wang, T.-X.; Shi, M.-M.; Jiang, J.-G. Bioassay-guided isolation and identification of anticancer and antioxidant compounds from *Gynostemma pentaphyllum* (Thunb.) Makino. *RSC Adv.* **2018**, 8, 23181–23190.

Wang, Z.; Wang, J.; Sun, Y.; Li, S.; Wang, H. Purification of caffeic acid, chlorogenic acid and luteolin from *Caulis Lonicerae* by high-speed counter-current chromatography. *Sep. Purif. Technol.* **2008**, 63, 721–724.

Wong-Paz, J.E.; Muñiz-Márquez, D.B.; Martínez-Ávila, G.C.G.; Belmares-Cerda, R.E.; Aguilar, C.N. Ultrasound-assisted extraction of polyphenols from native plants in the mexican desert. *Ultrason. Sonochem.* **2015**, 22, 474–481. https://doi.org/10.1016/j.ultsonch.2014.06.001.

Xu, J.Q.; Fan, N.; Yu, B.Y.; Wang, Q.Q.; Zhang, J. Biotransformation of quercetin by *Gliocladium deliquescens* NRRL 1086. *Chin. J. Nat. Med.*, **2017**, 15, 615–624.

Yahia, E.M.; Gutiérrez-Orozco, F.; Moreno-Pérez, M.A. Identification of phenolic compounds by liquid chromatography-mass spectrometry in seventeen species of wild mushrooms in Central Mexico and determination of their antioxidant activity and bioactive compounds. *Food Chem.* **2017**, 226, 14–22. https://doi.org/10.1016/j.foodchem.2017.01.044.

Yang, B.; Liu, H.; Yang, J.; Gupta, V.K.; Jiang, Y. New insights on bioactivities and biosynthesis of flavonoid glycosides. *Trends Food Sci. Technol.* **2018**, 79, 116–124.

Yang, J.; Zhang, Y. I-TASSER server: new development for protein structure and function predictions. *Nuc. Acids Res*. **2015**, 43, W174–W181.

Yeoman, C.J.; Han, Y.; Dodd, D.; Schroeder, C.M.; Mackie, R.I.; Cann, I.K.O. *Thermostable Enzymes as Biocatalysts in the Biofuel Industry.* 1st ed., Elsevier Inc.: Amsterdam, **2010**, 70. https://doi.org/10.1016/S0065-2164(10)70001-0.

Zhang, B.; Yang, R.; Zhao, Y.; Liu, C.-Z. Separation of chlorogenic acid from honeysuckle crude extracts by macroporous resins. *J. Chromatogr. B.* **2008**, 867, 253–258.

Zhang, C.; Freddolino, P.L.; Zhang, Y. COFACTOR: improved protein function prediction by combining structure, sequence and protein–protein interaction information. *Nuc. Acids Res.* **2017**, 45, W291–W299.

Zhang, G.; Hu, M.; He, L.; Fu, P.; Wang, L.; Zhou, J. Optimization of microwave-assisted enzymatic extraction of polyphenols from waste peanut shells and evaluation of its antioxidant and antibacterial activities in vitro. *Food Bioprod. Process.* **2013**, 91, 158–168. https://doi.org/10.1016/j.fbp.2012.09.003.

Zhang, L.; Li, Y.; Liang, Y.; Liang, K.; Zhang, F.; Xu, T.; Wang, M. Determination of phenolic acid profiles by HPLC-MS in vegetables commonly consumed in China. Food Chem. **2019**, 276, 538–546. https://doi.org/10.1016/j.foodchem.2018.10.074.

Zhang, Q.-W.; Lin, L.-G.; Ye, W.-C. Techniques for extraction and isolation of natural products: a comprehensive review. *Chin. Med.* **2018**, 13, 20.

Zhang, Y.; Skolnick, J. TM-align: a protein structure alignment algorithm based on TM-score. *Nuc. Acids Res.* **2005**, 33, 2302–2309.

Zhang, P.; Zhu, K.-L.; Zhang, J.; Li, Y.; Zhang, H.; Wang, Y. Purification of flavonoids from mulberry leaves via high-speed counter-current chromatography. *Processes* **2019**, 7, 1–11.

Zheng, W.; Zheng, Q.; Xue, Y.; Hu, J.; Gao, M.T. Influence of rice straw polyphenols on cellulase production by *Trichoderma reesei*. *J. Biosci. Bioeng.*, **2017**, 123, 731–738.

Zhu, Y.; Jia, H.; Xi, M.; Xu, L.; Wu, S.; Li, X. Purification and characterization of a naringinase from a newly isolated strain of *Bacillus amyloliquefaciens* 11568 suitable for the transformation of flavonoids. *Food Chem.* **2017**, 214, 39–46. https://doi.org/10.1016/j.foodchem.2016.06.108.

Index

Printed in the United States
by Baker & Taylor Publisher Services

Printed in the United States
by Baker & Taylor Publisher Services